PACIFIC SYMPOSIUM ON BIOCOMPUTING 2017

PACIFIC SYMPOSIUM ON BIOCOMPUTING 2017

Kohala Coast, Hawaii, USA,
4 – 8 January 2017

Edited by

Russ B. Altman
Stanford University, USA

A. Keith Dunker
Indiana University, USA

Lawrence Hunter
University of Colorado Health Sciences Center, USA

Marylyn D. Ritchie
Geisinger Health System, USA

Tiffany Murray
Stanford University, USA

Teri E. Klein
Stanford University, USA

World Scientific

NEW JERSEY • LONDON • SINGAPORE • BEIJING • SHANGHAI • HONG KONG • TAIPEI • CHENNAI • TOKYO

Published by

World Scientific Publishing Co. Pte. Ltd.
5 Toh Tuck Link, Singapore 596224
USA office: 27 Warren Street, Suite 401-402, Hackensack, NJ 07601
UK office: 57 Shelton Street, Covent Garden, London WC2H 9HE

ISSN: 2335-6936

British Library Cataloguing-in-Publication Data
A catalogue record for this book is available from the British Library.

BIOCOMPUTING 2017
Proceedings of the Pacific Symposium

ISBN 978-981-3207-80-6

Preface vii

COMPUTATIONAL APPROACHES TO UNDERSTANDING THE EVOLUTION OF MOLECULAR FUNCTION

Session Introduction 1
Yana Bromberg, Matthew W. Hahn, Predrag Radivojac

Identification and Analysis of Bacterial Genomic Metabolic Signatures 3
Nathan Bowerman, Nathan Tintle, Matthew DeJongh, Aaron A. Best

When should we NOT transfer functional annotation between sequence paralogs? 15
Mengfei Cao, Lenore J. Cowen

ProSNet: integrating homology with molecular networks for protein function prediction 27
Sheng Wang, Meng Qu, Jian Peng

On the power and limits of sequence similarity based clustering of proteins into families 39
Christian Wiwie, Richard Röttger

IMAGING GENOMICS

Session Introduction 51
Li Shen, Lee A.D. Cooper

Adaptive testing of SNP-brain functional connectivity association via a modular network analysis 58
Chen Gao, Junghi Kim, Wei Pan

Exploring Brain Transcriptomic Patterns: A Topological Analysis Using Spatial Expression Networks .. 70
Zhana Kuncheva, Michelle L. Krishnan, Giovanni Montana

Integrative Analysis for Lung Adenocarcinoma Predicts Morphological Features Associated with Genetic Variations 82
Chao Wang, Hai Su, Lin Yang, Kun Huang

Identification of Discriminative Imaging Proteomics Associations in Alzheimer's Disease via a Novel Sparse Correlation Model 94
Jingwen Yan, Shannon L. Risacher, Kwangsik Nho, Andrew J. Saykin, Li Shen

Enforcing Co-expression in Multimodal Regression Framework 105
Pascal Zille, Vince D. Calhoun, Yu-Ping Wang

METHODS TO ENSURE THE REPRODUCIBILITY OF BIOMEDICAL RESEARCH

Session Introduction 117
Konrad J. Karczewski, Nicholas P. Tatonetti, Chirag J. Patel, Arjun K. Manrai, C. Titus Brown, John P.A. Ioannidis

Exploring the Reproducibility of Probabilistic Causal Molecular Network Models 120

Ariella Cohain, Aparna A. Divaraniya, Kuixi Zhu, Joseph R. Scarpa, Andrew Kasarskis, Jun Zhu, Rui Chang, Joel T. Dudley, Eric E. Schadt

Reproducible Drug Repurposing: When Similarity Does Not Suffice 132
Emre Guney

Empowering Multi-Cohort Gene Expression Analysis to Increase Reproducibility 144
Winston A. Haynes, Francesco Vallania, Charles Liu, Erika Bongen, Aurelie Tomczak, Marta Andres-Terrè, Shane Lofgren, Andrew Tam, Cole A. Deisseroth, Matthew D. Li, Timothy E.Sweeney, and Purvesh Khatri

Rabix: An Open-Source Workflow Executor Supporting Recomputability and Interoperability of Workflow Descriptions 154
Gaurav Kaushik, Sinisa Ivkovic, Janko Simonovic, Nebojsa Tijanic, Brandi Davis-Dusenbery, Deniz Kural

Data Sharing and Reproducible Clinical Genetic Testing: Successes and Challenges 166
Shan Yang, Melissa Cline, Can Zhang, Benedict Paten and Stephen E. Lincoln

PATTERNS IN BIOMEDICAL DATA-HOW DO WE FIND THEM?

Session Introduction 177
Anna Okula Basile, Anurag Verma, Marta Byrska-Bishop, Sarah P. Pendergrass, Christian Darabos, H. Lester Kirchner

Learning Attributes of Disease Progression from Trajectories of Sparse Lab Values 184
Vibhu Agarwal, Nigam H. Shah

Computer Aided Image Segmentation and Classification for Viable and Non-Viable Tumor Identification in Osteosarcoma 195
Harish Babu Arunachalam, Rashika Mishra, Bogdan Armaselu, Ovidiu Daescu, Maria Martinez, Patrick Leavey, Dinesh Rakheja, Kevin Cederberg, Anita Sengupta, Molly Ni'suilleabhain

Missing Data Imputation in the Electronic Health Record Using Deeply Learned Autoencoders 207
Brett K. Beaulieu-Jones, Jason H. Moore, The Pooled Resource Open-Access ALS Clinical Trials Consortium

A Deep Learning Approach for Cancer Detection and Relevant Gene Identification 219
Padideh Danaee, Reza Ghaeini, David Hendrix

Development and Performance of Text-Mining Algorithms to Extract Socioeconomic Status from De-Identified Electronic Health Records 230
Brittany M. Hollister, Nicole A. Restrepo, Eric Farber-Eger, Dana C. Crawford, Melinda C. Aldrich, Amy Non

Genome-Wide Interaction with Selected Type 2 Diabetes Loci Reveals Novel Loci for Type 2 Diabetes in African Americans 242
Jacob M. Keaton, Jacklyn N. Hellwege, Maggie C. Y. Ng, Nicholette D. Palmer, James S. Pankow, Myriam Fornage, James G. Wilson, Adolfo Correa, Laura J. Rasmussen-Torvik, Jerome

I. Rotter, Yii-Der I. Chen, Kent D. Taylor, Stephen S. Rich, Lynne E. Wagenknecht, Barry I. Freedman, Donald W. Bowden

Deep Motif Dashboard: Visualizing and Understanding Genomic Sequences Using Deep Neural Networks 254
Jack Lanchantin, Ritambhara Singh, Beilun Wang, Yanjun Qi

Meta-Analysis of Continuous Phenotypes Identifies a Gene Signature that Correlates with COPD Disease Status 266
Madeleine Scott, Francesco Vallania, and Purvesh Khatri

Predictive Modeling of Hospital Readmission Rates Using Electronic Medical Record-Wide Machine Learning: A Case-Study Using Mount Sinai Heart Failure Cohort 276
Khader Shameer, Kipp W. Johnson, Alexandre Yahi, Riccardo Miotto, Li Li, Doran Ricks, Jebakumar Jebakaran, Patricia Kovatch, Partho P. Sengupta, Annetine Gelijns, Alan Moskovitz, Bruce Darrow, David L. Reich, Andrew Kasarskis, Nicholas P. Tatonetti, Sean Pinney, Joel T. Dudley

Learning Parsimonious Ensembles for Unbalanced Computational Genomics Problems 288
Ana Stanescu, Gaurav Pandey

Methods for Clustering Time Series Data Acquired from Mobile Health Apps 300
Nicole Tignor, Pei Wang, Nicholas Genes, Linda Rogers, Steven G. Hershman, Erick R. Scott, Micol Zweig, Yu-Feng Yvonne Chan, Eric E. Schadt

A New Relevance Estimator for the Compilation and Visualization of Disease Patterns and Potential Drug Targets 312
Modest von Korff, Tobias Fink, Thomas Sander

Network Map of Adverse Health Effects Among Victims of Intimate Partner Violence 324
Kathleen Whiting, Larry Y. Liu, Mehmet Koyuturk, Gunnur Karakurt

Discovery of Functional and Disease Pathways by Community Detection in Protein-Protein Interaction Networks 336
Stephen J. Wilson, Angela D. Wilkins, Chih-Hsu Lin, Rhonald C. Lua, Olivier Lichtarge

PRECISION MEDICINE: DATA AND DISCOVERY FOR IMPROVED HEALTH AND THERAPY

Session Introduction 348
Alexander A. Morgan, Dana C. Crawford, Josh C. Denny, Sean D. Mooney, Bruce J. Aronow, Steven E. Brenner

Opening the Door to the Large Scale Use of Clinical Lab Measures for Association Testing: Exploring Different Methods for Defining Phenotypes 356
Christopher R. Bauer, Daniel Lavage, John Snyder, Joseph Leader, J. Matthew Mahoney, Sarah A. Pendergrass

A Powerful Method for Including Genotype Uncertainty in Tests of Hardy-Weinberg Equilibrium 368

Andrew Beck, Alexander Luedtke, Keli Liu, Nathan Tintle

Temporal Order of Disease Pairs Affects Subsequent Disease Trajectories: The Case of Diabetes and Sleep Apnea..........380
Mette Beck, David Westergaard, Leif Groop and Soren Brunak

MicroRNA-Augmented Pathways (mirAP) and Their Applications to Pathway Analysis and Disease Subtyping..........390
Diana Diaz, Michele Donato, Tin Nguyen, Sorin Draghici

Frequent Subgraph Mining of Personalized Signaling Pathway Networks Groups Patients with Frequently Dysregulated Disease Pathways and Predicts Prognosis..........402
Arda Durmaz, Tim A.D. Henderson, Douglas Brubaker, Gurkan Bebek

Human Kinases Display Mutational Hotspots at Cognate Positions Within Cancer..........414
Jonathan Gallion, Angela D. Wilkins, Olivier Lichtarge

MUSE: A Multi-locus Sampling-based Epistasis Algorithm for Quantitative Genetic Trait Prediction..........426
Dan He, Laxmi Parida

ceRNA Search Method Identified a MET-activated Subgroup Among EGFR DNA Amplified Lung Adenocarcinoma Patients..........438
Halla Kabat, Leo Tunkle, Inhan Lee

Improved Performance of Gene Set Analysis on Genome-Wide Transcriptomics Data when Using Gene Activity State Estimates..........449
Thomas Kamp, Micah Adams, Craig Disselkoen, Nathan Tintle

methylDMV: Simultaneous Detection of Differential DNA Methylation and Variability with Confounder Adjustment..........461
Pei Fen Kuan, Junyan Song, Shuyao He

Identify Cancer Driver Genes Through Shared Mendelian Disease Pathogenic Variants and Cancer Somatic Mutations..........473
Meng Ma, Changchang Wang, Benjamin Glicksberg, Eric Schadt, Shuyu D. Li, Rong Chen

Identifying Cancer Specific Metabolic Signatures Using Constraint-Based Models..........485
André Schultz, Sanket Mehta, Chenyue W. Hu, Fieke W Hoff, Terzah M. Horton, Steven M. Kornblau, Amina A. Qutub

Differential Pathway Dependency Discovery Associated with Drug Response across Cancer Cell Lines..........497
Gil Speyer, Divya Mahendra, Hai J. Tran, Jeff Kiefer, Stuart L. Schreiber, Paul A. Clemons, Harshil Dhruv, Michael Berens, Seungchan Kim

A Methylation-to-Expression Feature Model for Generating Accurate Prognostic Risk Scores and Identifying Disease Targets in Clear Cell Kidney Cancer..........509
Jeffrey A. Thompson, Carmen J. Marsit

De Novo *Mutations in Autism Implicate the Synaptic Elimination Network*..........521

Guhan Ram Venkataraman, Chloe O'Connell, Fumiko Egawa, Dorna Kashef-Haghighi, Dennis Paul Wall

Identifying Genetic Associations with Variability in Metabolic Health and Blood Count Laboratory Values: Diving into the Quantitative Traits by Leveraging Longitudinal Data from an EHR 533
Shefali S. Verma, Anastasia M. Lucas, Daniel R. Lavage, Joseph B. Leader, Raghu Metpally, Sarathbabu Krishnamurthy, Frederick Dewey, Ingrid Borecki, Alexander Lopez, John Overton, John Penn, Jeffrey Reid, Sarah A. Pendergrass, Gerda Breitwieser, Marylyn D. Ritchie

Strategies for Equitable Pharmacogenomic-Guided Warfarin Dosing Among European and African American Individuals in a Clinical Population 545
Laura Wiley, Jacob VanHouten, David Samuels, Melinda Aldrich, Dan Roden, Josh Peterson, Joshua Denny

SINGLE-CELL ANALYSIS AND MODELLING OF CELL POPULATION HETEROGENEITY

Session Introduction 557
Nikolay Samusik, Nima Aghaeepour, Sean Bendall

Production of a Preliminary Quality Control Pipeline for Single Nuclei RNA-Seq and Its Application in the Analysis of Cell Type Diversity of Post-Mortem Human Brain Neocortex 564
Brian Aevermann, Jamison Mccorrison, Pratap Venepally, Rebecca Hodge, Trygve Bakken, Jeremy Miller, Mark Novotny, Danny N. Tran, Francisco Diez-Fuertes, Lena Christiansen, Fan Zhang, Frank Steemers, Roger S. Lasken, Ed Lein, Nicholas Schork, Richard H. Scheuermann

Tracing Co-Regulatory Network Dynamics in Noisy, Single-Cell Transcriptome Trajectories 576
Pablo Cordero, Joshua M. Stuart

An Updated Debarcoding Tool for Mass Cytometry with Cell Type-Specific and Cell Sample-Specific Stringency Adjustment 588
Kristin I. Fread, William D. Strickland, Garry P. Nolan, Eli R. Zunder

Mapping Neuronal Cell Types Using Integrative Multi-Species Modeling of Human and Mouse Single Cell RNA Sequencing 599
Travis Johnson, Zachary Abrams, Yan Zhang, Kun Huang

A Spatiotemporal Model to Simulate Chemotherapy Regimens for Heterogeneous Bladder Cancer Metastases to the Lung 611
Kimberly R. Kanigel Winner, James C. Costello

Scalable Visualization for High-dimensional Single-cell Data 623
Juho Kim, Nate Russell, Jian Peng

WORKSHOPS

Harnessing Big Data for Precision Medicine: Infrastructures and Applications 635
Kun-Hsing Yu, Steven N. Hart, Rachel Goldfeder, Qiangfeng Cliff Zhang, Stephen C. J. Parker, Michael Snyder

The Training of Next Generation Data Scientists in Biomedicine......640
Lana Garmire, Stephen Gliske, Quynh C Nguyen, Jonathan H. Chen, Shamim Nemati, John D. Van Horn, Jason H. Moore, Carol Shreffler, Michelle Dunn

No-Boundary Thinking in Bioinformatics......646
Jason H. Moore, Steven F. Jennings, Casey S. Greene, Lawrence E. Hunter, Andy D. Perkins, Clarlynda Williams-Devane, Donald C. Wunsch, Zhongming Zhao, Xiuzhen Huang

Open Data for Discovery Science......649
Philip R.O. Payne, Kun Huang, Nigam H. Shah, Jessica Tenenbaum

PACIFIC SYMPOSIUM ON BIOCOMPUTING 2017

2017 marks the 22nd Pacific Symposium on Biocomputing (PSB)! Biocomputing, biomedical informatics and data science have become high profile activities in recent years. The associated emphasis on the need for policies and technologies for effective data sharing has also received quite a bit of attention. In a series of editorials, the New England Journal of Medicine (Longo & Drazen, N Engl J Med 2016; 374:276-277. 1/21/2016. DOI: 10.1056/NEJMe1516564 and Drazen, N Engl J Med 2016; 374:e24. 5/12/2016. DOI: 10.1056/NEJMe1601087) discussed the particular challenges for data sharing in the context of clinical data, generating much discussion about the proper approaches for these activities. Most notable, however, was the introduction of the phrase "research parasite" to describe "people who had nothing to do with the design and execution of the study but use another group's data for their own ends, possibly stealing from the research productivity planned by the data gatherers, or even use the data to try to disprove what the original investigators had posited." Of course, this perfectly describes (although in somewhat more negative terms than are typically used by data scientists) the motivation behind data sharing in genomics and molecular biology—and the potential value of secondary analysis of public data sets, including the occasional disproof of an incorrect scientific finding or conclusion—thus saving time, resources and potentially lives. The depiction of secondary analyses and the potential refutation of primary analyses as a negative surprised many, but points to the special challenges of sharing sensitive clinical data. Nonetheless, "I am a research parasite" became a refrain among biomedical data analysts who felt strongly that these analyses are indeed quite positive and exactly what is needed. In that vein, the PSB organizers were approached with an idea to host an award to recognize those who have made substantial scientific contributions by analyzing data collected by others. These "Research Parasite Awards" quickly attracted financial support and will be presented for the first time at this meeting. To his credit, Dr. Jeffrey Drazen, the Editor-in-Chief of the New England Journal of Medicine, has also agreed to give a talk about the challenges in clinical data sharing, particularly from his vantage point as an editor of an influential clinical journal. Dr. Drazen has no connection with the Research Parasite Awards, but is fully aware that they have been created and will be awarded at the meeting this year.

The mission of PSB is to provide a forum for the best *emerging* science in Biocomputing, providing both formal and informal mechanisms for scientific communication. PSB depends on the community to define emerging areas in biomedical computation. Its sessions are usually conceived at the previous PSB meeting as people discuss trends and opportunities for new science. The typical program includes sessions that evolve over two to three years as well as entirely new sessions. This year we revisit new dimensions of precision medicine, ranging from single cell measurements to populations.

In addition to being published by World Scientific and indexed in PubMED, the proceedings from all PSB meetings are available online at http://psb.stanford.edu/psb-online/. PSB has published more than 800 papers. These papers are often cited in journal articles and represent early contributions in emerging subfields—many times before there is an established literature in more traditional journals; for this reason, many papers have garnered hundreds of citations. The Twitter handle PSB 2017 is @PacSymBiocomp and the hashtag this year will be #psb17.

The efforts of a dedicated group of session organizers have produced an outstanding program. The sessions of PSB 2017 and their hard- working organizers are as follows:

Computational approaches to understanding the evolution of molecular function
Yana Bromberg, Matthew Hahn, and Predrag Radivojac

Imaging Genomics
Li Shen and Lee Cooper

Methods to Ensure the Reproducibility of Biomedical Research
Konrad J. Karczewski, Nicholas P. Tatonetti, Chirag J. Patel, Arjun K. Manrai, C. Titus Brown, and John P.A. Ioannidis

Patterns in Biomedical Data - How do we find them?
Anurag Verma, Anna Okula Basile, Marta Byrska-Bishop, Christian Darabos, H. Lester Kirchner, and Sarah Pendergrass

Precision medicine: from genotypes and molecular phenotypes towards improved health and therapies
Bruce Aronow, Steven E. Brenner, Dana C. Crawford, Joshua C. Denny, Sean D. Mooney, and Alexander A. Morgan

Single-cell analysis and modelling of cell population heterogeneity
Nikolay Samusik, Sean Bendall, and Nima Aghaeepour

We are also pleased to present four workshops in which investigators with a common interest come together to exchange results and new ideas in a format that is more informal than the peer-reviewed sessions. For this year, the workshops and their organizers are:

Harnessing Big Data for Precision Medicine: Infrastructures and Applications
Kun-Hsing Yu, Steven Hart, Rachel Goldfeder, Qiangfeng Cliff Zhang, Stephen Parker, and Michael Snyder

The Making of Next Generation Data Scientists in Biomedicine
Lana Garmire, Shamim Nemati, John D. Van Horn, Jason Moore, Carole Shreffler and Michelle Dunn

No-Boundary Thinking in Bioinformatics
Xiuzhen Huang and Jason H. Moore

Open Data for Discovery Science
Philip R.O. Payne, Kun Huang, Nigam H. Shah, and Jessica Tenenbaum

We thank our keynote speakers Neil Risch (Science keynote) and David Magnus (Ethical, Legal and Social Implications keynote). We also thank Jeffrey Drazen for his talk.

Tiffany Murray has managed the peer review process and assembly of the proceedings since 2003, and also plays a key role in many other aspects of the meeting. We are grateful for the support of the The Penn Institute for Biomedical Informatics; Rxight Pharmacogenetics Program; and the Institute for Computational Biology, a collaborative effort of Case Western Reserve University, the Cleveland Clinic Foundation, and University Hospitals for their support of PSB 2017. We also thank the National Institutes of Health[1] and the International Society for Computational Biology (ISCB) for travel grant support. The research parasite awards benefit by support from: Jeff Stibel, GigaScience (Biomed Central), Nature Genetics, Scientific Data (Nature), the Gordon & Betty Moore Foundation, the Arnold Foundation, Tim Triche Jr., and Casey Greene.

We are particularly grateful to the onsite PSB staff Al Conde, Ryan Whaley, Georgia Hansen, BJ Morrison-McKay, Cynthia Paulazzo, Jackson Miller, Kasey Miller, and Paul Murray for their assistance. We also acknowledge the many busy researchers who reviewed the submitted manuscripts on a very tight schedule. The partial list following this preface does not include many who wished to remain anonymous, and of course we apologize to any who may have been left out by mistake.

[1] Funding for this conference was made possible (in part) by Grant # 5 R13 LM006766 – 20 from the National Library of Medicine. The views expressed in written conference materials or publications, and by speakers and moderators, does not necessarily reflect the official policies of the Department of Health and Human Services; nor does mention by trade names, commercial practices, or organizations imply endorsement by the U.S. Government.

We look forward to a great meeting once again. Aloha!

Pacific Symposium on Biocomputing Co-Chairs,
October 15, 2016

Russ B. Altman
Departments of Bioengineering, Genetics & Medicine, Stanford University

A. Keith Dunker
Department of Biochemistry and Molecular Biology, Indiana University School of Medicine

Lawrence Hunter
Department of Pharmacology, University of Colorado Health Sciences Center

Marylyn D. Ritchie
Department of Biomedical and Translational Informatics, Geisinger Health System

Teri E. Klein
Department of Genetics, Stanford University

Thanks to the reviewers...

Finally, we wish to thank the scores of reviewers. PSB aims for every paper in this volume to be reviewed by three independent referees. Since there is a large volume of submitted papers, paper reviews require a great deal of work from many people. We are grateful to all of you listed below and to anyone whose name we may have accidentally omitted or who wished to remain anonymous.

Nima Aghaeepour
Harindra Arachchi
Mohammad Arbabshirani
Bruce Aronow
Chloe-Agathe Azencott
Anna Basile
Kayhan Batmanghelich
Chris Bauer
Asa Ben-Hur
Sean Bendall
Tyler Burns
William Bush
Mariusz Butkiewicz
Fabien Campagne
Kevin Chen
Shuo Chen
Yin Hoon Chew
Moo Chung
Jessica Cooke Bailey
James Costello
Christian Darabos
Christophe Dessimoz
Lei Du
Todd Edwards
Arne Elofsson
Niclas Eriksson
Tilman Flock
Yi Gao
Brice Gaudilliere
Tian Ge
Jeff Gentry
Olivier Gevaert
Jesse Gillis
Anthony Gitter
Rachel Goldfeder
Casey Greene
Jake Hall
Xiaoke Hao
Yun Hao
Imran Haque
Jaroslaw Harezlak
Blanca Himes
Isaac Hodes
Emily Holzinger
Ting Hu
Junzhou Huang
Jake Hughey
Shaun Jackman
Ola Jacunski
Shuiwang Ji
Kipp Johnson
Konrad Karczewski
Jonathan Karr
Dokyoon Kim
Sungeun Kim
H. Lester Kirchner
Jun Kong
Linglong Kong
Mickey Kosloff
Irina Kufareva
Willaim Lai
Chirag Lakhani
Nicholas Larson
Roman Laskowski
Li Li
Ruowang Li
Nita Limdi
Jingyu Liu
Kefei Liu
Liang Liu
Colton Lloyd
Tal Lorberbaum
Jose Lugo-Martinez
Gang Luo
Emily Mallory
Elisabetta Manduchi
Arjun Manrai
Tyler Massaro
Brett McKinney
Andrew Michaels
Marghoob Mohiyuddin
Jason Moore
Yves Moreau
Spencer Muse
Kelly Nudelman
Randy Olson
Casey Overby
Bernhard Palsson
Gaurav Pandey
Chirag Patel
Vikas Pejaver
Sarah Pendergrass
Hanchuan Peng
Minoli Perera
Abhishek Pratap
Wei-Qi Qei
Marylyn Ritchie
Igor Rogozin
Joe Romano
Mert Sabuncu
Satya Sahoo
Erin Simonds
Marina Sirota
Johannes Soding
Sudeep Srivastava
Jason Stein
Timothy Sweeney
Suzanne Tamang
Haixu Tang
Nicholas Tatonetti
Shaolei Teng
Gregg Thomas
William Thompson
Stefano Toppo
Ryan Urbanowicz
Giorgio Valentini
Elise Valkanas
Sofie Van Gassen
Fabio Vandin
Shankar Vembu
Anurag Verma
Yogasudha Veturi
Bjarni Vihjalmsson
Susann Vorberg
Pei Wang
Qianghu Wang
Marquitta White
Chunlei Wu
Rong Xu
Ya Yang
Dmeliha Yetisgen
Pooya Zakeri
Daoqiang Zhang
Xiaobo Zhou
Chengsheng Zhu

COMPUTATIONAL APPROACHES TO UNDERSTANDING THE EVOLUTION OF MOLECULAR FUNCTION

Yana Bromberg

Department of Biochemistry and Microbiology, Rutgers University
New Brunswick, New Jersey, U.S.A.

Matthew W. Hahn[†], Predrag Radivojac

[†]Department of Biology, Indiana University
Department of Computer Science and Informatics, Indiana University
Bloomington, Indiana, U.S.A.

1. Introduction

Understanding the function of biological macromolecules and their interactions is a grand challenge of modern biology, and a key foundation for biomedical research.[1,2] It is now evident that the function of these molecules, in isolation or in groups, can be productively studied in the context of evolution.[3,4] Therefore, understanding how these molecules and their functions evolve is an important step in understanding the specific events that lead to observable changes in molecular and biological processes.

With the advent of high-throughput technologies and the rapid accumulation of molecular data over the past several decades, the evolution of molecular function can be systematically studied at multiple levels. This includes the evolution of protein structure, 3D organization and dynamics, protein and gene expression, as well as the higher-level organization of function contained within pathways.[5–11] New experiments using the latest gene-editing technologies (such as CRISPR-Cas9) have also made it possible to directly test hypotheses about function in almost any organism.[12] Combining these data with theory and computational tools taken from evolutionary biology and related fields has led to an explosion in the study of how function evolves.

2. Overview of Contributions

Our session includes four accepted papers covering a variety of the subjects in this field. The papers address biological questions from metabolic processes to the evolution of duplicated genes; they use computational methods ranging from learning functions on biological networks to the optimal way to choose clustering parameters to identify homologs. Bowerman et al. investigate a set of about one hundred fully sequenced bacterial species mapped onto a space of metabolic variants via a literature search. They subsequently use these data to learn metabolic signatures among these species, an approach that can ultimately lead to a predictive system of metabolic potential for any bacterial species. Cao and Cowen study protein function transfer within a single species and ask under what conditions it leads to accurate prediction. Several sequence, network, and evolutionary features were examined to conclude that the level of sequence divergence is the major determinant of accurate function transfer among within-species paralogs in yeast. The paper relates to several earlier studies addressing evolutionary relationships and functional similarity.[13–17] Wang et al. present and evaluate a new approach for protein function prediction. Their method is based on amino acid sequences and protein-protein interaction networks over multiple species, integrated into a single heterogeneous network. Network integration is often challenging to formalize considering practical problems such as missing data, sample selection bias, and noise in available protein-protein interactions. Nevertheless, the approach showed good performance upon data integration and provided the insight that the combination of data sources contributed to increased accuracy. Finally, Wiwie and Röttger study the behavior and performance of several clustering algorithms in the context of detecting protein families in similarity graphs. Protein clustering is difficult owing to the unequal sizes of homologous families and the sensitivity of clusters to the parameters of the algorithm. They show that the original data can, in principle, be used to predict clustering performance but also highlight difficulties in finding optimal clustering parameters.

References

1. R. Rentzsch and C. A. Orengo, *Trends Biotechnol* **27**, 210 (2009).
2. P. Radivojac *et al.*, *Nat Methods* **10**, 221 (2013).
3. J. A. Eisen, *Genome Res* **8**, 163 (1998).
4. M. Pellegrini *et al.*, *Proc Natl Acad Sci U S A* **96**, 4285 (1999).
5. N. V. Grishin, *J Struct Biol* **134**, 167 (2001).
6. E. V. Koonin, *Annu Rev Genet* **39**, 309 (2005).
7. D. A. Drummond *et al.*, *Proc Natl Acad Sci U S A* **102**, 14338 (2005).
8. C. Pal *et al.*, *Nat Rev Genet* **7**, 337 (2006).
9. M. E. Peterson *et al.*, *Protein Sci* **18**, 1306 (2009).
10. W. Qian *et al.*, *Proc Natl Acad Sci U S A* **108**, 8725 (2011).
11. C. Park *et al.*, *Proc Natl Acad Sci U S A* **110**, E678 (2013).
12. J. A. Doudna and E. Charpentier, *Science* **346**, p. 1258096 (2014).
13. S. Mika and B. Rost, *PLoS Comput Biol* **2**, p. e79 (2006).
14. R. A. Studer and M. Robinson-Rechavi, *Trends Genet* **25**, 210 (2009).
15. N. L. Nehrt *et al.*, *PLoS Comput Biol* **7**, p. e1002073 (2011).
16. A. M. Altenhoff *et al.*, *PLoS Comput Biol* **8**, p. e1002514 (2012).
17. G. Plata and D. Vitkup, *Nucleic Acids Res* **42**, 2405 (2014).

IDENTIFICATION AND ANALYSIS OF BACTERIAL GENOMIC METABOLIC SIGNATURES

NATHANIEL BOWERMAN
Department of Biology, Hope College, 35 E 12th St,
Holland, MI 49423 USA
nathaniel.bowerman@hope.edu

NATHAN TINTLE
Department of Mathematics and Statistics, Dordt College, 498 4th Ave NE
Sioux Center, IA 51250, USA
nathan.tintle@dordt.edu

MATTHEW DEJONGH
Department of Computer Science, Hope College, 27 Graves Place,
Holland, MI 49423 USA
dejongh@hope.edu

AARON A. BEST*
Department of Biology, Hope College, 35 E 12th St,
Holland, MI 49423 USA
best@hope.edu

With continued rapid growth in the number and quality of fully sequenced and accurately annotated bacterial genomes, we have unprecedented opportunities to understand metabolic diversity. We selected 101 diverse and representative completely sequenced bacteria and implemented a manual curation effort to identify 846 unique metabolic variants present in these bacteria. The presence or absence of these variants act as a metabolic signature for each of the bacteria, which can then be used to understand similarities and differences between and across bacterial groups. We propose a novel and robust method of summarizing metabolic diversity using metabolic signatures and use this method to generate a metabolic tree, clustering metabolically similar organisms. Resulting analysis of the metabolic tree confirms strong associations with well-established biological results along with direct insight into particular metabolic variants which are most predictive of metabolic diversity. The positive results of this manual curation effort and novel method development suggest that future work is needed to further expand the set of bacteria to which this approach is applied and use the resulting tree to test broad questions about metabolic diversity and complexity across the bacterial tree of life.

* To whom correspondence should be addressed.

1. Introduction

The metabolism of an organism relies on thousands of biochemical reactions, which comprise a network that allows the cell to grow, reproduce, and respond to changing environmental conditions. The set of metabolic reactions are defined by the genes the organism carries and dictate the metabolic properties of the organism. Developing an understanding of the metabolic reactions possible by an organism begins to coalesce into a coherent picture of the metabolic capability of the cell. With thousands of annotated genome sequences of microbial organisms available, it is now possible to analyze not only the metabolic properties of individual organisms, but also the patterns that are seen in metabolic networks across organisms. This includes analyses of the evolution of specific metabolic pathways [*e.g.*, 1,2], analyses based on network topology and properties [*e.g.*, 3–6], analyses of simulated metabolic networks [*e.g.*, 7,8], and combinations of flux balance analysis based modeling of metabolic networks within the context of phylogenies [9–11]. Such analyses can lead to a deeper understanding of the metabolic landscape represented by microbial diversity. Further, sequence-based taxonomic surveys and metagenomic analyses of diverse environments are beginning to allow the systematic exploration of relationships between microbial diversity, functional diversity and environment [12–16].

Accurate annotation of sequenced genomes is foundational to downstream analyses of genomes and metagenome communities. We have reviewed [17] the rapid and accurate subsystem approach to genome annotation implemented in the SEED [18] and RAST [19] frameworks. Achieving highly accurate automated annotations of genomes in RAST is predicated upon a core set of manually curated subsystems in which an expert has catalogued the functional elements of a biological process (*e.g.*, a metabolic pathway) and assigned genes to those functional elements for a large set of sequenced microbes. This ensures high quality annotation of each subsystem and the propagation of knowledge captured in the subsystem to all existing and newly sequenced genomes. One outcome of the subsystems approach is the declaration and discovery of metabolic variants, which are defined as different forms or combinations of forms of a functioning metabolic process [17,20,21]. By identifying patterns of genes comprising a variant, one can quickly assign an organism to a particular variant based on the pattern of genes found during the annotation process. Thus, an organism is assigned a variant code for each subsystem, which yields an abstraction of the metabolic capabilities and the forms of those metabolic functions. Further, a catalogue of functional variants that exist for a particular subsystem captures the diversity with which that biological process is performed among sequenced microbes. Such a catalogue represents a rich data set through which we can gain insight into the complexity and diversity of microbial metabolism.

To enable these types of inquiries and to provide consistent descriptions of metabolic variants among sequenced microbes, we selected a representative set of 101 microbial genomes that were used to manually define and annotate metabolic variants in 139 distinct subsystems covering much of known metabolism. We used this resource to (i.) generate a metabolic signature for each of the 101 organisms comprised of assigned variants for each of the 139 subsystems and (ii.) conduct comparative analyses of metabolic signatures of this diverse set of microbes. These variants and their definitions yield a set of high-confidence metabolic subsystems that have been used to aid the automated generation of genome-scale metabolic reconstructions [22], provide a framework for automated recognition and propagation of variants to newly sequenced genomes, and allow for comparative studies of metabolic variation observed in sequenced microbes.

2. Results

2.1 *Defining Metabolic Variants for Sequenced Bacteria*

A metabolic variant can be described as a particular version of a metabolic process performed by an organism [21]. We will use the synthesis of isoprenoids (terpenoids) to illustrate the concept of metabolic variants and how particular variants are assigned to an organism. Isoprenoids (*e.g.*, chlorophyll and cholesterol) are found in all organisms and are essential to survival. Key isoprenoid precursors, isopentenyl diphosphate (IPP) and dimethylallyl diphosphate (DMAPP) are produced via two known biosynthetic pathways, the so-called mevalonate and non-mevalonate (DOXP) pathways [1]. The reactions in each pathway are catalyzed by non-homologous proteins, and represent two distinct routes to IPP and DMAPP for organisms. In considering the simplest case of defining metabolic variants for this metabolic process, each of these routes represent separate variants – alternative ways to accomplish the same function of producing precursors to isoprenoids. A third variant exists in the case of an organism containing the necessary genes for both of these pathways. A fourth variant indicates absence of this function through known metabolic pathways in an organism. For each variant (defined in this case as A, B, C and -1, respectively), the possible patterns of metabolic steps involved in each variant is generated, and a brief verbal description of the variant is given. Assignment of any one organism to a known variant of the pathway is accomplished by identifying genes in the organism's genome that encode functions corresponding to the area of metabolism and matching the pattern of metabolic steps the organism is predicted to be capable of to one of the defined variants (see Supplemental Figure 1 for additional details).

We have implemented the approach of identifying variants, defining variants, and assigning variants to organisms in the framework of SEED subsystems [18]. This represents a significant, multi-year manual curation effort on the part of SEED annotators through the capture of known metabolic diversity described in the literature and the analysis of patterns seen in sequenced microbial genomes. We chose a set of 101 bacterial genomes, representing 14 bacterial divisions, and 139 subsystems in the SEED that maximized our coverage of metabolism represented in major metabolic databases (*e.g.*, KEGG) and that facilitated the automated generation of metabolic models for bacteria [22]. We characterized a total of 846 metabolic variants in these subsystems that our set of organisms are capable of based on known information of each subsystem and the annotated function of genes in each genome. The outcome of this curation effort is a metabolic variant catalogue comprising descriptions of naturally occurring variations of central and intermediate metabolism for a phylogenetically diverse group of bacteria. Supplemental Figure 2 and Supplemental Files 1-4 give detailed information on the organisms and variants selected and defined.

2.2 *Analyses of Bacterial Metabolic Signatures*

In order to gain a more thorough understanding of metabolic diversity and how metabolic functions are distributed throughout Bacteria, we devised a measure of the metabolic distance, D_{FM}, between two organisms based on the curated metabolic variant catalogue. For a given organism *i*, it is possible to summarize the metabolic capabilities as a binary vector, v_i, of 846 0's and 1's, representing the absence and presence, respectively, of each of the 846 metabolic variants. In effect, *v* is a metabolic signature (or barcode) describing the metabolic capabilities of an organism. D_{FM} measures the metabolic distance between two given organisms *i* and *j* by comparing the similarity of v_i and v_j to the likelihood of observing the similarity between the two vectors by chance. We utilized complete linkage hierarchical clustering of all pairwise D_{FM} of organisms in our dataset to produce a dendrogram summarizing the relationships of the organisms based on metabolic distances (Figure 1). We used a false discovery rate (FDR) of 1×10^{-15} to identify 5 distinct clusters of organisms (Clusters A through E in Figure 1). Each cluster represents a group of organisms with highly similar metabolic signatures. To assess the face validity of the resulting metabolic signature tree, we sought to confirm that the ordering seen in the tree met reasonable biological expectations. For instance, one would expect that closely related organism pairs are likely to be closely paired on the dendrogram – *E. coli* and *Salmonella* are nearest neighbors in the tree as are two representatives of the genus *Shewanella*. Furthermore, the four oxygenic photosynthetic organisms in the set form a tight cluster (FDR $<1 \times 10^{-60}$, Supplemental Figure 3a, organism names colored green). These observations, and many others not detailed here (for example, Supplemental Figures 3b and 3c), indicate that the metabolic distance metric reveals biologically meaningful patterns and gave us confidence that we could use the tree to address additional biological questions of interest.

2.3 *Contribution of Organism Characteristics to Bacterial Metabolic Signatures*

To provide a quantitative estimate of the ability of organism characteristics to explain the clustering observed in the metabolic signature tree, we produced a data set capturing 19 characteristics for each of the 101 organisms, covering attributes such as phylogenetic grouping, environment classification, and oxygen utilization (Supplemental File 1). We performed a multiple regression analysis, using the 19 phenotypic characteristics to predict metabolic distance. The variables in our data set were able to explain 50% of the variance of metabolic distance ($r^2 = 0.50$). The top four characteristics contributing to the clustering are genome size, metabolic mode, host association, and ability to survive in an intracellular environment, uniquely explaining 19.7%, 9.6%, 7.5% and 7.2% of the overall variation in metabolic distance, respectively. All other characteristics contribute to ~5% or less of the overall r^2. Phylogenetic distance ranked 11th of the 19 characteristics, indicating that only a small fraction of the metabolic distance variance could be attributed to phylogeny. A phylogenetic tree of the organisms in this study annotated with metabolic signature cluster membership shows the clear mixing of related organisms throughout the 5 clusters (Supplemental Figure 2). A follow-up analysis which removed 14 organisms with small genomes (Cluster B), showed that there is a slight decrease in the ability to explain the overall variation in metabolic distance ($r^2 = 0.48$) with the 19 phenotypes combined, and less predictive

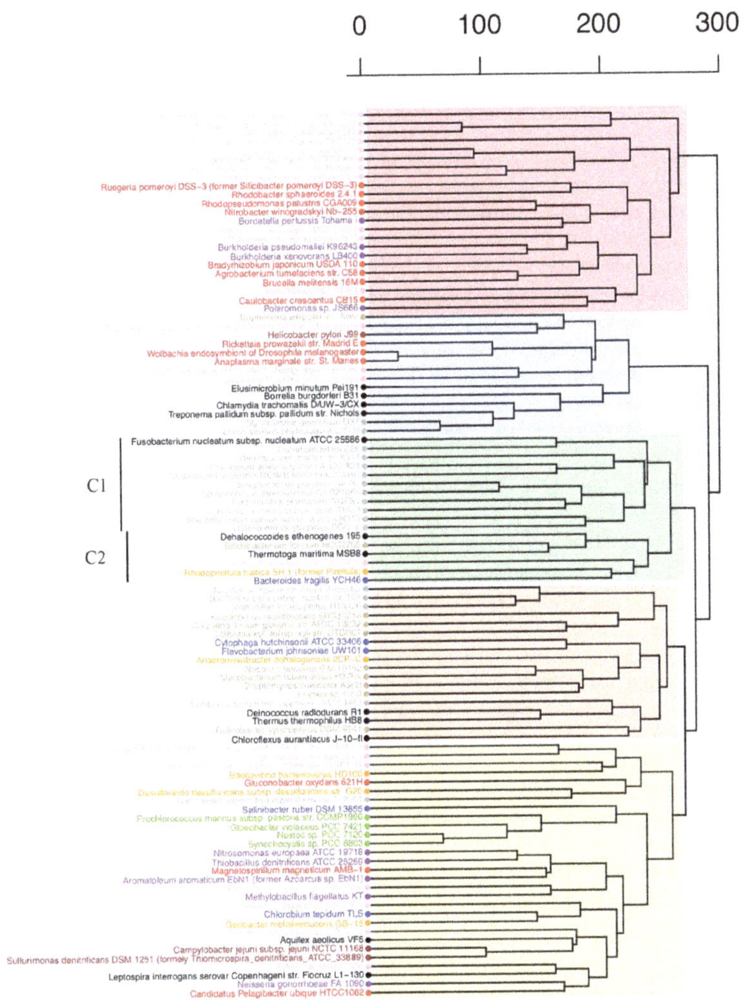

Figure 1. Metabolic Signature Tree from complete linkage hierarchical clustering of D_{FM} of organisms. Five clusters corresponding to an FDR of 1 x 10^{-15} are highlighted by shading – Clusters A-E; pink, blue, green, orange and yellow, respectively. Subclusters C1 and C2 are indicated by black bars. Organism names are colored according to phylogenetic classification: Actionomycetes, Tan; Firmicutes, Gray; Cyanobacteria, Green; Bacteroides/Chloribi, Blue; Other, Black; Proteobacteria: Alpha, Red; Beta, Purple; Delta, Orange; Epsilon, Brown; Gamma, Pink.

ability of genome size (from 19.7% to 6.7%). Full results are provided in Supplemental Figures 4a and 4b.

2.4 *Specific Phenotypes Associated with Individual Clusters*

In addition to characterizing the influence of phenotype on the global topology of the metabolic tree, it is possible to associate specific phenotypic characteristics with individual clusters of organisms in the metabolic tree. We assessed the distribution of each phenotypic characteristic within a cluster and compared this to the distribution of that phenotypic character in the other clusters to yield a statistical measure of the differential distribution of any one phenotypic trait among clusters (see Methods). Each of the clusters is characterized by a particular set of phenotypes as summarized in Table 1 that are over or underrepresented at a conservative measure of statistical confidence ($p < 0.0006$). As expected, the phenotypic characters with the lowest p-values for each cluster correspond to initial observations seen with the overlay of phenotypic characters on the metabolic tree, while providing more specificity to the observations and highlighting characters that may not be otherwise apparent. Cluster A consists completely of Gram negative organisms that also tend to have large genomes (5.2 Mb vs 3.1 Mb average for entire dataset). All organisms are phylogenetically related, being members of the α, β, and γ Proteobacteria. However, these taxonomic groups are not identified as statistically significant due to the broad distribution of other members of these taxonomic groups throughout the clusters (*i.e*, B and E). This result is consistent with the diverse habitats and lifestyles associated with Proteobacteria. Cluster B contains organisms that tend to have small genome sizes, are classified as intracellular and obligate host associated, and have a low GC%. Obligate intracellular parasites tend to have smaller genomes as they require fewer genes due to obtaining resources from the host cell and smaller genomes tend to have lower GC content to facilitate evolution through an increased mutation rate. Cluster C consists of organisms that tend to be in the phylum Firmicutes, families Bacillales or Lactobacillales, are Gram positive, and are anaerobic. Cluster D contains an over-representation of Actinomycetes, Gram positive bacteria, and sporulating bacteria. Cluster E contains many phylogenetically unrelated organisms, a majority of organisms that have preferred metabolic modes other than chemoheterotrophy, and also contains a disproportionate number of Gram negative bacteria.

2.5 *Metabolic Variants Associated with Specific Clusters*

As a complementary approach to exploring organism characteristics associated with specific clusters, it is also possible to explore whether particular metabolic variants are over- or under-represented in the specific metabolic clusters. As an example, we observed that Cluster C could be divided into two subgroups, C1 and C2. The organisms in Cluster C1 are low-GC Gram positive organisms in the phylum Firmicutes with the exception of *Fusobacterium*; the subcluster can be further divided by the oxygen requirement characteristic – the organisms in class Clostridia and *Fusobacterium* are all obligate anaerobes, whereas the organisms in class Bacilli are facultative (Figure 1, Supplemental File 1). We hypothesized that there should be specific metabolic variants (likely related to respiratory systems) that would distinguish these two groups. To investigate this and similar hypotheses, we used an approach that compared the frequencies of metabolic variants in two groups of organisms (*e.g.*, subgroups of Cluster C1) to highlight those variants that were the most different between the groups (see Methods for details). In the case of Cluster C1, there

Table 1. Over- and under-representation of characteristics by cluster

Cluster	Characteristic*	Present Inside Cluster	Present Outside Cluster	p-value
A	Genome Size	Mean = 5.2	Mean = 3.1	5.88×10^{-6}
	Gram Stain Negative	23/23 (100%)	43/78 (55%)	1.22×10^{-5}
	Gram Stain Positive	0/23 (0%)	26/78 (33%)	6.79×10^{-4}
B	Genome Size	Mean = 1.1	Mean = 4	3.71×10^{-24}
	Intracellular Survival - Obligate Intracellular	8/14 (57%)	0/87 (0%)	1.49×10^{-8}
	Free Living/Host Associated - Obligate Host Association	12/14 (86%)	10/87 (11%)	3.96×10^{-8}
	Host Type - Arthropod/Insect	8/14 (57%)	2/87 (2%)	5.94×10^{-7}
	GC Content	35.5	51.6	1.47×10^{-5}
	Free Living/Host Associated - Free Living	0/14 (0%)	50/87 (57%)	4.35×10^{-5}
	Intracellular Survival - Not Applicable	0/14 (0%)	50/87 (57%)	4.35×10^{-5}
	Habitat Outside Host - Soil	0/14 (0%)	38/87 (44%)	8.39×10^{-4}
C	Taxonomic Class - Mixed Firmicutes	10/17 (59%)	7/84 (8%)	1.17×10^{-5}
	Gram Stain - Positive	12/17 (71%)	14/84 (17%)	2.25×10^{-5}
	Oxygen Requirement - Aerobe	1/17 (6%)	47/84 (56%)	1.09×10^{-4}
	GC Content	Mean = 39.5	Mean = 51.4	1.24×10^{-4}
	Oxygen Requirement - Anaerobe	9/17 (53%)	8/84 (1%)	1.44×10^{-4}
	Gram Stain - Negative	4/17 (24%)	62/84 (74%)	1.47×10^{-4}
	Bacillales, Lactobacillales	6/17 (35%)	3/84 (4%)	5.98×10^{-4}
D	Taxonomic Class - Actinomycetes	8/18 (44%)	2/83 (2%)	7.96×10^{-6}
	Gram Stain - Positive	12/18 (67%)	14/83 (17%)	5.73×10^{-5}
	Sporulation - Sporulating	8/18 (44%)	5/83 (6%)	1.67×10^{-4}
	Gram Stain - Negative	5/18 (28%)	61/83 (73%)	5.86×10^{-4}
	Sporulation - Nonsporulating	10/18 (56%)	76/83 (92%)	6.77×10^{-4}
E	Prefered Metabolic Mode - Chemoorganoheterotroph	14/29 (48%)	69/72 (96%)	1.29×10^{-7}
	Prefered Metabolic Mode - Photolithoautotroph	6/29 (21%)	0/72 (0%)	3.75×10^{-4}
	Gram Stain - Positive	1/29 (3%)	25/72 (35%)	7.92×10^{-4}

*Genome Size given as average number of megabases in group; GC Content given as average percentage in group

are 12 metabolic variants that are unequally distributed between anaerobic and facultative organisms (p-value ≤ 0.05) within the cluster (Supplemental File 5). These 12 variants represent 7 unique subsystems associated with the synthesis of cofactors, vitamins, and isoprenoids. Three of these subsystems are associated with respiratory functions (heme and siroheme biosynthesis, sulfur related anaerobic respiratory reductases, and sodium translocating oxidoreductases). There is differential distribution between the anaerobic (4 of 5) and facultative (0 of 6) cluster members for the presence of sulfur reductases. Likewise, 4 of 5 anaerobic cluster members have an operon of *rnf* like genes encoding putative electron transport complexes associated with nitrogen fixation, whereas none of the facultative cluster members have this operon, which is consistent with the classical differentiation of *Clostridia* from *Bacilli* organisms in the low-GC Gram positive group.

3. Discussion

We have described a novel approach to examining the metabolic relationships among bacterial genomes that focuses on the collection of metabolic variants associated with an organism. The vector of metabolic variants succinctly describes the organism's metabolic capabilities and allows for statistical comparison of vectors between organisms that is scalable to thousands of genomes. In the current study, we have provided a proof of concept with a phylogenetically diverse set of 101 bacterial genomes, comprising 846 variants and covering much of known metabolism. The variant definitions are the result of a targeted manual curation effort in the framework of the SEED database [18], which breaks down bacterial metabolism into subsystems (defined as collections of functional roles necessary to perform a cellular function). In this study, 139 subsystems were individually examined to define the possible metabolic variants. The outcome of the manual curation effort is a set of curated metabolic variants that can be rapidly assigned to bacterial genomes and used to compare the metabolic capabilities present in the genomes.

Many of the approaches to understanding the breadth, conservation and evolution of metabolic networks found in the bacterial domain have focused on properties of network architecture such as scale, network path length, network motifs, centrality, modularity and connectedness [3,4,23]. Common themes are observed in that metabolic networks have been shown to be scale-free and highly modular for most organisms. It has been shown that the complexity of a metabolic network can be associated with particular lifestyles/habitats. For example, obligate symbionts that experience relatively stable environments have less complex networks than organisms that are free-living and exposed to many environments. These approaches are highly granular in that they connect networks on the level of individual reactions, compounds and enzymes. An extension to network based approaches was introduced by Mazurie *et al.* [5] that compares higher level functional units called networks of interacting pathways. These were used to classify organisms into phenotypic categories. They observed similar trends with respect to the nature of the networks as seen with other network-based approaches and were able to assign functional pathways to organisms of particular phenotypes. For instance, free-living and host-associated organisms differed with respect to frequency of observed carbohydrate and energy metabolism pathways; motile and non-motile organisms differed with respect to xenobiotic degradation pathways. More recently, Pearcy *et al.* [6] introduced a method that produced vectors for an organism whose elements described individual network motifs. They analyzed 3 and 4 node motifs that are abstractions of specific compound and reaction connections and identified network motifs that were enriched for organisms with different habitats/lifestyles, such as aerobic/facultative vs. anaerobic. By looking at the reactions and compounds that made up the enriched motifs, it was possible to identify specific metabolites associated with the different lifestyles. Patterns such as these supported the assertion that environmental conditions shape the properties of metabolic networks that occur in organisms. In a departure from analyzing network properties, Poot-Hernandez *et al.* [24] calculated linear enzymatic step sequences (ESS) found in metabolic maps in KEGG and defined core and peripheral metabolic pathways for 40 gamma proteobacteria species. An analysis of the relationships of ESS vectors among organisms was not conducted. Mithani *et al.* [4] analyzed the presence/absence of enzymatic reactions in pathways of *Pseudomonas* species based on KEGG map reaction mining. They found interesting patterns of gains and losses associated with niche specific adaptations to host association. Their approach is limited by the restriction to KEGG maps and boundary effects (reactions that appear in more than one map do not get connected). Further, the authors noted that other information such as genome context could improve understanding of evolutionary processes. The approach that we describe here is fundamentally different than those employed to date in that the unit being analyzed – variants associated with an organism – is non-network based; implicitly incorporates genome context, paralogs, isoenzymes and non-orthologous replacements through manual curation; and allows for coverage of metabolic capabilities across the modular nature of networks and their representation as disconnected metabolic maps. Further, each variant represents a functioning biological process, allowing the succinct assertion of organism capabilities (both positive and negative attribution). The analysis of variant vectors and the patterns observed therein give rise to clusters of metabolic forms comprised of the organisms and their individual variants. It is then possible to attribute the influence of phenotypic characters and phylogenetic relationships to these clusters through standard statistical approaches. It would be instructive to map individual variants to data types analyzed previously (e.g., networks of interacting pathways, individual network motifs, and ESS) to enable systematic comparison of each of these approaches to the variant approach.

We identified five main clusters of metabolically related organisms in our analysis (A-E in Fig. 1), each of which share some phenotypic traits (Table 1). We also described a complementary approach to evaluate which variants are most differentially distributed between clusters on the tree. These analyses yield patterns that are consistent with the approaches mentioned above. For instance, Cluster B is comprised of organisms that are host-associated and found in relatively stable environments; the 144 variants that are significantly differentially distributed ($p < 0.05$) include the absence of functions in amino acid, purine and pyrimidine, and vitamin/co-factor biosynthesis pathways (Supplemental File 5). There are other cases where there are hints at what drives the members of a cluster together in metabolic space (*e.g.*, *Neisseria*, *Pelagibacter*, *Xylella*, *Leptospira* – amino acid usage; *Gluconobacter*, *Desulfovibrio*, *Carboxydothermus* – extreme environments), but the current sampling of 101 organisms limits the statistical analysis of small clusters such as these.

These types of problems will become tractable with the inclusion of new genomes that begin to fill out metabolic signature clusters. Importantly, the fundamental structure of the dendrogram will not change as new genomes are added given a constant set of defined variants (*e.g.*, Cluster B will continue to contain organisms with small genomes/symbionts, Cluster C will contain most low GC Gram positive organisms, Cluster D will contain high GC organisms, and Cluster E will likely expand and subdivide as representation of metabolically diverse organisms increases). Organisms that do not follow these expectations may yield insight into novel combinations of metabolic capabilities; the current metabolic clusters represent a framework of hypotheses about relationships between suites of metabolic variants associated with any one organism. In contrast, as additional metabolic variants are identified, curated and assigned, the nature of the metabolic clusters may change. In short, as more well-annotated genomes are included, the statistical power for this type of analysis increases, enhancing our ability to examine the metabolic relationships between organisms and what factors impact these metabolic commonalities.

The proof of concept described in this work serves as a foundation for identifying metabolic signatures for all sequenced bacteria and associating those signatures with specific organism characteristics and metabolic variants. Analyses of correlations between metabolic variants observed across bacterial life will enhance our understanding of the nature of the metabolic space occupied by diverse organisms.

4. Methods

4.1 *Organisms and Features*

The 101 organisms chosen were representatives of 14 phylogenetic divisions of eubacteria (Supp Fig. 2), which provides a reasonable coverage of sequenced microbial diversity with complete genomes. Each of the 101 organisms were classified on 19 different phenotypic features based on information already present in the SEED and via literature review. The features considered here and summary statistics are provided as Supplemental File 1. In order to generate maximum likelihood phylogenetic distances for each pair of organisms, we selected a representative 16S rRNA sequence of each organism from the Silva SSU Reference Set Release 106 using the ARB environment [25]. RaxML 7.0.4 [26] was then used to generate a set of maximum likelihood pairwise distances. Pairwise phylogenetic distances are included as Supplemental File 2.

4.2 *Creating a Metabolic Distance Measure*

We calculated a measure of metabolic distance, D_{FM}, between organisms based on the vector v_i, where i is the ith organism, of 0's and 1's, indicating the presence/absence of the 846 subsystem variants. In general, the metabolic distance between organisms i and j, will be a function that measures the dissimilarity of vectors v_i and v_j. While there are numerous options for measuring dissimilarity or similarity between two vectors (e.g., Euclidean distance, Pearson correlation), we chose to use a novel method based on Fisher's exact test because of its robustness to the widely varying numbers of 0's and 1's observed in vector v, along with its ability to directly integrate a measure of statistical confidence into the distance measure, making D_{FM} an indirect measurement of the likelihood of two organisms possessing the observed degree of overlap in metabolism 'by chance.' To generate D_{FM}, first, for each of the 5050 $(101 * 100/2)$ pairs of organisms, a 2x2 cross tabulation table was created and a Fisher's exact test p-value was generated. The Fisher's exact test p-value (that is, the likelihood of observing pattern of metabolic consistency by chance) acts as a measure of metabolic similarity and is available for all pairs of organisms in Supplemental File 6. We transformed p-values using: D_{FM}=300+ln(p) to yield a metric of metabolic distance, D_{FM}, which is always greater than 0 in our dataset.

4.3 *Statistical Analyses*

Four main statistical analyses were performed on D_{FM}. First, hierarchical clustering with complete linkage was conducted on the 101 organisms using D_{FM} as computed between all 5050 pairs of organisms. A dendrogram was created and phenotypic features were overlaid on it to aid in interpretation of subsequent analyses. Clusters of interest on the dendrogram were determined using a false discovery rate (FDR) based on the Fisher's exact test p-values. Second, a multiple regression analysis was conducted to investigate the extent to which the metabolic distance, D_{FM}, could be explained by the 19 phenotype features. We used a dataset comprising 19 phenotype features and metabolic distance for each of the 5050 pairs of organisms (supplemental file #2). Models regressed metabolic distance on each of the 19 phenotype features. Third, we conducted analyses designed to answer the question "Which phenotypes explain why this cluster (on the dendrogram) exists?" After 'cutting' the dendrogram by looking at all of the mutually exclusive clusters for which all pairs of organisms within the cluster have a certain level of association, we wish to compare two mutually exclusive clusters of organisms to attempt to identify phenotypic differences in the clusters which are likely candidates for why the organisms separated into two mutually exclusive clusters. For categorical phenotypes, a Fisher's exact test is conducted which compares the proportion of organisms in cluster #1 with the phenotypic characteristic to the proportion of organisms in cluster #2 with the characteristic. For quantitative phenotypes, a two-sample t-test is used. Full results for all phenotypes and clusters A, B, C, D, E1 and E2 are provided in Supplemental File 7. Lastly, we conducted the same analysis as just described to answer the question "Which metabolic variants associate with specific clusters?" by using the Fisher's exact test approach on mutually exclusive clusters, evaluating association between metabolic variants and cluster memberships. Unless otherwise indicated, all analyses were conducted using R (www.r-project.org).

Supplemental Files

All supplemental files are available online at the following URL: http://homepages.dordt.edu/ntintle/metsig.zip

5. Acknowledgments

This work is supported by NSF MCB-1330734. We gratefully acknowledge discussions with Andrei Osterman, Ross Overbeek and other members of the Fellowship for the Interpretation of the Genome in early phases of this project.

References

[1] Y. Boucher and W. F. Doolittle, *Mol. Microbiol.* **37**, 703 (2000).
[2] G. Xie, C. A. Bonner, T. Brettin, R. Gottardo, N. O. Keyhani, and R. A. Jensen, *Genome Biol.* **4**, R14 (2003).
[3] A. Kreimer, E. Borenstein, U. Gophna, and E. Ruppin, *Proc. Natl. Acad. Sci.* **105**, 6976 (2008).
[4] A. Mithani, G. M. Preston, and J. Hein, *PLoS Comput. Biol.* **6**, (2010).
[5] A. Mazurie, D. Bonchev, B. Schwikowski, and G. A. Buck, *BMC Syst. Biol.* **4**, 59 (2010).
[6] N. Pearcy, J. J. Crofts, and N. Chuzhanova, *Mol. BioSyst.* **11**, 77 (2015).
[7] A. Barve and A. Wagner, *Nature* **500**, 203 (2013).
[8] J. Raymond, *Science* **311**, 1764 (2006).
[9] R. Braakman and E. Smith, *PLoS Comput. Biol.* **8**, e1002455 (2012).
[10] R. Braakman and E. Smith, *Phys. Biol.* **10**, 11001 (2013).

[11] R. Braakman and E. Smith, *PLoS One* **9**, e87950 (2014).
[12] J. Raes, I. Letunic, T. Yamada, L. J. Jensen, and P. Bork, *Mol. Syst. Biol.* **7**, 473 (2014).
[13] C. von Mering, P. Hugenholtz, J. Raes, S. G. Tringe, T. Doerks, L. J. Jensen, N. Ward, and P. Bork, *Science* **315**, 1126 (2007).
[14] T. A. Gianoulis, J. Raes, P. V. Patel, R. Bjornson, J. O. Korbel, I. Letunic, T. Yamada, A. Paccanaro, L. J. Jensen, M. Snyder, P. Bork, and M. B. Gerstein, *Proc. Natl. Acad. Sci.* **106**, 1374 (2009).
[15] S. Chaffron, H. Rehrauer, J. Pernthaler, and C. von Mering, *Genome Res.* **20**, 947 (2010).
[16] N. Fierer, J. W. Leff, B. J. Adams, U. N. Nielsen, S. T. Bates, C. L. Lauber, S. Owens, J. A. Gilbert, D. H. Wall, and J. G. Caporaso, *Proc. Natl. Acad. Sci.* **109**, 21390 (2012).
[17] C. S. Henry, R. Overbeek, F. Xia, A. A. Best, E. Glass, J. Gilbert, P. Larsen, R. Edwards, T. Disz, F. Meyer, V. Vonstein, M. Dejongh, D. Bartels, N. Desai, M. D'Souza, S. Devoid, K. P. Keegan, R. Olson, A. Wilke, J. Wilkening, and R. L. Stevens, *Biochim. Biophys. Acta* **1810**, 967 (2011).
[18] R. Overbeek, T. Begley, R. M. Butler, J. V Choudhuri, H. Y. Chuang, M. Cohoon, V. de Crecy-Lagard, N. Diaz, T. Disz, R. Edwards, M. Fonstein, E. D. Frank, S. Gerdes, E. M. Glass, A. Goesmann, A. Hanson, D. Iwata-Reuyl, R. Jensen, N. Jamshidi, L. Krause, M. Kubal, N. Larsen, B. Linke, A. C. McHardy, F. Meyer, H. Neuweger, G. Olsen, R. Olson, A. Osterman, V. Portnoy, G. D. Pusch, D. A. Rodionov, C. Ruckert, J. Steiner, R. Stevens, I. Thiele, O. Vassieva, Y. Ye, O. Zagnitko, and V. Vonstein, *Nucleic Acids Res.* **33**, 5691 (2005).
[19] R. K. Aziz, D. Bartels, A. A. Best, M. DeJongh, T. Disz, R. a Edwards, K. Formsma, S. Gerdes, E. M. Glass, M. Kubal, F. Meyer, G. J. Olsen, R. Olson, A. L. Osterman, R. a Overbeek, L. K. McNeil, D. Paarmann, T. Paczian, B. Parrello, G. D. Pusch, C. Reich, R. Stevens, O. Vassieva, V. Vonstein, A. Wilke, and O. Zagnitko, *BMC Genomics* **9**, 75 (2008).
[20] A. Osterman and R. Overbeek, Curr. Opin. Chem. Biol. **7**, 238 (2003).
[21] Y. Ye, A. Osterman, R. Overbeek, and A. Godzik, Bioinformatics **21**, i478 (2005).
[22] C. S. Henry, M. DeJongh, A. A. Best, P. M. Frybarger, B. Linsay, and R. L. Stevens, *Nat. Biotechnol.* **28**, 977 (2010).
[23] A.-L. Barabasi and Z. N. Oltvai, Nat Rev Genet **5**, 101 (2004).
[24] A. C. Poot-Hernandez, K. Rodriguez-Vazquez, and E. Perez-Rueda, *BMC Genomics* **16**, 957 (2015).
[25] W. Ludwig, O. Strunk, R. Westram, L. Richter, H. Meier, Yadhukumar, A. Buchner, T. Lai, S. Steppi, G. Jobb, W. Förster, I. Brettske, S. Gerber, A. W. Ginhart, O. Gross, S. Grumann, S. Hermann, R. Jost, A. König, T. Liss, R. Lüssmann, M. May, B. Nonhoff, B. Reichel, R. Strehlow, A. Stamatakis, N. Stuckmann, A. Vilbig, M. Lenke, T. Ludwig, A. Bode, and K.-H. Schleifer, *Nucleic Acids Res.* **32**, 1363 (2004).
[26] A. Stamatakis, *Bioinformatics* **22**, 2688 (2006).

WHEN SHOULD WE *NOT* TRANSFER FUNCTIONAL ANNOTATION BETWEEN SEQUENCE PARALOGS?

MENGFEI CAO and LENORE J. COWEN

Department of Computer Science, Tufts University,
Medford, MA 02155, USA
Email: mengfei.cao@tufts.edu and lenore.cowen@tufts.edu

Current automated computational methods to assign functional labels to unstudied genes often involve transferring annotation from orthologous or paralogous genes, however such genes can evolve divergent functions, making such transfer inappropriate. We consider the problem of determining when it is correct to make such an assignment between paralogs. We construct a benchmark dataset of two types of similar paralogous pairs of genes in the well-studied model organism *S. cerevisiae:* one set of pairs where single deletion mutants have very similar phenotypes (implying similar functions), and another set of pairs where single deletion mutants have very divergent phenotypes (implying different functions). State of the art methods for this problem will determine the evolutionary history of the paralogs with references to multiple related species. Here, we ask a first and simpler question: we explore to what extent any computational method with access only to data from a single species can solve this problem.

We consider divergence data (at both the amino acid and nucleotide levels), and network data (based on the yeast protein-protein interaction network, as captured in BioGRID), and ask if we can extract features from these data that can distinguish between these sets of paralogous gene pairs. We find that the best features come from measures of sequence divergence, however, simple network measures based on degree or centrality or shortest path or diffusion state distance (DSD), or shared neighborhood in the yeast protein-protein interaction (PPI) network also contain some signal. One should, in general, not transfer function if sequence divergence is too high. Further improvements in classification will need to come from more computationally expensive but much more powerful evolutionary methods that incorporate ancestral states and measure evolutionary divergence over multiple species based on evolutionary trees.

Keywords: protein function prediction, paralogs

1. Introduction

When new genes are sequenced and deposited into databases, a variety of manual and automated curation is involved in associating functional annotation to these genes. One of the most common practices is to transfer functions based on some threshold of sequence similarity.[1] However, when this sequence similarity threshold results in automatically transferring functional annotation between all pairs of orthologous and paralogous genes, this is deeply problematic because there are cases when the functions of the genes have diverged.[2]

In this paper, we consider the question of transfer of functional annotation *solely for paralogous genes.* It was widely believed that paralogs were more likely to acquire divergent functions than orthologs (the so-called *ortholog conjecture*),[3,4] but in recent years, this assumption has been the subject of spirited debate.[3–6] The present study requires neither a positive nor negative resolution of the ortholog conjecture, nor does it directly shed light on the conjecture itself, since it focuses only on a practical problem in the field of automatic function prediction artificially restricted to a single species: we ask whether any computational method *with access to information based only on the single species in which the paralogs reside*, can distinguish the pairs whose functional roles are similar from those where functional roles are diverged.

We construct a benchmark dataset of two types of paralogous pairs of genes in the well-studied model organism *S. cerevisiae:* one set of pairs where single deletion mutants have very similar phenotypes (implying similar function), and another set of pairs where single deletion mutants have very divergent phenotypes (implying different function). We are fortunate in that there exist data in *S. cerevisiae* where the similarity of phenotypes of deletion mutants has been categorized: in particular, the extensive phenotype data from Hillenmeyer et al.[7] who look at the phenotypes of homozygous single gene deletion knockouts under 418 different conditions such as depletion of certain amino acids or nutrients.

The Hillenmeyer et al data[7] allows us to construct a gold-standard benchmark dataset of paralogous yeast gene pairs, some with highly similar and some with highly dissimilar functions, as follows. We consider two different datasets of paralogous gene pairs in *S. cerevisiae.* The first dataset we construct from scratch by taking pairs of yeast genes with high sequence similarity. The second dataset is derived from the study of the putative whole genome duplication event for *S. cerevisiae* by Kellis et al.[8] who identify 450 paralogous gene pairs. For each of the paralog pairs in the two datasets we compute a co-fitness score[7] to represent to what extent the two gene deletion knockouts have similar phenotypes. We choose a subset of these paralogs with very high co-fitness score and a subset of these paralogs with very low co-fitness score. The subset with the high co-fitness score are our *same* or *conserved function* paralogs, and the subset with the low co-fitness score are our *divergent function* paralogs. Note that Hillenmeyer et al.[7] has already shown that when genes are clustered using such a co-fitness score, they find clusters that are consistent with shared Gene Ontology annotations for biological process and molecular functions. Gu et al.[9] also uses fitness effect data to study functional compensation among gene duplicates.

We note that a recent study of Plata and Vitkup[10] also considered the genetic robustness and functional evolution of gene duplicates in yeast, based on the same gene deletion knockout set of Hillenmeyer et al. However, they considered a measure that is different than our co-fitness score over the collection of gene deletion mutants. In particular, under the assumption that paralogs with similar function could mutually compensate for each other whereas paralogs with divergent function could not, they considered the average number of "sensitive" conditions (i.e. conditions where a growth defect was observed with a P value cutoff of 0.01) between paralog pairs. Paralogs with a small average number of conditions where there was a growth defect (also alternatively, with a small average fraction of conditions where there was data, to deal with missing data), they assumed meant that the paralogs were mutually able to compensate for one another in the deletion mutant. We discuss how well this measure correlates with our "similar function" co-fitness score below.

In addition to nucleotide and amino acid sequence similarity, we sought to investigate whether simple features of the PPI network would also help distinguish same function from divergent function paralogs. Mika and Rost[11] showed that PPI interactions were better conserved within species than across species: a sort of anti-ortholog conjecture for interlogs. Thus it is reasonable to think that the interaction partners of a gene will be more similar for genes with similar functions; the problem is of course complicated by the fact that existing PPI data is both noisy and also extremely incomplete. We consider some simple well-studied parameters of this network, namely degree, shortest path distance, shared neighborhood, as well as our diffusion-based DSD measure,[12,13] which has been shown to be especially robust to noise and missing data,[14] to find out to what extent these are informative features for our problem.

2. Related work

The most related paper to the current one is the previously mentioned work of Plata and Vitkup.[10] In addition, there have been some previous studies that have tried to place paralogous gene pairs into different functional categories based on a variety of information sources, including the recent SIFTER[2] which performed quite well in the past two Critical Assessment of protein Functional Annotation algorithm (CAFA) experiments[15] for automated function prediction. Unlike the present study, SIFTER assumes access to information from ancestral states, not just the species in which the paralogous gene pairs themselves reside, so they are able to leverage the power of evolutionary information. Other work[16–19] has used gene expression levels, the number of shared interacting partners, and shared Gene Ontology annotations, in order to predict or assess which pairs are instances of conserved function, subfunctionalization, and neofunctionalization. In each of these papers, ground truth for the predictions are assessed in different ways. Zeng and Hannenhalli[16] compare tissue specific gene expression levels from an ancestral gene (a single-copy gene from a closely related species) and the duplicated genes, where in the neofunctionalization case, for example, they assume the ancestral gene's expression level should be lower than that of both duplicates. In addition to this being a somewhat controversial assumption, noise in measuring expression levels can impact their conclusions. Nakhleh's group[17] uses the yeast PPI network to study the problem of categorizing different evolutionary fates of duplicate genes, but in their case, instead of using the structure of the PPI network to assist in predicting the categories, they used the network to *define* their ground truth gold standard for the categories. In particular, they define gene pairs as similar and divergent in function based on comparing the number of known interacting partners of the ancestral gene and the duplicated genes, a measure that will be very sensitive to noise and incomplete data even in the relatively well-studied yeast interactome.[20]

Our method of determining ground truth for same and divergent functions is less noise sensitive than either of these two other methods, but it is much more restrictive than the methods of previous studies. First, it presents only two categories of functional similarity and divergence. More importantly, it makes use of extensive phenotype data from single deletion mutants: a dataset available for yeast but unavailable for most other species at this time. Thus these other measures may be the only ones available in other species; conversely, if one accepts that the single deletion phenotype data is the best measure of ground truth when available for this problem, then the subject of this paper, namely, determining which *other* more easily obtainable sequence and network measures best correlate with this standard, might be the most important application to studying computational transfer of functional annotation standards in other organisms of interest.

Finally, the most common way in the field to determine if paralogs share the same function is simply to look up the curated functional annotations in a database based on a human-created ontology structure such as MIPS[21] and GO[22] to see if they are annotated with the same functional labels. However, we note that in many databases, paralogs with nearly identical or identical sequence are often annotated with the same functional labels, even if that annotation comes from experiments with only one of the paralogs.

3. Materials and Methods

3.1. *Physical interaction network*

We download all the 141,327 physical interactions compiled from 7601 publications by BioGRID, version 3.3.122 (date March 3rd, 2015), each interaction of which is experimentally verified and associated with one of the following experimental evidence codes: "Affinity Capture-Luminescence", "Affinity Capture-MS", "Affinity Capture-RNA", "Affinity Capture-Western", "Biochemical Activity", "Co-crystal Structure", "Co-fractionation", "Co-localization", "Co-purification", "Far Western", "FRET", "PCA", "Protein-peptide", "Protein-RNA", "Proximity Label-MS", "Reconstituted Complex", "Two-hybrid". While collecting these physical interactions, we adopt the scoring scheme from Cao et al.[13] and assign real value confidence scores in $(0, 1)$ as weights to interactions, where the scoring scheme weights interactions as higher confidence when they are verified by experiments from multiple publications, plus low-throughput experiments are deemed more reliable than high-throughput experiments. Since we only consider interactions associated for genes in the list of 5091 verified ORFs (open reading frames) from the *Saccharomyces* Genome Database (download date: April 11th, 2014), we exclude the interactions that are associated with non-verified ORFs. Using the data above, we build an undirected weighted graph where a node is a protein, a weighted edge between two nodes exists if and only if there is a physical interaction between the two nodes and the weight on each edge is calculated as the confidence score. As a result, we obtain a connected simple graph, involving 5043 nodes and 79594 edges, with diameter 5.

3.2. *Duplicated gene pairs*

We collect two sets of duplicated gene pairs in two ways. We construct the first set that we call the "SequenceCover" or "SC" set based on sequence similarity using the following process: We first collect the result from all against all BLAST searches over the 5043 proteins and then build a sequence similarity graph where a node is a protein and an edge between two proteins exists if and only if the sequence identity is at least 80% and the BLAST E-value is below 10^{-5}. We then find the maximal independent edge set using a naive heuristic algorithm from the graph which satisfies two conditions: (1) if an edge (a, b) is chosen, none of a or b's neighbors will be chosen; (2) no more edges can be added to the set without violating (1). Because these conditions together with the sequence similarity threshold settings are so strict, we will generate edges for protein pairs that have very high sequence similarity and any two of them in the set will not share a common node. Note that in order to analyze the gene pairs using their fitness data, we exclude from the edge set the edges that are incident to nodes that do not have fitness data available from Hillenmeyer et al.;[7] as a result, we are restricted to the 3732 genes out of 5043 genes have fitness data available. However, we may find different maximal independent edge sets if we choose different random seeds.

We randomly pick one of these independent maximal edge sets, which then defines our first duplicated gene pair set. For the second set of duplicated gene pairs, we download all 450 WGD gene pairs from Kellis et al.[8] The Kellis et al[8]'s gene pair set consists of gene pairs that are believed to be paralogs derived from the whole genome wide duplication event, inferred by both sequence mapping and gene locus. This data set has also been widely used by many other groups studying function of paralogous genes.[16–19,23] Again we restrict ourselves to the subset of the WGD gene pair set where both nodes have fitness data from Hillenmeyer et al.[7]

Note that by restricting the SC set to gene pairs with a relatively high degree of sequence similarity, it will miss some pairs of distant paralogs. However, since our focus is not on the behavior of the landscape of all paralogs, but rather on the scenario where one might computationally decide to transfer functional annotation based on sequence similarity, this is a reasonable threshold.

3.3. *Fitness profile*

We download all the 1,982,156 fitness defect log ratio scores (where 188,642 scores are missing) derived from the homozygous gene deletion experiments from Hillenmeyer et al.[7] Each log ratio score indicates the fitness defect for one of the 4769 homozygous gene deletion strains involving 4742 genes under one of the 418 different testing environments such as depletion of amino acids or nutrients. With respect to a strain for one gene, Hillenmeyer et al.[7] defines a fitness profile as the 418-dimensional vector where each entry is the log ratio score corresponding to each testing environment. Using the fitness profile, we then follow Hillenmeyer et al's[7] analysis and calculate a co-fitness score for each pair of genes that captures the phenotype similarity based on the growth defect under different testing environments. As a preprocessing step accounting for the missing entries, we impute the missing value as follows: 1) when only one gene has fitness values missing for a given environment, we use the same fitness value for both. 2) when both genes have their fitness values missing for a given environment, we use the mean value over all strains under that environment for both. We calculate the co-fitness scores between any two genes as the cosine distance between the fitness profile vectors as defined in Hillenmeyer et al.[7] In total, we obtain co-fitness scores for $4769 \times (4769 - 1)/2 = 11,369,296$ unique pairs.

In order to provide statistical analysis on the co-fitness scores for our targeted duplicated gene pairs, we define a z-score z_{cfs} as a normalized co-fitness score: $z_{cfs} = \frac{c_{ij}-\mu}{\sigma}$, where c_{ij} is the co-fitness score between gene i and gene j, μ is the mean co-fitness score over all pairs and σ is the standard deviation over all co-fitness scores. Therefore our empirical $p - value$ is computed as the probability that we will see a result at the normalized co-fitness score using the $t - distribution$ with $n - 1$ degrees of freedom where n is the number of distinct pairs of strains, namely 11,369,296. We report for each of our targeted duplicated gene pairs the co-fitness score and their $p - values$, of which a higher value will indicate that the pair of genes is less likely to share phenotype similarity and thus less likely to carry out the same biological function. Since the problem we are trying to solve is to distinguish between paralogous gene pairs with divergent functions and that with shared functions, we need to have two separate

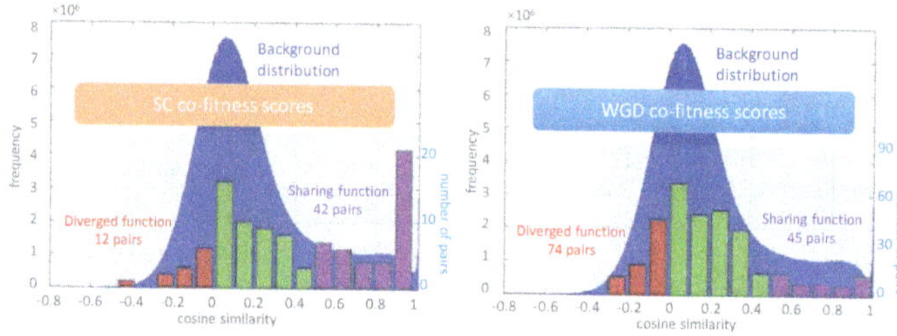

Fig. 1. Defining gene pairs with diverged functions and that with shared functions for the SC and WGD sets using the co-fitness scores.

sets of gene pairs, one set that with high confidence includes gene pairs with divergent functions and the other that with high confidence includes gene pairs with shared functions. We set the thresholds {0.00, 0.50} to define the gene pairs as with divergent functions if the co-fitness score is below 0.00 and the pairs to be with shared functions if the co-fitness score is above 0.50 as shown in Figure 1. Among the SC set where all genes in the 100 gene pairs have the fitness data available, we define 12 gene pairs with diverged functions and 42 gene pairs with shared functions; among the 450 WGD gene pairs where only 337 gene pairs have both genes' fitness data available, we define 74 gene pairs with diverged functions and 45 gene pairs with shared functions. The remaining unclassified gene pairs with co-fitness scores in the middle range, we decline to classify as "shared" or "divergent". Among the 54 classified gene pairs in the SC set, 78% pairs are considered as shared function pairs, while among the 119 classified gene pairs in the WGD set, only 38% pairs are considered as sharing function– this is not surprising as the WGD set includes gene pairs whose duplication event was in the very far past with a lot of evolutionary time to evolve mutations that could affect function.

Finally, we note that among the 119 pairs of paralogs that make up our WGD set, a total of 7 pairs lie on the same yeast chromosome. Among the 54 pairs of paralogs that make up our SC set, also a total of 7 pairs lie on the same yeast chromosome.

3.4. *Sequence similarity*

To measure amino acid similarity, for each of the classified gene pairs, we collect the BLAST bit-score, BLAST alignment length and the percentage identity as the protein sequence similarity measurements.

For nucleotide sequence similarity, for each of the classified gene pairs, we estimate the Ka score, the non-synonymous substitution rate, and the Ks score, synonymous substitution rate, as the nucleotide sequence divergence measurements.[24] More specifically, we compute the pairwise alignment for each gene pair using clustalw2,[25] then we translate the protein alignment to a codon alignment and estimate Ka and Ks scores using the KaKs_Calculator of Zhang et al.[26] with default parameters. In addition, we also compute the Ka/Ks ratio, which is commonly considered as an indicator of selective pressure acting on a protein-coding gene. We note that for the more distant pairs, these statistics do not give reliable indications of expected evolutionary divergence, however, we can still calculate the values: we just need to assume their correlation with true evolutionary divergence is weaker.

3.5. *PPI network based measures*

For each of the classified gene pairs, we compute a set of network based similarities (or distances): the number of shared interacting neighbors, the normalized shared neighborhood size, the normalized degree difference, the normalized betweenness-centrality score difference, the shortest-path distance and the diffusion state distance (as defined by Cao et al.[13]). In the case of the normalized shared neighborhood size, degree difference and betweenness-centrality score difference, we simply divide by the maximum of the quantities for each paralog to normalize: i.e. to compute the normalized degree difference of paralogs A and B, we simply take $|deg(A) - deg(B)|/\max(deg(A), deg(B))$.

3.6. *Problem formulation*

For each of the measures defined above, we can rank the paralogous gene pairs according to each measure. However, in order to appropriately set a cutoff for each measure, beyond which we predict "conserved function" or "divergent function," we need a training set of labeled examples. First we report the predictive power of each measure described above using a leave-one-out cross validation paradigm. Namely, we learn the optimal cutoffs for classifying pairs as conserved or divergent based on all the data except the held out pair, and then classify the held-out pair according to those thresholds, and report percentage accuracy.

Then we look at the power of some standard machine learning methods when given access to all the features. In particular, we consider: decision trees, naive Bayes, support vector machines (with linear kernel), K-nearest-neighbor (with K=1), logistic regression, random forest, multilayer perceptron, one rule method, and AdaBoost with decision tree, all implemented in WEKA,[27] and see their power on the task of distinguishing the same-function from the divergent-function pairs also in leave-one-out cross validation.

4. Results

4.1. *Classification using each individual similarity measurement*

We assess the predictive power of each individual similarity/divergence measurement using the leave-one-out cross validation paradigm. Specifically, per each measurement, for each paralogous gene pair, we learn a classification threshold based on all the other gene pairs where we will classify pairs above (or below, as appropriate) the threshold as "similar function" and below the threshold (or above) as "diverged function". We then count the percent of pairs that we classify correctly. This list is somewhat deceptive in measuring true performance because of the unbalanced class sizes: but we find the nucleotide sequence-based scores uniformly more informative than the protein sequence-based scores that we measure. Moreover the Ka and Ks scores remain good classifiers even if the thresholds are trained across the two datasets (see Table 2): for example when the Ks threshold for best classification is trained on WGD and tested on SC, and when the Ks threshold for best classification is trained on SC and tested on WGD, the percent accuracies become 88.89% and 80.67%, respectively. Figure 2 presents the scatter plot of the Ks score versus our co-fitness score.

None of the network measures perform as well as the sequence similarity measures, but the best performing network measures were related to shared neighborhood size.

Table 1. Accuracies for individual measurements

Performance for leave-one-out-cross-validation (% Accuracy)		SC	WGD
Protein sequence measurements	AA percent identity	74.07%	78.99%
	AA BLAST alignment length	87.04%	69.75%
	AA BLAST bit-score	81.48%	63.03%
	AA ClustalW length	83.33%	76.47%
Sequence measurements	Ka	90.74%	76.47%
	Ks	88.89%	79.83%
	Ka/Ks	79.63%	71.43%
Network measurements	degree-difference (normalized)	70.37%	61.34%
	bc-difference (normalized)	72.22%	63.03%
	shared neighborhood size (SNH)	75.93%	73.11%
	normalized SNH	87.04%	72.27%
	shortest path distance	81.48%	62.18%
	DSD	74.07%	69.75%

Table 2. Cross-training performance accuracy

	TrainOnWGDTestOnSC	trainOnSCTestOnWGD
Ka/Ks	64.81%	57.98%
Ka	87.04%	79.83%
Ks	88.89%	80.67%

4.2. *Common supervised learning methods for using all measurements*

Motivated by the observation above, we place all the 13 measurements into one feature vector for each gene pair and then try several common supervised learning methods. However, as shown in Table 3, none of the learning algorithms obtain better performance than the best performing individual measurement. Thus it remains an open question how to develop better algorithms that can separate the same-function from divergent function yeast paralogs in our set.

We also wondered whether our method of filling in missing data from the Hillenmeyer et al.[7] experiments contributed to the misclassification rates we saw. Recall that when data

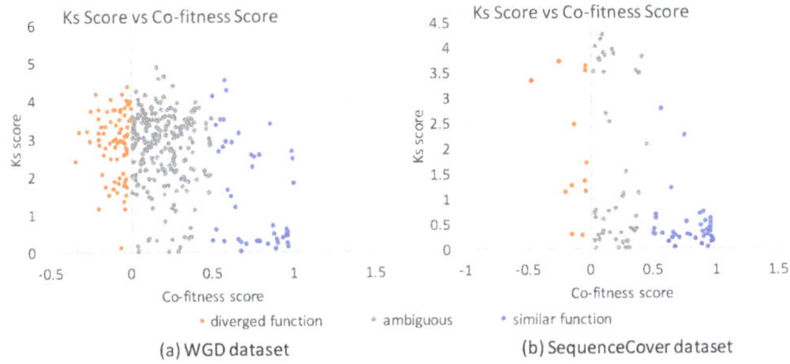

Fig. 2. Scatter plot of Ks score v.s. co-fitness scores for paralog gene pairs

Table 3. Accuracies for different learning algorithms		
Performance for leave-one-out-cross-validation (% Accuracy)	SC	WGD
Decision Tree	79.63%	78.15%
Naïve Bayes	90.74%	73.95%
Support Vector Machine (linear kernel)	83.33%	78.15%
k-nearest-neighbor	90.74%	68.07%
Logistic regression	85.19%	75.63%
Random forest	87.04%	78.15%
Multilayer perceptron	87.04%	68.91%
One-Rule	83.33%	75.63%
AdaBoost + Decision Tree	85.19%	70.59%

was missing from a phenotype experiment, we filled in artificial fitness values: if the value was missing in only one of the paralogs, we matched the other paralog; if it was missing for both paralogs, we utilized the mean fitness value over all the deletion experiments for that phenotype for both paralogs. This would make yeast paralogs that are in fact divergent be more likely to have co-fitness scores that would result in our classification as "same function", if at least one had many missing values.

This did, in fact, seem to underlie some of the bad classification results for the WGD dataset in particular. For example, for the SC dataset, among the 48/54 pairs for which the Ks feature results in the correct classification, the average missing ratio is .41, whereas among the 6/54 pairs where the Ks feature results in incorrect classification, the average missing ratio is .37, whereas, for the WGD dataset, among the 95/119 examples where the Ks feature is correct, the average missing ratio is .12, whereas for the 24/119 examples where the Ks feature is wrong, the average missing ratio is .45. Removing all examples with missing data will result in too small a benchmark set; it thus remains an open question to find better ways to deal with missing values in construction of the benchmark datasets.

5. Some example paralog pairs

We looked in more detail at some of the pairs we classified as paralogs with divergent function. Because *S. cerevisiae* is so well-studied, we thought that some of the paralog pairs that we classified as divergent function, might have support from functional annotations in the SGD database, or in the literature. We found the situation quite heterogeneous– for some of the pairs we found support for functional divergence in the literature, for others, there seems to be no annotation indicating that anyone has noted any functional divergence in the two paralogs.

For example, GPP1 and GPP2 are an example of a paralog pair where some functional divergence is known. In particular, GPP1 and GPP2 seem to behave very similarly under aerobic conditions but very differently under anerobic conditions.[28] Another paralog pair where functional divergence is documented is OAF1 and PIP2. Both genes are involved in fatty acid induction of the peroxisomal β-oxidation machinery involving regulation by the oleate response element, and form a heterodimer. But OAF1 binds fatty acids and PIP2 does not.[29] GDH1 and GDH3 are both involved in glutamate biosynthesis, but their regulation indicates that they are utilized under different growth conditions: expression of GDH3 is induced by ethanol and repressed by glucose, whereas GDH1 expression is high in either carbon source.[30] TPK1 and TPK3, together with a third gene, TPK2, are functionally redundant for cell viability, but they have differing protein targets, and also recognize and affect the transcription of different sets of gene targets.[31] In each of these cases, manual inspection implies that functional labels, at least at the top levels of MIPS or GO, would correctly transfer between paralogs, despite these pairs having documented different roles within these broad functional categories.

On the other hand, the majority of the paralog pairs which we mark as "functionally divergent" have no indication in the literature or SGD database that any functional divergence between the paralogs is yet documented. ALK1 and ALK2 is a typical case. Despite a low co-fitness score, this gene pair is currently annotated in a fashion very similar to "same function" pairs: the summary description of ALK2 reads: *Protein kinase; along with its paralog, ALK1, required for proper spindle positioning and nuclear segregation following mitotic arrest, proper organization of cell polarity factors in mitosis, proper localization of formins and polarity factors, and survival in cells that activate spindle assembly checkpoint; phosphorylated in response to DNA damage; ALK2 has a paralog, ALK1, that arose from the whole genome duplication; similar to mammalian haspins.* The description for ALK1 is identical except with the names interchanged.

6. Discussion

The problem of predicting when functional annotation terms should transfer between sequence homologs and paralogs is a difficult but urgent one in the field of automatic prediction of protein function. Here, we have done a very simple study in a single, well-studied species without leveraging the wealth of evolutionary information that is available in sequences. Clearly any reasonable solution will have to leverage this evolutionary information in order to make more accurate predictions.

One clue as to the difficulty of the task might come from the alternative definitions of Plata and Vitkup.[10] We sought to measure how our co-fitness score correlated with the genetic robustness measures of Plata and Vitkup:[10] for each duplicate pair, following their paper, the

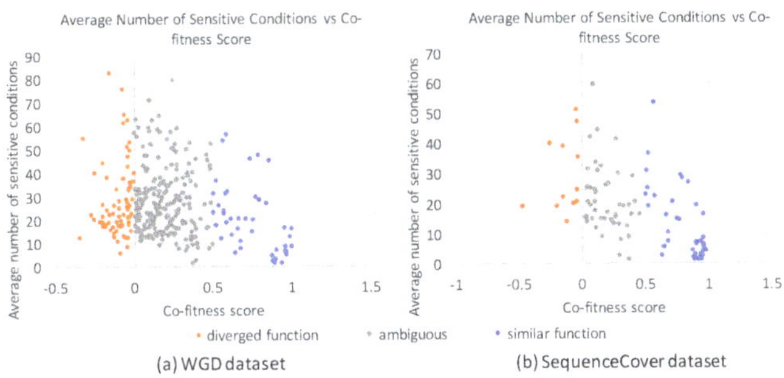

Fig. 3. Scatter plot of average number of sensitive conditions v.s. co-fitness scores for paralog gene pairs.

number of "sensitive" conditions is measured for each deletion mutant, where a "sensitive" condition is defined as a growth defect with $p < 0.01$. We report the number of sensitive conditions, averaged over the two paralogs in the pair, and plot its correlation with our co-fitness score. We find a negative correlation of -0.2046 ($P < 1.49E^{-04}$) (WGD pairs) and a negative correlation of -0.5484 ($P < 2.74E^{-09}$) (SC pairs). A scatter plot appears in Figure 3.

We notice that in both datasets, there are small but distinct set of pairs with both a very low number of average sensitive conditions, and a very high co-fitness score. For these pairs, it is hard to tell if the paralogs are similar function, or if no phenotype is frequently observed because there are third or fourth copy duplicate genes that can buffer both deletion mutants. Thus full understanding of both our co-fitness and their sensitive condition scores may require the consideration of higher order duplicates.

The SC and WGD benchmark datasets are available at bcb.cs.tufts.edu/paralogs

7. Acknowledgements

We thank the entire Tufts BCB group for helpful discussions.

References

1. I. Friedberg, *Briefings in Bioinformatics* **7**, 225 (2006).
2. S. M. Sahraeian, K. R. Luo and S. E. Brenner, *Nucleic Acids Research* , p. gkv461 (2015).
3. N. L. Nehrt, W. T. Clark, P. Radivojac and M. W. Hahn, *PLOS Comput Biol* **7**, p. e1002073 (2011).
4. R. A. Studer and M. Robinson-Rechavi, *Trends in Genetics* **25**, 210 (2009).
5. A. M. Altenhoff, R. A. Studer, M. Robinson-Rechavi and C. Dessimoz, *PLOS Comput Biol* **8**, p. e1002514 (2012).
6. X. Chen and J. Zhang, *PLOS Comput Biol* **8**, p. e1002784 (2012).
7. M. E. Hillenmeyer, E. Fung *et al.*, *Science* **320**, 362 (2008).
8. M. Kellis, B. W. Birren and E. S. Lander, *Nature* **428**, 617 (2004).
9. Z. Gu, L. M. Steinmetz, X. Gu, C. Scharfe, R. W. Davis and W.-H. Li, *Nature* **421**, 63 (2003).
10. G. Plata and D. Vitkup, *Nucleic Acids Research* **42**, 2405 (2014).

11. S. Mika and B. Rost, *PLOS Comput Biol* **2**, p. e79 (2006).
12. M. Cao, H. Zhang, J. Park, N. M. Daniels, M. E. Crovella, L. J. Cowen and B. Hescott, *PLOS One* **8**, p. e76339 (2013).
13. M. Cao, C. M. Pietras, X. Feng, K. J. Doroschak, T. Schaffner, J. Park, H. Zhang, L. J. Cowen and B. Hescott, *Bioinformatics* **30**, i219 (2014).
14. I. Fried, A. Cannistra, C. Casey, A. Piel, M. Crovella and B. Hescott, *ISMB Late Breaking Research* (2015).
15. P. Radivojac, W. T. Clark, T. R. Oron, A. M. Schnoes, T. Wittkop, A. Sokolov, K. Graim, C. Funk, K. Verspoor, A. Ben-Hur *et al.*, *Nature Methods* **10**, 221 (2013).
16. J. Zeng and S. Hannenhalli, *BMC Genomics* **14**, p. 1 (2013).
17. Y. Zhu, Z. Lin and L. Nakhleh, *G3: Genes— Genomes— Genetics* **3**, 2049 (2013).
18. A. Baudot, B. Jacq and C. Brun, *Genome Biology* **5**, p. 1 (2004).
19. L. Hakes, J. Pinney, S. Lovell, S. Oliver and D. Robertson, *Genome Biology* **8**, p. 1 (2007).
20. J.-F. Rual, K. Venkatesan *et al.*, *Nature* **437**, 1173 (2005).
21. A. Ruepp, A. Zollner, D. Maier *et al.*, *Nucleic Acids Research* **32**, 5539 (2004).
22. M. Ashburner, C. A. Ball *et al.*, *Nature Genetics* **25**, 25 (2000).
23. C. Roth, S. Rastogi, L. Arvestad, K. Dittmar, S. Light, D. Ekman and D. A. Liberles, *Journal of Experimental Zoology Part B: Molecular and Developmental Evolution* **308**, 58 (2007).
24. Z. Yang and R. Nielsen, *Molecular Biology and Evolution* **17**, 32 (2000).
25. M. A. Larkin, G. Blackshields *et al.*, *Bioinformatics* **23**, 2947 (2007).
26. Z. Zhang, J. Li, X.-Q. Zhao, J. Wang, G. K.-S. Wong and J. Yu, *Genomics, Proteomics & Bioinformatics* **4**, 259 (2006).
27. M. Hall, E. Frank, G. Holmes, B. Pfahringer, P. Reutemann and I. H. Witten, *ACM SIGKDD Explorations Newsletter* **11**, 10 (2009).
28. A.-K. Påhlman, K. Granath, R. Ansell, S. Hohmann and L. Adler, *Journal of Biological Chemistry* **276**, 3555 (2001).
29. A. Gurvitz and H. Rottensteiner, *Biochimica et Biophysica Acta (BBA)-Molecular Cell Research* **1763**, 1392 (2006).
30. A. DeLuna, A. Avendaño, L. Riego and A. González, *J. of Biol. Chem.* **276**, 43775 (2001).
31. L. S. Robertson and G. R. Fink, *PNAS* **95**, 13783 (1998).

PROSNET: INTEGRATING HOMOLOGY WITH MOLECULAR NETWORKS FOR PROTEIN FUNCTION PREDICTION

SHENG WANG, MENG QU, AND JIAN PENG

Department of Computer Science,
University of Illinois at Urbana-Champaign,
Champaign, IL, USA
**E-mail: jianpeng@illinois.edu*

Automated annotation of protein function has become a critical task in the post-genomic era. Network-based approaches and homology-based approaches have been widely used and recently tested in large-scale community-wide assessment experiments. It is natural to integrate network data with homology information to further improve the predictive performance. However, integrating these two heterogeneous, high-dimensional and noisy datasets is non-trivial. In this work, we introduce a novel protein function prediction algorithm ProSNet. An integrated heterogeneous network is first built to include molecular networks of multiple species and link together homologous proteins across multiple species. Based on this integrated network, a dimensionality reduction algorithm is introduced to obtain compact low-dimensional vectors to encode proteins in the network. Finally, we develop machine learning classification algorithms that take the vectors as input and make predictions by transferring annotations both within each species and across different species. Extensive experiments on five major species demonstrate that our integration of homology with molecular networks substantially improves the predictive performance over existing approaches.

Keywords: protein function prediction, homology, molecular networks, dimensionality reduction, data integration

1. Introduction

Comprehensively annotating protein function is crucial in illustrating activities of millions of proteins at molecular level, which can further advance basic biological research and biomedical sciences.[1] Although massive annotations have been curated, such as popular Gene Ontology (GO) annotations,[2] current experimental approaches are infeasible to fully exploring protein function annotations. As a result, computational approaches have become a more accessible way to annotate protein function[3,4] and help biologists prioritize their experiments.

Computational prediction of protein function has been extensively studied in the context of molecular evolution. Homologous proteins have most likely evolved from a common ancestor. They often carry out similar protein functions, because functions are generally conserved during molecular evolution. Consequently, computational approaches can predict the function of query proteins by transferring those of their annotated homologs. In addition to automatic annotations based on orthology or domain information or pre-existing cross-references and keywords,[5] a variety of machine learning algorithms[6–12] have been proposed to extract annotations based on sequence similarity-detection tools such as BLAST, PSI-BLAST,[13] and phylogenetic analysis.[14,15] Despite the success of homology-based approaches, their major constraint arises from a lack of annotated sequences.[16] In fact, among over 65 million protein sequences in publicly accessible databases,[17] only 2 million of them are manually curated.[18] Consequently, the predictive power of homology-based methods has been limited due to the scarcity of annotations. Furthermore, reliable homology relationships are sparse between distantly related species, thus posing computational and statistical challenges when making faithful predictions.

Fortunately, the rapidly growing interactome data from high-throughput experimental techniques allows us to extract patterns from neighbors in molecular networks[19–21] in addition to homologous proteins. This idea is supported by the established "guilt-by-association" principle, which states that proteins that are associated or interacting in the network are more likely to be functionally related.[22] Recently, this "guilt-by-association" principle has become the foundation of many network-based function prediction algorithms.[23–30] Among them, GeneMANIA[31] and clusDCA[32] are state-of-the-art network-based function prediction approaches. In addition to incorporating network topology, clusDCA also leverages the similarity between GO labels and obtains substantial improvement on sparsely annotated functions. GeneMANIA uses a label propagation algorithm on an integrated network specifically constructed for each functional label, and is currently available as a state-of-the-art web interface for gene function prediction for many organisms.

Intuitively, integrating homology data with molecular networks can synergistically improve function prediction results. On one hand, it enables us to transfer annotations from functionally well-characterized neighbors in the molecular network as well as from homologous proteins with conserved similar functions. On the other hand, homology data can further mitigate the incomplete and noisy nature of molecular networks through interologs,[33] which states that a conserved interaction occurs between a pair of proteins that have interacting homologs in another organism.[34]

Nevertheless, integrating homology data with molecular networks is both computationally and statistically challenging. Since they are heterogeneous data sources, it is likely sub-optimal to integrate them in an additive way which simply averages the prediction results of either of these two data sources. Moreover, we also need an efficient algorithm that scales to hundreds of thousands of proteins from multiple species. One way to integrate these two heterogeneous data sources seamlessly is to construct a multiple species heterogeneous network in which both nodes and edges are associated with different types. With this network, we can predict functions for query proteins based on annotations extracted from both their homologs and their neighbors in molecular networks. Furthermore, information can also be transferred between two proteins that are neither homologs nor neighbors in molecular networks. Notably, the only previous attempt to integrate these two heterogeneous data sources is using multi-view learning.[35] However, it does not scale to multiple species. In addition, they formulated protein function prediction as a structured-output hierarchical classification problem whose performance for sparsely annotated functional labels is far from satisfactory.[32]

In this work, we introduce **ProSNet**, a novel **Pro**tein function prediction algorithm which efficiently integrates **S**equence data with molecular **Net**work data across multiple species. Specifically, an integrated heterogeneous network is first constructed to include all molecular networks of multiple species, in which homologous proteins across multiple species are also linked together. Based on this integrated network, a novel dimensionality reduction algorithm is applied to obtain compact low-dimensional vectors for proteins in the network. Proteins that are topologically close in the molecular networks and/or have similar sequences are co-localized in this low-dimensional space based on their vectors. These low-dimensional vectors are then used as input features to two classifiers which utilize annotations from molecular networks and homologous proteins, respectively. In addition, ProSNet is inherently parallelized, which further promises scalability. When compared to the state-of-the-art methods that only use homology data or molecular networks, ProSNet substantially improves the function prediction performance on five major species.

2. Methods

As an overview, ProSNet first constructs a heterogeneous biological network by integrating homology data with molecular network data of multiple species. It then performs a novel dimensionality reduction algorithm on this heterogeneous network to optimize a low-dimensional vector representation for each protein. The vectors of two proteins will be co-localized in the low-dimensional space if the proteins are close to each other in the heterogeneous biological network. A key computational contribution is that ProSNet obtains low-dimensional vectors through a fast online learning algorithm instead of the batch learning algorithm used by previous work.[23,32] In each iteration, ProSNet samples a path from the heterogeneous network and optimizes low-dimensional vectors based on this path instead of all pairs of nodes. Therefore, it can easily scale to large networks containing hundreds of thousands or even millions of edges and nodes. After finding low-dimensional vector representation for each node, ProSNet calculates an intra-species affinity score and an inter-species affinity score by transfering annotations within the same species and across different species, respectively. Finally, ProSNet predicts functions for a query protein by averaging these scores and picking the function(s) with the highest score(s).

2.1. *Heterogeneous biological network*

Definition 1. Heterogeneous Biological Networks (HBNs) are biological networks where both nodes and edges are associated with different types. In an HBN $G = (V, E, R)$, V is the set of typed nodes (i.e., each node has its own type), R is the set of edge types in the network, and E is the set of typed edges. An **edge** $e \in E$ in a heterogeneous biological network is an ordered triplet $e = \langle u, v, r \rangle$, where $u \in V$ and $v \in V$ are two typed nodes associated with this edge and $r \in R$ is the edge type.

Definition 2. In an HBN $G = (V, E, R)$, a **heterogeneous path** is a sequence of compatible edge types $\mathcal{M} = \langle r_1, r_2, \ldots, r_L \rangle$, $\forall i,\ r_i \in R$. The outgoing node type of r_i should match the incoming node type of r_{i+1}. Any path $\mathcal{P}_{e_1 \rightsquigarrow e_L} = \langle e_1, e_2, \ldots, e_L \rangle$ connecting node u_1 and u_{L+1} is a **heterogeneous path instance** following $\mathcal{M}$, iff $\forall i,\ e_i$ is of type r_i.

In particular, any edge type r is a length-1 heterogeneous path $\mathcal{M} = \langle r \rangle$. We show a toy example of an HBN under our function prediction framework in Fig. 1.

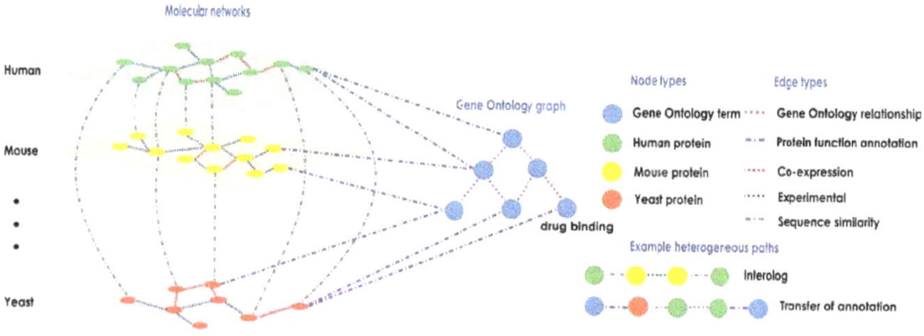

Fig. 1. An example of the heterogeneous biological network under our function prediction framework. The node set V consists of four types, {"*Human protein*", "*Yeast protein*", "*Mouse protein*", and "*Gene Ontology term*"}. The edge type set R consists of five types, {"*Sequence similarity*", "*Protein function annotation*","*Gene Ontology relationship*","*Experimental*", and "*Co-expression*" }. This HBN explicitly captures *interolog* and *transfer of annotation* through heterogeneous paths across different species.

2.2. *Low-dimensional vector learning in the heterogeneous biological network*

ProSNet finds the low-dimensional vector for each node through first sampling a large number of heterogeneous path instances according to the HBN. It then finds the optimal low-dimensional vector so that nodes that appear together in many instances turn to have similar vector representations. We first define the conditional probability of node v connected to node u by a heterogeneous path $\mathcal{M}$ as:

$$Pr(v|u,\mathcal{M}) = \frac{\exp(f(u,v,\mathcal{M}))}{\sum_{v' \in V} \exp(f(u,v',\mathcal{M}))}, \tag{1}$$

where f is a scoring function modeling the relevance between u and v conditioned on $\mathcal{M}$. Inspired from the previous work,[36] we define the following scoring function:

$$f(u,v,\mathcal{M}) = \mu_{\mathcal{M}} + \mathbf{p}_{\mathcal{M}}{}^T \mathbf{x_u} + \mathbf{q}_{\mathcal{M}}{}^T \mathbf{x_v} + \mathbf{x_u}{}^T \mathbf{x_v}. \tag{2}$$

Here, $\mu_{\mathcal{M}} \in \mathbb{R}$ is the global bias of the heterogeneous path $\mathcal{M}$. $\mathbf{p}_{\mathcal{M}}$ and $\mathbf{q}_{\mathcal{M}} \in \mathbb{R}^d$ are local bias d dimensional vectors of the heterogeneous path $\mathcal{M}$. $\mathbf{x_u}$ and $\mathbf{x_v} \in \mathbb{R}^d$ are low-dimensional vectors for nodes u and v respectively. Our framework models different heterogeneous paths differently by using $\mathbf{p}_{\mathcal{M}}$ and $\mathbf{q}_{\mathcal{M}}$ to weight different dimensions of node vectors according to the heterogeneous path $\mathcal{M}$.

For a heterogeneous path instance $\mathcal{P}_{e_1 \rightsquigarrow e_L} = \langle e_1 = \langle u_1, v_1, r_1 \rangle, \ldots, e_L = \langle u_L, v_L, r_L \rangle \rangle$ following $\mathcal{M} = \langle r_1, r_2, \ldots, r_L \rangle$, we propose the following approximation.

$$Pr(\mathcal{P}_{e_1 \rightsquigarrow e_L}|\mathcal{M}) \propto C(u_1, 1|\mathcal{M})^{\gamma} \times Pr(\mathcal{P}_{e_1 \rightsquigarrow e_L}|u_1, \mathcal{M}), \tag{3}$$

where $C(u,i|\mathcal{M})$ represents the count of path instances following $\mathcal{M}$ with the ith node being u. $C(u,i|\mathcal{M})$ can be efficiently computed through a dynamic programming algorithm. γ is a widely used parameter to control the effect of overly-popular nodes, which is set to 0.75 in previous work.[37] We assume that each node on the path only depends on its previous node. Then we have

$$Pr(\mathcal{P}_{e_1 \rightsquigarrow e_L}|u_1, \mathcal{M}) = \prod_{i=1}^{L} Pr(v_i|u_i, r_i). \tag{4}$$

Given the conditional distribution defined in Eq. (1) and (3), the maximum likelihood training is tractable but expensive because computing the gradient of the log-likelihood takes time linear in the number of nodes. Following the noise-contrastive estimation (NCE),[38] we reduce the problem of density estimation to a binary classification, discriminating between samples from path instances following the heterogeneous path and samples from a known noise distribution. In particular, we assume these samples come from the following mixture.

$$\frac{1}{\theta+1} Pr^{+}(\mathcal{P}_{e_1 \rightsquigarrow e_L}|\mathcal{M}) + \frac{\theta}{\theta+1} Pr^{-}(\mathcal{P}_{e_1 \rightsquigarrow e_L}|\mathcal{M}), \tag{5}$$

where θ is the negative sampling weight and $Pr^{+}(\mathcal{P}_{e_1 \rightsquigarrow e_L}|\mathcal{M})$ denotes the distribution of path instances in the HBN following the heterogeneous path $\mathcal{M}$. $Pr^{-}(\mathcal{P}_{e_1 \rightsquigarrow e_L}|\mathcal{M})$ is a noise distribution, and for simplicity we set

$$Pr^{-}(\mathcal{P}_{e_1 \rightsquigarrow e_L}|\mathcal{M}) \propto \prod_{i=1}^{L+1} C(u_i, i\,|\mathcal{M})^{\gamma}. \tag{6}$$

We further assume noise samples are θ times more frequent than positive path instance samples. The posterior probability that a given sample D came from positive path instance samples following the given heterogeneous path is

$$Pr(D=1|\mathcal{P}_{e_1 \rightsquigarrow e_L}, \mathcal{M}) = \frac{Pr^{+}(\mathcal{P}_{e_1 \rightsquigarrow e_L}|\mathcal{M})}{Pr^{+}(\mathcal{P}_{e_1 \rightsquigarrow e_L}|\mathcal{M}) + \theta \cdot Pr^{-}(\mathcal{P}_{e_1 \rightsquigarrow e_L}|\mathcal{M})}, \tag{7}$$

where $D \in \{0,1\}$ is the label of the binary classification. Since we would like to fit $Pr(\mathcal{P}_{e_1 \rightsquigarrow e_L}|\mathcal{M})$ to $Pr^{+}(\mathcal{P}_{e_1 \rightsquigarrow e_L}|\mathcal{M})$, we simply maximize the following expectation.

$$\begin{aligned} \mathcal{L}_{\mathcal{M}} =& \mathbb{E}_{Pr^{+}}\left[\log \frac{Pr(\mathcal{P}_{e_1 \rightsquigarrow e_L}|\mathcal{M})}{Pr(\mathcal{P}_{e_1 \rightsquigarrow e_L}|\mathcal{M}) + \theta \cdot Pr^{-}(\mathcal{P}_{e_1 \rightsquigarrow e_L}|\mathcal{M})}\right] \\ &+ \theta \cdot \mathbb{E}_{Pr^{-}}\left[\log \frac{\theta \cdot Pr^{-}(\mathcal{P}_{e_1 \rightsquigarrow e_L}|\mathcal{M})}{Pr(\mathcal{P}_{e_1 \rightsquigarrow e_L}|\mathcal{M}) + \theta \cdot Pr^{-}(\mathcal{P}_{e_1 \rightsquigarrow e_L}|\mathcal{M})}\right]. \end{aligned} \tag{8}$$

The loss function can be derived as

$$\begin{aligned} \mathcal{L}_{\mathcal{M}} \approx & \sum_{\mathcal{P}_{e_1 \rightsquigarrow e_L} \text{ following } \mathcal{M}} \log \sigma(\sum_{i=1}^{L} f(u_i, v_i, r_i)) + \\ & \sum_{j=1}^{\theta} \mathbb{E}_{\mathcal{P}^{j}_{e_1 \rightsquigarrow e_L} \sim Pr^{-}|u_1, \mathcal{M}}\left[\log\left(1 - \sigma(\textstyle\sum_{i=1}^{L} f(u_i^j, v_i^j, r_i))\right)\right], \end{aligned} \tag{9}$$

where $\sigma(\cdot)$ is the sigmoid function. Note that when deriving the above equation we used $\exp(f(u, v, \mathcal{M}))$ in place of $Pr(v|u, \mathcal{M})$, ignoring the normalization term in Eq. (1). We can do this because the NCE objective encourages the model to be approximately normalized and recovers a perfectly normalized model if the model class contains the data distribution.[38] Following the idea of negative sampling,[37] we also replaced $\sum_{i=1}^{L} f(u_i, v_i, r_i) - \log\left(\theta \cdot Pr^{-}(\mathcal{P}_{e_1 \rightsquigarrow e_L}|\mathcal{M}\right)$ with $\sum_{i=1}^{L} f(u_i, v_i, r_i)$ for ease of computation. We optimize parameters $\mathbf{x_u}, \mathbf{x_v}, \mathbf{p_r}, \mathbf{q_r}$, and $\mu_\mathbf{r}$ based on Eq. (9).

2.3. *Runtime improvements through online learning*

Like diffusion component analysis,[23] the number of pairs of nodes $\langle u, v\rangle$ that are connected by some path instances following at least one of the paths is $O(|V|^2)$ in the worst case. This is too large for storage or processing when $|V|$ is at the order of hundreds of thousands. Therefore, sampling a subset of path instances according to their distribution is the most feasible choice when optimizing, instead of going through every path instance per iteration. Thus, our method is still very efficient for networks containing large numbers of edges. Based on Eq. (3), we can sample a path instance by sampling the nodes on the heterogeneous path one by one. Once a path instance has been sampled, we use gradient descent to update the parameters $\mathbf{x_u}, \mathbf{x_v}, \mathbf{p_r}, \mathbf{q_r}$, and $\mu_\mathbf{r}$ based on Eq. (9). As a result, our sampling-based framework becomes a stochastic gradient descent framework. The derivations of these gradients are trivial and thus are omitted. Moreover, since stochastic gradient descent can generally be parallelized without locks, we can further optimize via multi-threading. Decomposing a heterogeneous network with more than sixty thousand nodes and ten million edges into a 500-dimensional vector space takes less than 30 minutes on a 12-core 3.07GZ Intel Xeon CPU through this online learning framework.

2.4. *Function prediction*

After using the above framework to find the low-dimensional vector for each protein in the HBN, ProSNet transfers annotations both within the same species and across different species to predict for a query protein.

To transfer annotations within the same species, ProSNet first uses diffusion component analysis[23] on the Gene Ontology graph[2] to find low-dimensional vector $\mathbf{y_i}$ for each functional label i. It then uses a transformation matrix $\mathbf{W}$ to project proteins from the protein vector space to the function vector space, which allows us to match proteins to functions based on geometric proximity. Let $\mathbf{y_i'}$ be the projection of the protein vector $\mathbf{x_i}$:

$$\mathbf{y_i'} = \mathbf{x_i}\mathbf{W}. \tag{10}$$

We define the intra-species affinity score z_{ij} between gene i and function j to be used for function prediction as:

$$z_{ij} = \mathbf{x_i}\mathbf{W}\mathbf{y_j^T}. \tag{11}$$

A larger z_{ij} indicates that gene i is more likely to be annotated with function j. We follow clusDCA[32] to find the optimal $\mathbf{W}$.

Since proteins from different species are located in the same low-dimensional vector space, ProSNet is able to use the annotations across different species as well. Instead of using the annotations from all the other proteins, ProSNet only considers the k most similar proteins based on the cosine similarity between their low-dimensional vectors. It then calculates the inter-species affinity score s_{ij} between gene i and function j as:

$$s_{ij} = \sum_{g \in B_i} \cos(\mathbf{x_i}, \mathbf{x_g}) \cdot \mathbb{1}(g \in T_j), \tag{12}$$

where B_i is the set of k most similar proteins of i and T_j is the set of genes that are annotated to function j in the training data.

After obtaining the intra-species affinity score $\mathbf{z}$ and inter-species affinity score $\mathbf{s}$, ProSNet normalizes them by z-scores. It predicts functions for a query protein by averaging these two normalized affinity scores and picking the function(s) with the highest score(s)

3. Experimental results

3.1. *Construction of heterogeneous biological network for function prediction*

To construct the heterogeneous biological network (HBN), we obtained six molecular networks for each of five species, including human (*Homo sapiens*), mouse (*Mus musculus*), yeast (*Saccharomyces cerevisiae*), fruit fly (*Drosophila melanogaster*), and worm (*Caenorhabditis elegans*) from the STRING database v10.[20] These six molecular networks are built from heterogeneous data sources, including high-throughput interaction assays, curated protein-protein interaction databases, and conserved co-expression data. We excluded text mining-based networks to avoid potential confounding. Each edge in the molecular networks has been associated with a weight between 0 and 1 representing the confidence of interaction. Next, we obtained protein-function annotations and the ontology of functional labels from the GO Consortium.[2] We only used annotations that have experimental evidence codes including EXP, IDA, IPI, IMP, IGI, and IEP. As a result, annotations that are based on an *in silico* analysis of the gene sequence and/or other data are removed to avoid potential leakage of labels. We built a directed acyclic graph of GO labels from all three categories [biological process (BP), molecular function (MF) and cellular component (CC)] based on "*is a*" and "*part of*" relationships. This graph has 13,708 functions and 19,206 edges. We set all edge weights of protein-function links to 1 and all edge weights between GO labels to 1. Finally, we extracted amino acid sequences of all proteins in our five-species network from the STRING database and the Universal Protein Resource (Uniprot).[17] To construct homology edges, we performed all-vs-all BLAST[13] and excluded edges with E-value larger than 1e-8. We then used the negative logarithm of the E-values as the edge weights and rescaled them into $[0, 1]$. We showed the statistics of our HBN in Tab. 1. For simplicity, all edges are undirected. Note that we excluded the protein-function annotation edges that are in the hold-out test set in the following experiments for rigorous comparisons. Our heterogeneous network is similar to the example network in Fig. 1, except that our network has five species and six different types of molecular networks.

3.2. *Experimental setting*

We used 3-fold cross-validation to evaluate the methods of interest. For a given species for evaluation, we randomly split proteins of the species into three equal-size subsets. Each time, the GO annotations of proteins in one subset were held out for testing, and the annotations of the other two subsets were used for intra-species classification training. For inter-species training, we used all experimental GO annotations from the other four species, ensuring no leakage of label information in the training data. To evaluate the predictive performance, we

Table 1. Statistics of our heterogeneous network

	Human	Mouse	Yeast	Fruit fly	Worm
#proteins	16,544	16,649	6,307	11,261	13,469
#co-expression edges	1,319,562	1,406,572	628,014	2,466,234	2,774,840
#co-occurrence edges	28,334	29,472	5,328	17,962	14,678
#database edges	275,860	347,406	66,972	116,748	69,948
#experimental edges	492,548	672,326	439,956	380,046	298,684
#fusion edges	2,678	3,994	2,722	4,026	4,336
#neighborhood edges	78,440	77,962	91,220	69,934	49,890
#human homology edges	0	525,221	55,884	202,993	159,481
#mouse homology edges	525,221	0	52,916	188,729	151,408
#yeast homology edges	55,884	52,916	0	26,950	28,269
#fruit fly homology edges	202,993	188,729	26,950	0	75,831
#worm homology edges	159,481	151,408	28,269	75,831	0
#annotations	77,950	66,238	28,668	32,259	21,655

measured the extent to which the predicted ranked list was consistent with the ground truth ranked list by computing the receiver operating characteristic curve (AUROC). We used the macro-AUROC as the evaluation metric following previous work.[31,32] The macro-AUROC is calculated by separately averaging the area under the curves for each label. We set the vector dimension $d = 500$, the number of nearest neighbors $k = 2000$, and the negative sampling weight $\theta = 5$ in our experiment. We observed that the performance of our algorithm is quite stable with different d, k, and θ values. We included all edge types in the predefined heterogeneous path set. Additionally, we added "*transfer of annotation*" to the predefined heterogeneous path set (Fig. 1).

To show the improvement from integrating homology data with molecular networks of multiple species, we compared our method with three existing state-of-the-art function prediction methods: GeneMANIA,[31] clusDCA,[32] and BLAST.[13] GeneMANIA and clusDCA integrate protein molecular networks within a given species. Neither of them is able to integrate information across different species. We used the latest released code and the suggested parameter settings for these two methods. BLAST uses bit score to rank annotations from significant hits by BLAST. We used the same datasets (i.e. annotations, proteins, and networks) and the same evaluation scheme for every method we tested.

3.3. *Molecular network data and homology data are complementary in function prediction*

We first studied whether information extracted from homology and from molecular networks are complementary. We compared the predictive performance of three different data sources: 1) molecular networks, 2) homology, 3) both molecular network and homology (integrated). We used clusDCA to predict function annotations based on molecular networks. We used BLAST to make predictions of function annotation based on homology. We summarized how many functions can be accurately annotated (AUROC>0.9) by each data source (Fig. 2). We notice that there are many functions that can only be accurately predicted by homology or network. For example, on mouse MF with 3-10 labels, 9% of functions (difference between

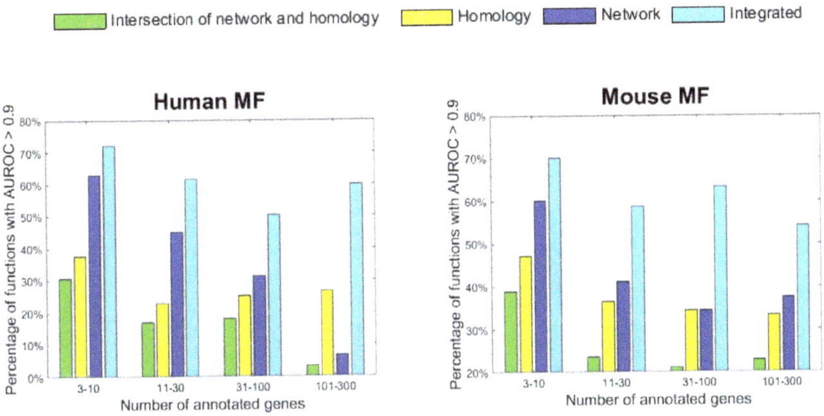

Fig. 2. Comparison of using different data sources for function prediction

yellow bar and green bar) can be accurately predicted only by homology but not by network. In the same category, another 21% of functions (difference between blue bar and green bar) can be accurately predicted only by network but not by homology. This suggests that these two data sources are complementary, and integrating them can synergistically improve the function prediction results. To this end, we integrated homology and network data by simply taking average of the z-scores of predicted annotations from these two data sources. We found that the predictive performance using both molecular network data and homology data is significantly better than only using one in all categories on both human and mouse. For example, on human MF with 101-300 labels, using both network data and homology data accurately annotates 60% of functions, which is much higher than 4% of only using network data and 26% of only using homology data. Notably, we only use the homology data from five species here. When including homology data from more species in the future, homology data may further boost the function prediction performance.

3.4. *ProSNet substantially improves function prediction performance*

We performed large-scale function prediction on all five species to compare our method to other state-of-the-art function prediction approaches. The results are summarized in Fig. 3 and Supplementary Fig. 1 (Supplementary Data). It is clear that our approach achieved the best overall results in all five species. When comparing with homology-based methods, we found that ProSNet significantly outperforms BLAST on both sparsely annotated and densely annotated labels (data not shown). For example, ProSNet achieves 0.8690 AUROC on human BP labels with 3-10 annotations, which is much higher than the 0.6326 AUROC by BLAST.

Furthermore, we compared ProSNet to existing state-of-the-art network-based methods, in-

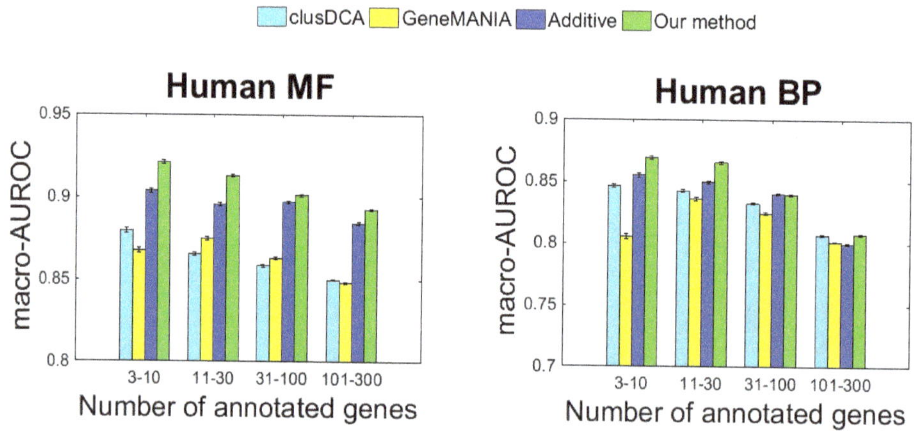

Fig. 3. Comparison of different methods

cluding clusDCA and GeneMANIA, which only integrate molecular networks of single species. We found that the overall performance of our approach is substantially higher than that of both of these methods. For instance, in human, our method achieved 0.9211 AUROC on MF labels with 3-10 annotations, which is much higher than 0.8673 by GeneMANIA and 0.8794 by clusDCA. In mouse, our method achieved 0.8523 AUROC on BP labels with 31-100 annotations, which is much higher than 0.8078 AUROC by GeneMANIA and 0.8299 AUROC by clusDCA.

To evaluate the integration of homology and network data, we developed a baseline approach that simply merges predictions made from homology data and sequence data, separately. This additive approach takes the average z-scores of the annotation score of clusDCA and BLAST to rank functional labels for each protein. We note that this baseline approach outperformes both GeneMANIA and clusDCA, indicating that integrating homology with molecular networks can substantially improve the function prediction performance. We then compared this additive approach to our method. We found that ProSNet also outperforms the additive approach. For instance, in human, our method achieves 0.9129 AUROC on MF labels with 11-30 labels, which is higher than 0.8956 AUROC by the additive approach. The improvement of our method in comparison to the additive approach demonstrates a better data integration by constructing a heterogeneous network and finding low-dimensional vector representations for each node in this network.

The improvement of ProSNet over existing network-based approaches is more pronounced on sparsely annotated functions. Since very few proteins are annotated to these functions, it is very easy to overfit any classification algorithm if we only use the data from a single species. With the integrated heterogeneous biological network, ProSNet successfully transfers annotations from other species to have a more robust and improved predictive performance on sparsely annotated functions.

4. Conclusion

In this paper, we have presented ProSNet, a novel protein function prediction method which seamlessly integrates homology data and molecular network data. ProSNet constructs a heterogeneous network to include molecular networks from all species and homology links across different species. We have designed an efficient dimensionality reduction approach which only takes 30 minutes to decompose a heterogeneous network containing hundreds of thousands of proteins. We have demonstrated that ProSNet outperforms state-of-the-art network-based approaches and homology-based approaches on five major species. Furthermore, ProSNet has achieved improved performance over an additive integration approach that simply adds predictions from network and homology data. This result supports our hypothesis that constructing a heterogeneous network and then finding low-dimensional vector representations for each node in this network is a better data integration approach. In the future, we plan to study how to annotate proteins of species that have very sparse molecular networks or even no molecular network. In addition, we plan to pursue further improvement by integrating networks and homology data from a complete spectrum of reference species.

Supplementary Data:
`http://web.engr.illinois.edu/~swang141/PSB/ProSNetSupp.pdf`

Funding
Jian Peng is supported by Sloan Research Fellowship. This research was partially supported by grant 1U54GM114838 awarded by NIGMS through funds provided by the trans-NIH Big Data to Knowledge initiative. The content is solely the responsibility of the authors and does not necessarily represent the official views of the National Institutes of Health.

References

1. B. Rost, P. Radivojac and Y. Bromberg, *FEBS Letters* (2016).
2. M. Ashburner, C. A. Ball, J. A. Blake, D. Botstein, H. Butler, J. M. Cherry, A. P. Davis, K. Dolinski, S. S. Dwight, J. T. Eppig, M. A. Harris, D. P. Hill, L. Issel-Tarver, A. Kasarskis, S. Lewis, J. C. Matese, J. E. Richardson, M. Ringwald, G. M. Rubin and G. Sherlock, *Nat. Genet.* **25**, 25 (May 2000).
3. P. Radivojac, W. T. Clark, T. R. Oron, A. M. Schnoes, T. Wittkop, A. Sokolov, K. Graim, C. Funk, K. Verspoor, A. Ben-Hur *et al.*, *Nature methods* **10**, 221 (2013).
4. Y. Jiang, T. R. Oron, W. T. Clark, A. R. Bankapur, D. D'Andrea, R. Lepore, C. S. Funk, I. Kahanda, K. M. Verspoor, A. Ben-Hur *et al.*, *arXiv preprint arXiv:1601.00891* (2016).
5. S. Burge, E. Kelly, D. Lonsdale, P. Mutowo-Muellenet, C. McAnulla, A. Mitchell, A. Sangrador-Vegas, S.-Y. Yong, N. Mulder and S. Hunter, *Database* **2012**, p. bar068 (2012).
6. Y. Loewenstein, L. Yaniv, R. Domenico, O. C. Redfern, W. James, F. Dmitrij, L. Michal, O. Christine, T. Janet and T. Anna, *Genome Biol.* **10**, p. 207 (2009).
7. W. T. Clark and P. Radivojac, *Proteins: Structure, Function, and Bioinformatics* **79**, 2086 (2011).

8. J. Gillis and P. Pavlidis, *BMC bioinformatics* **14**, p. 1 (2013).
9. R. Rentzsch and C. A. Orengo, *BMC bioinformatics* **14**, p. 1 (2013).
10. D. Cozzetto, D. W. Buchan, K. Bryson and D. T. Jones, *BMC bioinformatics* **14**, p. S1 (2013).
11. D. Lee, O. Redfern and C. Orengo, *Nature Reviews Molecular Cell Biology* **8**, 995 (2007).
12. G. Yachdav, E. Kloppmann, L. Kajan, M. Hecht, T. Goldberg, T. Hamp, P. Hönigschmid, A. Schafferhans, M. Roos, M. Bernhofer *et al.*, *Nucleic acids research* , p. gku366 (2014).
13. S. F. Altschul, T. L. Madden, A. A. Schäffer, J. Zhang, Z. Zhang, W. Miller and D. J. Lipman, *Nucleic acids research* **25**, 3389 (1997).
14. B. E. Engelhardt, M. I. Jordan, K. E. Muratore and S. E. Brenner, *PLoS Comput Biol* **1**, p. e45 (2005).
15. B. E. Engelhardt, M. I. Jordan, J. R. Srouji and S. E. Brenner, *Genome research* **21**, 1969 (2011).
16. Y. Jiang, W. T. Clark, I. Friedberg and P. Radivojac, *Bioinformatics* **30**, i609 (2014).
17. U. Consortium *et al.*, *Nucleic acids research* , p. gku989 (2014).
18. R. P. Huntley, T. Sawford, P. Mutowo-Meullenet, A. Shypitsyna, C. Bonilla, M. J. Martin and C. O'Donovan, *Nucleic acids research* **43**, D1057 (2015).
19. A. Chatr-Aryamontri, B.-J. Breitkreutz, S. Heinicke, L. Boucher, A. Winter, C. Stark, J. Nixon, L. Ramage, N. Kolas, L. ODonnell *et al.*, *Nucleic acids research* **41**, D816 (2013).
20. D. Szklarczyk, A. Franceschini, S. Wyder, K. Forslund, D. Heller, J. Huerta-Cepas, M. Simonovic, A. Roth, A. Santos, K. P. Tsafou, M. Kuhn, P. Bork, L. J. Jensen and C. von Mering, *Nucleic Acids Res.* **43**, D447 (January 2015).
21. T. Rolland, M. Taşan, B. Charloteaux, S. J. Pevzner, Q. Zhong, N. Sahni, S. Yi, I. Lemmens, C. Fontanillo, R. Mosca *et al.*, *Cell* **159**, 1212 (2014).
22. S. Oliver, *Nature* **403**, 601 (10 February 2000).
23. H. Cho, B. Berger and J. Peng, Diffusion component analysis: unraveling functional topology in biological networks, in *RECOMB*, 2015.
24. E. Sefer, S. Emre and K. Carl, Metric labeling and semi-metric embedding for protein annotation prediction, in *Lecture Notes in Computer Science*, 2011 pp. 392–407.
25. T. Milenkovic, V. Memisevic, A. K. Ganesan and N. Przulj, *J. R. Soc. Interface* **7**, 423 (6 March 2010).
26. M. Cao, C. M. Pietras, X. Feng, K. J. Doroschak, T. Schaffner, J. Park, H. Zhang, L. J. Cowen and B. J. Hescott, *Bioinformatics* **30**, i219 (2014).
27. A. K. Wong, A. Krishnan, V. Yao, A. Tadych and O. G. Troyanskaya, *Nucleic acids research* **43**, W128 (2015).
28. R. Sharan, I. Ulitsky and R. Shamir, *Molecular systems biology* **3**, p. 88 (2007).
29. E. Nabieva, K. Jim, A. Agarwal, B. Chazelle and M. Singh, *Bioinformatics* **21**, i302 (2005).
30. S. Navlakha and C. Kingsford, *Bioinformatics* **26**, 1057 (2010).
31. S. Mostafavi and Q. Morris, *Bioinformatics* **26**, 1759 (2010).
32. S. Wang, H. Cho, C. Zhai, B. Berger and J. Peng, *Bioinformatics* **31**, i357 (2015).
33. H. Yu, N. M. Luscombe, H. X. Lu, X. Zhu, Y. Xia, J.-D. J. Han, N. Bertin, S. Chung, M. Vidal and M. Gerstein, *Genome Res.* **14**, 1107 (June 2004).
34. A. J. Walhout, *Science* **287**, 116 (2000).
35. A. Sokolov, S. Artem and B.-H. Asa, Multi-view prediction of protein function, in *BCB '11*, 2011.
36. J. Pennington, R. Socher and C. D. Manning, Glove: Global vectors for word representation., in *EMNLP*, 2014.
37. T. Mikolov, I. Sutskever, K. Chen, G. S. Corrado and J. Dean, Distributed representations of words and phrases and their compositionality, in *NIPS*, 2013.
38. M. U. Gutmann and A. Hyvärinen, *The Journal of Machine Learning Research* **13**, 307 (2012).

ON THE POWER AND LIMITS OF SEQUENCE SIMILARITY BASED CLUSTERING OF PROTEINS INTO FAMILIES

CHRISTIAN WIWIE and RICHARD RÖTTGER

Department of Mathematics and Computer Science, University of Southern Denmark, Odense, Fyn, Denmark

E-mail: {wiwiec, roettger}@imada.sdu.dk

Over the last decades, we have observed an ongoing tremendous growth of available sequencing data fueled by the advancements in wet-lab technology. The sequencing information is only the beginning of the actual understanding of how organisms survive and prosper. It is, for instance, equally important to also unravel the proteomic repertoire of an organism. A classical computational approach for detecting protein families is a sequence-based similarity calculation coupled with a subsequent cluster analysis. In this work we have intensively analyzed various clustering tools on a large scale. We used the data to investigate the behavior of the tools' parameters underlining the diversity of the protein families. Furthermore, we trained regression models for predicting the expected performance of a clustering tool for an unknown data set and aimed to also suggest optimal parameters in an automated fashion. Our analysis demonstrates the benefits and limitations of the clustering of proteins with low sequence similarity indicating that each protein family requires its own distinct set of tools and parameters. All results, a tool prediction service, and additional supporting material is also available online under `http://proteinclustering.compbio.sdu.dk`.

Keywords: Protein Classification, Protein Evolution, Clustering

1. Introduction

With current wet-lab technology, we are producing a vast amount of genomic data at an ever increasing pace.[1] The knowledge of the very sequence of the organism is only one part of the complex puzzle of how organisms survive, reproduce and adopt to changing environmental conditions.[2] In order to benefit from the genomic data of an organism the data needs to be analyzed in an efficient and automated manner.

Of fundamental importance is the identification and classification of protein families fostering insights in the functional diversity of homologous proteins allowing to investigate the evolutionary history of the proteins.[3,4] Several, hand-curated databases exist providing information on protein family classification, e.g., SCOP[5] or PFAM.[6] Even though these databases are impressive in size, the number of known protein families is still growing with every sequenced organism.[7] Therefore, it is of importance to have reliable and automated means of classifying proteins in families, which can generally be separated into three groups:[8,9] pairwise alignment algorithms, generative models, and discriminative classifiers. Here, we are focusing on the common approach of pairwise alignments using NCBI BLAST[10] followed by a cluster analysis. There exists a myriad of clustering tools, all of them require different parameters and can only be used efficiently with a profound understanding of the underlying algorithm. Furthermore, as every clustering approach uses a different way of determining its optimal clustering, there is no universal best performer suiting all data sets equally well.[11]

There have been several studies comparing the performance of various clustering approaches for this task, discussing the problem from various points of view. For example, the study of Chan *et al.*[12] compares the performance of two clustering tools on three different genomes in order to assess the sensitivity of these tools towards the C+G content. The main limitation of this study is the small number of data sets and tools utilized. In a different study by Bernardes *et al.*[3] a larger-scale attempt was taken to compare the general performance of four different clustering approaches on data sets similar to our setting. The main focus of the paper was to demonstrate the limitations of sequence-based similarity functions compared to their novel profile based similarity function. Nevertheless, this work applied the tools in question only to the entire SCOP data set (with various levels of sequence identities) and clustered them into families and superfamilies. This approach neglects the variety within the protein families but gives a good overview of the general performance of the tested tools.

In contrast to previous works, we create several hundred data sets comprising smaller subsets of the SCOP data set in order to strategically assess the variance of the different protein families and their consequences to the different clustering tools. Further, we clustered each of our hundreds of data sets with extensive parameter training (1,000 parameters per data set per tool) using seven popular clustering approaches which have already demonstrated to work well on protein data sets.[11] This approach allows for a more detailed evaluation of the performances and limitations of the clustering tools. We further use the massive database of 100 of thousands clustering results generated during this work in order to conduct a meta learning approach, comparable to the work of De Souto *et al.*,[13] for the prediction of the expected clustering performance and thus a tool ranking. We also suggest the parameter settings for the tools, as we can identify the most similar data set in our database together with the best parameters.

To summarize, we present an in-depth analysis of protein clustering and the inherent variability of the data sets. We intensively investigated the performance of the tools on 202 different data sets with 1,000 different parameter settings each. We investigated the behavior of the tools and their parameters, reflecting the diversity of the different protein families. With a meta-learning approach we aim to predict the expected performance of the clustering tools on unseen data sets. We utilized intrinsic properties of the data sets (e.g., matrix rank or the cluster coefficient) and used them as features of a regression model for the prediction. We also provide the performance predictor as a web-service together with all results, the source code of the predictor, and additional information at `http://proteinclustering.compbio.sdu.dk`.

2. Materials

2.1. *Data sets*

We based our work on the Astral SCOPe 2.06 data set with less than 40% sequence identity.[5] This scenario is very challenging for clustering tools as the alignment scores fall into the so-called twilight zone when the sequence identity drops below 35%.[14] The data set provides a gold standard classification derived from the SCOP database which we utilize in order to assess the cluster quality. The Astral data set classifies each protein into a hierarchy of *class*, *fold*, *superfamily* and eventually *family*.

For our goal of predicting the expected performance of the clustering tools we require a multitude of data sets. Therefore, we have created sub-samples of the Astral data set by splitting it into classes and folds, i.e., we have created a single data set for each class, containing only the sequences of the one class and one data set for each fold in the same fashion. In the remainder we will refer to them as the *class data sets* and the *fold data sets*. This serves two purposes: (1) we received a sufficient number of data sets and (2) we were able to assess the diversity of the protein families and their impact on the clustering tools. We calculated pairwise BLAST[10] hits (E-value cut-off 100) between all protein sequences and converted them into similarities using the "Coverage BeH" method by Wittkop *et al.*[15] (coverage factor $f = 20$, cut-off $100,000$).

Given these data sets, we cluster each of them into the corresponding families, leading to the following two *scenarios*: *Class → Families* and *Fold → Families*. We performed a final filtering process by excluding all those data sets containing only one cluster, e.g., a fold containing only one family. We excluded them because they are trivial to cluster and would hugely distort the parameter prediction. After this final step we created seven class data sets and 195 fold data sets.

2.2. *Clustering Tools*

Table 1. Overview of the chosen clustering methods. We assign an abbreviation to each of the tools. We optimized the denoted parameters for each of the tools.

Abbreviation	Name	Optimized Parameter(s)
CDP	Clusterdp[16]	Kernel radius $dc \in [\wedge, \vee]$
HC(linkage)	Hierarchical Clustering[17]	Number of clusters $k \in [2, n]$
MCL	Markov Clustering[18]	Inflation $I \in [1.1, 10]$
PAM	Partitioning Around Medoids[19]	Number of clusters $k \in [2, n-1]$
TC	Transitivity Clustering[20]	$T \in [\wedge, \vee]$

We based our tool selection on the top performers (using the F1-score[21]) of a previous large-scale performance comparison of various clustering approaches,[11] summarized in Table 1. The F1-score is defined as the harmonic mean of precision and recall when comparing a cluster result with a gold standard. Generally, external validity indices (i.e., measures comparing against a gold standard) evaluate a result with regard to the purity of individual clusters and the completeness of the clusters.[11,21] In that context, the F1-score is a comprehensive measure that takes both of these into account by combining two external measures (precision and recall). The F1-score is the quasi-standard in clustering evaluation and has already proved useful in many biomedical contexts.[11,21] All considered clustering tools performed very well with an average F1-score of over 0.7 in the original study. We excluded tools which return overlapping clusters, as the F1-Score is undefined for such clusterings. We treat hierarchical clustering as three tools, depending on the linkage function used (single, complete, average).

3. Methods

3.1. *Data Statistics & Clustering*

For each data set, we calculated 25 data statistics (see Table 2). We selected these statistics to reflect a wide variety of properties of the data sets. Note, that some of the statistics are

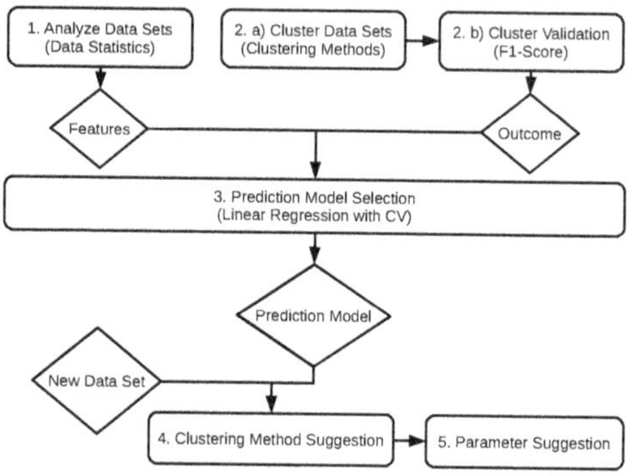

Fig. 1. Overview of the workflow of the presented method. **(1)** We calculate the features for the models, **(2a)** perform a clustering of all data sets and **(2b)** evaluate their quality. (1) and (2) are used to **(3)** train a regression model. **(4)** This model is used to predict the expected performance of each tool and suggests **(5)** the parameters.

correlated; this fact and the influence on the models is discussed in Section 3.2. The ranges of all statistics except *Minimal Similarity*, *Maximal Similarity* and *Number Samples* were normalized to $[0,1]$ to avoid biases in the trained regression models due to differences in the value ranges.

We utilized ClustEval[11] to execute each clustering tool with 1,000 different parameter sets as indicated in Table 1 and validated the results using the F1-Score. The maximal execution time of any tool per clustering was limited to 15 minutes as we occasionally observed degenerated execution times depending on the used parameters.

3.2. *Regression & Feature Selection*

For each clustering tool we selected an ordinary, Lasso and Ridge regression model. We used the R functions *lm*, *glmnet* ($\alpha = 1$) and *glmnet* ($\alpha = 0$) to train ordinary, Lasso and Ridge regression models respectively. The data set statistics used as features for the regression models are potentially correlated and thus might be troublesome for regression models. For this reason, we perform a feature selection for the ordinary linear regression. Lasso and Ridge regression already have an intrinsic feature selection, thus they were not subject to an additional feature selection.

We trained each of the three regression models per tool using the data statistics as feature variables. The outcome variables are either the best achieved F1-Score of each tool on each data set, or the parameter leading to the best result; depending on whether we want to predict the F1-Scores or the parameters. To assess the quality of the prediction, we used the mean absolute error (MAE) to measure error rates: $\text{MAE}(\hat{y}, y) = \frac{1}{N}\sum_{i=1}^{N} |\hat{y}_i - y_i|$ where $\hat{y}_i$ denotes the prediction, y_i the real value for data set i, and N the total number of data sets. Using MAE

Table 2. Overview of the calculated data statistics. The *Absolute Z-Score*, *Assortativity* and *Similarity Percentiles* are parameterized, i.e., we calculate the same statistic multiple times for different parameters. The brackets behind the statistic name denotes the number of parameters used.

Data Statistic Name	Description
Absolute Z-Scores (4)	The fraction of all object pairs having a similarity within {1,2,3,4} standard deviations from the mean.
Assortativity, un/weighted[22] (2)	The preference for vertices with same degree to connect to each other in the similarity graph.
Clustering Coefficient, avg.[23]	The ratio of fully connected triplets of nodes to connected triplets of nodes in the similarity graph.
Graph Adhesion[24]	The number of edges to remove such that the similarity graph falls into several connected components.
Graph Density[25]	The ratio of the number of edges and the number of possible edges in the similarity graph.
Graph Diversity, avg.[26]	The average scaled Shannon entropy of the weights of the incident edges on each vertex in the similarity graph.
Graph Min-Cut[25]	The sum of edge weights to remove such that the similarity graph falls into several connected components.
Matrix Rank	The number of independent rows in the similarity matrix.
Maximal Similarity	The largest similarity in the similarity matrix.
Minimal Similarity	The smallest similarity in the similarity matrix.
Number Samples	The number of objects in the input data set.
Similarity Percentiles (10)	The fraction of all object pairs having a similarity within the {[0-10],[10-20],...,[90-100]} similarity percentile.

allows for easy interpretation of the error-rate compared to other measures such as the root mean squared error (RMSE).[27]

3.2.1. *Cross Validation*

In order to estimate prediction errors for a trained model we utilize a 10-fold cross validation. We repeated the cross validations 100 times with different folds to minimize the influence of a single fold. Note that the Astral data set has only seven classes, thus when only using the class data sets, a Leave-one-out cross validation (LOOCV) was performed instead.

3.2.2. *Feature Selection for Ordinary Regression Models*

We utilized a greedy forward feature selection approach coupled with 10-fold cross validations to select features and thus models with small prediction error while trying to avoid overfitting. In each step of the process, we successively added that feature to the model which lead to the smallest cross validation prediction error estimate.

During this feature selection procedure, we generate models of increasing complexity, i.e., using more features. Thus, both training and testing errors of the cross-validation will decrease in the beginning. However, with increasing number of features, the model will overfit the training data which is indicated by a growing prediction error. The moment we observe a growing prediction error, we stop adding features and report the current model as the final model. A similar feature selection procedure was previously published in Pahikkala *et al.*[28]

4. Results & Discussion

4.1. *Data Statistics*

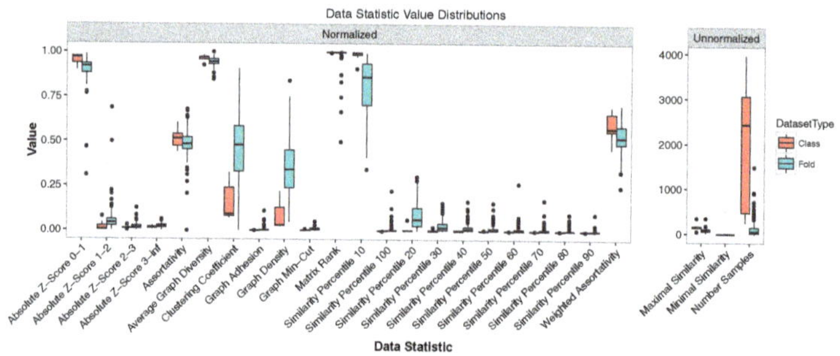

Fig. 2. The distributions of data statistic values for the class and fold data sets. We normalized data statistics using the theoretical maxima where available. The statistics *Minimal Similarity*, *Maximal Similarity* and *Number Samples* are not normalized.

Figure 2 summarizes the calculated data statistics for both class and fold data sets. Generally, some statistics such as *Graph Min-Cut*, *Graph Density* or *Clustering Coefficient* emphasize the sparsity of the pair-wise similarity matrix of the protein sequences. This is due to the fact that the proteins in the Astral data set do not have large sequence similarities resulting in many protein pairs without any significant BLAST hit. Further, we want to highlight two interesting observations:

(1) There is a clear difference in the statistical properties between the class and fold data sets. Again, this is due to the many protein pairs without any BLAST hit. The ratio of these pairs is larger in the class data sets which contain even more distantly related proteins. This is most clearly seen on Statistics such as *Average Graph Diversity*, *Clustering Coefficient*, *Graph Density*, *Similarity Percentile 10/20* and *Absolute Z-Score 0-1/1-2* which are very sensitive to this proportion.

(2) Even data sets of the same type (i.e., fold or class) vary hugely. This demonstrates the variety of the different protein families. This is even more pronounced in the fold data sets as they contain fewer families and thus are more susceptible to "outlier" families whereas in the class data sets, the variety of the different statistics is generally more balanced.

4.2. *Clustering Tool Performances*

We clustered all data sets into protein families using the clustering tools summarized in Table 1 to all previously mentioned class and fold data sets. The resulting F1-Scores are depicted in Figure 3. Generally, the selected clustering methods perform well on the data sets. HC(complete), MCL and PAM perform on average slightly worse than their competitors. The

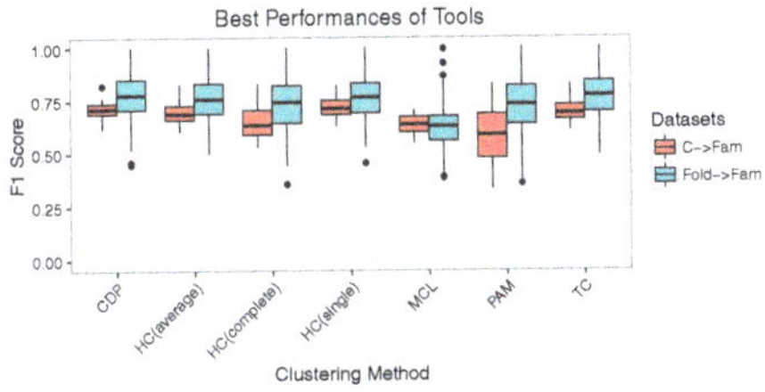

Fig. 3. The best tool performances as F1-Scores for the two scenarios: Clustering class (C) or fold (Fold) into families (Fam).

performance of PAM on class data sets might be due to our execution time limit of 15 minutes per clustering. For k-parameter values close to the real number of clusters in the classes, the algorithm does not finish in time. On the other hand, we only have seven of those data sets in this study, so the effect on the performance should be limited. None of the other methods were affected by the time limit. The general trend is that fold data sets can be clustered better (on average) than class data sets which can be explained by the fact that the class data sets are sparser. When ranking the tools by their F1-Score performance for each data set it shows that there is no best performer across all data sets, as expected. Rather, several tools alternate in taking the top ranks. The lack of a universal best performer and the variance in the rankings emphasize that performances and rankings are highly data set dependent. This further motivates the demand of a predictor based on data statistics.

4.3. *Clustering Tool Parameters*

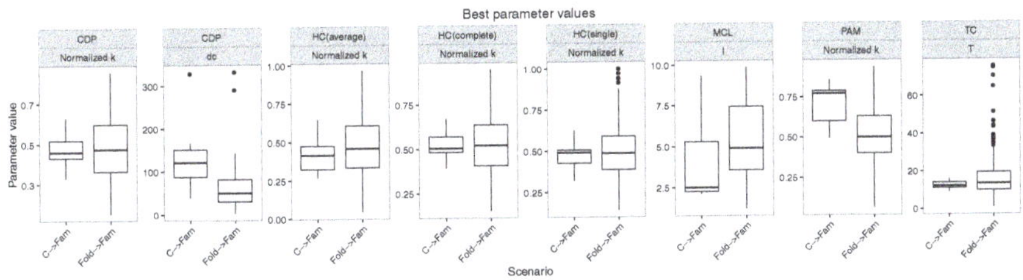

Fig. 4. The best performing parameter values of all tools. If a tool performed best with multiple parameter values we took the mean. Each tool was executed using 1,000 different parameter values. Because CDP has two main parameters we assessed both of them with an equal number of values: $32 * 32 = 1024$. Note that the parameter k is normalized by the number of objects.

We compared the best parameters of each tool for the two scenarios. Figure 4 summarizes our findings. Clearly, when clustering a fold data set we can observe a considerably larger variety for all tools. Parameters directly reflecting the desired number of clusters, i.e. k, have been normalized with the number of objects in the data set. Please note, that we cannot use the mean k parameter as a general "rule-of-thumb" as this value entirely depends on the average family size in the data set which is determined by the way we created the data sets. Nevertheless, the variance in the k parameters certainly demonstrate the variance in protein families. The only outlier with respect to the k parameter is PAM, again likely due to the runtime restriction.

Interestingly, the parameters of CDP and MCL have different means when clustering classes compared to clustering folds. This has practical implications, as for an unknown data set it is impossible to determine whether it is comprised of a class, a fold or a mixture. The threshold T of TC remains stable regardless of the data set type, with a larger variance for the fold data sets, including some significant outliers. Overall, this indicates that a naive parameter suggestion for arbitrary protein data sets is not feasible at least it does not do justice to the variety present in different protein families.

4.4. *Predicting Tool Performance*

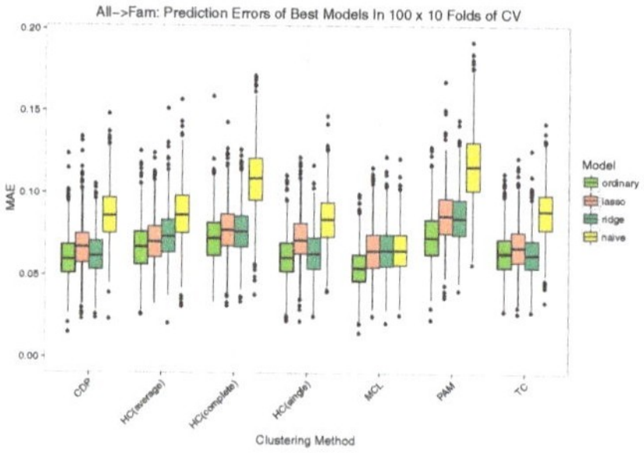

Fig. 5. The tool performance prediction errors for the final models of each tool when trained on all data sets. The prediction errors were estimated with 100×10-fold cross validations. The yellow boxes represent the performance of the naive model.

Figure 5 compares the tool performance prediction errors of the final models for all tools when trained on all data sets. We also calculate a naive predictor serving as baseline which predicts the average performance of each tool over all training data sets.

Generally, our final models outperform the naive models for all clustering tools except MCL (difference in MAE of ≥ 0.025). Note that prediction errors are relatively low for both kinds of models as all clustering tools performed well on the selected data sets. On average, the predictions of the naive models have an MAE ≈ 0.1, while those of our final models show an MAE ≈ 0.075. Ordinary models generally outperform Lasso and Ridge regression models in terms of MAE. The general trend is MAE(ordinary) $<$ MAE(lasso) $<$ MAE(ridge). However, the differences between ordinary, Lasso and Ridge regression are very small.

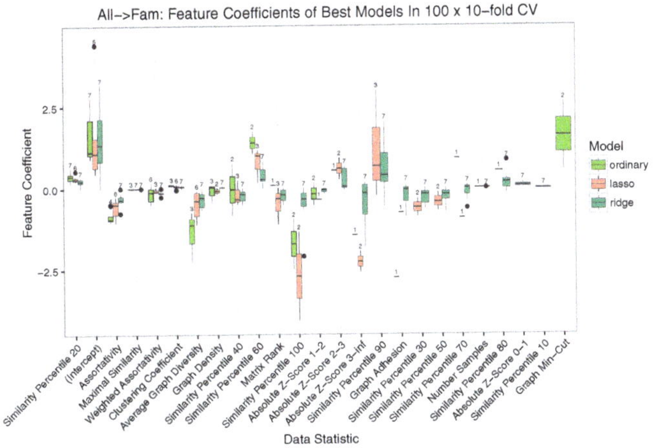

Fig. 6. This figure depicts the features and their average value in the different regression models for tool performance prediction. The features are sorted according to how often they have been selected by all models. We treated features with zero coefficient as not being in the model. The small number above each box indicates how often the feature was selected by the model represented by the box. Please note that the feature *Minimal Similarity* was never selected and thus is omitted from the figure.

Here, we want to point out the limitations of the models presented. A meaningful prediction is only possible in case the features of the unknown data set are in the same range as the features of the training data sets. We have chosen the ASTRAL data set with only up to 40% sequence similarity as we expected to observe here the most extreme feature distributions compared to data sets with higher sequence similarity.

Therefore, we have also tested the performance of the prediction with data sets not used for training. For that we have used the SCOP data set with proteins having 95% or less sequence identity; we proceeded as with the original data set and separated it also into the different classes. The error for the predicted F1 score with 0.083 for Lasso and 0.084 for Ridge regression was still remarkably small. Only the ordinary regression model showed a clear drop in performance with an average error of 0.191. This indicates that the ordinary regression is the most sensitive model with respect to unseen feature values. We will constantly update the model with new clustering results in order to further improve the quality and robustness of the models over time. To this point, the presented models should rather be regarded as a proof-of-concept.

Furthermore, we compared which data statistics have been chosen as features in the different types of models (see Figure 6). Features that have been chosen by all models clearly have predictive power for the tool performances. Examples for such features are the $[10, 20]$-*Similarity Percentile*, *Assortativity*, *Maximal Similarity* and *Weighted Assortativity*. The coefficients of the maximal similarity are very small compared to the other features, as this feature is not normalized and thus takes large values across the data sets.

The *Graph Diversity* measures whether a node in the similarity graph is very similar to only few other nodes (low diversity) or is equally similar to many nodes (high diversity). All model types chose this statistic as a predictor with negative impact on the tool performance. This might be explained by the fact that a very high diversity implies equal similarities between all nodes, leading to the lack of an actual cluster structure.

Interestingly, the selected *Similarity Percentile* statistics indicate that details of the similarity distribution have a large predictive power for the tool performance. For example, many pairwise similarities between the [10 – 20]-*Similarity Percentile* indicate a better tool performance while fewer pairwise similarities between the [90 – 100]-*Similarity Percentile* have the opposite effect.

Surprisingly, the *Clustering Coefficient* does not enter many models with a large coefficient. Equally surprising, given the performance difference between the class and fold data sets, is that the data set size is only very rarely chosen as a feature.

4.5. *Predicting Tool Parameters*

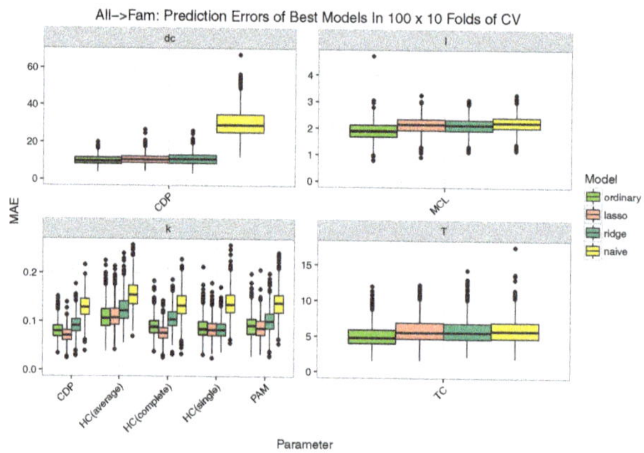

Fig. 7. The prediction performances for clustering tool parameters when trained on all data sets. Note that the various k parameters are summarized in one common plot and are normalized by the data set size.

As already previously discussed, a simple parameter suggestion valid for all data sets is not feasible due to the large variance in the protein families. Therefore, we applied the same pipeline as for the quality prediction to the parameters of the tools as well.

The results are summarized in Figure 7 and show a more mixed quality. We do not outperform the naive predictor for the threshold parameter T of TC and the Inflation parameter of MCL. We clearly outperform the naive predictor in the case of the *dc* parameter of CDP as well as the k parameters of all tools using such a parameter. Nevertheless, as discussed earlier, the k parameter is highly dependent on the way we have sampled the data sets, thus the predictive power has to be taken into account with care. Overall, the results indicate that an automated parameter prediction is not reliably possible with the presented simple models and may require more test data and more sophisticated models. In practice, the user has to resort to other methods for finding suitable parameters.[29]

5. Conclusion

With this work, we have thoroughly investigated the performance of seven well-known and established clustering tools and have particularly investigated the behavior of the tools' parameters. We have observed that all tools perform quite well on these data sets. Nevertheless, the good performance can only be reached when exhaustive parameter finding by means of a comparison against a gold standard is performed. In practice, such gold standards are not available and consequently the parameters need to be retrieved by different means. When investigating the behavior of the parameters, we cannot suggest the user a single parameter for all data sets due to the high variance of the protein families. Only TC shows a consistent behavior of a parameter which is not directly dependent on the number of clusters. Overall, a single fixed parameter cannot account for the potential variety in the data sets. Even though the k parameter also shows a consistent behavior, it is not suitable for any recommendations as this behavior results from the way we have sampled our data sets which cannot be expected in practice.

Given this massive repository of clustering results at hand, we utilized it for learning regression models for predicting the expected performance of the investigated tools on previously unseen data sets. The presented model does outperform the naive model. Especially when considering that all clustering tools performed constantly well, the achieved prediction accuracy is notable. We also tested the models on data sets which have not been part of the training process. This can be seen as a strong indicator that it is generally possible to identify data sets suitable for a particular tool in an automated fashion. We have created a web-service where the user can upload a data set and receive the expected performance of the different tools. Please be advised that the model might fail when presented with data sets whose feature values are outside of the range of values the model was trained on. The web service also presents the features of the most similar training data set for comparison. The service is available under `http://proteinclustering.compbio.sdu.dk`. We will constantly enhance the model with additional data in order to cover a broader variety of data set features and thus creating more reliable predictions.

More generally speaking, the study shows that state-of-the-art clustering tools, when presented only with sequence similarities, have limitations with capturing the high diversity of protein families and require a specific parameter for every data set which cannot be easily provided in practice. Nevertheless, the performance achieved by the tools is certainly good enough to render this approach a viable one; probably the biggest limitation is due to the rather simple similarity function only using sequence data. Fed with more sophisticated similarity functions, these tools might be able to capture the nature of the data set even better.

References

1. M. L. Metzker, *Nature reviews genetics* **11**, 31 (2010).
2. J. Baumbach, A. Tauch and S. Rahmann, *Briefings in bioinformatics* **10**, 75 (2009).
3. J. S. Bernardes, F. R. Vieira, L. M. Costa and G. Zaverucha, *BMC bioinformatics* **16**, p. 1 (2015).
4. S. Whelan and N. Goldman, *Molecular biology and evolution* **18**, 691 (2001).
5. N. K. Fox, S. E. Brenner and J.-M. Chandonia, *Nucleic Acids Research* **42**, D304 (dec 2013).
6. R. D. Finn, P. Coggill, R. Y. Eberhardt, S. R. Eddy, J. Mistry, A. L. Mitchell, S. C. Potter, M. Punta, M. Qureshi, A. Sangrador-Vegas *et al.*, *Nucleic acids research* **44**, D279 (2016).
7. V. Kunin, I. Cases, A. J. Enright, V. de Lorenzo and C. A. Ouzounis, *Genome biology* **4**, p. 1 (2003).
8. J. Chen, B. Liu and D. Huang, *BioMed Research International* **2016** (2016).
9. L. Liao and W. S. Noble, *Journal of computational biology* **10**, 857 (2003).
10. S. F. Altschul, W. Gish, W. Miller, E. W. Myers and D. J. Lipman, *Journal of Molecular Biology* **215**, 403 (oct 1990).
11. C. Wiwie, J. Baumbach and R. Röttger, *Nature Methods* **12**, 1033 (sep 2015).
12. C. X. Chan, M. Mahbob and M. A. Ragan, *BMC bioinformatics* **14**, p. 1 (2013).
13. M. C. De Souto, R. B. Prudencio, R. G. Soares, D. S. De Araujo, I. G. Costa, T. B. Ludermir and A. Schliep, Ranking and selecting clustering algorithms using a meta-learning approach, in *2008 IEEE International Joint Conference on Neural Networks (IEEE World Congress on Computational Intelligence)*, 2008.
14. B. Rost, *Protein engineering* **12**, 85 (1999).
15. T. Wittkop, J. Baumbach, F. P. Lobo and S. Rahmann, *BMC Bioinformatics* **8**, p. 396 (2007).
16. A. Rodriguez and A. Laio, *Science* **344**, 1492 (jun 2014).
17. R Core Team, *R: A Language and Environment for Statistical Computing.* R Foundation for Statistical Computing, Vienna, Austria, (2014).
18. S. Dongen, *A Cluster Algorithm for Graphs*, tech. rep. (Amsterdam, The Netherlands, The Netherlands, 2000).
19. M. Maechler, P. Rousseeuw, A. Struyf, M. Hubert and K. Hornik, *cluster: Cluster Analysis Basics and Extensions*, (2016). R package version 2.0.4 — For new features, see the 'Changelog' file (in the package source).
20. T. Wittkop, D. Emig, S. Lange, S. Rahmann, M. Albrecht, J. H. Morris, S. Böcker, J. Stoye and J. Baumbach, *Nature Methods* **7**, 419 (jun 2010).
21. J. Handl, J. Knowles and D. B. Kell, *Bioinformatics* **21**, 3201 (may 2005).
22. M. E. J. Newman, *Physical Review E* **67** (feb 2003).
23. A. Barrat, M. Barthlemy, R. Pastor-Satorras and A. Vespignani, *Proc Natl Acad Sci U S A* **101**, 3747 (Mar 2004).
24. D. R. White and F. Harary, *Sociological Methodology* **31**, 305 (2001).
25. R. Diestel, *Graph Theory (Graduate Texts in Mathematics)* (Springer, 2006).
26. N. Eagle, M. Macy and R. Claxton, *Science* **328**, 1029 (2010).
27. C. J. Willmott and K. Matsuura, *Climate research* **30**, 79 (2005).
28. T. Pahikkala, A. Airola and T. Salakoski, Speeding up greedy forward selection for regularized least-squares, in *2010 Ninth International Conference on Machine Learning and Applications*, (Institute of Electrical & Electronics Engineers (IEEE), dec 2010).
29. R. Röttger, P. Kalaghatgi, P. Sun, S. de Castro Soares, V. Azevedo, T. Wittkop and J. Baumbach, *Bioinformatics* , p. bts653 (2012).

IMAGING GENOMICS

LI SHEN

Center for Neuroimaging, Department of Radiology and Imaging Sciences
Center for Computational Biology and Bioinformatics
Indiana University School of Medicine
355 West 16th Street Suite 4100, Indianapolis, IN 46202
E-mail: shenli@iu.edu

LEE A.D. COOPER

Department of Biomedical Informatics, Emory University School of Medicine
Department of Biomedical Engineering, Georgia Institute of Technology
PAIS Building, 36 Eagle Row, 5th Floor South, Atlanta, GA 30322
E-mail: lee.cooper@emory.edu

Imaging genomics is an emerging research field, where integrative analysis of imaging and omics data is performed to provide new insights into the phenotypic characteristics and genetic mechanisms of normal and/or disordered biological structures and functions, and to impact the development of new diagnostic, therapeutic and preventive approaches. The Imaging Genomics Session at PSB 2017 aims to encourage discussion on fundamental concepts, new methods and innovative applications in this young and rapidly evolving field.

1. Introduction

Imaging genomics[1–9] is an emerging research field that arises with the recent advances in acquiring high throughput omics data and multimodal imaging data. Its major task is to perform integrative analysis of genomics data and structural, functional and molecular imaging data. Bridging imaging and genomic factors and exploring their connections have the potential to provide important new insights into the phenotypic characteristics and genetic mechanisms of normal and/or disordered biological structures and functions, which in turn will impact the development of new diagnostic, therapeutic and preventive approaches.

Binformatics strategies for imaging genomics, which is a relatively young field,[1–4] have been rapidly evolving. Early studies started with the simplest strategy to examine pairwise univariate associations[10,11] between genetic markers and imaging phenotypes. To identify more flexible associations involving multiple genetic markers and multiple imaging phenotypes, recent studies employed multiple regression and multivariate models,[12] sometimes coupled with powerful machine learning approaches[13] and valuable prior knowledge[14] to discover relevant imaging and genomic features. To increase statistical power and reduce false positives, meta-analysis studies[15,16] were performed to quantitatively synthesize imaging genomic findings from multiple independent analyses. To hunt for "missing heritability", epistatic studies[17] were performed to examine genetic interaction effects on imaging phenotypes. To identify biologically meaningful findings with increased statistical power, imaging genetic enrichment analysis[18] was proposed to mine set level associations in both imaging and genomic domains.

The topic of imaging genomics has recently been addressed in several medical imaging and bioinformatics conferences. The most focused one is the International Imaging Genetics

Conference (IIGC, http://www.imaginggenetics.uci.edu/), which is an annual meeting organized at the UC Irvine since 2005. The MICCAI Workshop on Imaging Genetics (MICGen, http://micgen.csail.mit.edu/) has been held twice in conjunction with the major medical image computing conference MICCAI in 2014 and 2015. An educational course on "Introduction to Imaging Genetics" has been offered at the annual meeting of the Organization for Human Brain Mapping (OHBM) since 2009. The topic of imaging genomics has also been covered in the following two events in the bioinformatics field: (1) ACM BCB 2015 Workshop on The Computational Pathology: Linking Tissue Phenotypes with Genomics and Clinical Outcomes, and (2) ICIBM 2015 Tutorial in Bioimage Informatics and Integrative Genomics.

As the field of imaging genomics contains a significant genomics (or omics in general) component in addition to biomedical imaging, we feel that it is timely for a major bioinformatics conference such as PSB to address this important, relevant and emerging topic. We believe that PSB offers an ideal and timely opportunity to bring together people with different expertise and shared interests in this rapidly evolving field. Specifically, the computational biology and bioinformatics expertise of the PSB and ISCB communities can provide important new perspective, complementary to the expertise of the IIGC, MICCAI, OHBM, ACM BCB and ICIBM communities, and thus can help contribute new concepts, methods, and applications to the analysis of emerging imaging and genomic data.

The scale and complexity of multidimensional imaging and omics data provide us unprecedented opportunities in enhancing mechanistic understanding of complex disorders such as neurological diseases[19–21] and cancers,[22,23] which can benefit public health outcomes by facilitating diagnostic and therapeutic progress. However, due to the extremely high dimensionality and complex structure of these data sets, this field is facing major computational and bioinformatics challenges. The technological advance in this field is urgently needed and has the potential to significantly contribute to multiple national health priority areas including *the Precision Medicine Initiative*,[24] *the Brain Initiative*,[25] and *the Big Data to Knowledge Initiative*.[26]

The objective of this Imaging Genomics Session at PSB 2017 is to encourage discussion on fundamental concepts, novel methods and innovative applications. We hope that this session will become a forum for researchers to exchange ideas, data, and software, in order to speed up the development of innovative technologies for hypothesis testing and data-driven discovery in Imaging Genomics.

2. Session Summary

This session includes an invited lecture and five accepted presentations with peer-reviewed papers. Three presentations will be delivered as platform talks and the other two as posters.

2.1. *Invited Talk*

Our invited lecture will be given by Dr. Paul Thompson, a world renowned pioneer in imaging genomics. Dr. Thompson is from the University of Southern California (USC). At USC, he is a Professor of Neurology, Psychiatry, Radiology, Pediatrics, Engineering, and Ophthalmology, the director of the USC Imaging Genetics Center, and the director of the ENIGMA

Center for Worldwide Medicine, Imaging & Genomics – an $11M NIH Center of Excellence in Big Data Computing. Dr. Thompson's major contributions to the field of imaging genomics and to the science in general can be summarized by the following text quoted from http://keck.usc.edu/faculty/paul-m-thompson/:

> Paul Thompson directs the ENIGMA Consortium, a global alliance of 307 scientists in 33 countries who conduct the largest studies of 10 major brain diseases – ranging from schizophrenia, depression, ADHD, bipolar illness and OCD, to HIV and addictions on the brain. ENIGMA's genomic screens of over 31,000 people's brain scans and genome-wide data (published in Nature Genetics, 2012; Nature, 2015) have brought together experts from 185 institutions to unearth genetic variants that affect brain structure, disease risk, and brain connectivity. Collaborating with imaging labs around the world, Dr. Thompson and his students have published over 1,300 publications (h-index: 116) describing novel mathematical and computational strategies for analyzing brain image databases, for detecting pathology in individual patients and groups, and for creating disease-specific atlases of the human brain.

2.2. *Papers*

In *Integrative analysis for lung adenocarcinoma predicts morphological features associated with genetic variations*, **Wang et al.** analyzed an imaging genomic data set downloaded from the TCGA portal, containing 201 patients with lung adenocarcinoma (LUAD). The data includes clinical information, mRNA expression profiles, and histopathologic whole slide images of the patients. On the imaging end, the authors calculated 283 morphological features from histopathologic images, and identified features strongly correlated with patient survival outcome. On the genomic end, the authors constructed the gene co-expression network and extracted gene co-expression clusters. To relate imaging with genomics, the authors regressed the outcome-relevant morphological feature on multiple co-expressed gene clusters using Lasso. The study identified gene clusters highly associated with DNA copy number variations. These observations may lead to new insight on lung cancer development, suggesting biological pathways from genetic variations, gene transcription, cancer morphology to survival outcome.

In *Identification of discriminative imaging proteomics associations in Alzheimer's disease via a novel sparse canonical correlation model*, **Yan et al.** analyzed an imaging proteomic data set downloaded from the Alzheimer's Disease Neuroimaging Initiative (ADNI) database. Participants include 42 healthy controls, 67 patients with mild cognitive impairment (MCI), and 67 patients with Alzheimer's disease (AD). The data includes clinical information, magnetic resonance imaging (MRI) scans, and expression data of 229 proteomic analytes (83 from cerebrospinal fluid and 146 from plasma). The authors developed a novel machine learning model, called discriminative sparse canonical correlation analysis (DSCCA), and applied it to the joint analysis of imaging, proteomic and diagnostic data. This analysis yielded a strong imaging proteomic association so that the identified imaging and proteomic components had also high discriminative power. Such an outcome-relevant imaging proteomic pattern has the potential to improve mechanistic understanding of the disease.

In *Enforcing co-expression in multimodal regression framework*, **Zille et al.** analyzed an imaging genomic data set collected by Mind Clinical Imaging Consortium (MCIC). Participants include 116 controls and 92 schizophrenia patients. The data includes clinical information, functional MRI (fMRI) scans, and genotyping data. The authors developed a new machine learning model, called MT-CoReg, by combining sparse regression with canonical correlation analysis; and applied it to the analysis of the MCIC data. The analysis identified imaging and genomic markers that not only induce a strong imaging genomic association but also can jointly predict the outcome.

In *Adaptive testing of SNP-brain functional connectivity association via a modular network analysis*, **Gao et al.** analyzed an imaging genomic data set downloaded from the ADNI database. Participants include 162 ADNI subjects: 73 with no *APOE* E4 allele, 67 with one copy of the *APOE* E4 allele, and 22 with two copies of the *APOE* E4 allele. The authors analyzed the resting-state fMRI data to identify modular structures in brain functional networks, using a weighted gene co-expression network analysis (WGCNA) framework, coupled with topological overlap matrix (TOM) elements in hierarchical clustering. After that, they employed an adaptive association test based on the proportional odds model to identify distinct modular structures in brain functional networks in relation to different *APOE* E4 groups.

In *Exploring brain transcriptomic patterns: a topological analysis using spatial expression networks*, **Kuncheva et al.** analyzed whole genome whole brain gene expression data downloaded from the Allen Human Brain Atlas (AHBA). Participants include six AHBA donors. The authors focused on 16,906 genes selected based on a previous study, and 105 brain regions where at least one measurement in all 6 brains were available. A Spatial Expression Network (SEN) was extracted for each gene to quantify co-expression patterns amongst several anatomical locations. After that, network similarity measures were computed and used to quantify the topological resemblance between pairs of SENs and identify naturally occurring clusters. The analysis identified three stable clusters, including one with genes specifically involved in the nervous system, and the other two representing immunity, transcription and translation.

2.3. *Discussion*

Most of these studies were facilitated by and conducted using the Big Data resources available in the open science domain, including TCGA analyzed in (**Wang et al.**), ADNI analyzed in (**Yan et al.** & **Gao et al.**), and AHBA analyzed in (**Kuncheva et al.**). The imaging data investigated by these studies ranged from histological whole slide images of cancer specimens in (**Wang et al.**), structural MRI scans in (**Yan et al.**), functional MRI scans in (**Zille et al.** & **Gao et al.**), to images of mRNA expression levels across the brain in (**Kuncheva et al.**). The omics data examined in these studies were also diverse, including DNA genotyping data in (**Zille et al.** & **Gao et al.**), mRNA expression profiles in (**Wang et al.** & **Kuncheva et al.**), and proteomic expression profiles in (**Yan et al.**).

These studies were performed to better understand the brain transcriptomic patterns in healthy controls (**Kuncheva et al.**), the brain imaging genomic or imaging proteomic patterns in Alzheimer's disease (**Yan et al.** & **Gao et al.**) or schizophrenia (**Zille et al.**), and biological pathways from gene transcription, tissue morphology to survival outcome in lung cancer

(**Wang et al.**). As to the bioinformatics strategies, a variety of machine learning methods were employed or newly developed in these studies, including network analysis and clustering models used in (**Wang et al.**, **Gao et al.** & **Kuncheva et al.**), regression models used in (**Wang et al.**), an adaptive association test used in (**Gao et al.**), an integrative regression and canonical correlation analysis model used in (**Zille et al.**), and an outcome-regularized sparse canonical correlation analysis model used in (**Yan et al.**).

While **Gao et al.** studied functional brain network as an innovative imaging phenotype, **Kuncheva et al.** aimed to identify gene clusters using whole brain spatial expression networks. The remaining three studies (**Wang et al.**, **Yan et al.** & **Zille et al.**) shared a common theme to examine the relationship among three levels (i.e., omics features, imaging phenotypes, and clinical outcomes). This suggests a promising future direction to integrate imaging genomics with systems biology, which attempts to model complex and interactive multilevel biological systems using multimodal imaging and multidimensional omics data sets.

3. Acknowledgements

We would like to thank all the authors for their high-quality submissions and excellent presentations. We would like to thank all the reviewers for taking their time and effort to evaluate the papers and provide valuable feedbacks. We would like to thank Dr. Paul Thompson of University of Southern California for giving an outstanding invited lecture. We would like to thank Dr. Kun Huang of Ohio State University for sharing his valuable experience on how to organize a successful PSB session. We would like to thank the PSB 2017 chairs and Tiffany Murray of Stanford University for their great help and support.

References

1. A. R. Hariri and D. R. Weinberger. Imaging genomics. *Br Med Bull*, 65:259–70, 2003.
2. V. S. Mattay and T. E. Goldberg. Imaging genetic influences in human brain function. *Curr Opin Neurobiol*, 14(2):239–47, 2004.
3. A. R. Hariri, E. M. Drabant, and D. R. Weinberger. Imaging genetics: perspectives from studies of genetically driven variation in serotonin function and corticolimbic affective processing. *Biol Psychiatry*, 59(10):888–97, 2006.
4. D. C. Glahn, T. Paus, and P. M. Thompson. Imaging genomics: mapping the influence of genetics on brain structure and function. *Hum Brain Mapp*, 28(6):461–3, 2007.
5. P. M. Thompson, N. G. Martin, and M. J. Wright. Imaging genomics. *Curr Opin Neurol*, 23(4):368–73, 2010.
6. L. Shen, P. M. Thompson, S. G. Potkin, L. Bertram, L. A. Farrer, T. M. Foroud, R. C. Green, X. Hu, M. J. Huentelman, S. Kim, J. S. Kauwe, Q. Li, E. Liu, F. Macciardi, J. H. Moore, L. Munsie, K. Nho, V. K. Ramanan, S. L. Risacher, D. J. Stone, S. Swaminathan, A. W. Toga, M. W. Weiner, A. J. Saykin, and Alzheimer's Disease Neuroimaging Initiative. Genetic analysis of quantitative phenotypes in AD and MCI: imaging, cognition and biomarkers. *Brain Imaging Behav*, 8(2):183–207, 2014.
7. M. G. ElBanan, A. M. Amer, P. O. Zinn, and R. R. Colen. Imaging genomics of Glioblastoma: state of the art bridge between genomics and neuroradiology. *Neuroimaging Clin N Am*, 25(1):141–53, 2015.
8. W. B. Pope. Genomics of brain tumor imaging. *Neuroimaging Clin N Am*, 25(1):105–19, 2015.

9. A. J. Saykin, L. Shen, X. Yao, S. Kim, K. Nho, S. L. Risacher, V. K. Ramanan, T. M. Foroud, K. M. Faber, N. Sarwar, L. M. Munsie, X. Hu, H. D. Soares, S. G. Potkin, P. M. Thompson, J. S. Kauwe, R. Kaddurah-Daouk, R. C. Green, A. W. Toga, M. W. Weiner, and Alzheimer's Disease Neuroimaging Initiative. Genetic studies of quantitative MCI and AD phenotypes in ADNI: Progress, opportunities, and plans. *Alzheimers Dement*, 11(7):792–814, 2015.
10. L. Shen, S. Kim, S. L. Risacher, K. Nho, S. Swaminathan, J. D. West, T. Foroud, N. Pankratz, J. H. Moore, C. D. Sloan, M. J. Huentelman, D. W. Craig, B. M. Dechairo, S. G. Potkin, Jr. Jack, C. R., M. W. Weiner, A. J. Saykin, and Alzheimer's Disease Neuroimaging Initiative. Whole genome association study of brain-wide imaging phenotypes for identifying quantitative trait loci in MCI and AD: A study of the ADNI cohort. *Neuroimage*, 53(3):1051–63, 2010.
11. J. L. Stein, X. Hua, S. Lee, A. J. Ho, A. D. Leow, A. W. Toga, A. J. Saykin, L. Shen, T. Foroud, N. Pankratz, M. J. Huentelman, D. W. Craig, J. D. Gerber, A. N. Allen, J. J. Corneveaux, B. M. Dechairo, S. G. Potkin, M. W. Weiner, P. Thompson, and Alzheimer's Disease Neuroimaging Initiative. Voxelwise genome-wide association study (vGWAS). *Neuroimage*, 53(3):1160–74, 2010.
12. D. P. Hibar, J. L. Stein, O. Kohannim, N. Jahanshad, A. J. Saykin, L. Shen, S. Kim, N. Pankratz, T. Foroud, M. J. Huentelman, S. G. Potkin, Jr. Jack, C. R., M. W. Weiner, A. W. Toga, P. M. Thompson, and Alzheimer's Disease Neuroimaging Initiative. Voxelwise gene-wide association study (vGeneWAS): multivariate gene-based association testing in 731 elderly subjects. *Neuroimage*, 56(4):1875–91, 2011.
13. M. Vounou, E. Janousova, R. Wolz, J. L. Stein, P. M. Thompson, D. Rueckert, G. Montana, and Alzheimer's Disease Neuroimaging Initiative. Sparse reduced-rank regression detects genetic associations with voxel-wise longitudinal phenotypes in Alzheimer's disease. *Neuroimage*, 60(1):700–16, 2012.
14. J. Yan, L. Du, S. Kim, S. L. Risacher, H. Huang, J. H. Moore, A. J. Saykin, L. Shen, and Alzheimer's Disease Neuroimaging Initiative. Transcriptome-guided amyloid imaging genetic analysis via a novel structured sparse learning algorithm. *Bioinformatics*, 30(17):i564–71, 2014.
15. J. L. Stein, S. E. Medland, A. A. Vasquez, D. P. Hibar, R. E. Senstad, A. M. Winkler, R. Toro, K. Appel, R. Bartecek, O. Bergmann, M. Bernard, A. A. Brown, D. M. Cannon, M. M. Chakravarty, A. Christoforou, M. Domin, O. Grimm, M. Hollinshead, A. J. Holmes, G. Homuth, J. J. Hottenga, C. Langan, L. M. Lopez, N. K. Hansell, K. S. Hwang, S. Kim, G. Laje, P. H. Lee, X. Liu, E. Loth, A. Lourdusamy, M. Mattingsdal, S. Mohnke, S. M. Maniega, K. Nho, A. C. Nugent, C. O'Brien, M. Papmeyer, B. Putz, A. Ramasamy, J. Rasmussen, M. Rijpkema, S. L. Risacher, J. C. Roddey, E. J. Rose, M. Ryten, L. Shen, E. Sprooten, E. Strengman, A. Teumer, D. Trabzuni, J. Turner, K. van Eijk, T. G. van Erp, M. J. van Tol, K. Wittfeld, C. Wolf, S. Woudstra, A. Aleman, S. Alhusaini, L. Almasy, E. B. Binder, D. G. Brohawn, R. M. Cantor, M. A. Carless, A. Corvin, M. Czisch, J. E. Curran, G. Davies, M. A. de Almeida, N. Delanty, C. Depondt, R. Duggirala, T. D. Dyer, S. Erk, J. Fagerness, P. T. Fox, N. B. Freimer, M. Gill, H. H. Goring, D. J. Hagler, D. Hoehn, F. Holsboer, M. Hoogman, N. Hosten, N. Jahanshad, M. P. Johnson, D. Kasperaviciute, Jr. Kent, J. W., P. Kochunov, J. L. Lancaster, S. M. Lawrie, D. C. Liewald, R. Mandl, M. Matarin, M. Mattheisen, E. Meisenzahl, I. Melle, E. K. Moses, T. W. Muhleisen, et al. Identification of common variants associated with human hippocampal and intracranial volumes. *Nat Genet*, 44(5):552–61, 2012.
16. D. P. Hibar, J. L. Stein, M. E. Renteria, A. Arias-Vasquez, S. Desrivieres, N. Jahanshad, R. Toro, K. Wittfeld, L. Abramovic, M. Andersson, B. S. Aribisala, N. J. Armstrong, M. Bernard, M. M. Bohlken, M. P. Boks, J. Bralten, A. A. Brown, M. M. Chakravarty, Q. Chen, C. R. Ching, G. Cuellar-Partida, A. den Braber, S. Giddaluru, A. L. Goldman, O. Grimm, T. Guadalupe, J. Hass, G. Woldehawariat, A. J. Holmes, M. Hoogman, D. Janowitz, T. Jia, S. Kim, M. Klein, B. Kraemer, P. H. Lee, L. M. Olde Loohuis, M. Luciano, C. Macare, K. A. Mather, M. Mattheisen, Y. Milaneschi, K. Nho, M. Papmeyer, A. Ramasamy, S. L. Risacher, R. Roiz-Santianez, E. J.

Rose, A. Salami, P. G. Samann, L. Schmaal, A. J. Schork, J. Shin, L. T. Strike, A. Teumer, M. M. van Donkelaar, K. R. van Eijk, R. K. Walters, L. T. Westlye, C. D. Whelan, A. M. Winkler, M. P. Zwiers, S. Alhusaini, L. Athanasiu, S. Ehrlich, M. M. Hakobjan, C. B. Hartberg, U. K. Haukvik, A. J. Heister, D. Hoehn, D. Kasperaviciute, D. C. Liewald, L. M. Lopez, R. R. Makkinje, M. Matarin, M. A. Naber, D. R. McKay, M. Needham, A. C. Nugent, B. Putz, N. A. Royle, L. Shen, E. Sprooten, D. Trabzuni, S. S. van der Marel, K. J. van Hulzen, E. Walton, C. Wolf, L. Almasy, D. Ames, S. Arepalli, A. A. Assareh, M. E. Bastin, H. Brodaty, K. B. Bulayeva, M. A. Carless, S. Cichon, A. Corvin, J. E. Curran, M. Czisch, et al. Common genetic variants influence human subcortical brain structures. *Nature*, 520(7546):224–9, 2015.

17. A. L. Zieselman, J. M. Fisher, T. Hu, P. C. Andrews, C. S. Greene, L. Shen, A. J. Saykin, and J. H. Moore. Computational genetics analysis of grey matter density in Alzheimer's disease. *BioData Min*, 7:17, 2014.
18. Xiaohui Yao, Jingwen Yan, Sungeun Kim, Kwangsik Nho, Shannon L. Risacher, Mark Inlow, Jason H. Moore, Andrew J. Saykin, and Li Shen. Two-dimensional enrichment analysis for mining high-level imaging genetic associations. *Brain Informatics*, pages 1–11, 2016.
19. M. W. Weiner, D. P. Veitch, P. S. Aisen, L. A. Beckett, N. J. Cairns, J. Cedarbaum, R. C. Green, D. Harvey, C. R. Jack, W. Jagust, J. Luthman, J. C. Morris, R. C. Petersen, A. J. Saykin, L. Shaw, L. Shen, A. Schwarz, A. W. Toga, J. Q. Trojanowski, and Alzheimer's Disease Neuroimaging Initiative. 2014 update of the Alzheimer's Disease Neuroimaging Initiative: A review of papers published since its inception. *Alzheimers Dement*, 11(6):e1–120, 2015.
20. R. J. Hodes and N. Buckholtz. Accelerating Medicines Partnership: Alzheimer's Disease (AMP-AD) knowledge portal aids alzheimer's drug discovery through open data sharing. *Expert Opin Ther Targets*, 20(4):389–91, 2016.
21. Parkinson Progression Marker Initiative. The Parkinson Progression Marker Initiative (PPMI). *Prog Neurobiol*, 95(4):629–35, 2011.
22. J. McCain. The cancer genome atlas: new weapon in old war? *Biotechnol Healthc*, 3(2):46–51B, 2006.
23. K. Clark, B. Vendt, K. Smith, J. Freymann, J. Kirby, P. Koppel, S. Moore, S. Phillips, D. Maffitt, M. Pringle, L. Tarbox, and F. Prior. The Cancer Imaging Archive (TCIA): maintaining and operating a public information repository. *J Digit Imaging*, 26(6):1045–57, 2013.
24. F. S. Collins and H. Varmus. A new initiative on precision medicine. *N Engl J Med*, 372(9):793–5, 2015.
25. M. McCarthy. US to launch major brain research initiative. *BMJ*, 346:f2156, 2013.
26. L. Ohno-Machado. NIH's Big Data to Knowledge initiative and the advancement of biomedical informatics. *J Am Med Inform Assoc*, 21(2):193, 2014.

ADAPTIVE TESTING OF SNP-BRAIN FUNCTIONAL CONNECTIVITY ASSOCIATION VIA A MODULAR NETWORK ANALYSIS

CHEN GAO, JUNGHI KIM and WEI PAN*, for the Alzheimer's Disease Neuroimaging Initiative*
Division of Biostatistics, School of Public Health, University of Minnesota
**E-mail: weip@biostat.umn.edu*

Due to its high dimensionality and high noise levels, analysis of a large brain functional network may not be powerful and easy to interpret; instead, decomposition of a large network into smaller subcomponents called modules may be more promising as suggested by some empirical evidence. For example, alteration of brain modularity is observed in patients suffering from various types of brain malfunctions. Although several methods exist for estimating brain functional networks, such as the sample correlation matrix or graphical lasso for a sparse precision matrix, it is still difficult to extract modules from such network estimates. Motivated by these considerations, we adapt a weighted gene co-expression network analysis (WGCNA) framework to resting-state fMRI (rs-fMRI) data to identify modular structures in brain functional networks. Modular structures are identified by using topological overlap matrix (TOM) elements in hierarchical clustering. We propose applying a new adaptive test built on the proportional odds model (POM) that can be applied to a high-dimensional setting, where the number of variables (p) can exceed the sample size (n) in addition to the usual $p < n$ setting. We applied our proposed methods to the ADNI data to test for associations between a genetic variant and either the whole brain functional network or its various subcomponents using various connectivity measures. We uncovered several modules based on the control cohort, and some of them were marginally associated with the APOE4 variant and several other SNPs; however, due to the small sample size of the ADNI data, larger studies are needed.

Keywords: aSPU test; brain functional connectivity; functional MRI; proportional odds model; single nucleotide polymorphism; weighted gene co-expression network analysis; WGCNA.

1. Introduction

Resting-state functional magnetic resonance imaging (rs-fMRI) is gaining popularity in studies of brain functional connectivity with applications to detection of subtle network reorganizations in Alzheimer's disease.[1] Disruption of connectivity in the brain functional network is related to many pathological conditions in the brain, such as Alzheimer's disease,[2] schizophrenia,[3] or autism.[4] This necessitates the development of methods for modelling the brain functional network its statistical inference.

A network is comprised of nodes and edges connecting the nodes. Based on functional MRI data, a popular choice of nodes are brain regions of interest (ROIs) while the edges are connectivities reflecting statistical dependencies between ROIs. An important network model, the scale-free network,[5] assumes that most nodes in a network are sparsely connected with the exception of a few "hub" nodes that are densely connected with other nodes. In the scale-free network model, new connections are more likely to occur for those hub nodes with already-high

*Data used in preparation of this article were obtained from the Alzheimer's Disease Neuroimaging Initiative (ADNI) database (`adni.loni.usc.edu`). As such, the investigators within the ADNI contributed to the design and implementation of ADNI and/or provided data but did not participate in analysis or writing of this report. A complete listing of ADNI investigators can be found at: `http://adni.loni.usc.edu/wp-content/uploads/how_to_apply/ADNI_Acknowledgement_List.pdf`

connectivity. There has been empirical evidence supporting this model for brain functional networks,[6] though it is still debatable. In addition, the scale-free network model also admits a modular topological structure, which can be extracted for more efficient analyses for human brains.

Methods for drawing statistical inference to distinguish brain connectivity for different groups of subjects are still under development. The first question encountered is how to define brain functional connectivity. Ref. 7 discussed the choice between Pearson's marginal correlation coefficient and partial correlation coefficient as a network connectivity measure, though other measures are possible and it is yet unclear which one is best. To reduce dimensionality and to reach sparseness, graphical lasso is often used for estimating networks for different groups. Since an estimated network with the imposed sparsity penalty may not demonstrate modular structures, a better approach is to directly discover the modules in a network. A general framework for estimating scale-free networks and detecting modules is proposed in Ref. 8 for gene network analysis, which has gained tremendous popularity in genomics.[9] It starts by defining a similarity measure between two nodes in a network, called adjacency, using the marginal correlation coefficient. Soft-thresholding is then applied, leading to a weighted network. The soft-thresholded adjacency is further transformed to a topological overlap matrix (TOM) element, which is converted to a dissimilarity measure for hierarchical clustering, grouping closely connected nodes together as modules in the network. The above framework not only provides multiple network connectivity measures, but also carries out modular structure identification. The connectivity measures and identified modules in the brain functional network may help statistical inference and offer biological insights.[9]

In this paper, for the first time, we adapt the use of WGCNA for gene expression data to rs-fMRI data, constructing weighted brain functional networks and identifying their subnetworks or modules using the Alzheimer's Disease Neuroimaging Initiative (ADNI) data. We explored using the adjacency matrix element and TOM element, in addition to the marginal correlation or covariance, to characterize connectivity in brain functional networks. Taking advantages of detected network modules, we conduct association analysis of genetic variants with not only the whole brain functional network, but also its various subcomponents, including its modules, which aims to not only improve statistical power, but also offer better biological interpretation. We propose applying a new adaptive association test based on a proportional odds model (POM) accounting for the ordinal nature of the SNP genotype. We found evidence of associations between several network modules and the APOE4 variant, which is by far the most significant genetic risk factor for Alzheimer's disease.

This paper is organized as follows. We first review the method of WGCNA, including its module identification, then introduce the adaptive test based on a POM. We demonstrate the application of our methods to the ADNI data before summarizing our findings and future research directions in the discussion section.

2. Methods

2.1. *Module detection via weighted gene co-expression network analysis*

In this section, we briefly review the work in Ref. 8 on the weighted gene-coexpression network analysis (WGCNA) framework for network construction and module identification.

2.1.1. *Adjacency matrix*

The first step of the WGCNA framework is to define a similarity measure between gene expression profiles; in the current context, we use the BOLD signals in each of multiple ROIs from one or more subjects to calculate a similarity between any two ROIs. The similarity measure is required to take values between 0 and 1. A typical choice of this similarity measure is the absolute value of the Pearson correlation coefficient $s_{uv} = |cor(u,v)|$, for nodes u and v. Another choice, which preserves the sign of correlation, is defined as $s_{uv} = [1 + cor(u,v)]/2$. We refer the first one as unsigned similarity measure, and the second one as the signed similarity measure. From our experience of applications to the ADNI data, the identified modules have negligible differences using either unsigned or signed similarity measure. We used the unsigned similarity measure throughout this paper.

Once the similarity measure is computed, the next step is to transform the similarity matrix $S = [s_{uv}]$ into an adjacency matrix using an adjacency function. Hard thresholding is often used to yield a binary or unweighted network with a 0/1 adjacency indicating no-connection/connection and thus possible loss of information, though a more efficient multi-scale approach with multiple thresholds yielding a set of binary networks has been proposed.[10] Soft thresholding is a simple and popular alternative with more flexibilities. One choice is the power adjacency function

$$a_{uv} = power(s_{uv}, \delta) \equiv |s_{uv}|^{\delta} \tag{1}$$

with parameter δ, which is chosen as the smallest integer such that the scale-free network model fitting is above a certain threshold.

2.1.2. *Topological overlap matrix*

Instead of using only the adjacency matrix, Ref. 11 advocated a topological overlap matrix $\Omega = [\omega_{uv}]$ with its element as a potentially more useful measure that reflects the relative interconnectedness of two nodes u and v after accounting for their shared neighbors. The topological overlap matrix element is defined as

$$\omega_{uv} = \frac{l_{uv} + a_{uv}}{\min\{k_u, k_v\} + 1 - a_{uv}} \tag{2}$$

with $k_u = \sum_v a_{uv}$ and $l_{uv} = \sum_q a_{uq}a_{qv}$. For a binary network with $a_{uv} = 0$ or 1, k_u is the connectivity of node u representing the number of its direct neighbors, while l_{uv} equals the number of nodes that connect both nodes u and v; $\omega_{uv} = 0$ if the nodes u and v are not connected and they are not connected to the same neighbors; in contrast, $\omega_{uv} = 1$ if the nodes u and v are connected and the neighbors of the node with fewer edges are also connected to the one with more edges. For any network, $0 \leq a_{uv} \leq 1$ implies $0 \leq \omega_{uv} \leq 1$.

2.1.3. *Module identification*

To identify modules in a network, we need to have a dissimilarity or distance measure. An intuitive way is to convert a similarity measure. Based on the topological overlap matrix element ω_{uv}, we can simply define the dissimilarity measure as $d^{\omega}_{uv} = 1 - \omega_{uv}$. The TOM-based dissimilarity d^{ω}_{uv} is used as the input for average linkage hierarchical clustering. The output from hierarchical clustering is a dendrogram composed of branches and leaves. In a brain functional network, each leaf corresponds to a ROI. The hierarchical clustering algorithm groups the closest ROIs and forms the branches. By cutting the branches of the dendrogram, closely related ROIs are identified as a module. Among the several methods for cutting the branches of the dendrogram, the default used in the WGCNA framework is Dynamic Tree Cut from the R package `dynamicTreeCut`.

Once modules are identified, one can calculate an intramodular connectivity

$$\omega.in_u = \sum_{v \in M} \omega_{uv} \tag{3}$$

for each node u in its module M. Ref. 8 pointed out that intramodular connectivities $\omega.in$ may represent important features of the nodes (i.e. ROIs).

2.2. *An adaptive association test based on the proportional odds model*

Let $Y_i = 0, 1, 2$ denote the count of the minor allele for subject i for a given SNP of interest, then Y_i has $J = 3$ ordered categories. The logistic regression model cannot be applied in this situation, because it only allows the response variable to be binary. A popular choice for ordinal data is the proportional odds model (POM),[12] which we will briefly describe here.

Suppose subject i has p network connectivities denoted by $X_i = (x_{i1}, \ldots, x_{ip})$ and l covariates denoted by $Z_i = (z_{i1}, \ldots, z_{il})$. For the proportional odds model, we define the regression coefficients $\beta = (\beta_1, \ldots, \beta_p)'$ for the network connectivities and $\delta = (\delta_1, \ldots, \delta_l)'$, and a vector of intercepts $\alpha = (\alpha_0, \ldots, \alpha_{J-2})'$. The proportional odds model is

$$\text{logit}[Pr(Y_i \leq j)] = \alpha_j + Z_i\delta + X_i\beta, \qquad j = 0, 1. \tag{4}$$

The likelihood for equation Eq. 4 can be derived based on the multinomial distribution for the categorical variable Y_i, from which maximum likelihood estimates and statistical inference can be obtained as implemented in R package `MASS` or `VGAM`. However, numerical issues such as non-convergence arise when p, the dimension of β, is relatively large as compared to the sample size n.

Here we propose applying a class of tests that are applicable to the high-dimensional setting with $p > n$, from which an adaptive test is constructed to summarize information across the tests. No that most existing tests cannot be applied to the case $p > n$. To test the null hypothesis $H_0 : \beta = (\beta_1, \beta_2, \ldots, \beta_p)' = 0$, we can use the score vector derived in Ref. 13,

$$U_\beta = \sum_{i=1}^{n} \sum_{j=0}^{J-2} (1 - \hat{r}_{i(j-1)} - \hat{r}_{ij}) \cdot I(Y_i = j) \cdot X_i \tag{5}$$

where $\hat{r}_{ij} = \exp(\hat{\alpha} + Z_i\hat{\delta})/[1 + \exp(\hat{\alpha} + Z_i\hat{\delta})]$ comes from the fitted null model of Eq. 4 (i.e. with

$\beta = 0$); $\hat{\alpha}$ and $\hat{\delta}$ are estimated by the `polr` function in the R package `MASS`. Let U_k denote the kth component of the score vector $U_\beta = (U_1, \ldots, U_p)'$. The SPU($\gamma$) test statistic is defined as

$$T_{SPU(\gamma)} = \sum_{k=1}^{p} U_k^\gamma, \tag{6}$$

where $\gamma \geq 1$ is an integer. As the parameter γ increases, a connectivity with a larger absolute value of the score gains a higher weight. In the extreme situation, when $\gamma \to \infty$ as an even integer, $SPU(\infty)$ takes only the maximum component of the score vector, i.e., $T_{SPU(\infty)} = \max_{k=1}^{p} |U_k|$.

The p-values of the SPU tests are computed by permuting the residuals from the null model B times, and the p-value can be calculated as

$$P_{SPU(\gamma)} = \frac{(\sum_{b=1}^{B} I[|T_{SPU(\gamma)}^{(b)}| \geq |T_{SPU(\gamma)}|] + 1)}{(B+1)}, \tag{7}$$

where $T_{SPU(\gamma)}^{(b)}$ is the SPU(γ) statistic based on the bth set of permuted residuals. Since the value of γ that yields highest power cannot be determined a priori, an adaptive SPU (aSPU) test is introduced to combine the evidence across multiple SPU tests,

$$T_{aSPU} = \min_{\gamma \in \Gamma} P_{SPU(\gamma)}, \tag{8}$$

where $P_{SPU(\gamma)}$ is the p-value of $SPU(\gamma)$ test statistics and Γ is a set of integers for the power of aSPU test. In the numerical examples throughout this paper, we chose γ from the set $\Gamma = \{1, 2, \ldots, 8, \infty\}$. To calculate the p-value of T_{aSPU}, we can use the same permutation scheme as used for calculating the p-values of T_{SPU}'s. For each permuted residual set b, after calculating $T_{SPU(\gamma)}^{(b)}$ and its p-value $p_\gamma^{(b)} = (\sum_{b_1 \neq b} I[T_{SPU(\gamma)}^{(b_1)} \geq T_{SPU(\gamma)}^{(b)}] + 1)/B$. Then we can obtain $T_{aSPU}^{(b)} = \min_{\gamma \in \Gamma} p_\gamma^{(b)}$, and the p-value of T_{aSPU} is

$$P_{aSPU} = \frac{(\sum_{b=1}^{B} I[T_{aSPU}^{(b)} \leq T_{aSPU}] + 1)}{(B+1)}. \tag{9}$$

A step-wise procedure is used to gradually increase B if needed. We can start with $B = 10^3$ initially, then increase to $B = 10^5$ (or bigger) if a p-value is smaller than 5×10^{-3} (or smaller). The test is implemented in R package `POMaSPU` to be available on CRAN.

3. Results

3.1. *ADNI Data*

Data used in the preparation of this article were obtained from the Alzheimer's Disease Neuroimaging Initiative (ADNI) database (`adni.loni.usc.edu`). We included all subjects from the normal and Alzheimer's disease (AD) groups in the ADNI data. We applied motion correction and global signal regression to reduce noises.

Here we used the power adjacency function $a_{uv} = power(s_{uv}, \beta) = |s_{uv}|^\beta$ (equation (1)). β was selected as the smallest β such that the scale-free model fitting R^2 was above a pre-set threshold 0.85.

3.2. *Distinct modular structures in brain functional networks based on APOE4 SNP genotype scores*

For the ADNI data, we grouped the subjects based on the APOE4 SNP (rs429358) minor allele counts $(0, 1, 2)$. APOE4 plays a major role in the pathogenesis of Alzheimer's disease.[14,15] The APOE4 variant is a major risk factor for both early- and late-onset Alzheimer's disease.[14,15] We removed those subjects with a missing rs429358 value, resulting in a total of 162 subjects. Among them, 73 subjects have no minor allele at rs429358, whereas 67 subjects have one minor allele and 22 subjects have two. In order to establish possible modular structures in brain functional networks in the normal condition, we first applied the WGCNA framework to the rs-fMRI data of the control subjects only. Specifically, for each ROI, we concatenated the BOLD time series of all the control subjects, which were used to calculate the similarity between any two ROIs (i.e. the absolute value of Pearson's correlation between any two BOLD time series), then conducting the subsequent analyses in the WGCNA framework. At the end, we identified four modules based on the data from the control cohort (Figure 1).

Based on the modules identified, we continued to explore them for each APOE4 SNP genotype group. To measure the network connectivities, we used the correlation matrix, covariance matrix, and the topological overlap matrix (TOM). The rows and columns are ordered in the same way as in Figure 1. Distinct modular structures seem to be present in the correlation, covariance and TOM plots across the APOE4 genotype groups (Figure 2).

3.3. *Adaptive testing for SNP-module associations*

Using the APOE4 SNP (rs429358) minor allele counts as the response in a POM, we tested the association between the APOE4 SNP and the network connectivities. Covariates including age, gender and years of education were adjusted. Using the aSPU test, we found that the covariance matrix elements were marginally associated with the APOE4 SNP ($P = 0.033$, Table 1). We further decomposed the whole network connectivities into two exclusive subsets: connectivities within the four modules and those between the modules. Both the between-modular covariance and TOM were associated with the APOE4 SNP with $P < 0.05$.

Next we focused on the network connectivities in each individual module, and tested their association with the APOE4 SNP (Table 2). The network connectivities defined by the correlations in the yellow module showed evidence of association with the APOE4 SNP ($P = 0.017$). In addition, the network connectivities defined by covariance matrix elements in the blue and yellow modules were also associated with the APOE4 SNP ($P = 0.034, P = 0.011$).

Finally we tested for association between each module-specific intramodular connectivity $\omega.in$ and the APOE4 SNP. Only the yellow module showed a significant association with $P = 0.007$.

There are 30 and 19 ROIs in the blue and yellow modules, respectively. The ROIs identified in the yellow modules includes left/right sides of posterior cingulate cortex, angular gyrus, superior frontal cortex, middle frontal cortex, and inferior frontal cortex. For comparison, Ref. 13 identified 18 nodes related to the default mode network (DMN), including left/right sides of superior frontal cortex, medial prefrontal cortex, ventral anterior cingulate cortex, posterior cingulate cortex, parahipppocampal cortex, inferior parietal cortex, angular, middle

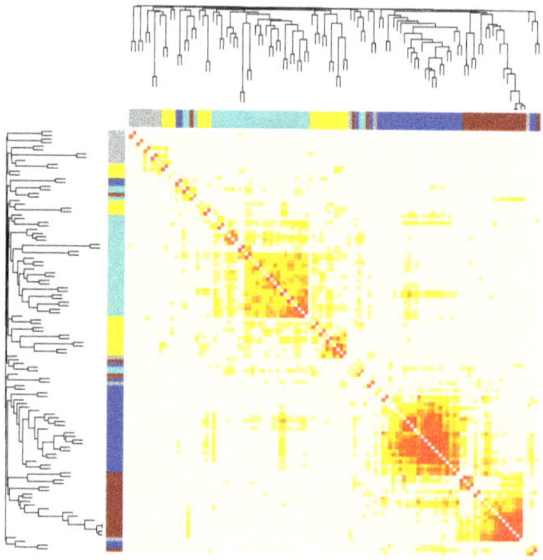

Fig. 1. TOM plot of the whole brain functional network and its modules for normal subjects. The rows and columns are the ROIs, ordered by their distance in the tree.

temporal gyrus, and inferior temporal cortex.[16–18] We found that 15 ROIs in the yellow module are also related to the 18 nodes in the DMN. For example, the posterior cingulate cortex plays a pivotal role in the default mode network of the brain.[19,20] The posterior cingulate cortex is linked to cognitive functions such spatial memory, configural learning, and maintenance of discriminative avoidance learning and.[21,22] It is shown in the DMN that Alzheimer's disease affects the posterior cingulate cortex.[20] Angular gyrus is another region found in both DMN and the yellow module. Loss of grey matter volume in angular gyrus has been associated with dementia and progression to Alzheimer's disease.[23] The association between the APOE4 variant and the network connectivity measures in the yellow module also uncovers some key brain regions in DMN that were found to be affected in Alzheimer's disease.

The ROIs in the blue module includes the left/right sides of hippocampus, lingual gyrus, cuneus, calcarine fissure and superior occipital gyrus, cerebellum and vermis. Hippocampus is well known for its key role in memory.[24] Hippocampal neuronal loss and structural change have been connected with Alzheimer's disease.[25,26] Alzheimer's disease patients have also demonstrated neuronal and glial loss and structural changes in cerebellum and vermis.[27] Lingual gyrus, cuneus, calcarine fissure and superior occipital gyrus are located in the occipital lobe, which are mainly related to vision processing.[28] In addition, lingual gyrus plays an important role in the identification and recognition of words.[29] The association between the APOE4

SNP and the network connectivity measures may reflect the pathological changes of the brain functional network in Alzheimer's disease.

3.4. *GWAS scan with individual modules*

We tested for associations of the SNPs across the whole genome with the functional connectivity measures in the yellow and blue modules respectively. For genotype data, we included all SNPs with a minor allele frequency (MAF) ≥ 0.05, genotyping rate $\geq 90\%$, and passing the Hardy-Weinberg equilibrium test with a p-value > 0.001. After filtering with the above criteria, we obtained 579,382 SNPs.

The genome-wide scan showed that among the SNPs associated with the network connectivities (measured by Pearson's correlation) in the yellow module, rs17114690 on chromosome 14 was the only SNP that had a p-value smaller than 10^{-3}. Three SNPs were founded to be associated with the network connectivities (correlations) in the blue module, with p-values smaller than 10^{-3}. They are located on chromosome 1 (rs7536105, rs11265187) and chromosome 2 (rs17498117). rs7536105 is located in the chromatin interactive region, while rs11265187 is located in the enhancer region of gene olfactory receptor family 10 subfamily J member 9 pseudogene (OR10J9P).

The genome-wide scan also identified 5 SNPs associated with the intramodular network connectivity $\omega.in$ for the yellow module, with $P < 10^{-5}$. They are located on chromosome 1 (rs6656071, rs12043216), chromosome 7 (rs1178127, rs12674460), and chromosome 13 (rs2819239). SNP rs1178127 is a missense variant in gene histone deacetylase 9 (HDAC9),[30] an important gene with function in transcriptional regulation and cell cycle in the Wnt signalling pathway.

4. Discussion

In this paper we adapted WGCNA for network construction and module detection to rs-fMRI data. Based on the identified modules, we also proposed applying a new adaptive association test for single SNP association with the connectivities of the whole network or its components in a proportional odds model. While the whole network was not associated, some module-based connectivities were significantly associated with the APOE4 SNP rs429358. Given the major role of APOE4 in the pathogenesis of Alzheimer's disease, our finding seems plausible, suggesting its possible use for genome-wide scans to detect SNP variants associated with altered brain networks and AD. Although none of the associations was highly or genome-wide significant, it was perhaps due to a too small sample size; larger studies are needed. Our use of modules, with either various ROI-to-ROI connectivities (e.g. TOM in addition to standard correlations) or some module-based node measures (such as intramodular connectivity), not only may reduce the dimension and thus improve the statistical power, but also can enhance result interpretation, highlighting where is the association if any. In particular, we found that intramodular connectivities showed more significant associations with more SNPs, possibly due to their lower dimensions (i.e. p_1 in a module with p_1 ROIs as compared to $p_1(p_1-1)/2$ of ROI-to-ROI connectivities) and/or higher information contents.

The multiple traits used in this paper, including various network connectivity measures in the whole network or its various subcomponents, differ from most of the previous neuroimaging studies,[31] in which the focus was on some direct measures on ROIs, not their connectivities as shown here. These phenotypes are often high dimensional with dimension exceeding the sample size. Many software packages cannot handle such a situation with $p > n$, which limits their use. The adaptive association test used in this paper can be applied to such high-dimensional traits. It can be a useful and powerful method for identifying associations between high-dimensional neuroimaging traits and SNPs. In this paper, we have focused on the study of the association between neuroimaging phenotypes and SNP genotype scores; however, other ordinal outcomes such as a disease status (e.g. normal, MCI and AD in the ADNI data) can be tested for their associations with neuroimaging and other endophenotypes.

Acknowledgment

This research was supported by NIH grants R01GM113250, R01HL105397 and R01HL116720, and by the Minnesota Supercomputing Institute.

Data collection and sharing for this project was funded by the Alzheimer's Disease Neuroimaging Initiative (ADNI) (National Institutes of Health Grant U01 AG024904) and DOD ADNI (Department of Defense award number W81XWH-12-2-0012). ADNI is funded by the National Institute on Aging, the National Institute of Biomedical Imaging and Bioengineering, and through generous contributions from the following: Alzheimer's Association; Alzheimer's Drug Discovery Foundation; BioClinica, Inc.; Biogen Idec Inc.; Bristol-Myers Squibb Company; Eisai Inc.; Elan Pharmaceuticals, Inc.; Eli Lilly and Company; F. Hoffmann-La Roche Ltd and its affiliated company Genentech, Inc.; GE Healthcare; Innogenetics, N.V.; IXICO Ltd.; Janssen Alzheimer Immunotherapy Research Development, LLC.; Johnson Johnson Pharmaceutical Research Development LLC.; Medpace, Inc.; Merck & Co., Inc.; Meso Scale Diagnostics, LLC.; NeuroRx Research; Novartis Pharmaceuticals Corporation; Pfizer Inc.; Piramal Imaging; Servier; Synarc Inc.; and Takeda Pharmaceutical Company. The Canadian Institutes of Health Research is providing funds to support ADNI clinical sites in Canada. Private sector contributions are facilitated by the Foundation for the National Institutes of Health (`www.fnih.org`). The grantee organization is the Northern California Institute for Research and Education, and the study is coordinated by the Alzheimer's Disease Cooperative Study at the University of California, San Diego. ADNI data are disseminated by the Laboratory for Neuro Imaging at the University of California, Los Angeles. This research was also supported by NIH grants P30 AG010129 and K01 AG030514.

References

1. Y. I. Sheline and M. E. Raichle, *Biological Psychiatry* **74**, 340 (2013).
2. K. Supekar, V. Menon, D. Rubin, M. Musen and M. D. Greicius, *PLoS Computational Biology* **4**, e1000100 (2008).
3. A. Zalesky, A. Fornito, M. L. Seal, L. Cocchi, C.-F. Westin, E. T. Bullmore, G. F. Egan and C. Pantelis, *Biological Psychiatry* **69**, 80 (2011).
4. M. K. Belmonte, G. Allen, A. Beckel-Mitchener, L. M. Boulanger, R. A. Carper and S. J. Webb, *The Journal of Neuroscience* **24**, 9228 (2004).
5. A.-L. Barabási and R. Albert, *Science* **286**, 509 (1999).

6. C. C. Hilgetag and A. Goulas, *Brain Structure and Function* , 1 (2015).
7. J. Kim, W. Pan and the Alzheimer's Disease Neuroimaging Initiative, *NeuroImage: Clinical* **9**, 625 (2015).
8. B. Zhang and S. Horvath, *Statistical Applications in Genetics and Molecular Biology* **4** (2005).
9. L. Zhu, J. Lei, B. Devlin and K. Roeder, *arXiv preprint arXiv:1606.00252* (2016).
10. H. Lee, H. Kang, M. K. Chung, B.-N. Kim and D. S. Lee, *IEEE transactions on medical imaging* **31**, 2267 (2012).
11. E. Ravasz, A. L. Somera, D. A. Mongru, Z. N. Oltvai and A.-L. Barabási, *Science* **297**, 1551 (2002).
12. P. McCullagh, *Journal of the Royal Statistical Society. Series B (Methodological)* , 109 (1980).
13. J. Kim, W. Pan and the Alzheimer's Disease Neuroimaging Initiative, *Unpublished* (2016).
14. J. Kim, J. M. Basak and D. M. Holtzman, *Neuron* **63**, 287 (2009).
15. E. Genin, D. Hannequin, D. Wallon, K. Sleegers, M. Hiltunen, O. Combarros, M. J. Bullido, S. Engelborghs, P. De Deyn, C. Berr *et al.*, *Molecular Psychiatry* **16**, 903 (2011).
16. M. D. Greicius, G. Srivastava, A. L. Reiss and V. Menon, *Proceedings of the National Academy of Sciences of the United States of America* **101**, 4637 (2004).
17. L. Q. Uddin, A. Clare Kelly, B. B. Biswal, F. Xavier Castellanos and M. P. Milham, *Human Brain Mapping* **30**, 625 (2009).
18. S. Passow, K. Specht, T. C. Adamsen, M. Biermann, N. Brekke, A. R. Craven, L. Ersland, R. Grüner, N. Kleven-Madsen, O.-H. Kvernenes *et al.*, *Human Brain Mapping* **36**, 2027 (2015).
19. P. Fransson and G. Marrelec, *Neuroimage* **42**, 1178 (2008).
20. R. L. Buckner, J. R. Andrews-Hanna and D. L. Schacter, *Annals of the New York Academy of Sciences* **1124**, 1 (2008).
21. R. J. Maddock, A. S. Garrett and M. H. Buonocore, *Neuroscience* **104**, 667 (2001).
22. R. Leech and D. J. Sharp, *Brain* **137**, 12 (2014).
23. G. Karas, J. Sluimer, R. Goekoop, W. Van Der Flier, S. Rombouts, H. Vrenken, P. Scheltens, N. Fox and F. Barkhof, *American Journal of Neuroradiology* **29**, 944 (2008).
24. L. R. Squire, *Psychological Review* **99**, 195 (1992).
25. B. T. Hyman, G. W. Van Hoesen, A. R. Damasio and C. L. Barnes, *Science* **225**, 1168 (1984).
26. M. J. West, P. D. Coleman, D. G. Flood and J. C. Troncoso, *The Lancet* **344**, 769 (1994).
27. M. Sjöbeck and E. Englund, *Dementia and Geriatric Cognitive Disorders* **12**, 211 (2001).
28. R. Malach, J. Reppas, R. Benson, K. Kwong, H. Jiang, W. Kennedy, P. Ledden, T. Brady, B. Rosen and R. Tootell, *Proceedings of the National Academy of Sciences* **92**, 8135 (1995).
29. A. Mechelli, G. W. Humphreys, K. Mayall, A. Olson and C. J. Price, *Proceedings of the Royal Society of London B: Biological Sciences* **267**, 1909 (2000).
30. C. Fernandez-Rozadilla, L. De Castro, J. Clofent, A. Brea-Fernandez, X. Bessa, A. Abuli, M. Andreu, R. Jover, R. Xicola, X. Llor *et al.*, *PLoS One* **5**, e12673 (2010).
31. L. Shen, S. Kim, S. L. Risacher, K. Nho, S. Swaminathan, J. D. West, T. Foroud, N. Pankratz, J. H. Moore, C. D. Sloan, M. J. Huentelman, D. W. Craig, B. M. DeChairo, S. G. Potkin, C. R. Jack Jr, M. W. Weiner, A. J. Saykin and the Alzheimer's Disease Neuroimaging Initiative, *Neuroimage* **53**, 1051 (2010).

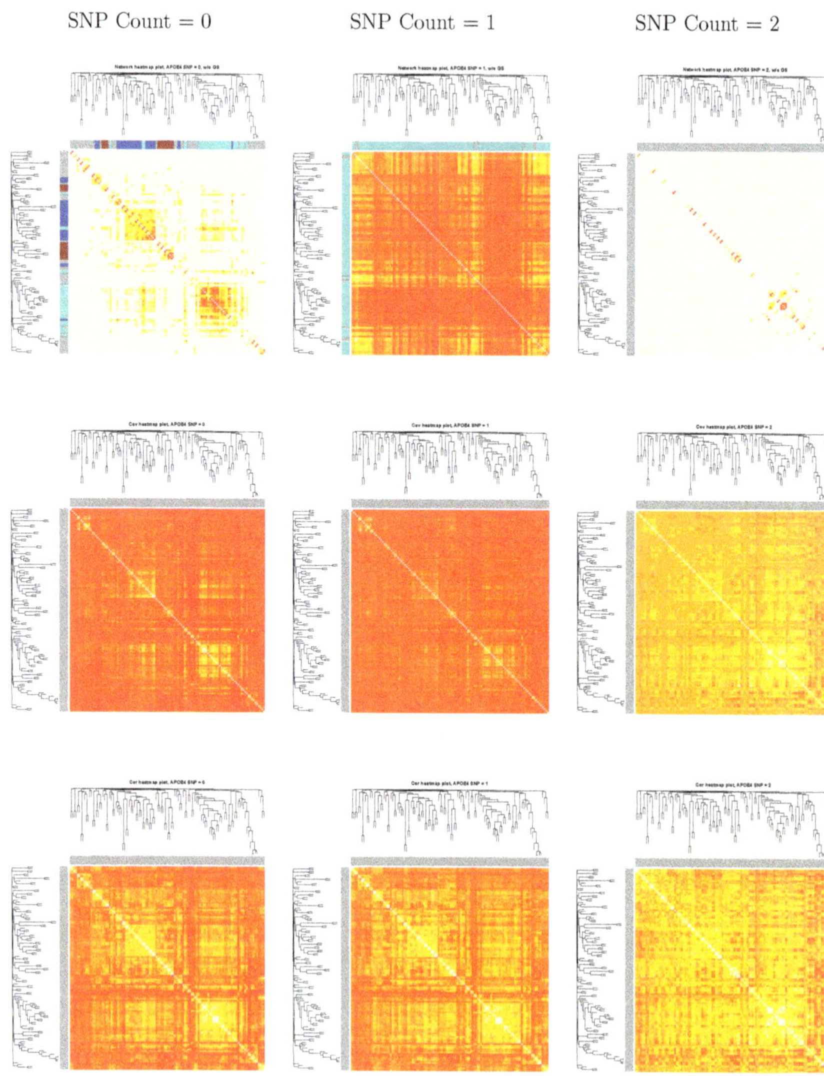

Fig. 2. TOM plot (top), covariance matrix plot (middle) and correlation matrix plot (bottom) of the brain functional networks for the three genotype groups based on APOE4 SNP (rs429358) (with its minor allele counts equal to 0, 1 or 2 from left to right).

Table 1. P-values of the tests for SNP-whole network associations using the correlation, covariance, TOM or adjacency matrix elements as the network connectivity measure respectively. W-mod and Btw-mod stand for within-modular and between-modular, respectively.

Test	Correlation			Covariance			TOM			Adjacency		
	All	W-mod	Btw-mod	All	W-mod	Btw-mod	All	W-mod	Btw-mod	All	W-mod	Btw-mod
SPU(1)	0.477	0.365	0.526	0.052	0.052	0.087	0.530	0.637	0.109	0.477	0.527	0.479
SPU(2)	0.161	0.154	0.207	0.012	0.016	0.008	0.099	0.250	0.014	0.477	0.515	0.487
SPU(3)	0.323	0.202	0.377	0.018	0.025	0.010	0.817	0.902	0.224	0.472	0.482	0.498
SPU(4)	0.150	0.122	0.197	0.019	0.009	0.004	0.325	0.424	0.172	0.463	0.434	0.516
SPU(5)	0.248	0.137	0.299	0.130	0.066	0.004	0.892	0.987	0.317	0.442	0.402	0.528
SPU(6)	0.141	0.101	0.209	0.120	0.008	0.003	0.444	0.533	0.330	0.416	0.365	0.554
SPU(7)	0.216	0.111	0.267	0.429	0.052	0.004	0.657	0.890	0.393	0.381	0.348	0.568
SPU(8)	0.137	0.089	0.225	0.188	0.009	0.004	0.498	0.603	0.410	0.348	0.334	0.583
SPU(∞)	0.122	0.079	0.356	0.210	0.009	0.007	0.463	0.706	0.655	0.263	0.239	0.460
aSPU	0.208	0.146	0.296	0.033	0.025	0.007	0.181	0.384	0.039	0.356	0.328	0.585

Table 2. P-values of the tests for SNP-individual network module associations using the correlation, covariance, TOM or adjacency matrix elements as the network connectivity measure.

Module	Test	Correlation		Covariance		TOM		Adjacency	
		W-mod	Btw-mod	W-mod	Btw-mod	W-mod	Btw-mod	W-mod	Btw-mod
Blue	SPU(1)	0.140	0.272	0.018	0.012	0.946	0.093	0.504	0.488
	SPU(2)	0.090	0.077	0.025	0.001	0.440	0.067	0.515	0.481
	SPU(3)	0.099	0.144	0.028	0.001	0.793	0.396	0.502	0.455
	SPU(4)	0.101	0.071	0.033	0.001	0.528	0.265	0.496	0.423
	SPU(8)	0.139	0.075	0.050	0.011	0.566	0.458	0.464	0.264
	SPU(∞)	0.267	0.070	0.069	0.015	0.571	0.554	0.458	0.180
	aSPU	0.160	0.119	0.034	0.003	0.654	0.141	0.581	0.244
Turquoise	aSPU	0.648	0.172	0.277	0.011	0.139	0.192	0.605	0.309
Brown	aSPU	0.260	0.182	0.016	0.015	0.459	0.219	0.249	0.327
Yellow	SPU(1)	0.040	0.500	0.005	0.084	0.083	0.172	0.357	0.510
	SPU(2)	0.024	0.219	0.005	0.012	0.060	0.155	0.323	0.509
	SPU(3)	0.020	0.350	0.005	0.015	0.147	0.458	0.297	0.514
	SPU(4)	0.016	0.220	0.010	0.006	0.188	0.445	0.262	0.522
	SPU(8)	0.008	0.271	0.100	0.004	0.317	0.690	0.177	0.500
	SPU(∞)	0.006	0.525	0.200	0.008	0.383	0.848	0.129	0.411
	aSPU	0.017	0.354	0.011	0.010	0.106	0.289	0.183	0.524

EXPLORING BRAIN TRANSCRIPTOMIC PATTERNS: A TOPOLOGICAL ANALYSIS USING SPATIAL EXPRESSION NETWORKS

ZHANA KUNCHEVA

Department of Mathematics
Imperial College London, UK
E-mail: z.kuncheva12@imperial.ac.uk

MICHELLE L. KRISHNAN

Perinatal Imaging and Health
King's College London, UK
E-mail: michelle.krishnan@kcl.ac.uk

GIOVANNI MONTANA

Biomedical Engineering Department
King's College London, UK
E-mail: giovanni.montana@kcl.ac.uk

Characterizing the transcriptome architecture of the human brain is fundamental in gaining an understanding of brain function and disease. A number of recent studies have investigated patterns of brain gene expression obtained from an extensive anatomical coverage across the entire human brain using experimental data generated by the Allen Human Brain Atlas (AHBA) project. In this paper, we propose a new representation of a gene's transcription activity that explicitly captures the pattern of spatial co-expression across different anatomical brain regions. For each gene, we define a Spatial Expression Network (SEN), a network quantifying co-expression patterns amongst several anatomical locations. Network similarity measures are then employed to quantify the topological resemblance between pairs of SENs and identify naturally occurring clusters. Using network-theoretical measures, three large clusters have been detected featuring distinct topological properties. We then evaluate whether topological diversity of the SENs reflects significant differences in biological function through a gene ontology analysis. We report on evidence suggesting that one of the three SEN clusters consists of genes specifically involved in the nervous system, including genes related to brain disorders, while the remaining two clusters are representative of immunity, transcription and translation. These findings are consistent with previous studies showing that brain gene clusters are generally associated with one of these three major biological processes.

Keywords: Spatial gene expressions; Biological networks

1. Introduction

The human brain is a complex interconnected structure controlling all elementary and high-level cognitive tasks[1]. This complexity is a result of the cellular diversity distributed across hundreds of distinct brain anatomical structures[2,3]. One of the main tasks of the neuroscience community in the past decade has been to connect the underlying genetic information of the anatomical structures to their underlying biological function[3–5]. A useful data source for such studies is the Allen Human Brain Atlas (AHBA)[3], which provides microarray expression profiles of almost every gene of the human genome with emphasis on an extensive anatomical coverage across the entire human brain.

In this paper, we make use of the experimental data provided by the AHBA project to study the spatial microarray variability at the single gene level. Analyzing the complete transcription architecture of the human brain in this way may be informative of the impact of genetic disorders on different brain regions that would otherwise not be apparent due to the coarse resolution.

To gain new insights into the expression patterns of the human brain and identify potentially important biomarkers, many studies involving the AHBA data explore gene to gene relationships[3,4]. Each gene is represented by its expression levels across anatomical locations. Genes with correlated expression profiles are grouped together based on an appropriate similarity measure. The analysis of the resulting gene co-expression networks provides evidence that transcriptional regulation relates to anatomy and brain function[2–4]. There are also studies that consider the genetic similarity between pairs of regions, and show that transcriptional regulation varies enormously with anatomic location[3,4,6,7]. These findings indicate the necessity to adopt a new representation of a gene's transcription activity that explicitly captures the pattern of spatial co-expression across different anatomical brain regions.

We propose a new and unexplored way to model the spatial variability at the single gene level. For each gene, we create a spatial expression network, or SEN. Each node of the network corresponds to a pre-defined brain region for which we have sufficient transcriptomic data, and each edge weight represents the similarity in gene expression levels, for that gene, between two brain regions. Applying this procedure to genes that have been found to be stably expressed across specimens gives rise to a population of approximately $17,000$ gene networks, each one representing a brain-wide spatial pattern of gene expression. Using this representation, we investigate whether the topological similarity of the SENs reflects the biological similarity of genes through an integrative analysis based on network clustering and gene ontologies. Our hypothesis is that, if clusters of topologically similar SENs can be identified, the corresponding genes within each cluster may also share similar biological properties.

A robust cluster analysis of all SENs has indicated the presence of three large and stable clusters of SENs, each one having significantly different topological features as well as different biological function. In particular, one of the clusters has been found to be uniquely enriched for brain-related terms, neurological diseases and genes with enriched expression in neurons. Overall, our analysis provides evidence supporting the notion that topological proximity of spatial gene networks is indicative of similar biological function.

2. Materials and Methods

2.1. *Spatial Expression Networks (SENs)*

The Allen Human Brain Atlas (AHBA)[3,8] is a publicly available atlas of the human brain with microarray-based genome-wide transcriptional profiling of specific brain regions spanning all major anatomical structures of the adult brain. The data set includes transcriptional profiling data from more than 3500 samples comprising approximately 200 brain regions in 6 clinically unremarkable adult human brains. The Agilent 4×44 Whole Human Genome platform was used for gene expression extraction. Two donors contributed samples representing approximately 1000 structures across the whole brain, while the other four approximately 500 samples from the left hemisphere. Our analyses is based on 16,906 pre-selected genes from a previous study[5]. We use the normalized expression levels, which were normalized across samples and across different brains as in previous analyses[9].

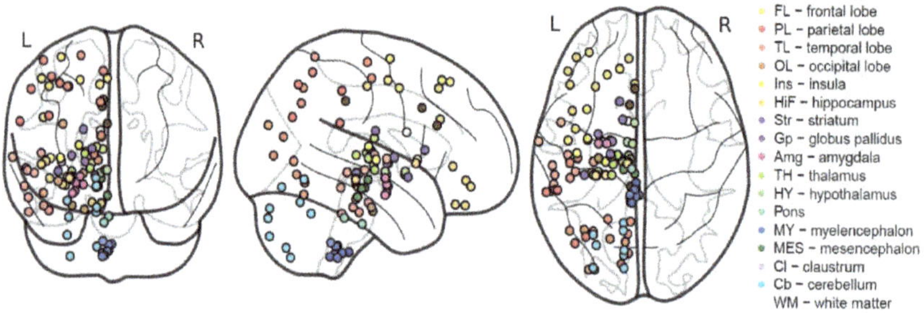

Fig. 1. Anatomical maps of the 105 brain regions used to construct the SENs. The maps show the brain regions as seen from inferior, lateral and superior views, from left to right. All regions are in the left hemisphere and they are located in the Thalamus, Cerebellum, Pons, Midbrain, Medulla and Cerebral cortex. Coloring of the regions is consistent with anatomical tissue and is obtained from AHBA ontology atlas.[8]

For each of the 16,906 genes, we constructed an individual spatial expression network (SEN) representing patterns of expression variability in the brain. Only brain regions with at least one measurement in all 6 brains were included in the analysis resulting in a total of $N = 105$ regions from the left hemisphere, as shown in Fig. 1.

The mean expression level for a gene in brain region i is denoted by g_i. The distribution of the mean and median values for each brain region over all genes were not found statistically different (Kolmogorov-Smirnov test[10]; all $p >> 0.05$). Furthermore, for more than 97% of all region samples across all genes, the standard deviation of the expression values is less than 20% of the mean value, indicating that the mean can be taken as representative of the expression values at a given region for a given gene.

Formally, we define a SEN as a fully connected network $G = (V, E)$ with node set $V = \{i : i = 1, 2, ..., N\}$ indicating the brain regions and weighted edge set $E = \{E_{ij} : i, j = 1, 2, ..., N; i \neq j\}$. Each edge weight $E_{ij} \in [0, 1]$ quantifies the similarity in gene expression between regions i and j. The maximum value is reached when the mean expression levels in the two brain regions are equal. We impose that E_{ij} monotonically decreases with an increasing absolute difference between mean expression levels; accordingly, the edge weights are defined as

$$E_{ij} := \frac{1}{1 + |g_i - g_j|}.$$

This network representation allows us to capture the interconnected variability of gene expression across the brain at the gene level.

2.2. *Clustering SENs*

In order to address our hypothesis that topological similarity may reflect biological similarity, initially we set out to explore whether SENs form naturally occurring clusters. For this we first required an appropriate measure of topological dissimilarity between pairs of SENs. We first mapped each SEN G to a N-dimensional feature vector $\boldsymbol{d} = (d_1, d_2, ..., d_N)$ with each elements representing the node degree, i.e. $d_i = \sum_{j=1}^{N} E_{i,j}$. The degree for each node captures the global transcriptomic similarity of the corresponding brain region to all other brain regions for a given gene. If the node degrees for two SENs are very different, then the corresponding genes have very different global transcriptomic patterns. The dissimilarity between two SENs, G_l and G_k, was taken to be the Euclidean distance between the corresponding feature vectors, $\boldsymbol{d}_l$ and $\boldsymbol{d}_k$.

Three different clustering algorithms were used – partitioning around medoids (PAM)[11], k-means[12] and fuzzy C-means[13] – all providing a partition of all the SENs into k different clusters. To determine an appropriate number of clusters k using each one of these algorithms we performed a stability analysis[12]. The k clusters are deemed "stable" if random changes in the SEN configurations generate almost identical k clusters. To introduce random changes in the networks, we use a randomization strategy by which the observed networks in network space Γ are perturbed slightly. For this analysis we used two different randomization procedures: (a) vertex permutations, i.e. we permuted the node labels of a random subset of networks so as to preserve the node degrees but not their order, (b) edge perturbation, i.e. we perturbed the edge weights of a random subset of networks so as to make the cluster robust against white noise.

To obtain a measure of cluster instability, we use the following steps: First, we generate perturbed versions Γ_b $(b = 1, 2, ..., b_{\max})$ of Γ, and cluster the networks in Γ_b into k clusters thus obtaining $\mathcal{C}_b(k)$. In addition, we randomize the cluster assignments[14] in $\mathcal{C}_b(k)$ to obtain random clustering $\mathcal{C}_{b,\text{rand}}(k)$. Second, for $b, b' = 1, 2, ..., b_{\max}$, we compute the pairwise distances $[1 - NMI(\mathcal{C}_b(k), \mathcal{C}_{b'}(k))]$ between the clusterings $\mathcal{C}_b(k)$ and $\mathcal{C}_{b'}(k)$, and between the randomized clusterings $\mathcal{C}_{b,\text{rand}}(k)$ and $\mathcal{C}_{b',\text{rand}}(k)$. The normalized mutual information (NMI) is used as a similarity measure between partitions[15]. The cluster instability index is defined as the mean distance between clusterings $\mathcal{C}_b(k)$, i.e.

$$I(k) = \frac{1}{b_{\max}^2} \sum_{b,b'=1}^{b_{\max}} [1 - NMI(\mathcal{C}_b(k), \mathcal{C}_{b'}(k))] \,. \tag{1}$$

We use the normalized instability index, $I_{\text{norm}}(k) := I(k)/I_{\text{rand}}(k)$, which corrects for a scaling[14] of $I(k)$ with an increasing number of clusters k. We choose number of clusters k that gives the lowest $I_{\text{norm}}(k)$.

2.3. *Topological characterization of SEN clusters*

To characterize the topological properties of SENs in each cluster, we use global topological measures that capture different aspect of the network such as its density, the tendency of its nodes to cluster and form communities, the presence of central and hub nodes. Overall, we use eight such different measures: average node degree[16], average closeness centrality[16], weighted diameter[17], global clustering coefficient for weighted networks[17], number of non-overlapping communities, average authority score[18], the number of nodes with authority score > 0.95, and the number of nodes with authority score < 0.05. All measures were computed for all SENs within each cluster. To test for statistically significant differences in network topology across clusters, we performed a multivariate ANOVA test[19].

Furthermore, for each SEN we derived a measure of community structure[20]. In our context, the presence of a community in a given SEN indicates that there is a set of highly interconnected brain regions whose gene expression similarity is higher compared to the rest of the network. For this analysis we used the Fast Greedy algorithm[21], which is based on the optimization of the modularity function that sums the edge weights within a community and corrects for the expected edge weights by chance. The algorithm is discriminative of small edge weight differences and can yield sensitive separation of brain regions into communities. Genes with similar community structures indicate the presence of similar local coherent transcriptomic patterns for groups of brain regions.

For each cluster, we quantify the similarity of a pair of brain regions using the communities detected in all the SENs by counting the number of times the two regions fall within the same community. This count is then divided by the total number of SENs in the cluster in order to obtain an index lying in the $[0, 1]$ range, which we call the "coherence index". Values close to 1 indicate high coherency between the two brain regions, i.e. the average tendency to fall within communities of highly interconnected brain regions.

2.4. *Biological characterization of SEN clusters*

In order to investigate whether naturally occurring clusters formed by SENs can be related to distinct biological function, we require a procedure which assigns representative biological terms to each cluster. For this purpose we use a Gene Ontology (GO) enrichment analysis pipeline which first collects broad GO information for the biological context of genes in each of the main clusters, and then reduces this information to representative GO terms for final interpretation of the clusters.

Each SEN cluster was first annotated for significantly enriched Biological Process (BP) terms using a standard hypergeometric test for over-represented terms ($p < 0.001$) implemented in the GOstats R package[22]. Using a clustering methodology implemented in the tool REVIGO[23], we group semantically similar GO terms based on the established *SimRel* measure. The algorithm finds a representative term for each group based on the enrichment p-values, with a bias away from very general parent GO terms. The size of the resulting summary list is controlled by setting the threshold for the *SimRel* similarity measure at 0.5. Results are summarized by retaining the cluster representatives for each GO term that can reveal underlying function of these clusters.

Genes in each of the clusters were also annotated for disease enrichment using the WebGestalt tool[24], which interfaces with the GLAD4U platform[25] to retrieve and prioritize disease-gene links from publications, using a hypergeometric test with multiple testing correction and the genome as background.

3. Experimental results

3.1. *Topologically different SEN clusters*

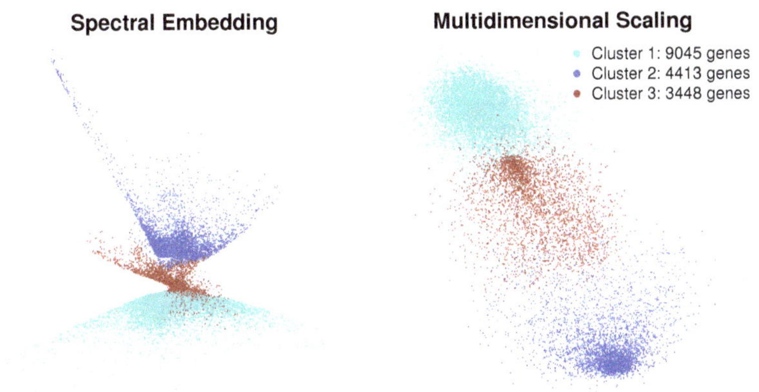

Fig. 2. Two-dimensional visualization of all SENs using two different dimensionality reduction algorithms: spectral embedding[26] (left) and multidimensional scaling[27] (right). The color scheme indicates the cluster membership as determined by the PAM algorithm. Both visualizations indicate three main clusters.

All SENs were clustered into up to six clusters using the procedures outlined in Sec. 2.2. The two instability analyses were each performed using $b_{\max} = 500$. Using the first randomization scheme, 5% of networks were randomly sampled for node permutation, while in the second procedure 20% of networks were randomly sampled and white noise was introduced by adding ±20% to each edge weight. The results for all three clustering procedures, Tab. 1, show that PAM clustering has the lowest instability followed by fuzzy C-means. Furthermore, for all three clustering methods grouping data into two and three clusters leads to the lowest instabilities.

Table 1. Different stability analyses for three different clustering algorithms using two randomization strategies (vertex and edge permutation).

	Vertex permutation			Edge perturbation		
$I_{\text{norm}}(k)$	PAM	Cmeans	k-means	PAM	Cmeans	k-means
$I_{\text{norm}}(2)$	0.016	0.020	0.065	0.009	0.015	0.065
$I_{\text{norm}}(3)$	**0.018**	**0.023**	**0.076**	**0.010**	**0.016**	**0.071**
$I_{\text{norm}}(4)$	0.023	0.031	0.092	0.019	0.033	0.180
$I_{\text{norm}}(5)$	0.026	0.038	0.171	0.025	0.030	0.191
$I_{\text{norm}}(6)$	0.031	0.080	0.187	0.027	0.086	0.208

The PAM algorithm was chosen to generate the final partitions as it yields the lowest instability index. As an additional validation to support the choice of three PAM clusters, we used three internal validation measures: the Sillhouette width[28], the Dunn index[13] and the within-cluster variance[29]. The Dunn index and Silhouette width support the presence of two to three clusters, see Tab. 2. However, the intra-cluster variance, which is known to be more sensitive to the existence of sub-clusters[30], shows that grouping data into two clusters leads to high within-cluster variability compared to a higher number of clusters. By taking all these criteria into account, we have chosen to consider $k = 3$ since this leads to the lowest instability and within-cluster variability whilst having as high as possible Dunn and Sillhouette scores.

Table 2. Cluster validation measures for clustering SENs into k clusters using PAM.

k	Dunn	Silhouette	Within-cluster Variance
2	2.20	0.66	0.276
3	**1.20**	**0.44**	**0.225**
4	0.61	0.30	0.215
5	0.63	0.23	0.223
6	0.48	0.18	0.211

In an attempt to visually assess whether this choice seems appropriate, we used a distance-preserving projection of all 16906 SENs into a 2D-dimensional space using two different dimensionality reduction procedures: spectral embedding[26] and multidimensional clustering[27]. The resulting projections can be found in Fig. 2. All three clusters – 1 (turquoise), 2 (blue) and 3 (brown) – appear well-separated.

3.2. *Topological differences amongst SEN clusters*

To validate that the three SEN clusters have distinct topological structure, we used the eight global network measures outlined in Sec. 2.3. The frequency distribution of the topological measures for each cluster is summarized in Fig. 3 where a clear mean difference can be observed for each individual measure across clusters. Using a MANOVA test, we reject the null hypothesis of equality of topological features across clusters ($p < 2.2e-16$; Wilk's $\Lambda = 0.3589$).

We have found that Cluster 1 mostly consists of SENs with the highest node degree, centrality measures, diameter, authority score and number of nodes with high authority score, while there are only a few number of communities and few nodes with low authority score. These properties imply coherent expression levels across all brain regions. On the other hand, Cluster 2 comprises of SENs with the lowest node degree, centrality measures, diameter, authority score and number of nodes with high authority scores, and the highest number of communities and nodes with low authority score. This indicates that most SENs within this cluster are sparse, and that there is high variability between expression levels across brain regions. Finally, Cluster 3 consists of SENs with medium ranged values for all network

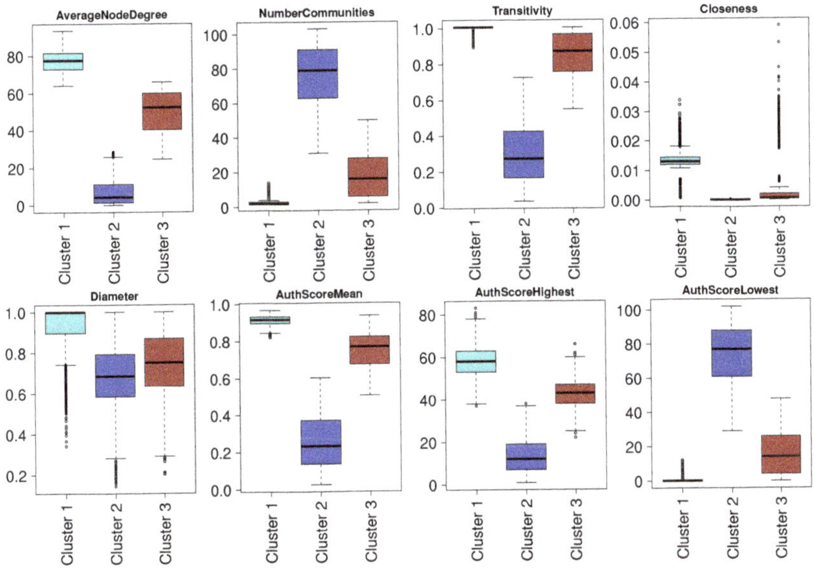

Fig. 3. Distribution of topological network measures in the three clusters obtained using the PAM algorithm. High node degrees imply high edge weights with fewer low-weighted shortest paths and fewer discrepancies in edge values. This leads to high transitivity and closeness values, simultaneously reducing the number of communities SENs are partitioned into. Higher node degrees lead to more nodes having high authority scores thus increasing both the average authority scores and the number of nodes with high authority. Low node degrees signify sparseness of the SENs and more low-weighted shortest paths. This results in more nodes being grouped in their own communities, in addition to low closeness and transitivity. Sparse networks and low node degrees result in lower authority scores and fewer nodes with high authority score.

measures, implying moderate variability between expression levels across brain regions.

3.3. *Biological differences amongst SEN clusters*

We investigated the local transcriptomic patterns within each of the three clusters using the "coherence index" defined in Sec. 2.3. The three clusters have different transcriptomic patterns, Fig. 4, and comparing heatmaps of the three clusters to one for all 16906 genes shows that Cluster 1 is closest to the genome-wide global patterning, while Cluster 2 and Cluster 3 are carriers of imposed heterogeneity. The patterns of the 16906 genes are also consistent with existing work, and largely replicate previous findings[3,4,6]. In particular, homogeneity within the Neocortex and Cerebellum, and increased heterogeneity in the Basal Ganglia, have been previously reported. Cluster 2 has few coherency patterns in the Basal Ganglia regions and Cerebellum. Cluster 1 exhibits high homogeneity within the Cerebellum and the Neocortex, and between subdivisions of the subcortical structure and the Hippocampus. Cluster 3 appears to have coherent patterns in the Cerebellum and the Neocortex but increased variability in the Basal Ganglia.

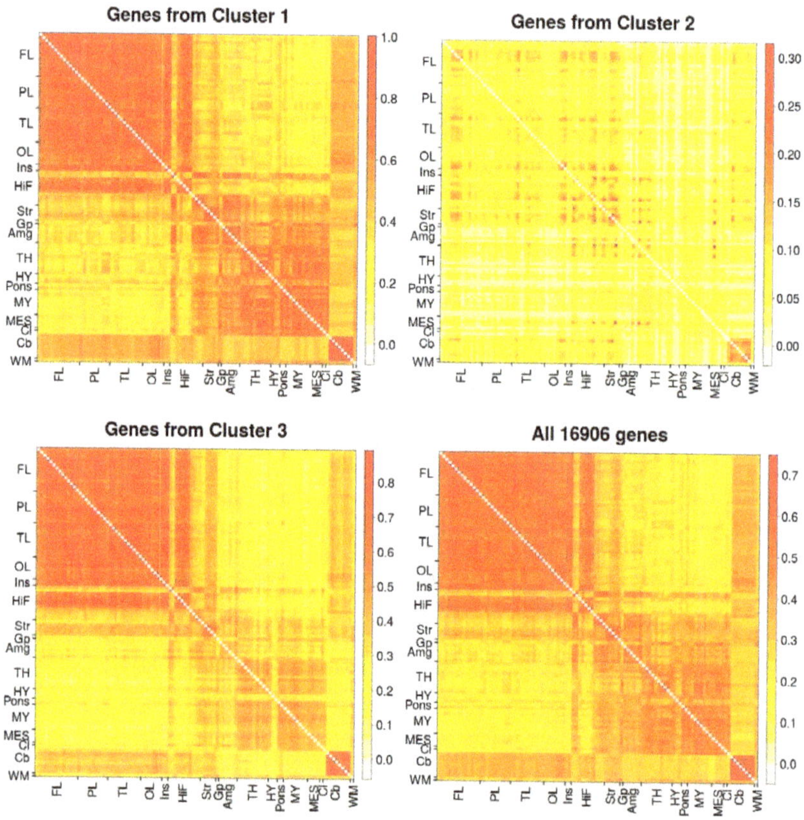

Fig. 4. Heatmaps representing the "coherence index" between pairs of brain regions in each of the three SEN clusters and across all 16906 genes. Each pixel on the heatmap is the "coherence index" between the two corresponding brain regions. Each heatmap is accompanied by a color key, where higher values indicate high homogeneity of expression levels and lower values indicate heterogeneous expression levels. The 105 brain regions are mapped to 17 major brain structures using the AHBA ontology atlas[8] and abbreviated as indicated in Fig. 1.

Obtaining detailed annotation as described in Sec. 2.4 revealed that all three clusters are significantly enriched ($p < 0.001$) for a variety of GO BP terms. We reduced these large sets of GO terms to smaller non-redundant sets by applying REVIGO[23].

The BP representative terms selected on the basis of enrichment p-values and semantic similarity indicate that Cluster 1 genes can be described primarily by "RNA processing" and "ribonucleoprotein complex biogenesis". Cluster 2 genes are predominantly involved in immunity including "immune system process", "leukocyte proliferation" and "G-protein coupled receptor signaling pathway" terms primarily associated to the immune system, whereas Cluster 3 genes are uniquely involved in "behavior", "metal ion transport" and "nervous system development". On closer inspection of Cluster 3, these representative terms comprise several linked biological processes specific to the Nervous System, and which are not found on either Cluster 1 or 2, such as "synaptic transmission" and "dendrite extension".

The significant disease enrichment (adjusted $p < 0.001$) also supported the functional distinctiveness of the three clusters, with Cluster 1 being enriched for Mitochondrial disease, Cluster 2 being significantly enriched for genes involved with Immune System and Inflammatory disease, and Cluster 3 being principally involved in Nervous system disorders. Given the observed functional differentiation between clusters, we investigated whether this might correspond to cell-type specialization. We obtained lists of neuron- and microglia-enriched genes in a repository of detailed RNA-sequencing and splicing data from purified cell cultures[31] , and computed significant intersections using the SuperExactTest[32]. This showed that genes in Cluster 3 have significant overlap with neuron- and microglia-specific genes ($p < 0.05$). Cluster 2, on the other hand, has a unique association to microglia-specific genes only ($p < 0.05$).

4. Discussion

Analyzing the transcriptome architecture of the human brain is a challenging task due to the high-dimensionality and biological complexity of the data. This is compounded by technical factors related to sample acquisition and measurement error that can influence the results. We addressed the issue of anatomical variability in gene expression by proposing to model each gene's spatial co-expression pattern across anatomical regions as an individual spatial network, or SEN. To explore whether topological similarity of gene expression as captured by SENs is related to biological similarity, we used network dissimilarity to obtain clusters of genes with similar patterns of spatial co-expression. We aimed to gain additional insights into the biological interpretation of regional anatomical specialization of the brain.

We demonstrated that there is evidence to support the presence of three topologically distinct clusters of SENs, with each cluster being characterised by particular network properties. Furthermore, investigating the community structure of the SENs, we identified possible anatomical basis for the difference in the topological properties in the three clusters. The differences between clusters are mainly due to the heterogeneity of expression levels in the Basal Ganglia, and between the Neocortex and Cerebellum.

We also found these three topologically distinct clusters to have biologically distinct properties. On closer inspection we find Cluster 3 to be specific to the nervous system, while Cluster 2 appears to be involved with immunity and Cluster 1 with transcription and translation. These associations are in line with previous results on the AHBA data set[3,4], where the majority of clusters obtained using WGCNA[33], a well-known gene clustering procedure, were also associated to immunity, nervous system or transcription and translation.

To gain an insight into possible cellular contributions to these differences, we included cell-type specific data and observe that the overlap of neuron- and microglia- specific genes in Cluster 3 is in keeping with current hypotheses regarding the significant interactions between these two cell-types, including the possible modulatory activity of microglia in synaptic pruning and cell communication beyond purely immune functions[34].

We found significant disease associations for all three clusters, implying the high biological impact of the genes involved and the utility of our modular clustering approach for the identification of therapeutic targets. There is a preponderance of neurological and neuropsychiatric conditions linked to Cluster 3 genes, and immune disorders linked to Cluster 2, reflecting their biological functions as described above and supporting those annotations.

One important concern was whether the above results were specific to using node degrees or they could be reproduced using other feature vectors. Thus we constructed two different sets of feature vectors based on node centrality as captured by the authority score and based on the raw edges of the SEN. Based on each new set of feature vectors, results not included in this paper demonstrated evidence to support the presence of three topologically distinct clusters of SENs. For both feature vectors, the three clusters were again marked by different topological properties although there were shifts in the distributions of those properties. Even so, in both cases the three clusters were uniquely associated to the immune system, nervous system or transcription and translation.

For comparison purposes, we used WGCNA on the gene expression values of the 16906 genes for the 105 brain regions. Results not included in this paper showed that WGCNA did not assign a cluster membership to the majority of genes in Cluster 2 due to the sparseness of their expression levels. More and smaller clusters were discovered with higher instability. The advantage of our method compared to WGCNA is that the structure of SENs allows us to use a number of clustering procedures to detect stable gene clusters, whose validation could be achieved using both topological and biological measures. We determine the biological function of a cluster using the gene ontology of the entire set of genes in the cluster, which is robust to slight changes in the cluster membership.

A next step in the analysis of SENs should consider additional clusters to detect more specialized biological functions. Furthermore, it is well known that gene expressions in the cerebellum, subcortical and cortical regions differ significantly from each other based on their composition of different cell types[3,4]. Future work in this direction will include an analysis where only neocortex regions are used to construct SENs.

5. Conclusion

An important and challenging task in studying the brain transcriptional architecture is integrating and modelling the high dimensionality of the gene expression across the brain. To the best of our knowledge, our work is the first to perform a region-wise comprehensive profiling of gene-specific co-expression patterns across the human brain. By modelling gene expression as SENs and employing network embeddings, we identified distinct clusters of genes associated to specific biological functions, topological properties and cell-types, with potential implications for neuropsychiatric disease. Modelling genes as SENs across brain regions could be used for future studies in helping to identify genes with particular co-expression patterns across a set of spatial brain locations of interest, enabling the identification of genes that act in spatially contextualized clusters with high biological impact. As more microarray gene expression data become available at higher spatial resolution and cell-type specificity, modelling gene co-expression across the brain will be increasingly important to understanding the brain transcriptome architecture at a microstructural scale.

References

1. M. C. Oldham, G. Konopka *et al.*, *Nat. Neurosci.* **11**, 1271 (nov 2008).
2. M. Hawrylycz, L. Ng *et al.*, *Neural Netw.* **24**, 933 (nov 2011).
3. M. J. Hawrylycz, E. S. Lein *et al.*, *Nature* **489**, 391 (sep 2012).
4. M. Hawrylycz, J. A. Miller *et al.*, *Nat. Neurosci.* **18**, 1832 (nov 2015).
5. J. Richiardi, A. Altmann *et al.*, *Science* **348**, 1241 (jun 2015).
6. A. Mahfouz, M. van de Giessen *et al.*, *Methods* **73**, 79 (mar 2015).
7. P. Goel, A. Kuceyeski *et al.*, *Hum. Brain Mapp.* **35**, 4204 (aug 2014).
8. Allen Institute for Brain Science., Allen Human Brain Atlas (2014).
9. Allen Human Brain Atlas, *Technical White Paper: Microarray Data Normalization*, tech. rep., Allen Institute (2013).
10. G. Marsaglia, W. W. Tsang *et al.*, *J. Stat. Softw.* **8**, 1 (2003).
11. L. Kaufman and P. J. Rousseeuw, *Finding Groups in Data: An Introduction to Cluster Analysis* (John Wiley & Sons Ltd, 1990).
12. U. Von Luxburg, *Clustering Stability: An Overview* (Now Publishers Inc., 2010).
13. J. C. Dunn, *J. Cybern.* **3**, 32 (1973).
14. T. Lange, V. Roth *et al.*, *Neural Comput.* **16**, 1299 (2004).
15. M. Meila, *J. Multivar. Anal.* **98**, 873 (may 2007).
16. T. Opsahl, F. Agneessens *et al.*, *Soc. Networks* **32**, 245 (jul 2010).
17. A. Barrat, M. Barthélemy *et al.*, *Proc. Natl. Acad. Sci. U. S. A.* **101**, 3747 (mar 2004).
18. J. M. Kleinberg, *J. ACM* **46**, 604 (sep 1999).
19. S. Scheiner, *Des. Anal. Ecol. Exp.* , 94 (2001).
20. S. Fortunato, *Phys. Rep.* **486**, 75 (2010).
21. A. Clauset, M. Newman *et al.*, *Phys. Rev. E* **70**, p. 066111 (dec 2004).
22. S. Falcon and R. Gentleman, *Bioinformatics* **23**, 257 (jan 2007).
23. F. Supek, M. Bošnjak *et al.*, *PLoS One* **6**, p. e21800 (jan 2011).
24. B. Zhang, S. Kirov *et al.*, *Nucleic Acids Res.* **33**, W741 (jul 2005).
25. J. Jourquin, D. Duncan *et al.*, *BMC Genomics* **13 Suppl 8**, p. S20 (jan 2012).
26. U. V. Luxburg, *A Tutorial on Spectral Clustering*, tech. rep., Max Planck Institute for Biological Cybernetics (2007).
27. I. Borg and P. J. F. Groenen, *Modern Multidimensional Scaling: Theory and Applications* (Springer Science & Business Media, 2005).
28. P. J. Rousseeuw, *J. Comput. Appl. Math.* **20**, 53 (nov 1987).
29. M. Halkidi and M. Vazirgiannis, Clustering Validity Asessment: Finding the Optimal Partitioning of a Data Set, in *Proc. 2001 IEEE Int. Conf. Data Min.*, (IEEE Comput. Soc, 2001).
30. Y. Liu, Z. Li *et al.*, Understanding of Internal Clustering Validation Measures, in *2010 IEEE Int. Conf. Data Min.*, (IEEE, dec 2010).
31. Y. Zhang, K. Chen *et al.*, *J. Neurosci.* **34**, 11929 (sep 2014).
32. M. Wang, Y. Zhao *et al.*, *Sci. Rep.* **5**, p. 16923 (jan 2015).
33. S. Horvath, *Weighted Network Analysis: Applications in Genomics and Systems Biology* (Springer, 2011).
34. M.-È. Tremblay, R. L. Lowery *et al.*, *PLoS Biol.* **8**, p. e1000527 (jan 2010).

INTEGRATIVE ANALYSIS FOR LUNG ADENOCARCINOMA PREDICTS MORPHOLOGICAL FEATURES ASSOCIATED WITH GENETIC VARIATIONS*

CHAO WANG

Electrical and Computer Engineering, The Ohio State University
Columbus, Ohio, 43210, USA
Email: wang.2031@osu.edu

HAI SU

Biomedical Engineering, University of Florida
Gainsville, Florida, 32611, USA
Email: hai.su@bme.ufl.edu

LIN YANG

Biomedical Engineering, University of Florida
Gainsville, Florida, 32611, USA
Email: lin.yang@bme.ufl.edu

KUN HUANG

Biomedical Informatics, The Ohio State University
Columbus, Ohio, 43210, US
Email: kun.huang@osumc.edu

Lung cancer is one of the most deadly cancers and lung adenocarcinoma (LUAD) is the most common histological type of lung cancer. However, LUAD is highly heterogeneous due to genetic difference as well as phenotypic differences such as cellular and tissue morphology. In this paper, we systematically examine the relationships between histological features and gene transcription. Specifically, we calculated 283 morphological features from histology images for 201 LUAD patients from TCGA project and identified the morphological feature with strong correlation with patient outcome. We then modeled the morphology feature using multiple co-expressed gene clusters using Lasso-regression. Many of the gene clusters are highly associated with genetic variations, specifically DNA copy number variations, implying that genetic variations play important roles in the development cancer morphology. As far as we know, our finding is the first to directly link the genetic variations and functional genomics to LUAD histology. These observations will lead to new insight on lung cancer development and potential new integrative biomarkers for prediction patient prognosis and response to treatments.

1. Introduction

Lung cancer is one the most deadly cancers in the world. Among lung cancers, lung adenocarcinoma (LUAD) is a subtype of the non-small cell lung cancer (NSCLC) and is the most common histological type of lung cancers (*1*). However, despite the fact that it is a sub-classification of lung cancer, LUAD is a heterogeneous group of tumors with a highly variable prognosis and responses to treatment (*2*).

The high-throughput sequencing technologies are making targeted therapies possible for LUAD (*3*). The advance of these technologies allows molecular diagnostic biomarkers for the detection of lung cancer in addition to computed tomography (CT) screening (*4–7*). For example, the utility of epidermal growth factor receptor (EGFR) mutation testing is strongly recommended (*8*) in clinical practice. However, although EGFR-mutant lung cancers are

* This work is supported by UK-OSU Joint CCTS grant, NCI ITCR 1U01CA188547-01A1 grant, the OSU Pelotonia Fellowship, and the Ohio Supercomputer Center.

sensitive to EGFR tyrosine kinase inhibitors (TKIs), they develop resistance (*9*). Therefore, novel biomarkers for for LUAD are needed for enhanced personalized treatment.

Lung cancer diagnosis and classification have been traditionally based on imaging approaches, such as CT and histopathology (*10, 11*). For instance, five distinct histologic subtypes and radiologic patterns have been reported recently. Traditionally, histopathology images serve as a golden standard for lung cancer diagnosis. Cellular and inter-cellular level morphology are usually used by the pathologists for making diagnosis decisions. However, the current pathology diagnosis is commonly based on individual pathologists' interpretations of the samples which are subject to large inter-observer variations and low throughput analysis. Unbiased quantitative pathology methods are showing promise by offering more cellular information (*12–14*). Recently, pathology informatics study on lung cancer has attracted more interests. In one study (*15*), the diagnostic significance of nuclear features in differentiating small cell lung cancer (SCLC) and non-small cell lung cancer (NSCLC) was investigated. Edwards et.al.(*16*) showed that adenocarcinoma diagnosis is more challenging compared to squamous carcinoma. An early automatic pathology analysis system was proposed in (*17*). In the study by Mijović et al. (*18*), diagnostic values of seven Karyometric variables are examined for diagnosis of major histological types of lung carcinoma. In Zhang et al's study (*19*), an image classification system is proposed to differentiate lung adenocarcinoma and squamous carcinoma. The work by Yao et al (*20*) developed topological features for lung cancer diagnosis. Compared to genomic biomarkers, advanced imaging may provide more clinically relevant information.

In order to take advantage of both the richness of histopathological information and molecular profiles, we aim to develop an integrative computational pipeline that exploits diagnostic images and mRNA expression. A related work on lung cancer was recently published on integrating histopathologcal images with genetic data for outcome prediction (*21*). The pipeline allowed us to discover the associations between cellular level and molecular level phenotypes, and thus novel biomarkers can be unveiled. In this paper, we extracted 283 histopathological features from LUAD tissue slides and initially attempted to identify co-expression gene clusters that have high correlation with these image features. Such approach in other cancers has led to new insight on cancer biology and new potential biomarkers (*22*). However, as shown in this paper, the morphology of LUAD is much more complicated and it turned out that the morphological features have low correlations with gene expression profiles. Figure 1 shows a 'highly-correlated' pairs between the imaging features and gene clusters. It is thus plausible that the LUAD morphology is regulated by any particular group of genes; instead a specific morphological characteristic is a manifestation of a combined effect from multiple groups of genes. Based on these quantitative experiments, we assert that a multivariate model is needed.

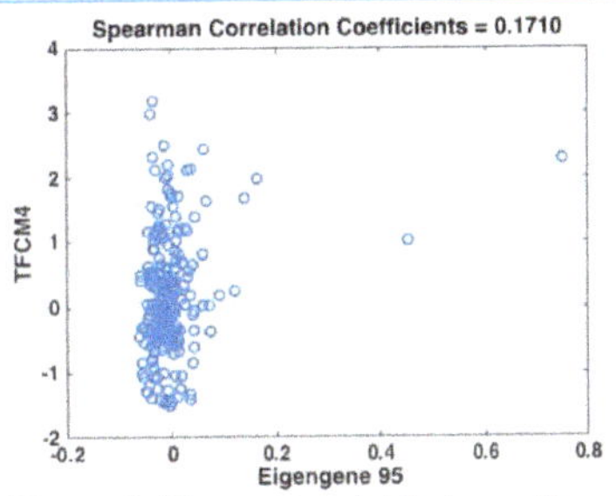

Figure 1: The scatter plot between the eigengene 95 and the *tfcm4*, which have the highest SCC between all eigengenes with tfcm4 (SCC = 0.170).

Therefore in this paper, we demonstrate that the morphological characteristics of LUAD can be explained by a combination of multiple gene clusters identified using sparse modeling based on the Lasso algorithm. In addition, we found that many of the gene clusters are associated with putative copy number variations, implying that genetic variations play important roles in the development cancer morphology. As far as we know, our finding is the first to directly link the genetic variations and functional genomics to LUAD histology. These observations will lead to new insight on lung cancer development and potential new integrative biomarkers for prediction patient outcome and response to treatments.

2 Methods and Materials

Our analysis involve molecular and histological analysis based on data from The Cancer Genome Atlas (TCGA) LUAD project. The data we use include mRNA profiling, histological images and clinical data including survival information.

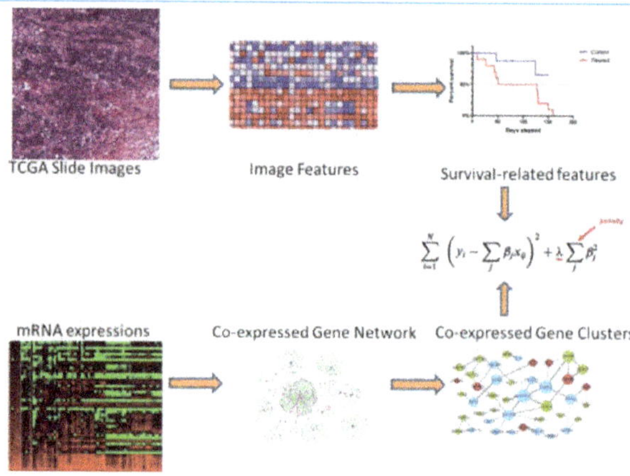

Figure 2: Schematic overview of constructing gene co-expression networks and analyzing the relationships between gene networks and morphological features.

2.1 Integrated Analysis Pipeline

We collected matched diagnostic images and gene expression data for a discovery dataset of 201 LUAD patients from the TCGA. The integrative analysis workflow is shown in Figure 2. Our automatic imaging processing pipeline detected cell nuclei and extracted predefined features evaluating staining variations. To select imaging features with clinical relevance, survival-related imaging features were identified. At molecular level, gene expression profiles (mRNA levels) were filtered and clustered using our co-expression network analysis algorithm. Strongly co-expressed gene clusters were represented by eigengenes. Then, we built a lasso regression model to select gene clusters that regulate the image feature that has the strongest association with survival times. By finding the co-expression patterns that are associated with the selected imaging feature, we can discover biological processes and genetic variations associated with cancer histology.

2.2 Image and Genomic Data Collection

We focus on LUAD patients with clinical information, genomic information, and histopathologic whole slide images. The data were downloaded from TCGA (The Cancer Genome Atlas) Data Portal. Data for 201 LUAD patients with all the three data types are downloaded for the experiments in 2014. For each patient, a representative image patch of size 1712 x 952 without damage or artifact is cropped from the tumor region. Expression profiles of 20,530 unique genes were investigated in the 201 patients (*23*).

2.3 Data Preprocessing and Imaging Feature Extraction

2.3.1 Imaging features

We adopt the cell detection and segmentation methods proposed in (*24*). In the cell detection stage, a radial voting scheme with Gaussian pyramid is employed (*25*). For each image, a Gaussian pyramid is created. A single–pass voting is applied to each layer. The voting region receives scores weighted by a distance transform. Therefore, such weighted voting encourages the pixels closer to the cell center accumulates higher voting scores. The final voting score is obtained by summing up the voting scores from different layers. In the segmentation stage, a marker based active contour with a repulsive term is applied to the images using the detection results as the markers. An initial contour associated with each detected marker is created first. The contours evolve through an iterative procedure to reach the real boundaries of the cells. The repulsive term serves to prevent the contours from crossing and merging with each other.

Group 1: Geometry Features. Based on the segmentation results, five geometry features are calculated for each lung cancer cell to capture the cell shape information, including cell area, contour perimeter, circularity, major-minor axis ratio, and contour solidity. Contour solidity is defined as the ratio of the area of a cell region over the convex hull defined by the segmentation boundary.

Group 2: Pixel Intensity Statistics. Pixel intensity statistics features are used to capture the color of the segmented cells. This group of features are calculated based on the intensity of the pixels within the segmented cells, including intensity mean, standard deviation, skewness, kurtosis, entropy, and energy. *Lab* color space is used for a better color representation.

Group 3: Texture Features: Texture is an important feature found to be closely related to cancer diagnosis in radiomics. This is rooted in the fact that texture patterns are linked to difference in protein expressions (*26*). This group of features consists of co-occurrence matrix (*27*), center symmetric auto-correlation (CSAC) (*28*), local binary pattern (LBP) (*29*), texture feature coding method (TFCM) (*30*). The co-occurrence matrix (*27*) computes an estimation of the joint probability distribution of the intensity of two neighboring pixels. CSAC is a measure of the local patterns with symmetrical structure. These patterns are characterized by a series of local auto-correlation and covariance introduced in (*28*), including symmetric texture covariance (SCOV), variance (SVR), and within-pair variance (WVAR), and between-pair variance (BVAR). 3×3 pixel unit of each channel is considered. LBP (*29*) feature measures the local textures by assigning a binary code to a pixel with respect to its intensity and those of its neighboring pixels. A histogram of the generated binary codes reveals the distribution of the present repeated local patterns. Similar to LBP, in TFCM (*30*), a texture feature number (TFN) is assigned to each pixel by comparing this pixel with its neighbors in four directions: 0°, 45°, 90°, and 135°. A histogram is calculated based on the TFNs of one image patch.

2.3.2 Gene transcriptome data

The expression profiles of 201 samples with primary lung cancer adenocarcinoma from TCGA LUAD project were downloaded from TCGA data portal in January 2014. Specifically, RNA-seq data for the tumor samples were obtained using Illumina sequencing and processed as described in (*6*). The mapping results were converted to RPKM (read per kilobase per million reads) values for 20,530 genes. Genes with low expression levels (with no data in the top 15 percentile) and low variance (in the lowest 10 percentile) were removed resulting in 9,179 genes.

2.4 Gene co-expression network analysis and summarization

While our goal is to establish the relationships between gene expression levels and the imaging features, we first carry out gene co-expression network analysis (GCNA) to cluster genes into co-expressed clusters. There are multiple reasons for carrying out GCNA before associating them with the imaging features. First, there is a large number of genes. If the association between every pair of gene and imaging feature is calculated and tested for significance, then more than half a million tests will be carried out which leads to low statistical power. In addition, since we will explore the association beyond univariate relationships using sparse analysis, the large number of genes (which are not always independent), pose serious computing challenges to the sparse modeling algorithms such as Lasso. Thus we first group genes with highly correlated expression profiles into co-expression clusters using GCNA then summarize the expression profiles within each cluster as an "eigengene" using the protocol described in (*31*). Essentially the expression profiles of each gene are first centralized (by subtracting the mean for each gene) and then standardized to have norm one. After the processing steps, singular value decomposition is applied to obtain the *eigengene* as the principal vector in the direction with the largest variance among the samples. Another advantage of the GCNA approach is that the highly co-expressed gene clusters are usually highly enriched in specific biological processes, regulatory factors or structural variations (e.g., copy number variations) (*32*), making the interpretation of the results straightforward.

While there are many algorithms for performing GCNA including the well known WGCNA package, we use a weighted network mining algorithm called local maximum quasi-clique merging (lmQCM) algorithm we recently developed (*32*). Unlike WGCNA which uses hierarchical clustering and does not allow overlaps between clusters, our algorithm is a greedy approach allowing genes to be shared among multiple clusters, in consistent with the fact the genes often participate in multiple biological processes. In addition, we have shown that lmQCM can find smaller co-expressed gene clusters which are often associated structual mutations such as copy number variations in cancers. The lmQCM algorithm has four parameters γ, α, t, and β. Among these parameters, γ is the most influencial, it decides if a new cluster can be initiated by setting the weight threshold for the first edge of the cluster as a subnetwork. In our GCN analysis, we directly use the absolute values of the Spearman correlation coefficients between expression profiles of genes as weights for which we have shown to be effective in previous studies.

2.4 Associations between Morphology and Transcriptomes

2.4.1 Correlation analysis

We first examined the correlation between the imaging features and the eigengenes for the gene clusters identified using lmQCM by calculating the Spearman correlation coefficients between them. However, as shown in the Results, the correlations between imaging features and eigengenes are not strong (none of them is significant if Bonferroni correction is applied for multiple test compensation). While this is different from the case in breast cancer, it suggests that the tissue morphology development is a complicated process involving in multiple processes and genetic factors. Thus in order to explain the morphology development, we need to resort to multi-variate modeling methods such as lasso regression.

2.4.2 Sparse modeling using Lasso regression

We model imaging features as manifestations of gene expression. Given the data availability, we focus on transcriptome data. Lasso regression model minimizes the residual sum of squares while at the same time enforcing sparsity of the model by adding a penalty term of the L_1-norm of the model coefficients.

Consider the linear regression model: we have (x_i, y_i), $i = 1,2,\ldots,N$, where $x_i = (x_{i1},\ldots,x_{ip})^T$ and y_i are eigen-gene expression and image feature value for the i th observation(patient sample), respectively. With regular regression model, the least square estimates are obtained by minimizing the residual squared error. However, in feature selection models to predict biomarkers, only imperative transcriptomes contribute to biological functions and processes, requiring more stringent and interpretable features. With large number of features, we would like to determine a small subset of them that can predict strong correlations. Let $\beta = (\beta_1, ..., \beta_p)^T$ and β_0 to be a scalar. The lasso model estimate (β, β_0) by solving the following problem

$$\min_{\beta, \beta_0} (\frac{1}{2N} \sum_{i=1}^{N} (y_i - \beta_0 - x_i^T \beta)^2 + \lambda \sum_{j=1}^{p} |\beta_j|), \quad (1)$$

where λ is nonnegative regularization parameter giving the weight for the model complexity term. As λ increases, the number of nonzero components of β decreases, leading to smaller numbers of predictors.

2.5 Identification of Survival-Related Image Features

Univariate Cox Proportional Hazard models are used to identify morphological features and genes that have expression related significantly to survival. Morphological features that have p-values less than 0.05 are recorded.

Table 1: Prognostic values of various image features in discover dataset. The features are listed by their significance in the survival model.

Feature Names	p-value		Feature Names	p-value
tfcm4	0.00456904		contrast1	0.01210092
tfcm9	0.00532429		tfcm12	0.01247155
tfcm3	0.00563955		tfcm11	0.01361604
tfcm1	0.0064998		csac23	0.01754474
tfcm2	0.00657692		tfcm7	0.0178572
tfcm10	0.00685436		fourier15	0.0178766
contrast2	0.0082282		csac5	0.01896244
tfcm8	0.0093341		entry4	0.01995154

2.6 Validating the Identified Genes in Other Data Set

For genes associated with the imaging features with the highest potential for predicting patient survival, we also test them on another publicly available dataset obtained from the NCBI Gene Expression Omnibus. The dataset GSExxxx contains transcriptome data of 149 non-small cell lung cancer patients, among which 90 are unique lung adenocarcinoma patients with clinical outcome (survival time and status). We use the genes to be tested as features to separate the 90 patients into two groups using K-means algorithms (K=2, Euclidean distance, average linkage, and 10 replicates). The survival times of the two groups are then visualized using Kaplan-Meier curves and compared using Cox Proportional Hazard regression.

2.7 Enrichment analysis of gene clusters

To interpret the biological meaning of the identified gene clusters, enrichment analysis tools such as TOPPGene (https://toppgene.cchmc.org/enrichment.jsp) are used. In addition, information about the genes are extracted from cBioPortal (http://www.cbioportal.org/).

3 Results

3.1 Image Feature Calculation

As shown in Figure 3 Left, the images reveal clear heterogeneity of the tumors among the patients. We calculated 283 image features from the images. As described in Section 2.3.1, there are multiple types of features and many features are strongly correlated (Figure 3 Right) such as part of the TCFM family (the block of 211 to 222). In this paper, we analyze each feature individually, but some of the highly features can be combined in future analysis.

3.2 Survival-related Image Features and Gene Cluster

Using a univariate Cox proportional hazards regression model, we assessed the image features related risk score in the prediction of the LUAD patient survival. Significant

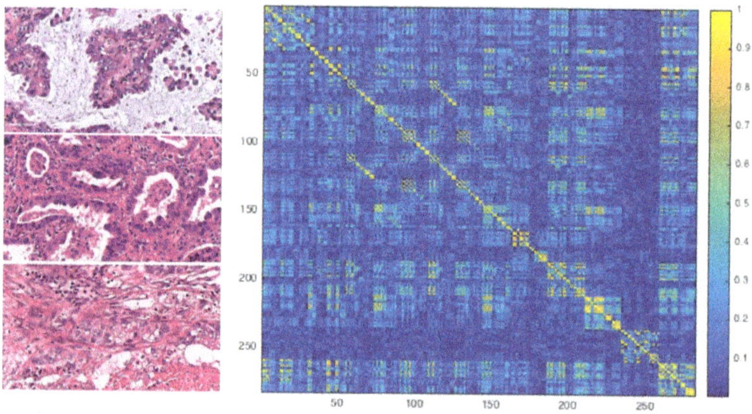

Figure 3. **Left**: Examples of the image patches from different patients. **Right**: Heatmap of the correlation (Spearman) matrix for the 283 image features.

morphological features are listed in Table 1. Among the six categories of imaging features, the *tfcm* category shows the most significant prognostic power, indicating texture features in lung adenocarcinoma have a strong potential for predicting patients' outcomes. In fact, all of the top six survival-related imaging features are in the *tfcm* category. Other features that capture prognosis are *contrast2*, *contrast*, *csac23*, *fourier15*, *csac5* and *entry4*.

3.2 Gene Co-Expression Network Analysis

As mentioned in Section 2.3.2, *9,179* genes were kept for analysis. The absolute value of the Spearman rank correlation coefficients were used for cluster detection using lmQCM algorithm. We allow the smallest gene clusters to have five genes. Then we found with γ = 0.75, t = 1, α = 1, and β = 0.4 the algoirthm yielded co-expressed gene clusters with balanced sizes. Specifically, it led to 95 clusters ranging from 5 to 120 genes. Many of the gene clusters are consistent to the ones frequently found in cancers. Most of these clusters involved in hallmark cancer biological processes such as cell cycle/genome stability (cluster 1), immune respones (cluster 2), translation / protein synthesis (cluster 3), and extracellular matrix development (cluster 7). However, some of them are more associated with specific cytobands (e.g., chr19p13), implying potential CNV sites.

3.3 Correlations between Image Features and Gene Clusters

The image analysis pipeline allowed us to quantify tumor characteristics on cellular level and associate these tumor characteristics with patient outcomes. In this study, we calculated 283 imaging features for the 201 patients and correlated with the 95 eigengenes. The correlation coefficient with the large absolute value is -0.2990 (p=1.7728e-05). In Table 2, we list the strongest correlation between eigengenes and the top five imaging features (in

Table 1) with the most significant power for predicting patient outcome. It is clear from the table that none of such correlations is statistically significant (after multiple test compensation), suggesting that complex phenomena such as cell and tissue morphology in lung cancers can only be explained by multiple molecular and genetic factors.

Table 2. Imaging features and the eigengenes with the strongest correlations with them.

Imaging feature	Eigengene (cluster)	SCC/p-value	Enrichment
tfcm4	95	0.1710/0.0153	18q12.1 (p=1.175e-9), all five genes on 18q12
tfcm9	59	0.1677/0.0174	16p11.2 (p=1.364e-10), all seven genes on 16p11
tfcm3	59	-0.1658/0.0188	
tfcm1	59	0.1704/0.0157	
tfcm2	59	0.1508/0.0327	

3.4 Lasso Regression Model for Imaging Features Using Eigengenes

Since the imaging features with prognostic power do not have strongly correlated gene clusters, we resort to multivariate models to explain the cell and tissue morphology using molecular data. Specifically, we built a lasso regression model. The lasso model selects a sparse set of eigengenes to explain the selected imaging feature. We rank the importance of image features by their significance in survival analysis. The top 10 image features in Table 1 belong to only two categories – TFCM and Contrast. Features within each category are highly correlated (for the eight TFCM features, the smallest of the absolute value of the SCC is 0.6840, the two SCC between the two Contrast features is 0.9923). Since eight out of 10 top image features are from the TFCM family, we chose one feature from for our modeling, namely *tfcm4*.

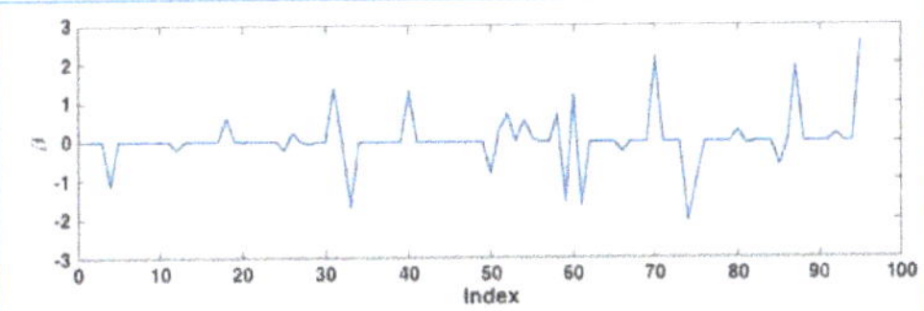

Figure 4. Coefficients (β values) of the lasso regression for *tfcm4*.

For *tfcm4*, it is found that the lowest MSE is found at λ = 0.0371 for the cost function in Eq.(1). Figure 4 shows the values of the coefficients β. Among the 95 eigengenes, 28 have non-zero coefficients among which 18 are larger than 0.5 and 12 are larger than 1. For the analysis of genes, we collected 185 genes from the 18 clusters with absolute value of coefficients larger than 0.5. In addition, Figure 5 shows the correlation between the combined eigengenes using the calculated β values with the *tfcm4* values in contrast to the correlation between the 95th eigengene (as listed in Table 2) and *tfcm4* (Figure 1).

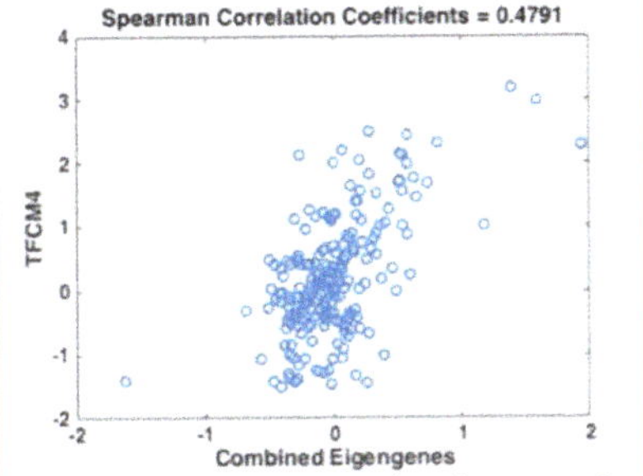

Figure 5. The scatter plot between the combined eigengenes using the lasso coefficients and the *tfcm4* (SCC = 0.4791).

3.5 Functional and Genetic Analysis of Gene Clusters Associated with Imaging Features

In order to understand the functional roles of the gene clusters associated with *tfcm4*, enrichment analysis was carried out using TOPPGene and the results for the 18 gene clusters are shown in Table 2. Among the gene clusters whose eigengenes are associated with tfcm4, the largest cluster is the cluster #4, consisting of 59 genes and is highly enriched with ribosomal genes and thus protein translation function. Other related biological processes including immune response (response to virus, cluster #18), response to steroid hormone, negative regulation of epithelial cell proliferation, and mitochondrial ATP synthesis.

Interestingly, 14 out of the 18 gene clusters are highly enriched on specific cytobands. It has been previously noticed that many of the co-expressed clusters in cancers are associated with copy number variations (CNVs) in specific cytobands (*32*). CNVs are common genetic variations playing important roles cancer initiation and development. Functional CNVs usually lead to changes in expression levels of genes on that region due to the "dose effect", which also leads to co-expression of the transcribed genes. Figure 6 Left shows an example of the RPRD1A gene in cluster #95, whose mRNA level has a strong correlation with its copy number measurement and it shows a strong co-expression relationship with the ELP2 genes on the same cytoband.

Table 2: Gene clusters showing strong correlation with texture image feature tfcm, and their Gene Ontology terms and enriched cytobands.

Gene Cluster (size)	beta	GO Biological Process/p-values	Cytobands/p-values	Notes:
4 (59)	-1.1558	GO:0006614 SRP-dependent cotranslational protein targeting to membrane / 9.105E-98		
18 (14)	0.6328	GO:0009615response to virus/ 9.965E-15		
31 (10)	1.3894			Genes down-regulated in nsopharyngeal carcinoma relative to the normal tissue (p = 5.074e-19, all 10 genes)
33 (10)	-1.7213		19q13.42/5.525e-6	All 10 genes on 19q13.3-4
40 (8)	1.2977		8q24.13/3.263e-5	Seven genes on 8q21-24, one on 8q13
50 (8)	-0.8343	GO:0048545response to steroid hormone / 2.290E-8		
52 (8)	0.7075		7q33/4.800E-5	All eight genes on 7q21-36
54 (7)	0.5669	GO:0006413 translational initiation /1.096E-5	Yq11/2.305E-6, Xq13.2/2.856E-5	Four genes on Yq11, two on Xq13.2, one on Yp11.3
58 (7)	0.6952		8p21.1/ 6.631E-6	Five genes on 8p21, two on 8p12
59 (7)	-1.5729		16p11.2/1.364e-10	All seven genes on 16p11
60 (7)	1.2103		Xq28/1.982e-13	All seven genes on Xq27-28
61 (7)	-1.6639		6p21.1/4.436e-7	Six genes on 6p21-22, one on 6p12
70 (6)	2.1783		17q21.31/5.532e-	All six genes are on 17q21

			10	
74 (6)	-2.0544		8p11.2/1.048e-9	All six genes are on 8p11.2
75 (6)	-1.0093	GO:0050680 negative regulation of epithelial cell proliferation / 3.290E-6	17q11.2/6.880e-7	All six genes are on 17q11-12
85 (5)	-0.6095	GO:0042776 mitochondrial ATP synthesis coupled proton transport / 6.311E-9	21q22.11/3.344E-5	Four genes on 21q21-22
87 (5)	1.9569		19q13.2/1.131e-6	All five genes on 19q13
95 (5)	2.5783		18q12.1/1.175e-9	All five genes on 18q12

3.4 Prognostic Validation

Validation on heterogeneous external data sets allows for evaluation of the generalizability. To test the importance of cilium-related genes, we further performed survival analysis on a publicly available dataset with 90 LUAD patients. Among the 185 genes correlated with the image feature category tfcm, 118 of the gene symbols can be matched exactly to the external dataset. In the validation dataset, lung adenocarcinoma patients were stratified into two groups using K-means based on their expression levels of the 118 genes. In both datasets, a statistically significant group of patients with worse outcomes were differentiated (n = 44 and n = 46, respectively). The difference between the two groups is significant (Cox hazard proportional model p-value 2.1285e-6). Figure 7 shows the Kaplan-Meier curves of the two patient cohorts.

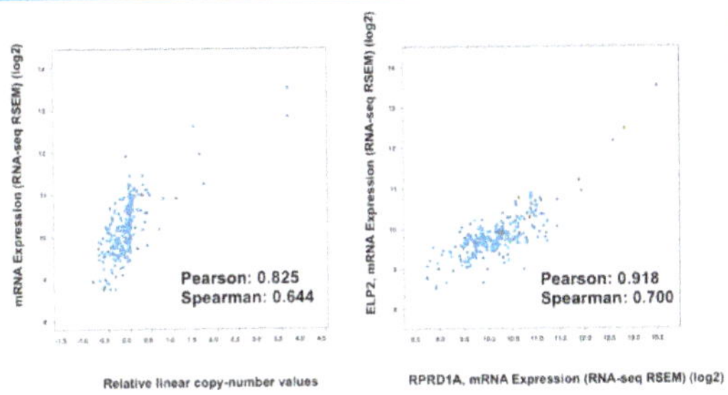

Figure 6. **Left**: Gene RPRD1A shows strong correlation between mRNA level and copy number values. **Right**: RPRD1A and ELP2 (both in cluster #95) are strongly co-expressed.

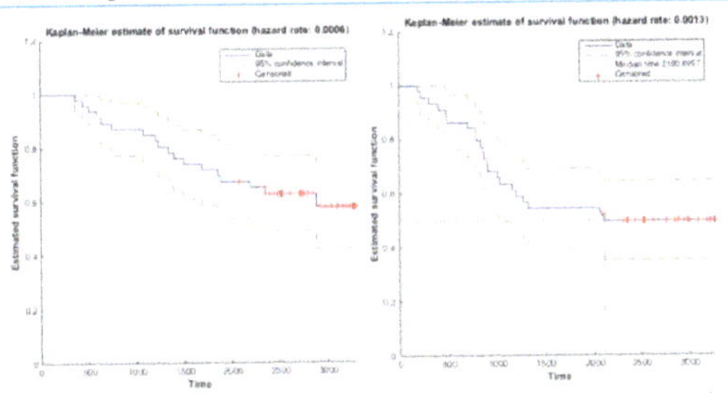

Figure 7: Kaplan–Meier survival curves of prognostic model on two clusters from the lung adenocarcinoma dataset. Survival curves of two groups (group 1 with 44 patients on the left and group 2 with 46 patients on the right). The difference of survival time between the two groups is significant (Cox hazard proportional model p-value 2.1285e-6).

4 Discussion and Conclusion

Our integrative analysis

pipeline allows us to find survival related textural features of lung adenocarcinoma. In addition to the image features, we also demonstrated that modeling of the histology at cellular and tissue levels using "omics" data may involve multiple groups of genes. Interestingly, our results showed that that the histological phenotype may be manifestations of multiple genetic variations, especially copy number variations. Specifically, many of the enriched cytobands we identified have been previously associated with lung cancer development including 19q13 (*33, 34*) , 8q24 (*33*), 7q21-36 (*35*), 8p21 (*33*), 16p11 (*36*), Xq27-28, 6p21 (*34*), 17q21 (*34*), 21q22 (*35*), and 18q12 (*33*). While there is no report on the association of Xq27-28 with lung cancer, Xq26 has been shown to be associated with lung cancers (*36*), suggesting that the genetic variations should be further explored to identify potential "driver" genes for lung cancer. We also showed that the genes in the clusters can indeed predict patient prognosis, which leads to discovery of potential biomarkers. While our study is focused on patient prognosis, the process can be repeated for patient treatment response prediction with appropriate data. Overall we demonstrated that the morphology is a complex phenomenon and its development may involve multiple groups of genes. In cancers, this process is even more complex as the genetic variations also contribute significantly to this process. Our findings indeed support this notion.

References

1. S. Couraud, G. Zalcman, B. Milleron, F. Morin, P.-J. Souquet, *Eur. J. Cancer.* **48**, 1299–311 (2012).
2. P. A. Russell *et al.*, *J. Thorac. Oncol.* **6**, 1496–504 (2011).
3. E. Conde *et al.*, *Clin. Transl. Oncol.* **15**, 503–8 (2013).
4. D. Hokka *et al.*, *Lung Cancer.* **79**, 77–82 (2013).
5. X. Li *et al.*, *Neoplasma.* **59**, 500–7 (2012).
6. E. A. Collisson *et al.*, *Nature.* **511**, 543–50 (2014).
7. C. Camps, Jantus-Lewintre, Usó, Sanmartin, *Lung Cancer Targets Ther.*, 21 (2012).
8. P. F. Robert T. Adamson, *Am. J. Manag. Care.* **19** (2013).
9. J. Chmielecki *et al.*, *J. Thorac. Oncol.* **7**, 434–42 (2012).
10. J. H. M. Austin *et al.*, *Radiology.* **266**, 62–71 (2013).
11. L. M. Solis *et al.*, *Cancer.* **118**, 2889–99 (2012).
12. Y. Yuan *et al.*, *Sci. Transl. Med.* **4**, 157ra143–157ra143 (2012).
13. a. H. Beck *et al.*, *Sci. Transl. Med.* **3**, 108ra113–108ra113 (2011).
14. H. Wang, F. Xing, H. Su, A. Stromberg, L. Yang, *BMC Bioinformatics.* **15**, 1–12 (2014).
15. F. B. Thunnissen *et al.*, *Pathol. Res. Pract.* **188**, 531–5 (1992).
16. S. L. Edwards *et al.*, *J. Clin. Pathol.* **53**, 537–40 (2000).
17. K. Kayser, D. Radziszowski, P. Bzdyl, R. Sommer, G. Kayser, *Rom. J. Morphol. Embryol.* **47**, 21–8 (2006).
18. M. Mijovic, Zaklina; Mihailovic, Dragan; Kostov, *Med. Biol.* **15**, 28 – 32 (2008).
19. X. Zhang, L. Yang, W. Liu, H. Su, S. Zhang, in *Medical Image Computing and Computer-Assisted Intervention – MICCAI 2014* (2014), pp. 479–486.
20. J. Yao *et al.*, in *Proceedings of the 6th International Workshop on Machine Learning in Medical Imaging - Volume 9352* (Springer-Verlag New York, Inc., 2015; http://link.springer.com/10.1007/978-3-319-24888-2_35), pp. 288–295.
21. X. Zhu *et al.*, in *2016 IEEE 13th International Symposium on Biomedical Imaging (ISBI)* (IEEE, 2016; http://ieeexplore.ieee.org/document/7493475/), pp. 1173–1176.
22. C. Wang *et al.*, *J. Am. Med. Inform. Assoc.* **20**, 680–7.

23. TCGA, The Cancer Genome Atlas - Data Portal.
24. F. Xing, L. Yang, in *2013 IEEE 10th International Symposium on Biomedical Imaging* (IEEE, 2013), pp. 386–389.
25. X. Qi, F. Xing, D. J. Foran, L. Yang, *IEEE Trans. Biomed. Eng.* **59**, 754–65 (2012).
26. P. Lambin *et al.*, *Eur. J. Cancer.* **48**, 441–6 (2012).
27. R. M. Haralick, K. Shanmugam, I. Dinstein, *IEEE Trans. Syst. Man. Cybern.* **3**, 610–621 (1973).
28. K. Laws, in *24th Annual Technical Symposium*, T. F. Wiener, Ed. (International Society for Optics and Photonics, 1980), pp. 376–381.
29. T. Ojala, M. Pietikainen, T. Maenpaa, *IEEE Trans. Pattern Anal. Mach. Intell.* **24**, 971–987 (2002).
30. M.-H. Horng, Y.-N. Sun, X.-Z. Lin, *Comput. Med. Imaging Graph.* **26**, 33–42 (2002).
31. P. Langfelder, S. Horvath, *BMC Bioinformatics.* **9**, 559 (2008).
32. J. Zhang, K. Huang, *Cancer Inform.* **1**, 1 (2016).
33. B. R. Balsara *et al.*, *Cancer Res.* **57**, 2116–20 (1997).
34. P. P. Medina *et al.*, *Hum. Mol. Genet.* **18**, 1343–52 (2009).
35. F. Li, L. Sun, S. Zhang, *Oncol. Rep.* **34**, 1701–7 (2015).
36. N. A. Levin *et al.*, *Cancer Res.* **54**, 5086–91 (1994).

IDENTIFICATION OF DISCRIMINATIVE IMAGING PROTEOMICS ASSOCIATIONS IN ALZHEIMER'S DISEASE VIA A NOVEL SPARSE CORRELATION MODEL

JINGWEN YAN*

Department of BioHealth Informatics, Indiana University,
Center for Computational Biology and Bioinformatics, School of Medicine, Indiana University
Indianapolis, 46202, USA
E-mail: jingyan@iupui.edu

SHANNON L. RISACHER, KWANGSIK NHO, ANDREW J. SAYKIN

Department of Radiology and Imaging Sciences, School of Medicine, Indiana University,
Indianapolis, 46202, USA
E-mail:{srisache,knho,asaykin}@iupui.edu

LI SHEN*

Department of Radiology and Imaging Sciences, School of Medicine, Indiana University,
Center for Computational Biology and Bioinformatics, School of Medicine, Indiana University
Indianapolis, 46202, USA
E-mail:shenli@iu.edu

FOR THE ALZHEIMER'S DISEASE NEUROIMAGING INITIATIVE†

Brain imaging and protein expression, from both cerebrospinal fluid and blood plasma, have been found to provide complementary information in predicting the clinical outcomes of Alzheimer's disease (AD). But the underlying associations that contribute to such a complementary relationship have not been previously studied yet. In this work, we will perform an imaging proteomics association analysis to explore how they are related with each other. While traditional association models, such as Sparse Canonical Correlation Analysis (SCCA), can not guarantee the selection of only disease-relevant biomarkers and associations, we propose a novel discriminative SCCA (denoted as DSCCA) model with new penalty terms to account for the disease status information. Given brain imaging, proteomic and diagnostic data, the proposed model can perform a joint association and multi-class discrimination analysis, such that we can not only identify disease-relevant multimodal biomarkers, but also reveal strong associations between them. Based on a real imaging proteomic data set, the empirical results show that DSCCA and traditional SCCA have comparable association performances. But in a further classification analysis, canonical variables of imaging and proteomic data obtained in DSCCA demonstrate much more discrimination power toward multiple pairs of diagnosis groups than those obtained in SCCA.

Keywords: Imaging genomics; Alzheimer's disease; Proteomics; Canonical correlation analysis; Multi-class discrimination.

---

*To whom correspondence should be addressed

†Data used in preparation of this article were obtained from the Alzheimer's Disease Neuroimaging Initiative (ADNI) database (adni.loni.usc.edu). As such, the investigators within the ADNI contributed to the design and implementation of ADNI and/or provided data but did not participate in analysis or writing of this report. A complete listing of ADNI investigators can be found at: http://adni.loni.usc.edu/wp-content/uploads/how_to_apply/ADNI_Acknowledgement_List.pdf.

1. Introduction

Alzheimer's disease (AD) has been well known as one of the most common brain dementia, a major neurodegenerative disorder that has been characterized by gradual memory loss and brain behavior impairment. According to the latest report,[1] more than 5 million Americans are living with Alzheimer's and it has been officially listed as the 6th leading cause of death. Also, due to the significant decline of self-care capabilities during disease, it is not only the patients who suffer, but also the family members, friends, communities and the whole society considering the time-consuming daily care and high health care expenditures needed. In the past decade, deaths attributed to Alzheimer's disease has increased 68 percent, while deaths attributed to the number one cause, heart disease, has decreased 16 percent. And all of these situations will continue to deteriorate as the population ages during the next several decades. To prevent such health care crisis, substantial efforts have been made to help cure, slow or stop the progression of the disease.

In the last few years, many efforts have been dedicated to explore whether the combination of multi-modal measures, e.g. brain atrophy measured by magnetic resonance imaging (MRI), hypometabolism measured by functional imaging and quantification of proteins, can better predict the clinical outcomes of AD, such as disease status and cognitive outcomes.[19] In many of these works, it has been found that brain imaging and protein expression, from both cerebrospinal fluid (CSF) and blood plasma, hold some complementary information.[12,18] But how they are related with each other still remains elusive.

In this work, we will explore the relationships between brain imaging and protein expression using bi-multivariate association models. Sparse Canonical Correlation Analysis (SCCA)[11,16] is a typical example that has been widely used for associative analysis in both real[8,15] and simulated[3] -omics data sets.[2,11,17] But it can not guarantee the selection of disease-relevant biomarkers and therefore the associations generated in SCCA are not necessarily related to a specific disease either, unless the input features are already prefiltered disease-related biomarkers.[5] On the other hand, most existing SCCA algorithms use the soft threshold strategy for solving the Lasso[11,16] regularization terms, which assumes the independence structure of data features. Unfortunately, this independence assumption does not hold in neither imaging nor proteomics data, and will inevitably limit the capability of yielding optimal solutions.

To overcome these limitations, we propose a novel discriminative SCCA (DSCCA) model, coupled with a new algorithm to eliminate the independence assumption, to explore the imaging and proteomic associations. Given imaging, proteomic and diagnostic data, the proposed model can perform a joint association and multi-class discrimination analysis. As such, we can not only identify disease-relevant multimodal biomarkers, but also reveal strong association between them. We perform an empirical comparison between the proposed DSCCA algorithm and a widely used SCCA implementation in the PMA software package (http://cran.r-project.org/web/packages/PMA/).[16] The results show that DSCCA and SCCA have comparable association performances. But in a further classification analysis, canonical variables of imaging and proteomic data obtained in DSCCA demonstrate much more discrimination power toward diagnosis groups than those obtained in SCCA.

2. Discriminative SCCA (DSCCA)

Throughout this section, we denote vectors as boldface lowercase letters and matrices as boldface uppercase ones. For a given matrix $\mathbf{M} = (m_{ij})$, we denote its i-th row and j-th column to $\mathbf{m}^i$ and $\mathbf{m}_j$ respectively. Let $\mathbf{X} = \{x_1, ..., x_n\} \subseteq \Re^p$ be the imaging data and $\mathbf{Y} = \{y_1, ..., y_n\} \subseteq \Re^q$ be the protein data, where n is the number of participants, p and q are the number of brain regions and proteins respectively.

Canonical correlation analysis (CCA) is a bi-multivariate method that explores the linear transformations of variables $\mathbf{X}$ and $\mathbf{Y}$ to achieve the maximal correlation between $\mathbf{Xu}$ and $\mathbf{Yv}$, which can be formulated as:

$$\max_{\mathbf{u},\mathbf{v}} \mathbf{u}^T\mathbf{X}^T\mathbf{Y}\mathbf{v} \qquad s.t.\ \mathbf{u}^T\mathbf{X}^T\mathbf{X}\mathbf{u} = 1, \mathbf{v}^T\mathbf{Y}^T\mathbf{Y}\mathbf{v} = 1 \tag{1}$$

where $\mathbf{u}$ and $\mathbf{v}$ are canonical loadings or weights, reflecting the significance of each feature in identified associations.

However, the power of CCA in biomedical applications is quite limited due to 1) its requirement on the relatively large number of observations n which is expected to exceed the combined dimension of $\mathbf{X}$ and $\mathbf{Y}$, and 2) its nonsparse outputs $\mathbf{u}$ and $\mathbf{v}$ which make the ultimate pattern hard to interpret. To address this concerns, sparse CCA (SCCA) method was later proposed, where two penalty terms on both weight vectors $P_1(\mathbf{u}) \leq c_1$ and $P_2(\mathbf{v}) \leq c_2$ were introduced to help generate sparse results.

A widely used SCCA implementation, PMA package,[16] applied L_1 norm penalty for both P_1 and P_2. But without diagnosis information, its capability in identifying disease-relevant biomarkers is quite limited. Thus the ultimate association relationships are not necessarily related to a specific disease either. Another limitation of PMA is that it takes the soft threshold strategy in the solution, which requires the input data to have an linear independence design $\mathbf{X^TX} = \mathbf{I}$ and $\mathbf{Y^TY} = \mathbf{I}$ (see Section 10 in[14]). Unfortunately, this independence assumption does not hold in both imaging and proteomics data (e.g., correlated voxels in an ROI, correlated protein expressions), and will inevitably limit the capability of identifying meaningful imaging proteomics associations.

To overcome these limitations, we propose a novel discriminative SCCA (denoted as DSCCA) algorithm to not only take into account the diagnosis information but also eliminate the independence assumption. Inspired by the application of locality preserving projection (LPP) in linear discriminative analysis,[10] we add two new constraints as P_1 and P_2 for multi-class discrimination.

$$\begin{aligned} P_1(\mathbf{u}) &= ||\mathbf{u}||_D = \alpha\mathbf{u}^T\mathbf{X}^T\mathbf{L}_w\mathbf{X}\mathbf{u} - (1-\alpha)\mathbf{u}^T\mathbf{X}^T\mathbf{L}_b\mathbf{X}\mathbf{u}, \\ P_2(\mathbf{v}) &= ||\mathbf{v}||_D = \alpha\mathbf{v}^T\mathbf{Y}^T\mathbf{L}_w\mathbf{Y}\mathbf{v} - (1-\alpha)\mathbf{v}^T\mathbf{Y}^T\mathbf{L}_b\mathbf{Y}\mathbf{v}, \end{aligned} \tag{2}$$

Here, we construct two graphs $\mathbf{G}_w$ and $\mathbf{G}_b$ to account for the diagnosis groups, where each vertex indicates one subject (Fig. 1). In $\mathbf{G}_w$, only subjects within the same diagnosis group have connections to each other. In other words, we build a complete graph for all the subjects belonging to the same diagnosis group. In $\mathbf{G}_b$, only subjects from different diagnosis

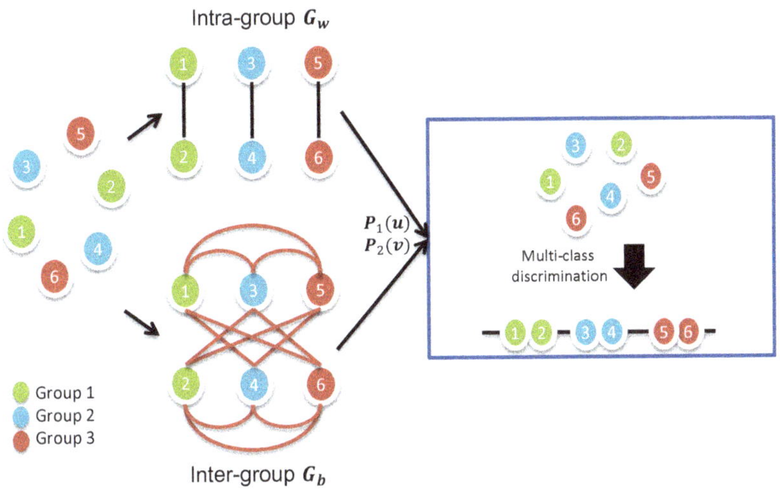

Fig. 1. Illustration of within- and between-group graphs $\mathbf{G}_w$ and $\mathbf{G}_b$. Each circle indicates one subject and subjects from the same diagnosis group are colored the same.

groups have connections. $\mathbf{L}_w$ and $\mathbf{L}_b$ are the Laplacian graphs of $\mathbf{G}_w$ and $\mathbf{G}_b$ respectively. While the traiditonal L_1 norm helps ascertain the sparsity of selected imaging and protein biomarkers, the new penalty term $||\cdot||_D$ encourages the closeness between subjects within the same diagnosis groups and distance between subjects from different diagnosis groups after projection. α is a trade off parameter that help balance the within- and between-group constraints. Since canonical variables $\mathbf{Xu}$ and $\mathbf{Yv}$ have the exact same length, we use the same α for both penalties P_1 and P_2.

The final objective function of DSCCA can be written as follows:

$$\max_{\mathbf{u},\mathbf{v}} \mathbf{u}^T\mathbf{X}^T\mathbf{Y}\mathbf{v} - \frac{\beta_1}{2}P_1(\mathbf{u}) - \frac{\beta_2}{2}P_2(\mathbf{v}) \tag{3}$$

$$s.t.\ \ \mathbf{u}^T\mathbf{X}^T\mathbf{X}\mathbf{u} = 1, \mathbf{v}^T\mathbf{Y}^T\mathbf{Y}\mathbf{v} = 1, ||\mathbf{u}||_1 \le c_1, ||\mathbf{v}||_1 \le c_2$$

Using Lagrange multipliers, Eq. (3) can be reformulated as follows:

$$\max_{\mathbf{u},\mathbf{v}} \mathbf{u}^T\mathbf{X}^T\mathbf{Y}\mathbf{v} - \frac{\gamma_1}{2}||\mathbf{Xu}||_2^2 - \frac{\gamma_2}{2}||\mathbf{Yv}||_2^2 - \frac{\beta_1}{2}P_1(\mathbf{u}) - \frac{\beta_2}{2}P_2(\mathbf{v}) - \lambda_1||\mathbf{u}||_1 - \lambda_2||\mathbf{v}||_1 \tag{4}$$

Eq. (4) is known as a bi-convex problem, which can be easily solved using an alternating algorithm as discussed in.[16] By fixing $\mathbf{u}$ and $\mathbf{v}$ respectively, we will have the following two minimization problems shown in Eq. (5) and (6).

$$\min_{\mathbf{u}} -\mathbf{u}^T\mathbf{X}^T\mathbf{Y}\mathbf{v} + \frac{\gamma_1}{2}\mathbf{u}^T\mathbf{X}^T\mathbf{X}\mathbf{u} + \frac{\beta_1}{2}P_1(\mathbf{u}) + \lambda_1||\mathbf{u}||_1, \quad (5)$$

$$\min_{\mathbf{v}} -\mathbf{u}^T\mathbf{X}^T\mathbf{Y}\mathbf{v} + \frac{\gamma_2}{2}\mathbf{v}^T\mathbf{Y}^T\mathbf{Y}\mathbf{v} + \frac{\beta_2}{2}P_2(\mathbf{v}) + \lambda_2||\mathbf{v}||_1, \quad (6)$$

Both objective functions can be efficiently solved using the Nesterovs accelerated proximal gradient optimization algorithm.[9] Algorithm 2.1 summarizes the optimization procedure. The convergence is based on the value changes of the objective function and we use 10^{-6} as stop criteria. Five-fold nested cross-validation was applied to automatically tune the parameters β_1, β_2, λ_1 and λ_2. According to,[2] the learned pattern and performance are insensitive to γ_1 and γ_2 settings. Therefore in this paper we set both of them to 1 for simplicity. The optimization method used in steps 3 and 4 is similar to that proposed in.[9]

Algorithm 2.1 Discriminative SCCA (DSCCA)

Require:

$\mathbf{X} = \{x_1, ..., x_n\}$, $\mathbf{Y} = \{y_1, ..., y_n\}$, $\mathbf{L}_w \subseteq \Re^{n \times n}$, $\mathbf{L}_b \subseteq \Re^{n \times n}$

Ensure:

Canonical vectors $\mathbf{u}$ and $\mathbf{v}$.

1: $t = 1$, Initialize $\mathbf{u}_t \in \Re^{p\times 1}$, $\mathbf{v}_t \in \Re^{q\times 1}$;
2: **while** not converge **do**
3: Solve Eq. (5) using Nesterov's method and obtain $\mathbf{u}$;
4: Solve Eq. (6) using Nesterov's method and obtain $\mathbf{v}$;
5: Scale $\mathbf{u}$ so that $\mathbf{u}^T\mathbf{u} = 1$
6: Scale $\mathbf{v}$ so that $\mathbf{v}^T\mathbf{v} = 1$
7: $t = t + 1$.
8: **end while**

3. Results

3.1. *Data and Experimental Setting*

The MRI data, quantification of proteins in CSF and blood plasma were downloaded from the Alzheimer's Disease Neuroimaging Initiative (ADNI) database. The primary goal of ADNI has been to test whether serial MRI, PET, other biological markers, and clinical and neuropsychological assessment can be combined to measure the progression of mild cognitive impairment (MCI) and early AD. For up-to-date information, see adni.loni.usc.edu.

We totally extracted 246 subjects with all MRI, CSF and plasma proteomic data available. To balance the diagnostic groups, we randomly removed some mild cognitive impairment (MCI) participants. Finally, 176 subjects (67 AD, 67 MCI and 42 healthy control (HC)), were included in this study (Table 1). For each baseline MRI scan, FreeSurfer (FS) V4 was employed to extract 73 cortical thickness measures and 26 volume measures, as well as to extract the intracranial volume (ICV). CSF and blood plasma samples were evaluated by Rules Based Medicine, Inc. (RBM) proteomic panel and 229 proteomic analytes survived the quality control process, with 83 from CSF and 146 from plasma. Using the regression weights from HC participants, all the MRI, CSF and blood plasma proteomic measures were pre-adjusted for the baseline age, gender, education, and handedness, with ICV as an additional covariate for MRI only.

Table 1. Participant characteristics

	HC	MCI	AD
Number	67	67	42
Gender(M/F)	38/29	45/22	22/20
Handedness(R/L)	64/3	64/3	38/4
Age(mean±std)	75.15±7.68	74.28±7.25	75.93±5.82
Education(mean±std)	15.12±3.01	15.96±2.92	15.88±2.77

3.2. *Experimental Results*

Both DSCCA and PMA were performed on the normalized FS and proteomic measures. To avoid the over-fitting problem, 5-fold nested cross-validation was applied, which also helped to optimally tune the parameters. Table 2 shows 5-fold cross-validation canonical correlation results. It is observed that proposed DSCCA and PMA have comparable performances in identifying imaging proteomic associations, whereas DSCCA is slightly better in performance stability.

Next, we examined the discriminative power of canonical variables $\mathbf{Xu}$ and $\mathbf{Yv}$ generated by DSCCA and PMA. Area under ROC curve (AUC) was calculated for each single canonical variable of five folds. Both imaging and proteomic canonical variables of PMA and imaging canonical variable of DSCCA were found to have little discrimination power in all HC vs MCI, HC vs AD and MCI vs AD cases. Proteomic canonical variable $\mathbf{Yv}$ of DSCCA has the best performance, with an averaged AUC around 0.7 for all three cases. Shown in Fig. 2 is an example plot of $\mathbf{Xu}$ against $\mathbf{Yv}$ in one fold. Dot colors represent different diagnostic groups. Compared to one single canonical variable, we observe that combination of two canonical variables generated in DSCCA demonstrated much more discrimination power than PMA. In Fig. 2(a) three diagnosis groups are all very well separated, whereas in Fig. 2(b) subjects are mixing together.

To further validate our results, a follow up classification analysis was performed using both imaging and proteomic canonical variables as predictors. Canonical loadings learned in the training data set are applied to both training and test data to calculate the training and test canonical variables respectively. The LIBSVM toolbox was employed to implement the SVM using a linear kernel under default settings. Three pair-wise binary classification analyses were performed between HC vs MCI, HC vs AD, and MCI vs AD respectively. Shown in Table. 3 are the classification performance comparison between DSCCA and PMA. The results are very encouraging. Canonical variables of DSCCA significantly outperformed those of PMA in terms of the overall accuracy in almost all the cases. The resulting best prediction rates for HC vs AD (92.1%), HC vs MCI (75.3%) and MCI vs AD (70.3%) were competitive with prior multi-modal studies,[6,19] especially considering that it is under default parameter settings.

All five-fold experiments generated similar sparse results in terms of selection of imaging and proteomic markers. Fig. 3 shows the imaging and proteomic markers commonly identified across all folds using DSCCA, where the color represents the weights of corresponding brain regions. Top brain regions identified include entorhinal cortex, amygdala volume, hippocampal volume, etc. (Fig. 3(a)), which are all aligned with previous AD findings.[12,19] In terms of proteomic markers, expression levels of 12 proteins from CSF and 19 proteins from blood plasma were found to be strongly associated with those brain regions. According to the STRING database (http://string-db.org/), these proteins are highly interconnected with each other, as shown in Fig. 3(b). Edges are colored based on the evidence of the connection, such as experimental interaction, co-expression or co-occurrence in the literature. The more edges two proteins have, the more confident their connection will be.

In particular, four proteins, apolipoprotein E (*APOE*), AXL receptor tyrosine kinase(*AXL*), interleukin 6 receptor (*IL6R*) and vascular endothelial growth factor (*VEGF*), were identified in both CSF and blood plasma. *APOE* is the top risk gene of AD. *AXL* is a member of the Tyro3-Axl-Mer (TAM) receptor tyrosine kinase subfamily, which has been previously reported to be involved in Amyloidogenic APP Processing and β-Amyloid Deposition in AD.[20] For growth factor VEGF, both its variants and expression changes are found to be associated with AD.[4,13] *IL6R* is less explored in terms of its relationship with dementia. But in a recent study it was reported to have significant associations with proteins involved in amyloid processing and inflammation.[7] These findings suggest the existence of certain connections between brain and blood biomarkers. Thus, more accessible fluid biomarkers from blood should have potential to provide extra insights of AD and guidance for future therapeutic intervention activities.

Table 2. Five-fold cross validation canonical correlation results

		f1	f2	f3	f4	f5	mean
DSCCA	Train	0.796	0.670	0.820	0.680	0.636	0.720
	Test	0.424	0.476	0.281	0.392	0.312	0.377
PMA	Train	0.529	0.629	0.505	0.524	0.504	0.538
	Test	0.410	0.095	0.324	0.201	0.460	0.298

4. Discussion

We performed an integrative analysis of brain imaging and protein expression data to jointly identify AD related biomarkers and their associations using a new sparse learning model DSCCA. The overall association performance of DSCCA is better than SCCA. the combination of its two canonical variables are much more powerful in discriminating multiple diagnostic groups simultaneously. Using both imaging and proteomic canonical variables in DSCCA as predictors, we obtained very promising prediction performances: HC vs AD (92.1%), HC vs MCI (75.3%) and MCI vs AD (70.3%), which were competitive with prior multi-modal studies. Since the classification was done under default parameter settings and the sample size is

Table 3. Five-fold cross validation classification performances (%) using canonical variables **Xu** and **Yv**. HC vs MCI, MCI vs AD, and HC vs AD are performed as three tasks separately.

		Train			Test		
		HC vs MCI	HC vs AD	MCI vs AD	HC vs MCI	HC vs AD	MCI vs AD
DSCCA	f1	97.17	100.00	94.19	75.00	91.30	60.87
	f2	86.79	96.51	84.88	85.71	95.65	60.87
	f3	96.23	100.00	94.19	85.71	91.30	86.96
	f4	93.40	95.35	75.58	57.14	100.00	78.26
	f5	72.32	82.61	69.57	72.73	82.35	64.71
	mean	89.18	94.89	83.68	75.26	92.12	70.33
PMA	f1	60.38	77.91	65.12	71.43	86.96	73.91
	f2	66.98	84.88	74.42	71.43	95.65	60.87
	f3	66.04	80.23	63.95	50.00	86.96	60.87
	f4	68.87	80.23	59.30	42.86	82.61	78.26
	f5	65.18	77.17	60.87	31.82	64.71	64.71
	mean	65.49	80.09	64.73	53.51	83.38	67.72

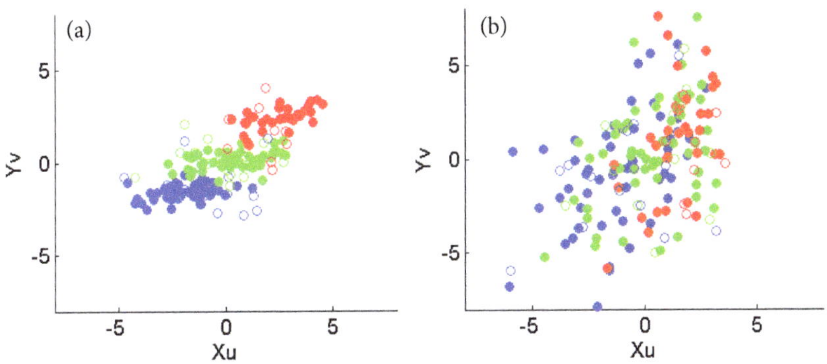

Fig. 2. Plot of canonical variables **Xu** and **Yv**. Left: DSCCA; Right: PMA; Red: AD; Green: MCI; Blue: HC; Solid: Training; Circle: Test.

very limited, we expect improved performances with more advanced parameter optimization strategies and/or larger sample sizes.

In real applications, many identified proteomic markers are found to be interconnected, but the underlying mechanisms still warrant further investigation. Replication in independent large samples will be important to confirm these findings. Further pathway enrichment analysis could be performed as a future direction to identify underlying biological pathways of relevant genes and proteins. Considering the ever increasing data volume and diversity in many complex diseases, another potential future topic is to investigate whether DSCCA can help identify valuable complementary information between new -omics features and further improve the classification performance.

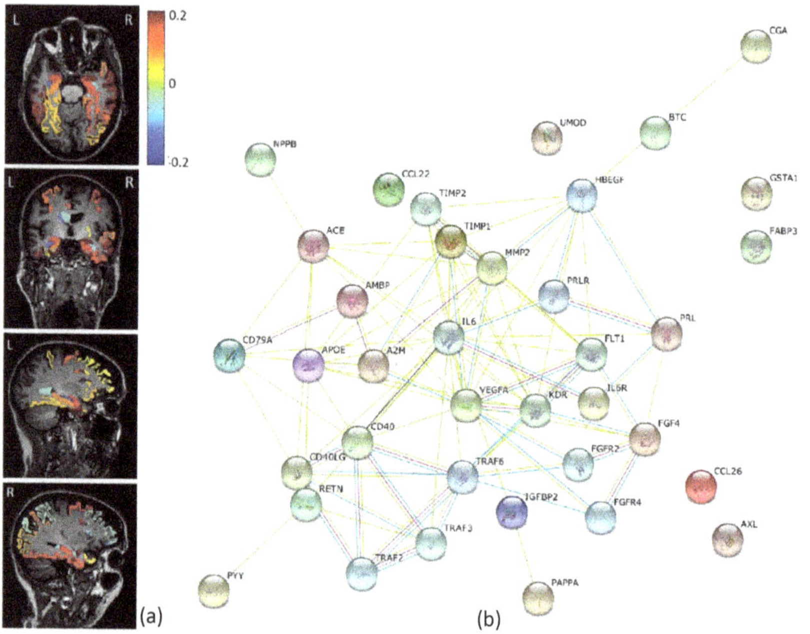

Fig. 3. Common imaging and proteomic markers across 5-fold cross-validation. (a): Mapping of imaging canonical loadings onto the brain; (b): Known interactions between identified protein biomarkers from STRING database.

Acknowledgement

This work was supported by NIH R01 EB022574, R01 LM011360, U01 AG024904, R01 AG19771, P30 AG10133, UL1 TR001108, K01 AG049050 and R00 LM011384; DOD W81XWH-14-2-0151, W81XWH-13-1-0259, and W81XWH-12-2-0012; and NCAA 14132004 at Indiana University.

Data collection and sharing for this project was funded by the Alzheimer's Disease Neuroimaging Initiative (ADNI) (National Institutes of Health Grant U01 AG024904). ADNI is funded by the National Institute on Aging, the National Institute of Biomedical Imaging and Bioengineering, and through generous contributions from the following: Abbott; Alzheimer's Association; Alzheimer's Drug Discovery Foundation; Amorfix Life Sciences Ltd.; AstraZeneca; Bayer HealthCare; BioClinica, Inc.; Biogen Idec Inc.; Bristol-Myers Squibb Company; Eisai Inc.; Elan Pharmaceuticals Inc.; Eli Lilly and Company; F. Hoffmann-La Roche Ltd and its affiliated company Genentech, Inc.; GE Healthcare; Innogenetics, N.V.; Janssen Alzheimer Immunotherapy Research & Development, LLC.; Johnson & Johnson Pharmaceutical Research & Development LLC.; Medpace, Inc.; Merck & Co., Inc.; Meso Scale Diagnostics, LLC.; Novartis Pharmaceuticals Corporation; Pfizer Inc.; Servier; Synarc Inc.; and Takeda Pharmaceutical Company. The Canadian Institutes of Health Research is providing funds to support ADNI clinical sites in Canada. Private sector contributions are facilitated by the Foundation for the National Institutes of Health (www.fnih.org). The grantee organization is the Northern California Institute for Research and Education, and the study is coordinated by the Alzheimer's Disease Cooperative Study at the University of California, San Diego. ADNI data are disseminated by the Laboratory for Neuro Imaging at the University of Southern California.

References

1. Alzheimers-Association: Alzheimers disease facts and figures. Alzheimers and Dementia 12, 4 (2016)
2. Chen, X., Liu, H., Carbonell, J.G.: Structured sparse canonical correlation analysis. In: International Conference on Artificial Intelligence and Statistics (2012)
3. Chi, E., Allen, G., et al.: Imaging genetics via sparse canonical correlation analysis. In: Biomedical Imaging (ISBI), 2013 IEEE 10th Int Sym on. pp. 740–743 (2013)
4. Del Bo, R., Ghezzi, S., Scarpini, E., Bresolin, N., Comi, G.: Vegf genetic variability is associated with increased risk of developing alzheimer's disease. Journal of the neurological sciences 283(1), 66–68 (2009)
5. Du, L., Yan, J.W., Kim, S., Risacher, S.L., Huang, H., Inlow, M., Moore, J.H., Saykin, A.J., Shen, L., Initia, A.D.N.: A novel structure-aware sparse learning algorithm for brain imaging genetics. Medical Image Computing and Computer-Assisted Intervention - Miccai 2014, Pt Iii 8675, 329–336 (2014)
6. Hinrichs, C., Singh, V., Xu, G., Johnson, S.C.: Predictive markers for ad in a multi-modality framework: an analysis of mci progression in the adni population. Neuroimage 55(2), 574–89 (2011)
7. Kauwe, J., Bailey, M., Ridge, P., Perry, R., Wadsworth, M., Hoyt, K., Ainscough, B.: Genome-wide association study of csf levels of 59 alzheimer's disease candidate proteins: significant associations with proteins involved in amyloid processing and inflammation. Plos Genetics 10(10), e1004758 (2014)
8. Lin, D., Calhoun, V.D., Wang, Y.P.: Correspondence between fMRI and SNP data by group sparse canonical correlation analysis. Med Image Anal (2013)
9. Liu, J., Ji, S., Ye, J.: Multi-task feature learning via efficient l2,1-norm minimization. In: In Proceedings of the twenty-fifth conference on uncertainty in artificial intelligence. pp. 339–348. AUAI Press (2009)
10. Lu, K., Ding, Z.M., Ge, S.: Sparse-representation-based graph embedding for traffic sign recognition. Ieee Transactions on Intelligent Transportation Systems 13(4), 1515–1524 (2012)
11. Parkhomenko, E., Tritchler, D., Beyene, J.: Sparse canonical correlation analysis with application to genomic data integration. Statistical Applications in Genetics and Molecular Biology 8, 1–34 (2009)
12. Shen, L., Kim, S., Qi, Y., Inlow, M., Swaminathan, S., Nho, K., Wan, J., Risacher, S.L., Shaw, L.M., Trojanowski, J.Q., Weiner, M.W., Saykin, A.J., Adni: Identifying neuroimaging and proteomic biomarkers for mci and ad via the elastic net. Multimodal Brain Image Analysis 7012, 27–34 (2011)
13. Tarkowski, E., Issa, R., Sjgren, M., Wallin, A., Blennow, K., Tarkowski, A., Kumar, P.: Increased intrathecal levels of the angiogenic factors vegf and tgf- in alzheimers disease and vascular de-

mentia. Neurobiology of aging 23(2), 237–243 (2002)
14. Tibshirani, R.: Regression shrinkage and selection via the lasso. Journal of the Royal Statistical Society. Series B (Methodological) 58(1), 267–288 (1996)
15. Wan, J., Kim, S., et al.: Hippocampal surface mapping of genetic risk factors in AD via sparse learning models. MICCAI 14(Pt 2), 376–83 (2011)
16. Witten, D.M., Tibshirani, R., Hastie, T.: A penalized matrix decomposition, with applications to sparse principal components and canonical correlation analysis. Biostatistics 10(3), 515–34 (2009)
17. Yan, J., Du, L., Kim, S., Risacher, S.L., Huang, H., Moore, J.H., Saykin, A.J., Shen, L.: Transcriptome-guided amyloid imaging genetic analysis via a novel structured sparse learning algorithm. Bioinformatics 30(17), i564–71 (2014)
18. Yan, J., H, H., Kim, S., Moore, J., Saykin, A., Shen, L., Initia, A.D.N.: Joint identification of imaging and proteomics biomarkers of alzheimer's disease using network-guided sparse learning. In: In Proceedings of the twenty-fifth conference on uncertainty in artificial intelligence. pp. 665–668. IEEE (2014)
19. Zhang, D.Q., Wang, Y.P., Zhou, L.P., Yuan, H., Shen, D.G., Initia, A.D.N.: Multimodal classification of alzheimer's disease and mild cognitive impairment. Neuroimage 55(3), 856–867 (2011)
20. Zheng, Y., Wang, Q., Xiao, B., Lu, Q., Wang, Y., Wang, X.: Involvement of receptor tyrosine kinase tyro3 in amyloidogenic app processing and -amyloid deposition in alzheimer's disease models. Plos One 7(6), e39035 (2012)

ENFORCING CO-EXPRESSION IN MULTIMODAL REGRESSION FRAMEWORK

PASCAL ZILLE[1], VINCE D. CALHOUN[2] and YU-PING WANG[1,*]
[1] *Biomedical Engineering Department, Tulane University.*
[2] *The Mind Research Network, University of New Mexico.*
[*] *E-mail: wyp@tulane.edu*

We consider the problem of multimodal data integration for the study of complex neurological diseases (e.g. schizophrenia). Among the challenges arising in such situation, estimating the link between genetic and neurological variability within a population sample has been a promising direction. A wide variety of statistical models arose from such applications. For example, Lasso regression and its multitask extension are often used to fit a multivariate linear relationship between given phenotype(s) and associated observations. Other approaches, such as canonical correlation analysis (CCA), are widely used to extract relationships between sets of variables from different modalities. In this paper, we propose an exploratory multivariate method combining these two methods. More Specifically, we rely on a 'CCA-type' formulation in order to regularize the classical multimodal Lasso regression problem. The underlying motivation is to extract discriminative variables that display are also co-expressed across modalities. We first evaluate the method on a simulated dataset, and further validate it using Single Nucleotide Polymorphisms (SNP) and functional Magnetic Resonance Imaging (fMRI) data for the study of schizophrenia.

Keywords: Multimodal Analysis, Collaborative Regression, CCA, Sparse Models, Schizophrenia.

1. Introduction

An increasing amount of high-dimensional biomedical data such as micro arrays (mRNA, SNP) or brain imaging sequences (MRI, PET) is collected every day. Classical unimodal analysis often ignore the potential joint effects that may exist, for example, between genes and specific brain regions for diseases such as Schizophrenia, Alzheimer, etc. By harnessing these joint effects across modalities, we might be able to identify new mechanisms that uni-modal methods may fail to capture. Imaging genomics is an emerging field whose aim is precisely to leverage the wealth of biomedical knowledge provided by genomic and brain imaging data. Integrating such multimodal data sets is critical to extract meaningful bio-markers, improve clinical outcome prediction or identify potential associations across modalities. Unfortunately, as mentioned by Lin[1], such studies using genomic and brain imaging data often run into two limitations: The first one is an average small sample size, which may result in over fitting issues. In order to address such constraint, many authors relied on the use of sparse models. One classical method introduced by Tibshirani[2] is the Lasso regression. The second limitation is poor biomarker reproducibility across studies. Although this issue remains an open problem, one may hope that using appropriate priors over the solution will lead to an improved consistency of the result across different studies.

1.1. *Motivation: the study of Schizophrenia*

Schizophrenia is a serious neurological disorder that affects around 1% of the general population. It is regarded as the result of various factors including genetic variants, brain development abnormalities and environmental effects. Identifying critical genes or SNPs related to schizophrenia[3,4] has been a challenging issue. Many studies relied as well on brain imaging techniques[5,6] to pinpoint functional abnormalities in brain regions for schizophrenia patients. Multimodal analysis (e.g. using both genomic and brain imaging) often improve generalization in situations in which many irrelevant features are present. In their recent paper, Cao et al.[7] proposed a sparse representation based variable selection (SRVS) algorithm relying on sparse regression model to integrate both SNP and fMRI in order to perform biomarker selection for the study of schizophrenia. Lin[8] proposed a group sparse canonical correlation analysis (CCA) method based on SNP and fMRI data to extract correlation between genes and brain regions. Le Floch et al.[9] combined univariate filtering and Partial Least Squares (PLS) to identify SNPs covarying with various neuroimaging phenotypes. It appears that both regression and CCA methods display promising behaviors when combining SNP and fMRI data for the study of schizophrenia. In this work, we will try to merge these two methods in order to make the most out of both formulations.

The rest of this paper is organized as follows: we introduce in Section 2 some of the relevant methods as well as the motivation for this work. A novel approach to multivariate regression problems is then proposed in Section 3. Such method is then evaluated on both synthetic and real datasets in Section 4, followed by some discussions and concluding remarks in Section 5.

2. Methods

2.1. *Learning with L_1 penalty*

We consider $M \in \mathbb{N}^+$ distinct (i.e. from different modalities) datasets with n samples and $p_m \in \mathbb{N}^+$ $(m = 1, .., M)$ variables each. The m-th dataset is represented by a matrix $\mathbf{X}_m \in \mathbb{R}^{n \times p_m}$. Additionally, each sample is assigned a class label (e.g. case/controls) $y_i \in \{-1, 1\}$, $i = 1, .., n$. Our goal is to look for a linear link between those class labels and the M data matrices. Let us consider the following regression model:

$$\min_{\beta} \sum_{m=1}^{M} ||\mathbf{y} - \mathbf{X}_m \beta_m||_2^2 + \lambda ||\beta||_1 \tag{1}$$

The model described by Eq. 1 performs both variable selection and regularization. It often improves the prediction accuracy and interpretability of the results compared to the use of classical ℓ_2 norm regularization terms, especially when the number of variables is far greater than the number of observations. In some situations, we have several output vectors $\mathbf{y}_m, \forall m = 1, .., M$ and the m datasets are from the same modality: multi-task Lasso[10] was proposed to capture shared structures among the various regression vectors. We consider the following model:

$$\min_{\beta} \sum_{m=1}^{M} ||\mathbf{y}_m - \mathbf{X}_m \beta_m||_2^2 + \lambda \sum_{p=1}^{P} ||\beta^p||_2 \tag{2}$$

where P is the dimension of the problem and β^p is the p-th row of the matrix such that $\beta = [\beta_1, .., \beta_m]$ (i.e. the β_m are stacked horizontally). Such norm is also referred to as the ℓ_1/ℓ_2 norm, and is used to both enforce joint sparsity across the multiple β_m and estimate only a few non-zero coefficients. Enforcing regularity within a modality[11,12] (and across tasks) has been an active aspect of regression models, and has proven to increase reliability and results. However, since often pair-wise closeness is looked for in the common subspace, such methods will often fail to capture relationships across modalities.

2.2. *Collaborative learning*

Collaborative (or Co-regularized) methods[13] are based on the optimization of measures of agreement and smoothness across multi-modal datasets. Smoothness across modalities is enforced through a joint regularization term. Their general model can be expressed as follows:

$$J(\beta) = \sum_{m=1}^{M} ||\mathbf{y} - \mathbf{X}_m\beta_m||_2^2 + \gamma \sum_{m,q=1}^{M} ||\mathbf{U}_m\beta_m - \mathbf{U}_q\beta_q||_2^2 + \lambda||\beta||_1 \quad (3)$$

where the $\mathbf{U}_m$, $m = 1, .., M$ are arbitrary matrices whose roles are to control the cross-view joint regularization between each pair of vectors (β_m, β_q), $m, q = 1, .., M$. Scalar parameter $\gamma \geq 0$ controls the influence of such cross-regularization term. Notice that if $\gamma = 0$, we fall back on the original Lasso formulation. Collaborative learning is an interesting extension of Eq.1 allowing the user to explicitly enforce regularization across modalities. In this work, we rely on a special case of collaborative methods (introduced later in section 3) to address the following aspects: (*i*) Extend the regularization idea across modalities; (*ii*) Assume that relationships between variable are not available as a prior knowledge (as opposed, e.g., to Xin[11]); (*iii*) Define links between components using correlation measure. To do so, we first briefly introduce in the next section some of the classical methods to extract meaningful relationships between variables across modalities.

2.3. *Extracting relationship between datasets*

A wide variety of problems amount to the joint analysis of multimodal datasets describing the same set of observations. Often, a mean to perform such analysis is to learn projection subspaces using paired samples such that structures of interest appear more clearly. Some of these methods are for example: Canonical correlation analysis[14] (CCA), Partial least squares[9] (PLS) or cross-modal factor analysis (CFA). Among them, CCA is probably the most widely used. Its goal is to extract linear combinations of variables with maximal correlation between two (or more) datasets. Using similar notations as in the previous section,and assuming $M = 2$, one formulation of CCA is expressed as follow:

$$\underset{\beta_1,\beta_2}{\operatorname{argmin}} \; J_{cca}(\beta_1, \beta_2) = ||\mathbf{X}_1\beta_1 - \mathbf{X}_2\beta_2||^2 \quad (4)$$

to which a constraint on the norm of canonical vectors β_1, β_2 is added to avoid the trivial null solution. In recent years, CCA has been widely applied to genomic data analysis. As a consequence, many studies on sparse versions of CCA (sCCA) have been proposed[8,15–18] to cope with the high dimension but low sample size problem. In the next section, we will rely on a CCA term to measure co-expression between variables from different modalities.

3. Enforcing cross-correlation in regression problems

3.1. *MT-CoReg formulation*

As discussed in Section 1, several methods have been proposed to: (*i*) Associate a phenotype and datasets while enforcing prior over solution; (*ii*) Extract relationships between coupled or co-expressed datasets. In the present study, we propose to associate both the regression and CCA frameworks in the case of $M = 2$ datasets. Our motivation is to extract informative features that also display a significant amount of correlation across modalities. A simple way to combine Lasso and sparse CCA would be a weighted combination of Eq.(1) and Eq.(4):

$$\min_{\beta} J(\beta) = (1-\gamma)\sum_{m=1}^{2} \|\mathbf{y} - \mathbf{X}_m\beta_m\|_2^2 + \gamma\|\mathbf{X}_1\beta_1 - \mathbf{X}_2\beta_2\|^2 + \lambda\|\beta\|_1 \tag{5}$$

where $\gamma \in [0,1]$ is a weight parameter. Notice that Eq.(5) can be expressed within the collaborative framework introduced in Section 2.2. If we take a look at Eq.(3) with $M = 2$, $\mathbf{U}_1 = \mathbf{X}_1$ and $\mathbf{U}_2 = \mathbf{X}_2$, we fall back on Eq.(5). Let us call this model CoReg for *Collaborative Regression*. Interestingly, a similar model has been considered before by Gross[19] to perform prediction using breast cancer data. However, to our opinion, such formulation might prove to be too constraining. It essentially amounts to force each component of the β_m's to fit both the regression term and the CCA one. We illustrate such behaviour using a toy dataset later in Section 3.4. Since our goal is to perform feature selection, we may allow the model to be slightly more flexible. We thus propose an alternative formulation by first duplicating each β_m into two components such that:

$$\beta_m = [\alpha_m, \theta_m] \ , \ \forall m = 1,2 \tag{6}$$

where α_m, θ_m are vectors from $\mathbb{R}^{p_m}$. As a consequence, the β_m's are now matrices such that $\beta_m \in \mathbb{R}^{p_m \times 2}$ $\forall m = 1,2$. We then propose the following MT-CoReg formulation:

$$\min_{\beta} J(\beta) = (1-\gamma)\sum_{m=1}^{2} \|\mathbf{y} - \mathbf{X}_m\alpha_m\|_2^2 + \gamma\|\mathbf{X}_1\theta_1 - \mathbf{X}_2\theta_2\|^2 + \lambda\sum_{m=1}^{2}\sum_{i=1}^{p_m}\|\beta_m^i\|_2 \tag{7}$$

where β_m^i is the i-th row of β_m, i.e. $\beta_m^i = [\alpha_m(i), \theta_m(i)] \in \mathbb{R}^2$. The third term of Eq.(3.3) is simply the ℓ_1/ℓ_2 norm of each of the β_m. As we can observe from looking at Eq.(3.3), each 'component' (i.e. column of β_m) will be involved in separate parts of the functional J: (*i*) components α_m are the fit to the regression term of Eq.(3.3); (*ii*) components θ_m are the fit of the CCA term of Eq.(3.3). Each pair (α_m, θ_m) and $m = 1,2$ is coupled through the use of the ℓ_1/ℓ_2 norm from the third term in Eq.(3.3). Although their values are different, shared sparsity patterns are encouraged within each pair (α_m, θ_m). As a consequence, we allow the method to be significantly more flexible in terms of solutions: different values can be taken to simultaneously fit the Regression and CCA parts. We hope that such framework will encourage the selection of features that are discriminative (via the regression part) but also co-expressed across modalities (via the CCA part). Note that when $\gamma = 0$, criterion (3.3) essentially reduces to the initial regression problem of Eq.(1), while setting $\gamma = 1$ amounts to solving a conventional sparse CCA problem. A schematic view of the MT-CoReg pipeline can be seen in Fig.(1). In the next section, we briefly explain how to solve the problem described in Eq.(3.3).

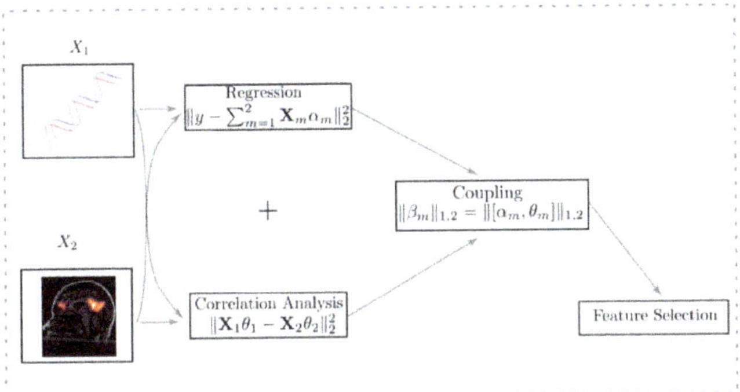

Fig. 1. Schematic view of the MT-CoReg pipeline. From two different datasets $X1$ and $X2$ from different modalities (here SNP and fMRI respectively), we fit both a regression and CCA terms and couple the resulting components (α_m, θ_m) using the $\ell 1/\ell_2$ norm denoted $\|\cdot\|_{1,2}$ here. The ultimate goal is to find discriminative SNP and brain regions that are also co-expressed across modalities.

3.2. *Optimization*

We solve the problem from Eq.(3.3) by optimizing the β_m's alternatively over iterations until convergence, in a similar fashion to Wilms[20] et al. formulation of sCCA. Suppose we have an initial value β_1^* for β_1, and want to estimate β_2. Updating matrix β_2 can be recast into a problem of the following form:

$$\min_{\tilde{\beta}_2} J(\tilde{\beta}_2|\beta_1^*) = \|\tilde{\mathbf{y}}_2 - \tilde{\mathbf{X}}_2\beta_2\|_F^2 + \lambda \sum_{i=1}^{p_2} \|\beta_2^i\|_2 \tag{8}$$

where

$$\tilde{\mathbf{y}}_2 = [\sqrt{(1-\gamma)}\mathbf{y}, \sqrt{\gamma}\mathbf{X}_1\theta_1^*], \quad \tilde{\mathbf{X}}_2 = [\sqrt{(1-\gamma)}\mathbf{X}_2, \sqrt{\gamma}\mathbf{X}_2] \tag{9}$$

Obviously, Eq.(8) is a classical group-lasso regression problem[10] (cf. Eq.(2)). It is easy to show that updating β_1 reduces to solving a similar problem. As a consequence, solving our mixed Lasso/CCA problem from Eq.(3.3) can be briefly summarized as:

1 Initialization: estimate initial values for α_1, β_1, α_2, β_2 using ridge regression and ridge CCA.
2 Assume β_1's value fixed, and update β_2 using Eq.(8).
3 Assume β_2's value fixed, and update β_1 using the adapted version of Eq.(8).
4 Go back to step 2. until convergence

3.3. *Parameter selection*

Solving problem from Eq.(3.3) requires the estimation of two parameters, λ and γ, which respectively control the weights of the sparsity and the co-expression regularization terms.
The choice of sparsity parameter λ for this type of problems is known to display a high sensitivity[21]. In order to make the searching process more robust, we chose to let the sparsity level of the solution control the tuning parameter value[22,23]. Consider a column vector $\beta \in \mathbb{R}^p$ (e.g. a column of β from Eq.): let us denote $|\beta|_\kappa$ the κ-th ($\kappa \in \mathbb{N}^+$) largest absolute magnitude of β. We can define a correspondence between λ and κ by making sure that for each iteration, we have $\lambda \in [|\beta|_\kappa, |\beta|_{\kappa+1}]$. The selection can be looked for around the sample size (i.e. $\kappa = n$ for the entire estimation process), which helps drastically stabilize the estimation process in practice.
As for the estimation of γ, we chose to rely on a technique introduced by Sun et al.[24] based on variable selection stability. Its main goal is to select a given tuning parameter so that the associated variable selection method (in our case, the model from Eq.(3.3)) is stable in terms of the features it selects. In this framework, the training set is split in two halves using resampling (bootstrap resampling in our case). The variable selection method is then applied to each of the subsamples along a grid of candidate values for the parameter. Kappa selection criterion[25] is then used to measure the degree of agreement between the two sets of variables obtained for a given parameter value. This process is then repeated a number of times, and an approximated measure of selection consistency is derived. The parameter value for which this consistency is the highest (after correction for the number of non-zeros elements retained) is the one kept for the estimation.

3.4. *MT-CoReg VS. CoReg*

As mentioned earlier in Section 3.1, in their CoReg model from Eq.(5) Gross et al.[19] did not separate the solution vectors β_m into two components. We then propose to illustrate the behavior of both models (Eq.(5) and Eq.(3.3)) on a toy dataset.
We generated $M = 2$ data matrices $\mathbf{X}_1, \mathbf{X}_2$ such that $p_1 = p_2 = 30$ and $n = 50$ observations. We used a latent variable model to simulate cross-correlated components so that columns $p = [1, ..5] \cup [10, ..15]$ of $\mathbf{X}_1, \mathbf{X}_2$ are mutually co-expressed. We further use columns $p = [10, ..15] \cup [20, ..25]$ to generate a phenotype vector $\mathbf{y}$ such that $\mathbf{y}_i \in \{-1; 1\}$. With such setup, columns $p = [10, ..15]$ correspond to both non-zeros values in the true regression and canonical coefficients. Furthermore, let us point out that these non-zero values are diferent (canonical coefficients' amplitude is lower than the regression ones). This setup can be seen in the first row of Fig.(2, *Truth*), where the blue and red curves are the values taken by the canonical and regression coefficients respectively. Resulting estimates for sCCA, Lasso, CoReg[19] as well as proposed method MT-CoReg can also be seen in Fig.(2). In such scenario, while CoReg model assumes that regression and canonical coefficients have identical values, MT-CoReg has a wider scope and allows a finer joint estimation of both components types.

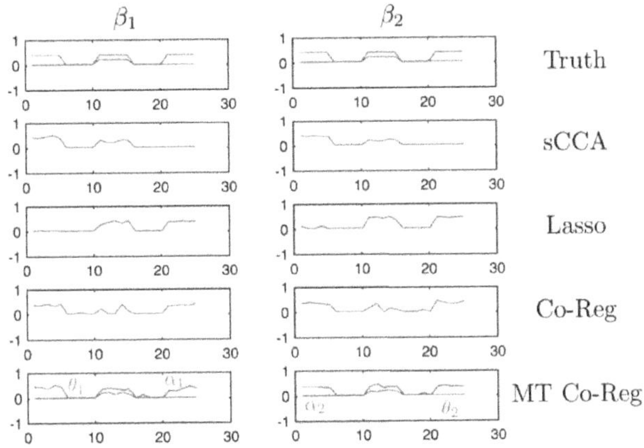

Fig. 2. Resulting estimates β_1, β_2 on the toy dataset. (*Truth*) blue and red curves are the values taken by the true canonical and regression coefficients respectively. Solutions obtained with sCCA, Lasso, CoReg[19] and proposed method MT-CoReg are displayed. Notice that columns columns $p = [10, ..15]$ correspond to both non-zeros values in the true regression and canonical coefficients, although their amplitudes are different. By relaxing the assumption that regression and canonical coefficients have identical values, MT-CoReg allows a finer joint estimation of both components types compared to CoReg.

4. Experiments

In this section, we evaluate the proposed estimator from Eq.3.3. Performances will be assessed in terms of feature selection relevance on both simulated and real data.

4.1. *Results on synthetic data*

For our first test, we simulate both fMRI and SNP datasets. Similar to the toy dataset from Section 3.4, we start by generating explanatory variables $\alpha_1^*, \alpha_2^* \in \mathbb{R}^{900}$ for both genomic and brain imaging data. The first 100 components of α_1^*, α_2^* are drawn from Normal distribution, while the rest is set to zero. The total number of observations is set to $n = 200$. Genomic values are coded as 0 (no minor allele), 1 (one minor allele), and 2 (two minor allele). We first define a minor allele frequency η drawn from a uniform distribution $\mathcal{U}([0.2, 0.4])$. The i-th SNP is then generated from a binomial distribution $\mathcal{B}(2, \eta_i)$. For the imaging data, voxels values were drawn from a Gaussian distribution $\mathcal{N}(0, I_p)$. Finally, binary phenotype $\mathbf{y}$ data are generated from $\mathcal{B}(1, d_i)$, where $d_i = \dfrac{exp(5\sum_{m=1}^{M} \mathbf{X}_m \alpha_m^*)}{1 + exp(5\sum_{m=1}^{M} \mathbf{X}_m \alpha_m^*)}$. Furthermore, we add 100 additional variables to the problem that will play the role of cross-correlated variables. Two canonical vectors $\theta_1^*, \theta_2^* \in \mathbb{R}^{100}$ are drawn from Normal distribution. Cross-correlated SNP are drawn from $\mathcal{B}(2, \text{logit}^{-1}(-a_i + \text{logit}(\eta_i)))$ where a is issued from $\mathcal{N}(\theta_1^* \mathbf{y}, I_{100})$, while cross-correlated voxels are drawn from $\mathcal{N}(\theta_2^* \mathbf{y}, I_{100})$. The final dataset is made of $n = 200$ observations of $p = 1000$ variables for both SNP and fMRI. Each of these datasets is made of explanatory and cross-correlated components. A common way to assess the performance of a model when it comes to feature selection is to measure the true positive rate (TPR) and false positive rate (FPR). TPR reflects the proportion of variables that are correctly identified, while FDR reflects the proportion of variables that are incorrectly selected by the model. We apply MT-CoReg to 100 random generation of the dataset described above. The tuning parameter γ from Eq.(3.3) that weights the CCA term against the regression one is optimized through a grid search over $\{[0] \cup [10^{-1+\ell/20}]; \ell = 0, .., 20\}$. We plotted TPR values against FDR ones in Fig.(3) for two different cases. In the first (left) subfigure are displayed TPR/FDR values relative to non-zero components of α_1^*, α_2^* for $\gamma = 0$ (i.e. classical Lasso), $\gamma = \gamma(\text{C.S.})$ where the weight value is determined using consistency selection (C.S.) scheme described in Section 3.3, and $\gamma = 1$ (i.e. classical sCCA). We can observe that although classical regression seems to perform slightly better for really low FDR values, MT-CoReg is quickly catching up around $FDR \approx 0.15$. sCCA, on the other hand, has a low selection power. The second (bottom) figure displays TPR/FDR values relative to non-zero components of θ_1^*, θ_2^*, i.e. the cross-correlated components. We can observe that MT-CoReg performs as well as sCCA, while Lasso is unable to properly select the components of interest. It is encouraging to see that MT-CoReg takes the best of both methods and seems to properly select the components we are interested in. It seems to confirm our hypothesis that using a mix of both terms may lead to an improved feature selection accuracy. In the next section, we apply the same method to a real dataset of fMRI and SNP data.

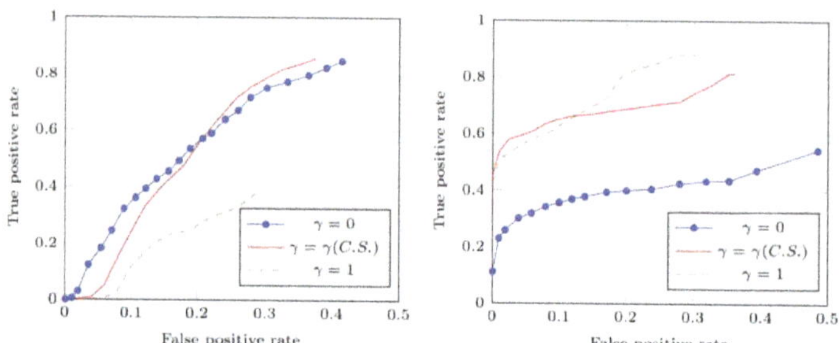

(a) (b)

Fig. 3. TPR against FDR values averaged over 100 simulations for different γ values. Fixing $\gamma = 0$ amounts to using Lasso regression, while $\gamma = 1$ is equivalent to classical sparse CCA. $\gamma(\text{C.S.})$ is the ROC curve obtained while using consistency selection (C.S.) scheme described in section 3.3 to automatically estimate γ. (a) values for the selection of first 100 components (i.e. the explanatory components) only (b) values for the selection of the last 100 components (i.e. the cross-correlated components). It can be seen that a non-trivial weight combinaison for γ seems to be taking the best of the two methods that are Lasso ($\gamma = 0$) and CCA ($\gamma = 1$).

4.2. *Results on real imaging genetics data*

4.2.1. *Data acquisition*

Both SNP and fMRI acquisition were conducted by the Mind Clinical Imaging Consortium (MCIC) for 214 subjects, including 92 schizophrenia patients (age: 34 ± 11, 22 females) and 116 controls (age 32 ± 11, 44 females). Schizophreniac were diagnosed based on DSM-IV-TR citeria. Controls were free of any medical, neurological of psychiatric illnesses.

fMRI were acquired during a sensor motor task with auditory simulation. Data were pre-processed with SPM5, spatially normalized and resliced, smoothed, and analyzed by multiple regression considering the stimulus and their temporal derivatives plus an intercept term as regressors. For each patient, a stimulus-on vs. stimulus-off contrast image was extracted. 116 ROIs were extracted based on the aal brain atlas, which resulted in 41236 voxels left for analysis. SNP data were obtained from blood sample using Illumina Infinium HumanOmni1-Quad array covering 1,140,419 SNP loci. After standard quality control procedures using PLINK software package [a], a final dataset spanning $777,635$ SNP loci was available. Each SNP was categorized into three clusters based on their genotype and was represented with discrete numbers: 0 (no minor allele), 1 (one minor allele) and 2 (two minor alleles). SNPs with $> 20\%$ missing data were deleted and missing data were further imputed. SNPs with minor allele frequency $< 5\%$ were removed. This procedure yielded a final set of $129,145$ SNPs.

4.2.2. *Significance analysis*

In order to achieve a stable feature selection process, we follow Lin[8] and perform $N = 100$ random samplings out of the 214 total subjects, where for each time 80% are used for training and parameter selection, while the remaining 20% are used for evaluation. At the $k - th$ random sampling, we can calculate a set of solution vectors $\hat{\beta}_m^k, m \in \{1,2\}$. It is then possible to define a measure of relevance p_m^i for the i-th feature in the m-th dataset such that: $p_m^i = \frac{1}{N}\sum_{k=1}^{N} I(\hat{\beta}_m^k(i) \neq 0)$ where $i = 1,..,d_m$ is the feature index and $I(\cdot)$ is the indicator function. We can then rank each SNP and voxel based on their associated relevance measure and apply a cut-off threshold of 0.3 (c.f. Lin[8]). After applying this significance test, we were left with a subset of 43 SNP spanning 30 genes and 6 ROI with a number of selected voxels over 5.
We display in Table.1 the list of each of the 43 selected SNP, as well as their associated genes. Some of them have been identified by other similar studies[8,26,27] such as CNTNAP2, GLI2, GRIK3, NOTCH4, SUCLG2, GABRG2. Others have been identified from well-known databases[28] such as GRIK4 or HTR4. We display in Table.2 the list of the selected ROI as well as the corresponding voxel count for each one of them. ROI for which less than 5 voxels were selected where dismissed. Once again, it is encouraging to note that each of the selected ROI (3, 7, 11, 40, 51, 100 from aal.) have been identified in similar studies[8,29] on the same dataset. Other studies pointed out both functionnal or structural differences in the middle occipital gyrus[30] and the parahippocampal gyrus[31] for schizophrenic patients. Finally, a detailled slice view of the selected voxels can be seen in Fig.(4).

[a]http://pngu.mgh.harvard.edu/ purcell/plink

Table 1. List of selected SNP and their associated genes.

SNP ID	Gene name	SNP ID	Gene name	SNP ID	Gene name	SNP ID	Gene name
rs3856465	ATP6V1C2	rs11607732	GRIK4	rs815533	CACNA2D3	rs10748732	HPSE2
rs12333931	CNTNAP2	rs12332417	HTR4	rs2373347	CNTNAP2	rs13359903	HTR4
rs2407264	CYSLTR2	rs7725785	HTR4	rs9535112	CYSLTR2	rs11875988	LIPG
rs6567629	DHRSX	rs12454370	LIPG	rs858341	ENPP1	rs9787820	LRRC4C
rs16842460	EPHB1	rs17819648	MAML2	rs11927660	FGF12	rs3134797	NOTCH4
rs17599845	FHIT	rs3134799	NOTCH4	rs10926254	FMN2	rs394657	NOTCH4
rs4659573	FMN2	rs1009708	PDE2A	rs11060822	FZD10	rs7111188	PDE2A
rs12824777	FZD10	rs17016738	RARB	rs2963094	GABRG2	rs12101383	SMAD6
rs10831614	GALNTL4	rs7030433	SMARCA2	rs7602673	GLI2	rs573700	SPRY2
rs6753202	GPD2	rs9849270	SUCLG2	rs1392744	GRIK3	rs1105880	UGT1A6
rs10502240	GRIK4	rs17863787	UGT1A6				

Table 2. List of selected ROI (from aal.) and associated voxel count.

ROI ID (aal.)	ROI name	voxels nb.
51	Left middle occipital gyrus	13
7	Left middle frontal gyrus	11
11	Left middle frontal gyrus, orbital part	9
100	Right lobule VI of cerebellar hemisphere	9
3	Left superior frontal gyrus	8
40	Right parahippocampal gyrus	7

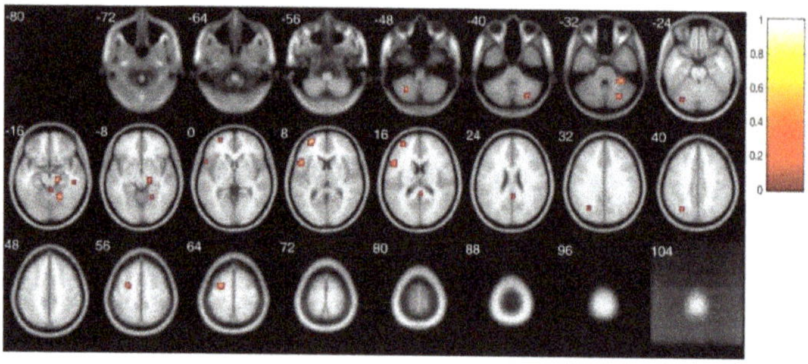

Fig. 4. Slice view of the selected voxels (without thresholding using cluster size) and their significance.

4.2.3. *Quantitative analysis*

In this section, we try to analyze the results of MT-CoReg using some quantitative metrics. We can first turn our attention to the Sum of Squared Errors (SSE) values obtained on the testing set during our tests. Histograms of SSE distributions for different γ values (i.e.

Lasso, MT-CoReg and sCCA) can be seen in Fig.(5,left): unsurprisingly, Lasso and MT-CoReg produce the lowest RSS values, while sCCA does not fit the phenotype. If we now look at Fig.(5,middle) where distributions of Pearson's correlation on the testing set are displayed for the same 3 strategies, we can see that MT-CoReg produces a better selection than Lasso in terms of cross-correlation. This seems to confirm our intuition that MT-CoReg makes the best of both Lasso and CCA by producing a solution that is good fit to the phenotype while selecting co-expressed features across modalities.
Distribution of γ values produced by the consistency selection scheme described in Section 3.3 can be seen in Fig.(5,right). Most of these values fall into the range $[0; 0.4]$, with a peak in $[0.2; 0.3]$. It does appear, at least in term of feature consistency selection, that a non-zero weight for the CCA term in Eq.(3.3) leads to improved performances.

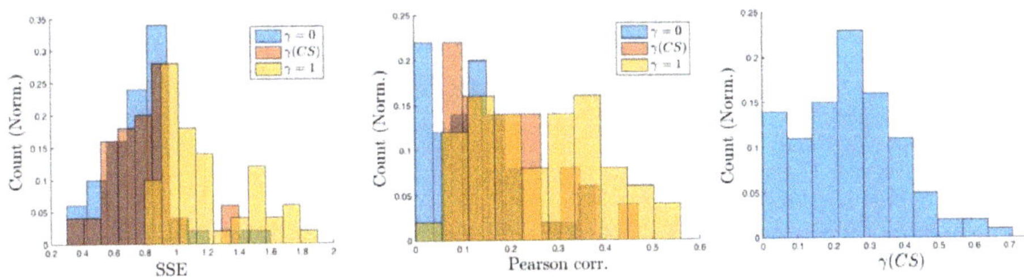

Fig. 5. Frequency distribution of RSS values (on the test set) for $N = 100$ sub-sampling of the original set of observations.

5. Conclusions

The main contributions of this paper can be summarized as follows. First, we proposed a novel variable selection approach using a CCA-like regularization term in order to enforce co-expression between modalities. Secondly, we present an efficient algorithm to solve this problem, as well as strategies to estimate the tuning parameters. On top of that, a series of experiments on both synthetical and real datasets were conducted, allowing us to evaluate the performances of the proposed method. We identified two sets of SNP and voxels in which a number of them have been previously reported to have potential relationship with the risk of schizophrenia. Further exploration of the optimization scheme (alternate estimations) as well as the selection of regularization parameter λ (see Section 3.3) will be needed in the future.

6. Acknowledgments

The authors wish to thank the NIH (NSF EPSCoR#1539067) for their partial support.

References

1. D. Lin, J. Zhang, J. Li, H. He, H.-W. Deng and Y.-P. Wang, *Multi-omic Data Integration* , p. 126 (2015).

2. R. Tibshirani, *Journal of the Royal Statistical Society. Series B (Methodological)* , 267 (1996).
3. C. M. Lewis, D. F. Levinson, L. H. Wise, L. E. DeLisi, R. E. Straub, I. Hovatta, N. M. Williams, S. G. Schwab, A. E. Pulver, S. V. Faraone *et al.*, *The American Journal of Human Genetics* **73**, 34 (2003).
4. S. R. Sutrala, D. Goossens, N. M. Williams, L. Heyrman, R. Adolfsson, N. Norton, P. R. Buckland and J. Del-Favero, *Schizophrenia research* **96**, 93 (2007).
5. M. E. Shenton, C. C. Dickey, M. Frumin and R. W. McCarley, *Schizophrenia research* **49**, 1 (2001).
6. S. A. Meda, M. Bhattarai, N. A. Morris, R. S. Astur, V. D. Calhoun, D. H. Mathalon, K. A. Kiehl and G. D. Pearlson, *Schizophrenia research* **104**, 85 (2008).
7. H. Cao, J. Duan, D. Lin, Y. Y. Shugart, V. Calhoun and Y.-P. Wang, *Neuroimage* **102**, 220 (2014).
8. D. Lin, V. D. Calhoun and Y.-P. Wang, *Medical image analysis* **18**, 891 (2014).
9. É. Le Floch, V. Guillemot, V. Frouin, P. Pinel, C. Lalanne, L. Trinchera, A. Tenenhaus, A. Moreno, M. Zilbovicius, T. Bourgeron *et al.*, *Neuroimage* **63**, 11 (2012).
10. M. Yuan and Y. Lin, *Journal of the Royal Statistical Society: Series B (Statistical Methodology)* **68**, 49 (2006).
11. B. Xin, Y. Kawahara, Y. Wang, L. Hu and W. Gao, *ACM Transactions on Intelligent Systems and Technology (TIST)* **7**, p. 60 (2016).
12. B. Jie, D. Zhang, B. Cheng and D. Shen, *Human brain mapping* **36**, 489 (2015).
13. U. Brefeld, T. Gartner, T. Scheffer and S. Wrobel, 137 (2006).
14. H. Hotelling, *Biometrika* **28**, 321 (1936).
15. D. M. Witten and R. J. Tibshirani, *Statistical applications in genetics and molecular biology* **8**, 1 (2009).
16. J. Chen, F. D. Bushman, J. D. Lewis, G. D. Wu and H. Li, *Biostatistics* **14**, 244 (2013).
17. L. Du, H. Huang, J. Yan, S. Kim, S. L. Risacher, M. Inlow, J. H. Moore, A. J. Saykin, L. Shen, A. D. N. Initiative *et al.*, *Bioinformatics* , p. btw033 (2016).
18. Springer, *A novel structure-aware sparse learning algorithm for brain imaging genetics* 2014.
19. S. M. Gross and R. Tibshirani, *Biostatistics* **16**, 326 (2015).
20. I. Wilms and C. Croux, *Biometrical Journal* **57**, 834 (2015).
21. E. Parkhomenko, D. Tritchler and J. Beyene, *Statistical Applications in Genetics and Molecular Biology* **8**, 1 (2009).
22. J. Duan, J.-G. Zhang, H.-W. Deng and Y.-P. Wang, *PloS one* **8**, p. e59128 (2013).
23. Z. Xu, X. Chang, F. Xu and H. Zhang, *IEEE Transactions on neural networks and learning systems* **23**, 1013 (2012).
24. W. Sun, J. Wang and Y. Fang, *Journal of Machine Learning Research* **14**, 3419 (2013).
25. J. Cohen, *Psychological bulletin* **70**, p. 213 (1968).
26. D. Lin, H. He, J. Li, H.-W. Deng, V. D. Calhoun and Y.-P. Wang, 9 (2013).
27. J. Sun, P.-H. Kuo, B. P. Riley, K. S. Kendler and Z. Zhao, *American Journal of Medical Genetics Part B: Neuropsychiatric Genetics* **147**, 1173 (2008).
28. P. Jia, J. Sun, A. Guo and Z. Zhao, *Molecular psychiatry* **15**, 453 (2010).
29. D. Lin, H. Cao, V. D. Calhoun and Y.-P. Wang, *Journal of neuroscience methods* **237**, 69 (2014).
30. S. Singh, S. Modi, S. Goyal, P. Kaur, N. Singh, T. Bhatia, S. N. Deshpande and S. Khushu, *Journal of biosciences* **40**, 355 (2015).
31. M. J. Escartí, M. de la Iglesia-Vayá, L. Martí-Bonmatí, M. Robles, J. Carbonell, J. J. Lull, G. García-Martí, J. V. Manjón, E. J. Aguilar, A. Aleman *et al.*, *Schizophrenia research* **117**, 31 (2010).

METHODS TO ENSURE THE REPRODUCIBILITY OF BIOMEDICAL RESEARCH

KONRAD J. KARCZEWSKI
Massachusetts General Hospital, Boston, MA; Broad Institute, Cambridge, MA
Email: konradjkarczewski@gmail.com

NICHOLAS P. TATONETTI
Columbia University, New York, NY
Email: nick.tatonetti@columbia.edu

ARJUN K. MANRAI
Harvard Medical School, Boston, MA
Email: manrai@post.harvard.edu

CHIRAG J. PATEL
Harvard Medical School, Boston, MA
Email: chirag_patel@hms.harvard.edu

C. TITUS BROWN
University of California , Davis, CA
Email: ctbrown@ucdavis.edu

JOHN P. A. IOANNIDIS
Stanford University, Stanford, CA
Email: jioannid@stanford.edu

Science is not done in a vacuum – across fields of biomedicine, scientists have built on previous research and used data published in previous papers. A mainstay of scientific inquiry is the publication of one's research and recognition for this work is given in the form of citations and notoriety -- ideally given in proportion to the quality of the work. Academic incentives, however, may encourage individual researchers to prioritize career ambitions over scientific truth. Recently, the *New England Journal of Medicine* published a commentary calling scientists who repurpose data "research parasites" who *mis*use data generated by others to demonstrate alternative hypotheses[1]. In our opinion, the concept of data hoarding not only runs contrary to the spirit of, but also hinders scientific progress. Scientific research is meant to seek objective truth, rather than promote a personal agenda, and the only way to do so is through maximum transparency and reproducibility, no matter who is using the data.

To maintain the integrity of the scientific process, it is necessary to cultivate practices that ensure reproducibility, especially as large and public heterogeneous databases proliferate. Many of these paradigms can be likened to open-source practices already adopted by much of the computer science community. These include, but are not limited to, version control, code review, and containerization. There are many benefits to improving reproducibility: aside from the general benefit to science through increased transparency, releasing code enables additional peer review and is educational and efficient as it reduces duplications of efforts. Of course, these approaches require additional time for investigators to document and clean up code and data for release, which is the top reason for not sharing data and code[2] (in addition to managing the intricacies of tools for version control, for example). Various incentive structures have been proposed to improve reproducibility rates across scientific fields, including creation of requirements by funding agencies or establishment of reward systems[3]. Additionally, like many computational skills, these require some initial effort, but have long-term benefits and will eventually become ingrained. Finally, public release of code can enable public code review, which improves programming habits: efforts such as Software Carpentry have been established to teach these skills and have met with recent success[4].

Reproducibility can take a number of forms and the desired extent of reproducibility has been debated in other fora: whatever the ideal solution, there is room for improvement in ensuring that research is reproducible. A growing number of researchers have begun to share their code and processed data, where possible. For instance, the ENCODE project released a virtual machine image that contained the code and data to reproduce the figures in their manuscript[5] [http://encodeproject.org/ENCODE/integrativeAnalysis/VM]. Similarly, the ExAC consortium deposited the figure generating code for their recent papers[6,7] on Github [https://github.com/macarthur-lab/exac_papers; https://github.com/ericminikel/prnp_penetrance]. Some have gone even further as to publicly release a full manuscript under version control[8,9] and document the process for others to do so [http://ivory.idyll.org/blog/2014-our-paper-process.html].

In this session, we feature five papers that explore research on the topic of reproducibility. This year, we required submissions to strive for reproducibility by depositing data and code on public repositories. The authors have stepped up to the challenge and are practicing what they preach: where possible, they have released applicable code and/or data to make their own research as reproducible as possible.

Session Contributions

Cohain, Divaraniya, and colleagues[10] address an important challenge for reproducibility of Bayesian networks. While frequentist approaches can rely on p-values to predict replication, the construction of a Bayesian network is a data-dependent and heuristic process, and consistency between multiple analyses has not been rigorously performed. This paper explores the replication of Bayesian networks, particularly in relation to key driver nodes and hubs, as well as edge reproducibility.

Hundreds of studies have used publicly available data to predict adverse drug reactions and drug indications and have reported seemingly exceptional predictive accuracy: Guney[11] investigates the issue of performance overestimation for drug side effect and indication, and finds that major assumption of these methods (independence) is violated, which overestimates their performance. Haynes et al[12] present a pipeline for expression meta-analysis, which fills an unmet need for systematic processing and visualization of results from such analyses. Kaushik and colleagues[13] describe a workflow engine that uses graph theory approaches to optimize and ensure reproducible data analyses. Finally, Yang et al[14] provide a detailed look on the reproducibility of clinical genetics data: concordance across variant classifications is reasonably high, but more work will be required to resolve differences and accurately classify all variants as pathogenic or benign. In summary, these exemplar papers demonstrate how to enhance research reproducibility across a variety of biomedical domains critical in this era of "big data" and precision medicine.

References

1. Longo, D. L. & Drazen, J. M. Data Sharing. *http://dx.doi.org.ezp-prod1.hul.harvard.edu/10.1056/NEJMe1516564* **374,** 276–277 (2016).
2. Stodden, V. The Scientific Method in Practice: Reproducibility in the Computational Sciences. *SSRN Journal* (2010). doi:10.2139/ssrn.1550193
3. Ioannidis, J. P. A. How to Make More Published Research True. *PLoS Med* **11,** e1001747 (2014).
4. Wilson, G. Software Carpentry: lessons learned. *F1000Research* **3,** (2014).
5. ENCODE Project Consortium *et al.* An integrated encyclopedia of DNA elements in the human genome. *Nature* **489,** 57–74 (2012).
6. Lek, M. *et al.* Analysis of protein-coding genetic variation in 60,706 humans. *Nature* **536,** 285–291 (2016).
7. Minikel, E. V. *et al.* Quantifying prion disease penetrance using large population control cohorts. *Science Translational Medicine* **8,** 322ra9–322ra9 (2016).
8. Pell, J. *et al.* Scaling metagenome sequence assembly with probabilistic de Bruijn graphs. *Proc Natl Acad Sci USA* **109,** 13272–13277 (2012).
9. Zhang, Q., Pell, J., Canino-Koning, R., Howe, A. C. & Brown, C. T. These Are Not the K-mers You Are Looking For: Efficient Online K-mer Counting Using a Probabilistic Data Structure. *PLoS ONE* **9,** e101271 (2014).
10. Cohain A, Divaraniya AA, Zhu K, Zhu J, Chang R, Dudley JT, Schadt EE. "Exploring the reproducibility of probabilistic causal molecular network models" *Pac. Symp Biocomput* (2017).
11. Guney E. "Reproducible Drug Repurposing: When Similarity Does Not Suffice" *Pac. Symp Biocomput* (2017).
12. Haynes WA, Vallania F, Liu C, Bongen E, Tomczak A, Andres-Terrè M, Lofgren S, Tam A, Deisseroth CA, Li MD, Sweeney TE, Khatri P. "Empowering Multi-Cohort Gene Expression Analysis to Increase Reproducibility" *Pac. Symp Biocomput* (2017).
13. Kaushik G, Ivkovic S, Simonovic J, Tijanic N, Davis-Dusenbery B, Kural D. "Graph Theory Approaches For Optimizing Biomedical Data Analysis Using Reproducible Workflows" *Pac. Symp Biocomput* (2017).
14. Yang S, Cline M, Zhang C, Paten B, Lincoln SE. "Data Sharing and reproducible Clinical genetic testing: successes and challenges" *Pac. Symp Biocomput* (2017)

EXPLORING THE REPRODUCIBILITY OF PROBABILISTIC CAUSAL MOLECULAR NETWORK MODELS

ARIELLA COHAIN*, APARNA A. DIVARANIYA*, KUIXI ZHU, JOSEPH R. SCARPA, ANDREW KASARSKIS, JUN ZHU, RUI CHANG, JOEL T. DUDLEY, ERIC E. SCHADT†

Icahn Institute and Department of Genetics and Genomics, Icahn School of Medicine at Mount Sinai, One Gustave L. Levy Place, Box 1498,
New York, NY, 10029, USA
Email: eric.schadt@mssm.edu

Network reconstruction algorithms are increasingly being employed in biomedical and life sciences research to integrate large-scale, high-dimensional data informing on living systems. One particular class of probabilistic causal networks being applied to model the complexity and causal structure of biological data is Bayesian networks (BNs). BNs provide an elegant mathematical framework for not only inferring causal relationships among many different molecular and higher order phenotypes, but also for incorporating highly diverse priors that provide an efficient path for incorporating existing knowledge. While significant methodological developments have broadly enabled the application of BNs to generate and validate meaningful biological hypotheses, the reproducibility of BNs in this context has not been systematically explored. In this study, we aim to determine the criteria for generating reproducible BNs in the context of transcription-based regulatory networks. We utilize two unique tissues from independent datasets, whole blood from the GTEx Consortium and liver from the Stockholm-Tartu Atherosclerosis Reverse Network Engineering Team (STARNET) study. We evaluated the reproducibility of the BNs by creating networks on data subsampled at different levels from each cohort and comparing these networks to the BNs constructed using the complete data. To help validate our results, we used simulated networks at varying sample sizes. Our study indicates that reproducibility of BNs in biological research is an issue worthy of further consideration, especially in light of the many publications that now employ findings from such constructs without appropriate attention paid to reproducibility. We find that while edge-to-edge reproducibility is strongly dependent on sample size, identification of more highly connected key driver nodes in BNs can be carried out with high confidence across a range of sample sizes.

1. Introduction

Biological networks provide a graphical framework for organizing complex relationships among many thousands of variables in ways that can reveal coherent structures. These structures reveal knowledge and improve the understanding of molecular processes linked to higher order functioning of living systems. Vast arrays of data are being generated in numerous areas of biomedical research such as large-scale multi-'omic' studies across many cell types, comprehensive characterizations of microbiota living in and around us, advanced imaging data, and deep clinical characterizations of populations to name a few. This upsurge of big data has forced the life and biomedical sciences to rapidly turn to the use of network constructs. One such organizing framework for integrating data comes in the form of probabilistic network models that seek to capture the regulatory states of a system and their association to complex phenotypes such as disease. A particular class of probabilistic causal networks being applied to model the complexity and causal structure of biological data is Bayesian networks (BNs).

---

* Co-first Authors
† Corresponding Author

BNs are increasingly used in the field of genetics to describe and predict gene, metabolite, and protein level interactions. These networks are able to infer causal relationships among variables by employing mutual information or conditional independence measures based on Bayes Theorem. Since 2000 when this method was first applied to understand gene regulation[1], numerous studies have showcased the advantage of using such methods to uncover biological insights that are not easily captured through descriptive methods such as hierarchical clustering or coexpression network analysis. Whether predicting regulatory genetic drivers of complex phenotypes such as human diseases or enabling identification of novel drug target interactions and adverse side effects, BNs have helped uncover the individual genes and biological processes involved in a broad range of human conditions, including cancer, diabetes and obesity, asthma and COPD, cardiovascular disease, and Alzheimer's disease[2–9]. For example, BNs generated from ileal pediatric samples identified a causal gene resulting in a predictor for adult-onset inflammatory bowel disease[10]. As sample sizes increase, it can be envisioned that more groups will use BNs to predict individual response to treatment and it will enable fine-tuning for precision medicine[11].

Constructing a BN structure from data is an NP-hard problem with the complexity equaling $O(n^n)$, where n is the number of nodes in the structure. Many heuristic approaches are applied in searching for an optimal structure from the given data. However, these heuristic methods may find many local sub-optimal structures with no guarantee of finding a global optimal structure. To achieve high accuracy BNs, especially with respect to edge direction, large sample sizes or "big data" are required[12,13]. With the number of large datasets for which BN reconstruction algorithms could be applied growing at an exponential rate, the application of BN algorithms face a similar trend regarding the number of networks being constructed to derive data-driven hypotheses. However, assessing the reproducibility of BNs in the context of gene regulatory networks has not kept pace, with there being no studies to our knowledge systematically exploring this issue. Thus, we thought it crucial to test the conservation and reproducibility of BN constructions as a way to gain confidence in the methods currently used in the field. While significant work has been carried out to assess the construction methods that perform best across different types of biological data[14–16], these types of comparisons do not explicitly address the reproducibility of any given BN.

Perhaps among the gravest concerns in the field of biomedical research today is the lack of reproducibility. It is estimated that over $28 billion of research money, or roughly 50% of life-science research, is not reproducible[17]. The scientific method is rooted with principles of reproducibility giving credence to hypotheses only if they can withstand the scrutiny of many groups trying to reproduce them. In the current era of big data biology, the number of hypotheses generated in even a single publication can number in the hundreds (e.g., GWAS study on a complex trait). These hypotheses are difficult to validate across multiple groups, as the number of groups to rigorously pursue every hypothesis generated is limited. While intuition may argue that the large sample sizes and the robustness of the models may inherently address issues relating to reproducibility compared to traditional biological studies, recent claims indicate that about one quarter (25.5%) of studies not reproduced are due to data-analysis and reporting issues[17]. We therefore focused our study on the reproducibility of individual directed edges and key driver nodes of BNs, as these are generally considered targets for biological validation studies.

2. Study design

Two different gene expression datasets and a simulated dataset were used in this reproducibility study. The first gene expression dataset was obtained from the GTEx Consortium where RNA was extracted from multiple tissues from deceased, healthy individuals. Here, we used data from whole blood, which had a large sample size (N = 379)[18,19]. The second gene expression dataset was comprised of atherosclerosis patients undergoing Coronary Artery Bypass Grafting (CABG) surgery, at which time multiple tissues were extracted and RNA sequenced from the Stockholm-Tartu Atherosclerosis Reverse Network Engineering Team (STARNET)[20] We chose to utilize the liver tissue (N=545), which contained the strongest eQTL signal[20], a prior in the BN reconstruction algorithm we employed that helps reduce the search space and resolve true causal relationships. By leveraging these real-world datasets, we are able to capture the complex correlation structures that derive from gene expression data measured in populations. RNA levels are high fidelity sensors of the state of the system and of technical noise, where the many different variance components (technical, genetic, micro- and macro-environment) form a complex covariance structure that is difficult to reproduce in simulated datasets. In addition, these two biological datasets represent not only two distinct tissues, but also reflect different states of disease and wellness (Table 1).

Table 1. Overview of datasets used. This table provides details on the two datasets used in this study.

	GTEx	STARNET
Tissue	Whole Blood	Liver
Patient Status	Deceased - Healthy	Living - Undergoing CABG
# Samples	379	545
# Genes Used	455	385
Priors	cis eQTLs	cis eQTLs + Causal Inference Priors

To assess and compare networks in a thorough manner, we restricted attention to a subset of genes (N=465) that have been previously identified as highly informative for inflammatory diseases and associated with immune and inflammation response[2,5,8,21–24]. By selecting this set of genes to use in the analysis, we reduced the computational time and cost required to generate each network.

In order to assess the reproducibility of BNs, we subsampled from the complete datasets to generate datasets reflecting different sample sizes under identical conditions. Towards this end, we subsampled the data in three ways: 1) a subsampling of 50% of the samples (referred to as the subsampled-50 networks), 2) a subsampling of 80% of the samples (referred to as the subsampled-80 networks), and 3) a subsampling of 90% of the samples (referred to as the subsampled-90 networks) (Fig 1). All subsampling divisions were replicated five times. The first scenario was intended to mimic the situation in which an initial study producing a BN is followed by an equivalent replication study producing a confirmatory BN, while the second and third scenarios represent incremental data releases, as happens in the context of large studies where data freezes are employed. The same process was used with the simulated dataset, however, here we were able to control the power and increased our sample size (N=1000) to the point of reaching near perfect reproducibility. For the simulated datasets, we subsampled at 50%, 80%, and 90%, with five replicates generated at each level. We also generated the simulated data at a subsampling of 10% to represent how data with limited noise is reproduced at a small sample size (N=100).

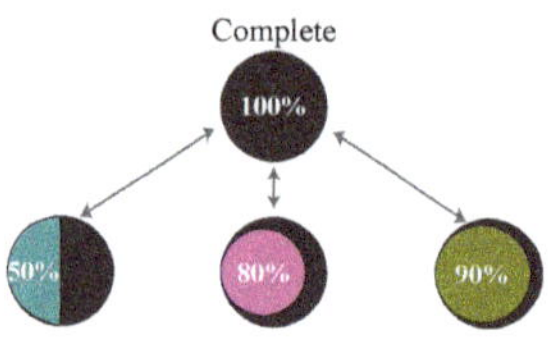

Figure 1. Schematic of the study design.

For all datasets, networks were generated using the Reconstructing Integrative Molecular Bayesian Networks (RIMBANet) algorithm[25,26] as the output has been validated extensively (see methods). When available, eQTL data as well as previous information regarding the causal association between several genes (nodes) in the network were used as structural priors[5,9,20,25,26]. With BNs, the predominant method for assessing confidence of an edge is based on the posterior probability associated with that edge. This is computed either directly from the network model or is empirically estimated by generating a distribution of models and computing summary statistics across the networks comprising the distribution. We utilized the latter scenario where the posterior probability is approximated by computing the number of networks that contain a particular edge and dividing this number by the total number of networks generated. In this study, we considered nine different posterior probability thresholds (0.1 to 0.9 in 0.1 increments) to explore the reproducibility of edges across different confidence levels. Thus, for each dataset, we generated nine networks for the complete and each of the subsampled datasets.

3. Results

3.1. *Exploring edge-to-edge reproducibility*

Comparing BN's is a multifaceted task in itself as they are complex representations of high-dimensional data. To provide a more intuitive comparison consistent with how BNs are used in practice in the life sciences and biomedical research spaces, we compared networks in two ways: 1) by evaluating the confidence levels of individual edges and 2) by evaluating the higher-level topology of the network.

Table 2: Overlap of five replications of complete BN. For each posterior probability, all combinations of replicates were looked at to calculate the percentage overlap divided by the total edges of each replicate. Here we report the mean percentage and standard deviation.

Posterior Probability	0.1	0.2	0.3	0.4	0.5	0.6	0.7	0.8	0.9
GTEx	99% (± 0.008)	99% (± 0.008)	99% (± 0.007)	99% (± 0.008)	99% (± 0.008)	99% (± 0.006)	99% (± 0.004)	98% (± 0.01)	97% (± 0.02)
STARNET	99% (± 0.01)	99% (± 0.01)	99% (± 0.01)	99% (± 0.01)	98% (± 0.01)	99% (± 0.02)	99% (± 0.01)	98% (± 0.02)	96% (± 0.02)

Given the stochastic search employed in the BN construction process, we first compared five networks generated on the complete dataset (includes all samples) for each cohort to characterize the degree of variability. As depicted in Table 2, at a posterior probability of 0.1, both datasets have a mean edge overlap of 99%. While the edges with high confidence (at a posterior probability >0.9) are found on average 97% in other replicates in GTEx and 96% in STARNET, we observe that 100% of these edges are present in other replicates when the posterior probability is > 0.5.

As the stochasticity of the BN reconstruction process does not seem to affect the reproducibility of the BNs, we next calculated the Jaccard index with respect to all network pairs within a given subsampled set (Table 3). The Jaccard index is a measure commonly used when comparing sets, and ranges from 0, for completely unrelated sets, to 1, for highly similar sets. In our case, the edge counts between replicates are comparable when the number of samples and posterior probability are the same (see standard deviations in Table 4), thus the maximum Jaccard index should be close to 1 (complete reproducibility). The Jaccard index had a mean of 0.27 when comparing edges from the subsampled-50 networks across the different posterior probability thresholds within the replicates or to the complete network within each cohort (Table 3).

Interestingly, the Jaccard index achieved values close to 0.5 for edges from the subsampled-90 networks (Table 3), which is very different from the values we saw when comparing the replicates of the complete networks (mean >0.95 in both at a posterior probability >0.1). These results suggest that even with 90% overlap of samples, the edge-set overlap can still be different, highlighting significant reproducibility issues even among highly comparable sample sets. The data suggests that statistical power in resolving network relationships may be primarily responsible for the lower than expected reproducibility, an issue that can be experimentally addressed by increasing the sample size.

Table 3. Jaccard index values. We calculated the Jaccard index (intersection divided by union) for the edges found in the networks at each posterior probability threshold. We compared the subsampling networks to their respective replicates and to the complete BN at the same posterior probability threshold. Standard deviation ranges from 0.01-0.04 in all cases.

		GTEx		STARNET	
Sub-sampling	**Posterior Probability**	**To Other Replicate**	**To Complete**	**To Other Replicate**	**To Complete**
50%	0.1	0.23	0.26	0.22	0.27
	0.5	0.23	0.26	0.21	0.27
	0.9	0.20	0.20	0.16	0.19
80%	0.1	0.34	0.40	0.37	0.43
	0.5	0.34	0.40	0.36	0.43
	0.9	0.29	0.33	0.26	0.35
90%	0.1	0.44	0.51	0.43	0.52
	0.5	0.43	0.53	0.43	0.52
	0.9	0.33	0.39	0.36	0.42
Complete	0.1	0.98	---	0.97	---
	0.5	0.98		0.97	
	0.9	0.95		0.93	

The number of edges in a BN is at least partially a function of power, given that as sample size increases, an increase in the number of edges in the BN is observed (Table 4). Thus, a more applicable measure for assessing reproducibility among networks is by looking at the number of overlapping edges between a subsampled network and the complete network, divided by the number of edges in the subsampled network. This measure relates to precision or positive predictive value, given here we accepted as truth the complete network (in the context of the simulated data, true and false positives are known with certainty). The flip side of precision is recall, or sensitivity, defined by dividing the overlap number of edges by the total number of edges in the complete network (Fig 2A).

For both GTEx and STARNET, when comparing the subsampled and the complete network at the same posterior probability cutoff, we found that on average 44% of GTEx and 38%

Table 4. Number of edges in each network. We calculated the number of edges present in each subsampled network. Displayed are the mean and standard deviation for number of edges at select posterior probabilities.

	GTEx			STARNET			Simulation		
Sub-sampling	**0.1**	**0.5**	**0.9**	**0.1**	**0.5**	**0.9**	**0.1**	**0.5**	**0.9**
10%	---	---	---	---	---	---	209 (± 4.637)	192 (± 3.391)	47.4 (± 5.030)
50%	297.8 (± 8.349)	257.2 (± 3.271)	89 (± 9.055)	291.4 (± 5.683)	262.4 (± 4.336)	113.6 (± 11.393)	345.2 (± 3.493)	329 (± 2.550)	149.8 (± 7.396)
80%	390.8 (± 5.586)	343.6 (± 5.459)	136.6 (± 5.459)	373.2 (± 8.349)	329.8 (± 2.775)	135 (± 8.337)	380.6 (± 4.722)	368.6 (± 1.342)	149 (± 4.000)
90%	414.4 (± 6.465)	365 (± 5.099)	135 (± 4.950)	395 (± 6.205)	350.6 (± 3.647)	144.6 (± 3.782)	385.2 (± 1.095)	373.8 (± 3.493)	185.4 (± 8.081)
Complete	441.6 (± 0.894)	388.4 (± 1.517)	138.2 (± 2.280)	396.8 (± 4.382)	364.2 (± 4.025)	151 (± 1.225)	393.2 (± 0.447)	379.8 (± 0.447)	189.8 (± 0.837)

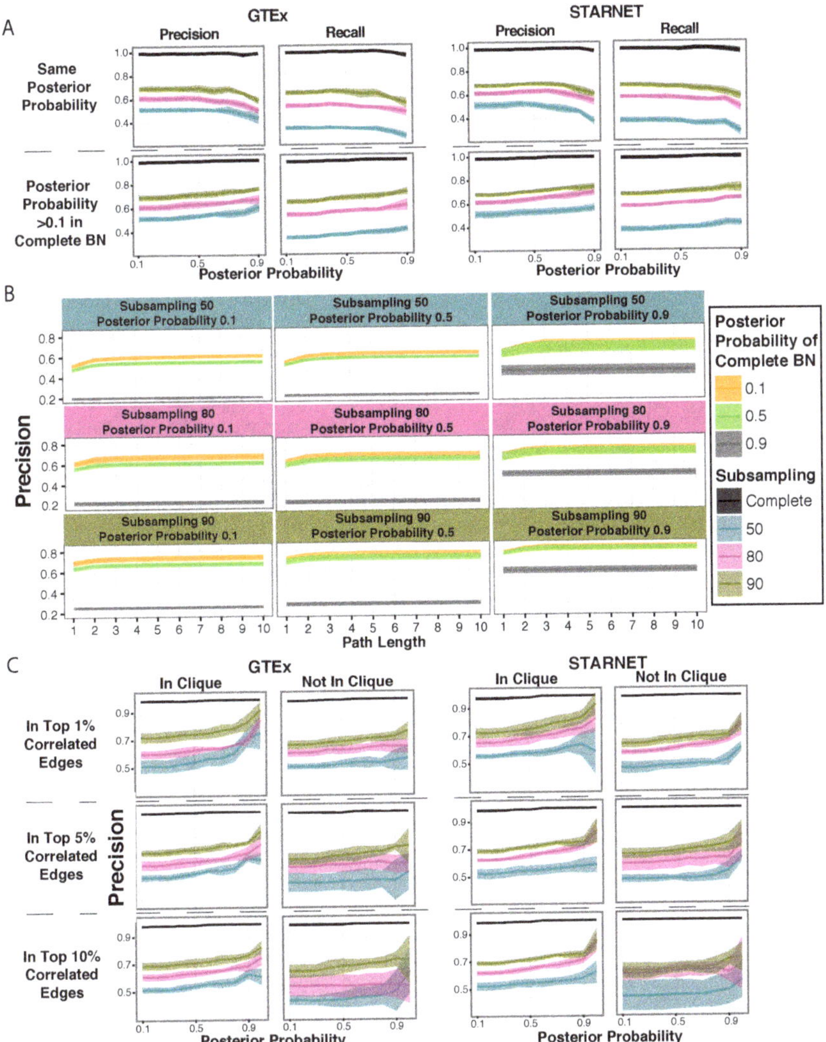

Figure 2. Edge reproducibility rate. In panel A, we compared the number of edges present in the complete BN to the subsampling network at the same posterior probability (top half) and by fixing the threshold for the subsampling networks but allowing any edge for the complete BN (posterior probability >0.1) as seen in the bottom half. In panel B, we show the results from the GTEx data as we allow for edges to be considered reproduced if there is a connection in the complete BN between those two nodes at a path length up to 10. In panel C, we illustrate the percision of edges depending on if the nodes are in the same correlation clique or not. For all panels coloring depicts the subsampling networks and the complete BN.

STARNETs' most confident edges (posterior probability >0.9) in the subsampled-50 networks were reproduced, and this increases to 58% in GTEx and 61% in STARNET for the subsampled-90 networks (Fig 2A). We observed a trend of the precision increasing as the posterior probability increased to 0.4-0.5, but then observed a decrease as the confidence in the edges increased (Fig 2A). This is most likely due to a decrease in the number of edges in the BNs as the posterior probability increases (Table 4). We further evaluated the precision by relaxing the posterior probability for edges in the complete network to >0.1 (Fig 2A). In this case, on average 61% in GTEx and 57% in STARNET of the most confident edges (posterior probability >0.9) were reproduced in the subsampled-50 networks whereas for the subsampled-90 networks 76% in GTEx and 75% in STARNET were reproduced (Fig 2A).

The above definitions of precision at the edge level require the presence of the exact same edge, whereas causal relationships in one network may also be reflected in a different network via intermediary nodes. For example, in one network an edge might be present from A → B (path length=1) and in a second network it may appear as A→C→B, where there is a path from A to B, but via C (path length=2). We hypothesized that this may explain some portion of the edges that failed to reproduce. To test this, we further evaluated if two connected nodes from the subsampled networks were connected in the complete network within a path length of ten. For the GTEx BNs, we saw that in the subsampled-50 networks, the precision increased to an average rate of 67% (up from 61%) at a path length of five for the most confident edges (posterior probability >0.9), while in the subsampled-90 networks, the precision increased to an average rate of 81% (up from 76%) at a path length of three (similar results were seen for STARNET as well). The precision increased with both the path length and sample size (Fig 2B). It should be noted that after a path length of 3, the precision plateaus, providing confidence that increasing the path length further would not have added any new information in the context of our networks.

BNs reflect complex correlation structures or rich substructures in which the expectancy of certain nodes to be more or less connected may be contained within the network. Higher-order correlation structures have been informative for the underlying biology from large datasets[13,27]. To explore whether the correlation structure of the data affected edge reproducibility, we examined whether genes in clique structures (groups of highly interconnected genes) were more or less likely to be reproduced, compared to the average precision of the network. For each data set, we computed the correlation matrix and took the top 1%, 5% and 10% most correlated values to build an undirected, correlation-based network. We focused on the most stringent correlation criteria to define edges, which was the top 1%. From these networks we were able to call all clique communities using the program COS (https://sourceforge.net/projects/cosparallel/). This enabled us to determine if both nodes of an edge were included in the same clique. We found that the precision was further improved in edges whose nodes were found in the same clique (Fig 2C). In the STARNET subsampled-90 networks, the most confident edges (posterior probability >0.9) present in a clique obtained using the top 1% correlated values had a mean precision measure of 85% compared to 71% for edges in which both nodes were not found in the same clique (whereas all edges had a mean precision of 75%). In the subsampled-50 networks, the edges in a clique had a precision rate of 65% versus 53% for edges comprised of nodes that did not both fall within the same clique (whereas all edges had a mean precision of 57%). The GTEx dataset provided similar results, showing that we were able to improve the precision of edges by incorporating correlation clique information.

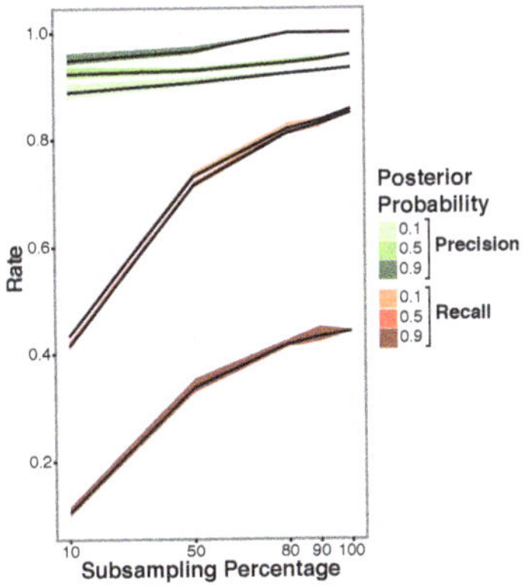

Figure 3. Simulation data precision and recall. We simulated a BN for 300 nodes, 1000 samples with discrete data and looked at the precision and recall for the subsampling at 10%, 50%, 80%, 90%, and 100%. The color scale represents the posterior probability threshold. We show the mean and standard deviation for the five replicates.

Precision and recall trends with the simulated datasets were similar to those observed in the biological datasets. This confirmed not only that our simulated data was reflective of the biological datasets, but also that by increasing sample size we could address the edge-level precision and improve recall (Fig 3). Thus, as larger datasets are generated, the issue of reproducibility of networks should be addressed.

3.2. *On the reproducibility of key driver nodes*

Another important aspect of BNs is their higher order topology. Not all nodes in a BN are equivalent, but rather some are more connected having a substantial causal impact on many more nodes in the network (referred to here as key driver nodes, or KD nodes). One way to assess reproducibility of these types of important topological features is by examining the reproducibility of KDs. KD nodes are important and commonly inferred from networks as they help elucidate the regulatory states of complex systems, and are crucial from a diagnostic and drug discovery standpoint[2,5,28]. Thus, we decided to assess the reproducibility of the detection of these types of nodes.

We calculated the KDs for each network built at each posterior probability threshold and assessed the precision of the KDs in the same manner applied to the edges (see methods). First, we evaluated the overlap of KDs between the complete and subsampled networks when they were built at the same fixed posterior probability. To see if a difference between the ranking of KDs and their precision could be measured, we defined the top KDs as being in the 97.5 – 100 percentile and bottom KDs as being in the 95 – 97.5 percentile. When evaluating the KDs of the network built from the most confident edges (posterior probability >0.9), we found that the top KDs from the subsampled-90 networks were reproduced at an average rate of only 49% while the bottom KDs were reproduced at an average rate of 54% in GTEx. In STARNET, the top KDs were reproduced at an average rate of 85% while the bottom KDs were reproduced at an average rate of 43% (Fig 4). To see if we could improve the reproducibility rate, we relaxed the threshold for the complete BN and allowed for the KD to be present at any posterior probability (similar to what was done with the edges). This drastically improved the reproducibility of the KDs. In GTEx, the top KDs from the subsampled-90 networks built on the most confident edges (posterior probability >0.9) were reproduced at an average rate of 87% while the bottom KDs were reproduced at an average rate of 77%. A similar evaluation of the STARNET results showed the top KDs were reproduced on average 93%, while the bottom KDs were reproduced at 60%. We saw in the subsampled-50 networks, at a posterior probability >0.5 that while the edge-overlap was on average 54% in GTEx and 53% in STARNET, the KD overlap was 58% in GTEx and 66% in STARNET. In the subsampled-90 networks, where the edge-overlap was on average 72% in GTEx and 71% in STARNET, the KD overlap increased to 76% in GTEx and 87% in STARNET. The KDs performed as well if not better than the edges, indicating that the KDs of BNs are more conserved than edges. Since the networks with fewer samples have fewer edges present, it could help explain why we see such low precision in the subsampled-50 networks. These results further support that a larger sample size, or increased power, will lead to more reproducible KDs.

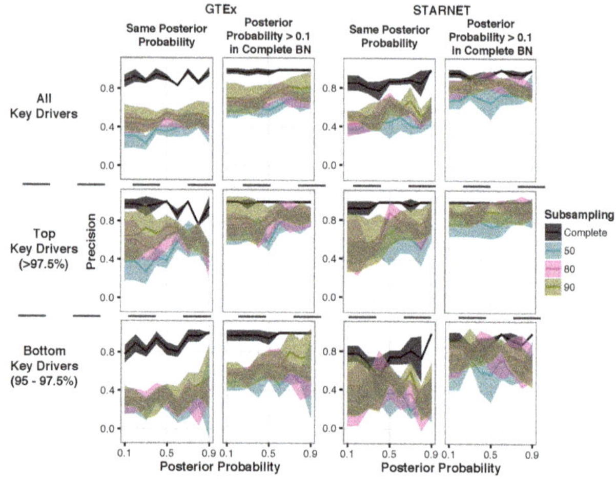

Figure 4. Precision of key driver (KDs). Precision is the % KDs of the subsampling network present in the complete BN (at either the same posterior probability threshold or at any). Left panel shows all KDs; Middle panel shows Top KDs (top 97.5% based on the weighted number of connections, see methods); Right panel, shows bottom KDs (95 – 97.5%). Mean and standard deviation for the five replicates are displayed, and color depicts subsampling.

As the KDs take into account the shortest path to reach all nodes, we thought to additionally assess nodes with the highest number of first-degree downstream targets, hub nodes. These nodes have the most local and direct impact on other nodes. Here we took the top 10% of nodes based on their total number of out edges and applied the same analysis pipeline defined above for KDs. We found that when the posterior probability >0.1 for the complete network, the hub nodes were more reproduced in the subsampled networks, as can be seen by the subsampled-90 networks reaching an average rate of 78% in GTEx and 83% in STARNET at a posterior probability threshold of 0.5 (Fig 5). However, if we hold the posterior probabilities constant in both the complete and subsampled networks, the precision fluctuates in the GTEx dataset but appears to perform better in the STARNET dataset. This could be explained by the larger sample size of the STARNET dataset.

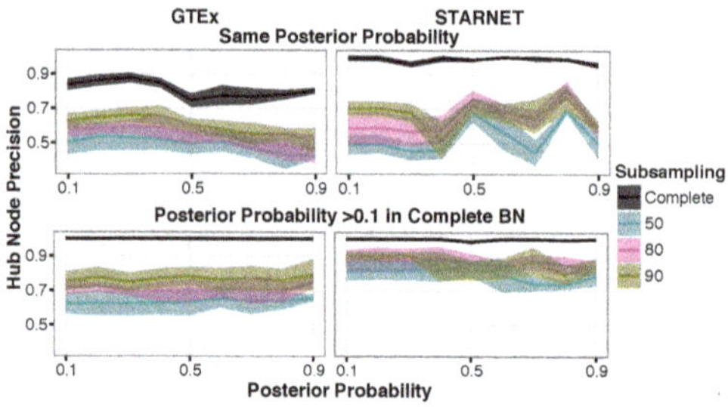

Figure 5. Hub nodes precision. We define hub nodes as nodes in the 90th percentile based on the number of first degree out edges. The top half illustrates the precision when the posterior probability is the same in both the subsampling and the complete BN. The bottom half illustrates the precision when the posterior probability in the subsampling network is fixed but the hub node in the complete BN can be at any posterior probability. The mean and standard deviation for the five replicates is displayed and color depicts subsampling.

4. Discussion

In this study on the reproducibility of BNs in the context of regulatory gene networks, while we found a high degree of reproducibility at the edge and key driver node levels, we also noted that a large proportion of edges and key driver nodes were not reproduced. Given the rate at which edges and key driver nodes did not reproduce in networks constructed from a moderate number of samples, caution should be exercised when interpreting specific features of a network. Validating hypotheses generated from networks is critical to ensure the accuracy of network predictions. However, we also observed that the lack of reproducibility might be attributed to power issues, which can be straightforwardly addressed by increasing sample sizes for network reconstructions. As obtaining large sample size is difficult and expensive, our results stress the need to assess the reproducibility of methods being deployed in the field. We must be aware of limitations so we can strive to improve them.

While we restricted attention to a coherent subset of several hundred genes to contain computational costs, we have observed similar trends in BNs built on 10,000 or more genes using the GTEx whole blood samples, suggesting that the subset of genes used was a good proxy for how larger networks of genes would behave. Ideally, we would have run our analysis on a completely validated BN from a biological dataset. However, at the time of this study, such a validated network was not available. Instead, we complemented our study of networks constructed from gene expression datasets with examination of simulated datasets containing discretized data for a comparable number of genes.

We used structural priors to generate the BNs, which could bias the structure of the resulting networks. However, we saw a decrease in precision and recall when priors were not used, further demonstrating the importance of high-confidence priors. We chose to include priors as this is typically done in practice today and their use has shown to increase accuracy of networks based on smaller sample sizes[26].

The reproducibility of KDs was of particular interest, given the role they play in current biological investigations of complex systems. KDs represent central information flow points in the network that are identified in disease studies as potential targets of therapeutic intervention or as features that may be critical as biomarkers of disease. We observed that KDs were more reproduced than edges. This suggests that while the edges may be less conserved due to nonlinear interactions or stochasticity, the overall structure of the network may still be well conserved, explaining the increased confidence in key driver node predictions. In particular, the top KDs, which are most connected and predicted to significantly impact network states, were reproduced at exceptionally high rates.

As biomedical and life sciences research gravitates toward network-based constructs, issues of reproducibility will come front and center. It is critical to characterize network reconstruction methods from the standpoint of what is required to lead to reproducible structures that in turn, lead to high-confidence hypotheses. Our analysis shows that well-powered Bayesian networks are highly reproducible. Since high power is not always possible to achieve because samples are scarce and assays are expensive, our results provide guidance on interpreting and using Bayesian networks. In cases of diminished power, it is critical to realize that key drivers, in particular the strongest key drivers, and hub nodes are more robustly reproduced than individual edges.

5. Methods

Bayesian Network Construction: RIMBANet was used to construct all Bayesian Networks[9,12,26]. Continuous data was used for calculating partial priors, which are then used as priors in the network construction. Additional priors included genes that are *cis* eQTLs and the results from the causal inference test of *cis gene* → *trans gene* (for STARNET only)[20,29]. For the eQTL priors, if a gene also has a strong eQTL associated with it in *cis*, such a gene can be considered as a parent node, given the genotype cannot be the effect of a gene expression change. The data was discretized into 3 states for each gene: high expression levels, low expression levels and unexpressed. This is done by first normalizing the values for each gene to ensure a normal distribution. Then, k-means clustering (k=3) is used with the option of dropping groups should there not be enough members to fill it to assign the values for each sample. In a case where there are only two clusters they would be classified as high and low[30]. For the sake of quicker run times, when looking for the parents of each gene, the other genes were sorted by their mutual information and only the top 80% were considered as candidates. Also, the maximum number of parent nodes that were allowed for any given node was set to 3. After running successfully 1,000 reconstructions, the networks were pooled together. Finally, because a BN is a directed acyclic graph (DAG) by definition, the consensus network was obtained by searching for the shortest cycle and then the edge with the weakest weight (the smallest number of times it occurs in 1,000 reconstructions) was removed. This process was repeated until no cycles were present and the resulting network was a DAG.

Generation of Simulated Dataset: To generate the synthetic true network, we used the SynTRen software v1.2[31]. We extracted a subnetwork with 300 nodes from the background source network "DAG1_clean.sif" with default settings. We limited the node selection to 300 nodes to reduce the computational time required to generate all of the networks and to mimic the size of the biological datasets used in this study. Next, to generate the synthetic discretized data from the known network structure, we utilized Bayes Net Toolbox (BNT) for Matlab [https://code.google.com/archive/p/bnt/]. The conditional probability was customized so that we could discretize the data into three bins, similar to RIMBAnet. Given the configuration of parent node, the child nodes were skewed towards one of the three discretized states with a probability between 0.8 and 0.9, therefore, ensuring assignment to a given bin with high confidence.

Key Driver Node Detection: Key driver nodes (KDs) were detected by calculating the shortest downstream path length between each pair of nodes in the network. For each candidate key driver node, we took the inverse of path length between the candidate key driver node and every other node in the network. We then summed the inverse path lengths to obtain a final score per node. Based on this calculation, we defined nodes in the 95th percentile as KDs[5]. We define top KDs as nodes in the 97.5 - 100 percentile and bottom KDs as nodes in the 95 - 97.5 percentile.

Code and data can be found at https://github.com/divara01/PSB2017_ReproducibilityOfBNs/ and http://research.mssm.edu/integrative-network-biology/Software.html

Acknowledgements

Funding for this project was provided by National Institute of Health (NIH) grants U54CA189201, R01DK098242, 5U01AG046170, and 1R01MH109897 and Leducq Foundation grant 12CVD02.

References

1. Friedman, N., Linial, M., Nachman, I. & Pe'er, D. Using Bayesian Networks to Analyze Expression Data. *J. Comput. Biol.* **7**, 601–620 (2000).
2. Jostins, L. *et al.* Host-microbe interactions have shaped the genetic architecture of inflammatory bowel disease. *Nature* **491**, 119–24 (2012).
3. Korucuoglu, M., Isci, S., Ozgur, A. & Otu, H. H. Bayesian Pathway Analysis of Cancer Microarray Data. *PLoS One* **9**, e102803 (2014).

4. Schwartz, S. M., Schwartz, H. T., Horvath, S., Schadt, E. & Lee, S.-I. A systematic approach to multifactorial cardiovascular disease: causal analysis. *Arterioscler. Thromb. Vasc. Biol.* **32,** 2821–35 (2012).
5. Zhang, B. *et al.* Integrated systems approach identifies genetic nodes and networks in late-onset Alzheimer's disease. *Cell* **153,** 707–20 (2013).
6. Kidd, B. a, Peters, L. a, Schadt, E. E. & Dudley, J. T. Unifying immunology with informatics and multiscale biology. *Nat. Immunol.* **15,** 118–27 (2014).
7. Schadt, E. E. Molecular networks as sensors and drivers of common human diseases. *Nature* **461,** 218–23 (2009).
8. Greenawalt, D. M. *et al.* A survey of the genetics of stomach, liver, and adipose gene expression from a morbidly obese cohort. *Genome Res.* **21,** 1008–16 (2011).
9. Zhu, J. *et al.* Integrating large-scale functional genomic data to dissect the complexity of yeast regulatory networks. **40,** 854–861 (2008).
10. Li, Q. *et al.* Variants in TRIM22 That Affect NOD2 Signaling Are Associated With Very-Early-Onset Inflammatory Bowel Disease. *Gastroenterology* **150,** 1196–207 (2016).
11. Uzilov, A. V. *et al.* Development and clinical application of an integrative genomic approach to personalized cancer therapy. *Genome Med.* **8,** 62 (2016).
12. Zhu, J. *et al.* An integrative genomics approach to the reconstruction of gene networks in segregating populations. *Cytogenet. Genome Res.* **105,** 363–74 (2004).
13. Langfelder, P. & Horvath, S. WGCNA: an R package for weighted correlation network analysis. *BMC Bioinformatics* **9,** 559 (2008).
14. Marbach, D., Schaffter, T., Mattiussi, C. & Floreano, D. Generating realistic in silico gene networks for performance assessment of reverse engineering methods. *J. Comput. Biol.* **16,** 229–39 (2009).
15. Saez-Rodriguez, J. *et al.* Crowdsourcing biomedical research: leveraging communities as innovation engines. *Nat. Rev. Genet.* **17,** 470–486 (2016).
16. Hill, S. M. *et al.* Inferring causal molecular networks: empirical assessment through a community-based effort. *Nat. Methods* **13,** 310–318 (2016).
17. Freedman, L. P., Cockburn, I. M. & Simcoe, T. S. The Economics of Reproducibility in Preclinical Research. *PLoS Biol.* **13,** e1002165 (2015).
18. Lonsdale, J., Thomas, J., Salvatore, M. & Phillips, R. The genotype-tissue expression (GTEx) project. *Nat. ...* **45,** 580–5 (2013).
19. Ardlie, K. G. *et al.* The Genotype-Tissue Expression (GTEx) pilot analysis: Multitissue gene regulation in humans. *Science (80-.).* **348,** 648–660 (2015).
20. Franzén, O. *et al.* Cardiometabolic risk loci share downstream cis- and trans-gene regulation across tissues and diseases. *Science* **353,** 827–30 (2016).
21. Chen, Y. *et al.* Variations in DNA elucidate molecular networks that cause disease. *Nature* **452,** 429–35 (2008).
22. Emilsson, V. *et al.* Genetics of gene expression and its effect on disease. *Nature* **452,** 423–8 (2008).
23. Wang, I.-M. *et al.* Systems analysis of eleven rodent disease models reveals an inflammatome signature and key drivers. *Mol. Syst. Biol.* **8,** 594 (2012).
24. Yang, X. *et al.* Validation of candidate causal genes for obesity that affect shared metabolic pathways and networks. *Nat. Genet.* **41,** 415–23 (2009).
25. Zhu, J. *et al.* Stitching together multiple data dimensions reveals interacting metabolomic and transcriptomic networks that modulate cell regulation. *PLoS Biol.* **10,** e1001301 (2012).
26. Zhu, J. *et al.* Increasing the Power to Detect Causal Associations by Combining Genotypic and Expression Data in Segregating Populations. **3,** (2007).
27. Song, W.-M. *et al.* Multiscale Embedded Gene Co-expression Network Analysis. *PLOS Comput. Biol.* **11,** e1004574 (2015).
28. Dudley, J. T. *et al.* Computational repositioning of the anticonvulsant topiramate for inflammatory bowel disease. *Sci. Transl. Med.* **3,** 96ra76 (2011).
29. Millstein, J., Zhang, B., Zhu, J. & Schadt, E. E. Disentangling molecular relationships with a causal inference test. *BMC Genet.* **10,** 23 (2009).
30. Zhu, J. *et al.* Complexity of Yeast Regulatory Networks. **40,** 854–861 (2009).
31. Van den Bulcke, T. *et al.* SynTReN: a generator of synthetic gene expression data for design and analysis of structure learning algorithms. *BMC Bioinformatics* **7,** 43 (2006).

REPRODUCIBLE DRUG REPURPOSING: WHEN SIMILARITY DOES NOT SUFFICE

EMRE GUNEY*

Joint IRB-BSC-CRG Program in Computational Biology, Institute for Research in Biomedicine
c/ Baldiri Reixac 10-12, Barcelona, 08028, Spain
**E-mail: emre.guney@irbbarcelona.org*

Repurposing existing drugs for new uses has attracted considerable attention over the past years. To identify potential candidates that could be repositioned for a new indication, many studies make use of chemical, target, and side effect similarity between drugs to train classifiers. Despite promising prediction accuracies of these supervised computational models, their use in practice, such as for rare diseases, is hindered by the assumption that there are already known and similar drugs for a given condition of interest. In this study, using publicly available data sets, we question the prediction accuracies of supervised approaches based on drug similarity when the drugs in the training and the test set are completely disjoint. We first build a Python platform to generate reproducible similarity-based drug repurposing models. Next, we show that, while a simple chemical, target, and side effect similarity based machine learning method can achieve good performance on the benchmark data set, the prediction performance drops sharply when the drugs in the folds of the cross validation are not overlapping and the similarity information within the training and test sets are used independently. These intriguing results suggest revisiting the assumptions underlying the validation scenarios of similarity-based methods and underline the need for unsupervised approaches to identify novel drug uses inside the unexplored pharmacological space. We make the digital notebook containing the Python code to replicate our analysis that involves the drug repurposing platform based on machine learning models and the proposed disjoint cross fold generation method freely available at `github.com/emreg00/repurpose`.

Keywords: Drug repurposing; Machine learning; Drug similarity; Stratified disjoint cross validation.

1. Introduction

Computational drug repurposing has gained popularity over the past decade, offering a possibility to counteract the increasing costs associated with the conventional drug development pipelines. Several studies have focused on training similarity-based predictors (also known as knowledge-based or guilt-by-association-based methods) using drug chemical, target and side effect similarity between drugs (see Refs. 1–3 for recent reviews). These studies often combine various features including but not limited to chemical 2D fingerprint similarity, overlap or interaction network closeness of drug targets and correlation between drug side effects and build a machine learning model based on different algorithms, such as support vector machines, random forests and logistic regression classifiers.[4–11] The proposed models are then compared in a cross validation setting, in which a portion of the known drug-disease associations are hidden during training and used for the validation afterwards. The areas under reciver operating characteristic (ROC) curves in the cross validation analysis reported for these models range between 75-95%, suggesting that some of these models can accurately identify novel drug-disease associations. Nevertheless, in reality, the applicability of these methods for discovery of novel drug-disease associations has been limited due to "the reliance on data existing nearby in pharmacological space" as highlighted by Hodos et *al.*[2] Moreover, Vilar and colleagues alert the community about the potential "upstream bias introduced with the information provided in the construction of the similarity measurement" in similarity-based predictors.[12] Yet, since many studies do not provide the data and code used to build the models for repurposing, it is often cumbersome to validate, reproduce and reuse the underlying methodology.

In this study, first, we provide a Python-based platform for reproducible similarity-based drug repurposing and then seek to quantify the effect of the assumptions on the existing data nearby in pharmacological space. Following similar works evaluating various cross validation approaches for drug-target and protein-protein interaction prediction,[13,14] we adopt a stratified disjoint cross validation strategy for splitting drug-disease pairs, where none of the drugs in the training set appear in the test set. We show that, although a simple logistic regression classifier can achieve good performance on the data set under a conventional cross validation setting, it performs poorly when it faces with drugs it has never seen before.

Overall, our results suggest that the prediction accuracies reported by existing supervised methods are optimistic, failing to represent what one would expect in a real-world setting. We believe that the platform provided in this study could be useful for prospective studies to perform benchmarking in a unified manner.

2. Results

2.1. *A Python platform for reproducible similarity-based drug repurposing*

To incentivize reproducibility in computational drug repurposing research, we provide a Python-based platform[a] encapsulating several machine learning algorithms available in Python Scikit-learn package[15] available both as stand alone code and Jupyter notebook. The platform consists of methods to (*i*) parse a publicly available data set containing drug chemical substrucuture, target, side effect information, (*ii*) calculate drug similarity using a combination of the three features provided in the data set, (*iii*) balance data such that the drug-disease pairs have the same proportion of positive and negative instances, (*iv*) apply cross validation, and (*v*) build classifiers (Fig. 1).

The platform facilitates access to several machine learning algorithms and cross validation utilities available in Scikit-learn. By changing the configuration values, the user can build a classifier using default parameters based on logistic regression, k-nearest neighbor classifier, support vector machine, random forest, and gradient boosting classifier. We note, however, these methods are provided as is and the user still has to take the necessary steps for parameter optimization for these methods. The user can also adjust the proportion of the positive and negative pairs within each fold by changing the parameter file. Furthermore, the platform is easily customizable, allowing the user to define her own data balancing, cross validation and classifier building methods.

2.2. *Evaluating similarity-based drug repurposing via cross validation*

Next, we show the utility of the platform by building a logistic regression based drug repurposing classifier that incorporates drug chemical, target, and side effect similarity, a simplified

[a]Available at `github.com/emreg00/repurpose`

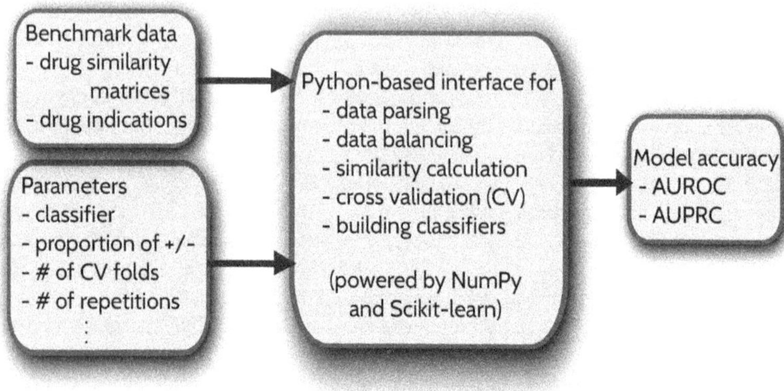

Fig. 1. Overview of the reproducible similarity-based repurposing platform.

version of the classifier suggested in a seminal paper by Gottlieb and colleagues.[4] Our model uses three drug-drug similarity based features compared to the combination of five drug-drug similarity (similarity of targets in terms of gene ontology functions and protein interaction network closeness in addition to the drug chemical, target, and side effect similarity) and two disease-disease similarity-based features (ten in total) proposed by Gottlieb and colleagues. We also incorporate the k-nearest-neighbor approach used by Zhang and coworkers,[7] who recently, built a classifier based on similarity to the 20 most similar drugs and compared it to Gottlieb and colleagues. We build our model on the same data set[b] used by Zhang and coworkers. We calculate the Pearson correlation between drugs using each of the three features mentioned above. For each feature, we assign a score corresponding to the likelihood of a given drug to be indicated for a disease based on the similarity scores and labels of the most similar 20 drugs. These scores are then combined in a logistic regression model and coefficients of the model is derived using a cross validation scheme (see Methods).

We test the prediction accuracy of the classifier under a ten fold cross validation scheme, where we split the available data set into ten groups, leave one group for testing the accuracy of the classifier and use the remaining groups to train the classifier. We repeat the cross validation analysis ten times to get estimates on the mean and standard deviation of the areas under ROC curves (AUC) and report these values in Table 1. We find that the AUC of the classifier is 84%, comparable to 87% reported by Zhang and coworkers. The slight discrepancy between the values can be explained by *(i)* the original study using imputation on the feature set and/or *(ii)* the authors reporting the AUC value from a single run instead of the mean

[b]Made publicly available by the Zhang *et al.* at `http://astro.temple.edu/~tua87106/drugreposition.html`

over multiple cross validation runs (due to the random subsampling of the data, the AUC values in consequent runs might vary slightly).

Table 1. Areas under ROC and Precision-Recall curves (AUC and AUPRC) under various validation schemes averaged over ten runs of ten-fold cross validation (S.d.: Standard deviation).

Disjoint folds	Mean AUC (%)	S.d. AUC (%)	Mean AUPRC (%)	S.d. AUPRC (%)
No	84.1	0.3	83.7	0.3
Yes	65.6	0.5	62.8	0.5

2.3. *Revisiting cross validation using disjoint folds*

Existing studies often assume that the drugs that are in the test set will also appear in the training set, a rather counter-intuitive assumption as, in practice, one is often interested in predicting truly novel drug-disease associations (i.e. for drugs that have no known indications previously). We challenge this assumption by evaluating the effect of having training and test sets in which none of the drugs in one overlaps with the drugs in the other. Accordingly, we implement a disjoint cross validation fold generation method that ensures that the drug-disease pairs are split such that none of the drugs in the training set appear in the test set (Fig. 2, see Methods for details).

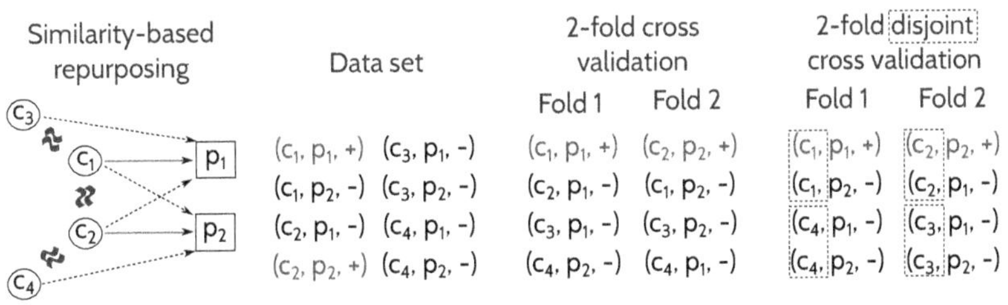

Fig. 2. Schematic representation of similarity-based repurposing and cross validation strategy. On a toy data set consisting of four compounds c_1, c_2, c_3, c_4 and two phenotypes p_1, p_2, the similarity-based drug repurposing approach is illustrated. c_1 and c_2 are indicated for p_1 and p_2, respectively. For instance, c_3 can be inferred to be useful for p_1 due to its similarity to c_1. Conventionally, k-fold cross validation randomly splits the data into k groups preserving the overall proportion of the labels in the data. We propose a disjoint cross validation scheme for paired data, such as drug-disease pairs in drug repurposing studies, that does not only preserve the proportion of the labels but also ensures that none of the drugs from the pairs in one fold are in the other folds. We demonstrate this on the toy data for k = 2 (two-fold cross validation).

In fact, several studies aim to investigate the prediction performance when the drugs in the test set are dissimilar to those in the training data set. Nonetheless, they usually do not guarantee that the trained models are unbiased with respect to unseen data. For instance,

Luo *et al.*[11] use an independent set of drug-disease associations, yet, 95% of the drugs in the independent set are also in the original data set (109 out of 115). On the other hand, Gottlieb *et al.*[4] create the folds such that 10% of the drugs are hidden instead of 10% of the drug-disease pairs, but they do not ensure that the drugs used to train the model are disjoint from the drugs in the test set.

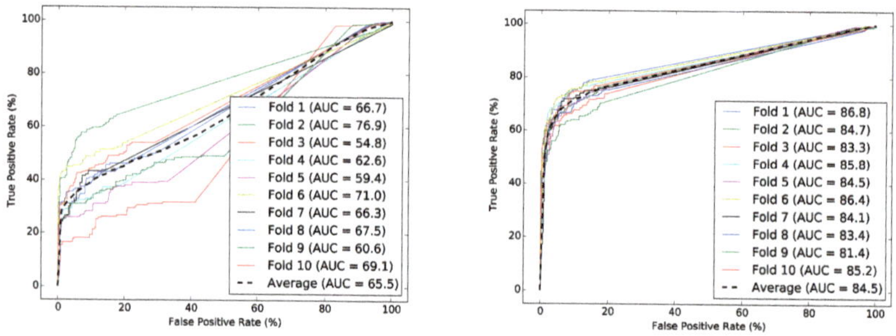

Fig. 3. ROC curves for each fold with and without disjoint cross validation (in a single run).

2.4. *Effect of the cross validation strategy on classifier performance*

We use the drug-wise disjoint cross validation strategy to study its effect on the classifier performance. We observe that the AUC drops significantly from 84% to 66% ($P = 6.9 \times 10^{-23}$, assessed by two-sided t-test) when the classifier is trained with drug-disease associations coming from the drugs that do not exist in the test data set (Table 1).

We suspect that this is due to the limited information within the test set from which the similarity-based drug-disease associations are calculated (using 20 most similar drugs) before they are fed to the classifier. To verify this, we repeat the analysis using two-fold, five-fold and 20-fold cross validation and show that the number of folds does indeed have an effect on the classifier performance (Table 2). In the two-fold disjoint cross validation scheme, the classifier accuracy is almost as good as the ten-fold cross validation accuracy without using disjoint folds, probably due to the number of drug-disease pairs within the test fold being large enough to capture the similarity relationships between drugs. Conversely, in the 20-fold disjoint cross validation scheme, the AUC drops to 59%, emphasizing the effect of the test set size due to the increased number of folds.

We next turn to the ROC curve of each cross validation fold under the two different strategies to examine the consistency among different folds (Fig. 3). We recognize that the variance between the ROC curves is higher when the folds are drug-wise disjoint compared to when drugs are shared among folds. As a result, the standard deviation over the corresponding AUC values is larger in the drug-wise disjoint case (6.0% in disjoint vs 1.5% in non-disjoint),

Table 2. Areas under ROC and Precision-Recall curves under disjoint k-fold cross validation scheme for $k = 2, 5, 10$ and 20 averaged over ten runs.

Number of folds	Mean AUC (%)	S.d. AUC (%)	Mean AUPRC (%)	S.d. AUPRC (%)
2	80.7	0.3	79.3	0.3
5	73.6	0.7	71.9	0.7
10	65.6	0.5	62.8	0.5
20	59.1	0.6	57.0	0.3

suggesting that the predictions are less robust against the partitioning of the drugs in disjoint cross validation.

Compiled mainly via text mining, the drug side effect information in SIDER is prone to a high number of false positives. Given the reduced number of drugs with high similarity, the effect of false positive associations might be more pronounced in the disjoint cross validation than the non-disjoint scenario. Thus, to inspect whether the observed decline in the AUC can be attributed to one of the features used in the classifier –such as side effect based similarity–, we check the contribution of each feature under the disjoint cross validation scheme (Fig. 4). We confirm that this is not the case. In fact, the feature based on side effect similarity is slightly more predictive than the rest (AUC=65% for side effect similarity vs 62% and 61% for chemical and target similarity, respectively), corroborating the promise of side effect profiles to describe similarities between drugs,[4,7,16,17] despite potential noise in the annotations.

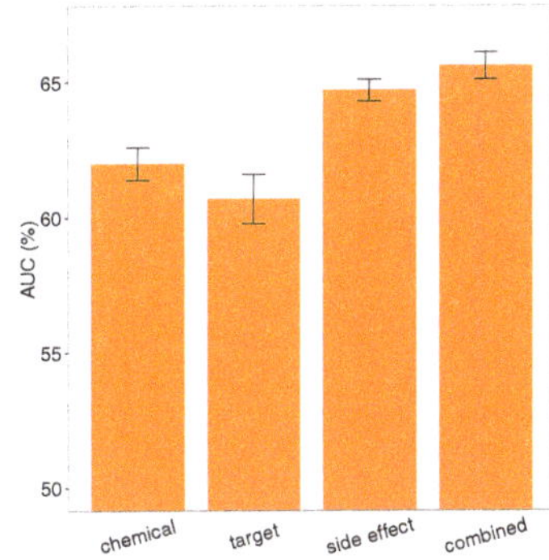

Fig. 4. Prediction accuracy (AUC) when each similarity feature used individually in disjoint cross validation. Error bars show standard deviation of AUC over ten runs of ten-fold cross validation.

2.5. *When similarity does not suffice*

The drop in the AUC confirms that many drug-disease associations are missed when the drugs in the test set have not been seen while training the classifier. For instance, the gold standard data contains several lipid lowering agents indicated for hypercholesterolemia: cholesterol absorption inhibitors (ezetimibe); fibrates (clofibrate, fenofibrate, gemfibrozil); and statins (atorvastatin, fluvastatin, lovastatin, pravastatin, simvastatin). We observe that most of these drugs can be predicted for hypercholesterolemia due to their chemical, target, and side effect based similarity to the other drugs within the same family when drugs are allowed to overlap across cross validation folds. However, when the classifier is trained using disjoint cross validation, most of these drug-disease associations can not be predicted correctly. Likewise, the drugs used for juvenile rheumatoid arthritis (diclofenac, ibuprofen, methotrexate, naproxen, oxaprozin, sulfasalazine, toletin) fail to manifest similarity to other drugs in the cross validation fold, hence missed by the classifier. We also note a similar trend for acute myeloid leukemia drugs (cyclophosphamide, daunorubicin, etoposide, idarubicin, mitoxantrone). In Table 3, we highlight the similarity-based scores of the drug to the other drugs and the probability calculated by the logistic regression classifier in a cross validation fold for several of these drug-disease associations.

Table 3. Similarity scores and logistic regression based probabilities for several known drug-disease associations missed using disjoint cross validation.

	Non-disjoint cross validation				Disjoint cross validation			
Drug	Chemical score	Target score	Side effect score	Probability	Chemical score	Target score	Side effect score	Probability
Hypercholesterolemia drugs								
fenofibrate	0.76	0.71	1.10	0.82	0.57	0	0.46	0.36
lovastatin	1.93	1.97	2.92	0.99	0	0	0	0.14
Juvenile rheumatoid arthritis drugs								
ibuprofen	0.82	3.50	1.08	1.00	0	0.50	0.43	0.43
sulfasalazine	1.39	1.99	0.43	0.96	0	0.50	0.43	0.43
Acute myeloid leukemia drugs								
daunorubicin	1.77	1.50	0	0.87	0	0	0	0.15
idarubicin	0.78	2.00	0.81	0.97	0	0	0	0.14

3. Methods

3.1. *Data sets*

We have retrieved the data set Zhang *et al.* curated for the analysis of the drug repurposing classifier they proposed.[7] They collected 1,007 approved drugs and their targets from DrugBank,[18] the chemical structure information of these drugs from PubChem[19] and the side effect information from SIDER.[20] The drugs were represented by a combination of 775 targets extracted from DrugBank and 881 substructures in PubChem. They were able to map side effects of 888 out of 1,007 drugs using SIDER, covering 61,102 drug-side effect associations coming from 1,385 side effects. The known drug-disease indications span 3,250 associations between 799 drugs and 719 diseases and were extracted from the National Drug File - Reference Terminology (NDF-RT) as suggested in a previous study by Li and Lu.[21] The data set is publicly available online at `http://astro.temple.edu/~tua87106/drugreposition.html`. We used the 536 drugs that were common among chemical, target, side effect, and indication data, corresponding to 2,229 drug-disease associations covering 578 diseases and 40,455 drug-side effect associations covering 1,252 side effects.

3.2. *Drug similarity definitions*

We used the data sets described above to build a drug-drug similarity matrix for each one of the three feature types: chemical substructures, targets, side effects. For each feature type, the drug i was defined by a binary vector $X_i = [x_1, x_2, \ldots, x_n]^T$, corresponding to the existence of the feature for that drug (1 if exists, 0 otherwise). The Pearson product-moment correlation coefficient between two drugs i and j was then calculated using $\rho_{ij} = \frac{C_{ij}}{\sqrt{C_{ii} * C_{jj}}}$, where C_{ij} given by

$$C_{ij} = \text{cov}(X_i, X_j) = \text{E}[(X_i - \text{E}(X_i))(X_j - \text{E}(X_j))]$$

The *corrcoef* function implemented in NumPy Python package was used to calculate correlation coefficients for each drug-drug pair.

3.3. *Similarity-based logistic regression classifier*

We trained a logistic regression model to predict the drug-disease associations based on the drug-drug similarities defined by the targets, chemical substructures, and side effects combined for the 20 most similar drugs to the drug in concern. Therefore, the probability of observing an association between the drug i and the disease j is

$$P(Y_{ij} = 1 | s_{ij}^{\text{chemical}}, s_{ij}^{\text{target}}, s_{ij}^{\text{side effect}}) = \frac{1}{1 + e^{-(\beta_0 + \beta_1 * s_{ij}^{\text{chemical}} + \beta_2 * s_{ij}^{\text{target}} + \beta_3 * s_{ij}^{\text{side effect}})}}$$

where for each feature $f \in$ { chemical, target, and side effect }, the similarity-based drug-disease score s_{ij}^f is defined as

$$s_{ij}^f = \sum_{k \in \text{NN}(i)} \text{sim}^f(i,k) * X(k,j)$$

with $sim^f(i,k)$ being the similarity between two drugs i and k (calculated via Pearson product-moment correlation coefficient as explained above), NN(i) is the set of 20 most similar drugs to drug i (nearest neighbors in the similarity space), and $X(k,j)$ being an indicator function with values 1 if drug k is a known indication for disease j, and 0 otherwise.

We used the *LogisticRegression* function in Scikit-learn Python package with the L2 regularization option and the default values (inverse regularization strength of 1 and stopping tolerance of 0.0001).

3.4. *Prediction accuracy evaluation*

To assess the prediction performance of the logistic regression classifier, we calculated the area under ROC curve (AUC) using k-fold cross validation scheme (e.g., $k = 2, 5, 10, 20$). We used 2,229 known drug-disease associations as the positive instances and marked all remaining possible associations between 536 drugs and 578 diseases ($536 \times 578 - 2,229 = 307,579$ associations) as negative instances. Following the previous studies, we balanced the data set such that it contained twice as many negative instances as positives.[4,7] Thus, in a k-fold cross validation run, we created k groups containing $2,229/k$ positive instances and $2 \times 2,229/k$ negative instances that were randomly chosen among all negative instances. Each fold was used as the test set once, in which all the remaining folds were used to train the classifier. In order to get robust estimates of the AUC, we repeated the cross validation procedure ten times and recorded the mean and the standard deviation of the AUC values over these runs. Note that, the classifier we built relies on both the similarity and the labels of the training drug-disease associations, as we calculate a drug-disease association score using the most similar 20 drugs and their indication information. We made sure not to use the training information in the test phase and calculated the drug-disease association scores within the training and test folds separately. We used the *roc_curve* and *auc* functions in Scikit-learn Python package to first get the true and false positive rates at various cutoffs and then to calculate the AUC using the trapezoidal rule.

3.5. *Stratified disjoint cross validation for defining non-overlapping drug groups*

To investigate the robustness of the drug-disease association classifier in the case of unseen data, we used a disjoint cross validation scheme, in which none of the drugs in one fold appear in another fold. We created cross validation folds such that all the drugs with the same name were in the same fold by first converting the drug's name into an integer value and then taking the modulo (k) of this value (for k-fold cross validation). To produce different groupings at each run, we added a random integer to the integer value of the drug calculated based on its name. The details of the algorithm are as follows:

D: data set containing drug-disease pairs, c: drug, p: disease,
l: label (1 if c is known to be indicated for p, 0 otherwise), k: number of cross validation folds,
$fold$: dictionary containing the fold index of each drug-disease pair
$i :=$ random([0, 100])
$fold := \{\}$
for each $(c, p, l) \in D$ **do**
  $sum := 0$
  for each $x \in$ characters(c) **do**
    $sum := sum +$ to_integer(x)
  $fold(c, p) :=$ modulo($sum + i, k$)
return $fold$

To preserve the balance between positive and negative instances (stratified cross validation), we first grouped the data set into positive ($D^{l=1}$) and negative ($D^{l=0}$) pairs and applied the proposed disjoint fold generation algorithm above to each group.

4. Conclusions

Many recent similarity-based drug repurposing studies reported stunningly high prediction performances, suggesting that drugs can be predicted for novel uses almost with perfect accuracy. Yet, there has not been an observable improvement in the drug discovery in the pharma industry over the past years. We suspect this could be *(i)* because similarity-based methods do not provide insights on the mechanism of action of drugs, failing to explain clinical failures due to the lack of efficacy and safety and/or *(ii)* the reported accuracies being unrealistic due to the underlying validation scheme.

To look into various validation schemes and toward increasing the reproducibility in computational drug repurposing research, we provide a Python-based platform encapsulating machine learning algorithms available in Python Scikit-learn package and propose a disjoint cross fold generation method. This platform allows us to easily evaluate the prediction performance of a logistic regression classifier built using drug chemical, target, and side effect similarity under various experimental settings. Using this platform, we investigate the role of the experimental settings in similarity-based drug repurposing studies in producing optimistic prediction accuracies. In particular, we seek to validate the drug repurposing model when it has never seen the drug beforehand. To test this idea, we use a cross validation approach in which the data is split such that none of the drugs in the test set are in the training set. We show that the high success rate of the model drops sharply under such cross validation setting.

Indeed, in many computational biology problems dealing with paired data, such as predicting drug targets, side effects, drug-drug interactions, protein-protein interactions, functional annotations, and disease-genes, researches aim to leverage machine learning methods using similarity between biomolecules. Our findings suggest that failure to take into account the parity in such data sets can produce optimistic prediction accuracies, supporting earlier studies on drug-target and protein-protein interaction prediction.[13,14] We particularly point out the effect of the training set size when the drugs in the training and test sets do not overlap. Hence, we argue that, though useful in highlighting potential unknown drug-disease pairs, similarity-based methods are likely to be ineffective to explore drugs that are not in the nearby pharmacological space, i.e. the drugs with low chemical similarity or for which target and side effect data are not abundant.

Alternatively, systems-level drug discovery approaches can offer insights on the mechanism of action of the drugs by matching gene expression signatures upon drug treatment to compensate the genomic changes caused by the disease[22,23] or exploiting the paths from drug targets to the genes perturbed in the diseases to explain the efficacy of treatments given the interaction network.[24] Nonetheless, these approaches are still at their infancy and their accuracies remain modest,[24,25] leaving room for improvement.

Acknowledgments

The author is grateful to Dr. Patrick Aloy for hosting EG who is supported by EU-cofunded Beatriu de Pinós incoming fellowship from the Agency for Management of University and

Research Grants (AGAUR) of Government of Catalunya. I also would like to thank Zhang and colleagues for making the data used in their analysis online.

References

1. G. Jin and S. T. C. Wong, Toward better drug repositioning: prioritizing and integrating existing methods into efficient pipelines, *Drug Discovery Today* **19**, 637 (May 2014).
2. R. A. Hodos, B. A. Kidd, K. Shameer, B. P. Readhead and J. T. Dudley, In silico methods for drug repurposing and pharmacology, *Wiley Interdisciplinary Reviews: Systems Biology and Medicine* **8**, 186 (May 2016).
3. S. Vilar and G. Hripcsak, The role of drug profiles as similarity metrics: applications to repurposing, adverse effects detection and drug-drug interactions, *Briefings in Bioinformatics* , p. bbw048 (June 2016).
4. A. Gottlieb, G. Y. Stein, E. Ruppin and R. Sharan, PREDICT: a method for inferring novel drug indications with application to personalized medicine, *Mol Syst Biol* **7**, p. 496 (June 2011).
5. J. Li and Z. Lu, A new method for computational drug repositioning using drug pairwise similarity, in *2012 IEEE International Conference on Bioinformatics and Biomedicine (BIBM)*, October 2012.
6. F. Napolitano, Y. Zhao, V. M. Moreira, R. Tagliaferri, J. Kere, M. D'Amato and D. Greco, Drug Repositioning: A Machine-Learning Approach through Data Integration, *Journal of Cheminformatics* **5**, p. 30 (2013).
7. P. Zhang, P. Agarwal and Z. Obradovic, Computational Drug Repositioning by Ranking and Integrating Multiple Data Sources, in *Machine Learning and Knowledge Discovery in Databases*, eds. H. Blockeel, K. Kersting, S. Nijssen and F. ZeleznyLecture Notes in Computer Science (Springer Berlin Heidelberg, September 2013) pp. 579–594. DOI: 10.1007/978-3-642-40994-3_37.
8. M. Oh, J. Ahn and Y. Yoon, A Network-Based Classification Model for Deriving Novel Drug-Disease Associations and Assessing Their Molecular Actions, *PLOS ONE* **9**, p. e111668 (October 2014).
9. Z. Liu, F. Guo, J. Gu, Y. Wang, Y. Li, D. Wang, L. Lu, D. Li and F. He, Similarity-based prediction for Anatomical Therapeutic Chemical classification of drugs by integrating multiple data sources, *Bioinformatics* **31**, 1788 (June 2015).
10. W. Dai, X. Liu, Y. Gao, L. Chen, J. Song, D. Chen, K. Gao, Y. Jiang, Y. Yang, J. Chen and P. Lu, Matrix Factorization-Based Prediction of Novel Drug Indications by Integrating Genomic Space, *Computational and Mathematical Methods in Medicine* **2015**, p. e275045 (May 2015).
11. H. Luo, J. Wang, M. Li, J. Luo, X. Peng, F.-X. Wu and Y. Pan, Drug repositioning based on comprehensive similarity measures and Bi-Random walk algorithm, *Bioinformatics* , p. btw228 (May 2016).
12. S. Vilar, E. Uriarte, L. Santana, T. Lorberbaum, G. Hripcsak, C. Friedman and N. P. Tatonetti, Similarity-based modeling in large-scale prediction of drug-drug interactions, *Nature Protocols* **9**, 2147 (September 2014).
13. T. Pahikkala, A. Airola, S. Pietil, S. Shakyawar, A. Szwajda, J. Tang and T. Aittokallio, Toward more realistic drugtarget interaction predictions, *Briefings in Bioinformatics* **16**, 325 (March 2015).
14. Y. Park and E. M. Marcotte, A flaw in the typical evaluation scheme for pair-input computational predictions, *Nature methods* **9**, 1134 (December 2012).
15. F. Pedregosa, G. Varoquaux, A. Gramfort, V. Michel, B. Thirion, O. Grisel, M. Blondel, P. Prettenhofer, R. Weiss, V. Dubourg, J. Vanderplas, A. Passos, D. Cournapeau, M. Brucher, M. Perrot and E. Duchesnay, Scikit-learn: Machine Learning in Python, *Journal of Machine Learning Research* **12**, p. 28252830 (October 2011).

16. M. Campillos, M. Kuhn, A.-C. Gavin, L. J. Jensen and P. Bork, Drug Target Identification Using Side-Effect Similarity, *Science* **321**, 263 (July 2008).
17. L. Yang and P. Agarwal, Systematic Drug Repositioning Based on Clinical Side-Effects, *PLOS ONE* **6**, p. e28025 (December 2011).
18. D. S. Wishart, C. Knox, A. C. Guo, D. Cheng, S. Shrivastava, D. Tzur, B. Gautam and M. Hassanali, DrugBank: a knowledgebase for drugs, drug actions and drug targets, *Nucleic Acids Research* **36**, D901 (January 2008).
19. Y. Wang, J. Xiao, T. O. Suzek, J. Zhang, J. Wang and S. H. Bryant, PubChem: a public information system for analyzing bioactivities of small molecules, *Nucleic Acids Research* **37**, W623 (July 2009).
20. M. Kuhn, M. Campillos, I. Letunic, L. J. Jensen and P. Bork, A side effect resource to capture phenotypic effects of drugs, *Molecular Systems Biology* **6**, p. 343 (2010).
21. J. Li, X. Zhu and J. Y. Chen, Building disease-specific drug-protein connectivity maps from molecular interaction networks and PubMed abstracts, *PLoS computational biology* **5**, p. e1000450 (July 2009).
22. J. Lamb, E. D. Crawford, D. Peck, J. W. Modell, I. C. Blat, M. J. Wrobel, J. Lerner, J.-P. Brunet, A. Subramanian, K. N. Ross, M. Reich, H. Hieronymus, G. Wei, S. A. Armstrong, S. J. Haggarty, P. A. Clemons, R. Wei, S. A. Carr, E. S. Lander and T. R. Golub, The Connectivity Map: Using Gene-Expression Signatures to Connect Small Molecules, Genes, and Disease, *Science* **313**, 1929 (September 2006).
23. M. Sirota, J. T. Dudley, J. Kim, A. P. Chiang, A. A. Morgan, A. Sweet-Cordero, J. Sage and A. J. Butte, Discovery and preclinical validation of drug indications using compendia of public gene expression data, *Science Translational Medicine* **3**, p. 96ra77 (August 2011).
24. E. Guney, J. Menche, M. Vidal and A.-L. Barabási, Network-based in silico drug efficacy screening, *Nature Communications* **7**, p. 10331 (February 2016).
25. J. Cheng, L. Yang, V. Kumar and P. Agarwal, Systematic evaluation of connectivity map for disease indications, *Genome Medicine* **6** (December 2014).

EMPOWERING MULTI-COHORT GENE EXPRESSION ANALYSIS TO INCREASE REPRODUCIBILITY

WINSTON A HAYNES[1,2,3], FRANCESCO VALLANIA[1], CHARLES LIU[1,4], ERIKA BONGEN[1], AURELIE TOMCZAK[1,3], MARTA ANDRES-TERRÈ[1], SHANE LOFGREN[1], ANDREW TAM[1], COLE A DEISSEROTH[1,4], MATTHEW D LI[1], TIMOTHY E SWEENEY[1,3], and PURVESH KHATRI[1,3,*]

[1]*Stanford Institute for Immunity, Transplantation, and Infection, Stanford University*
[2]*Biomedical Informatics Training Program, Stanford University*
[3]*Stanford Center for Biomedical Informatics Research, Stanford University*
[4]*Stanford Institutes of Medicine Research Program, Stanford University*
Stanford, CA 94305 USA
[*]*E-mail: pkhatri@stanford.edu*

A major contributor to the scientific reproducibility crisis has been that the results from homogeneous, single-center studies do not generalize to heterogeneous, real world populations. Multi-cohort gene expression analysis has helped to increase reproducibility by aggregating data from diverse populations into a single analysis. To make the multi-cohort analysis process more feasible, we have assembled an analysis pipeline which implements rigorously studied meta-analysis best practices. We have compiled and made publicly available the results of our own multi-cohort gene expression analysis of 103 diseases, spanning 615 studies and 36,915 samples, through a novel and interactive web application. As a result, we have made both the process of and the results from multi-cohort gene expression analysis more approachable for non-technical users.

Keywords: Multi-cohort Analysis; Meta-Analysis; Gene Expression; Reproducibility; Web Application; Software

1. Introduction

Prior to translation of the results of a biological experiment into clinical practice, they must be replicated and validated in multiple independent cohorts. However, the majority of findings fail to validate, leading to a 'reproducibility crisis' in science.[1,2] One of the factors in this lack of reproducibility is that traditional, single cohort studies do not represent the heterogeneity observed in the real world patient population.[3] As a result, observed and reported effects are often specific to a population subset instead of generalizable across the population.

More than two million publicly available gene expression microarrays present novel opportunities to incorporate the real-world heterogeneity observed in patient populations into analysis.[4,5] However, the biological (tissue, treatment, demographics) and technical (experimental protocol, microarray) heterogeneity present in such data poses a daunting challenge for their integration and reuse. Consequently, many tools, which allow reuse of these data, are unable to combine evidence across multiple data sets and place that burden on the end user, leading to under-utilization of these datasets.[6,7]

Previously, we have described a novel multi-cohort analysis framework for integrating multiple heterogeneous datasets to identify robust and reproducible signatures by leveraging the biological and technical heterogeneity in these datasets. We have repeatedly demonstrated the utility of our framework for identifying novel diagnostic and prognostic biomarkers, drug targets, and repurposing FDA-approved drugs in diverse diseases, including organ transplantation, cancer, infection, and neurodegenerative diseases.[8-16] In each of these analyses, we analyzed more than a thousand human samples from more than 10 independent cohorts to generate and validate data-driven hypotheses. Many of these results also been further validated in experimental settings.[8,11,16] These results have further demonstrated the ability of our framework to create "Big Data" by combining multiple smaller studies that are collectively representative of the real word patient population heterogeneity.

We recently published a systematic comparison of gene expression meta-analysis to evaluate existing tools, including GeneMeta, MAMA, MetaDE, ExAtlas, rmeta, and metafor,[17-21] and described best practice guidelines for gene expression meta-analysis.[22] While these existing packages perform both generic and gene expression meta-analysis, none provide coverage of the entire gene expression meta-analysis workflow: downloading data from public repositories, rigorously implementing gene expression meta-analysis best practices, and providing visualizations of the final results.

2. Multi-Cohort Gene Expression Analysis with MetaIntegrator

Despite its demonstrated utility in identifying robust, reproducible, and biologically as well as clinically relevant disease signatures, our multi-cohort analysis framework has previously required manual dataset download, pipeline set up, and visualization generation. To lower this barrier to entry, we have developed MetaIntegrator, an R package that automates most of the multi-cohort analysis framework. Our package guides the user from data download to execution of statistical analysis to evaluation of the results [Figure 1].

2.1. *Data Processing*

The first step in the multi-cohort analysis is downloading the requisite experimental information, notably the class labels (case or control), the gene expression data, and any interesting phenotypic information about the samples. Since we have found that most users will download data from the NCBI's Gene Expression Omnibus (GEO), we have integrated an automatic downloading and processing of GEO data into our analysis pipeline. MetaIntegrator will automatically download the expression data and all available annotations, perform sanity checks that the data have been appropriately normalized, and compile the data into the MetaIntegrator object format.

2.2. *Multi-cohort Analysis*

2.2.1. *Combining effect sizes*

Our meta-analysis approach computes an Hedges g effect size for each gene in each dataset defined as:

$$g = J\frac{\bar{X}_1 - \bar{X}_0}{\sqrt{\frac{(n_1-1)S_1^2+(n_0-1)S_0^2}{n_1+n_0-2}}} \tag{1}$$

where $\bar{X}_1$ and $\bar{X}_0$ are the average expression for cases and controls, S_1 and S_0 are the standard deviations for cases and controls, and n_1 and n_0 are the number of cases and controls.[8,23] J is

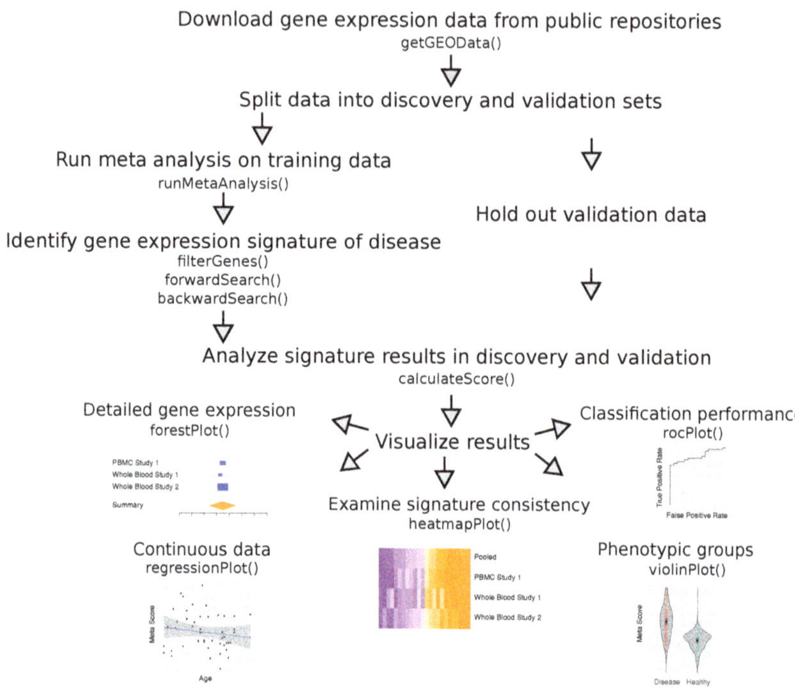

Fig. 1. Gene expression meta-analysis workflow with MetaIntegrator utility functions.

the Hedges' g correction factor, which is computed as:

$$J = 1 - \frac{3}{4df - 1} \tag{2}$$

where *df* are the degrees of freedom.

To pool these effect sizes across datasets, the summary effect size g_s is computed using a random effect model as:

$$g_s = \frac{\sum_i^n W_i g_i}{\sum_i^n W_i} \tag{3}$$

where n is the number of studies, g_i is the Hedges' g of that gene within dataset i, W_i is a weight equal to $1/(V_i + T^2)$, V_i is the variance of that gene within a given dataset i, and T^2 is the inter-dataset variation as estimated by the DerSimonian-Laird method.[23,24] The standard error for the summary effect size is $SE_{g_s} = \sqrt{\frac{1}{\sum_i^n W_i}}$. Given g_s and SE_{g_s}, we calculate a p-value based on a standard normal distribution and perform a Benjamini-Hochberg FDR correction for multiple hypothesis testing across all genes.[25]

2.2.2. *Heterogeneity of effect size*

We calculate Cochrane's Q value for evaluating heterogeneity of effect size estimates between studies:

$$Q = \sum_{i=1}^{n} W_i \left(g_i - g_s\right)^2 \tag{4}$$

where W_i, g_i, and g_s are the same as above.[23] The p-value of Cochrane's Q is calculated against a chi-squared distribution and adjusted for multiple hypothesis testing using the Benjamini-Hochberg FDR method across all genes.[25] A statistically significant Cochrane's Q indicates significant heterogeneity of effect sizes between studies.

2.2.3. *Combining p-values*

We use Fisher's method for combining p-values across studies.[26] We calculate the log sum of p-values that each gene is up-regulated as:

$$F_{\text{up}} = -2 \sum_{i=1}^{n} \log(p_i) \tag{5}$$

where n is the number of studies and p_i is the t-test p-value that the gene of interest is up-regulated in study i. Similarly, we calculate F_{down} as the log-sum of p-values that each gene is down-regulated.

For each gene, we calculate the p-value of F_{up} and F_{down} under a chi-squared distribution and perform a Benjamini-Hochberg FDR correction across all genes.[25]

2.3. *Signature Selection*

Once meta-analysis is performed, a subset of genes must be identified as the disease signature. MetaIntegrator allows the user to identify these genes by varying the filtering parameters based on gene effect size, effect size false discovery rate, Fisher's method false discovery rate, heterogeneity of effect size, and the number of studies in which the gene was present. In order to avoid disproportionate influence of a single study, MetaIntegrator allows the user only include genes which were similarly significant across all leave-one-dataset-out analyses. By varying these criterion, the user may control whether they identify a larger set of genes, which may be ideal for understanding molecular pathogenesis and identifying drug targets, or a smaller set of genes, which may be optimal developing a parsimonious clinical diagnostic.

For users that are particularly interested in developing a powerful diagnostic, we have integrated forward and backward search, which reduce gene set size to optimize the area under the receiver operating characteristic curve on the training data.[10]

2.4. *Score Calculation*

For a set of signature genes, a signature score can be computed for every sample, i, as:

$$S_i = \left(\prod_{\text{gene}\in\text{pos}} x_i(\text{gene})\right)^{\frac{1}{\|\text{pos}\|}} - \left(\prod_{\text{gene}\in\text{neg}} x_i(\text{gene})\right)^{\frac{1}{\|\text{neg}\|}} \tag{6}$$

where pos and neg are the sets of positive and negative genes, respectively, and $x_i(\text{gene})$ is the expression of any particular gene in sample i (a positive score indicates an association with cases and a negative score with controls). This score S_i is normalized to a z-score to center the samples for each study around zero.

2.5. *Visualization*

With scores calculated for each sample, we are able to visualize comparisons of cases vs. controls, regression of continuous variables against the score, and consistency of gene expression across datasets. Some of the built in visualizations, in counter-clockwise order from Figure 1:

- **Forest plots.** Examine the effect sizes and standard errors for a single gene across studies, including the summary effect size.
- **Regression plots.** Evaluate the relationship of the signature score with continuous variables like clinical severity and time.
- **Heatmap plots.** Observe consistency of differential expression for all signature genes across studies.
- **Violin plots.** Compare signature scores across categorical variables like disease subtypes, treatment protocols, and demographic groups.
- **ROC plots.** Evaluate classification performance for signature score on a single dataset in terms of specificity and sensitivity.

3. Data-Driven Biological Hypotheses with MetaSignature

We have created MetaSignature (`http://metasignature.stanford.edu`), a web application which empowers researchers to generate data-driven hypotheses by enabling access to the results of our multi-cohort gene expression analysis framework. We focused on enabling intuitive data access for researchers with specific interest in either a disease, a gene, or several genes, while requiring little or no analytic background.

3.1. *Data*

Thus far, we have aggregated 615 gene expression studies composed of more than 35,000 human samples with approximately 1.5 billion data points from 103 diseases, a number which we will continue to grow. For each disease, we applied our multi-cohort analysis approach to compute the gene expression differences between the manually curated cases and controls. To perform these multi-cohort analyses, we searched for relevant studies in GEO, identified cases and controls in every study, and calculated disease effect sizes using the MetaIntegrator R

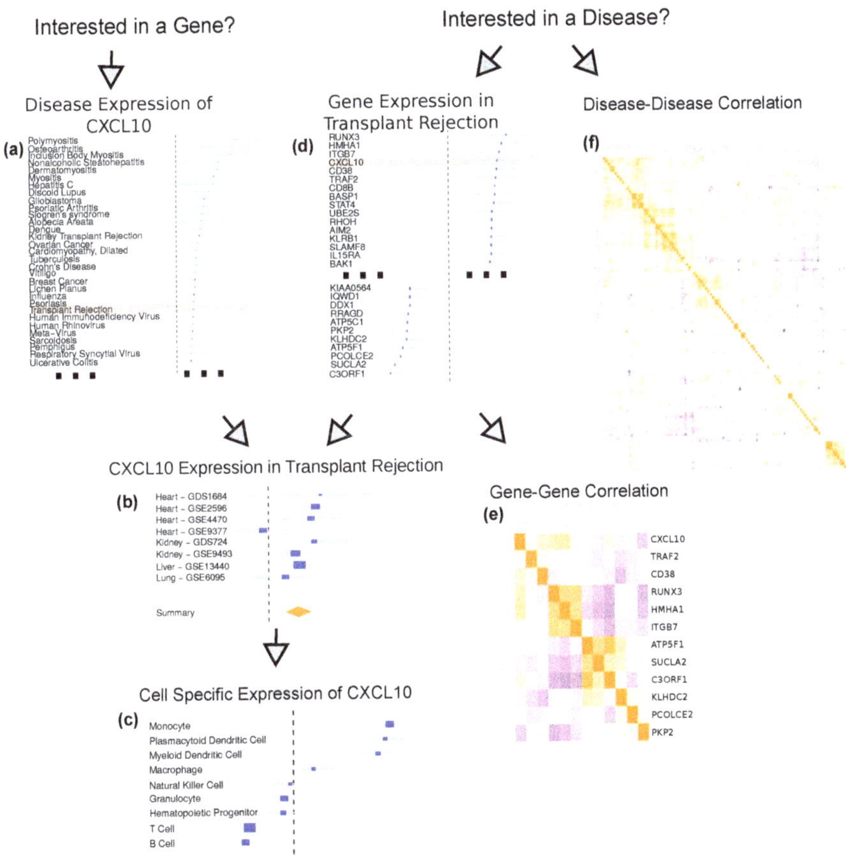

Fig. 2. Diagram of the MetaSignature web application.

package. We stored the multi-cohort analysis results in a MySQL database for rapid retrieval. As more studies are incorporated into our database, we recalculate the disease summary effect sizes.

3.2. *Gene-centric Analysis*

For researchers that are interested in the expression of a particular gene, MetaSignature provides visualizations that allow researchers to quickly identify the diseases in which specified gene is most differentially expressed [Figure 2a], study-level data of the gene expression in particular diseases [Figure 2b], and cell type-specific gene expression patterns [Figure 2c].

For instance, consider a researcher who has developed a drug, such as atorvastatin, that effectively reduces plasma levels of *CXCL10*, and seeks to identify the most promising clinical applications. Using MetaSignature, she determines *CXCL10* is significantly up-regulated in transplant rejection [Figure 2a]. A drilldown further identifies eight separate studies that have measured *CXCL10* in transplant rejection, indicating a highly positive effect size in all except one of these studies [Figure 2b]. The researcher further observes that *CXCL10* is up-regulated in monocytes, compared to other immune cell types. [Figure 2c]. Taken together, these findings would motivate a clinical investigation of the use of a *CXCL10* inhibitor, such as atorvastatin, in monocytes of patients at risk for transplant rejection. We have already verified this data-driven hypothesis in mouse models and patient electronic health records, where, in both cases, atorvastatin increases graft survival.[8]

Beyond single gene analysis, MetaSignature empowers users to examine gene sets in terms of correlation of those genes based on their disease effect sizes [Figure 2e] and correlation of diseases based on expression of that set of genes [similar to Figure 2f]. These visualizations enable dissection of positively- and negatively-correlated members of gene families.

3.3. *Disease-centric Analysis*

If a researcher is more interested in a particular disease, then MetaSignature enables identification of genes that are most up- or down-regulated in that disease [Figure 2d] and exploration of that disease's relationship to other diseases based on gene expression [Figure 2f]. When we compute disease-disease correlation based on gene expression data, we observe clustering patterns that map to established disease categories.

To follow our example from the gene-centric analysis, consider a researcher who is interested in improving transplant rejection outcomes. To gain a global understanding of transplant rejection, the researcher observes that transplant rejection falls into a cluster of inflammatory diseases, including discoid lupus, Crohn's disease, and interstitial cystitis [Figure 2f]. By examining the transplant rejection expression data in MetaSignature, he or she would recognize that *CXCL10*, a chemokine important in inflammatory response, is one of the most up-regulated genes in transplant rejection [Figure 2d].[27] After verifying that this observation is consistent across studies [Figure 2b], the researcher identifies that *CXCL10* is a reasonable target for therapeutic inhibition in transplant rejection. Looking at other genes which are up- and down-regulated in transplant rejection, he or she recognizes that CXCL10 expression is in a positively correlated with several other genes, including *TRAF2* and *CD38* [Figure 2e]. Collectively from these observations, the researcher has learned that transplant rejection is related to inflammatory diseases, which is consistent with the observed up-regulation of CXCL10, an inflammatory chemokine. As noted in the gene-centric analysis above, we have observed increased graft survival through administration of atorvastatin.[8]

4. Discussion

The reproducibility crisis in biomedical research has led to erroneous conclusions and wasted resources. Here, we present a vertically integrated platform that can both assist with gene expression multi-cohort analysis (MetaIntegrator) and provide aggregated results for users who wish to rapidly test hypotheses (MetaSignature). By leveraging the growing public data available for study, this new resource can drastically reduce the time and effort for biological hypothesis testing across numerous studies and diseases. While many software packages exist for similar analyses,[17–21] ours offers simple, custom software for plotting and analysis, automated downloading of data from GEO, and integration to the MetaSignature database.

Our package is complementary to the recently published OMiCC platform, which enables curation and meta-analysis of GEO studies.[28] OMiCC relies on the RankProd package for performing meta-analysis using rank-based statistics for identifying differentially expressed genes.[29] While others have provided thorough comparisons of the different meta-analysis methods, the most notable difference between RankProd and MetaIntegrator is that rank-based statistics fail to produce a summary effect size across multiple studies.[30,31] By leveraging our MetaIntegrator package, OMiCC could produce differential gene expression profiles across multiple studies instead of internal to single studies.

Although MetaIntegrator is currently applicable to microarray gene expression data, we plan to expand the MetaIntegrator package to handle the count data which is generated by RNAseq experiments. Additionally, we intend to enable download from additional data repositories including ArrayExpress and, once RNAseq processing is implemented, Sequence Read Archive.[?,5]

Our work promises to increase reproducibility of research for both data analysts and wet lab researchers. For data analysts, we have made multi-cohort gene expression analysis publicly available through a straightforward R package. By performing integrative, multi-cohort analyses, these analysts will generate more reproducible results. For wet lab researchers, we are empowering data-driven hypotheses prior to experimentation. Rather than performing broad assays to identify disease related genes, researchers can focus on performing targeted experiments on genes which are reproducible across cohorts.

5. Package and Source Code Distribution

The MetaIntegrator R package, including an introductory vignette, may be installed using the following command in R:

```
install.packages("MetaIntegrator")
```

The source code for MetaIntegrator is available at:
https://cran.rstudio.com/web/packages/MetaIntegrator/
MetaSignature was developed using R and Shiny and is hosted at:
http://metasignature.stanford.edu/

6. Acknowledgements

We thank Alex Schrenchuk for computer support. WAH is funded by the National Science Foundation Graduate Research Fellowship under Grant No. DGE-114747. FV is funded by the National Institute of Health K12 Career Award 5K12HL120001-02. MAT is funded by La Caixa Foundation. EB is funded by Gabilan Fellowship. PK is funded by the National

Institute of Allergy and Infectious Diseases grants 1U19AI109662, U19AI057229, U54I117925, and U01AI089859.

References

1. J. P. A. Ioannidis, *PLoS medicine* **2**, p. e124 (August 2005).
2. M. Baker, *Nature* , 452 (2016).
3. J. P. Ioannidis, E. E. Ntzani, T. A. Trikalinos and D. G. Contopoulos-Ioannidis, *Nature Genetics* **29**, 306 (November 2001).
4. R. Edgar, *Nucleic Acids Research* **30**, 207 (January 2002).
5. A. Brazma, H. Parkinson, U. Sarkans, M. Shojatalab, J. Vilo, N. Abeygunawardena, E. Holloway, M. Kapushesky, P. Kemmeren, G. G. Lara, A. Oezcimen, P. Rocca-Serra and S.-A. Sansone, *Nucleic Acids Research* **31**, 68 (January 2003).
6. J. M. Engreitz, R. Chen, A. A. Morgan, J. T. Dudley, R. Mallelwar and A. J. Butte, *Bioinformatics* **27**, 3317 (December 2011).
7. R. Petryszak, T. Burdett, B. Fiorelli, N. A. Fonseca, M. Gonzalez-Porta, E. Hastings, W. Huber, S. Jupp, M. Keays, N. Kryvych, J. McMurry, J. C. Marioni, J. Malone, K. Megy, G. Rustici, A. Y. Tang, J. Taubert, E. Williams, O. Mannion, H. E. Parkinson and A. Brazma, *Nucleic Acids Research* **42**, D926 (January 2014).
8. P. Khatri, S. Roedder, N. Kimura, K. De Vusser, A. A. Morgan, Y. Gong, M. P. Fischbein, R. C. Robbins, M. Naesens, A. J. Butte and M. M. Sarwal, *The Journal of experimental medicine* **210**, 2205 (October 2013).
9. R. Chen, P. Khatri, P. K. Mazur, M. Polin, Y. Zheng, D. Vaka, C. D. Hoang, J. Shrager, Y. Xu, S. Vicent, A. J. Butte and E. A. Sweet-Cordero, *Cancer Research* **74**, 2892 (May 2014).
10. T. E. Sweeney, A. Shidham, H. R. Wong and P. Khatri, *Science Translational Medicine* **7**, p. 287ra71 (May 2015).
11. M. Andres-Terre, H. M. McGuire, Y. Pouliot, E. Bongen, T. E. Sweeney, C. M. Tato and P. Khatri, *Immunity* **43**, 1199 (December 2015).
12. M. D. Li, T. C. Burns, A. A. Morgan and P. Khatri, *Acta neuropathologica communications* **2**, p. 93 (January 2014).
13. P. K. Mazur, N. Reynoird, P. Khatri, P. W. T. C. Jansen, A. W. Wilkinson, S. Liu, O. Barbash, G. S. Van Aller, M. Huddleston, D. Dhanak, P. J. Tummino, R. G. Kruger, B. A. Garcia, A. J. Butte, M. Vermeulen, J. Sage and O. Gozani, *Nature* **advance on** (May 2014).
14. P. K. Mazur, A. Herner, S. S. Mello, M. Wirth, S. Hausmann, F. J. Sánchez-Rivera, S. M. Lofgren, T. Kuschma, S. A. Hahn, D. Vangala, M. Trajkovic-Arsic, A. Gupta, I. Heid, P. B. Noël, R. Braren, M. Erkan, J. Kleeff, B. Sipos, L. C. Sayles, M. Heikenwalder, E. Heß mann, V. Ellenrieder, I. Esposito, T. Jacks, J. E. Bradner, P. Khatri, E. A. Sweet-Cordero, L. D. Attardi, R. M. Schmid, G. Schneider, J. Sage and J. T. Siveke, *Nature Medicine* **21**, 1163 (September 2015).
15. T. E. Sweeney, L. Braviak, C. M. Tato and P. Khatri, *The Lancet Respiratory Medicine* **4**, 213 (2016).
16. T. E. Sweeney, H. R. Wong and P. Khatri, *Science translational medicine* **8**, p. 346ra91 (July 2016).
17. L. Lusa, R. Gentleman and M. Ruschhaupt, *GeneMeta: MetaAnalysis for High Throughput Experiments.*
18. I. Ihnatova., *MAMA: Meta-Analysis of MicroArray*, (2013).
19. T. Lumley, *rmeta: Meta-analysis*, (2012).
20. X. Wang, D. D. Kang, K. Shen, C. Song, S. Lu, L. C. Chang, S. G. Liao, Z. Huo, S. Tang, Y. Ding, N. Kaminski, E. Sibille, Y. Lin, J. Li and G. C. Tseng, *Bioinformatics* **28**, 2534 (2012).

21. A. A. Sharov, D. Schlessinger and M. S. H. Ko, *Journal of Bioinformatics and Computational Biology* **13**, p. 1550019 (2015).
22. T. E. Sweeney, W. A. Haynes, F. Vallania, J. P. Ioannidis and P. Khatri, *Nucleic acids research* , p. gkw797 (September 2016).
23. M. Borenstein, L. V. Hedges, J. P. T. Higgins and H. R. Rothstein, *Introduction to Meta-Analysis* 2009.
24. R. DerSimonian and R. Kacker, *Contemporary Clinical Trials* **28**, 105 (2007).
25. Y. Benjamini and Y. Hochberg, *Journal of the Royal Statistical Society. Series B (Methodological)* **57**, 289 (1995).
26. R. Fisher, *Statistical methods for research workers*, 1925).
27. L. F. Neville, G. Mathiak and O. Bagasra, *Cytokine & Growth Factor Reviews* **8**, 207 (September 1997).
28. N. Shah, Y. Guo, K. V. Wendelsdorf, Y. Lu, R. Sparks and J. S. Tsang, *Nature Biotechnology* (June 2016).
29. F. Hong, R. Breitling, C. W. McEntee, B. S. Wittner, J. L. Nemhauser and J. Chory, *Bioinformatics (Oxford, England)* **22**, 2825 (November 2006).
30. L.-C. Chang, H.-M. Lin, E. Sibille and G. C. Tseng, *BMC bioinformatics* **14**, p. 368 (2013).
31. A. Ramasamy, A. Mondry, C. C. Holmes and D. G. Altman, *PLoS medicine* **5**, p. e184 (September 2008).

RABIX: AN OPEN-SOURCE WORKFLOW EXECUTOR SUPPORTING RECOMPUTABILITY AND INTEROPERABILITY OF WORKFLOW DESCRIPTIONS

GAURAV KAUSHIK[†]
Seven Bridges Genomics
1 Main Street, Cambridge, MA 02140, USA
Email: gaurav@sevenbridges.com

SINISA IVKOVIC
Seven Bridges Genomics
Omladinskih brigada 90g, Belgrade, Republic of Serbia
Email: sinisa.ivkovic@sevenbridges.com

JANKO SIMONOVIC
Seven Bridges Genomics
Omladinskih brigada 90g, Belgrade, Republic of Serbia
Email: janko.simonovic@sevenbridges.com

NEBOJSA TIJANIC
Seven Bridges Genomics
Omladinskih brigada 90g, Belgrade, Republic of Serbia
Email: boysha@sevenbridges.com

BRANDI DAVIS-DUSENBERY
Seven Bridges Genomics
1 Main Street, Cambridge, MA 02140, USA
Email: brandi@sevenbridges.com

DENIZ KURAL
Seven Bridges Genomics
1 Main Street, Cambridge, MA 02140, USA
Email: deniz.kural@sevenbridges.com

As biomedical data has become increasingly easy to generate in large quantities, the methods used to analyze it have proliferated rapidly. Reproducible and reusable methods are required to learn from large volumes of data reliably. To address this issue, numerous groups have developed workflow specifications or execution engines, which provide a framework with which to perform a sequence of analyses. One such specification is the Common Workflow Language, an emerging standard which provides a robust and flexible framework for describing data analysis tools and workflows. In addition, reproducibility can be furthered by executors or workflow engines which interpret the specification and enable additional features, such as error logging, file organization, optim[1]izations to computation and job scheduling, and allow for easy computing on large volumes of data. To this end, we have developed the Rabix Executor[a], an open-source workflow engine for the purposes of improving reproducibility through reusability and interoperability of workflow descriptions.

[1]This project has been funded in whole or in part with Federal funds from the National Cancer Institute, National Institutes of Health, Department of Health and Human Services, under Contract No. HHSN261201400008C.
[†] Corresponding author
[a]The Rabix Executor is available on GitHub: http://github.com/rabix/bunny

1. Introduction

Reproducible analyses require the sharing of data, methods, and computational resources.[1] The probability of reproducing a computational analysis is increased by methods that support replicating each analysis and the capability to reuse code in multiple environments. In recent years, the practice of organizing data analysis via computational workflow engines or accompanying workflow description languages has surged in popularity as a way to support the reproducible analysis of massive genomics datasets.[2,3] Robust and reliable workflow systems share three key properties: flexibility, portability, and reproducibility. Flexibility can be defined as the ability to gracefully handle large volumes of data with multiple formats. Adopting flexibility as a design principle for workflows ensures that multiple versions of a workflow are not required for different datasets and a single workflow or pipeline can be applied in many use cases. Together, these properties reduce the software engineering burden accompanying large-scale data analysis. Portability, or the ability to execute analyses in multiple environments, grants researchers the ability to access additional computational resources with which to analyze their data. For example, workflows highly customized for a particular infrastructure make it challenging to port analyses to other environments and thus scale or collaborate with other researchers. Well-designed workflow systems must also support reproducibility in science. In the context of workflow execution, computational reproducibility (or recomputability) can be simply defined as the ability to achieve the same results on the same data regardless of the computing environment or when the analysis is performed. Workflows and the languages that describe them must account for the complexity of the information being generated from biological samples and the variation in the computational space in which they are employed. Without flexible, portable, and reproducible workflows, the ability for massive and collaborative genomics projects to arrive at synonymous or agreeable results is limited.[4,5]

Biomedical or genomics workflows may consist of dozens of tools with hundreds of parameters to handle a variety of use cases and data types. Workflows can be made more flexible by allowing for transformations on inputs during execution or incorporating metadata, such as sample type or reference genome, into the execution. They can allow for handling many use cases, such as dynamically generating the appropriate command based on file type or size, without needing to modify the workflow description to adjust for edge cases. Such design approaches are advantageous as they alleviate the software engineering burden and thus the accompanying probability of error associated with executing extremely complex workflows on large volumes of data. In addition, as the complexity of an individual workflow increases to handle a variety of use cases or criteria, it becomes more challenging to optimally compute with it. For example, analyses may incorporate nested workflows, business logic, memoization or the ability to restart failed workflows, or require parsing of metadata -- all of which compound the challenges in optimizing workflow execution.

As a result of the increasing volume of biomedical data, analytical complexity, and the scale of collaborative initiatives focused on data analysis, reliable and reproducible analysis of biomedical data has become a significant concern. Workflow descriptions and the engines that interpret and execute them must be able to support a plethora of computational environments and ensure reproducibility and efficiency while operating across them. It is for this reason that we have developed the Rabix Executor (on GitHub as Project "Bunny")[a], an open-source workflow engine designed to support computational reproducibility/recomputability through the use of standard workflow descriptions, a software model that supports metadata integration, provenance over file organization, the ability to reuse workflows efficiently, and which combines an array of optimizations used separately in existing workflow execution methods.[6–12]

For the 1.0 release of the Rabix Executor (or Rabix), we've focused on supporting the Common Workflow Language (CWL), an open, community-driven specification for describing tools and workflows with a focus on features that support reproducibility.[2] The Common Workflow Language is used to describe individual "processes" or "applications", which can be either a single tool or an entire workflow. Workflows are described as a series of "steps," each of which is a single tool or another, previously-described workflow. Each step in the workflow has a set of "ports" which represent data elements that are either inputs or outputs of the tool. A single port represents a specific data element that is required for execution of the tool or is the result of its execution. For data elements that are passed between applications, there must be an output port from the upstream application and a complementary input port on the downstream application.

CWL is designed to be extensible, so the specification may grow based on the community's needs. However, the software model for Rabix was designed with an abstract workflow execution model to anticipate support for additional workflow languages or syntax used by other workflow engines.

2. Software model used by Rabix to interpret and compute workflows

The Rabix Executor allows users to execute applications described by a workflow description language. First, the workflow description is submitted to the engine. Then, the Rabix engine interprets the workflow description and translates it into discrete computational processes or "jobs." Finally, the jobs are queued to a backend or computational infrastructure, such as a local machine, cluster, or cloud instances, for scheduling and execution. Each component of the executor (frontend, bindings, engine, queue, backend) is abstracted from each other to enable complete modularity; Developers are able to design custom frontends (e.g. command line or graphical user interface), bindings for the engine to parse different workflow languages, use the queuing protocol of their choice, and submit computational jobs to different backends. This flexible software model means that Rabix can be modified to perform data analysis on many different infrastructures as desired by the user or developer and achieve identical results or incorporate tools described by different languages or syntaxes into a single workflow.

3. Abstract representation of data analysis workflows in Rabix

Computational workflows are frequently understood as a directed acyclic graph (DAG)[3,13,14], a kind of finite graph which contains no cycles and which must be traversed in a specific direction. In this representation, each node is either an individual executable command, a "nested" workflow, or a set of commands that can be executed in parallel. The edges in the DAG represent execution variables (data elements such as files or parameters) which pass from upstream nodes to downstream ones.

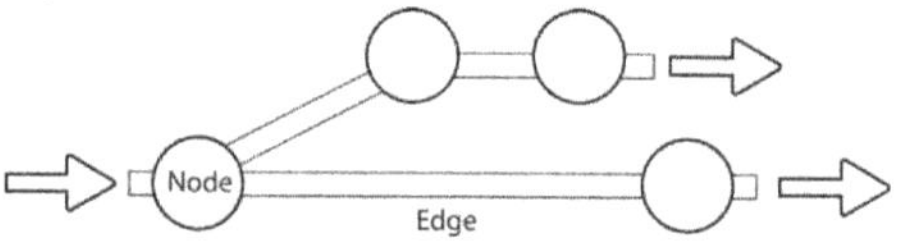

Figure 1. Illustration of a directed acyclic graph (DAG). The DAG may be traversed from left-to-right, moving from node-to-node along the edges that connect them.

Workflows can be described as machine-readable serialized data objects in either a general-purpose programming language (GPL), domain-specific language (DSL), or serialized object models for workflow description.[2,9,15] For example, an object model-based approach may describe the steps in a workflow in JSON format with a custom syntax. This workflow description can then be parsed by an engine or executor to create the DAG representation of the workflow. The executor may then translate the directions for workflow execution to actionable jobs in which data is analyzed on a computational infrastructure, such as a cloud computing instance, a high-performance computing cluster, or a personal computer.

A primary design constraint of the Rabix executor is to abstract components of a workflow to a data model that is comprehensive enough to allow for mapping the syntax of different workflow systems, whether they are DSLs or serialized data objects. In this way, tools and workflows from different systems can be used together in a single workflow.

3.1. *General structure of a workflow execution*

There are three general steps in preparing a workflow for execution: interpretation of a machine-readable workflow description, generation of the workflow DAG, and finally decomposition into individual jobs that can be scheduled for execution. At the beginning of execution, a workflow engine or interpreter is provided with the workflow description and the required inputs for execution of the workflow, such as parameters and file paths (Fig. 2a). The workflow description object is then parsed and a DAG is created (Fig. 2b), which contains the initial set of nodes and edges required for computation.

In addition to representing the steps in the workflow as a DAG (Fig. 2c), certain workflow ontologies model computational jobs as a composite (tree) pattern in which there are “parent nodes” (workflows), which can contain multiple executables or “leaf nodes” or other "parent" nodes (Fig. 2d).[16–20] The Rabix engine extends this model by generalizing “parent” nodes to include groups of jobs, such as when parallelization is possible at that node. It is important to note that the “parent-child” terminology is also applied to relations between individual workflow nodes by the Toil project, an executor which can also interpret Common Workflow Language.[10] However, Rabix uses these terms to refer to computational "jobs" and "subjobs", e.g. a “nested” workflow node is a child of a workflow and can be decomposed into an array of “subjobs”. The engine handles the "execution" or parsing of these parent jobs, while leaves are queued for scheduling and execution on a backend. This model allows for more efficient resolution of DAG features such as nodes in which steps can be parallelized or are nested. It also maintains a one-to-one mapping between the internal DAG representation and the workflow description supplied by the author.

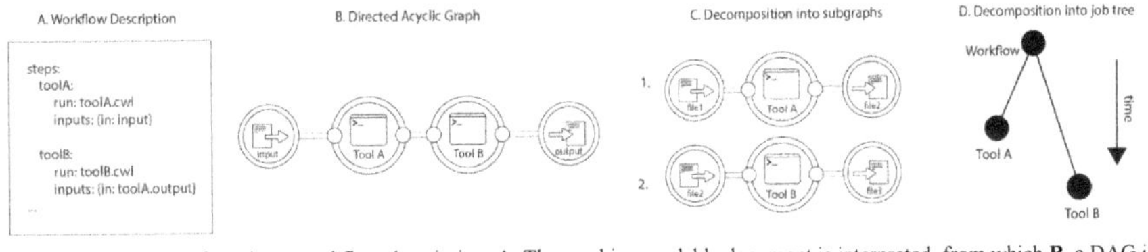

Figure 2. The process of parsing a workflow description. **A.** The machine-readable document is interpreted, from which **B.** a DAG is produced. From the DAG, **C.** subgraphs representing computational jobs that can be sent to backends for scheduling/execution and **D.** a job tree is resolved, which identifies “parent” and “leaf” nodes. Each leaf represents an individual job.

4. Optimization of CWL workflows via DAG transformations

The Rabix Executor began its development by examining how to interpret Common Workflow Language and interoperate on different versions or earlier drafts, in a way that is extensible to future versions and other workflow syntaxes. Rabix currently supports tools and workflows described in CWL Draft 2, Draft 3, and version 1.0, either individually or in combination.

When a CWL workflow is represented as a DAG, applications become nodes and edges indicate the flow of data elements between ports of linked tools. In the case of a simple workflow, there are no possible transformations of the DAG; each node represents a single command line execution and all data elements are simply passed from tool-to-tool as-is (Fig. 3).

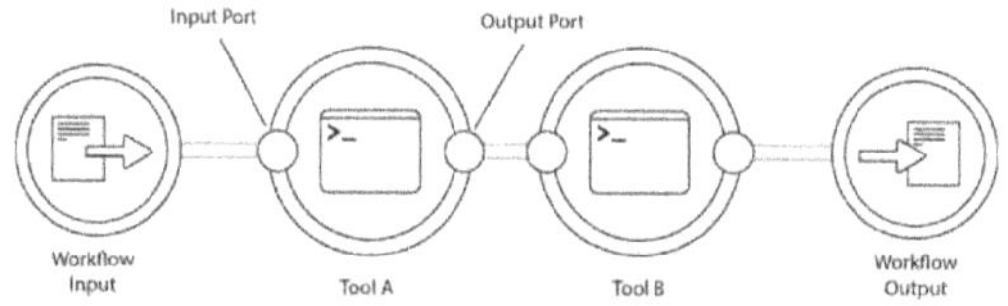

Figure 3. A DAG created from a workflow described by the Common Workflow Language which contains two tools (**A, B**). Tools have input and output ports, which define discrete data elements that are passed downstream along the edges of the DAG.

Additionally, CWL workflows can be designed such that data elements and the execution itself can be transformed during runtime. Developers are given several options for describing workflows which can enhance their utility and flexibility in handling biomedical data analysis:

1. The ability to generate "dynamic expressions" or transformations on data elements, inputs, outputs, and other command line arguments.
2. The ability to perform "scatter/gather I/O (input/output)", also known as vectored I/O, in which execution of the input data can be parallelized based on specific criteria. A common genomics use case for this is performing an analysis per chromosome, in which the set of chromosomes is delivered to a node as an array (e.g. [1, 2, 3, X]).
3. The ability to nest workflows within workflows, which allows for rapid composition of complex workflows and the ability to quickly reuse existing code.

4.1 *Rabix uses a custom data model and port-level inspection for workflow execution*

Though CWL provides a specification for how to describe the execution of tools and workflows, the exact way in which these features are implemented is left entirely to the execution engine that is interpreting it. Therefore, the Rabix engine has been designed to handle CWL descriptions with two optimizations:

1. Reacting to "port ready" events rather than "job done" events. "Port ready" is a state triggered by the evaluation of data elements produced by a port, whereas "job done" refers to *all* ports of a node being evaluated. In this approach, possible downstream executions are triggered if the edges leading to it are resolved. This allows further dynamic transformations of the DAG to optimize for when all prerequisites for downstream jobs are ready.
2. Reacting to "port ready" events from dynamically created subjobs and rewiring them to their final destinations, possibly creating and running subjobs before their parent fully evaluated (referred to as "look-ahead" method).

These functionalities enable the Rabix engine to create additional edges and nodes as needed, in order to decompose the workflow DAG as early as possible, allowing downstream jobs to be scheduled as soon as actual prerequisites are met.

The workflow DAG is stored in three tables, Variables, Jobs, and Links, which are accessed when a port value is updated. The Variables table contains the ports and their explicit values. The Jobs table stores each node of the workflow and a counter for the inputs and outputs that have been evaluated at that node. The Links table stores the edges in the DAG that is traversed.

As compared to other CWL execution models[2,10], computational events are triggered by "port" events instead of "job" events. In other words, when a port is evaluated, this triggers the executor to scan or update these tables in the following order: Variables, Jobs, Links. Any node for which all input ports are now evaluated is then executed.

Suppose for example, Rabix is executing the workflow in Figure 4. The engine will first parse the workflow description as a workflow DAG with two variables (W.I, W.O; Fig. 4a), which are yet to be evaluated. Additionally, there are two ports (#In, #Out), an input and an output. Next, the engine inspects the contents of the workflow (Fig. 4b) and is able to see the following steps: Tool A, Tool B, each of their ports, and the link between each step within scope.

After this, any known values are curried downstream through their links. The input for the workflow (W.I) is curried to Tool A through the link that has been identified between the two (W.I → W.I.A). The input job counter (#In) for Tool A is decremented to 0, thereby triggering an input event where a job (execution of Tool A with value1) is distributed to a backend for computation. The engine now waits for an event in which the output of Tool A (W.A.O) is reported as value.

Once the output for the job is evaluated and reported to the engine (value2), an output event is triggered. The output port for W.A is decremented to 0, the link from W.A.O to W.B.I is traversed, and W.B.I is evaluated as value2. This reduces the #In counter for W.B to 0 in the Jobs table and triggers a job, the execution of Tool B with its input (Fig. 4c). The execution finally concludes until the input port counter for W reaches 0 and W.O is evaluated (Fig. 4d).

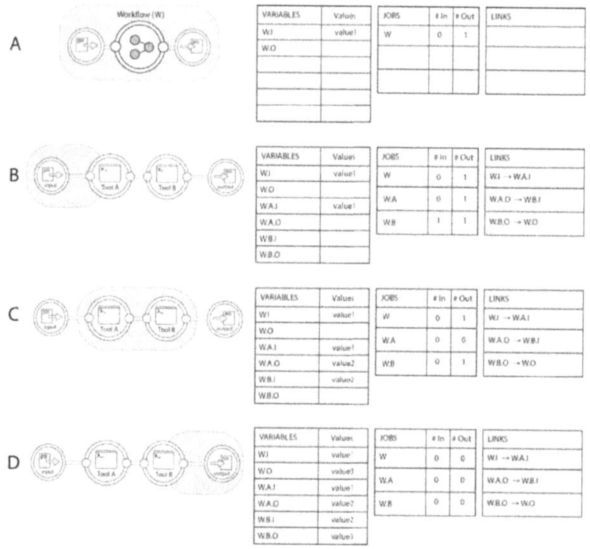

Figure 4. The algorithm as it is traversed. **A.** The engine interprets the top-level of the workflow description and **B.** inspects the contents of the workflow node and determines the DAG structure and links between each step (edges). The currying of value1 from the workflow input to the input of Tool A triggers an input event, where a job (analysis of Tool A with its inputs) is sent to a backend node. **C.** The execution continues and the engine traverses the DAG. **D.** The workflow is completed when the output of the final tool (W.B.O., value3) is curried to the overall workflow output (W.O). The port counters allow the engine to track when nodes are ready to be executed even if upstream jobs are only partially completed.

In the case where the engine is traversing a portion of the workflow that maps to a parent node beneath the root parent node, each output update event will generate an additional output update event. This strategy allows the engine to "look-ahead" towards future executions and apply optimizations to dynamic portions of the DAG, as outlined in the following sections.

4.2. *DAG transformations: parallelization with scatter/gather*

By evaluating workflows through this port-counter and trigger system, Rabix is capable of rewiring parallelizable nodes in the DAG when upstream jobs are only partially completed. Suppose we have a workflow where a data file and an array are inputs for a single tool, which then produces an output file (Fig. 5a). In this case, the tool is capable of being scattered over an array of variables (e.g. [1, 2, 3]). Normally, these executions will be performed sequentially on a single core, or on multiple threads if the tool allows it. However, on a workflow level, additional parallelization can be enabled by scattering the data over three separate executions of the tool based on the values in the array (Fig. 5b), thus allowing the jobs to be distributed to separate computational instances as needed.

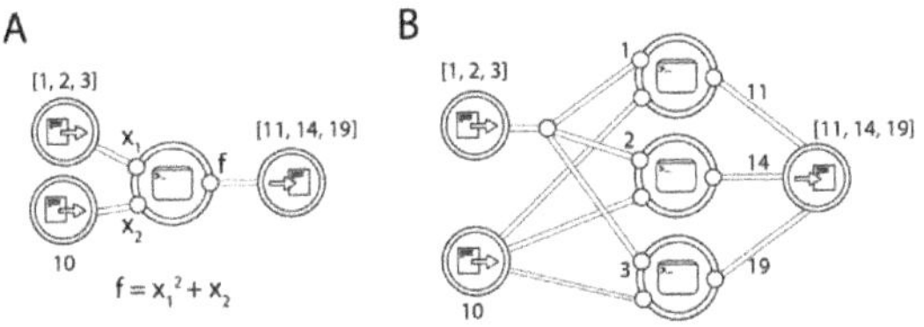

Figure 5. Graph transformations when performing parallelization. In this workflow, a function is performed on two inputs, an *int* and an *array of ints*. **B.** The flattened DAG created by the engine. Each value of the array is scattered as a single process to reduce computation time.

The advantages of the transformation approach is further demonstrated by another use case, in which there are two sequential, parallelizable jobs (Fig. 6a). Rabix employs a "look-ahead" strategy (Fig. 6b) which can mark downstream jobs as ready even though not all sub-jobs (leaves) are done from the upstream parent job.

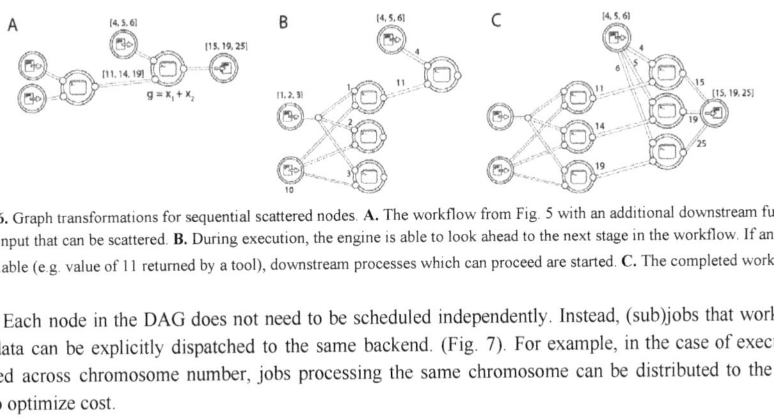

Figure 6. Graph transformations for sequential scattered nodes. **A.** The workflow from Fig. 5 with an additional downstream function with an input that can be scattered. **B.** During execution, the engine is able to look ahead to the next stage in the workflow. If any input is available (e.g. value of 11 returned by a tool), downstream processes which can proceed are started. **C.** The completed workflow.

Each node in the DAG does not need to be scheduled independently. Instead, (sub)jobs that work with same data can be explicitly dispatched to the same backend. (Fig. 7). For example, in the case of executions scattered across chromosome number, jobs processing the same chromosome can be distributed to the same node to optimize cost.

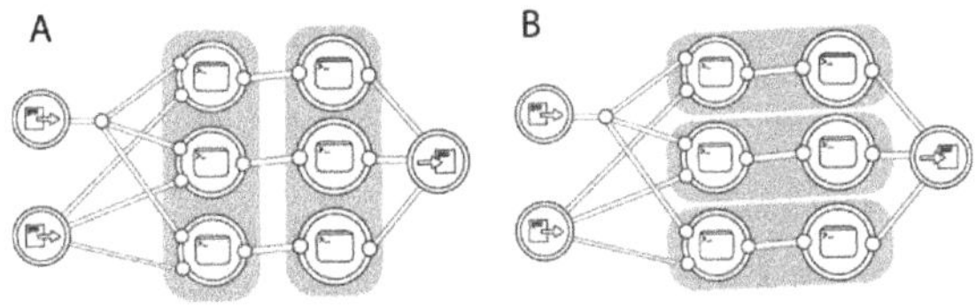

Figure 7. Jobs can be grouped (grey background) for execution on a backend node from criteria set by the workflow or tool author.

Figure 7 demonstrates two possible job group assignments. In the case of Figure 7a, the first tool can be executed simultaneously for each chunk of the data on a single backend node. Once any single job in the first group is finished, the second group of jobs can begin execution on a second node. In the case of Figure 7b, each chunk of data is parallelized across three nodes and the final output is gathered at the end.

The engine is also able to send information to the backends about upcoming jobs, which allows a backend scheduler to pre-allocate resources for them. When executing CWL workflows, both of these optimizations are enabled through the "hints" feature.

Whether these optimizations can be used are sometimes dependent on how the workflow is constructed. For example, a workflow author can make use of optimizations in Fig. 6 by grouping nodes that can be scattered into a nested workflow. This optimization can be especially useful when combined with nested workflow optimizations described in the next section, and allows for reusability of previously made workflows, as encouraged by CWL.

4.3. *Graph transformations: nested workflows*

CWL developers have the ability to reuse existing code and import previously-described workflows into other workflows. This feature means that it is possible to reuse code for additional workflows in lieu of refactoring and potentially introducing errors that break reproducibility. However, the ability to nest workflows presents a challenge to interpretation and optimization by the engine. If no DAG transformations are applied and nested workflows are only executed recursively, this can lead to unnecessarily prolonged execution time and cost.

Suppose a developer has described a workflow that takes two inputs and produces two outputs from two tools (Fig. 8a). In this workflow, one of the outputs is created by the upstream tool and one from the downstream tool. Later, the developer wishes to reuse this workflow description in another workflow, where the output of the upstream tool is passed to another tool for further analysis (Fig. 8b). As with sequentially scattered tools, the engine is capable of passing values from the nested workflow, once they're produced, to steps downstream using the "look-ahead" strategy. Commonly, the tool outside the nested workflow is blocked from execution until all outputs from the nested workflow are produced, leading to increased computation time and cost.

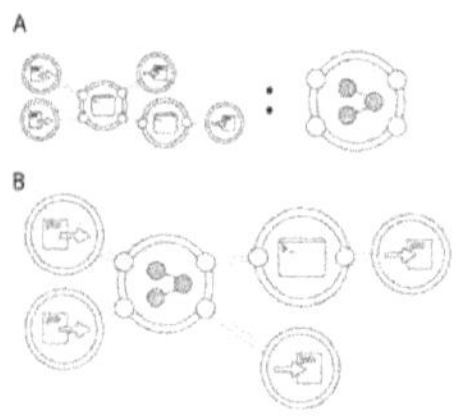

Figure 8. Graph transformations of nested workflows to optimize total execution time. **A.** Workflow consisting of two tools. **B.** Workflow in Fig. 8a. extended with third tool. The engine allows the downstream tool to start executing once the necessary inputs are ready, even if the upstream workflow has yet to produce all of its outputs. No code refactoring from the workflow in 8a is required.

4.4. Benefits to Logging, Orchestration, and Computation

The model used by the Rabix engine allows for improved optimization of data analysis at the workflow level. Further, it provides the ability to implement additional optimizations or features to enhance orchestration of jobs and computation, regardless of whether such features are supported by a workflow description language or specification.

Rabix keeps track of all jobs executed from the workflow and caches results. In addition, each parameter of a job is recorded and automatically logged for the researcher. These include the explicit command line arguments used, the files/paths, attributes of the data, metadata attributes, and any logs associated with the execution. In addition, a snapshot of the application is stored, along with the explicit values used in the execution. All of this is done at the job level, allowing for granular replication of subsets of a workflow. If the workflow contains a job that has previously been executed and the outputs are still available, the engine can reuse them even if the job was part of a different workflow run. Importantly, even if cached results are not available, the engine will look ahead in the DAG and may encounter cached downstream jobs which do have these files available, and so can resume failed or modified workflow jobs. This makes the caching mechanism comparable to declarative workflow description such as GNU Make.[8]

Additional business logic outside of a workflow specification can also be implemented. For example, CWL does not yet allow for conditional workflows, in which the entirety of DAG is not necessarily traversed but only paths based on checkpoints during the execution. Additionally, though a DAG is acyclic, Rabix could in principle enable loops for a tool or workflow which use iterative operations.

4.5. *Caveats to Graph Transformations and Possible Solutions*

An important caveat for these optimizations are external transformations in which the structure of data elements is modified before execution and thus cannot be anticipated by the engine. For example, CWL and other workflow description languages allow for modifications of input types before tool execution. In certain cases, such as for a tool which can be scattered, the data type may change or the length of the array that is being scattered cannot be known ahead of time. If the engine is unable to anticipate the length of an array that must be scattered upon execution, it is impossible for it to re-wire the DAG before evaluation. However, such hurdles can be overcome by allowing users to either define a mapping for individual array items or declaratively specifying the method of combining multiple ports before scattering (cross-product or dot-product). In these cases, the engine can still maintain its look-ahead optimizations.

4.6. *Furthering reproducibility by extending CWL to execution descriptions*

Workflows described using the Common Workflow Language require two objects for execution: the description of an application and an input object specifying the explicit values of the required inputs. Recording a task that has been previously executed is not, however, within the scope of CWL. However, an analyst may want to reinspect a prior analysis, reuse a workflow with a specific set of parameters on new data, or reanalyze the same data with a different workflow version. It is for these reasons that we have enabled an additional layer of task description and annotation within Rabix, alleviating the burden of logging the workflow execution.

Following the execution of a workflow, additional outputs and logs are produced by Rabix as a matter of course. The explicit command line execution, an object describing the output of the execution, and a description of the workflow execution are all recorded. From these objects, it is directly possible to reproduce a prior analysis or reanalyze additional data with the exact same parameters as previous. Rabix allows for replication of a previous execution or reproduction an exact workflow on new data with a single command line call. In this way, it is possible for an analyst to not only publish a workflow but also the explicit tasks as plain text files. These functionalities can be extended with new modules or plugins to enable a variety of use cases centered on reproducibility.

5. Rabix in the context of existing workflow models and engines

The primary design guideline for Rabix was to support Common Workflow Language in a way which will allow for supporting additional workflow languages, whether they are domain-specific languages or object-based. Further, "tools" or workflows described in different syntaxes should be interoperable such that a single workflow may be comprised of tools and workflows from a variety of syntaxes. In effect, certain optimizations described in Rabix above have been implemented in other systems, but not yet in a single executor capable of supporting emerging standards.

Most of the focus in this paper was on port-level inspection, an abstract data model for tools and workflows, and how they can enable additional optimizations when used in conjunction. However, certain features described here are also used by existing workflow systems,[6,7,10–12] most notably the support for multiple infrastructures. Additionally, there are certain features not yet implemented in Rabix but which are seen in other systems, such as conditional steps in a workflow, as seen in Toil. Though the Rabix model allows for conditional operations (e.g. for, if, while), we chose to focus on features supporting reusability and interoperability and computational optimizations for this manuscript.

6. Conclusions

The Rabix Executor is an open-source project designed to enable scalable and reproducible analysis of portable workflows, which is available on GitHub (http://github.com/rabix/bunny). Computational reproducibility, the ability to replicate a prior analysis or reuse prior workflows on new data, is required for accurately judging scientific claims or enabling large-scale data analysis initiatives in which synonymous results can be compared.[4,5,21] The Rabix engine additionally aims to optimize workflow executions by intelligently interpreting and handling complex workflows. This is achieved through a composite model in which workflows can be more fully decomposed. Finally, additional logic can be applied to optimize for user-defined variables, such as cost or execution time, regardless of the workflow description language being interpreted.

References

1. Peng, R. D. Reproducible Research in Computational Science. *Science* **334,** 1226–1227 (2011).
2. Amstutz, P. *et al.* Common Workflow Language v1.0. *FigShare* (2016). doi:10.6084/m9.figshare.3115156.v2
3. Leipzig, J. A review of bioinformatic pipeline frameworks. *Brief. Bioinform.* (2016). doi:10.1093/bib/bbw020
4. Kanwal, S., Lonie, A., Sinnott, R. O. & Anderson, C. Challenges of Large-Scale Biomedical Workflows on the Cloud -- A Case Study on the Need for Reproducibility of Results. in *2015 IEEE 28th International Symposium on Computer-Based Medical Systems* 220–225 (IEEE). doi:10.1109/CBMS.2015.28
5. Alioto, T. S. *et al.* A comprehensive assessment of somatic mutation detection in cancer using whole-genome sequencing. *Nat. Commun.* **6,** 10001 (2015).
6. Jason P Kurs, Manuele Simi, Fabien Campagne. NextflowWorkbench: Reproducible and Reusable Workflows for Beginners and Experts. (2016). doi:10.1101/041236
7. Cromwell: Workflow Execution Engine using WDL. Available at: https://github.com/broadinstitute/cromwell. (Accessed: 2016)
8. *GNU Make: A Program for Directing Recompilation : GNU Make Version 3.79.1.* (Free Software Foundation, 2002).
9. Köster, J. & Rahmann, S. Snakemake--a scalable bioinformatics workflow engine. *Bioinformatics* **28,** 2520–2522 (2012).
10. John Vivian, Arjun Rao, Frank Austin Nothaft, Christopher Ketchum, Joel Armstrong, Adam Novak, Jacob Pfeil, Jake Narkizian, Alden D. Deran, Audrey Musselman-Brown, Hannes Schmidt, Peter Amstutz, Brian Craft, Mary Goldman, Kate Rosenbloom, Melissa Cline, Brian O'Connor, Megan Hanna, Chet Birger, W. James Kent, David A. Patterson, Anthony D. Joseph, Jingchun Zhu, Sasha Zaranek, Gad Getz, David Haussler, Benedict Paten. Rapid and efficient analysis of 20,000 RNA-seq samples with Toil. *bioRxiv* (2016). doi:10.1101/062497
11. Wolstencroft, K. *et al.* The Taverna workflow suite: designing and executing workflows of Web Services on the desktop, web or in the cloud. *Nucleic Acids Res.* **41,** W557–61 (2013).
12. Goecks, J., Nekrutenko, A., Taylor, J. & Galaxy Team. Galaxy: a comprehensive approach for supporting accessible, reproducible, and transparent computational research in the life sciences. *Genome Biol.* **11,** R86 (2010).
13. Deelman, E. *et al.* Pegasus, a workflow management system for science automation. *Future Gener. Comput. Syst.* **46,** 17–35 (2015/5).
14. Guo, F., Yu, L., Tian, S. & Yu, J. A workflow task scheduling algorithm based on the resources' fuzzy clustering in cloud computing environment. *Int. J. Commun. Syst.* **28,** 1053–1067 (2015).
15. Workflow Description Language - Specification and Implementations. Available at: https://github.com/broadinstitute/wdl. (Accessed: 2016)
16. Belhajjame, K. *et al.* Using a suite of ontologies for preserving workflow-centric research objects. *Web Semantics: Science, Services and Agents on the World Wide Web* **32,** 16–42 (2015/5).
17. Terstyanszky, G. *et al.* Enabling scientific workflow sharing through coarse-grained interoperability. *Future Gener. Comput. Syst.* **37,** 46–59 (2014/7).
18. Gamma, E., Helm, R., Johnson, R. & Vlissides, J. *Design Patterns: Elements of Reusable Object-Oriented Software with Applying Uml and Patterns:An Introduction to Object-Oriented Analysis and Design and the Unified Process.* (Addison Wesley, 2003).
19. PROVO-O: The PROV Ontology. (2013). Available at: https://www.w3.org/TR/prov-o. (Accessed: 2016)
20. Hettne, K. M. *et al.* Structuring research methods and data with the research object model: genomics workflows as a case study. *J. Biomed. Semantics* **5,** 41 (2014).
21. Sandve, G. K., Nekrutenko, A., Taylor, J. & Hovig, E. Ten simple rules for reproducible computational research. *PLoS Comput. Biol.* **9,** e1003285 (2013).

DATA SHARING AND REPRODUCIBLE CLINICAL GENETIC TESTING: SUCCESSES AND CHALLENGES

SHAN YANG

Invitae
San Francisco, California, USA
Email: shan.yang@invitae.com

MELISSA CLINE

University of California Santa Cruz
Santa Cruz, California, USA
Email: cline@soe.ucsc.edu

CAN ZHANG

University of California Santa Cruz
Santa Cruz, California, USA
Email: mollyzhang@soe.ucsc.edu

BENEDICT PATEN

University of California Santa Cruz
Santa Cruz, California, USA
Email: benedict@soe.ucsc.edu

STEPHEN E. LINCOLN

Invitae
San Francisco, California, USA
Email: steve.lincoln@me.com

Open sharing of clinical genetic data promises to both monitor and eventually improve the reproducibility of variant interpretation among clinical testing laboratories. A significant public data resource has been developed by the NIH ClinVar initiative, which includes submissions from hundreds of laboratories and clinics worldwide. We analyzed a subset of ClinVar data focused on specific clinical areas and we find high reproducibility (>90% concordance) among labs, although challenges for the community are clearly identified in this dataset. We further review results for the commonly tested *BRCA1* and *BRCA2* genes, which show even higher concordance, although the significant fragmentation of data into different silos presents an ongoing challenge now being addressed by the BRCA Exchange. We encourage all laboratories and clinics to contribute to these important resources.

1. Background

1.1. *Clinical genetic testing*

Clinical genetic tests of germline DNA are routinely used to direct patient care in oncology, cardiology, neurology, pediatrics, obstetrics, and other clinical specialties. Excitement surrounds the future of medical genetics, which will likely involve routine and proactive sequencing of patient genomes or exomes. However, even today genetics is used pervasively: over one million clinical genetic tests will be performed in 2016 to inform various pressing medical decisions facing doctors and patients. This number is considerably larger if tests for infectious disease and tumors (somatic testing) are included. Such testing is regulated, often paid for by private insurance and public health systems, and written into many current clinical care guidelines established by payers and medical professional societies.

It is not glib to say that many of these tests are ordered in life-or-death situations. One example is *BRCA1* and *BRCA2* (collectively, *BRCA1/2*) tests, where erroneous results can have substantial deleterious consequences for patients. With a false positive, a radical preventative procedure such as prophylactic bilateral oophorectomy may be indicated, thereby causing an otherwise healthy woman to enter premature menopause and to face the multiple health risks associated with that procedure and with the hormone replacement therapy that often follows. Prophylactic chemotherapy (specifically, tamoxifen) is another option offered to some healthy *BRCA1/2* carriers, with significant side effects. Conversely a false negative could eliminate the chance to prevent a fatal early-onset carcinoma. Such errors are either analytic (reporting a variant to be present in a patient when it is not, or vice versa) or interpretive (concluding that a variant is pathogenic [disease causing] when it is not, or vice versa). This paper focuses on the latter subject.

1.2. *Clinical variant interpretation*

In response to concerns about reproducibility among laboratories, the American College of Medical Genetics (ACMG) and the Association for Molecular Pathology (AMP) jointly developed revised guidelines for clinical variant interpretation [Richards 2015]. These guidelines require laboratory directors to scrutinize the literature and all other available evidence for each variant observed in a patient. The guidelines provide a structured framework for which evidence is weighed in final interpretations. Under these guidelines, variants are classified as pathogenic (P), likely pathogenic (LP), variants of uncertain significance (VUS), likely benign (LB), or benign (B). Despite the significant improvement in standardization that these new guidelines represent compared with their predecessor, laboratory directors must still use a significant degree of expert judgment, which can result in different classifications from different laboratories for the same variant. Date also matters: classifications that pre-date availability of an important piece of evidence should indeed be different than those that post-date it.

1.3. *Data sharing and clinical genetics*

Of course, the first step toward achieving reproducibility is measuring reproducibility, which requires data to be shared among clinical labs. The sharing of genetic data from research projects has long been accepted and encouraged (despite being incompletely implemented). Unfortunately, the open sharing of de-identified clinical genetic data has been far less common owing to a combination of informed consent issues, the commercial interests of certain healthcare providers, and the lack of a community mechanism for doing so.

Recently, the National Institutes of Health established ClinVar, "a freely available archive for interpretations of clinical significance of variants for reported conditions" [Landrum 2016]. By storing only individual variants and classifications, the re-identification of patients whose genotypes are submitted to ClinVar becomes essentially impossible, at least without an independent test of the same variant in the same patient for comparison (in which case, the patient's genotype is already known). Thus, fully de-identified clinical genetic data can be disclosed publicly under US laws and regulations. The American Medical Association (AMA) and National Society of Genetic Counselors (NSGC), among others, have issued recommendations urging laboratories to share such data.

Some commercial and academic laboratories have, unfortunately, declined to participate. Most famously, Myriad Genetics, the largest *BRCA1/2* testing laboratory in the world, has maintained its large genetic database as a proprietary asset [Cook-Deegan 2013]. Moreover, Myriad claims that by leveraging this database, it can deliver superior variant classifications compared to other labs [Angrist 2014]. This stands in sharp contrast with the American Medical Association and the National Society for Genetic Counselors recommendations. It also is inconsistent with accepted practice in many non-genetics medical fields in which data sharing is common. Thankfully thousands of de-identified Myriad reports have been submitted to ClinVar by ordering clinicians through the Sharing Clinical Reports Project [SCRP website].

2. ClinVar

Since its inception in 2013, ClinVar has grown rapidly, and as of August 2016 contains more than 186,000 records from 560 submitters, most of which are clinical genetic testing laboratories [ClinVar website]. Importantly, three of the top eight submitters to ClinVar are commercial laboratories (GeneDx, Invitae, and Ambry). Another three are large academic laboratories (Harvard Partners Laboratory for Molecular Medicine, Emory Genetics Laboratory, and the University of Chicago Genetic Services Laboratories), and two are academic efforts that aggregate literature-based information (OMIM and GeneReviews) These submitters account for more than half of the data in ClinVar, although the many smaller submitters provide key data as well. This high degree of industry–academic collaboration is encouraging and critical given the degree of privatization in the American healthcare system.

2.1. *Data set used for analysis*

We extracted variant classifications from ClinVar (May 2016 XML download, which remains archived online [ClinVar website]). We included data for genes in six different clinical specialties that our laboratory (Invitae) offered for clinical testing at the time and with which we were thus familiar (Supplemental Data). For simplicity, when one gene may be tested by multiple specialties, we used the most common one. Because variant-phenotype assertions are inconsistently populated in ClinVar these were ignored. We further limited our data set to classifications of germline (not somatic) variants from licensed clinical diagnostic laboratories. Thus data submitted by literature curation efforts (e.g. OMIM), expert panels (e.g., ENIGMA, InSiGHT) and research were also excluded, as these do not reflect actual clinical test reports provided to physicians. Finally, we required that variant classifications be on the 5-class ACMG system and be asserted by at least two submitters. Our data set contained 9875 variants in 409 genes (Table 1, Supplemental Data). We note that many of these classifications pre-date the 2015 ACMG guidelines mentioned above.

	Variants	*Genes*	*Classifications*	*Variants/Gene*	*Classifications/Variant*
Cancer	4802	55	12,703	87.3	2.7
Cardiology	3289	163	7611	20.2	2.3
Epilepsy	739	58	1659	12.7	2.2
Metabolic	383	56	850	6.8	2.2
Neurology	662	77	1376	8.6	2.1
Total	**9875**	**409**	**24,199**	**24.1**	**2.5**

Table 1. ClinVar-based data set used in this analysis.

Overall, variants considered benign (B or LB) by most or all submitters composed the largest group (44.5%). Pathogenic variants (P or LP) made up 17.9% of the data set. Many variants (26.9%) were considered VUS, and 10.7% had no consensus (as defined below) for any category. This distribution varied significantly by clinical area (Figure 1).

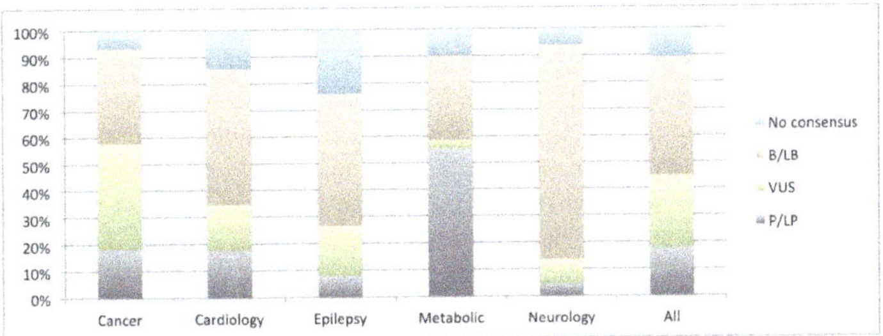

Figure 1. Fraction of variants in ClinVar for each clinical area by consensus pathogenicity.

2.2. *Rarity of clinically observed variants*

Because our data set was limited to variants from two or more submitters, it was naturally biased away from the rarest of variants. Nevertheless, this data set was predominantly composed of rare variants (Figure 2). Most (62%) of the ClinVar variants that also appear in ExAC [Lek, 2016] had population allele frequencies less than 0.001, and for 36%, that frequency was less than 0.0001. Another 22.8% of the ClinVar variants were not in ExAC at all, either because they are very rare or because they lie outside of ExAC's well-covered regions. This rarity also manifests itself in the number of submitters who have classified each variant: Most variants had been classified by only two or three of the 23 submitters in this data set (Table 1). Even in the case of *BRCA1/2*, one of the most common clinically tested genes, the average was only 2.9 classifications per variant. Rare variants comprise an even larger fraction of ClinVar overall, particularly variants with only a single submitter which were excluded from this data set.

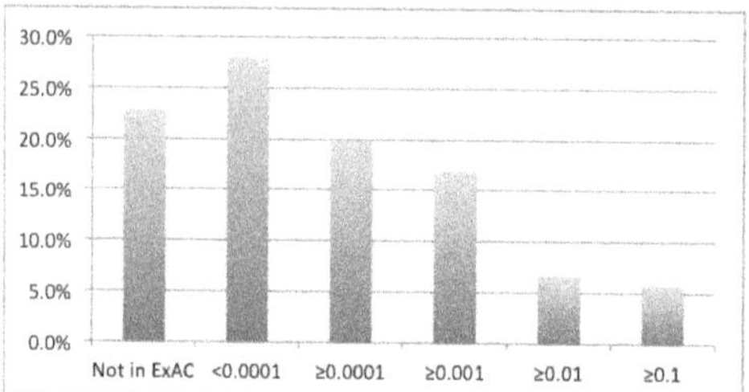

Figure 2. Histogram of allele frequency in ExAC for all ClinVar variants in our analysis regardless of pathogenicity. Note that the vast majority of ClinVar variants are in or near exons.

2.3. *Concordance of variant classifications*

We compared variant classifications in ClinVar to assess the degree of agreement among clinical testing laboratories (Figure 3). We first focused on differences between positive (P or LP) classifications, which are potentially clinically actionable, as opposed to findings that are not actionable (VUS, B, or LB). We refer to this analysis as the P-NP (positive versus not positive) comparison. Counting each of the 9875 variants as a data point, concordance among laboratories was high: 96.1% of variants agreed across all (two or more) submitters. For an additional 0.9% of variants, there was a consensus among a majority of the submitters. We defined consensus as agreement in two-thirds of the submissions (i.e., consensus required two of two submissions to agree, or 2/3, 3/4, 4/5, 4/6, etc.). In 3% of variants, there were only two submitters who disagreed, and only one variant had four submitters with a 2–2 tie. Clinical care guidelines generally state that patients with only VUS should be managed according to their personal and family histories and not their genetic test results [e.g. NCCN 2016]. Thus the P-NP comparisons correlate most with the impact of interpretation discordance on patient care decisions.

When the comparison was performed on a different basis—not combining VUS with B/LB classifications—concordance was, of course, lower. We refer to this analysis as the P-V-B (pathogenic versus VUS versus benign) comparison. In this evaluation, only 83% of variants agreed among all submitters. A further 6% achieved consensus but with some submitter(s) in dissent. This much lower rate indicates that the criteria for discriminating between VUS and B/LB variants varies among laboratories, more so than criteria for establishing pathogenicity.

Concordance varied considerably among clinical areas. On a P-NP basis, variants in cardiology and metabolic genes had concordances lower that those in the other areas, although in all cases concordance was greater than 90%. On a P-V-B basis, epilepsy genes fared the worst, followed by cardiology. The gap between P-NP and P-V-B is particularly large in epilepsy genes, suggesting that evidence against pathogenicity is used quite inconsistently by labs. Cursory analysis suggests that classification date, as expected, plays a significant role in discordance (Supplemental Data). A detailed analysis of the basis for discordance is important future work.

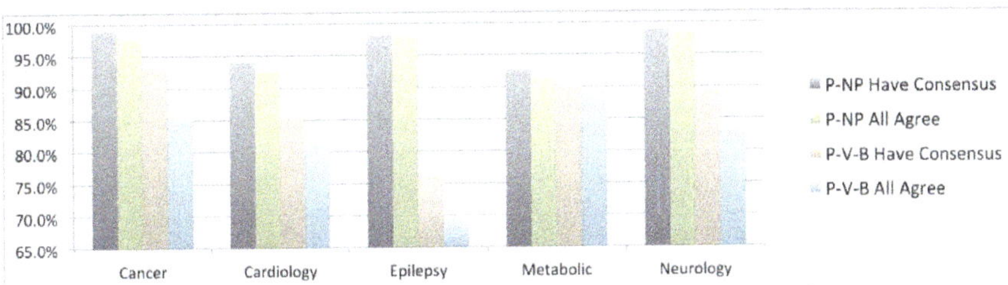

Figure 3. Concordance among labs measured in different ways. See text.

3. *BRCA1/2*

BRCA1/2 had the largest number of variants of any gene(s) in our ClinVar data set (1771 combined) for several reasons: *BRCA1* and *BRCA2* are not only among the most commonly tested genes in clinical practice today, but also have been clinically tested for more than 20 years. Moreover, a significant international effort has focused on adding *BRCA1/2* variants to ClinVar, whereas data sharing efforts for some other commonly tested genes center on previously established databases (e.g. the CFTR2 database for cystic fibrosis). Finally, compared with most human genes, *BRCA1/2* have relatively large coding sequences and thus can harbor an atypically large number of variants. Thus, the "long tail" of *BRCA1/2* variants is particularly long, and new variants needing classification are continually uncovered, as shown by our own data (Figure 4). This conclusion is consistent with unpublished reports from Myriad Genetics, which claims to encounter >50 new variants per week despite offering testing for 20 years [Myriad 2015]

3.1. *Concordance among BRCA1/2 variant classifications*

In a separate study, we performed a much more detailed comparison of ClinVar data for *BRCA1/2* using a ClinVar data set of more than 2000 comparable variants [Lincoln 2016]. This analysis considered only classifications from clinical labs with significant experience (as evidenced by submitting 200 or more variants to ClinVar) and excluded submitters where most classifications were >5 years old. On a P-NP basis, 98.5% of variants showed no disagreement among submitters—a concordance higher than that observed in ClinVar overall. This previous study also showed that variants with classification discordance were rare (allele frequencies were always less than 0.0005 and usually were immeasurably low). Although they are numerous, rare variants by definition appear in very few patients: less than 15% of the 30,000 patients studied carried any rare variants in *BRCA1* or *BRCA2*, and most of those were concordantly classified. In this prior study, concordance per patient (not per variant) was thus estimated to be 99.8%.

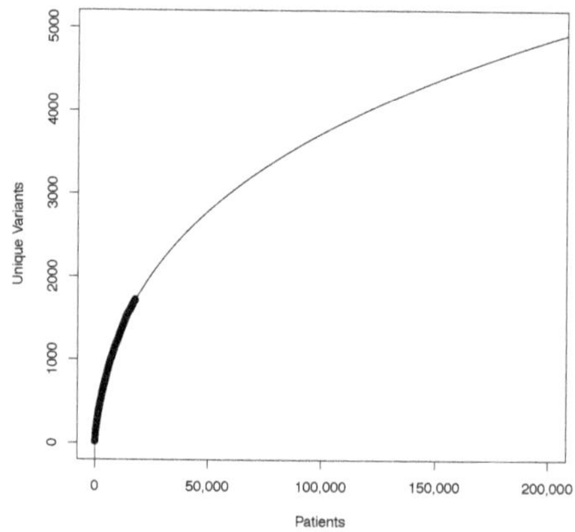

Figure 4. The relationship of number of unique BRCA1/2 variants to number of patients tested at Invitae (dark curve). The extrapolation (light curve) was fit in R using the formula poly(log(Patients), 3). We chose the polynomial degree empirically by minimizing the Akaike Information Criteria [Sakamoto 1986].

3.2. *Variants of Uncertain Significance (VUS) in BRCA1/2*

VUS can present a challenge in day-to-day clinical decision-making, and the most prevalent type of VUS are rare missense changes. VUS rates are traditionally defined as the fraction of patients with one or more VUS and no positive findings. Major U.S. laboratories report VUS rates in the range of 3–5% for *BRCA1/2*, although this rate varies considerably with ethnic mix and with the fraction of cancer-affected versus unaffected patients [Lincoln 2015]. On a per-variant (rather than per-patient) basis, the VUS rate is much higher: 31.4% of *BRCA1/2* variants in our data set (Table 1) were VUS, although most are very rare and thus appear in very few patients.

The evidence suggests that the majority of VUS are actually benign variants that have inadequate evidence to demonstrate that fact. This is supported by our own experience that most

VUS, when reclassified, are "downgraded" to LB or B. We also observed this in a sequential analysis of ClinVar releases from the past 2 years (available at [ClinVar website]) in which roughly 95% of *BRCA1/2* VUS reclassifications were downgrades. Others have also observed this in Myriad data [Murray 2011]. In terms of clinical impact, a rough approximation is that if 4% of patients have a VUS, and if 5% of those findings are truly pathogenic variants lacking evidence of pathogenicity, then 1/500 *BRCA1/2*-positive patients may currently be missed.

BRCA1/2 tests are increasingly being replaced by multi-gene panels that assay additional genes that significantly increase the risk of various cancers. By virtue of testing more genes, the VUS rate in these panels is substantially larger. For example, VUS rates of roughly 40% have been reported by 25-29 gene panels [Lincoln 2015; Desmond 2015; Tung 2015], although again, experience suggests that the majority of these VUS will ultimately be classified as benign.

3.3. *The BRCA Exchange*

As of August 2016, ClinVar contains more than 9000 variants in *BRCA1/2*, many of which are either unclassified or are considered VUS. Most of these variants have been reported by only a single submitter. These data still represent only a fraction of the known human variation in *BRCA1/2*, much of which is either not submitted to ClimVar or is not appropriate for ClinVar (yet is useful to have linked). In an effort to collect a more comprehensive view of *BRCA1/2* variation, the BRCA Exchange project has been initiated under the auspices of the Global Alliance for Genomics and Health's BRCA Challenge. European laboratory data, coordinated by the Leiden Open Variation Database (LOVD), population databases, and other data sources are being combined with ClinVar in this *BRCA1/2*-specific public database. In its current preliminary form, the BRCA Exchange describes more than 13,000 variants, many of which originate from only a single source database (Figure 5). Not only is the BRCA exchange database open, but the code that populates it is open source. Future analyses of the type described in this paper could and should leverage this code in order to further improve reproducibility of such research.

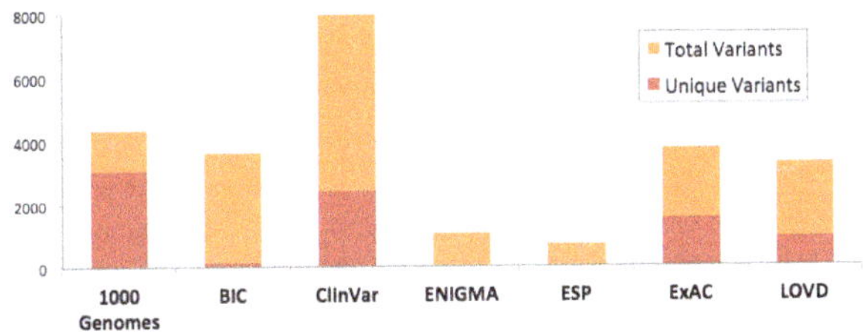

Figure 5. Sources of data in the pre-release BRCA exchange. Many variants not in ClinVar and indeed many are unique to a single database. For details and references see brcaexchange.org.

4. Discussion

4.1. *Summary: Most variant classifications agree, but . . .*

In the analysis described above, we examined nearly 10,000 variants from ClinVar in more than 400 genes across six clinical areas and found generally high (>90%) variant classification concordance among clinical laboratories in terms of potential effect on clinical management (our P-NP comparison). In separate prior studies, we examined *BRCA1/2* in particular detail found higher concordance on both a per-variant (99.0%) and a per-patient (99.8%) basis than is seen in the broader gene list. It is reassuring, at least for geneticists, to note that this level of concordance is higher than that observed among pathologists reading breast biopsies or radiologists reading mammograms [Elmore 2015(a,b); Elmore 2016; Sprague 2016]. Nevertheless, resolving differences in variant classification is critical to doctors and patients. Moreover, many variants (30.1%) have classifications that are concordantly VUS and much more work is required to classify these variants definitively even though laboratories agree.

Public databases such as ClinVar play critical roles in the identification of both disagreements and uncertainties, and these databases can facilitate collaborative interactions that will resolve many such issues. The value of such collaboration in improving variant classifications has recently been demonstrated by multiple groups [Amendola 2016]. Efforts are now organized into disease-specific working groups by the ClinGen consortium [Rehm 2015; Pfimister 2015] and support pre-existing efforts such as ENIGMA and InSiGHT. Those with interest and expertise in these areas should certainly consider joining and contributing.

Public databases can also play a critical role in laboratory quality control by allowing detailed independent peer scrutiny of all variant classifications by the global community. In our opinion, no laboratory could (or probably would) mount such an effort alone, and publication peer review processes can not provide this type of ongoing quality assessment. In our opinion, laboratory directors who are both confident in their quality yet continually working to improve should have no reservations about unrestricted public data submission of their data.

4.2. *Considerations when using public clinical databases*

Our analysis highlights important considerations users must keep in mind when accessing public databases such as ClinVar. Foremost is that it is a fallacy to say, for example, "ClinVar says that variant X is pathogenic." ClinVar itself generates no assertions; it only collects them from submitters. Database users must pay careful attention to the original source of each classification, which may be a reputable clinical laboratory rigorously following accepted classification guidelines, or it may not be. Dates are important, as submissions to these databases can become outdated, which results in false discrepancies.

It is important that users understand the biology and medical practice considerations for each gene they examine in a public database. Consider three examples of the rates of variant pathogenicity (Figure 1) which we find unsurprising: (a) In some genes (e.g., most hereditary cancer genes) loss-of-function variants are pathogenic, and nature provides many means of disabling genes or their proteins. In other cases (e.g., some neurology and cardiology genes), gain-of-function mutations are clinically more important, and these, by their very nature, are less numerous, reducing the fraction of pathogenic variants in ClinVar. (b) The large fraction of pathogenic variants and small fraction of VUS in metabolic genes reflect the fact that experimental confirmation of pathogenicity (e.g., through blood chemistry and urinalysis) is relatively straightforward and standard clinical practice. However, the relatively low concordance in metabolic genes (Figure 3) suggests that these procedures are imperfect. (c) In cardiology, complexities in both phenotyping and penetrance are well known to increase the complexity of variant classification [Van Driest].

Deliberate (and not nefarious) submission biases also affect ClinVar. Notably, laboratory policies vary as to whether and when B/LB variants are reported to patients/physicians or to ClinVar (even though benign polymorphisms are frequently observed). Similarly, practices for the detection and reporting of non-coding variants vary. Although many routine tests detect copy number variants, these variants are less commonly reported to ClinVar for logistical reasons (a situation we hope will change). Furthermore, a test may or may not be sensitive to complex alterations such as copy-neutral inversions, Alu insertions, or variants in low complexity or highly conserved regions. Although ClinVar can record the observed prevalence of any variant, this field is rarely filled in. Finally, ClinVar submissions generally represent laboratory patient series, which are subject to many undocumented ascertainment biases. For these reasons, ClinVar cannot be used to evaluate the spectrum of disease-causing or benign variation in any gene.

4.3. *Whither data sharing*

Although sharing of clinical genetic data has been successful, and clearly impactful, challenges remain. For example, during our various analyses of ClinVar, we uncovered a number of out of date and erroneous submissions, which are an obvious concern. A bigger problem is the multiple laboratories who do not contribute. In addition to not contributing, Myriad Genetics has updated its terms of service to, in theory, prohibit ordering clinicians from sharing data with ClinVar [Robinson 2016]. A further challenge is the fragmentation of data into multiple silos. Although the BRCA Exchange aims to address this problem for *BRCA1/2*, this is a considerable effort and only applies to these two genes, not the many others of clinical relevance.

In environmental policy, the term "greenwashing" has emerged to describe the characterization of various activities as environmentally friendly when in fact they are not. Activities can occur in our field that one might perhaps call "sharewashing". For example two large commercial labs (Labcorp and Quest) currently contribute variants only to BRCAShare, a database whose terms effectively prohibit either incorporation of the data into a common repository (like the BRCA Exchange) or its use in comparisons such as those described here. We hope this changes, but at present these data are not available in unrestricted form. The BRCAShare terms also prohibit use of the data by other commercial labs without paying a significant fee (unlike ClinVar). Separately, Myriad has tried to argue that its participation in the PROMPT patient registry comprises data sharing. PROMPT is indeed valuable, but serves a very different purpose than

ClinVar. We encourage all groups to support and contribute to open, unrestricted, public databases, particularly ClinVar.

5. References

Amendola LM, et al. Am J Hum Genet 2016; 10.1016/j.ajhg.2016.03.024.
Angrist M and Cook-Deegan R. Appl Transl Genom 2014;3(4):124-127.
ClinVar website: www.clinvar.com
Cook-Deegan R, et al. Eur J Hum Genet 2013;21(6):585-8.
Desmond A, et al. JAMA Oncol 2015;1(7):943-51.
ENIGMA website: enigmaconsortium.org
Elmore 2015(a): Elmore JG, et al. JAMA 2015;313(11):1122-32.
Elmore 2015(b): Elmore JG, et al.. JAMA 2015;314(1):83-4.
Elmore 2015(c): Elmore JG, et al. Ann Intern Med 2016;164(10):649-55.
Genereviews website: www.ncbi.nlm.nih.gov/books/NBK1116/
InSiGHT website: insight-group.org
Landrum MJ, et al. Nucleic Acids Res 2016;44(D1):D862-8.
Lek M, et al. Nature 2016; 536:285–291
Lincoln 2016: Lincoln SE, ACMG 2016 platform presentation, copy available at www.invitae.com; manuscript in review.
Lincoln 2015: Lincoln SE, et al. J Mol Diagn 2015;17(5):533-44.
Murray ML, et al. Genet Med 2011;13(12):998-1005.
Myriad Analyst Day Presentation, September 2015 from www.myriad.com
NCCN (National Comprehensive Cancer Network). NCCN Practice Guidelines in Oncology. Genetic/Familial High Risk Assessment: Breast and Ovarian, Version 2.2016. www.nccn.org
OMIM website: omim.org
Phimister EG. New Engl J Med 2015;372(23):2227-8.
Richards S, et al. Genet Med 2015;17(5):405-24.
Rehm HL, et al. New Engl J Med 2015;372(23):2235-42.
Robinson 2016: L. Robinson, genetic counselor, UT Southwestern; personal communication.
Sakamoto Y, et al. 1986. Akaike Information Criterion Statistics. (D. Reidel Publishing)
SCRP website: www.clinicalgenome.org/data-sharing/sharing-clinical-reports-project-scrp/
Sprague BL, et al. Ann Intern Med 2016; 10.7326/M15-2934.
Tung N, et al. Cancer 2015, 121:25e33
Van Driest SL, et al. JAMA 2016;315(1):47-57.

6. Supplement

The dataset upon which this analysis is based is available at:
https://drive.google.com/drive/folders/0B79LNgCdve9BSWN0VHhodFFsMmM

PATTERNS IN BIOMEDICAL DATA-HOW DO WE FIND THEM?

ANNA O. BASILE
The Pennsylvania State University, Department of Biochemistry and Molecular Biology
328 Innovation Blvd Ste 210
State College, PA 16803
azo121@psu.edu

ANURAG VERMA
Geisinger Health System
The Pennsylvania State University, Huck Institutes of the Life Sciences
328 Innovation Blvd Ste 210
State College, PA 16803
averma@geisinger.edu

MARTA BYRSKA-BISHOP
Geisinger Health System
328 Innovation Blvd Ste 210
State College, PA 16803
mbyrskabishop@geisinger.edu

SARAH A. PENDERGRASS
Geisinger Health System, Biomedical and Translational Informatics
122 Weis Center for Research
Danville, PA 17822
spendergrass@geisinger.edu

CHRISTIAN DARABOS
Dartmouth College, Research Computing Services
HB 6129
Hanover, NH 03755
christian.darabos@dartmouth.edu

H. LESTER KIRCHNER
Geisinger Health System, Biomedical and Translational Informatics
100 N. Academy Ave
Danville, PA 17822-4400
hlkirchner@geisinger.edu

Given the exponential growth of biomedical data, researchers are faced with numerous challenges in extracting and interpreting information from these large, high-dimensional, incomplete, and often noisy data. To facilitate addressing this growing concern, the "Patterns in Biomedical Data-How do we find them?" session of the 2017 Pacific Symposium on Biocomputing (PSB) is devoted to exploring pattern recognition using data-driven approaches for biomedical and precision medicine applications. The papers selected for this session focus on novel machine learning techniques as well as applications of established methods to heterogeneous data. We also feature manuscripts aimed at addressing the current challenges associated with the analysis of biomedical data.

1. Introduction

With great technological advances and numerous 'big data' initiatives targeted at generating and acquiring large amounts of biomedical information, there has been an astonishing growth in the volume of data in recent years [1]. Considering sequencing data alone, the size of data has approximately doubled every six months in the last decade [2]. Continuing at this rate, we can expect to reach a zettabyte of sequencing data generated per year by 2025 [2].

Thus, the age of big data is upon us, and with its arrival comes the potential to revolutionize many aspects of our lives. Decisions previously made using carefully constructed, simulated models of reality can now be made using measured data. While the term 'big data' is not well defined, it will be used herein to describe a situation where the amount of information far exceeds that which has been previously available [3]. Big data analyses impact many areas of society, culture, and research. To combat crime, law enforcement officials are employing seismology-like models to predict areas of high crime, and intervene to prevent them from occurring [3]. With large scale surveys, such as the Two Micron All-Sky Survey, which contains a petabyte of data, astronomers can now focus their efforts illuminating structures and exploring potential connections and hypotheses [4]. In the area of public health and precision medicine, large-scale efforts have been made to create datasets aimed at elucidating the genetic underpinnings of various traits as a means of disease prevention and development of effective treatment. For example, the Precision Medicine Initiative Cohort Program announced by President Obama plans to enroll one million participants spanning a multitude of age and race groups within the US [5]. Other large-scale genome projects include the UK 100,000 Genomes Project [6], and the Geisinger MyCode Community Health Initiative which unites Geisinger Health System and Regeneron Genetics Center in a collaboration aimed at bio-banking and whole-exome sequencing more than 200,000 patients [7]. Likewise, public datasets, such as The Cancer Genome Atlas (TCGA), which provides molecular characterization of cancer genomes, continue to provide a wealth of data to researchers with the hope of one day improving clinical patient care.

While these potentials are truly revolutionary, there are a number of challenges that can impede the promises of big data and make it difficult to extract the true value of this information. The sheer volume of available data and the rate at which it is being generated is overwhelming the majority of industries, many of which do not yet have the proper management, storage and analytical means of assessing this information [8]. Additionally, while small sample sizes are often prohibitive in research, the large sample sizes provided by big data initiatives may not be a panacea. Large sample sizes may be of little value if they are not representative of the population being assessed, are missing information (especially if missingness is nonrandom or important data is completely missing), or contain sampling biases [9]. Machine-learning approaches in this data-driven space will require an integration of different generated data types. In a biomedical setting, this may include clinical measurements, drug usage data, mRNA expression levels, and environmental exposures. These informatics methods must also be robust to incompleteness and variable sparsity, as well as heterogeneity which can present mixtures of categorical and numerical data. Further considerations that will need to be made include scalability and dealing with a feature space that far exceeds the number of samples.

The collection of papers presented in this session demonstrates a diversity of data-driven, pattern recognition approaches and challenges within the biomedical and precision health setting. These manuscripts span a wide range of categories from applications of well-studied informatics methods to novel pattern recognition techniques as well as approaches of overcoming big data challenges.

2. Session Contributions:

2.1 Machine Learning and Deep Learning Approaches

Machine learning and deep learning have recently received a great deal of attention due to their potentially transformative applications to big data. Machine learning refers to a class of algorithms that can learn from and also make predictions on data [10], while deep learning describes a branch of machine learning that models data using multiple levels of representation and abstraction. These methods do not require explicit rules as they rely on the data, and generally speaking, the more data, the better the outcome of these techniques. While the use of data-driven approaches is not new, this is an expanding area of biomedical research that is gaining momentum due to algorithmic sophistication, computational advancement, and the growth in volume and variety of available data.

Shameer et al. describe a data-driven feature selection and machine learning approach to predict hospital readmission in heart failure (HF) patients from electronic health records (EHR) in "*Predictive Modeling of Hospital Readmission Rates using Electronic Medical Record-wide Machine Learning: A Cased-Study Using Mount Sinai Health Cohort*". Several data domains were extracted from the EHR including diagnoses, medications, laboratory measurements, procedures, and vitals. Separate models were generated from the data domains using the Naïve Bayes algorithm and then combined. Feature selection was performed using a correlation-based method. Their approach was contrasted to using logistic regression, and it performed well over all existing predictive models in HF.

In the manuscript "*Missing data imputation in the electronic health record using deeply learned autoencoders*" **Beaulieu-Jones** et al. tackle the important issue of dealing with missing data, commonly encountered in the context of EHR. Specifically, the authors use the Pooled Resource Open-Access Amyotrophic Lateral Sclerosis (ALS) Clinical Trial Database (PRO-ACT) to evaluate missing data imputation performance of a machine learning approach, namely deeply learned autoencoders, and compare it to the performance of several established imputation strategies, such as mean, median, K-nearest neighbors, or Singular Value Decomposition (SVD). They show that autoencoders outperform other methods in imputation of data missing completely at random (MCAR), as well as data missing not at random (MNAR). Furthermore, they used data imputed by different methods to predict ALS progression and identify the most important predictors of ALS.

One of the challenges associated with applying machine learning approaches to biological problems is the interpretation of the models that arise from them. In the manuscript titled "*DG-Dashboard: Visualizing and Understanding Genomic Sequences Using Deep Neural Networks*", **Lanchantin** et al. present a visualization toolkit called the Deep Genomic Dashboard (DG-Dashboard), which facilitates interpretation of deep neural network models in the context of predicting transcription factor binding sites (TFBS) along genomic DNA. In particular, DG-Dashboard offers three strategies: saliency maps, temporal output scores, and class optimizations, which enable visualization of nucleotide importance within a particular motif, critical positions along a DNA sequence, as well as class-specific motif patterns for a particular TF based on predictions obtained from convolutional neural networks (CNNs), recurrent neural networks (RNNs), as well as convolutional-recurrent neural networks (CNN-RNNs). In addition to facilitating interpretation of the three deep neural network architectures, Lanchantin et al. demonstrate that CNN-RNNs outperform CNN and RNN in classification of TFBSs.

2.2 Pattern recognition applications in EHR, Medical Imaging, and Mobile Health data

Applications of machine learning approaches are widespread in the biomedical sector. EHRs, biomedical images, and mobile health apps are just a few of the many sources researchers are mining to advance human health. Data-driven approaches can leverage the wealth of information in these sources and extract meaningful knowledge which can then be utilized to study disease progression and symptom patterns, classify patient subgroups, and inform clinical practice and decision-making.

One such application is digital image analysis that was implemented to classify the bone cancer in "*Large Scale Image Segmentation and Classification for Viable and non-viable Tumor Identification in Osteosarcoma*". **Arunachalam** et al. demonstrate a high-throughput approach to classify the tumor region from images of Hematoxylin and eosin (H&E) stain slides from bone cancer patients. They proposed a multi-tier approach where they used pixel and object based approach to color and classify different histopathological regions of cancer cells in the digital stain images. Further, they used a combination of multiple clustering algorithms to define viable and non-viable tumors.

In "*Development and Performance of Text-Mining Algorithms to Extract Socioeconomics Status from DE-identified Electronic Health Records*", **Hollister** et al. describe a data mining approach, where they developed an algorithm to define a phenotype status from variety of structured and unstructured free text in EHR. In order to investigate socioeconomics status (SES) they developed seven different algorithms predictive of SES like Education, Occupation, Insurance Status, Retirement, Medicaid, and Homelessness. Their work addresses an important question associated with health outcomes and the socioeconomic status extracted from various semantic categories. They provide performance metric of seven algorithms, but also highlight many shortcoming and challenges that potentially affect phenotype algorithm development in current EHR systems.

In "*Methods for Clustering Time Series Data Acquired from Mobile Health Apps*", **Tignor** and colleagues present a method to cluster individuals with asthma using data collected from a mobile health app. The data represent a time series of daily asthma symptoms which exhibit non-ignorable missingness. Their work focuses on developing a novel probabilistic imputation method, and combined with a consensus clustering algorithm, is used to identify distinct symptom patterns. Variations on the algorithm implementation are devised and compared.

Studying the heterogeneous patterns of disease manifestation and progression is important for the clinical treatment and management of a condition. In "*Learning Attributes of Disease Progression from Trajectories of Sparse Lab Values*", **Agarwal** et al. use the Functional Clustering Model (FCM) to cluster sparse clinical lab measures from patients with Chronic Kidney Disease (CKD) from the Stanford Health Care (SHC) system. The authors hypothesize that using data-driven approaches on trajectories of sparse lab values can create clinically meaningful clusters that highlight alternate disease progression patterns in CKD. Irregularity and sparsity in longitudinal EHR data creates high variance in trajectory estimates and often leads to unstable clusters. The FCM approach addresses this challenge by treating curve coefficients as random effects, and then projecting the curve into a subspace where the cluster centers now represent the probability of cluster membership. Using this approach, the authors cluster creatinine trajectories of CKD patients to create two patient groupings which feature distinct clinical attributes.

2.3 Public Data Mining

The extraction and identification of higher level relationships from high-throughput data and data repositories is an important area of research. For example, with the ever increasing amount of study information existing within PubMed, it is a challenge to integrate that much information to gain higher level insights over trends that have been found for genes and diseases. The information gained from effectively integrating comprehensive data together in novel ways could ultimately result in the "sum being greater than the parts", providing new insights for further research and discovery.

In "*A new relevance estimator for compilation and visualization of disease patterns and potential drug targets*", **von Korff** et al. describe a tool, the Disease Relevance Miner (DDRelevanceMiner), which was developed using the concept of second order co-occurrence which takes advantage of calculating the similarity between two words that do not co-occur frequently, but co-occur with the same neighboring word. The authors used the basis of this approach but with the advancement of a relevance estimator. Using the DDRelevance Miner, the authors used HUGO gene identifiers, and then linked them to PubMed in order to extract relevant records for each gene, where each publication record in turn was searched with disease MeSH terms. Linking together these data along with a metric of relevance, provided detailed disease-gene and disease-disease associations which could be further explored. This includes the identification of gene drug targets that had indications of being highly specific to single diseases.

Wilson et al. evaluate the performance of four community detection algorithms to automatically determine groups of genes from protein-protein interaction networks using experimental data in "*Discovery of Functional and Disease Pathways by Community Detection in Protein-Protein Interaction Networks*". To date, biological pathway information has been based on experimentally gained understanding. The various pathway repositories that exist are incredibly important resources, a testament to how much has been learned of the underlying structure of biology. These resources contribute to a greater understanding of gene expression and genetic association results, as well as identification of genetic interaction candidates. High throughput computational approaches could help fast track the evaluation of new potential pathways. Determining communities of biological networks could shed new light on groupings of genes with common biological functions or features. With the reliance of many analyses based on gene and pathway information, such as the Gene Set Enrichment Analysis (GSEA) [11], Pathway Analysis by Randomization Incorporating Structure (PARIS) [12], and other tools like Biofilter [13], further identification of pathways could support new hypothesis generation for experimental validation. In the manuscript by Wilson et al., several possible community detection methods were tested using a STRING protein-protein interaction network [14]. Communities obtained were then compared to curated biological pathways, over multiple metrics. Both known pathways were re-identified and possibly novel pathways were identified, the authors carefully characterized other features of these networks as well, highlighting the utility of community detection methods in identifying new pathways for further study.

References:

1. Bourne PE, Bonazzi V, Dunn M, Green ED, Guyer M, Komatsoulis G, et al. The NIH Big Data to Knowledge (BD2K) initiative. J. Am. Med. Inform. Assoc. 2015;22:1114–1114.

2. Stephens ZD, Lee SY, Faghri F, Campbell RH, Zhai C, Efron MJ, et al. Big Data: Astronomical or Genomical? PLoS Biol. 2015;13:e1002195.

3. Hvistendahl M. Can "predictive policing" prevent crime before it happens? Sci. Mag. [Internet]. 2016; Available from: http://www.sciencemag.org/news/2016/09/can-predictive-policing-prevent-crime-it-happens

4. Zhang Y, Zhao Y. Astronomy in the Big Data Era. Data Sci. J. 2015;14:11.

5. PMI in the News [Internet]. Natl. Inst. Health NIH. 2015 [cited 2016 Sep 23]. Available from: https://www.nih.gov/node/19706/draft

6. Rabes T, 1 ratanaAug, 2014, Pm 12:30. U.K.'s 100,000 Genomes Project gets £300 million to finish the job by 2017 [Internet]. Sci. AAAS. 2014 [cited 2016 Sep 23]. Available from: http://www.sciencemag.org/news/2014/08/uks-100000-genomes-project-gets-300-million-finish-job-2017

7. Carey DJ, Fetterolf SN, Davis FD, Faucett WA, Kirchner HL, Mirshahi U, et al. The Geisinger MyCode community health initiative: an electronic health record-linked biobank for precision medicine research. Genet. Med. 2016;18:906–13.

8. Kaplan RM, Chambers DA, Glasgow RE. Big Data and Large Sample Size: A Cautionary Note on the Potential for Bias. Clin. Transl. Sci. 2014;7:342–6.

9. DeRouen TA. Promises and Pitfalls in the Use of "Big Data" for Clinical Research. J. Dent. Res. 2015;94:107S – 109S.

10. Vadrevu S. Understanding the Promise and Pitfalls of Machine Learning [Internet]. Data Inf. 2015 [cited 2016 Sep 30]. Available from: http://data-informed.com/understanding-promise-pitfalls-machine-learning/

11. Subramanian A, Tamayo P, Mootha VK, Mukherjee S, Ebert BL, Gillette MA, et al. Gene set enrichment analysis: A knowledge-based approach for interpreting genome-wide expression profiles. Proc. Natl. Acad. Sci. 2005;102:15545–50.

12. Butkiewicz M, Cooke Bailey JN, Frase A, Dudek S, Yaspan BL, Ritchie MD, et al. Pathway analysis by randomization incorporating structure-PARIS: an update. Bioinforma. Oxf. Engl. 2016;32:2361–3.

13. Pendergrass SA, Frase A, Wallace J, Wolfe D, Katiyar N, Moore C, et al. Genomic analyses with biofilter 2.0: knowledge driven filtering, annotation, and model development. BioData Min. 2013;6:25.

14. Franceschini A, Szklarczyk D, Frankild S, Kuhn M, Simonovic M, Roth A, et al. STRING v9.1: protein-protein interaction networks, with increased coverage and integration. Nucleic Acids Res. 2013;41:D808–15.

LEARNING ATTRIBUTES OF DISEASE PROGRESSION FROM TRAJECTORIES OF SPARSE LAB VALUES

VIBHU AGARWAL

Biomedical Informatics Training Program, Stanford University
Stanford, CA 94305, USA
Email: vibhua@stanford.com

NIGAM H SHAH

Center for Biomedical Informatics Research, Shah Lab, Stanford University
Stanford, CA 94305, USA
Email: nigam@stanford.com

There is heterogeneity in the manifestation of diseases, therefore it is essential to understand the patterns of progression of a disease in a given population for disease management as well as for clinical research. Disease status is often summarized by repeated recordings of one or more physiological measures. As a result, historical values of these physiological measures for a population sample can be used to characterize disease progression patterns. We use a method for clustering sparse functional data for identifying sub-groups within a cohort of patients with chronic kidney disease (CKD), based on the trajectories of their Creatinine measurements. We demonstrate through a proof-of-principle study how the two sub-groups that display distinct patterns of disease progression may be compared on clinical attributes that correspond to the maximum difference in progression patterns. The key attributes that distinguish the two sub-groups appear to have support in published literature clinical practice related to CKD.

1. Introduction

It is common knowledge that diseases manifest differently in different people. Knowing the alternative progression patterns of a disease for a given population, as well as the clinical attributes associated with the patterns, is therefore of interest to patients, doctors as well as researchers[1]. Knowing what to expect, empowers patients to make informed choices about their treatment options as well as plan a judicious acquisition of healthcare resources in the future. Furthermore, the ability to spot the unusual, and initiate a clinical evaluation in case the observed symptoms are anomalous with respect to known progression attributes, has the potential to improve the care delivery process. From the perspective of the care provider, knowing the attributes of the different paths of disease progression is essential for investigating risk factors associated with progression[2].

For a healthcare system preparing to care for an aging population, an understanding of disease paths as the basis for planning treatment can have a profound impact on the patient's wellness goals. For instance, it has been seen that classification of end-stage functional decline into four groups explains the observed patterns in a sample of older medicare decedents[3]. Insight into the most likely course of progression and the "signature" attributes, can prove invaluable to healthcare professionals. Finally, a knowledge of progression patterns is essential for discovering treatment options that alter disease progression. For instance, the stage duration as well as progression rates between normal aging and severe dementia, assessed via the Global Deterioration Scale in patients with Alzheimer's, show high heterogeneity[4]. A clinical evaluation of prospective therapies that seek to slow the cognitive decline in patients with Alzheimer's would need to be carried out in individuals with similar progression trends.

The general problem is of discovering patterns of clinical events associated with stages of progression and then classes of such sequential patterns. Generally, disease progression modelling efforts first learn a state transition model using comorbidity patterns and later infer the comorbidities that drive progression based on the observed symptoms[5]. However, for many diseases, the disease status can be reliably summarized by recording one or more physiological measures. Univariate measures such as Glycosylated Hemoglobin (Diabetes), Predicted Forced Vital Capacity (Scleroderma) and Estimated Glomerular Filtration Rate (Kidney Disease) are used routinely in medical practice. These measurements are typically recorded irregularly, and usually after long intervals, making the recorded trajectories sparse. For example, out of 18,342 patients with Type 2 Diabetes in our extract of patient data from the Stanford Clinical Data warehouse, only 8231 patients had two or more HbA1c measurements. The mean number of observations per patient was 7.49. An estimate of the disease progression path based on an observed trajectory of such measurements will have high variance. As a consequence, clusters derived from such path estimates are likely to be unstable.

We hypothesize that it is possible to learn clinically meaningful clusters of disease paths from sparse and irregular trajectories of lab values. In our earlier work[2] we have described a generative model for simultaneously modelling stages of progression in a cohort of Chronic Kidney Disease (CKD) patients, as well as discovering clusters of distinct progression sequence. In related work, there are prior efforts in creating finite dimensional representation of a trajectory captured by dense measurements, and cluster the trajectories using an appropriate similarity metric. Example of successful path estimation with methods employing Gaussian Process regression often involve measurements in post-operative care or the intensive care unit, where physiological measurements are recorded regularly and relatively few observations are missing[6–8].

In order to meaningfully cluster paths estimated from sparse measurement trajectories, it is possible to borrow support from other trajectories provided a large number of trajectories have been recorded for the full time grid. The Functional Clustering Model (FCM) proposed by James and Sugar[9] models sparse trajectories as random effects, after fitting natural cubic splines to observations from each trajectory. In the work presented here, we cluster creatinine measurements from patients with Chronic Kidney Disease using the FCM. We then compare the distribution of clinical features between patients in different clusters, by defining a time window around the region of maximum discrimination between the clusters. Finally, we examine the features whose distribution is significantly different between clusters, and interpret the differences in the light of published literature on the management of Chronic Kidney Disease. Figure 1 illustrates our approach and overall workflow.

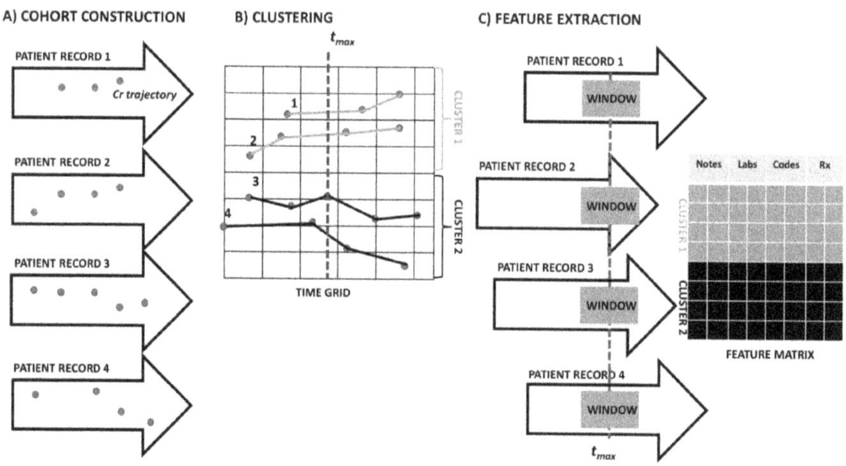

Figure 1A) Records for patients in the CKD cohort B) Clustering sparse trajectories of creatinine values. The time point at which the clustered trajectories are most discriminable is t_{max} C) Text concepts, lab results, ICD9 codes and prescriptions from a window centered at t_{max} in the patient records

2. Data

The patient dataset was extracted from the Stanford clinical data warehouse (SCDW), which integrates data from Stanford Children's Health (SCH) and Stanford Health Care (SHC). The extract comprises 2 million patients, with 49 million encounters, 35 million coded diagnoses and procedures, 204.8 million laboratory tests, 14 million medication orders as well as pathology, radiology, and transcription reports totaling over 27 million clinical notes. Our extract of the de-identified patient data from 01/1994 through 06/2013 from SCH and SHC is stored in a structured and indexed form within a MySQL relational database.

2.1. *Cohort selection*

In order to select patients with a diagnosis of Chronic Kidney Disease (CKD), we used the presence of the ICD-9 code 585.00 as our filtering criterion. We identified 959 CKD patients through this method, and kept 792 that had 3 or more creatinine measurements. We henceforth refer to this set of 792 patients as our cohort. Examining the sequence of creatinine measurements over time or "trajectories" from our cohort revealed significant sparsity in the creatinine measurements. Figure 2 shows the distribution of of the number of per patient measurements in our cohort, with a mean of 25.98 and median 13.

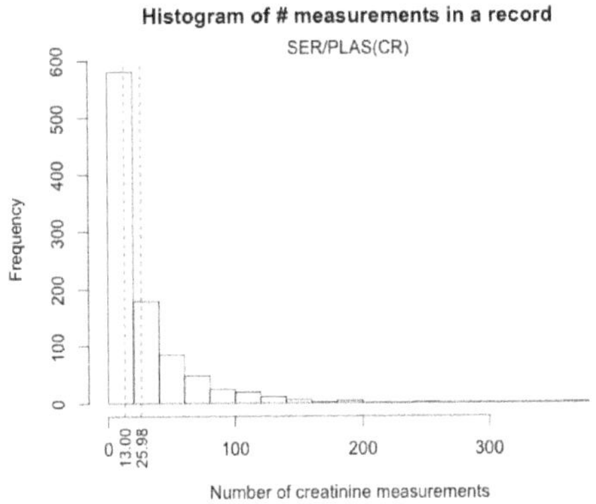

Figure 2 Distribution of number of measurements of creatinine per patient with the mean and median.

3. Methods

3.1. *Functional Clustering Model*

In order to cluster sparse observations in patient histories, we make use of the Functional Clustering Model (FCM)[9] which solves the problems of high variance due to sparsity as well as that of unequal variances due to irregular time instants. While a full description of the functional clustering approach and model fitting procedure is described by James and Sugar, the intuition behind the procedure is to project each curve onto a finite dimensional space using a natural cubic spline basis and cluster the resulting coefficients. However, instead of treating the basis coefficients as parameters James and Sugar model these are random effects that are ascertained by the clustering algorithm, via a global optimization over all curves. Doing so allows us to borrow strength across curves, allowing the method to work with sparsely or irregularly sampled curves, provided that the total number of observations is large enough.The model is fit to the data using an Expectation Maximization like procedure that iteratively updates the FCM parameters. As shown in Figure 3, each track is represented as a vector of measurements $\boldsymbol{Y}_i$ such that

$$Y_i = g_i + \epsilon_i \quad (1)$$

where g_i represents the true (unobserved) measurements at the same instants and ϵ_i is the vector of random observation errors $\epsilon_i \sim N(0, \sigma^2 I)$. The observation errors are assumed to be uncorrelated with each other and with g_i. We assume membership in one among G clusters. The true observations are modelled by using natural cubic splines as basis functions $s(t)$ so that

$$g_i(t) = s(t)^T \eta_i \quad (2)$$

where η_i is the vector of spline coefficients. The FCM uses a random effects model for the η_i, assuming a normal distribution of values around a cluster mean as follows

$$\eta_i = \mu_{z_i} + \gamma_i \quad (3)$$

where μ_{z_i} represents the mean of the ith cluster of tracks and $\gamma \sim N(0, \Gamma)$. An additional parameterization step allows for a low dimensional representation of η_i given by

$$\eta_i = \lambda_0 + \Lambda\alpha_{z_i} + \gamma_i \quad (4a)$$

$$\text{where } \mu_k = \lambda_0 + \Lambda\alpha_k \quad (4b)$$

Thereafter the FCM can be represented as below

$$Y_i = Si^T(\lambda_0 + \Lambda\alpha_{z_i} + \gamma_i) + \epsilon_i \quad (5)$$

where Si is the matrix of basis expansions of all time points for track i, α_{z_i} is a h vector representing the mean of the ith cluster of trajectories in a low dimensional space. The low dimensional representation is achieved via $\mu_k = \lambda_0 + \Lambda\alpha_k$ where both μ_k, λ_0 are vectors in $\mathbb{R}^p$, Λ is a p x h matrix and $h \leq min(p, G-1)$. To ensure a unique solution for μ_k, λ_0 and Λ, we require

$$\sum \alpha_i = 0 \quad (6a)$$

$$\Lambda^T S^T (\sigma^2 I + S\Gamma S^T)^{-1} S\Lambda = I \quad (6b)$$

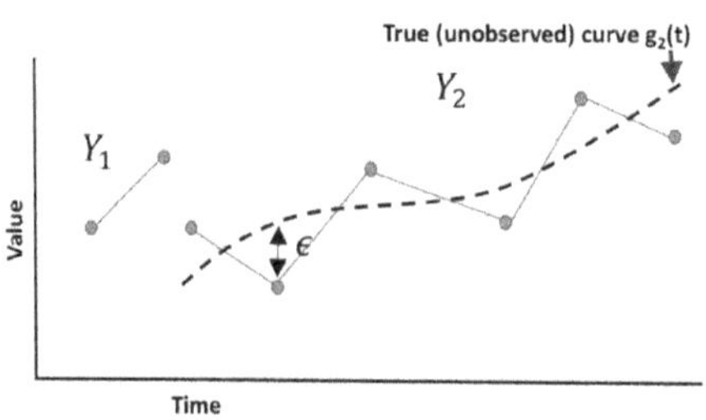

Figure 3 Modeling functional data

The form of (6b) ensures that $Cov(\alpha_i) = I$ for all i, when the observations for each track are measured at the same time points. The above formulation allows us to project every trajectory into h dimensional space to obtain the corresponding $\widehat{\alpha_i}$, such that the proximity between $\widehat{\alpha_i}$ and α_k represents the likelihood that the track i belongs to cluster k. Since $\hat{\alpha} = \Lambda^T S^T (\sigma^2 I + S\Gamma S^T)^{-1}(Y - S\lambda_0)$, the vector $\Lambda^T S^T(\sigma^2 I + S\Gamma S^T)^{-1}$ may be thought of as weights that determine the proximity to α_k, with the highest weight having the most influence on cluster assignment. This allows us to use $\Lambda^T S^T(\sigma^2 I + S\Gamma S^T)^{-1}$ as a discriminant function, for determining the time points that provide maximum cluster discrimination.

We provide here only an overview of the FCM and point the reader to a full description of the model, constraints and the fitting algorithm[9]. Since intuitively one expects to see a small number of distinct trajectory patterns, we clustered the creatinine measurements using small values of G. We obtained two nearly indistinguishable clusters with G=3, which suggests that G=2 may be an appropriate number of clusters for the creatinine trajectories in our cohort. After fitting the FCM with G=2 and making cluster assignments for every trajectory, we plotted the discriminant function for the two clusters in order to distinguish the time at which the two groups of trajectories most differ from each other.

3.2. *Feature Engineering and Analysis*

For each patient record in our cohort, we construct a one-year time window centered around the maximum discriminating time point as identified by the discriminant function described earlier. Within the time window, we represent the structured and unstructured data within a patient record as features from four categories – terms (or concepts), prescriptions, laboratory test results and diagnosis codes. Prescriptions, laboratory test results and diagnosis codes were taken from the structured record whereas terms were extracted from free text. We normalize terms into concepts in the same manner as in our earlier studies involving text mining on clinical notes—essentially using UMLS term-to-concept maps with suppression rules to weed out ambiguous mappings as described by Jung et al[10]. Such mapping reduces the total number of features as well as reduces the number of correlated features since synonyms get mapped to the same concept.

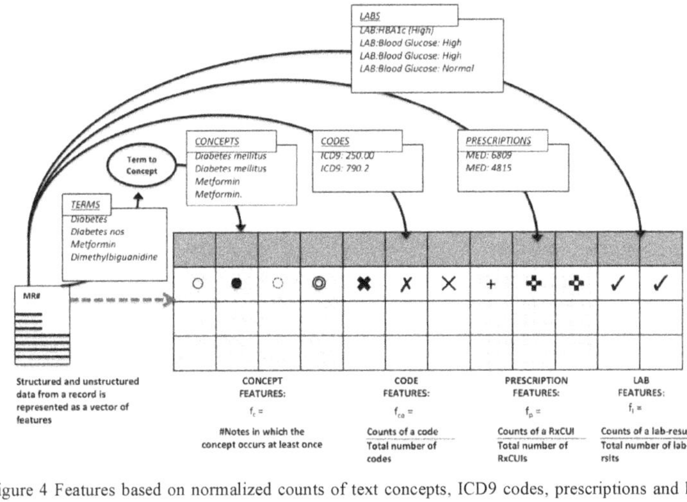

Figure 4 Features based on normalized counts of text concepts, ICD9 codes, prescriptions and lab results from the 1 year window around the maximum discriminating time point in the patient record

For concepts, we used the number of distinct notes in which the concept occurs at least once (note frequency) as the feature representation. For prescriptions and diagnostic codes, we used the normalized counts of the active ingredient for each medication (RxNorm concept unique identifier) and the normalized counts of each International Classification of Diseases, revision 9 (ICD9) code as the respective features. For laboratory test results, we utilized the categorical result status for each ordered test (high/ normal/low or normal/abnormal) as recorded in the Electronic Health Record (EHR) and calculated a feature based on the normalized counts for each test-result instance in the record. Our feature construction method is illustrated in Figure 4 which depicts our feature matrix along with the four categories (concepts, diagnosis codes, prescriptions and laboratory results) of data elements within the patient record from which the features are sourced. Finally, we performed an enrichment analysis on the feature matrix for the two clusters using Fisher's exact test, setting a false discovery rate of 5% to adjust for multiple testing. All analysis was performed in R using the Aphrodite[11] and the fclust[12] APIs.

4. Results

Figure 5 shows a randomly selected subset of 50 trajectories from each of the two clusters, indicating the overall progression pattern in each cluster. The thick lines corresponding to each cluster mean show progression patterns that the respective cluster represents. In case of cluster 1, the mean trajectory indicates that creatinine levels begin to rise around the age of 65 years, peak at around 72 years and then decrease. The trajectories in cluster 2 suggest an overall better control on creatinine levels.

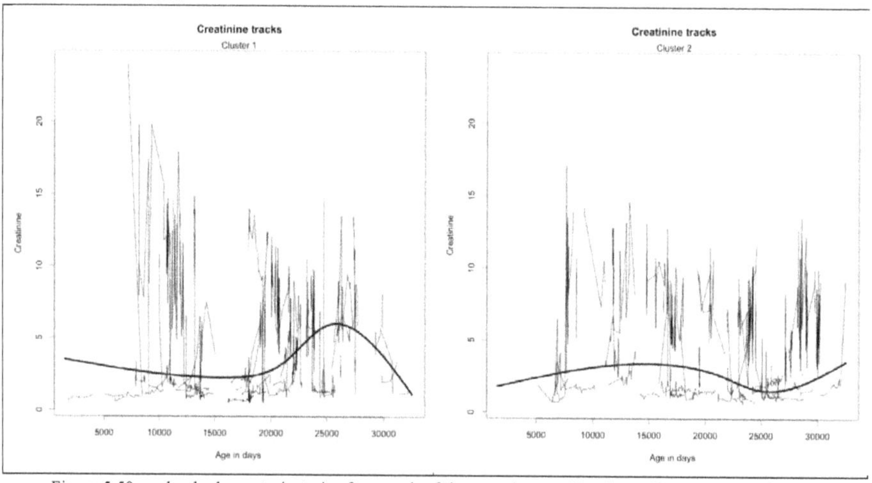

Figure 5 50 randomly drawn trajectories from each of the two clusters. The time offsets represent the age (in days) at which the measurement was taken. The heavy lines are the respective mean trajectories

The maximum value of the discriminant function occurs at a time offset t_{dmax} = 26,383 days (72 years), corresponding to the peak in the mean Creatinine trajectory in cluster 1. Drawing concepts, ICD9 codes, prescriptions and lab results from the EHR records of patients in each of the two clusters for the 1-year time window centered on t_{dmax} yields 1046 features. The rate of progression to end stage renal disease depends on a number of risk factors, such as the presence of CKD-associated illnesses, nutrition issues and adherence to medication[13] and we expect features from 1 year window to represent events associated with clinically significant decline in renal function in advanced-stage patients[14].

Table 1 Top ranked significant features

Rank	Feature ID (p)	Names	Rank	Feature ID (p)	Names
1	lab.3023314.0 (3.27e-05)	Hematocrit [Volume Fraction] of Blood by Automated count:	16	obs.4274025 (7.86e-04)	Disease
2	lab.3000963.0 (4.06e-05)	Hemoglobin:	17	obs.4322976 (8.01e-04)	Procedure
3	lab.3000905.0 (5.92e-05)	Leukocytes [#/volume] in Blood by Automated count:	18	obs.4229881 (8.46e-04)	Weight loss
4	obs.4187395 (7.19e-05)	Reflux	19	obs.46233416 (9.53e-04)	Assessment
5	obs.4187458 (1.46e-04)	Review of systems	20	obs.4143467 (1.09e-03)	Chief complaint
6	obs.4243768 (1.78e-04)	Auscultation	21	lab.3009261.0 (1.22e-03)	Glucose [Presence] in Urine by Test strip:
7	obs.4118663 (2.06e-04)	Related	22	obs.4209224 (1.51e-03)	Cyst
8	obs.31967 (3.23e-04)	Nausea	23	obs.441408 (1.67e-03)	Vomiting
9	lab.3013682.0 (3.25e-04)	Urea nitrogen serum/plasma:	24	obs.4147571 (1.82e-03)	Follow-up
10	obs.442985 (3.85e-04)	Male	25	obs.4267147 (1.85e-03)	Platelet count
11	lab.3014051.0 (4.03e-04)	Protein [Presence] in Urine by Test strip:	26	lab.3022621.0 (1.85e-03)	pH of Urine by Test strip:
12	obs.254761 (4.21e-04)	Cough	27	obs.77670 (1.86e-03)	Chest pain
13	obs.4077953 (4.88e-04)	Therapy	28	lab.3035350.0 (1.86e-03)	Ketones urine dipstick:
14	obs.4099313 (5.19e-04)	Urinalysis	29	lab.3004501.45876384 (1.96e-03)	Glucose lab:High
15	obs.4329041 (7.37e-04)	Pain	30	obs.4303558 (1.98e-03)	Touch

A Fisher's exact test of enrichment for each feature with respect to the two clusters, using a false discovery rate of 5% to adjust for multiple testing, identified 133 enriched features, of which 30 top ranked features along with their respective p-values are presented in Table 1.

All of the top 30 features are found to be enriched in cluster 1 and an examination of the features suggests concordance with the known attributes of disease severity. For example the top 2 features (lab orders for hematocrit and hemoglobin) suggest the presence of Anemia and Thrombocytopenia respectively, which are two common comorbidities in advanced CKD that are thought to occur as a result of reduced erythropoietin secretion[15]. An observation related to platelet counts (feature rank 25) corroborates this view. Similarly, automated blood count is routinely measured in CKD patients, particularly in those patients requiring management of Anemia. Recently studies show that spikes in granulocyte and monocyte count in CKD patients are associated with progression to end stage renal disease[16]. A finding of Proteinuria on dipstick urinalysis is amongst the early signs of kidney disease, however high levels of protein in the urine is an indicator of nephritic syndrome and is associated with edema, increased cholesterol levels and other comorbidities that increase the risk of CKD progression[17]. Lab orders for pH and ketone measurement in urine (feature rank 26 and 28 respectively) suggest diabetic ketoacidosis which is an uncommon but life threatening complication in chronic kidney disease. Poor glucose regulation (feature rank 29) further supports the view that several of the distinguishing attributes have etiological linkages with diabetic complications.

5. Discussion

Longitudinal patient data from a large sample of patients offers an opportunity to characterize the variability in how phenotypes progress over time. However, discovering patterns of progression for chronic diseases is challenging because of the irregularity and sparsity in the observations. Trajectory estimates derived from irregularly sampled and sparse observations have high variance which leads to unstable cluster definitions. The irregular sampling also poses an additional challenge – the variance of the estimated curve coefficients are different for each trajectory. The FCM described in the methods section addresses the problem by treating the curve coefficients as random effects and by projecting each curve into a subspace, such that the covariance normalized distance from the cluster center in this subspace, represents the probability of cluster membership. Using the FCM to cluster creatinine trajectories of CKD patients results in two clusters with distinct mean trajectories. Features based on counts of clinical attributes, from a windowed segment of the patients' EHR records at the point of maximal cluster separation, show a significantly different distribution in the two clusters; and many are supported by medical literature on CKD. However, several of our significant features refer to general CKD attributes and do not appear to have an obvious connection with disease severity or progression. Further, reducing the false discovery rate to 1% did not give any statistically significant features.

We acknowledge limitations of our approach. We made use of the ICD9 code for CKD for defining our cohort. Such a method could have a positive predictive value (PPV) as low as 53%[18] if relying solely on the ICD9 code. In which case, the the selected patients may not have the clinical indicators for the disease, while creatinine measurements could still be available since creatinine may be ordered routinely as part of a basic metabolic panel. We mitigate this issue by requiring at least three creatinine measurements for members of the analysis cohort. Further still, creatinine values are known to be altered through processes that are independent of renal function and standard practice requires that the estimated Glomerular Filtration Rate be used for assessing kidney health[19]. The FCM also implicitly assumes that the unobserved time points are missing at random. Given that our data comes from a referral facility, it is likely that there exists a "disease severity bias" in the missing observations.

The alternative to using ICD9 codes for cohort identification is to use a robust algorithm that has been validated to achieve a high PPV[20]. Patient records extracted from claims data may possibly provide better longitudinal coverage compared to EHR data from a tertiary care facility. Addressing our study's limitations through the aforementioned remedial measures appears feasible and we anticipate doing so in follow up work.

6. Conclusion

Being able to account for individual variability in the progression of diseases is of value to the practitioner, the patient as well as the researcher. For chronic diseases, learning the clinical attributes of the disease progression paths is possible by using a method for clustering irregularly sampled, sparse trajectories of disease markers, by defining a time window in which the clusters are most discriminable, and identifying discriminating features based on that time window. Our results from clustering creatinine trajectories of CKD patients demonstrate the feasibility of the approach.

7. Acknowledgements

We would like to thank our colleagues Sarah Poole and Jassi Pannu for their contributions to this study. This work was funded by NLM R01 LM011369.

References

1. Jensen PB, Jensen LJ, Brunak S. Mining electronic health records: towards better research applications and clinical care. *Nat Rev Genet.* 2012;13(6):395-405. doi:10.1038/nrg3208.
2. Yang J, McAuley J, Leskovec J, LePendu P, Shah N. Finding progression stages in time-evolving event sequences. *Proc 23rd Int Conf World wide web.* 2014:783-794. doi:10.1145/2566486.2568044.
3. Lunney JR, Lynn J, Hogan C. Profiles of Older Medicare Decedents. *J Am Geriatr Soc.* 2002;50(6):1108-1112. doi:10.1046/j.1532-5415.2002.50268.x.
4. Komarova NL, Thalhauser CJ. High degree of heterogeneity in Alzheimer's disease progression patterns. *PLoS Comput Biol.* 2011;7(11):e1002251. doi:10.1371/journal.pcbi.1002251.
5. Wang X, Wang F. Unsupervised Learning of Disease Progression Models. 2014. doi:10.1145/2623330.2623754.
6. Lasko TA, Denny JC, Levy MA. Computational phenotype discovery using unsupervised feature learning over noisy, sparse, and irregular clinical data. *PLoS One.* 2013;8(6):e66341. doi:10.1371/journal.pone.0066341.
7. Pimentel M, Clifton D, Clifton L, Tarassenko L. Modelling Patient Time-Series Data from Electronic Health Records using Gaussian Processes. *Adv neural Inf Process Syst Work Mach Learn Clin Data Anal.* 2013:1-4.
8. Pimentel, MAF, Clifton DA TL. Gaussian process clustering for the functional characterisation of vital-sign trajectories. In: *Machine Learning for Signal Processing*

(MLSP), 2013 IEEE International Workshop On. ; 2013:1-6. doi:10.1109/MLSP.2013.6661947.

9. James GM, Sugar C a. Clustering for Sparsely Sampled Functional Data. *J Am Stat Assoc.* 2003;98(462):397-408. doi:10.1198/016214503000189.
10. Jung K, LePendu P, Iyer S, Bauer-Mehren A, Percha B, Shah NH. Functional evaluation of out-of-the-box text-mining tools for data-mining tasks. *J Am Med Inform Assoc.* 2015;22(1):121-131. doi:10.1136/amiajnl-2014-002902.
11. OHDSI. Aphrodite. https://github.com/OHDSI/Aphrodite. Accessed October 3, 2016.
12. Gareth J. Fclust. http://www-bcf.usc.edu/~gareth/research/fclustdoc.pdf. Accessed October 3, 2016.
13. Thomas R, Kanso A, Sedor JR. Chronic kidney disease and its complications. *Prim Care.* 2008;35(2):329-344, vii. doi:10.1016/j.pop.2008.01.008.
14. Zhang A-H, Tam P, LeBlanc D, et al. Natural history of CKD stage 4 and 5 patients following referral to renal management clinic. *Int Urol Nephrol.* 2009;41(4):977-982. doi:10.1007/s11255-009-9604-3.
15. Akimoto T, Ito C, Kotoda A, et al. Challenges of caring for an advanced chronic kidney disease patient with severe thrombocytopenia. *Clin Med Insights Case Rep.* 2013;6:171-175. doi:10.4137/CCRep.S13238.
16. Agarwal R, Light RP. Patterns and prognostic value of total and differential leukocyte count in chronic kidney disease. *Clin J Am Soc Nephrol.* 2011;6(6):1393-1399. doi:10.2215/CJN.10521110.
17. Mehdi U, Toto RD. Anemia, diabetes, and chronic kidney disease. *Diabetes Care.* 2009;32(7):1320-1326. doi:10.2337/dc08-0779.
18. Cipparone CW, Withiam-Leitch M, Kimminau KS, Fox CH, Singh R, Kahn L. Inaccuracy of ICD-9 Codes for Chronic Kidney Disease: A Study from Two Practice-based Research Networks (PBRNs). *J Am Board Fam Med.* 28(5):678-682. doi:10.3122/jabfm.2015.05.140136.
19. Samra M, Abcar AC. False estimates of elevated creatinine. *Perm J.* 2012;16(2):51-52.
20. Nadkarni GN, Gottesman O, Linneman JG, et al. Development and validation of an electronic phenotyping algorithm for chronic kidney disease. *AMIA Annu Symp Proc.* 2014;2014:907-916.

COMPUTER AIDED IMAGE SEGMENTATION AND CLASSIFICATION FOR VIABLE AND NON-VIABLE TUMOR IDENTIFICATION IN OSTEOSARCOMA

HARISH BABU ARUNACHALAM[1]*, RASHIKA MISHRA[1], BOGDAN ARMASELU[1], DR. OVIDIU DAESCU[1] and MARIA MARTINEZ[2]

[1]*Department of Computer Science,*
[2]*Department of Biomedical Engineering,*
University of Texas at Dallas, Richardson, TX
**Email: harishb@utdallas.edu*

DR. PATRICK LEAVEY, DR. DINESH RAKHEJA, DR. KEVIN CEDERBERG, DR. ANITA SENGUPTA and MOLLY NI'SUILLEABHAIN

University of Texas Southwestern Medical Center, Dallas, TX

ABSTRACT: Osteosarcoma is one of the most common types of bone cancer in children. To gauge the extent of cancer treatment response in the patient after surgical resection, the H&E stained image slides are manually evaluated by pathologists to estimate the percentage of necrosis, a time consuming process prone to observer bias and inaccuracy. Digital image analysis is a potential method to automate this process, thus saving time and providing a more accurate evaluation. The slides are scanned in Aperio Scanscope, converted to digital Whole Slide Images (WSIs) and stored in SVS format. These are high resolution images, of the order of 10^9 pixels, allowing up to 40X magnification factor. This paper proposes an image segmentation and analysis technique for segmenting tumor and non-tumor regions in histopathological WSIs of osteosarcoma datasets. Our approach is a combination of pixel-based and object-based methods which utilize tumor properties such as nuclei cluster, density, and circularity to classify tumor regions as viable and non-viable. A K-Means clustering technique is used for tumor isolation using color normalization, followed by multi-threshold Otsu segmentation technique to further classify tumor region as viable and non-viable. Then a Flood-fill algorithm is applied to cluster similar pixels into cellular objects and compute cluster data for further analysis of regions under study. To the best of our knowledge this is the first comprehensive solution that is able to produce such a classification for Osteosarcoma cancer. The results are very conclusive in identifying viable and non-viable tumor regions. In our experiments, the accuracy of the discussed approach is 100% in viable tumor and coagulative necrosis identification while it is around 90% for fibrosis and acellular/hypocellular tumor osteoid, for all the sampled datasets used. We expect the developed software to lead to a significant increase in accuracy and decrease in interobserver variability in assessment of necrosis by the pathologists and a reduction in the time spent by the pathologists in such assessments.

Keywords: Image segmentation, Otsu thresholding, Osteosarcoma, SVS image analysis

1. Introduction

Pathology Informatics, one of the fastest growing fields in medical informatics, deals with mining information from medical pathology data and images. It involves the use of computational methods and analytical processes to make informed decisions that serve as assistive tools in clinical diagnosis. Due to the complexity of medical data and given the expert knowledge required for such analyses, it is often difficult to replicate the work of pathologists and physicians.[1] Though there is substantial literature published in the area of tumor research[2,3] the main challenge in the field is that all the methods are tumor specific which makes the development of one common method, that is applicable for all kinds of tumor, an arduous task. This necessitates the creation of ad hoc methods tied to each requirement, that consider signature of each tumor sample and incorporate tumor specific information such as the tumor spread, contextual information etc. Each tumor detection method utilizes specific information about the tumor and therefore one tumor identification approach may not be applicable for another. Hence it becomes a challenge to apply existing methods that work well for other types of tumors for Osteosarcoma detection, the tumor that is used in this study.

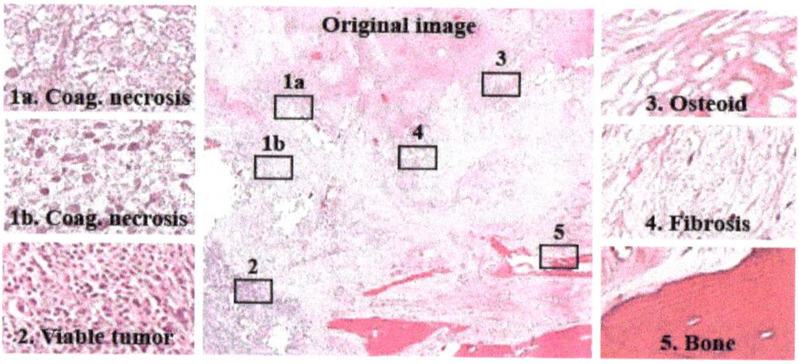

Fig. 1. WSI with different regions enlarged and their location in the image with figure 1a and 1b representing color and shape variations in coagulative necrosis regions for the same WSI. The numbered boxes represent the locations of histologically distinct regions in the image.

Osteosarcoma is the most common type of bone cancer that occurs in adolescents in the age of 10 to 14 years. The tumor usually arises in the long bones of the extremities in the metaphyses next to the growth plates.[4] What makes osteosarcoma analysis inherently challenging is that there is a high degree of histologic variability within the tumor (Figure 1) which is accentuated after therapy. In order to accurately identify tumor occurrence and estimate the treatment response, it is necessary to consider histologically distinct regions that include dense clusters of nuclei, fibrous tissues, blood cells, calcified bone segments, marrow cells, adipocytes, osteoblasts, osteoclasts, haemorrhagic tumor, cartilage, precursors, growth plates and osteoid (tumor osteoid and reactive osteoid) with and without cellular material. Each of these regions have different characteristic features that differ in color, shape, size, density, texture and area of occurrence. They also have significant differences in their biological features such as background stroma, presence or absence of certain cellular material, neighboring regions etc. There are also multiple color variations within the same dataset representing the same type of regions (Figure 1a and 1b), which makes segmentation based only on color ineffective. Due to the variable properties of the images, there is no one method that with certainty, can accurately segment regions and classify them. The literature available for osteosarcoma data image analysis/digital pathology is minimal which makes it critical to come up with methods that are applicable for this type of tumor analysis.

The goal of this paper is to present an approach to segment H&E[5] stained images into viable tumor and non-viable tumor regions using a combination of techniques (color segmentation, Otsu thresholding, and Flood-fill). It employs pixel-based and object-based segmentation by using color, shape and density parameters. The first step locates tumor and non-tumor regions and subsequent steps distinguish tumor into viable and non-viable regions. Color quantization in the approach accounts for variance in color distribution while shape properties such as area and circularity measures are utilized to accurately locate tumor. A further analysis performed on computed cluster data on resulting images characterize different regions in the image. The approach is shown to be robust and has a high accuracy overall in the datasets considered.

1.1. *Background and Setup*

We have assembled an investigative team of clinical scientists at University of Texas Southwestern Medical Center, Dallas and computer scientists at University of Texas at Dallas. Archival samples for 50 patients treated at Children's Medical Center, Dallas, between 1995 and 2015, have been identified. The treatment effect for each patient is estimated after surgical resection, by physically cutting the region of interest from the resected bone into pieces. These pieces are de-calcified, treated with H&E stain and converted to slides to be analyzed under microscope. Each patient case is represented by a single H&E stained slide at the time of biopsy when available and 8-50 H&E stained slides per case at time of resection when necrosis is determined. Each slide consists of 1-2cm X 1-2cm sections of the tumor in its widest coronal plane. Slides are scanned using an Aperio Scanscope at a magnification of upto 40X and stored in SVS format.[6] Each SVS WSI has a size between 150MB and 1.5GB and spans an order of 10^9 pixels. The experimental dataset consists of a subset of images from the above available patient datasets, manually annotated by pathologists.

1.2. *Related Work*

The Tumor identification/isolation problem has been well studied by researchers in the field of digital pathology[7–11] . The most common methods include image segmentation (region classification), image analysis (pattern analysis), regions of interest(ROI) identification, statistical analysis (such as number of clusters, mean size of groups) etc.[7,8] The methods used in all the studies include color based (pixel level), shape based (object level) and contextual information based methods.[8] Normally, pixel level methods, that make use of pixel level processing, form some of the basic approaches because they are the simplest but they are not the most efficient. Researchers have tried to analyze the pathological images based on quantitative metrics representing the spatial structure of histopathology imagery and include identifying structures such as nuclei, glands and lymphocytes etc. These spatial feature utilization[12] has become the backbone of histopathology image analysis techniques, as these are the prominent metrics that can yield maximum information. More advanced than the pixel based methods are the object based methods that make use of region growing and object identification utilizing shape properties. These methods provide better segmentation results than their pixel counterparts[2,13] however they are expensive in terms of the computational resources they use. Multi-level thresholding approach using Otsu segmentation promises good accuracy but when there is a lot of noise, this method alone may not fare well. Pattern Recognition Image Analysis(PRIA) describes a pattern recognition method based on a genetic algorithm that evolves over multiple iterations and compares the results with GeNIE,[14] a bio-image analysis tool from Aperio and manual segmentation by pathologists. A recent work on Lung cancer[11] identifies 9879 image features and uses regularized classifiers to estimate patient prognosis.

1.3. *Challenges*

A majority of pathological images are in proprietary format and there is no common standard for these images, which makes it difficult for researchers collaborating towards common goals to share information. Openslide[15] and VIPS[16] are image processing libraries that help to narrow down the gap by a small margin, however, the main problem of handling different imaging formats remains the same.

Pathology is a relatively subjective field dependent on the opinions of trained pathologists resulting in discrepancy in the accuracy of different image analysis approaches available for pathological images. A study on renal cell carcinoma[10] performed analysis on pathologists and found a high degree of subjectivity in their evaluation. Therefore, a standard objective procedure is recommended, which is important not only from a clinical standpoint but also for developing quality research application that will be reliable and independent of varying views of pathologists.[7] The size of images generated in these studies is large and as a result the algorithms for smaller images may not scale up due to memory issues.[17] Hence, there is a need for a standard approach that will process images one tile at a time and at the same time can scale up without loss of accuracy. Most of the tools that are developed by researchers in academia cater to a subset of problems. ImageJ[18] has many inbuilt image processing algorithms, however, is limited in its use to process proprietary formats and large files. PRIA by Webster et.al[1] is an advanced method, but it fails to perform well in identifying necrosis, which is one of the main tasks in this study. CellProfiler,[19] a tool for high throughput image analysis is good at identifying cellular objects and calculating their properties. However, the results of our trials with CellProfiler are inconclusive in identifying cellular objects in Osteosarcoma. Object based methods[2,13] work well on images with well-defined shapes but need pre-configured training sets. Machine learning approaches such as Bayesian classifier, Support Vector Machines[9] etc. are effective but would need a large annotated training data and the training phase is very time consuming.[7] Some of the related works[9–11] in lung and breast cancer focus on identifying properties and features of nuclei. Necrotic regions in this study do not necessarily have nuclei and hence the above works address only a part of the problem. The presence of a high degree of variability in the shapes, in Osteosarcoma datasets, makes the above methods unlikely to perform well. Another issue with WSI images is the color variance between different features,[9] which causes active tumor cells to have different color signatures, and thus segmentation based only on color is less accurate.

Given these challenges, in the next few sections we describe our approach explaining the algorithm and our results.

2. The Approach

We illustrate in Figure 2, the complete procedure, which includes K-means color segmentation, multi-level Otsu segmentation, Flood-fill clustering and statistical analysis. Based on inputs from pathologists, we define the different tumor regions with the following properties which are then quantified in the segmentation and classification approach.

(1) **Viable Tumor:** Nuclei densely aggregated together

(2) **Non-Viable Tumor:** Cells and tissues in the stage of recovery or dead

  (a) *Coagulative Necrosis:* disintegrated nuclei but with less color density than viable tumor.

  (b) *Fibrosis:* Fibrous collagen (protein) produced by fibroblasts (benign cells).

  (c) *Acellular/Hypocellular Tumor Osteoid (subsequently designated as Osteoid in this paper):* Eosinophilic/pink extracellular protein matrix produced by tumor cells.

The viable tumor and coagulative necrosis regions resemble each other in terms of high color density, closer to blue while fibrosis and osteoid have brighter color shade, closer to pink. The above regions are grouped together into two intermediary classes Ψ_1 and Ψ_2 based on high intra-class similarities, as follows:

(1) $\Psi_1 = \{Viable\ tumor, Coagulative\ necrosis\}$

(2) $\Psi_2 = \{Fibrosis,\ Osteoid\}$

The images in Ψ_1 are analyzed in terms of shape and density properties to classify them as viable tumor and non-viable tumor, while those in Ψ_2 are by default classified as non-viable tumor.

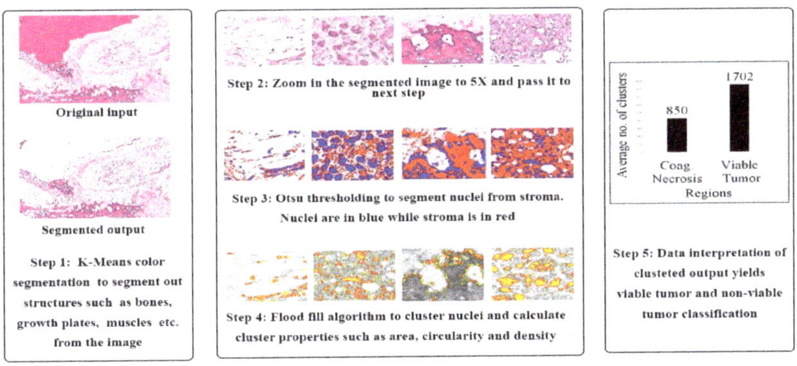

Fig. 2. **Algorithm pipeline**

2.1. *The Algorithm Pipeline*

The following is the general algorithm pipeline.

Input: Unprocessed SVS image

Output: Color segmented image and mapped regions identified as tumor (viable and non-viable) and non-tumor

Steps

For each SVS image given as input, at the eye fit (scaling factor 1X) zoom level, do the following:

(1) Run K-means color segmentation with K=3.
(2) For each of the tumor regions identified in step 1, increase the scaling factor to 5X.
(3) On a window size of 512 * 512, on the original 5X scaled images, do the following:
   (a) Generate red-blue segmentation using 2-level Otsu thresholding.
   (b) Compute the percentage of red pixels and blue pixels in each image.
   (c) Images with higher percentage of blue pixels fall into the Ψ_1 class
   (d) Images with higher percentage of red pixels fall into the Ψ_2 class
   (e) Create a tumor map of the pixel information by including pixel location, color values and class label
(4) For each entry in the tumor map,
   (a) Run Flood-fill algorithm to identify boundaries and group pixels in the cluster.
   (b) Remove clusters that are smaller than minimum cluster size and larger than maximum cluster size to remove false positives.
   (c) Output the remaining pixels in the original image along with cluster-mapped pixels.
(5) Run data analysis and classify images as viable and non-viable tumor based on the cluster data output.

2.2. *Color Segmentation*

At eye-fit level(scaling factor 1X), a 3-means color segmentation process is used to distinguish the given image into tumor and non-tumor image. Since all the WSIs are H&E stained, the pixels are made up of variants of Red and Blue channels. Hence it is imperative that more focus is given to these two channels. Given an Image I, of width I_w and height I_h, made up of $I_w * I_h$ pixels, each pixel P^i is represented by $\{P^i_r, P^i_g, P^i_b\}$, where P^i_r, P^i_g and P^i_b denote the red, blue and green channel values of ith pixel. Let the set $C \in (C^w, C^p, C^b)$ represent the cluster centers of white, pink and blue regions. These color values are taken from the empirical analysis of the stained images. Each P^i in the image is assigned to one of the cluster centers C^k by calculating distances between the pixel and the centroids. The distance is given by subtracting the color channel differences between the pixel and centroid. If $\phi(P^i)$ represents the cluster value for pixel i, then

$$\phi(P^i) = \arg\min_{C^j} \ \delta(P^i, C^j) \tag{1}$$

where,

$$\delta(P^i, C^j) = \sqrt{(P^i_r - P^k_r)^2 + (P^i_g - P^k_g)^2 + (P^i_b - P^k_b)^2} \tag{2}$$

where P^k_r, P^k_g, P^k_b represent RGB color channel values of the pixel, of kth cluster centroid. The centroids are initialized with random values and each pixel P^i in I is classified. The pixel values of centroids are then updated as follows:

$$P_r^k = \frac{1}{N_k}\sum_j P_r^j;\ P_g^k = \frac{1}{N_k}\sum_j P_g^j;\ P_b^k = \frac{1}{N_k}\sum_j P_b^j; \tag{3}$$

Where N_k is the number of pixels classified under k^{th} centroid. The algorithm is run for Γ iterations until there are no more changes to the clusters. The clusters represented by centroids C^b and C^p are regions of potential tumor whereas C^w represents non-tumor. The blue and pink clusters are further investigated at a higher level of magnification for detailed classification. After K-means, the data is passed on to the next step, with the following values populated for each pixel P^i. Map M contains { m_{p_1}, m_{p_2}...} and each m_{p_i} =*(pixel-location, color value, label)*

2.3. *Otsu multi-level threshold segmentation*

A 2-level threshold segmentation is used in the next step. A window of 512 x 512 is considered and the color image is converted to 24 bit grayscale image, with more weight to blue channel. This is due to the fact that tumor regions have more blue channel values than non-tumor regions. The gray scale values for two level threshold are represented as [1,2,...t] and [t+1,....L] respectively and the weighted class variance is calculated as

$$\sigma_w^2(t) \ = \ q_1(t)\sigma_1^2(t) \ + \ q_2(t)\sigma_2^2(t) \tag{4}$$

where

$$q_1(t) \ = \sum_{i=1}^{t} P(i);\ q_2(t) \ = \ \sum_{i=t+1}^{L} P(i) \tag{5}$$

are the class probabilities and the intra-class variance is given by

$$\mu_1(t) = \sum_{i=1}^{t} \frac{i * P(i)}{q_1(t)} \tag{6}$$

Now the image contains two classes of pixels following a bi-modal histogram. We calculate the optimum threshold separating the two classes so that their combined spread (intra-class variance) is minimal. Otsu thresholding creates a red-blue color map where red signifies the non-viable tumor pixel and blue is the viable tumor or coagulative necrosis pixel. The data from this stage is exported by updating the values in map M as m_{p_i}=(pixel, red/blue value, original color value, label)

2.4. *Calculating clusters*

This stage calculates blue clusters and their properties. We run Flood-fill algorithm to identify boundaries and compute clusters. Viable tumor and coagulative necrosis always are in the blue region and contain cellular structures within them of non-uniform circularity. Along with each cluster, the size of the cluster in terms of area a, circularity c, and average color of cluster are calculated. Clusters that are less than minimum cluster size (50 pixels) and greater than maximum cluster size (300 pixels) are discarded as they represent false positives. The output of this stage is a map M' (Cluster label, Start Point, Centroid, Circularity, Area, List of Points, Color) and an image of clusters with red borders.

3. Results

The application was created using Openslide-Java for processing SVS images, ImageJ and Java Advanced Imaging for basic image processing tasks and C# .NET to perform Otsu segmentation and Flood-fill. The dataset included 120 images of 1160 x 640 resolution and all data samples were manually analyzed and classified by pathologists at UT Southwestern. The output of the application was compared with the classified images for verification and was validated by the team at UT Southwestern.

The choice of parameters such as window size in Otsu step (512*512), minimum and maximum cluster sizes (50 pixels and 300 pixels) etc. are selected based on empirical values for which the accuracy is maximized. Each figure (Figure 3 - Figure 6) consists of an original image I from the dataset, shown in (a), the output of Otsu method applied on I, (b) and clustering of cells by flood fill method,(c). Each cluster inside the red boundary in the images is defined in memory as a map data structure, Cluster(*Centroid,OriginalColor,Area,Perimeter,Circularity*).

3.1. *Viable Tumor*

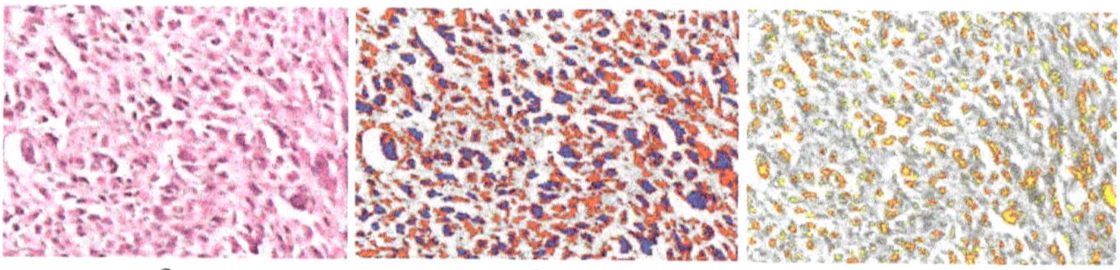

a b c

Fig. 3. Region: Viable tumor. (a) original image for viable tumor, (b) Otsu output showing more blue color, (c) Cell clustering using flood fill showing computed clusters.

Figure 3(a) shows that viable tumor has dense nuclei with more blue color. Otsu segmentation captures the nuclei with high accuracy represented by blue regions, as seen in 3(b). The percentage of blue pixels higher than the percentage of red pixels is a significant indicator for classifying this region into class Ψ_1. Clustered cellular information from Flood-fill 3(c) shows that there are more cells in the viable tumor region than the others. (see Figure 8(a)).

3.2. *Coagulative Necrosis*

Figure 4(a) shows coagulative necrosis containing cells with disintegrated nuclear matter, which makes the image appear brighter than viable tumor. Otsu segmentation step 4(b) yields higher percentage of blue pixels than red pixels, which is a key parameter in deciding regions belonging to class Ψ_1. Cluster properties in 4(c) show that the cell clusters are less dense and are distant from each other.

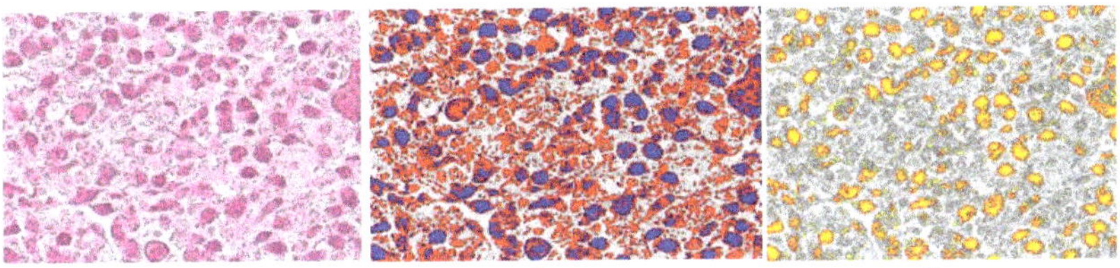

a b c

Fig. 4. Region: Coagulative necrosis. (**a**) original image for coagulative necrosis. (**b**) output of Otsu showing more blue than red, (**c**) Cell clustering using flood fill similar to viable tumor

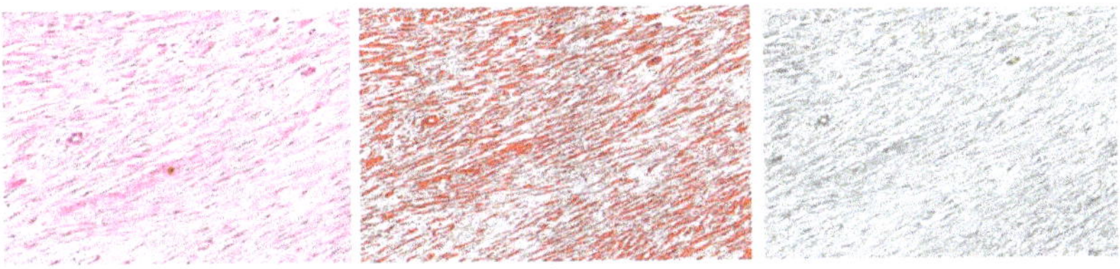

a b c

Fig. 5. Region: Fibrosis. (**a**) original image for fibrosis, (**b**) shows Otsu output for Fibrosis where there is a higher percentage of red pixels than blue, (**c**) cell clustering using flood fill, showing presence of fewer cells

3.3. *Fibrosis*

Figure 5(a) shows fibrosis region represented by strand like structures and absence of cells and nuclei. Due to this characterstic, Otsu in 5(b) produces higher percentage of red than blue pixels. This result distinguishes fibrosis from images in class Ψ_1. Flood-fill on this output, seen in 5(c), produces lesser number of clusters compared to viable tumor and coagulative necrosis.

3.4. *Osteoid*

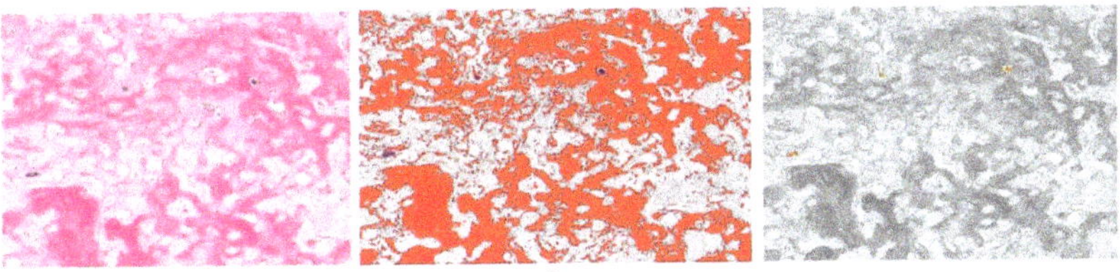

a b c

Fig. 6. Region: Osteoid. (**a**) original image for osteoid, (**b**) Otsu output for osteoid characterized by higher percentage of red than blue pixels. (**c**) Cell clustering flood fill output marked by absence of cells.

Figure 6(a) shows osteoid from the dataset, characterized by pink regions, background stroma and absence of cells and nuclei similar to fibrosis. Running Otsu on this image produces 6(b), with high percentage of red pixels due to absence of cells. This result makes osteoid to be grouped in Ψ_2 distinguishing them from images in class Ψ_1. Since osteoid is an extracellular protein matrix, it remains after tumor cells have undergone necrosis. However, there maybe interspersed cells found in the matrix. Flood-fill on this output, shown in 6(c), captures scattered cells that are less dense, unlike viable tumor and coagulative necrosis.

3.5. *Data Interpretation*

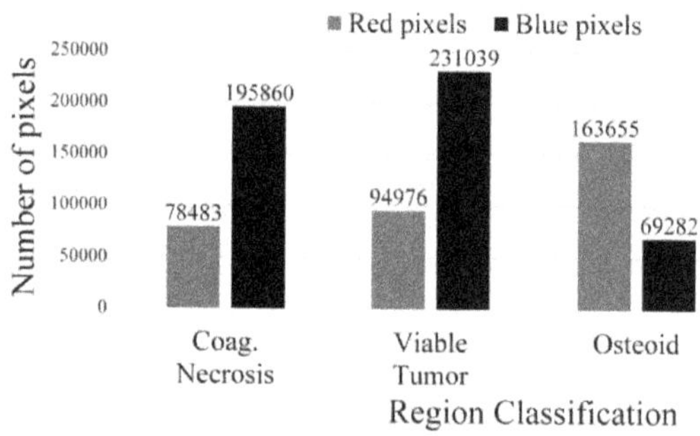

Fig. 7. Region wise average red and blue pixel count

A plot of average pixel count for classification regions shows that viable tumor and coagulative necrosis regions have more blue pixels, while fibrosis and osteoid regions have more red pixels. This result from Otsu step divides the image into prominent blue and red regions (see Figure 7). The regions that get classified under Ψ_1 have more cells than the regions under Ψ_2, and therefore have more blue pixels than red. Thus, images with viable tumor and coagulative necrosis regions can be classified into Ψ_1, while fibrosis and osteoid can be classified into Ψ_2.

A further analysis on average cell counts shows that viable tumor has 1702 cell clusters, while coagulative necrosis has 850 cells (see Figure 8(a)). This further distinguishes images in Ψ_1 into viable tumor and coagulative necrosis more accurately. We calculated the average density of cells in a 32x32 window as shown in Figure 8(b). It is observed that viable tumor has a cellular density of 2.4 while coagulative necrosis has 1.17. This important characteristic differentiates viable tumor from coagulative necrosis. The findings conclude that viable tumor is more dense and has closely aggregated cells than coagulative necrosis, the result of which has been used in classification.

It can be seen that fibrosis and osteoid regions have low cell clusters and high background stroma, hence concurring with the previous findings that these segmented images have less blue and more red pixels. Thus, the images in Ψ_2, that were identified as fibrosis and osteoid, have been categorized as non-viable tumor. Furthermore, in class Ψ_1, the cellular density distinguishes viable tumor from coagulative necrosis, which can be used to classify coagulative necrosis under non-viable tumor.

The accuracy of the method has been measured and is found in Table 1.

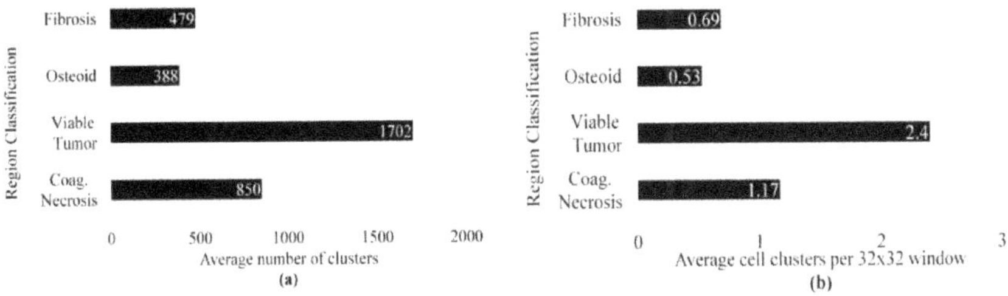

Fig. 8. (a) Region wise cellular count.(b) Region wise average cellular density count per 32x32 window.

Table 1. Quantitative metric comparison of classification regions

Region Type	Quantitative Metrics (Average)				
	red pixels count	blue pixels count	cell clusters	cell density (32x32 window)	Classification accuracy(%)
Viable Tumor	94976	231039	1702	2.4	100
Coagulative Necrosis	78483	195860	850	1.17	100
Osteoid	163655	69282	388	0.53	93
Fibrosis	140026	56313	479	0.69	89

4. Limitations and Future Improvements

The approach presented in this paper is limited to Osteosarcoma tumor identification. The current method works well in the given sampled datasets. We propose to extend it to all images in the dataset by incorporating contextual information that yield additional data. RGB color channels used in the experiments are affected by color variance in images and hence we would replace them with LAB colorspace. We plan to identify more relevant features from the images and subsequently use machine learning algorithms on them to improve classification accuracy.

Acknowledgments

This research was partially supported by NSF award IIP1439718 and CPRIT award RP150164. We would like to thank Dr.Lan Ma, University of Maryland and Dr.Riccardo Ziraldo, University of Texas at Dallas for their helpful discussions. We also would like to thank John-Paul Bach and Sammy Glick from UT Southwestern Medical Center for their help with datasets.

References

1. J. D. Webster, A. M. Michalowski, J. E. Dwyer, K. N. Corps, B.-R. Wei, T. Juopperi, S. B. Hoover, R. M. Simpson *et al.*, *Journal of pathology informatics* **3**, p. 18 (2012).
2. J. I. Zhongwu Wang, John R. Jensen, *Environmental Modelling and Software* **25**, 1149 (October 2010).
3. R. Harrabi and E. B. Braiek, *EURASIP Journal on Image and Video Processing* **2012**, 1 (2012).
4. G. Ottaviani and N. Jaffe, *The Epidemiology of Osteosarcoma*, in *Pediatric and Adolescent Osteosarcoma*, eds. N. Jaffe, S. O. Bruland and S. Bielack (Springer US, Boston, MA, 2010), Boston, MA, pp. 3–13.
5. A. H. Fischer, K. A. Jacobson, J. Rose and R. Zeller, *Cold Spring Harbor Protocols* **2008**, pdb (2008).
6. Aperio svs tiff format `http://openslide.org/formats/aperio/`.
7. M. N. Gurcan, L. E. Boucheron, A. Can, A. Madabhushi, N. M. Rajpoot and B. Yener, *IEEE Reviews in Biomedical Engineering* **2**, 147 (2009).
8. T. H. S. Sonal Kothari, John H Phan and M. D. Wang, *J Am Med Inform Assoc.* **20**, 1099 (November 2013).
9. A. N. Basavanhally, S. Ganesan, S. Agner, J. P. Monaco, M. D. Feldman, J. E. Tomaszewski, G. Bhanot and A. Madabhushi, *IEEE Transactions on Biomedical Engineering* **57**, 642 (March 2010).
10. T. J. Fuchs and J. M. Buhmann, *Computerized Medical Imaging and Graphics* **35**, 515 (2011).
11. K.-H. Yu, C. Zhang, G. J. Berry, R. B. Altman, C. Ré, D. L. Rubin and M. Snyder, *Nature Communications* **7** (2016).
12. K. L. Weind, C. F. Maier, B. K. Rutt and M. Moussa, Invasive carcinomas and fibroadenomas of the breast: comparison of microvessel distributions–implications for imaging modalities. *Radiology* **208**1998. PMID: 9680579.
13. T. Blaschke, G. J. Hay, M. Kelly, S. Lang, P. Hofmann, E. Addink, R. Q. Feitosa, F. van der Meer, H. van der Werff, F. van Coillie *et al.*, *ISPRS Journal of Photogrammetry and Remote Sensing* **87**, 180 (2014).
14. S. P. Brumby, N. R. Harvey, S. J. Perkins, R. B. Porter, J. J. Szymanski, J. P. Theiler and J. J. Bloch, Genetic algorithm for combining new and existing image processing tools for multispectral imagery, in *AeroSense 2000*,
15. A. Goode, B. Gilbert, J. Harkes, D. Jukic, M. Satyanarayanan *et al.*, Openslide: A vendor-neutral software foundation for digital pathology *Journal of pathology informatics* **4** (Medknow Publications, 2013).
16. K. Martinez and J. Cupitt, Vips-a highly tuned image processing software architecture, in *IEEE International Conference on Image Processing 2005*, 2005.
17. F. Wang, T. W. Oh, C. Vergara-Niedermayr, T. Kurc and J. Saltz, Managing and querying whole slide images, in *SPIE Medical Imaging*, 2012.
18. M. D. Abramoff, P. J. Magalhaes and S. J. Ram, *Biophotonics international* **11**, 36 (2004).
19. T. R. Jones, I. H. Kang, D. B. Wheeler, R. A. Lindquist, A. Papallo, D. M. Sabatini, P. Golland and A. E. Carpenter, *BMC bioinformatics* **9**, p. 1 (2008).

MISSING DATA IMPUTATION IN THE ELECTRONIC HEALTH RECORD USING DEEPLY LEARNED AUTOENCODERS*

BRETT K. BEAULIEU-JONES

Genomics and Computational Biology Graduate Group, Computational Genetics Lab, Institute for Biomedical Informatics, Perelman School of Medicine, University of Pennsylvania, 3700 Hamilton Walk, Philadelphia PA, 19104
Email: brettbe@med.upenn.edu

JASON H. MOORE

Computational Genetics Lab, Institute for Biomedical Informatics, University of Pennsylvania, 3700 Hamilton Walk, Philadelphia PA, 19104
Email: jhmoore@exchange.upenn.edu

THE POOLED RESOURCE OPEN-ACCESS ALS CLINICAL TRIALS CONSORTIUM†

Electronic health records (EHRs) have become a vital source of patient outcome data but the widespread prevalence of missing data presents a major challenge. Different causes of missing data in the EHR data may introduce unintentional bias. Here, we compare the effectiveness of popular multiple imputation strategies with a deeply learned autoencoder using the Pooled Resource Open-Access ALS Clinical Trials Database (PRO-ACT). To evaluate performance, we examined imputation accuracy for known values simulated to be either missing completely at random or missing not at random. We also compared ALS disease progression prediction across different imputation models. Autoencoders showed strong performance for imputation accuracy and contributed to the strongest disease progression predictor. Finally, we show that despite clinical heterogeneity, ALS disease progression appears homogenous with time from onset being the most important predictor.

* This work is supported by a Commonwealth Universal Research Enhancement (CURE) Program grant from the Pennsylvania Department of Health.

† Data used in the preparation of this article were obtained from the Pooled Resource Open-Access ALS Clinical Trials (PRO-ACT) Database. As such, the following organizations and individuals within the PRO-ACT Consortium contributed to the design and implementation of the PRO-ACT Database and/or provided data, but did not participate in the analysis of the data or the writing of this report: Neurological Clinical Research Institute, MGH, Northeast ALS Consoritium, Novartis, Prize4Life, Regeneron Pharmaceuticals, Sanofi, Teva Pharmaceutical Industries, Ltd.

1. Introduction

1.1. *Background*

Electronic health records (EHRs) are a core resource in genetic, epidemiological and clinical research providing phenotypic, patient progression and outcome data to researchers. Missing data presents major challenges to research by reducing viable sample size and introducing potential biases through patient selection or imputation[1,2].

Missing data is widely prevalent in the EHR for several reasons. EHRs are designed and optimized for clinical and billing purposes meaning data useful to research may not be recorded[2]. Outside of the design of EHRs, the reality of the clinic results in missing data. For example, clinicians must consider financial burden in ordering lab tests for patients and issue the minimum amount of testing and diagnostics to effectively treat their patients[3].

The various reasons data may be missing create different types of missing data: missing completely at random, missing at random and missing not at random[1,4,5]. This work focuses on data missing completely at random and data missing not at random. Data missing completely at random indicates that there is no systematic determination of whether a value is missing or present. The likelihood of response is independent of the data and any latent factors. Data missing not at random occurs when data is missing due to either observed values in the data or unobserved latent values. An example within the EHR would occur if a lab test is only issued based on a clinicians observation of the patient. Whether or not the test is issued provides insight into the patient's status.

Non-imputation approaches often exclude data from the analysis to allow for downstream analysis. One approach, complete-case analysis, throws out records with missing data. Within the EHR this would severely limit sample size, in addition if incomplete records have a systematic difference from complete records unintentional bias can be introduced[6]. As computational resources have increased, computationally complex imputation techniques have become feasible and are growing in popularity[1]. Computationally intensive techniques such as Singular Value Decomposition (SVD) based methods and weighted K-nearest neighbors (KNN) methods have joined less complex methods like mean and median imputation. Both SVD and KNN-based methods have been shown to perform effectively in microarray imputation[7]. Popular multiple imputation methods show particular challenges with data that are not missing at random[1].

Autoencoders are a variation of artificial neural networks that learn a distributed representation of their input[8]. They learn parameters to transform the data to a hidden layer and then reconstruct the original input. By using a hidden layer smaller than the number of input features, or "bottle-neck" layer the autoencoder is forced to learn the most important patterns in the data[9]. To prevent overreliance on specific features two techniques are commonly used. In a denoising autoencoder, noise is added to corrupt a portion of the inputs[9,10]. Alternatively, a technique called dropout in which random units and connections are removed from the network forcing it to learn generalizations[11]. Autoencoders were shown to generate useful higher representations in both simulated and real EHR data. Because autoencoders learn by reconstructing the original input from a corrupted version, imputation is a natural extension[12,13].

1.2. *ALS and the Pooled Resource Open-access Clinical Trials*

We evaluate each of the imputation methods on the ALS Pooled Resource Open-access Clinical Trials (PRO-ACT). Pooled clinical trial datasets present an ideal option for evaluating EHR imputation strategies because they include patients from differing environments with potential systematic biases. In addition, clinical trials represent the gold standard for data collection making it possible to spike-in missing data while maintaining enough signal to evaluate imputation techniques.

Prize4Life and the Neurological Clinical Research Institute (NCRI) at Massachusetts General Hospital created the Pooled Resource Open-Access ALS Clincial Trials (PRO-ACT) platform with funding from the ALS Therapy Alliance in and in partnership with the Northeast ALS Consortium. The PRO-ACT project was designed to empower translational ALS research and includes data from 23 clinical trials and 10,723 patients. In this work, we use the subset of 1,824 patients included in the Prize4Life challenges[14].

ALS is a progressive neurodegenerative disorder affecting both the upper and lower motor neurons causing muscle weakness, paralysis and leading to death[14]. ALS patients typically survive only 3 to 5 years from disease onset and show large degrees of clinical heterogeneity[15–18].

A common measure used to monitor an ALS patient's condition is the ALS functional rating scale (ALSFRS)[19,20]. The ALSFRS consists of 10 tests scored from 0-4 assessing patients' self-sufficiency in categories including: feeding, grooming, ambulation and communication. The change over time, or slope, is commonly used as a statistic to represent ALS progression.

2. Methods

We compare and evaluate a variety of methods to impute missing data in the EHR. We spiked-in missing data to the PRO-ACT dataset, and evaluated each approach's performance imputing known data. We also evaluated prediction accuracy using each of the imputation methods on the ALSFRS. Each of these is described in detail below and all analysis was run using freely available open source library packages, DAPS[12], FancyImpute[21], Keras[22] and Scikit-learn[23].

2.1. *Data preparation and standardization*

The PRO-ACT dataset includes patient demographic data, family history, concomitant medications, vital sign measurements, laboratory results, and patient history (disease onset etc.). PRO-ACT performed an initial data cleaning and quality assurance process. This process included extracting quantitative variables, merging laboratory tests with different names across trials, removable of indecipherable records and converting units. After processing the PRO-ACT dataset includes only quantitative values (continuous, binary, ordinal and categorical).

Our analysis encoded categorical variables using Sci-kit learn's OneHotEncoder[23]. Temporal or repeated measurements were encoded as the mean, minimum, maximum, count, standard deviation and slope across all measurements, creating 572 features for each samples. Additional measurements were standardized across scales (i.e. inches to cm). Non-numeric values in numeric measurements were coerced to numeric values. Where coercion failed they were replaced by NaN.

Input features were normalized and scaled to be between 0 and 1, with missing features remaining as NaN.

2.2. *Imputation Strategies*

2.2.1. *Imputing missing data with Autoencoders*

We constructed an autoencoder with a modified binary cross entropy cost between the reconstructed layer z and the input data x to better handle missing data as in Beaulieu-Jones and Greene (2016) (Formula 1)[12]. The modified function takes into account missing data, with m representing a "missingness" vector; m has a value of 1 where the data is present and 0 when the data is missing. By multiplying by m and dividing by the count of present features (sum of m) the result represents the average cost per present feature. The weights and biases of the autoencoder are trained only on present features and imputation does not need to be performed prior to training the autoencoder.

$$cost = -\sum_{k=1}^{d}[x_k \log(z_k)\ m_k + (1 - x_k)\log(1 - z_k)\ m_k] \ / \ count(m) \quad \text{(Formula 1)}$$

With the exception of the modified cost function autoencoders were trained as described by Vincent et al. with a 100 training epoch patience[10]. If a new minimum cost was not reached in 100 epochs, training was stopped. The autoencoder with dropout was implemented using the FancyImpute[21] and Keras[22] libraries with a Theano[24,25] backend.

We performed a parameter sweep to determine the hyperparameters for the autoencoder. In the sweep we included autoencoders of one to three hidden layers and each combination of 2, 4, 10, 100, 200, 500 and 1000 (over-complete representation) hidden nodes per layer. Autoencoders with two hidden layers made up of 500 nodes each are shown for all comparisons (Figure 1). Dropout levels of 5, 10, 20, 30, 40 and 50% were evaluated with 20% being shown for all comparisons.

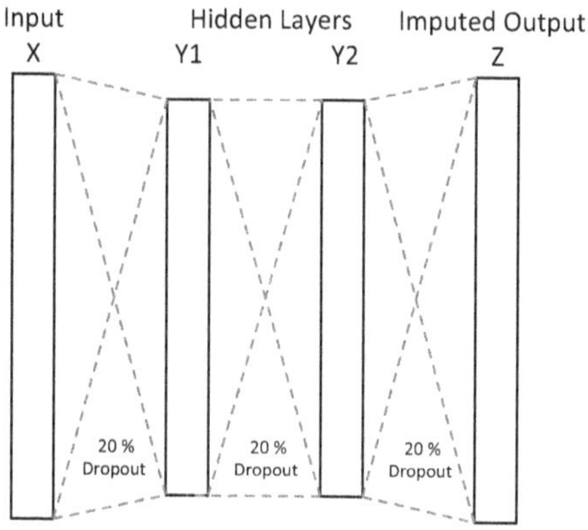

Figure 1. Schematic structure of the autoencoder used for evaluations, with two hidden layers and 20% dropout between each layer.

Binary cross entropy was used for training because it tends to be a better evaluator of quality when training neural networks[9,10,26,27]. We use a root mean squared error for comparison to other methods to prevent a bias in favor of autoencoders, as most other methods are not trained with cross entropy.

2.2.2. *Comparative imputation strategies*

We used the FancyImpute[21] libraries implementations for each of the other imputation strategies: 1) IterativeSVD, matrix completion by low rank singular value decomposition based on SVDImpute[7], 2) K-nearest neighbors imputation (KNNimpute), matrix completion by choosing the mean values of the K closest samples for features where both samples are present 3) SoftImpute[28], matrix completion by iterative replacement of missing values with values from a soft-thresholded singular value decomposition, 4) column mean filling and 5) column median filling. The standard implementations of the remaining algorithms in the FancyImpute library, MICE, Matrix Factorization and Nuclear Norm Minimization are known to be slow on large matrices and were impractically slow on this dataset[29–33].

We performed a parameter sweep for SVDimpute analyzing ranks of 5, 10, 20, 40 and 80. Ranks of 40 showed the strongest performance with this dataset and are shown for all comparisons. The parameter sweep for KNNimpute included 1, 3, 5, 7, 15 and 30 neighbors, k of 7 showed the strongest performance of the parameter sweep and is used for all comparisons.

2.3. *Missing Completely at Random Imputation Evaluation*

To evaluate imputation accuracy in a missing completely at random environment we performed trials replacing 10, 20, 30, 40 and 50% of known features at random with NaN. We performed each imputation strategy on the data with spiked-in missingness and evaluated the root mean squared error between the imputed estimates and the original data. We performed five trials for each amount of spiked-in data (Figure 2A). Performance was evaluated using the root mean squared error between the known value before spiking in missingness and the imputed value.

2.4. *Missing Not at Random Imputation Evaluation*

To perform a basic imputation simulation where data was missing not at random, varying percentages (10, 20, 30, 40, and 50%) of features were chosen at random. Half of the highest or lowest (randomly selected) quartile of values was replaced by NaN at random. Each imputation strategy was evaluated on five independent spike-in trails. Performance was evaluated using the root mean squared error between the imputed values and original values. This type of imputation could occur when the highest or lowest values represent the normal range and the clinician is able to determine a patient is normal through other factors. Alternatively the extreme values could represent a clear result where an additional is not needed to determine the result. Performance was evaluated using the root mean squared error between the known value before spiking in missingness and the predicted value.

2.5. *Progression Prediction Evaluation*

To predict disease progression as represented by the ALSFRS score slope, we first imputed the missing data using column mean averaging, column median averaging, SVDImpute, SoftImpute, KNNimpute, and an autoencoder with dropout. For prediction purposes we excluded all ALSFRS score and Forced Vital Capacity-related features.

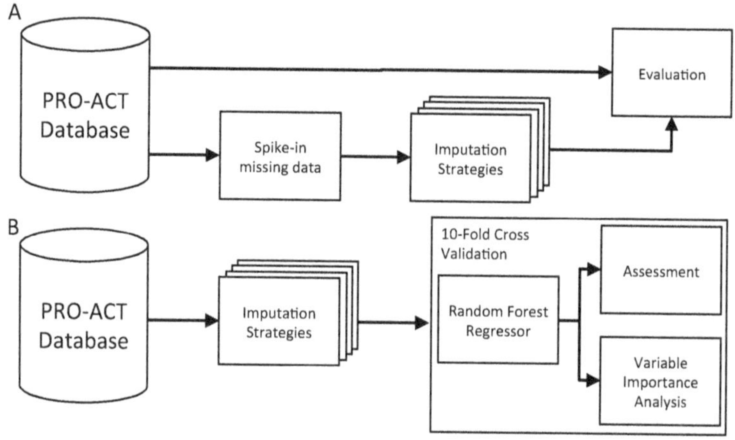

Figure 2. Evaluation outline **(a)** Imputation Evaluation. PRO-ACT patient data of 10,723 subjects has known data masked with spiked in missing data. Imputation strategies are performed in parallel and the RMSE is calculated between the masked input data and each strategies imputations. **(b)** Progression Prediction. PRO-ACT patients are imputed using each strategy. Ten-fold cross validation of a random forest regressor is performed on imputed patients.

We then used the scikit-learn implementation of a random forest regressor[23] to predict the ALSFRS score slope. The random forest regressor was chosen because four of the top six teams in the DREAM-Phil Bowen ALS Prediction Prize4Life challenge used variants of random forest regressors[14]. We also compare a random forest regressor modified to predict progression from the raw data without imputation[34]. Ten-fold cross validation was performed and the root mean squared error between the predicted slope and actual slope was calculated. We then extracted the top 10 most important features used in the trained model for analysis (Figure 2B).

3. Results

Most patients were missing approximately half of the features we extracted from the EHR (3A). The pooled aspect of the PRO-ACT data is particularly evident in the distribution of missing features as different clinical trials collected different amounts of data. Features tended to be observed in either less than 25% or in greater than 75% of patients (Figure 3B). Lab tests in particular demonstrated high variability of missingness among patients, with many present in small numbers of patients. It is impossible to determine the level of each type of missing data that exists, but it is clear that at least some of data missing is due to clinical factors (trial group etc.). The most complete features are demographics and family history information, information likely collected before entry into any of the clinical trials.

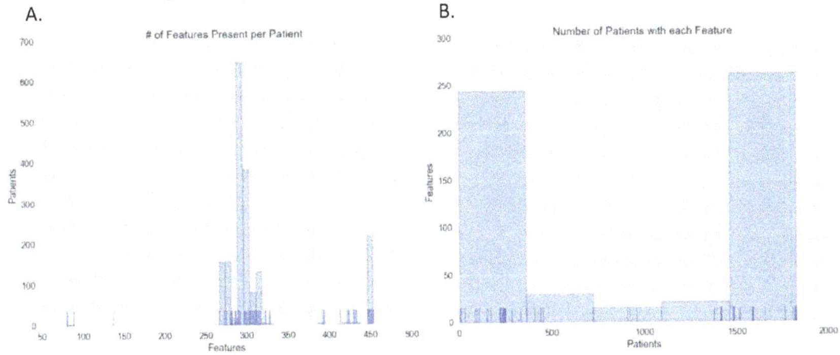

Figure 3. Histogram distribution and rug plot showing the number of patients each feature is present in. **(a)** The number of features each patient has. Ticks at the bottom indicate one patient with the count of features, bins indicate the number of patients in a range. **(b)** The number of patients having a recorded value for each feature. Ticks at the bottom indicate the number of patients a feature is present in, bins indicate the number of features in a range.

3.1. *Missing completely at random spike-in results*

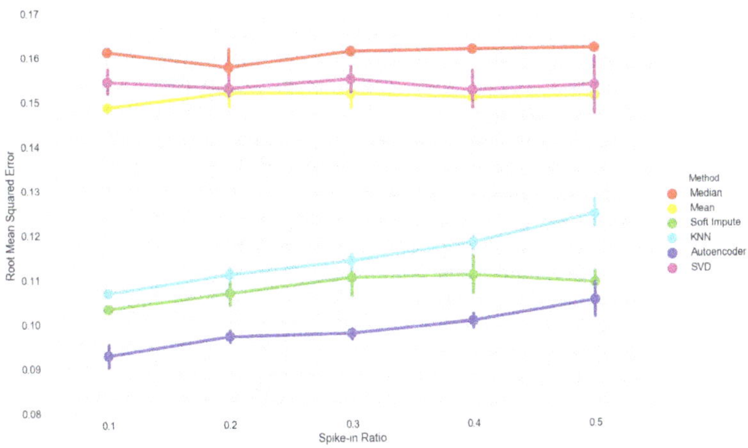

Figure 4. Effect of the amount of spiked-in missing data on imputation. Error bars indicate 5-fold cross validation score ranges.

Mean, Median and Singular Value Decomposition imputation perform poorly when data is missing completely at random. However, they do not appear to degrade as the spike-in ratio increase (Figure 4). This is not surprising for mean and median imputation because missing data is chosen completely at random and is unlikely to have a large effect on statistical averages. The autoencoder had the highest imputation performance despite increasing as the spike-in ratio increased.

3.2. *Not missing at random spike-in results*

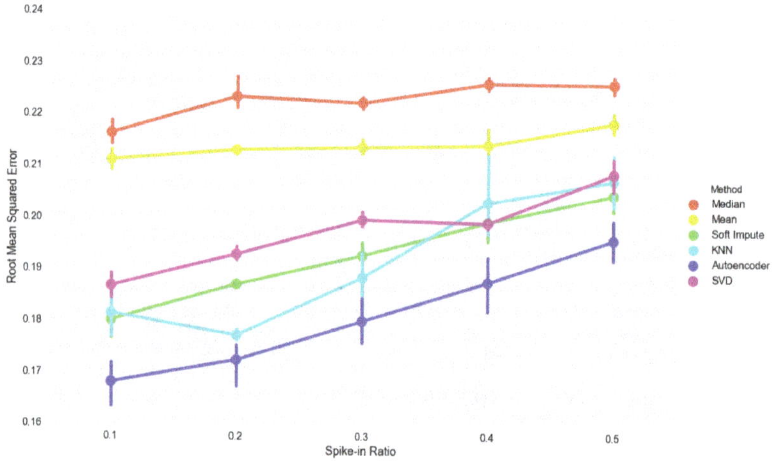

Figure 5. Effect of non-random spiked-in missing data on imputation (measured in root mean squared error). Autoencoder w/Dropout (2 layer 500 nodes each), SVD – SVDImpute with rank of 40, KNN - KNNimpute with 7 neighbors, Mean – Column Mean Averaging, Median – column median averaging, SI – SoftImpute.

The trends seen in the missing completely at random experiment largely repeat when the data is missing not at random. The autoencoder approach shows strong performance but is closely followed by the KNN, Softimpute and SVD approaches (Figure 5). KNN works by finding the k-nearest neighbors for shared values and taking the mean for the missing feature. Autoencoders work by learning the optimal network for reconstruction. Similar input values will have similar hidden node values. This similarity could explain the relatively even performance between the two methods. In addition to recognizing similar samples, autoencoders have been shown to perform well when there is dependency or correlation between variables[35]; this is the scenario when data is missing not at random. When spike-in ratios increase to high levels the methods begin to converge to the performance of mean and median imputation. This is likely because too much of the signal is lost as missing data to learn the correlation structure.

3.3. *ALS disease progression*

Imputation strategy has a modest but statistically significant impact on the root mean squared error of ALS disease progression prediction, but the autoencoder approach is the strongest performing

(Figure 6). Despite showing poor performance in the imputation accuracy exercises Singular Value Decomposition does approximately as well as k-nearest neighbors and SoftImpute in this experiment. A random forest regressor applied to the raw data is the worst performing, but is not significantly worse than any of the methods other than the Autoencoder. In terms of ALS disease progression, imputation does not appear to have a large effect on prediction, but can be vital to allow the use of other algorithms (prediction, clustering etc.) without modification.

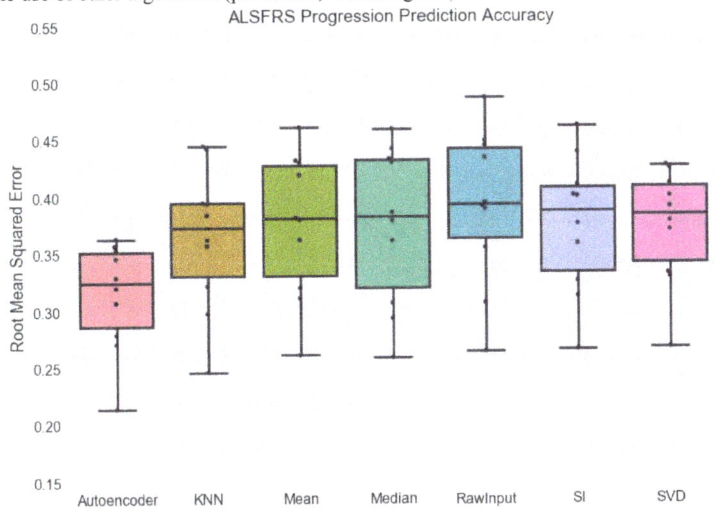

Figure 6. ALS Functional Rating Scale prediction accuracy shown for an autoencoder, k-nearest neighbors, mean averaging, median averaging, the raw input including missing values, soft impute and singular value decomposition. The box indicates inner quartiles with the line representing the median; the whiskers indicate outer quartiles excluding outliers.

3.4. *ALS progression predictive indicat*

Nine out of the top ten most important features in the autoencoder-imputed random forest regressor were among the top fifteen identified in the DREAM Prediction challenge (Figure 7A). The amount of time using Riluzole was not among the top fifteen previously identified. Riluzole is the only FDA approved medication for ALS treatment but it is believed to have a limited effect on survival[36–38]. The finding that Riluzole is protective of ALS slope indicates some level of efficacy.

Of the top ten most important features, five are missing in more than 50% of patients in the data set. This is a possible explanation for the improvement shown by Autoencoders, SVDimpute and KNNimpute over mean imputation.

By far the most important feature for prediction is the time from onset and several of the most important features are highly correlated with time from onset. ALSFRS slopes resemble a normal distribution (Figure 7B). When including the entire PRO-ACT dataset, the Kolmogorov-Smirnov test score is 0.05 for patients with negative slopes. This indicates the progression of the disease is similar to a truncated normal distribution. We exclude positive slopes because ALS patients do not typically get better, and signs of doing so are likely the result of measurement error. Despite presenting in clinically heterogeneous manners, ALS progression as defined by the ALSFRS appears to be largely homogenous. Patients fall within a relatively normal distribution and have increasingly negative slopes the longer they ALS.

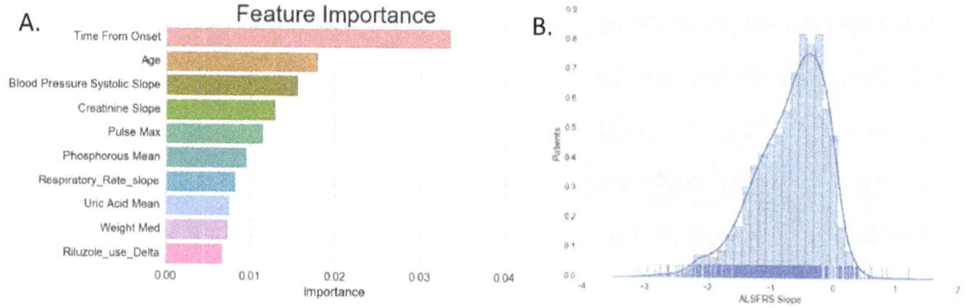

Figure 7. Prediction feature importance. **(a)** Importance levels of the top 10 features to the random forest regressor with autoencoder imputed data. **(b)** Histogram distribution of patient ALSFRS slope levels.

4. Discussion and Conclusions

In this study, we compared the performance of an autoencoder approach with popular imputation techniques in ALS EHR data. A multi-layer autoencoder with dropout showed robust imputation performance across a variety of spiked-in missing data experiments designed to be both completely at random and not at random. Furthermore, we found that imputation accuracy may not strictly correlate with predictive performance but the most accurate imputer provided the most accurate predictor. The importance of imputation is demonstrated by five of the top ten most important features for prediction being missing in more than 50% of patients.

Increased deterioration of imputation performance for KNNimpute and SVDimpute with increased missing data is at odds with previous research of imputation in microarrays[7]. Possible explanations include either reaching a threshold of missing data where the burden is too high for these methods to accurately impute or that a confounding systematic bias is introduced from the different clinical trials.

This work is a promising first step in utilizing deep learning techniques for missing data imputation in the EHR but challenges remain. Autoencoders are computationally intensive, but less so than imputation techniques like MICE, Matrix Factorization and Nuclear Norm Minimization. With GPU resources, autoencoders train in similar amounts of time to both KNN and SVD methods for these clinical trials. As data increases, autoencoder training time increases linearly in line with the number of samples. Methods like KNN require computing a distance matrix, which increases in exponential time. In addition, further examination is necessary to determine whether the strong performance shown by autoencoders is a result of the structure of this pooled clinical trial dataset. The subset of 1,800 patients is relatively small and methods may differ in performance increases with more patients.

This work offers promising results but has several limitations especially because it specifically analyzes pre-processed pooled clinical trial data. Clinical trials have more complete and cleaner data than raw EHR. Follow up work should be performed with other diseases and in the general patient population. These methods have also only been evaluated for quantitative values; in raw EHR data there will be an additional extraction step for raw text and qualitative observations that was not necessary due to PRO-ACT's preprocessing.

Additional future work will be concentrated on developing tools to better understand and interpret the structure of the trained autoencoder networks. We anticipate being able to recognize patterns in the trained weights to see correlation between input features. Understanding correlation will empower new clustering and visualization opportunities. Spike-in evaluations can provide a supervised context to otherwise unsupervised learning problems; further analysis should be performed on the higher-level learned features in the hidden layers of the autoencoders. We suspect these features may be useful in patient outcome classification and regression problems.

5. Acknowledgments

This work is supported by the Commonwealth Universal Research Enhancement (CURE) Program grant from the Pennsylvania Department of Health to JHM. The authors would like to thank Dr. Casey S Greene for helpful discussions. The authors acknowledge the support of the NVIDIA Corporation for the donation of a TitanX GPU used for this research.

References

1. Sterne JJ a C, White IRI, Carlin JJB, et al. Multiple imputation for missing data in epidemiological and clinical research: potential and pitfalls. *BMJ*. 2009;338(July):b2393. doi:10.1136/bmj.b2393.
2. Wells BJ, Chagin KM, Nowacki AS, Kattan MW. Strategies for handling missing data in electronic health record derived data. *EGEMS (Washington, DC)*. 2013;1(3):1035. doi:10.13063/2327-9214.1035.
3. McClatchey KD. *Clinical Laboratory Medicine*. Lippincott Williams & Wilkins; 2002.
4. Little R, Rubin D. *Statistical Analysis with Missing Data*. John Wiley & Sons; 2014.
5. Marlin B. Missing data problems in machine learning. 2008. http://www-devel.cs.ubc.ca/~bmarlin/research/phd_thesis/marlin-phd-thesis.pdf. Accessed August 7, 2016.
6. Gelman A, Hill J. *Data Analysis Using Regression and Multilevel/hierarchical Models*.; 2006. https://books.google.com/books?hl=en&lr=&id=c9xLKzZWoZ4C&oi=fnd&pg=PR17&dq=data+analysis+using+regression+and+multilevel/hierarchical+models&ots=baT3R3Mnng&sig=KpLzVOFtUseaK8_IhUfPLM2Y7fU. Accessed August 10, 2016.
7. Troyanskaya O, Cantor M, Sherlock G, et al. Missing value estimation methods for DNA microarrays. *Bioinformatics*. 2001;17(6):520-525. doi:10.1093/bioinformatics/17.6.520.
8. Bengio Y. Learning Deep Architectures for AI. *Found Trends® Mach Learn*. 2009;2(1):1-127. doi:10.1561/2200000006.
9. Vincent P, Larochelle H, Lajoie I, Bengio Y, Manzagol P-A. Stacked Denoising Autoencoders: Learning Useful Representations in a Deep Network with a Local Denoising Criterion. *J Mach Learn Res*. 2010;11(3):3371-3408. doi:10.1111/1467-8535.00290.
10. Vincent P, Larochelle H, Bengio Y, Manzagol P-A. Extracting and composing robust features with denoising autoencoders. *Proc 25th Int Conf Mach Learn - ICML '08*. 2008:1096-1103. doi:10.1145/1390156.1390294.
11. Srivastava N, Hinton GE, Krizhevsky A, Sutskever I, Salakhutdinov R. Dropout : A Simple Way to Prevent Neural Networks from Overfitting. *J Mach Learn Res*. 2014;15:1929-1958. doi:10.1214/12-AOS1000.
12. Beaulieu-Jones BK, Greene CS. Semi-Supervised Learning of the Electronic Health Record with Denoising

Autoencoders for Phenotype Stratification. *bioRxiv*. February 2016:39800. doi:10.1101/039800.
13. Miotto R, Li L, Kidd BA, et al. Deep Patient: An Unsupervised Representation to Predict the Future of Patients from the Electronic Health Records. *Sci Rep*. 2016;6:26094. doi:10.1038/srep26094.
14. Küffner R, Zach N, Norel R, et al. Crowdsourced analysis of clinical trial data to predict amyotrophic lateral sclerosis progression. *Nat Biotechnol*. 2015;33(1):51-57. doi:10.1038/nbt.3051.
15. Kollewe K, Mauss U, Krampfl K, Petri S, Dengler R, Mohammadi B. ALSFRS-R score and its ratio: A useful predictor for ALS-progression. *J Neurol Sci*. 2008;275(1-2):69-73. doi:10.1016/j.jns.2008.07.016.
16. Beghi E, Mennini T, Bendotti C, et al. The heterogeneity of amyotrophic lateral sclerosis: a possible explanation of treatment failure. *Curr Med Chem*. 2007;14(30):3185-3200. http://www.ncbi.nlm.nih.gov/pubmed/18220753. Accessed August 7, 2016.
17. Sabatelli M, Conte A, Zollino M. Clinical and genetic heterogeneity of amyotrophic lateral sclerosis. *Clin Genet*. 2013;83(5):408-416. doi:10.1111/cge.12117.
18. Ravits JM, La Spada AR. ALS motor phenotype heterogeneity, focality, and spread: deconstructing motor neuron degeneration. *Neurology*. 2009;73(10):805-811. doi:10.1212/WNL.0b013e3181b6bbbd.
19. Cedarbaum JM, Stambler N. Performance of the amyotrophic lateral sclerosis functional rating scale (ALSFRS) in multicenter clinical trials. In: *Journal of the Neurological Sciences*. Vol 152. ; 1997. doi:10.1016/S0022-510X(97)00237-2.
20. Cedarbaum JM, Stambler N, Malta E, et al. The ALSFRS-R: A revised ALS functional rating scale that incorporates assessments of respiratory function. *J Neurol Sci*. 1999;169(1-2):13-21. doi:10.1016/S0022-510X(99)00210-5.
21. Rubinsteyn A, Feldman S. fancyimpute: Version 0.0.9. March 2016. doi:10.5281/zenodo.47151.
22. Chollet F. Keras. *GitHub Repos*. 2015.
23. Pedregosa F, Varoquaux G, Gramfort A, et al. Scikit-learn: Machine Learning in Python. *… Mach Learn …*. 2012;12:2825-2830. http://dl.acm.org/citation.cfm?id=2078195\nhttp://arxiv.org/abs/1201.0490.
24. Bastien F, Lamblin P, Pascanu R, et al. Theano: new features and speed improvements. *arXiv Prepr arXiv …*. 2012:1-10. http://arxiv.org/abs/1211.5590.
25. Bergstra J, Breuleux O, Bastien F, et al. Theano: a CPU and GPU math compiler in Python. In: *9th Python in Science Conference*. ; 2010:1-7. http://www-etud.iro.umontreal.ca/~wardefar/publications/theano_scipy2010.pdf.
26. Socher R, Pennington J, Huang E, Ng A. Semi-supervised recursive autoencoders for predicting sentiment distributions. *Proc*. 2011. http://dl.acm.org/citation.cfm?id=2145450. Accessed August 8, 2016.
27. Hinton G, Salakhutdinov R. Reducing the dimensionality of data with neural networks. *Science (80-)*. 2006. http://science.sciencemag.org/content/313/5786/504.short. Accessed August 8, 2016.
28. Mazumder R, Hastie T, Edu H, Tibshirani R, Edu T. Spectral Regularization Algorithms for Learning Large Incomplete Matrices. *J Mach Learn Res*. 2010;11:2287-2322.
29. Buuren S van. *Flexible Imputation of Missing Data*.; 2012. doi:10.1201/b11826.
30. Royston P. Multiple imputation of missing values: update of ice. *Stata J*. 2005. https://www.researchgate.net/profile/James_Cui2/publication/23780230_Buckley-James_method_for_analyzing_censored_data_with_an_application_to_a_cardiovascular_disease_and_an_HIVAIDS_study/links/53d5866d0cf228d363ea0b7a.pdf#page=59. Accessed August 10, 2016.
31. Kim J, Park H. Toward Faster Nonnegative Matrix Factorization: A New Algorithm and Comparisons.
32. Lin C-J. Projected Gradient Methods for Non-negative Matrix Factorization.
33. Hsieh C-J, Olsen PA. Nuclear Norm Minimization via Active Subspace Selection.
34. Breiman L, Cutler A. Random Forests. 2004. *URL http//stat-www berkeley edu/users/breiman/RandomForests/cc home htm*. 2014.
35. Nelwamondo F V., Mohamed S, Marwala T. Missing Data: A Comparison of Neural Network and Expectation Maximisation Techniques. April 2007. http://arxiv.org/abs/0704.3474. Accessed September 30, 2016.
36. Zoccolella S, Beghi E, Palagano G, et al. Riluzole and amyotrophic lateral sclerosis survival: a population-based study in southern Italy. *Eur J Neurol*. 2007;14(3):262-268. doi:10.1111/j.1468-1331.2006.01575.x.
37. Traynor BJ, Alexander M, Corr B, Frost E, Hardiman O. An outcome study of riluzole in amyotrophic lateral sclerosis. *J Neurol*. 2003;250(4):473-479. doi:10.1007/s00415-003-1026-z.
38. Czaplinski A, Yen AA, Appel SH. Forced vital capacity (FVC) as an indicator of survival and disease progression in an ALS clinic population. *J Neurol Neurosurg Psychiatry*. 2006;77(3):390-392. doi:10.1136/jnnp.2005.072660.

A DEEP LEARNING APPROACH FOR CANCER DETECTION AND RELEVANT GENE IDENTIFICATION

PADIDEH DANAEE*, REZA GHAEINI

School of Electrical Engineering and Computer Science, Oregon State University, Corvallis, OR 97330, USA
**E-mail: danaeep@oregonstate.edu and ghaeinim@oregonstate.edu*

DAVID A. HENDRIX

School of Electrical Engineering and Computer Science, Department of Biochemistry and Biophysics, Oregon State University, Corvallis, OR 97330, USA E-mail: david.hendrix@oregonstate.edu

Cancer detection from gene expression data continues to pose a challenge due to the high dimensionality and complexity of these data. After decades of research there is still uncertainty in the clinical diagnosis of cancer and the identification of tumor-specific markers. Here we present a deep learning approach to cancer detection, and to the identification of genes critical for the diagnosis of breast cancer. First, we used Stacked Denoising Autoencoder (SDAE) to deeply extract functional features from high dimensional gene expression profiles. Next, we evaluated the performance of the extracted representation through supervised classification models to verify the usefulness of the new features in cancer detection. Lastly, we identified a set of highly interactive genes by analyzing the SDAE connectivity matrices. Our results and analysis illustrate that these highly interactive genes could be useful cancer biomarkers for the detection of breast cancer that deserve further studies.

Keywords: Cancer Detection; RNA-seq Expression; Deep Learning; Dimensionality Reduction; Stacked Denoising Autoencoder; Classification.

1. Introduction

The analysis of gene expression data has the potential to lead to significant biological discoveries. Much of the work on the identification of differentially expressed genes has focused on the most significant changes, and may not allow recognition of more subtle patterns in the data.[1–6] Tremendous potential exists for computational methods to analyze these data for the discovery of gene regulatory targets, disease diagnosis and drug development.[7–9] However, the high dimension and noise associated with these data presents a challenge for these tasks. Moreover, the mismatch between the large number of genes and typically small number of samples presents the challenge of a "dimensionality curse". Multiple algorithms have been used to distinguish normal cells from abnormal cells using gene expression.[10–13] Although there has been a lot of research into cancer detection from gene expression data, there remains a critical need to improve accuracy, and to identify genes that play important roles in cancer.

Machine learning methods for dimensionality reduction and classification of gene expression data have achieved some success, but there are limitations in the interpretation of the most significant signals for classification purposes.[14,15] Recently, there have been efforts to use single-layer, nonlinear dimensionality reduction techniques to classify samples based on gene expression data.[16] In similar studies of computer vision, unsupervised deep learning methods have been successfully applied to extract information from high dimensional image data.[17]

Similarly, one can extract the meaningful part of the expression data by applying such techniques, thereby enabling identification of specific subsets of genes that are useful for biologists and physicians, with the potential to inform therapeutic strategies.

In this work, we used stacked denoising autoencoders (SDAE) to transform high-dimensional, noisy gene expression data to a lower dimensional, meaningful representation.[18] We then used the new representations to classify breast cancer samples from the healthy control samples. We used different machine learning (ML) architectures to observe how the new compact features can be effective for a classification task and allow the evaluation of the performance of different models. Finally, we analyzed the lower-dimensional representations by mapping back to the original data to discover highly relevant genes that could play critical roles and serve as clinical biomarkers for cancer diagnosis. The performance of these methods affirm that SDAEs could be applied to cancer detection in order to improve the classification performance, extract both linear and nonlinear relationships in the data, and perhaps more important, to extract a subset of relevant genes from deep models as a set of potential cancer biomarkers. The identification of these relevant genes deserves further analysis as it potentially can improve methods for cancer diagnosis and treatment.

2. Background

Classification and clustering of gene expression in the form of microarray or RNA-seq data are well studied. There are various approaches for the classification of cancer cells and healthy cells using gene expression profiles and supervised learning models. The self-organizing map (SOM) was used to analyze leukemia cancer cells.[19] A support vector machine (SVM) with a dot product kernel has been applied to the diagnosis of ovarian, leukemia, and colon cancers.[11] SVMs with nonlinear kernels (polynomial and Gaussian) were also used for classification of breast cancer tissues from microarray data.[10]

Unsupervised learning techniques are capable of finding global patterns in gene expression data. Gene clustering represents various groups of similar genes based on similar expression patterns. Hierarchical clustering and maximal margin linear programming are examples of this learning and they have been used to classify colon cancer cells.[20,21] K-nearest neighbors (KNN) unsupervised learning also has been applied to breast cancer data.[12]

Due to the large number of genes, high amount of noise in the gene expression data, and also the complexity of biological networks, there is a need to deeply analyze the raw data and exploit the important subsets of genes. Regarding this matter, other techniques such as principal component analysis (PCA) have been proposed for dimensionality reduction of expression profiles to aid clustering of the relevant genes in a context of expression profiles.[22] PCA uses an orthogonal transformation to map high dimensional data to linearly uncorrelated components.[23] However, PCA reduces the dimensionality of the data linearly and it may not extract some nonlinear relationships of the data.[24] In contrast, other approaches such as kernel PCA (KPCA) may be capable of uncovering these nonlinear relationships.[25]

Similarly, researchers have applied PCA to a set of combined genes of 13 data sets to obtain the linear representation of the gene expression and then apply a autoencoder to capture nonlinear relationships.[26] Recently, a denoising autoencoder has been applied to extract a feature set from breast cancer data.[16] Using a single autoencoder may not extract all the useful representations from the noisy, complex, and high-dimensional expression data. However, by reducing the dimensionality incrementally, the multi-layered architecture of an SDAE may extract meaningful patterns in these data with reduced loss of information.[27]

3. Materials and Methods

We have applied a deep learning approach that extracts the important gene expression relationships using SDAE. After training the SDAE, we selected a layer that has both low-dimension and low validation error compared to other encoder stacks using a validation data set independent of both our training and test set.[28] As a result, we selected an SDAE with four layers of dimensions of 15,000, 10,000, 2,00, and 500. Consequently we used the selected layer as input features to the classification algorithms. The goal of our model is extracting a mapping that possibly decodes the original data as closely as possible without losing significant gene patterns.

We evaluated our approach for feature selection by feeding the SDAE-encoded features to a shallow artificial neural network (ANN)[29] and an SVM model.[30] Furthermore, we applied a similar approach with PCA and KPCA as a comparison.

Lastly, we used the SDAE weights from each layer to extract genes with strongly propagated influence on the reduced-dimension SDAE-encoding. These selected "deeply connected genes" (DCGs) are further tested and analyzed for pathway and Gene Ontology (GO) enrichment. The results from our analysis showed that in fact our approach can reveal a set of biomarkers for the purpose of cancer diagnosis. The details of our method are discussed in the following subsections, and the work-flow of our approach is shown in Fig 1.

3.1. *Gene Expression Data*

For our analysis, we analyzed RNA-seq expression data from The Cancer Genome Atlas (TCGA) database for both tumor and healthy breast samples.[31] These data consist of 1097 breast cancer samples, and 113 healthy samples. To overcome the class imbalance of the data, we used synthetic minority over-sampling technique (SMOTE) to transform data into a more balanced representation for pre-training.[32] We used the `imbalanced-learn` package for this transformation of the training data.[33] Furthermore, we removed all genes that had zero expression across all samples.

3.2. *Dimensionality Reduction Using Stacked Denoising Autoencoder*

An autoencoder (AE) is a feedforward neural network that produces the output layer as close as possible to its input layer using a lower dimensional representation (hidden layer). The autoencoder consists of an encoder and a decoder. The encoder is a nonlinear function, like a sigmoid, applied to an affine mapping of the input layer, which can be expressed as $f_\theta(X) = \sigma(Wx+b)$ with parameters $\theta = \{W, b\}$. The matrix W is of dimensions $d' \times d$ to go from a larger dimension of gene expression data d to a lower dimensional encoding corresponding to d'. The bias vector b is of dimension d'. This input layer encodes the data to generate a

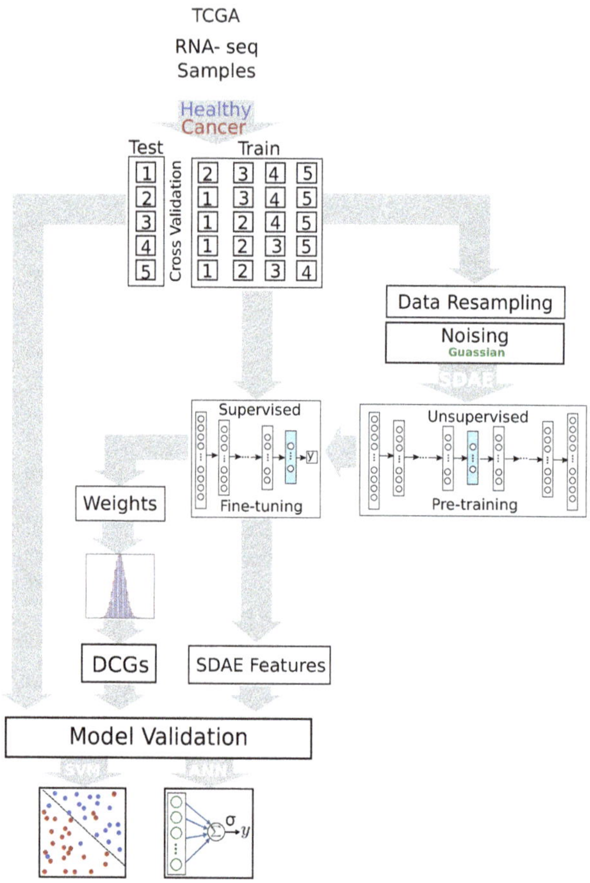

Fig. 1. The pipeline representing the stacked denoising autoencoder (SDAE) model for breast cancer classification and the process of biomarkers extraction.

hidden or latent layer. The decoder takes the hidden representations from the previous layer and decodes the data as closely as possible to the original inputs, and can be expressed as $z = g_{\theta'}(y) = \sigma(W'y + b')$. In our implementation, we imposed tied weights, with $W' = W^T$. We can refer to the weight matrix W and bias b as $\theta = \{W, b\}$ and similarly $\theta' = \{W', b'\}$.

A SDAE can be constructed as a series of AE mappings with parameters $\theta_1, \theta_2, ..., \theta_n$ and the addition of noise to prevent overfitting.[18] In order to get a good representation for each layer, we maximize the information gain between the input layer (modeled as a random variable X from an unknown distribution $q(X)$) and its higher level stochastic representation (random variable Y from a known distribution $p(X|Y;\theta')$). For layer i, we then learned a set of parameters θ_i and θ'_i from a known distribution p where $q(Y|X) = p(Y|X;\theta_i)$ and also $q(X|Y) = p(X|Y;\theta'_i)$ that maximize the mutual information.[18]

This maximization problem corresponds to minimizing the reconstruction error of the input layer using hidden representation. In this construction, the hidden layer contains the compressed information of the data by ignoring useless and noisy features. In fact, the autoencoder extracts a set of new representations which encompass the complex relationships between input variables. The reconstruction error of the input layer using this new representation is non-zero, but can be minimized. In practice, the weights of the model are learned through the stochastic gradient descent (SGD) algorithm.[34,35]

Autoencoders extract both linear and nonlinear relationships inherent in the input data, making them powerful and versatile. The encoder of the SDAE decreases the dimensionality of the gene expression data stack-by-stack, which leads to reduced loss of information compared to reducing the dimension in one step.[27] In contrast, the decoder increases the dimensionality to eventually achieve the full reconstruction of the original input as close as possible. In this procedure, the output of one layer is the input to the next layer. For this implementation, we used the `Keras` library with `Theano` backend running on an Nvidia Tesla K80 GPU.[36] Although it is difficult to estimate the time complexity of the deep architecture of the SDAE, with batch training and highly parallelizable implementation on GPUs, training takes a few minutes and testing of a sample is performed in a few seconds.

It is proven in practice that pre-training the parameters in a deep architecture leads to a better generalization on a specific task of interest.[18] Greedy layer-wise pre-training is an unsupervised approach that helps the model initialize the parameters near a good local minimum and convert the problem to a better form of optimization.[27] Therefore, we considered the pre-training approach as supposed to achieve smoother convergence and higher overall performance in cancer classification. After starting with the initial parameters resulting from the pre-training phase, we used supervised fine-tuning on the full training set to update the parameters.

To avoid overfitting in the learning phase (both pre-training and fine-tuning) of the SDAE, we utilized a dropout regularization factor, which is a method of randomly excluding fractions of hidden units in the training procedure by setting them to zero. This method prevents nodes from co-adapting too much and consequently avoids overfitting.[37] For the same purpose, we provided partially corrupted input values to the SDAE (denoising). The SDAE is robust, and its accuracy does not change upon introducing noise at a low rate. In fact, SDAE with denoising and dropout can find a better representation from the noisy data. Fig 2 shows the SDAE encoded, decoded, and denoised representations on the subset of genes.

3.3. *Differentially Expressed Genes*

We used significantly differentially expressed genes as a comparison to our SDAE features for cancer classification. First, we computed the log fold change comparing the median expression

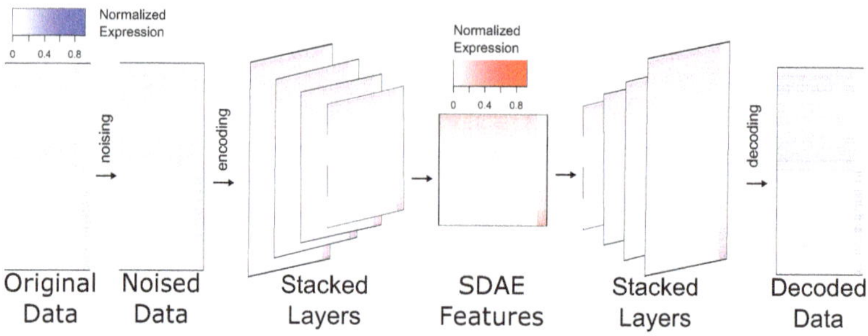

Fig. 2. SDAE representation using the enriched genes in the TCGA breast cancer. In this depiction for illustrative purposes, the top 500 genes with median expression across cancer samples enriched above health samples, and the top 500 genes with reduced median expression across cancer samples is shown.

in cancer tissue samples to that of healthy tissue samples. We then computed a two-tailed p-value using a Gaussian fit, followed by a Benjamini-Hochberg (BH) correction.[38] We identified two sets of differentially expressed genes. The first, DIFFEXP0.05 was the 206 genes, 98 up-regulated and 118 down-regulated, that were significant at an FDR of 0.05. The second set, DIFFEXP500, contains the top 500 most significant differentially expressed genes (the same dimension as the SDAE features) using the same 2-tailed p-values, containing 244 up-regulated and 256 down-regulated genes.

3.4. *Dimensionality Reduction Using Principal Component Analysis*

As a second level of comparison, we extracted features using linear PCA to provide a baseline for the performance of linear dimensionality reduction algorithms for our ML models. The same reduced dimensionality of 500 was used. In addition, we used KPCA with an RBF kernel to extract features that by default are of the same dimension as the number of training input samples. For both PCA and KPCA we used an implementation in the `scikit-learn` package.[39]

4. Results and Discussion

4.1. *Classification Learning*

In order to evaluate the effectiveness of our autoencoder-extracted features, we used two different supervised learning models to classify cancer samples from healthy control samples. First, we considered a single-layer ANN with input nodes directly connected to output layers without any hidden units. If we consider the input units as $X = (x_1, x_2, ..., x_n)$, the output values are calculated as $y = \sigma(\sum_i w_i x_i + b)$. Second, we considered both an SVM with a linear kernel and with a radial basis function kernel (SVM-RBF). We applied 5-fold cross-validation for to exhaustively split the data into train and test sets to estimate the accuracy of each model without overfitting. In each split, the model was trained on 4 partitions and tested on the 5th, ensuring that training and testing are performed on non-overlapping subsets.

5. Comparison of Different Models

To assess the effectiveness of the SDAE features, we compared their performance in classification to differentially expressed genes and to principal components for different machine learning models. The performance of the SDAE features for classification is summarized in Table 1. The best method varies depending on the performance metric, but on these data the SDAE features performed best on three of the five metrics we considered. The highest accuracy was attained using SDAE features applied to SVM-RBF classification. This method also had the highest F-measure. The highest sensitivity was found for SDAE features as well, but using the ANN classification model. KPCA features applied to an SVM-RBF had higher specificity and precision.

Table 1. Comparison of different feature sets using three classification learning models.

Features	Model	Accuracy	Sensitivity	Specificity	Precision	F-measure
	ANN	96.95	**98.73**	95.29	95.42	0.970
SDAE	SVM	98.04	97.21	99.11	99.17	0.981
	SVM-RBF	**98.26**	97.61	99.11	99.17	**0.983**
	ANN	63.04	60.56	70.76	84.58	0.704
DIFFEXP500	SVM	57.83	64.06	46.43	70.42	0.618
	SVM-RBF	77.391	86.69	71.29	67.08	0.755
	ANN	59.93	59.93	69.95	84.58	0.701
DIFFEXP0.05	SVM	68.70	82.73	57.5	65.04	0.637
	SVM-RBF	76.96	87.56	70.48	65.42	0.747
	ANN	96.52	98.38	95.10	95.00	0.965
PCA	SVM	96.30	94.58	98.61	98.75	0.965
	SVM-RBF	89.13	83.31	99.47	99.58	0.906
	ANN	97.39	96.02	99.10	99.17	0.975
KPCA	SVM	97.17	96.38	98.20	98.33	0.973
	SVM-RBF	97.32	89.92	**99.52**	**99.58**	0.943

6. Deep Feature Extraction and Deeply Connected Genes

Going beyond classification, there is potential biological significance in understanding what subsets of genes are involved in the new feature space that makes it an effective set for the cancer detection. Previous work on cancer detection using a single-layer autoencoder has evaluated the importance of each hidden node.[16] Here, we analyzed the importance of genes by considering combined effect of each stack of the deep architecture. To extract these genes, we utilized a strategy of computing the product of the weight matrices for each layer of our SDAE. The result is a $500 \times G$ dimensional matrix W, where G is the number of genes in the expression data, computed for an n-layer SDAE by

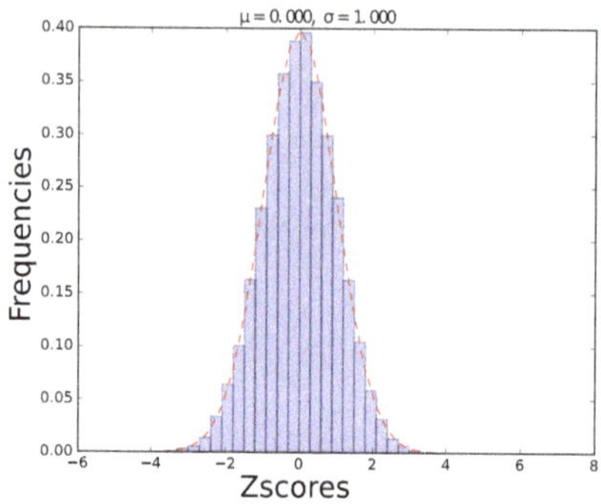

Fig. 3. Histogram of z-Scores from the dot product matrix of the weights connectivity of the SDAE.

$$W = \prod_{i=1}^{n} W_i.$$

Although the weights of each layer of the SDAE are computed with a nonlinear model, the matrix W is a linearization of the compounded effect of each gene on the SDAE features. Genes with the largest weights in W are the most strongly connected to the extracted and highly predictive features, so we called these genes DCGs. We found that the terms of matrix W were strongly normally distributed (Fig 3). We identified the subset of genes with the most statistically significant impact on the encoding by fitting the distribution of these values in W to a normal distribution, computing a p-value using this fit,and applying a BH correction with an FDR of 0.05.

6.1. *Gene Ontology*

We examined the functional enrichment of the DCGs through a GO term and Panther pathway analysis. Table 2 presents the statistically-enriched GO terms under "biological process", and having a Bonferroni-corrected p-value of less than 1e-10. Many of the most significant terms are related to mitosis, suggesting a large number of genes with core functionality that is relevant to cell proliferation. In addition, an analysis of the enrichment of Panther pathways led to a single enriched term, p53 pathway, where we observe 10 genes when 1.34 are expected, giving a p-value of 2.21E-04. P53 is known to be an important tumor-suppressor gene[40–42] , and this finding suggests a role of tumor suppressor function in many of the DCGs.

6.2. *Classification Learning*

Finally, we used the expression of the DCGs as features for the ML models previously mentioned. These genes served as useful features for cancer classification, achieving 94.78% accuracy (Table 3). Although these features performed a few percentage points below that of the

Table 2. Enriched GO terms associated with DCGs in breast cancer data from TCGA.

GO biological process	Total	Observed	Expected	Enrichment	P-value
cell cycle process (GO:0022402)	1079	100	16.46	6.07	1.12E-45
cell cycle (GO:0007049)	1311	108	20	5.4	3.28E-45
mitotic cell cycle process (GO:1903047)	741	85	11.31	7.52	1.06E-44
mitotic cell cycle (GO:0000278)	760	85	11.6	7.33	7.33E-44
nuclear division (GO:0000280)	470	63	7.17	8.78	1.52E-35
organelle fission (GO:0048285)	492	64	7.51	8.53	1.99E-35
mitotic nuclear division (GO:0007067)	357	56	5.45	10.28	1.34E-34
cell division (GO:0051301)	477	58	7.28	7.97	3.72E-30
chromosome segregation (GO:0007059)	274	46	4.18	11	5.46E-29
sister chromatid segregation (GO:0000819)	176	36	2.69	13.41	7.62E-25
nuclear chromosome segregation (GO:0098813)	230	38	3.51	10.83	3.97E-23
mitotic cell cycle phase transition (GO:0044772)	249	35	3.8	9.21	8.10E-19
mitotic prometaphase (GO:0000236)	99	25	1.51	16.55	1.56E-18
cell cycle phase transition (GO:0044770)	255	35	3.89	9	1.72E-18
regulation of cell cycle (GO:0051726)	943	62	14.39	4.31	2.48E-18
chromosome organization (GO:0051276)	984	63	15.01	4.2	4.15E-18
DNA metabolic process (GO:0006259)	768	52	11.72	4.44	2.67E-15
organelle organization (GO:0006996)	3133	112	47.8	2.34	4.27E-15
mitotic cell cycle phase (GO:0098763)	211	29	3.22	9.01	7.67E-15
cell cycle phase (GO:0022403)	211	29	3.22	9.01	7.67E-15
biological phase (GO:0044848)	215	29	3.28	8.84	1.25E-14
sister chromatid cohesion (GO:0007062)	113	22	1.72	12.76	1.18E-13
cellular resp. to DNA damage stimu. (GO:0006974)	719	48	10.97	4.38	1.27E-13
regulation of cell cycle process (GO:0010564)	557	42	8.5	4.94	2.53E-13
mitotic sister chromatid segregation (GO:0000070)	90	20	1.37	14.56	3.01E-13
cell cycle checkpoint (GO:0000075)	196	25	2.99	8.36	1.09E-11
M phase (GO:0000279)	173	23	2.64	8.71	6.55E-11
mitotic M phase (GO:0000087)	173	23	2.64	8.71	6.55E-11
regulation of mitotic cell cycle (GO:0007346)	461	35	7.03	4.98	1.10E-10
single-organism process (GO:0044699)	12451	253	189.98	1.33	4.67E-10
DNA replication (GO:0006260)	213	24	3.25	7.38	5.74E-10
anaphase (GO:0051322)	154	21	2.35	8.94	6.13E-10
mitotic anaphase (GO:0000090)	154	21	2.35	8.94	6.13E-10
cellular component organization (GO:0016043)	5133	139	78.32	1.77	7.74E-10

SDAE features, they still have advantage of being more readily interpreted. Future work is needed to improve the extraction of DCGs to enhance their utility as features for classification.

Table 3. Cancer classification results using deeply connected genes (DCGs).

Features	Model	Accuracy	Sensitivity	Specificity	Precision	F-measure
DCGs	ANN	91.74	98.13	87.15	85.83	0.913
	SVM	91.74	88.80	97.50	97.25	0.927
	SVM-RBF	94.78	93.04	97.5	97.20	0.951

6.3. *Conclusion*

In conclusion, we have used a deep architecture, SDAE, for the extraction of meaningful features from gene expression data that enable the classification of cancer cells. We were able to use the weights of this model to extract genes that were also useful for cancer prediction, and have potential as biomarkers or therapeutic targets.

One limitation of deep learning approaches is the requirement for large data sets, which may not be available for cancer tissues. We expect that as more gene expression data becomes available, this model will improve in performance and reveal more useful patterns. Accordingly, deep learning models are highly scalable to large input data.

Future work is needed to analyze different types of cancer to identify cancer-specific biomarkers. In addition, there is potential to identify cross-cancer biomarkers through the analysis of aggregated heterogeneous cancer data.

References

1. E. Kettunen, S. Anttila, J. K. Seppänen, A. Karjalainen, H. Edgren, I. Lindström, R. Salovaara, A.-M. Nissén, J. Salo, K. Mattson *et al.*, *Cancer genetics and cytogenetics* **149**, 98 (2004).
2. C. G. A. R. Network *et al.*, *Nature* **499**, 43 (2013).
3. J. Xu, J. A. Stolk, X. Zhang, S. J. Silva, R. L. Houghton, M. Matsumura, T. S. Vedvick, K. B. Leslie, R. Badaro and S. G. Reed, *Cancer research* **60**, 1677 (2000).
4. H. Li, B. Yu, J. Li, L. Su, M. Yan, J. Zhang, C. Li, Z. Zhu and B. Liu, *PloS one* **10**, p. e0125013 (2015).
5. T. Zhou, Y. Du and T. Wei, *Biophysics Reports* **1**, 106 (2015).
6. J. S. Myers, A. K. von Lersner, C. J. Robbins and Q.-X. A. Sang, *PloS one* **10**, p. e0145322 (2015).
7. M. Maienschein-Cline, J. Zhou, K. P. White, R. Sciammas and A. R. Dinner, *Bioinformatics* **28**, 206 (2012).
8. E. E. Schadt, J. Lamb, X. Yang, J. Zhu, S. Edwards, D. GuhaThakurta, S. K. Sieberts, S. Monks, M. Reitman, C. Zhang *et al.*, *Nature genetics* **37**, 710 (2005).
9. K. Shabana, K. A. Nazeer, M. Pradhan and M. Palakal, *BMC bioinformatics* **16**, p. 1 (2015).
10. S. Reddy, K. T. Reddy, V. V. Kumari and K. V. Varma, *International Journal of Computer Science and Information Technologies* **5**, 5901 (2014).
11. T. S. Furey, N. Cristianini, N. Duffy, D. W. Bednarski, M. Schummer and D. Haussler, *Bioinformatics* **16**, 906 (2000).
12. S. A. Medjahed, T. A. Saadi and A. Benyettou, *International Journal of Computer Applications* **62** (2013).
13. A. C. Tan and D. Gilbert (2003).
14. J. A. Cruz and D. S. Wishart, *Cancer informatics* **2** (2006).
15. K. Kourou, T. P. Exarchos, K. P. Exarchos, M. V. Karamouzis and D. I. Fotiadis, *Computational and structural biotechnology journal* **13**, 8 (2015).
16. C. C. e. a. Tan J, Ung M, Unsupervised feature construction and knowledge extraction from genome-wide assays of breast cancer with denoising autoencoders (2015).
17. H. Lee, R. Grosse, R. Ranganath and A. Y. Ng, Convolutional deep belief networks for scalable unsupervised learning of hierarchical representations, in *Proceedings of the 26th Annual International Conference on Machine Learning*, 2009.
18. P. Vincent, H. Larochelle, I. Lajoie, Y. Bengio and P.-A. Manzagol, *The Journal of Machine Learning Research* **11**, 3371 (2010).

19. T. R. Golub, D. K. Slonim, P. Tamayo, C. Huard, M. Gaasenbeek, J. P. Mesirov, H. Coller, M. L. Loh, J. R. Downing, M. A. Caligiuri *et al.*, *science* **286**, 531 (1999).
20. U. Alon, N. Barkai, D. A. Notterman, K. Gish, S. Ybarra, D. Mack and A. J. Levine, *Proceedings of the National Academy of Sciences* **96**, 6745 (1999).
21. J. Li, H. Liu, S.-K. Ng and L. Wong, *Bioinformatics* **19**, ii93 (2003).
22. K. Y. Yeung and W. L. Ruzzo, *Bioinformatics* **17**, 763 (2001).
23. S. Wold, K. Esbensen and P. Geladi, *Chemometrics and intelligent laboratory systems* **2**, 37 (1987).
24. A. Gupta, H. Wang and M. Ganapathiraju, Learning structure in gene expression data using deep architectures, with an application to gene clustering, in *Bioinformatics and Biomedicine (BIBM), 2015 IEEE International Conference on*, 2015.
25. B. Schölkopf, A. Smola and K.-R. Müller, Kernel principal component analysis, in *International Conference on Artificial Neural Networks*, 1997.
26. R. Fakoor, F. Ladhak, A. Nazi and M. Huber, Using deep learning to enhance cancer diagnosis and classification, in *Proceedings of the ICML Workshop on the Role of Machine Learning in Transforming Healthcare. Atlanta, Georgia: JMLR: W&CP*, 2013.
27. Y. Bengio, P. Lamblin, D. Popovici, H. Larochelle *et al.*, *Advances in neural information processing systems* **19**, p. 153 (2007).
28. G. E. Hinton and R. R. Salakhutdinov, *Science* **313**, 504 (2006).
29. S.-C. Wang, Artificial neural network, in *Interdisciplinary Computing in Java Programming*, (Springer, 2003) pp. 81–100.
30. C. Cortes and V. Vapnik, *Machine learning* **20**, 273 (1995).
31. J. N. Weinstein, E. A. Collisson, G. B. Mills, K. R. M. Shaw, B. A. Ozenberger, K. Ellrott, I. Shmulevich, C. Sander, J. M. Stuart, C. G. A. R. Network *et al.*, *Nature genetics* **45**, 1113 (2013).
32. N. V. Chawla, K. W. Bowyer, L. O. Hall and W. P. Kegelmeyer, *Journal of artificial intelligence research* **16**, 321 (2002).
33. G. Lemaître, F. Nogueira and C. K. Aridas, *CoRR* **abs/1609.06570** (2016).
34. D. Saad, *Online Learning* .
35. O. Bousquet and L. Bottou, The tradeoffs of large scale learning, in *Advances in neural information processing systems*, 2008.
36. F. Chollet, Keras `https://github.com/fchollet/keras`, (2015).
37. N. Srivastava, G. Hinton, A. Krizhevsky, I. Sutskever and R. Salakhutdinov, *The Journal of Machine Learning Research* **15**, 1929 (2014).
38. Y. Benjamini and Y. Hochberg, *Journal of the royal statistical society. Series B (Methodological)* , 289 (1995).
39. F. Pedregosa, G. Varoquaux, A. Gramfort, V. Michel, B. Thirion, O. Grisel, M. Blondel, P. Prettenhofer, R. Weiss, V. Dubourg, J. Vanderplas, A. Passos, D. Cournapeau, M. Brucher, M. Perrot and E. Duchesnay, *Journal of Machine Learning Research* **12**, 2825 (2011).
40. G. Matlashewski, P. Lamb, D. Pim, J. Peacock, L. Crawford and S. Benchimol, *The EMBO journal* **3**, p. 3257 (1984).
41. M. Isobe, B. Emanuel, D. Givol, M. Oren and C. M. Croce (1986).
42. S. E. Kern, K. W. Kinzler, A. Bruskin, D. Jarosz, P. Friedman, C. Prives and B. Vogelstein, *Science* **252**, 1708 (1991).

DEVELOPMENT AND PERFORMANCE OF TEXT-MINING ALGORITHMS TO EXTRACT SOCIOECONOMIC STATUS FROM DE-IDENTIFIED ELECTRONIC HEALTH RECORDS

BRITTANY M. HOLLISTER

Vanderbilt Genetics Institute, Vanderbilt University, 519 Light Hall, 2215 Garland Ave. South Nashville, TN, 37232, USA
Email: Brittany.M.Hollister@Vanderbilt.edu

NICOLE A. RESTREPO

Institute for Computational Biology, Department of Epidemiology and Biostatistics, Case Western Reserve University, Wolstein Research Building, 2103 Cornell Road, Cleveland, OH 44106, USA
Email: nrestrepo@case.edu

ERIC FARBER-EGER

Vanderbilt Institute for Clinical and Translational Research, Vanderbilt University Medical Center, 2525 West End Avenue, Suite 600, Nashville, TN 37203, UA
Email: eric.h.farber-eger@vanderbilt.edu

DANA C. CRAWFORD

Institute for Computational Biology, Department of Epidemiology and Biostatistics, Case Western Reserve University, Wolstein Research Building, 2103 Cornell Road, Suite 2527, Cleveland, OH 44106, USA
Email: dana.crawford@case.edu

MELINDA C. ALDRICH[†]

Department of Thoracic Surgery and Division of Epidemiology, Vanderbilt University Medical Center, 1313 21st Avenue South, 609 Oxford House, Nashville, TN 37232, USA
Email: melinda.aldrich@vanderbilt.edu

AMY NON[†]

Department of Anthropology, University of California, San Diego, 9500 Gilman Drive #0532 La Jolla, CA 92093, USA
Email: alnon@ucsd.edu

[†] Co-Senior authors

Socioeconomic status (SES) is a fundamental contributor to health, and a key factor underlying racial disparities in disease. However, SES data are rarely included in genetic studies due in part to the difficultly of collecting these data when studies were not originally designed for that purpose. The emergence of large clinic-based biobanks linked to electronic health records (EHRs) provides research access to large patient populations with longitudinal phenotype data captured in structured fields as billing codes, procedure codes, and prescriptions. SES data however, are often not explicitly recorded in structured fields, but rather recorded in the free text of clinical notes and communications. The content and completeness of these data vary widely by practitioner. To enable gene-environment studies that consider SES as an exposure, we sought to extract SES variables from racial/ethnic minority adult patients (n=9,977) in BioVU, the Vanderbilt University Medical Center biorepository linked to de-identified EHRs. We developed several measures of SES using information available within the de-identified EHR, including broad categories of occupation, education, insurance status, and homelessness. Two hundred patients were randomly selected for manual review to develop a set of seven algorithms for extracting SES information from de-identified EHRs. The algorithms consist of 15 categories of information, with 830 unique search terms. SES data extracted from manual review of 50 randomly selected records were compared to data produced by the algorithm, resulting in positive predictive values of 80.0% (education), 85.4% (occupation), 87.5% (unemployment), 63.6% (retirement), 23.1% (uninsured), 81.8% (Medicaid), and 33.3% (homelessness), suggesting some categories of SES data are easier to extract in this EHR than others. The SES data extraction approach developed here will enable future EHR-based genetic studies to integrate SES information into statistical analyses. Ultimately, incorporation of measures of SES into genetic studies will help elucidate the impact of the social environment on disease risk and outcomes.

1. Introduction

1.1. *Socioeconomic status and health*

Socioeconomic status (SES) is a major determinant of variation in health outcomes worldwide[1]. SES is typically defined as a combination of income or wealth, educational achievement, and occupation[2,3] and be can assessed at the individual, household, or neighborhood level. Health outcomes within the United States, ranging from cancer to hypertension, vary by socioeconomic levels, regardless of how they are measured[4]. Multiple measures of SES have been previously associated with health outcomes, including income[5], years of education[6,7], occupational prestige[2,8,9], insurance coverage[10], and homelessness[11].

SES likely affects health through various pathways including access to healthcare services, knowledge of health behaviors, exposure to environmental stressors and hazards, limited financial resources, and social support[2]. The relationship between SES and health is also highly entangled with race/ethnicity, as SES may covary with race and contribute in part to the existence of racial disparities in health[4,12]. Though these pathways are difficult to distinguish and could affect different populations to varying degrees, it is important to consider SES as a representation of these potential pathways in studies of human health.

Despite the overwhelming evidence that SES affects health outcomes, SES measures are often not included in genetic studies of disease. Neglect of SES data may be due to a lack of available SES information in existing cohorts, as well as the additional time and resources it takes to collect SES data in new studies. Despite these difficulties, it is vital to include measurements of SES in genetic association studies of racial disparities in health. In addition to the possible confounding that may occur due to the association of race/ethnicity with both SES and health[13], SES has the potential to modify the effect of genetic variants on health outcomes[14]. Therefore, the etiology of disease is likely to be misunderstood without the inclusion of SES data in association studies. Although prior genetic association studies have found some gene variants that may explain a small portion of racial disparities in disease prevalence and risk[15], SES factors are likely to play an even larger role in racial health disparities[6,7].

1.2. *SES data within electronic health records*

The use of electronic health records (EHRs) for research purposes is becoming increasingly prevalent. With the announcement of the Precision Medicine Initiative and its goal of recruiting one million participants with biological and EHR data, the research use of EHRs is anticipated to increase[16]. EHRs provide an attractive resource for biomedical researchers for many reasons, including their rich phenotypic and longitudinal data, as well as the lower cost of participant recruitment versus a traditional prospective cohort study. Additionally, clinical biobanks that contain biological samples linked to EHRs are becoming an invaluable resource for conducting genetic epidemiology studies. Despite the potential for EHRs in research settings, these clinical data repositories currently have noted deficits in the availability and completeness of important social and environmental data[17], including SES, that are known to contribute independently to health status and could modify genetic effects[18].

Recognizing the importance of formally and consistently capturing social and behavioral measures in the EHR, the Institute of Medicine (IOM) recently recommended SES measures, specifically educational attainment, financial resource strain, and neighborhood median household income be included in the EHR[19]. The committee also recommended that a plan be developed by the NIH to expand the research use of EHRs to include social and behavioral data[19]. Adoption of these recommendations will take time, and may not be universal across medical centers; therefore, there is a need to develop approaches and methods to access existing unstructured SES data within the EHR for research purposes. SES data are almost entirely found within the free text clinical notes from providers and the clinical communications between providers. Currently, there are no published algorithms available to extract SES data from EHRs. In this study, we developed an approach for extracting available SES information from the free text of a de-identified EHR. These algorithms will facilitate the immediate extraction of key SES information from clinical biobanks for incorporation into future biomedical research.

1.3. *BioVU*

BioVU is a DNA biobank of the Vanderbilt University Medical Center (VUMC) linked to de-identified EHRs. DNA samples are extracted from discarded blood samples drawn for routine clinical care[20]. DNA samples are linked to the Synthetic Derivative (SD), the de-identified version of the VUMC EHR, by a unique study ID. Medical records within the SD are scrubbed of all HIPAA identifiers such as names, locations, zip codes, and social security numbers. Dates within each SD record are shifted to prevent re-identification of the records. Date shifting is consistent within a single patient's record. As previously described[21], data from BioVU are de-identified in accordance with provisions of Title 45, Code of Federal Regulations, part 46 (45 CFT 46); consequently, this study is considered non-human subjects research by the Vanderbilt University Institutional Review Board.

2. Methods

2.1. *Population*

The study population included all racial/ethnic minority patients >18 years old participating in BioVU as of 2011[22]. The EHRs used for the development of the algorithms were updated in 2015 to include current information. Race/ethnicity is administratively reported in BioVU and strongly correlated with genetic ancestry[23,24]. The majority (81%) of patients in the dataset are black individuals with an average age of 50 years (Table 1). The mean number of clinic visits within a patient's EHR record is 40.45 visits, and the mean number of days between patients' first and last visit within the EHR is 2,340 days (Table 1).

Table 1. Vanderbilt BioVU racial/ethnic minority population characteristics

Characteristic	n= 9,977
Sex	
Male	3,568 (36%)
Female	6,409 (64%)
Race/ethnicity	
Black	8,078 (81%)
Hispanic	1,049 (10.5%)
Asian	850 (8.5%)
Age (mean, years ± SD)	49.8 ± 18.1
Number of clinic visits (mean ± SD)	40.5 ± 55.0
Number of days between visits (mean ± SD)	2,340 ± 1,793.1

2.2. *Development of algorithms*

We sought to develop algorithms to extract SES data from structured and unstructured data in the de-identified EHRs. We developed seven algorithms for the extraction of SES information including education level, occupation, unemployment, retirement, insurance status, Medicaid status, and homelessness (Table 2). The initial development of the SES algorithms began with a manual review of both structured and unstructured data within the de-identified EHR of 200 randomly selected patients within this minority population dataset to identify the following: 1) the categories of SES information most frequently mentioned, 2) where in the EHR this information is noted, and 3) the semantic language used by clinical providers for socioeconomic information (Figure 1). The manual review revealed that the SES data were found exclusively within the unstructured free text of the clinical notes, social history, and clinical communications of this EHR. It was also noted that the most frequently mentioned semantic categories were employment, education, insurance status, and homelessness, and thus these categories were chosen for extraction. Semantic tags for each category were selected if they appeared more than once within the 200 development records.

2.2.1. *Employment*

Employment information was extracted using three different algorithms designed to capture data on occupation, unemployment, and retirement. The occupation algorithm extracts the occupation

Table 2. Variables extracted by socioeconomic status (SES) algorithms applied to de-identified electronic health records

Semantic category	Format of algorithm output
Occupational prestige	0-100
Unemployment	Ever/never
Retirement	Ever/never
Education	-Never attended -Less than high school -High school graduate/GED -Associate's degree -Bachelor's degree -Master's degree -Professional degree -Doctoral degree
Uninsured	Ever/never
Medicaid	Ever/ never
Homelessness	Ever/never

mentioned in a patient's record and translates it to an occupational prestige score (scale of 0-100). This score represents how well-respected an occupation is within a society (i.e., subjective socioeconomic position). Occupational prestige scores were determined from a National Opinion Research Center (NORC) survey where respondents were asked to rank occupations according to their prestige[25]. The occupation tags utilized for the occupation algorithm were adopted from the most recent NORC report. The algorithm's occupation tags were shortened to 678 occupations from the original NORC list of 860 occupations given that some of the occupations were highly specific with repetitive occupational prestige scores. As an example, "teacher, elementary school" and "teacher, secondary school" were collapsed to "teacher."

We next used the occupation algorithm to search the unstructured data of the original 200 patients for the initial occupation tags. This search identified a large number of false positives, where the algorithm tagged occupation-related words that were not indicative of the patient's occupation. In order to filter these false positives, additional prefix language such as "works as," "is a/n," "employed" was added to a subset of occupations to filter out non-relevant terms, which greatly improved the algorithm. Unemployment data were extracted using semantic tags for unemployment (e.g., "unemployed," "does not work," "hasn't worked since"). The unemployment algorithm was then tested on the unstructured data from the 200 records used for development, and a high number of false positives were returned. These false positives were often in reference to medications. Therefore the tags "if this does not work" and "if that does not work" were excluded to filter false positives. Unemployment was classified as ever/never (Table 2). Retirement was also extracted from the EHR using the tag "retired" and classified as ever/never (Table 2).

2.2.2. *Education*

The education algorithm was designed to assign education level to a patient based on the highest education achieved and recorded in the EHR. Education levels were assigned to each relevant tag word or phrase found in the unstructured text of the EHR (Table 2). Sixty-two semantic tags were utilized and the highest level of education was determined for each patient. These tags were exclusive to an assigned education level. For example, the high school degree category of education level included tags such as "high school graduate" and "completed 12th grade," while the bachelor's degree category included terms such as "BS degree" and "completed college." The levels of education were based on U.S. census definitions with one modification such that all grade levels below high school graduate were collapsed into a "less than high school" category. We searched through the unstructured text of the 200 records used for development to determine if further filtering or modification was needed. Fifteen additional tags were used to filter false positive results related to types of medical education (e.g. "diet education," "dialysis education") and Vanderbilt Medical School students (e.g., "medical student," "pharmacy student," "student nurse").

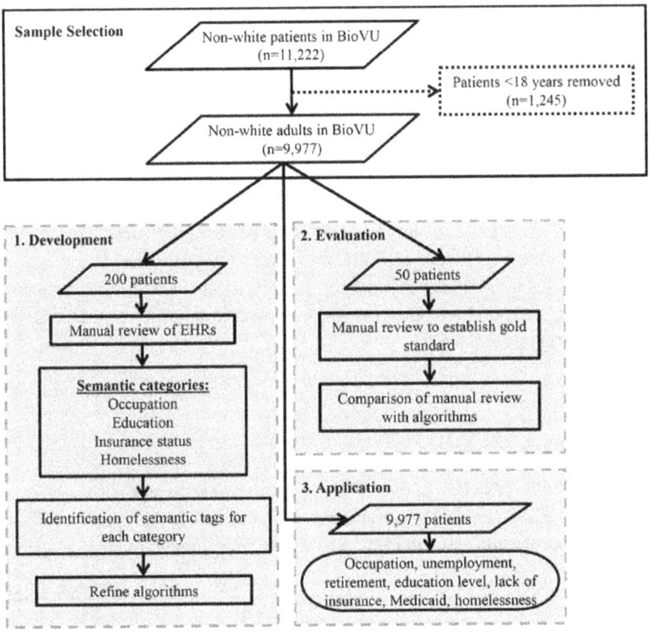

Figure 1. Overview of the development process for the SES algorithms

2.2.3. *Insurance status*

The extraction process for insurance status required two algorithms. The first algorithm was used to determine if there was any time point in the EHR when the patient did not have insurance based on the presence of five semantic tags (Table 2). These tags included "no insurance" and "does not have insurance" and excluded some language that was used in a standard discharge letter and therefore appeared frequently in the EHR. A second insurance algorithm extracted Medicaid information using specific phrases or keywords such as "Medicaid" and "TennCare" and was classified as ever/never in order to determine if a patient was ever on Medicaid in their EHR (Table 2).

2.2.4. *Homelessness*

Homelessness information was extracted using the tags "homeless" and "shelter" among the 200 development EHRs. After this search, several false positives were returned relating to patients who worked or volunteered at homeless shelters. Therefore, exclusion tags were added such as "volunteer at homeless shelter," "works at homeless shelter," "works with homeless," and "animal shelter." Homelessness was classified as ever/never (Table 2).

2.3. *Evaluation of algorithm performance*

To evaluate the performance of these SES algorithms, results were compared to findings from a manual review of 50 randomly selected patients. These 50 individuals were selected using random sampling without replacement. Two independent reviewers manually reviewed the clinical record of each patient and any discrepancies were resolved by discussion between the two reviewers. Comparison of results from the two independent reviewers was quantified using percent positive agreement, percent negative agreement, and kappa statistics for each of the seven categories and subcategories: education level, occupation, unemployment, retirement, uninsured, Medicaid, and homelessness. The manual review of 50 records was then compared to the algorithm results for each of the seven categories and subcategories. Sensitivity, specificity, and positive predictive value were estimated. The chi-square statistic was used to determine if the algorithms performed differently in different populations.

3. Results

3.1. *Population characteristics*

Among the total study population (n=9,977), we were able to extract at least one type of SES information from 8,282 (83.0%) individuals. We extracted education information for 3,780 individuals and occupation information for 7,296 individuals (Table 3). For the remaining

Table 3. Percent of records within the study population with algorithm-identified SES characteristics

Characteristics	Race			
	Black (n=8,078)	Hispanic (n=1,049)	Asian (n=850)	Total (n=9,977)
% with occupation	76.0	57.1	65.4	73.1
% unemployed	21.4	13.0	13.4	19.8
% retired	19.8	4.9	11.2	17.5
% with education	39.1	28.7	37.9	37.9
% uninsured	19.5	15.6	11.5	18.4
% on Medicaid	20.5	13.9	7.9	18.7
% homeless	3.7	1.3	1.0	3.2

categories, we were able to determine if an individual was unemployed, retired, uninsured, on Medicaid, or homeless at any point in his or her record. Of the total population for which we were able to extract SES data (n=8,282), 1,978 individuals were unemployed, 1,742 individuals were retired, 1,839 individuals were uninsured, 1,865 were on Medicaid, and 318 were homeless at least one time in their EHR (Table 3). For each of the seven categories, the algorithms returned SES information for a higher percentage of black patients than Hispanic or Asian patients ($p<0.00001$).

The five most frequently extracted occupations among those having occupation information (n=7,296) were manager, nurse, Army, manufacturer, and restaurant employee. Within the population with education information (n=3,780), the vast majority of individuals had a high school degree (n=2,066), followed by individuals without a high school degree (n=492), and individuals with a bachelor's degree (n=446).

3.2. *Algorithm Performance*

Prior to evaluating algorithm performance, the manual review results from the randomly selected records of 50 patients were compared between the two reviewers and any conflicts were resolved. The percent positive agreement between reviewers ranged from 98.0% to 100.0% and the percent negative agreement ranged from 94.7% to 100.0%. The Kappa statistic between reviewers ranged from 0.94 to 1.0.

Table 4. Comparison of manual review with algorithm results for each SES algorithm in a subset of randomly selected individuals (n=50)

Semantic Category	Records with SES information (%)	Sensitivity (%)	Specificity (%)	PPV (%)
Education level	48.0	66.7	84.5	80.0
Occupation	80.0	87.5	40.0	85.4
Unemployment	40.0	70.0	93.3	87.5
Retirement	14.0	100.0	90.7	63.6
Uninsured	8.00	75.0	78.3	23.1
Medicaid	18.0	100.0	95.1	81.8
Homelessness	2.00	100.0	95.9	33.3

Once all reviewer discrepancies were resolved, the manual review results were used as the gold standard and compared to the algorithm results. All the algorithms, with the exception of occupation, had very high specificity levels >78%. The lower specificity for occupation (40%) is due to six of the ten individuals who did not have occupation information (as identified by manual review) but were identified as having occupation information by the algorithm. All the algorithms also had high sensitivity levels (above 70%), with the exception of education level (66.7%) (Table 4). The lower sensitivity for education is driven by eight individuals who have an education level that was identified by manual review but not by the algorithm. The lower sensitivity for unemployment is due to the six individuals who were identified as unemployed by manual review but not by the algorithm. PPV values across the algorithms ranged from 23.1%-87.5%. The lower PPV for the retirement algorithm (63.6%) is due to the four individuals identified as retired by the algorithm but not retired by the manual review. (Table 4). The low PPV for the uninsured algorithm (23.1%) is due to the ten individuals who were identified as uninsured by the algorithm, but not by manual review. The low PPV for homelessness (33.3%) was a result of the fact that the manual review only identified one patient with homelessness in their record, whereas the algorithm misidentified two others.

3.2.1. *Missing data*

Of the total population (n=9,977), the algorithm was not able to extract any SES information for 1,695 individuals (17.0%). Of this group, there were 1,193 blacks, 309 Hispanics, and 193 Asians. Missing SES data were more common among Hispanic and Asian individuals, than among black individuals ($p<0.001$). The Hispanic and Asian populations represent 10.5% and 8.5% of the total dataset, respectively; however, these groups represent 18.2% and 11.4%, respectively, of the individuals with missing SES data. Males represent 35.8% of the study population and 28.0% of those without extracted SES data. The mean age for the total population is 49.9 years, and the mean age for the group without extracted SES information is 46.7 years.

4. Conclusion

Socioeconomic status is considered a fundamental cause of disease, because it affects so many proximate risk factors and disease outcomes[26]. SES has been consistently associated with health outcomes such as mortality, cancer, and cardiovascular disease[27,28]. Despite these consistent associations, SES data are typically not included in genetic studies of health outcomes. For studies that utilize biobanks, the lack of SES data is likely related to the difficulty in accessing these data within the EHR, where they are not usually recorded in structured fields. The algorithms described in this study are the first to extract these important data from EHRs for research purposes.

The SES algorithms described here focus on the extraction of data related to four semantic categories: occupation, education, insurance status, and homelessness. The occupation algorithms extracted and classified data as occupational prestige, unemployment (ever/never), and retirement (ever/never). The occupational prestige algorithm had a strong sensitivity and PPV; however it had a low specificity of 40%, reflective of the difficulty in filtering the occupation information. Although we took steps to remove false positives, it was difficult to completely eliminate all false positives without removing a large amount of accurate data. Our unemployment and retirement algorithms had high sensitivity (70% and 100%) and specificity (93.3% and 90.7%). While the unemployment algorithm had a high PPV, the retirement algorithm had a low PPV. Both unemployment and retirement were classified as ever/never because the EHR only captures a snapshot of time when the patient visits the clinic. It was not possible to accurately capture the length of time for unemployment or retirement as the patient's visits to the clinic may not reflect the length of time he or she was unemployed or retired. The sensitivity of the unemployment algorithm was affected by the varying language used to describe unemployment, which was identified in manual review but not consistently recognized by the algorithm ("does not work outside the home", "used to work in a restaurant"). The quality of the retirement algorithm was affected by false positives related to the identification of words related to retirement that were used in a context outside of the patient's retirement from an occupation.

The education algorithm identified the highest level of education that a patient achieved over the course of their EHR. This algorithm had a high specificity and PPV, but a low sensitivity. The low sensitivity was due to the inability of the algorithm to detect variations in education level compared with the manual review. The variation in language used by clinical providers made it difficult to include every mention of education while still maintaining some level of precision. For example, some of the Vanderbilt Medical School students were excluded ("medical student," "pharmacy student") because of the frequent mention of these terms in the EHR related to patient care, rather than education level. The reviewers were able to infer education level based on occupation and context clues as well as identify the medical school students, while the algorithm was not able to so. The algorithm that identified patients who were uninsured at some point in his or her record as well as the homelessness algorithm each had high sensitivity and specificity, but low PPV. Uninsured patients are the smallest portion of patients within VUMC, making up only 4.7% of the patient population in 2015[29]. The low PPV of these algorithms may be due to a low prevalence of uninsured patients and homeless individuals within the VUMC patient population. Within our randomly selected minority patient population used for evaluation, only four individuals were uninsured and one was homeless. These categories had the lowest prevalence within our evaluation dataset. The Medicaid algorithm was the highest performing algorithm, with a high sensitivity, specificity, and PPV.

The major challenges in utilizing EHR data in a research setting include missing data and the inconsistencies in the recording of SES data by clinical providers. While the majority of individuals within the study population had some SES information, a notable percentage of individuals did not have any SES information within their records (17.0%). The missing SES data could be a result of the lack of recording of information by the provider, either due to SES factors not being discussed in conversation with the patient, a low number of visits in the patient's EHR, or the willingness of the patient to provide SES information. Additionally, when variables are missing within a patient's record, we cannot distinguish whether it is due to negative data or just missing data. For example, if a patient does not have an occupation listed, we cannot assume that they are unemployed because it may have not been discussed with the provider or recorded by the provider. The higher level of missing data observed for Hispanic and Asian individuals in this dataset could be a reflection of the fact that the algorithms are optimized for the largest racial/ethnic population within the dataset (i.e., black patients).

The inconsistencies in the recording of the SES data are typical for social and environmental exposure data contained within free clinical text[17]. In the development of these algorithms, we noted that providers, in general, do not follow patterns when recording SES data within their notes in the EHR. The lack of consistent language and the numerous variations used to describe the SES information made extracting this information challenging. Furthermore, algorithms could also be limited by the accuracy of the selected filters and tags, rather than the information available within the EHR. While the aim of the algorithms was to include all possible semantic tags, there is a possibility that some information was missed by the algorithms or that information was captured inaccurately due to the limitations of the filtering process.

In addition to these general limitations, the algorithms developed here have specific limitations regarding portability. Even within the same dataset, we have noted a difference in tag retrieval for the SES categories queried across the three major racial/ethnic groups. Additional studies are required to improve the algorithms' performances and retrieval of semantic tags in multiple populations as well as within different study sites. Indeed, some of the tags developed here (such as "TennCare" in reference to Medicaid) are specific to Tennessee and will require modification to ensure portability regardless of the state in which the algorithms are deployed. Furthermore, these algorithms were created in a de-identified EHR, which required the development of a free text algorithm for insurance status, as the structured insurance information is considered identifying information. An identified EHR may have this insurance information within the structured text. However, the other categories of SES information are likely to only be found within the free text of an identified EHR.

Despite the many challenges faced with the extraction of SES data from the EHR, these algorithms were able to successfully extract a large amount of data not previously accessible for research purposes. The sensitivities, specificities, and PPVs for the algorithms were high considering the limitations of the SES data within the current EHR. Overall, these algorithms represent a first important step in incorporating SES data from EHRs into precision medicine research, as envisioned by the Institute of Medicine and others.

5. Resources

Semantic tag and filter lists for each algorithm can be found on the Vanderbilt University Medical Center TREAT Lung Cancer Research Program website (https://medschool.vanderbilt.edu/treat-lung-cancer-program/) and the Institute for Computational Biology website (http://www.icompbio.net/?page_id=1654). The code which was used to run the algorithms is available in GitHub.

6. Acknowledgements

This work was supported in part by NIH grants U01 HG004798 and its ARRA supplements (DCC) and 1K07CA172294 (MCA). BMH was supported by the NIH/NIGMS Genetics Predoctoral Research Training Program 5T32GM080178-07. The dataset(s) used for the analyses described were obtained from Vanderbilt University Medical Center's BioVU which is supported by institutional funding and by the Vanderbilt CTSA grant funded by the National Center for Research Resources, Grant UL1 RR024975-01, which is now at the National Center for Advancing Translational Sciences, Grant 2 UL1 TR000445-06.

References

1. *Poverty: Assessing the Distribution of Health Risks by Socioeconomic Position at National and Local Levels.* (2004).
2. T. Seeman *et al.*, *Social Science & Medicine* 66, 72-87 (2008).
3. P. Braveman *et al.*, *Public Health Reports* 129 Suppl 2, 19-31 (2014).
4. National Center for Health Statistics, *Health, United States, 2011: With Special Feature on Socioeconomic Status and Health* (2012).

5. V. Carrieri *et al.*, *Health Econ*, (2016).
6. A. L. Non *et al.*, *American Journal of Public Health* 102, 1559-1565 (2012).
7. M. C. Aldrich *et al.*, *American Journal of Public Health* 103, e73-80 (2013).
8. R. Hauser *et al.*, *Sociological Methodology* 27, 177-298 (1997).
9. K. Fujishiro *et al.*, *Social Science & Medicine* 71, 2100-2107 (2010).
10. in *Kaiser Commission on Medicaid and the Uninsured* T. H. J. K. F. Foundation, Ed. (Washington, D.C., 2012).
11. D. S. Morrison, *International Journal of Epidemiology* 38, 877-883 (2009).
12. National Center for Health Statistics *Health, United States, 2015: With Special Feature on Racial and Ethnic Health Disparities* (2016).
13. T. J. VanderWeele *et al.*, *Epidemiology* 25, 473-484 (2014).
14. S. Cakmak *et al.*, *Journal of Environmental Management* 177, 1-8 (2016).
15. J. S. Kaufman *et al.*, *American Journal of Epidemiology* 181, 464-472 (2015).
16. F. S. Collins *et al.*, *The New England Journal of Medicine* 372, 793-795 (2015).
17. I. S. Kohane, *Nature Reviews. Genetics* 12, 417-428 (2011).
18. J. Basson *et al.*, *American journal of hypertension* 27, 431-444 (2014).
19. IOM (Institute of Medicine), *Capturing Social and Behavioral Domains and Measures in Electronic Health Records: Phase 2* (2014).
20. D. M. Roden *et al.*, *Clinical Pharmacology and Therapeutics* 84, 362-369 (2008).
21. J. Pulley *et al.*, *Clinical and Translational Science* 3, 42-48 (2010).
22. D. C. Crawford *et al.*, *Human Heredity* 79, 137-146 (2015).
23. J. B. Hall *et al.*, *PloS one* 9, e99161 (2014).
24. L. Dumitrescu *et al.*, *Genetics in Medicine : official journal of the American College of Medical Genetics* 12, 648-650 (2010).
25. NORC, *Measuring Occupational Presitge on the 2012 General Social Survey* (2014).
26. B. G. Link *et al.*, *J Health Soc Behav* Spec No, 80-94 (1995).
27. T. N. Bethea *et al.*, *Ethnicity & Disease* 26, 157-164 (2016).
28. A. Rawshani *et al.*, *JAMA Internal Medicine*, (2016).
29. "2015 Financial Report " (Vanderbilt University, Nashville, TN. , 2015).

GENOME-WIDE INTERACTION WITH SELECTED TYPE 2 DIABETES LOCI REVEALS NOVEL LOCI FOR TYPE 2 DIABETES IN AFRICAN AMERICANS

JACOB M. KEATON[1,2,3], JACKLYN N. HELLWEGE[2,3], MAGGIE C. Y. NG[2,3], NICHOLETTE D. PALMER[2,3,4,5], JAMES S. PANKOW[6], MYRIAM FORNAGE[7], JAMES G. WILSON[8], ADOLFO CORREA[8], LAURA J. RASMUSSEN-TORVIK[9], JEROME I. ROTTER[10], YII-DER I. CHEN[10], KENT D. TAYLOR[10], STEPHEN S. RICH[11], LYNNE E. WAGENKNECHT[5,12], BARRY I. FREEDMAN[3,5,13], DONALD W. BOWDEN[2,3,4]

[1]*Molecular Genetics and Genomics Program, Wake Forest School of Medicine, Medical Center Blvd, Winston-Salem, NC, 27157, US*

[2]*Center for Genomics and Personalized Medicine Research, Wake Forest School of Medicine, Medical Center Blvd, Winston-Salem, NC, 27157, US*

[3]*Center for Diabetes Research, Wake Forest School of Medicine, Wake Forest School of Medicine, Medical Center Blvd, Winston-Salem, NC, 27157, US*

[4]*Department of Biochemistry, Wake Forest School of Medicine, Medical Center Blvd, Winston-Salem, NC, 27157, US*

[5]*Center for Public Health Genomics, Wake Forest School of Medicine, Medical Center Blvd, Winston-Salem, NC, 27157, US*

[6]*Division of Epidemiology and Community Health, University of Minnesota, Minneapolis, MN, 55455, US*

[7]*Institute of Molecular Medicine and Human Genetics Center, University of Texas Health Science Center at Houston, 7000 Fannin St #1200, Houston, TX, 77030, US*

[8]*University of Mississippi Medical Center, 2500 N State St, Jackson, MS, 39216, US*

[9]*Department of Preventive Medicine, Northwestern University Feinberg School of Medicine, 303 E Chicago Ave, Chicago, IL, 60611, US*

[10]*Institute for Translational Genomics and Population Sciences, Los Angeles BioMedical Research Institute, Harbor-UCLA Medical Center, 1000 W Carson St, Torrance, CA, 90502, US*

[11]*Center for Public Health Genomics, University of Virginia, Charlottesville, VA, 22904, US*

[12]*Division of Public Health Sciences, Wake Forest School of Medicine, Medical Center Blvd, Winston-Salem, NC, 27157, US*

[13]*Department of Internal Medicine - Section on Nephrology, Wake Forest School of Medicine, Medical Center Blvd, Winston-Salem, NC, 27157, US*

Type 2 diabetes (T2D) is the result of metabolic defects in insulin secretion and insulin sensitivity, yet most T2D loci identified to date influence insulin secretion. We hypothesized that T2D loci, particularly those affecting insulin sensitivity, can be identified through interaction with known T2D loci implicated in insulin secretion. To test this hypothesis, single nucleotide polymorphisms (SNPs) nominally associated with acute insulin response to glucose (AIR_g), a dynamic measure of first-phase insulin secretion, and previously associated with T2D in genome-wide association studies (GWAS) were identified in African Americans from the Insulin Resistance Atherosclerosis Family Study (IRASFS; n=492 subjects). These SNPs were tested for interaction, individually and jointly as a genetic risk score (GRS), using GWAS data from five cohorts (ARIC, CARDIA, JHS, MESA, WFSM; n=2,725 cases, 4,167 controls) with T2D as the outcome. In single variant analyses, suggestively significant ($P_{interaction} < 5\times10^{-6}$) interactions were observed at several loci including *DGKB* (rs978989), *CDK18* (rs12126276), *CXCL12* (rs7921850), *HCN1* (rs6895191), *FAM98A* (rs1900780), and *MGMT* (rs568530). Notable beta-cell GRS interactions included two SNPs at the *DGKB* locus (rs6976381; rs6962498). These data support the hypothesis that additional genetic factors contributing to T2D risk can be identified by interactions with insulin secretion loci.

1. Introduction

Although common variants examined in genome-wide association studies (GWAS) have identified ~80 loci associated with T2D risk, these variants explain only about 15% of T2D heritability[1,2]. A portion of the missing heritability may be explained by epistasis, which occurs when a genetic risk factor is modified by other factors in an individual's genetic background[3]. Epistasis, or gene-gene interaction, analyses may facilitate the detection of novel loci when non-additive effects exist, but may also provide novel insights illuminating biological mechanisms underlying complex diseases such as T2D[4].

T2D is characterized by impaired insulin secretion arising from pancreatic beta-cell dysfunction and insulin resistance in skeletal muscle, hepatic, and other peripheral tissues, leading to decreased plasma glucose uptake. However, documented T2D loci primarily map to genes influencing insulin secretion or other aspects of beta-cell biology[1]. Given the underlying bimodal pathophysiology, T2D may be a particularly well-suited disease model for hypothesis-driven investigation of epistatic interactions. Genetic insults to both insulin secretion and insulin sensitivity may jointly increase an individual's T2D risk in a non-additive manner. Considering the higher prevalence rate of T2D, insulin resistance, and obesity, African Americans are optimal for the study of genetic interactions that contribute to T2D risk.

In an effort to identify interactions contributing to T2D and to discover novel insulin sensitivity loci, we hypothesized that T2D risk loci, particularly those affecting insulin sensitivity, could be identified by interaction analyses with known T2D loci implicated in insulin secretion. In cross-sectional meta-analyses of five T2D studies (ARIC, CARDIA, JHS, MESA, and WFSM), we tested whether 5 SNPs from known T2D loci implicated in insulin secretion, or a genetic risk score summarizing these SNPs, modified genome-wide SNP associations with T2D risk.

2. Research Design and Methods

2.1 *Subjects*

Two sources of data were analyzed in this study. Primary inferences of association with insulin secretion were derived from African American participants (n=492 individuals from 42 families) in the Insulin Resistance Atherosclerosis Family Study (IRASFS), a metabolically well-characterized cohort[5]. Glucose homeostasis traits were measured by the frequently sampled intravenous glucose tolerance test (FSIGT)[5]. Briefly, a 50% glucose solution (0.3g/kg) and regular human insulin (0.03units/kg) were injected intravenously at 0 and 20 minutes, respectively. Blood was collected at −5, 2, 4, 8, 19, 22, 30, 40, 50, 70, 100, and 180 minutes for measurement of plasma glucose and insulin. AIR_g was calculated as the increase in insulin at 2–8 minutes above the basal (fasting) insulin level after the bolus glucose injection at 0-1 minute. Insulin sensitivity (S_I) was calculated by mathematical modeling using the MINMOD program (version 3.0 [1994])[6]. Disposition index (DI) was calculated as the product of S_I and AIR_g.

Inferences of genome-wide epistatic interaction with insulin secretion loci for T2D susceptibility were derived from African American participants from the Atherosclerosis Risk in Communities Study (ARIC; n = 955 T2D cases, 414 controls), Coronary Artery Risk Development in Young Adults (CARDIA; n = 94 T2D cases, 654 controls), Jackson Heart Study (JHS; n = 333 T2D cases, 1,450 controls), Multi-Ethnic Study of Atherosclerosis (MESA; n = 411 T2D cases, 793 controls), and the Wake Forest School of Medicine (WFSM; n = 932 T2D cases, 856 controls) cohorts for a total of 2,725 T2D cases and 4,167 controls[7–12]. T2D was diagnosed according to the American Diabetes Association criteria with at least one of the following: fasting glucose ≥126 mg/dL, 2-h oral glucose tolerance test glucose ≥200 mg/dL, random glucose ≥200 mg/dL, use of oral hypoglycemic agents and/or insulin, or physician diagnosed diabetes. Subjects diagnosed before 25 years of age were excluded. Normal glucose tolerance was defined as fasting glucose <100 mg/dL and 2-h oral glucose tolerance test glucose <140 mg/dL (if available) without reported use of diabetes medications. Control subjects <25 years of age were excluded.

IRB approval was obtained at all sites and all participants provided written informed consent. Descriptions of the T2D study cohorts are summarized in the Supplementary Methods.

2.2 *Genotyping, imputation, and quality control*

For the IRASFS samples, genotyping and quality control were performed at the Wake Forest Center for Genomics and Personalized Medicine Research using the Illumina Infinium HumanExome BeadChip v1.0 as previously described[13]. Briefly, the exome chip contained 247,870 variants (92% protein coding). In addition, the chip included 64 SNPs associated with T2D from previous GWAS in Europeans, many of which have been implicated in insulin secretion (exome chip design: http://genome.sph.umich.edu/wiki/Exome_Chip_Design). Sample and autosomal SNP call rates were ≥99%, and SNPs with poor cluster separation (<0.35) were excluded. Mendelian errors were identified using PedCheck[14] and resolved by removing conflicting genotypes. Hardy–Weinberg Equilibrium (HWE) was assessed in unrelated samples (n = 39) using PLINK (http://pngu.mgh.harvard.edu/purcell/plink)[15] to reduce biases introduced by familial allele frequencies. All variants were in accordance with HWE ($P > 1x10^{-5}$).

The T2D study samples were genotyped using the Affymetrix Genome-Wide Human SNP Array 6.0. For the ARIC, CARDIA, JHS, and MESA cohorts, genotyping and quality control were completed by the National Heart, Lung, and Blood Institute's (NHLBI's) Candidate Gene Association Resource (CARe) at the Broad Institute[16]. Genotyping for the WFSM study was performed at the Center for Inherited Disease Research (CIDR). For all T2D studies, imputation was performed using MACH with the function –mle (version 1.0.16, http://www.sph.umich.edu/csg/abecasis/MaCH/) to obtain missing genotypes and replace genotypes inconsistent with reference haplotypes as previously described[17]. SNPs with call rate ≥ 95% and minor allele frequency (MAF) ≥ 1% that passed study-specific quality control were used for imputation[16,18]. A 1:1 HapMap II (NCBI Build 36) CEU:YRI (European:African) consensus haplotype was used as reference. A total of 2,713,329 to 2,907,086 autosomal SNPs from each GWAS with call rate ≥95%, MAF ≥ 1%, and Hardy-Weinberg P-value ≥ 0.0001 for genotyped SNPs and MAF ≥ 1% and RSQ ≥ 0.5 for imputed SNPs were included in subsequent data analyses.

2.3 *Principal component analysis*

For IRASFS, admixture was estimated using principal components (PCs) from 39 ancestry informative markers (AIMs) and including HapMap CEU and YRI samples for comparison[19]. Only PC1 correlated with HapMap populations, and was thus used as a covariate in all analyses.

For the T2D studies, PCs were computed for each study using high-quality SNPs as previously described[13,16–18,20]. The first PC was highly correlated ($r^2 > 0.87$) with global African-European ancestry, as measured by ANCESTRYMAP[21], STRUCTURE[22], or FRAPPE[23]. The African American T2D study samples had an average of 80% African ancestry. By analyzing unrelated samples from all studies using SMARTPCA[20], only the first PC appeared to account for substantial genetic variation (data not shown), whereas the subsequent PCs may reflect sampling noise and/or relatedness in samples[21]. The first PC (PC1) was used as a covariate in all analyses to adjust for population substructure.

2.4 *Analysis of association with measures of glucose homeostasis in IRASFS*

To approximate a normal distribution, trait values were transformed by square root (AIR_g, DI) or natural logarithm plus a constant (S_I). Measured genotype association analyses of exome chip variants with AIR_g, S_I, and DI were performed under an additive model using the variance components method implemented in Sequential Oligogenic Linkage Analysis Routines (SOLAR)[24] with adjustment for age, gender, body mass index (BMI), and PC1.

2.5 *Genetic risk score construction*

We further explored our interaction approach by constructing genetic risk scores (GRS), both weighted and unweighted, summarizing the effects of SNPs associated with both T2D and insulin secretion (T2D-IS SNPs). The T2D-IS GRS was created using the T2D risk alleles for T2D-IS SNPs defined from the literature (Table 1). The unweighted risk score was calculated by summation of the number of risk alleles for each individual across all selected SNPs. The weighted T2D-IS GRS was calculated as the sum of risk alleles at each locus multiplied by the natural log of their T2D odds ratio (OR) defined from the literature[2,25–28]. Missing genotypes for a given SNP were imputed as the average number of risk alleles across all samples. The association of each GRS with both AIR_g and DI, a combinatorial measure of first-phase insulin secretion and insulin sensitivity, were evaluated in IRASFS using the variance components method implemented in SOLAR[24] adjusted for age, gender, and ancestry proportions.

Table 1. Characteristics and single-SNP AIR_g association results for T2D-IS SNPs in published GWAS and IRASFS

T2D-IS SNP	Chr	Position*	Gene	Published GWAS				IRASFS AIR_g			
				T2D Risk Allele	Other Allele	T2D OR†	PMID‡	RAF§	Beta	SE‖	P
rs7593730	2	161171454	RBMS1	T	C	1.11	20418489	0.39	-1.38	0.86	0.086
rs864745	7	28180556	JAZF1	T	C	1.10	18372903	0.72	-1.52	0.91	0.096
rs5215	11	17408630	KCNJ11	C	T	1.08	24509480	0.15	-2.60	1.18	0.033
rs1552224	11	72433098	ARAP1	A	C	1.14	20581827	0.06	-3.05	1.69	0.077
rs7119	15	77777632	HMG20A	C	T	1.24	22885922	0.52	-1.50	0.81	0.059

*NCBI build 37. †Reported odds ratio. ‡PubMed ID. §Risk allele frequency. ||Standard error.

2.6 *Analysis of interaction for T2D risk in the African American T2D case-control studies*

A logistic regression test for additive allelic interaction adjusted for age, gender, and PC1 was used for all interaction analyses with T2D as the outcome. Additional models included adjustment for BMI, and individuals with missing values were excluded (n = 110). In each study, genome-wide interaction tests were performed in PLINK between each SNP in the genome with each candidate SNP (i.e. insulin secretion SNP) and GRS (i.e. insulin secretion risk score). An example PLINK command is provided in the Supplementary Methods. Interaction results with extreme values (absolute β or SE > 10), primarily due to low cell counts, were excluded. Across interaction analyses with all SNPs and risk scores, the number of SNPs excluded as outliers ranged from 0 to 17,000. Interaction results were combined by fixed-effect inverse variance weighting for each candidate SNP or GRS in METAL (http://www.sph.umich.edu/csg/abecasis/metal/). Each meta-analysis contained results for 486,148 to 2,965,304 SNPs.

3. Results

3.1 *Candidate beta-cell function SNP selection*

The characteristics of IRASFS subjects are shown in Supplementary Table 1. Samples included 492 African Americans with mean age 41.2 years and mean BMI 29.1 kg/m^2. Average African ancestry proportion was 0.75. FSIGT was performed for all subjects without T2D (n = 492) to assess measures including insulin secretion (AIR_g), insulin sensitivity index (S_I), and disposition index (DI).

We identified 5 SNPs (Table 1) from established T2D risk loci from published GWAS[25–28] in which the T2D risk alleles were trending towards association ($P < 0.10$) with AIR_g in IRAS-FS (T2D-IS SNPs). Selected SNPs were identical to the published T2D GWAS index SNPs with the exception of rs7119 (*HMG20A*), which is in strong linkage disequilibrium with the GWAS index SNP rs7178572 in the current study ($r^2 \geq 0.73$ in all cohorts) and is suggestively associated with T2D ($P = 5.24\text{x}10^{-7}$) in individuals from Southeast Asia[29].

3.2 *Interaction analysis*

The selected SNPs were examined for genome-wide first order multiplicative interactions with 1) individual insulin secretion SNPs and 2) risk scores summarizing these insulin secretion SNPs. To maximize power, these analyses were performed in five studies (ARIC, CARDIA, JHS, MESA, and WFSM) including 2,725 T2D cases and 4,167 non-diabetic controls and results were meta-analyzed. Representative meta-analysis q-q plots are provided in Supplementary Figures 1 and 2. A flowchart summarizing experimental workflow is provided in Supplementary Figure 3.

The characteristics of T2D case (n = 2,725) and control subjects (n = 4,167) for each study cohort are shown in Supplementary Table 2. Mean age at examination ranged from 38.2 (CARDIA) to 67.6 (MESA) years. Mean age at diagnosis for T2D cases ranged from 35.0 (CARDIA) to 54.6 (MESA) years. In all cohorts except WFSM, BMI was >3 kg/m^2 higher in cases compared to controls.

3.3 *T2D-IS SNP interactions*

Five T2D-IS SNPs were tested for genome-wide interactions for T2D risk in the ARIC, CARDIA, JHS, MESA, and WFSM cohorts. Individual T2D-IS SNP results were meta-analyzed across cohorts. While no interactions were observed at a genome-wide significance level, a total of 21 SNP-pairs demonstrated suggestive evidence of interaction ($P_{\text{interaction}} < 5\text{x}10^{-6}$; Table 2). The most significant T2D-IS SNP interaction observed was between rs7119 at the *HMG20A* locus (T2D-IS SNP) and rs6487610 (interacting SNP; $P_{\text{interaction}} = 3.83\text{x}10^{-7}$). This interacting SNP is located in an intron of *SMCO2*, which encodes single-pass membrane protein with coiled-coil domains 2. Top interactions with T2D-IS SNPs overall were robust against BMI adjustment (Table 2), with similar p-values. Other notable interacting SNPs included rs978989 (*DGKB*), rs12126276 (*CDK18*), rs7921850 (*CXCL12*), rs6895191 (*HCN1*), rs1900780 (*FAM98A*), and rs568530 (*MGMT*).

Table 2. Top meta-analyzed interactions with T2D-IS SNPs regressed on T2D risk in ARIC, CARDIA, JHS, MESA, and WFSM

T2D-IS SNP (Gene)	Intxn SNP* (Gene)	Chr	Position†	MAF‡	β_{intxn}§	P_{intxn}§	P_{het}‖	$\beta_{intxn\_adj\_bmi}$¶	$P_{intxn\_adj\_bmi}$¶
rs5215 (*KCNJ11*)	rs3024370 (*F13A1*)	6	6250967	0.48	-0.52	3.01E-06	0.71	-0.56	2.32E-06
rs5215 (*KCNJ11*)	rs7842913 (*FUT10*)	8	33089041	0.07	-2.77	4.58E-06	1.00	-2.75	4.57E-06
rs7119 (*HMG20A*)	rs12121207 (*ATG4C*)	1	63232384	0.44	-0.29	2.68E-06	0.20	-0.28	1.43E-05
rs7119 (*HMG20A*)	rs1900780 (*FAM98A/MYADML*)	2	33901094	0.33	0.36	3.46E-06	0.76	0.37	6.92E-06
rs7119 (*HMG20A*)	rs978989 (*DGKB*)	7	14954759	0.27	0.33	2.72E-06	0.23	0.33	4.27E-06
rs7119 (*HMG20A*)	rs6487610 (*SMCO2*)	12	27628742	0.38	0.32	3.83E-07	0.42	0.32	8.45E-07
rs7119 (*HMG20A*)	rs7965793 (*ANKS1B*)	12	100175468	0.31	0.44	1.05E-06	0.76	0.47	7.74E-07
rs7119 (*HMG20A*)	rs1496811 (Intergenic)	18	38952563	0.49	0.27	4.95E-06	0.98	0.27	1.24E-05
rs7119 (*HMG20A*)	rs4812424 (Intergenic)	20	38654372	0.35	-0.47	4.68E-07	0.14	-0.46	1.51E-06
rs7119 (*HMG20A*)	rs6105151 (*ESF1*)	20	13691752	0.34	0.30	2.08E-06	0.42	0.32	7.23E-07
rs7593730 (*RBMS1*)	rs6895191 (*HCN1*)	5	45877674	0.28	0.32	2.80E-06	0.39	0.32	6.91E-06

rs7593730 (*RBMS1*)	rs4705321 (*SH3TC2/ABLIM3*)	5	148508860	0.31	0.30	4.13E-06	0.58	0.28	2.91E-05
rs7593730 (*RBMS1*)	rs16872382 (*ZFPM2*)	8	106108691	0.03	-0.97	7.34E-07	0.85	-0.99	8.49E-07
rs7593730 (*RBMS1*)	rs12865410 (Intergenic)	13	104785227	0.35	-0.30	9.69E-07	0.46	-0.32	6.44E-07
rs7593730 (*RBMS1*)	rs12863474 (Intergenic)	13	104784409	0.37	0.33	1.29E-06	0.89	0.36	4.48E-07
rs864745 (*JAZF1*)	rs12126276 (*CDK18*)	1	205494508	0.18	-0.92	1.31E-06	0.68	-0.92	2.98E-06
rs864745 (*JAZF1*)	rs12343907 (*GLT6D1*)	9	138498904	0.35	-0.34	1.44E-06	0.87	-0.34	2.04E-06
rs864745 (*JAZF1*)	rs7921850 (*CXCL12*)	10	44704401	0.37	-0.33	2.52E-06	0.56	-0.31	1.37E-05
rs864745 (*JAZF1*)	rs568530 (*MGMT*)	10	131018864	0.41	0.32	3.27E-06	0.30	0.32	1.03E-05
rs864745 (*JAZF1*)	rs16973790 (*WRD72/UNC13C*)	15	54188148	0.15	0.55	3.13E-06	0.27	0.51	3.09E-05
rs864745 (*JAZF1*)	rs12483006 (*SLC37A1*)	21	43953851	0.07	-0.66	1.95E-06	0.58	-0.64	8.17E-06

*SNP interacting with selected T2D-IS SNP. †NCBI build 37. ‡Minor allele frequency. §Meta-analyzed effect size and p-value from interaction models adjusted for age, gender, and PC1. ||Heterogeneity p-values across studies from interaction models adjusted for age, gender, and PC1. ¶ Meta-analyzed effect size and p-value from interaction models adjusted for age, gender, PC1, and BMI.

3.4 *GRS validation and interaction analysis*

Each GRS was tested for association with AIR_g and DI under an additive model using the variance components method with adjustment for age, gender, and PC1 in IRASFS (Supplementary Table 3). The weighted T2D-IS GRS was not associated with AIR_g; it was associated with DI with or without BMI adjustment ($P = 4.43\text{x}10^{-2}$ and $4.51\text{x}10^{-2}$, respectively). Since the weighted risk score was associated with measures of glucose homeostasis, analysis of this risk score was emphasized in the tests for genome-wide interaction in the ARIC, CARDIA, JHS, MESA, and WFSM cohorts.

Meta-analyzed estimates of genome-wide interactions with the weighted T2D-IS GRS are presented in Table 3. No interactions met conventional GWAS thresholds for significance. However, eight interactions with the weighted T2D-IS GRS reached a suggestive level of significance ($P_{interaction} < 5\text{x}10^{-6}$; Table 3). The most significant T2D-IS GRS interaction was with rs12434405 (Table 3, $P_{interaction} = 9.60\text{x}10^{-7}$). This is an intronic SNP in the gene *CEP128*, which encodes centrosomal protein 128kDa. Further, the T2D-IS GRS interaction analysis identified two SNPs at the *DGKB* locus, rs6976381 and rs6962498 ($r^2 \geq 0.75$ in all cohorts). This locus was identified in single variant interaction analyses with T2D-IS SNP rs7119 (*HMG20A*), though through a different interacting SNP (rs978989). Two SNPs at the *FAM98A* locus, rs6543772 and rs11687252, were also identified in this analysis. This locus was implicated in single variant analyses with T2D-IS SNP rs7119 (*HMG20A*) through the interacting SNP rs1900780. Top interactions with the T2D-IS GRS were also robust against BMI adjustment.

Table 3. Top meta-analyzed interactions with weighted T2D-IS GRS regressed on T2D risk in ARIC, CARDIA, JHS, MESA, and WFSM

Intxn SNP* (Gene)	Chr	Position†	MAF‡	β_{intxn}§	P_{intxn}§	P_{het}‖	$\beta_{intxn\_adj\_bmi}$¶	$P_{intxn\_adj\_bmi}$¶
rs6543722 (*FAM98A*)	2	33832523	0.39	-1.20	2.82E-06	0.79	-1.22	3.52E-06
rs11687252 (*FAM98A*)	2	33834496	0.38	-1.17	3.27E-06	0.68	-1.19	3.70E-06
rs6851672 (*DKK2*)	4	107907908	0.03	3.70	4.79E-06	0.82	3.63	9.62E-06
rs6976381 (*DGKB*)	7	15048814	0.18	-1.67	1.21E-06	0.73	-1.66	2.18E-06

rs6962498 (*DGKB*)	7	15050305	0.14	-1.77	3.71E-06	0.54	-1.77	6.65E-06
rs17082105 (*PCDH9*)	13	67685156	0.18	1.45	3.46E-06	0.86	1.51	2.65E-06
rs12434405 (*CEP128*)	14	81044614	0.12	-1.90	9.60E-07	0.12	-1.87	2.49E-06
rs16951940 (Intergenic)	16	80021664	0.03	3.40	2.29E-06	0.84	3.43	4.58E-06

*SNP interacting with the weighted T2D-IS GRS. †NCBI build 37. ‡Minor allele frequency. §Meta-analyzed effect size and p-value from interaction models adjusted for age, gender, and PC1. ||Heterogeneity p-values across studies from interaction models adjusted for age, gender, and PC1. ¶ Meta-analyzed effect size and p-value from interaction models adjusted for age, gender, PC1, and BMI.

4. Discussion

Meta-analyses of five African American T2D studies did not reveal genome-wide statistically significant ($P_{interaction} < 5\times10^{-8}$) first-order interactions with insulin secretion SNPs or composite risk scores. However, the observed interactions ($P_{interaction} < 5\times10^{-6}$) suggest that a candidate insulin secretion SNP/GRS interaction approach is a valid method for identifying insulin sensitivity and T2D risk loci. For example, analyses with the T2D-IS SNP rs864745 (*JAZF1*) revealed an interaction with rs7921850, an intergenic SNP downstream of the *CXCL12* gene encoding chemokine (C-X-C motif) ligand 12 (also known as stromal cell-derived factor 1). CXCL12 is an adipocyte-derived chemotactic factor that recruits macrophages and is required for the establishment of obesity-induced adipose tissue inflammation and systemic insulin resistance in mice[30].

Several genes related to pancreatic beta-cell function were also identified; suggesting interactions are not limited to insulin resistance as in our initial hypothesis. Evaluations of the T2D-IS SNP rs7119 (*HMG20A*) and the T2D-IS GRS identified interactions with rs978989 and rs6976381, respectively, intergenic SNPs downstream of the *DGKB* gene. Variants at *DGKB* have been associated with T2D, fasting glucose, and pancreatic islet beta-cell function as measured by HOMA-B[27,31]. Variants near *DGKB* disrupt islet-specific enhancer activity[32]. Several other variants detected in our analyses show interactions with similar biological relationships to insulin secretion and T2D.

Interestingly, we observed interactions discrete for individual loci. For example, analyses with rs864745 (*JAZF1*), a locus involved in transcriptional repression, showed an interaction with rs568530, an intergenic SNP upstream of *MGMT*, which encodes O-6-Methylguanine-DNA Methyltransferase. These observations may reflect different, input-dependent physiological characteristics of interaction results, and may lead to mechanistic insights about the underlying causes of T2D and defects in glucose homeostasis in expanded analyses.

Although results varied widely between interaction analyses, interactions with two loci, *DGKB* and *FAM98*, were replicated in multiple analyses. Functional characteristics of *FAM98* related to T2D and glucose homeostasis pathophysiology are not evident in the current literature.

Previous GWAS have largely ignored epistatic contributions to T2D risk due to the heavy multiple testing burden and computational challenges of exhaustive analytical approaches, and when they have considered this contribution, results have not been striking. For example, a recent genome-wide scan for two-locus interactions in the Wellcome Trust Case Control Consortium T2D GWAS data did not reveal any significant epistatic signals at a Bonferroni-corrected p-value threshold of 2.14×10^{-11} after adjusting for the main effects of the most strongly associated T2D locus, *TCF7L2*[33]. Further, Herold et al. estimated that analysis of all pairwise interactions among 550,000 SNPs in 1,200 samples on a 3 GHz computer would require a running time of 120 days[34]. The interaction analysis presented here overcomes the issue of a heavy multiple testing burden by using a candidate SNP approach. A recent study by Becker et al. demonstrated that a multiple test correction of 0.4m, where m is the number of SNP pairs tested, is sufficiently conservative for large-scale allelic interaction tests[35]. Further, Babron et al. show that a correction for the effective number of SNP pairs is equally sufficient[36]. Li et al. previously demonstrated that the effective number of SNPs for an imputed dataset is $\sim 10^6$. These findings suggest that a significance threshold of 1×10^{-8} is appropriate for this study.

We did not detect interactions even at the conventional GWAS threshold of 5×10^{-8} in the current study. In part, this likely reflects the challenge of inherently reduced power of interaction models due to the low frequency of compound genotypes[37]. Computational resources required for this study were equivalent to the requirements for running 12 GWAS (5 candidate insulin secretion SNPs plus a GRS, with and without BMI adjustment). This is a significant reduction compared to exhaustive approaches examining genome-wide interactions with all available SNP pairs.

In summary, our findings demonstrate that genome-wide interaction studies with selected insulin secretion variants is a powerful approach for the detection of T2D risk, insulin secretion, and insulin sensitivity loci. The use of a high-quality measure of first-phase insulin secretion, AIR_g, to identify candidate interaction SNPs yielded compelling associations. These results justify an expansion of the current study and further investigation of putative insulin sensitivity loci, namely *CXCL12*.

Acknowledgements

The authors would like to acknowledge the contributions of the involved research institutions, study investigators, field staff, and study participants of ARIC, CARDIA, JHS, MESA, and WFSM.

Genotyping services for the WFSM study were provided by CIDR. CIDR is fully funded through a federal contract from the National Institutes of Health (NIH) to The Johns Hopkins University (Contract HHSC268200782096C). The work at Wake Forest was supported by NIH grants K99-DK-081350 (N.D.P.), R01-DK-066358 (D.W.B.), R01-DK-053591 (D.W.B.), R01-HL-56266 (B.I.F.), and R01-DK-070941 (B.I.F.), and in part by the General Clinical Research Center of the WFSM Grant M01-RR-07122. This work was also supported by the NHLBI.

The following four parent studies have contributed parent study data, ancillary study data, and DNA samples through the Massachusetts Institute of Technology-Broad Institute (N01-HC-65226) to create this genotype/phenotype database for wide dissemination to the biomedical research community: ARIC, CARDIA, JHS, and MESA.

The Atherosclerosis Risk in Communities (ARIC) Study is carried out as a collaborative study supported by National Heart, Lung, and Blood Institute contracts (HHSN268201100005C, HHSN268201100006C, HHSN268201100007C, HHSN268201100008C, HHSN268201100009C, HHSN268201100010C, HHSN268201100011C, and HHSN268201100012C), R01HL087641, R01HL59367 and R01HL086694; National Human

Genome Research Institute contract U01HG004402; and National Institutes of Health contract HHSN268200625226C. The authors thank the staff and participants of the ARIC study for their important contributions. Infrastructure was partly supported by Grant Number UL1RR025005, a component of the National Institutes of Health and NIH Roadmap for Medical Research. The authors thank the staff and participants of the ARIC study for their important contributions.

The Coronary Artery Risk Development in Young Adults (CARDIA) Study is conducted and supported by the National Heart, Lung, and Blood Institute in collaboration with the University of Alabama at Birmingham (HHSN268201300025C & HHSN268201300026C), Northwestern University (HHSN268201300027C), University of Minnesota (HHSN268201300028C), Kaiser Foundation Research Institute (HHSN268201300029C), and Johns Hopkins University School of Medicine (HHSN268200900041C). CARDIA is also partially supported by the Intramural Research Program of the National Institute on Aging. Genotyping was funded as part of the NHLBI Candidate-gene Association Resource (N01-HC-65226) and the NHGRI Gene Environment Association Studies (GENEVA) (U01-HG004729, U01-HG04424, and U01-HG004446). This manuscript has been reviewed and approved by CARDIA for scientific content.

The Jackson Heart Study (JHS) is supported by contracts HHSN268201300046C, HHSN268201300047C, HHSN268201300048C, HHSN268201300049C, HHSN268201300050C from the National Heart, Lung, and Blood Institute and the National Institute on Minority Health and Health Disparities. The authors thank the participants and data collection staff of the Jackson Heart Study.

Multi-Ethnic Study of Atherosclerosis (MESA), and the MESA SHARe project are conducted and supported by the National Heart, Lung, and Blood Institute (NHLBI) in collaboration with MESA investigators. Support for MESA is provided by contracts N01-HC-95159, N01-HC-95160, N01-HC-95161, N01-HC-95162, N01-HC-95163, N01-HC-95164, N01-HC-95165, N01-HC-95166, N01-HC-95167, N01-HC-95168, N01-HC-95169, UL1-TR-001079, UL1-TR-000040, and DK063491. The MESA CARe data used for the analyses described in this manuscript were obtained through Genetics (CMP00068). Funding for CARe genotyping was provided by NHLBI Contract N01-HC-65226.

The views expressed in this manuscript are those of the authors and do not necessarily represent the views of the National Heart, Lung, and Blood Institute; the National Institutes of Health; or the U.S. Department of Health and Human Services.

Supplementary Material

Supplementary methods, tables, and figures can be found at http://csb.wfu.edu/SupplementaryData_online.docx.

References

1. Prasad, R. B. & Groop, L. Genetics of type 2 diabetes-pitfalls and possibilities. *Genes* **6,** 87–123 (2015).
2. Morris, A. P. *et al.* Large-scale association analysis provides insights into the genetic architecture and pathophysiology of type 2 diabetes. *Nat. Genet.* **44,** 981–990 (2012).

3. Cordell, H. J. Detecting gene-gene interactions that underlie human diseases. *Nat. Rev. Genet.* **10,** 392–404 (2009).
4. Moore, J. H. The ubiquitous nature of epistasis in determining susceptibility to common human diseases. *Hum. Hered.* **56,** 73–82 (2003).
5. Henkin, L. *et al.* Genetic epidemiology of insulin resistance and visceral adiposity. The IRAS Family Study design and methods. *Ann. Epidemiol.* **13,** 211–217 (2003).
6. Pacini, G. & Bergman, R. N. MINMOD: a computer program to calculate insulin sensitivity and pancreatic responsivity from the frequently sampled intravenous glucose tolerance test. *Comput. Methods Programs Biomed.* **23,** 113–122 (1986).
7. The ARIC Investigators. The Atherosclerosis Risk in Communities (ARIC) Study: design and objectives. The ARIC investigators. *Am. J. Epidemiol.* **129,** 687–702 (1989).
8. Friedman, G. D. *et al.* CARDIA: study design, recruitment, and some characteristics of the examined subjects. *J. Clin. Epidemiol.* **41,** 1105–1116 (1988).
9. Taylor, H. A. *et al.* Toward resolution of cardiovascular health disparities in African Americans: design and methods of the Jackson Heart Study. *Ethn. Dis.* **15,** S6-4–17 (2005).
10. Bild, D. E. *et al.* Multi-ethnic study of atherosclerosis: objectives and design. *Am. J. Epidemiol.* **156,** 871–881 (2002).
11. McDonough, C. W. *et al.* A genome-wide association study for diabetic nephropathy genes in African Americans. *Kidney Int.* **79,** 563–572 (2011).
12. Palmer, N. D. *et al.* A genome-wide association search for type 2 diabetes genes in African Americans. *PloS One* **7,** e29202 (2012).
13. Hellwege, J. N. *et al.* Genome-wide family-based linkage analysis of exome chip variants and cardiometabolic risk. *Genet. Epidemiol.* **38,** 345–352 (2014).
14. O'Connell, J. R. & Weeks, D. E. PedCheck: a program for identification of genotype incompatibilities in linkage analysis. *Am. J. Hum. Genet.* **63,** 259–266 (1998).
15. Purcell, S. *et al.* PLINK: a tool set for whole-genome association and population-based linkage analyses. *Am. J. Hum. Genet.* **81,** 559–575 (2007).
16. Lettre, G. *et al.* Genome-wide association study of coronary heart disease and its risk factors in 8,090 African Americans: the NHLBI CARe Project. *PLoS Genet.* **7,** e1001300 (2011).
17. Ng, M. C. Y. *et al.* Transferability and fine mapping of type 2 diabetes loci in African Americans: the Candidate Gene Association Resource Plus Study. *Diabetes* **62,** 965–976 (2013).
18. Hester, J. M. *et al.* Implication of European-derived adiposity loci in African Americans. *Int. J. Obes. 2005* **36,** 465–473 (2012).
19. Palmer, N. D. *et al.* Evaluation of DLG2 as a positional candidate for disposition index in African-Americans from the IRAS Family Study. *Diabetes Res. Clin. Pract.* **87,** 69–76 (2010).
20. Patterson, N., Price, A. L. & Reich, D. Population structure and eigenanalysis. *PLoS Genet.* **2,** e190 (2006).
21. Patterson, N. *et al.* Methods for high-density admixture mapping of disease genes. *Am. J. Hum. Genet.* **74,** 979–1000 (2004).
22. Pritchard, J. K., Stephens, M. & Donnelly, P. Inference of population structure using multilocus genotype data. *Genetics* **155,** 945–959 (2000).

23. Keene, K. L. *et al.* Exploration of the utility of ancestry informative markers for genetic association studies of African Americans with type 2 diabetes and end stage renal disease. *Hum. Genet.* **124,** 147–154 (2008).
24. Almasy, L. & Blangero, J. Multipoint quantitative-trait linkage analysis in general pedigrees. *Am. J. Hum. Genet.* **62,** 1198–1211 (1998).
25. Qi, L. *et al.* Genetic variants at 2q24 are associated with susceptibility to type 2 diabetes. *Hum. Mol. Genet.* **19,** 2706–2715 (2010).
26. Zeggini, E. *et al.* Meta-analysis of genome-wide association data and large-scale replication identifies additional susceptibility loci for type 2 diabetes. *Nat. Genet.* **40,** 638–645 (2008).
27. DIAbetes Genetics Replication And Meta-analysis (DIAGRAM) Consortium *et al.* Genome-wide trans-ancestry meta-analysis provides insight into the genetic architecture of type 2 diabetes susceptibility. *Nat. Genet.* **46,** 234–244 (2014).
28. Voight, B. F. *et al.* Twelve type 2 diabetes susceptibility loci identified through large-scale association analysis. *Nat. Genet.* **42,** 579–589 (2010).
29. Sim, X. *et al.* Transferability of type 2 diabetes implicated loci in multi-ethnic cohorts from Southeast Asia. *PLoS Genet.* **7,** e1001363 (2011).
30. Kim, D. *et al.* CXCL12 secreted from adipose tissue recruits macrophages and induces insulin resistance in mice. *Diabetologia* **57,** 1456–1465 (2014).
31. Dupuis, J. *et al.* New genetic loci implicated in fasting glucose homeostasis and their impact on type 2 diabetes risk. *Nat. Genet.* **42,** 105–116 (2010).
32. Pasquali, L. *et al.* Pancreatic islet enhancer clusters enriched in type 2 diabetes risk–associated variants. *Nat. Genet.* **46,** 136–143 (2014).
33. Bell, J. T. *et al.* Genome-wide association scan allowing for epistasis in type 2 diabetes. *Ann. Hum. Genet.* **75,** 10–19 (2011).
34. Herold, C., Steffens, M., Brockschmidt, F. F., Baur, M. P. & Becker, T. INTERSNP: genome-wide interaction analysis guided by a priori information. *Bioinforma. Oxf. Engl.* **25,** 3275–3281 (2009).
35. Becker, T., Herold, C., Meesters, C., Mattheisen, M. & Baur, M. P. Significance levels in genome-wide interaction analysis (GWIA). *Ann. Hum. Genet.* **75,** 29–35 (2011).
36. Babron, M.-C., Etcheto, A. & Dizier, M.-H. A New Correction for Multiple Testing in Gene-Gene Interaction Studies. *Ann. Hum. Genet.* (2015). doi:10.1111/ahg.12113
37. Lucas, G. *et al.* Hypothesis-Based Analysis of Gene-Gene Interactions and Risk of Myocardial Infarction. *PLoS ONE* **7,** (2012).

DEEP MOTIF DASHBOARD: VISUALIZING AND UNDERSTANDING GENOMIC SEQUENCES USING DEEP NEURAL NETWORKS

JACK LANCHANTIN, RITAMBHARA SINGH, BEILUN WANG, YANJUN QI

Department of Computer Science, University of Virginia
Charlottesville, VA 22903, USA
E-mail: {jjl5sw,rs3zz,bw4mw,yq2h}@virginia.edu

Deep neural network (DNN) models have recently obtained state-of-the-art prediction accuracy for the transcription factor binding (TFBS) site classification task. However, it remains unclear how these approaches identify meaningful DNA sequence signals and give insights as to why TFs bind to certain locations. In this paper, we propose a toolkit called the Deep Motif Dashboard (DeMo Dashboard) which provides a suite of visualization strategies to extract motifs, or sequence patterns from deep neural network models for TFBS classification. We demonstrate how to visualize and understand three important DNN models: convolutional, recurrent, and convolutional-recurrent networks. Our first visualization method is finding a test sequence's saliency map which uses first-order derivatives to describe the importance of each nucleotide in making the final prediction. Second, considering recurrent models make predictions in a temporal manner (from one end of a TFBS sequence to the other), we introduce temporal output scores, indicating the prediction score of a model over time for a sequential input. Lastly, a class-specific visualization strategy finds the optimal input sequence for a given TFBS positive class via stochastic gradient optimization. Our experimental results indicate that a convolutional-recurrent architecture performs the best among the three architectures. The visualization techniques indicate that CNN-RNN makes predictions by modeling both motifs as well as dependencies among them.

1. Introduction

In recent years, there has been an explosion of deep learning models which have lead to groundbreaking results in many fields such as computer vision,[1] natural language processing,[2] and computational biology.[3–8] However, although these models have proven to be very accurate, they have widely been viewed as "black boxes" due to their complexity, making them hard to understand. This is particularly unfavorable in the biomedical domain, where understanding a model's predictions is extremely important for doctors and researchers trying to use the model.

Aiming to open up the black box, we present the "Deep Motif Dashboard[a]" (DeMo Dashboard), to understand the inner workings of deep neural network models for a genomic sequence classification task. We do this by introducing a suite of different neural models and visualization strategies to see which ones perform the best and understand how they make their predictions.[b]

Understanding genetic sequences is one of the fundamental tasks of health advancements due to the high correlation of genes with diseases and drugs. An important problem within genetic sequence understanding is related to transcription factors (TFs), which are regulatory proteins that bind to DNA. Each different TF binds to specific transcription factor binding sites (TFBSs) on the genome to regulate cell machinery. Given an input DNA sequence, classifying whether or not there is a binding site for a particular TF is a core task of bioinformatics.[10]

For our task, we follow a two step approach. First, given a particular TF of interest and a dataset containing samples of positive and negative TFBS sequences, we construct three deep learning architectures to classify the sequences. Section 2 introduces the three different DNN structures that we use: a convolutional neural network (**CNN**), a recurrent neural network

[a]Dashboard normally refers to a user interface that gives a current summary, usually in graphic, easy-to-read form, of key information relating to performance.[9]
[b]We implemented our model in Torch, and it is made available at deepmotif.org

(**RNN**), and a convolutional-recurrent neural network (**CNN-RNN**).

Once we have our trained models to predict binding sites, the second step of our approach is to understand why the models perform the way they do. As explained in section 3, we do this by introducing three different visualization strategies for interpreting the models:

(1) Measuring nucleotide importance with **Saliency Maps**.
(2) Measuring critical sequence positions for the classifier using **Temporal Output Scores**.
(3) Generating class-specific motif patterns with **Class Optimization**.

We test and evaluate our models and visualization strategies on a large scale benchmark TFBS dataset. Section 4 provides experimental results for understanding and visualizing the three DNN architectures. We find that the CNN-RNN outperforms the other models. From the visualizations, we observe that the CNN-RNN tends to focus its predictions on the traditional motifs, as well as modeling long range dependencies among motifs.

2. Deep Neural Models for TFBS Classification

TFBS Classification. Chromatin immunoprecipitation (ChIP-seq) technologies and databases such as ENCODE[11] have made binding site locations available for hundreds of different TFs. Despite these advancements, there are two major drawbacks: (1) ChIP-seq experiments are slow and expensive, (2) although ChIP-seq experiments can find the binding site locations, they cannot find patterns that are common across all of the positive binding sites which can give insight as to why TFs bind to those locations. Thus, there is a need for large scale computational methods that can not only make accurate binding site classifications, but also identify and understand patterns that influence the binding site locations.

In order to computationally predict TFBSs on a DNA sequence, researchers initially used consensus sequences and position weight matrices to match against a test sequence.[10] Simple neural network classifiers were then proposed to differentiate positive and negative binding sites, but did not show significant improvements over the weight matrix matching methods.[12] Later, SVM techniques outperformed the generative methods by using k-mer features,[13,14] but string kernel based SVM systems are limited by expensive computational cost proportional to the number of training and testing sequences. Most recently, convolutional neural network models have shown state-of-the-art results on the TFBS task and are scalable to a large number of genomic sequences,[3,7] but it remains unclear which neural architectures work best.

Deep Neural Networks for TFBSs. To find which neural models work the best on the TFBS classification task, we examine several different types of models. Inspired by their success across different fields, we explore variations of two popular deep learning architectures: convolutional neural networks (CNNs), and recurrent neural networks (RNNs). CNNs have dominated the field of computer vision in recent years, obtaining state-of-the-art results in many tasks due to their ability to automatically extract translation-invariant features. On the other hand, RNNs have emerged as one of the most powerful models for sequential data tasks such as natural language processing due to their ability to learn long range dependencies. Specifically, on the TFBS prediction task, we explore three distinct architectures: (1) CNN, (2) RNN, and (3) a combination of the two, CNN-RNN. Figure 1 shows an overview of the models.

End-to-end Deep Framework. While the body of the three architectures we use differ, each implemented model follows a similar end-to-end framework which we use to easily compare and contrast results. We use the raw nucleotide characters (A,C,G,T) as inputs, where each character is converted into a one-hot encoding (a binary vector with the matching character entry being a 1 and the rest as 0s). This encoding matrix is used as the input to a convolutional,

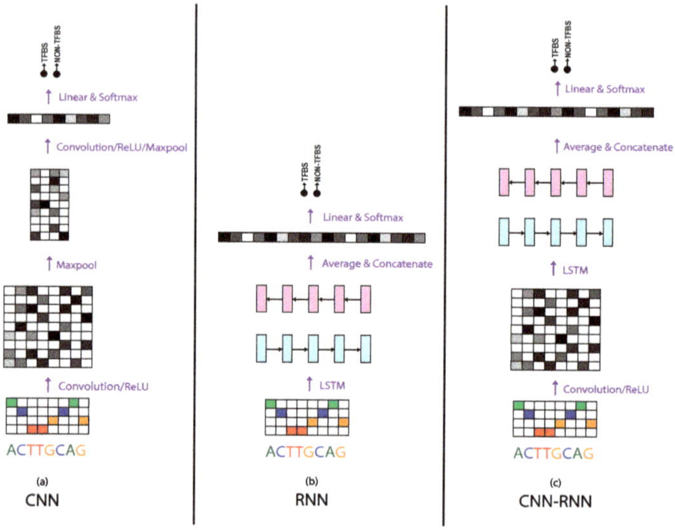

Fig. 1. **Model Architectures.** Each model has the same input (one-hot encoded matrix of the raw nucleotide inputs), and the same output (softmax classifier to make a binary prediction). The architectures differ by the middle "module", which are **(a)** Convolutional, **(b)** Recurrent, and **(c)** Convolutional-Recurrent.

recurrent, or convolutional-recurrent module that each outputs a vector of fixed dimension. The output vector of each model is linearly fed to a softmax function as the last layer which learns the mapping from the hidden space to the output class label space $C \in [+1, -1]$. The final output is a probability indicating whether an input is a positive or a negative binding site (binary classification task). The parameters of the network are trained end-to-end by minimizing the negative log-likelihood over the training set. The minimization of the loss function is obtained via the stochastic gradient algorithm Adam,[15] with a mini-batch size of 256 sequences. We use dropout[16] as a regularization method for each model.

2.1. *Convolutional Neural Network (CNN)*

In genomic sequences, it is believed that regulatory mechanisms such as transcription factor binding are influenced by local sequential patterns known as "motifs". Motifs can be viewed as the temporal equivalent of spatial patterns in images such as eyes on a face, which is what CNNs are able to automatically learn and achieve state-of-the art results on computer vision tasks. As a result, a temporal convolutional neural network is a fitting model to automatically extract these motifs. A temporal convolution with filter (or kernel) size k takes an input data matrix $\mathbf{X}$ of size $T \times n_{in}$, with length T and input layer size n_{in}, and outputs a matrix $\mathbf{Z}$ of size $T \times n_{out}$, where n_{out} is the output layer size. Specifically, $convolution(\mathbf{X}) = \mathbf{Z}$, where

$$\mathbf{z}_{t,i} = \sigma(\mathbf{B}_i + \sum_{j=1}^{n_{in}} \sum_{z=1}^{k} \mathbf{W}_{i,j,z} \mathbf{x}_{t+z-1,j}), \tag{1}$$

where $\mathbf{W}$ and $\mathbf{B}$ are the trainable parameters of the convolution filter, and σ is a function enforcing element-wise nonlinearity. We use rectified linear units (ReLU) as the nonlinearity:

$$\text{ReLU}(x) = \max(0, x). \tag{2}$$

After the convolution and nonlinearity, CNNs typically use maxpooling, which is a dimension reduction technique to provide translation invariance and to extract higher level features from a wider range of the input sequence. Temporal maxpooling on a matrix $\mathbf{Z}$ with a pooling size of m results in output matrix $\mathbf{Y}$. Formally, $maxpool(\mathbf{Z}) = \mathbf{Y}$, where

$$\mathbf{y}_{t,i} = \max_{j=1}^{m} \mathbf{z}_{m(t-1)+j,i} \tag{3}$$

Our CNN implementation involves a progression of convolution, nonlinearity, and maxpooling. This is represented as one convolutional layer in the network, and we test up to 4 layer deep CNNs. The final layer involves a maxpool across the entire temporal domain so that we have a fixed-size vector which can be fed into a softmax classifier.

Figure 1 (a) shows our CNN model with two convolutional layers. The input one-hot encoded matrix is convolved with several filters (not shown) and fed through a ReLU nonlinearity to produce a matrix of convolution activations. We then perform a maxpool on the activation matrix. The output of the first maxpool is fed through another convolution, ReLU, and maxpooled across the entire length resulting in a vector. This vector is then transposed and fed through a linear and softmax layer for classification.

2.2. *Recurrent Neural Network (RNN)*

Designed to handle sequential data, Recurrent neural networks (RNNs) have become the main neural model for tasks such as natural language understanding. The key advantage of RNNs over CNNs is that they are able to find long range patterns in the data which are highly dependent on the ordering of the sequence for the prediction task.

Given an input matrix $\mathbf{X}$ of size $T \times n_{in}$, an RNN produces matrix $\mathbf{H}$ of size $T \times d$, where d is the RNN embedding size. At each timestep t, an RNN takes an input column vector $\mathbf{x_t} \in \mathbb{R}^{n_{in}}$ and the previous hidden state vector $\mathbf{h_{t-1}} \in \mathbb{R}^d$ and produces the next hidden state $\mathbf{h_t}$ by applying the following recursive operation:

$$\mathbf{h}_t = \sigma(\mathbf{W}x_t + \mathbf{U}\mathbf{h}_{t-1} + \mathbf{b}), \tag{4}$$

where $\mathbf{W}, \mathbf{U}, \mathbf{b}$ are the trainable parameters of the model, and σ is an element-wise nonlinearity. Due to their recursive nature, RNNs can model the full conditional distribution of any sequential data and find dependencies over time, where each position in a sequence is a timestep on an imaginary time coordinate running in a certain direction. To handle the "vanishing gradients" problem of training basic RNNs on long sequences, Hochreiter and Schmidhuber[17] proposed an RNN variant called the Long Short-term Memory (LSTM) network (for simplicity, we refer to LSTMs as RNNs in this paper), which can handle long term dependencies by using gating functions. These gates can control when information is written to, read from, and forgotten. Specifically, LSTM "cells" take inputs $\mathbf{x}_t, \mathbf{h}_{t-1}$, and $\mathbf{c}_{t-1}$, and produce $\mathbf{h}_t$, and $\mathbf{c}_t$:

$$\begin{aligned}
\mathbf{i}_t &= \sigma(\mathbf{W}^i\mathbf{x}_t + \mathbf{U}^i\mathbf{h}_{t-1} + \mathbf{b}^i) \\
\mathbf{f}_t &= \sigma(\mathbf{W}^f\mathbf{x}_t + \mathbf{U}^f\mathbf{h}_{t-1} + \mathbf{b}^f) \\
\mathbf{o}_t &= \sigma(\mathbf{W}^o\mathbf{x}_t + \mathbf{U}^o\mathbf{h}_{t-1} + \mathbf{b}^o) \\
\mathbf{g}_t &= tanh(\mathbf{W}^g\mathbf{x}_t + \mathbf{U}^g\mathbf{h}_{t-1} + \mathbf{b}^g) \\
\mathbf{c}_t &= \mathbf{f}_t \odot \mathbf{c}_{t-1} + \mathbf{i}_t \odot \mathbf{g}_t \\
\mathbf{h}_t &= \mathbf{o}_t \odot tanh(\mathbf{c}_t)
\end{aligned}$$

where $\sigma(\cdot)$, $tanh(\cdot)$, and $\odot$ are element-wise sigmoid, hyperbolic tangent, and multiplication functions, respectively. $\mathbf{i}_t$, $\mathbf{f}_t$, and $\mathbf{o}_t$ are the input, forget, and output gates, respectively.

RNNs produce an output vector $\mathbf{h}_t$ at each timestep t of the input sequence. In order to use them on a classification task, we take the mean of all vectors $\mathbf{h}_t$, and use the mean vector $\mathbf{h}_{mean} \in \mathbb{R}^d$ as input to the softmax layer.

Since there is no innate direction in genomic sequences, we use a bi-directional LSTM as our RNN model. In the bi-directional LSTM, the input sequence gets fed through two LSTM networks, one in each direction, and then the output vectors of each direction get concatenated together in the temporal direction and fed through a linear classifier.

Figure 1 (b) shows our RNN model. The input one-hot encoded matrix is fed through an LSTM in both the forward and backward direction which each produce a matrix of column vectors representing the LSTM output embedding at each timestep. These vectors are then averaged to create one vector for each direction representing the LSTM output. The forward and backward output vectors are then concatenated and fed to the softmax for classification.

2.3. *Convolutional-Recurrent Network (CNN-RNN)*

Considering convolutional networks are designed to extract motifs, and recurrent networks are designed to extract temporal features, we implement a combination of the two in order to find temporal patterns between the motifs. Given an input matrix $\mathbf{X} \in \mathbb{R}^{T \times n_{in}}$, the output of the CNN is $\mathbf{Z} \in \mathbb{R}^{T \times n_{out}}$. Each column vector of $\mathbf{Z}$ gets fed into the RNN one at a time in the same way that the one-hot encoded vectors get input to the regular RNN model. The resulting output of the RNN $\mathbf{H} \in \mathbb{R}^{T \times d}$, where d is the LSTM embedding size, is then averaged across the temporal domain (in the same way as the regular RNN), and fed to a softmax classifier.

Figure 1 (c) shows our CNN-RNN model. The input one-hot encoded matrix is fed through one layer of convolution to produce a convolution activation matrix. This matrix is then input to the LSTM, as done in the regular RNN model from the original one-hot matrix. The output of the LSTM is averaged, concatenated, and fed to the softmax, similar to the RNN.

3. Visualizing and Understanding Deep Models

The previous section explained the deep models we use for the TFBS classification task, where we can evaluate which models perform the best. While making accurate predictions is important in biomedical tasks, it is equally important to understand why models make their predictions. Accurate, but uninterpretable models are often very slow to emerge in practice due to the inability to understand their predictions, making biomedical domain experts reluctant to use them. Consequently, we aim to obtain a better understanding of why certain models work better than others, and investigate how they make their predictions by introducing several visualization techniques. The proposed DeMo Dashboard allows us visualize and understand DNNs in three different ways: Saliency Maps, Temporal Output Scores, and Class Optimizations.

3.1. *Saliency Maps*

For a certain DNA sequence and a model's classification, a logical question may be: "which which parts of the sequence are most influential for the classification?" To do this, we seek to visualize the influence of each position (i.e. nucleotide) on the prediction. Our approach is similar to the methods used on images by Simonyan et al.[18] and Baehrens et al.[19] Given a sequence X_0 of length $|X_0|$, and class $c \in C$, a DNN model provides a score function $S_c(X_0)$. We rank the nucleotides of X_0 based on their influence on the score $S_c(X_0)$. Since $S_c(X)$ is a highly non-linear function of X with deep neural nets, it is hard to directly see the influence of each nucleotide of X on S_c. Mathematically, around the point X_0, $S_c(X)$ can be approximated by a linear function by computing the first-order Taylor expansion:

$$S_c(X) \approx w^T X + b = \sum_{i=1}^{|X|} w_i x_i + b \tag{5}$$

where w is the derivative of S_c with respect to the sequence variable X at the point X_0:

$$w = \left.\frac{\partial S_c}{\partial X}\right|_{X_0} = \textit{saliency map} \tag{6}$$

This derivative is simply one step of backpropagation in the DNN model, and is therefore easy to compute. We do a pointwise multiplication of the saliency map with the one-hot encoded sequence to get the derivative values for the actual nucleotide characters of the sequence (A,T,C, or G) so we can see the influence of the character at each position on the output score. Finally, we take the element-wise magnitude of the resulting derivative vector to visualize how important each character is regardless of derivative direction. We call the resulting vector a "saliency map[18]" because it tells us which nucleotides need to be changed the least in order to affect the class score the most. As we can see from equation 5, the saliency map is simply a weighted sum of the input nucleotides, where the each weight, w_i, indicates the influence of that nucleotide position on the output score.

3.2. *Temporal Output Scores*

Since DNA is sequential (i.e. can be read in a certain direction), it can be insightful to visualize the output scores at each timestep (position) of a sequence, which we call the temporal output scores. Here we assume an imaginary time direction running from left to right on a given sequence, so each position in the sequence is a timestep in such an imagined time coordinate. In other words, we check the RNN's prediction scores when we vary the input of the RNN. The input series is constructed by using subsequences of an input X running along the imaginary time coordinate, where the subsequences start from just the first nucleotide (position), and ends with the entire sequence X. This way we can see exactly where in the sequence the recurrent model changes its decision from negative to positive, or vice versa. Since our recurrent models are bi-directional, we also use the same technique on the reverse sequence. CNNs process the entire sequence at once, thus we can't view its output as a temporal sequence, so we use this visualization on just the RNN and CNN-RNN.

3.3. *Class Optimization*

The previous two visualization methods listed are representative of a specific testing sample (i.e. sequence-specific). Now we introduce an approach to extract a *class-specific* visualization for a DNN model, where we attempt to find the best sequence which maximizes the probability of a positive TFBS, which we call class optimization. Formally, we optimize the following equation where $S_+(X)$ is the probability (or score) of an input sequence X (matrix in our case) being a positive TFBS computed by the softmax equation of our trained DNN model for a specific TF:

$$\arg\max_X S_+(X) + \lambda \|X\|_2^2 \tag{7}$$

where λ is the regularization parameter. We find a locally optimal X through stochastic gradient descent, where the optimization is with respect to the input sequence. In this optimization, the model weights remain unchanged. This is similar to the methods used in Simonyan et al.[18] to optimize toward a specific image class. This visualization method depicts the notion of a positive TFBS class for a particular TF and is not specific to any test sequence.

3.4. *End-to-end Automatic Motif Extraction from the Dashboard*

Our three proposed visualization techniques allow us to manually inspect how the models make their predictions. In order to automatically find patterns from the techniques, we also propose methods to extract motifs, or consensus subsequences that represent the positive binding sites. We extract motifs from each of our three visualization methods in the following ways: (1) From each positive test sequence (thus, 500 total for each TF dataset) we extract a motif from the saliency map by selecting the contiguous length-9 subsequence that achieves the highest sum of contiguous length-9 saliency map values. (2) For each positive test sequence, we extract a motif from the temporal output scores by selecting the length-9 subsequence that shows the strongest score change from negative to positive output score. (3) For each different TF, we can directly use the class-optimized sequence as a motif.

3.5. *Connecting to Previous Studies*

Neural networks have produced state-of-the-art results on several important benchmark tasks related to genomic sequence classification,[3–5] making them a good candidate to use. However, *why* these models work well has been poorly understood. Recent works have attempted to uncover the properties of these models, in which most of the work has been done on understanding image classifications using convolutional neural networks. Zeiler and Fergus[20] used a "deconvolution" approach to map hidden layer representations back to the input space for a specific example, showing the features of the image which were important for classification. Simonyan et al.[18] explored a similar approach by using a first-order Taylor expansion to linearly approximate the network and find the input features most relevant, and also tried optimizing image classes. Many similar techniques later followed to understand convolutional models.[21,22] Most importantly, researchers have found that CNNs are able to extract layers of translational-invariant feature maps, which may indicate why CNNs have been successfully used in genomic sequence predictions which are believed to be triggered by motifs.

On text-based tasks, there have been fewer visualization studies for DNNs. Karpathy et al.[23] explored the interpretability of RNNs for language modeling and found that there exist interpretable neurons which are able to focus on certain language structure such as quotes. Li et al.[24] visualized how RNNs achieve compositionality in natural language for sentiment analysis by visualizing RNN embedding vectors as well as measuring the influence of input words on classification. Both studies show examples that can be validated by our understanding of natural language linguistics. Contrarily, we are interested in understanding DNA "linguistics" given DNNs (the opposite direction of Karpathy et al.[23] and Li et al.[24]).

The main difference between our work and previous works on images and natural language is that instead of trying to understand the DNNs given human understanding of such human perception tasks, we attempt to uncover critical signals in DNA sequences given our understanding of DNNs.

For TFBS prediction, Alipanahi et al.[3] was the first to implement a visualization method on a DNN model. They visualize their CNN model by extracting motifs based on the input subsequence corresponding to the strongest activation location for each convolutional filter (which we call convolution activation). Since they only have one convolutional layer, it is trivial to map the activations back, but this method does not work as well with deeper models. We attempted this technique on our models and found that our approach using saliency maps outperforms it in finding motif patterns (details in section 4). Quang and Xie[4] use the same visualization method on their convolutional-recurrent model for noncoding variant prediction.

Table 1. Variations of DNN Model Hyperparameters

Model	Conv. Layers	Conv. Size (n_{out})	Conv. filter Sizes (k)	Conv. Pool Size (m)	LSTM Layers	LSTM Size (d)
Small RNN	N/A	N/A	N/A	N/A	1	16
Medium RNN	N/A	N/A	N/A	N/A	1	32
Large RNN	N/A	N/A	N/A	N/A	2	32
Small CNN	2	64	9,5	2	N/A	N/A
Medium CNN	3	64	9,5,3	2	N/A	N/A
Large CNN	4	64	9,5,3,3	2	N/A	N/A
Small CNN-RNN	1	64	5	N/A	2	32
Medium CNN-RNN	1	128	9	N/A	1	32
Large CNN-RNN	2	128	9,5	2	1	32

4. Experiments and Results

4.1. *Experimental Setup*

Dataset. In order to evaluate our DNN models and visualizations, we train and test on the 108 K562 cell ENCODE ChIP-Seq TF datasets used in Alipanahi et al.[3] Each TF dataset has an average of 30,819 training sequences (with an even positive/negative split), and each sequence consists of 101 DNA-base characters (A,C,G,T). Every dataset has 1,000 testing sequences (with an even positive/negative split). Positive sequences are extracted from the hg19 genome centered at the reported ChIP-Seq peak. Negative sequences are generated by dinucleotide-preserving shuffle of the positive sequences. Due to the separate train/test data for each TF, we train a separate model for each individual TF dataset.

Variations of DNN Models. We implement several variations of each DNN architecture by varying hyperparameters. Table 1 shows the different hyperparameters in each architecture. We trained many different hyperparameters for each architecture, but we show the best performing model for each type, surrounded by a larger and smaller version to show that it isn't underfitting or overfitting.

Baselines. We use the "MEME-ChIP[25] sum" results from Alipanahi et al.[3] as one prediction performance baseline. These results are from applying MEME-ChIP to the top 500 positive training sequences, deriving five PWMs, and scoring test sequences using the sum of scores using all five PWMs. We also compare against the CNN model proposed in Alipanahi et al.[3] To evaluate motif extraction, we compare against the "convolution activation" method used in Alipanahi et al.[3] and Quang and Xie,[4] where we map the strongest first layer convolution filter activation back to the input sequence to find the most influential length-9 subsequence.

4.2. *TFBS Prediction Performance of DNN Models*

Table 2 shows the mean area under the ROC curve (AUC) scores for each of the tested models (from Table 1). As expected, the CNN models outperform the standard RNN models. This validates our hypothesis that positive binding sites are mainly triggered by local patterns or "motifs" that CNNs can easily find. Interestingly, the CNN-RNN achieves the best performance among the three deep architectures. To check the statistical significance of such comparisons, we apply a pairwise t-test using the AUC scores for each TF and report the two tailed p-values in Table 3. We apply the t-test on each of the best performing (based on AUC) models for each model type. All deep models are significantly better than the MEME baseline. The CNN is significantly better than the RNN and the CNN-RNN is significantly better than the CNN. In order to understand why the CNN-RNN performs the best, we turn to the dashboard visualizations.

Table 2. Mean AUC scores on the TFBS classification task

Model	Mean AUC	Median AUC	STDEV
MEME-ChIP[25]	0.834	0.868	0.127
DeepBind[3] (CNN)	0.903	0.931	0.091
Small RNN	0.860	0.881	106
Med RNN	0.876	0.905	0.116
Large RNN	0.808	0.860	0.175
Small CNN	0.896	0.918	0.098
Med CNN	0.902	0.922	0.085
Large CNN	0.880	0.890	0.093
Small CNN-RNN	0.917	0.943	0.079
Med CNN-RNN	**0.925**	**0.947**	**0.073**
Large CNN-RNN	0.918	0.944	0.081

Table 3. AUC pairwise t-test

Model Comparison[c]	p-value
RNN vs MEME	5.15E-05
CNN vs MEME	1.87E-19
CNN-RNN vs MEME	4.84E-24
CNN vs RNN	5.08E-04
CNN-RNN vs RNN	7.99E-10
CNN-RNN vs CNN	4.79E-22

4.3. *Understanding DNNs Using the DeMo Dashboard*

To evaluate the dashboard visualization methods, we first manually inspect the dashboard visualizations to look for interpretable signals. Figure 2 shows examples of the DeMo Dashboard for three different TFs and positive TFBS sequences. We apply the visualizations on the best performing models of each of the three DNN architectures. Each dashboard snapshot is for a specific TF and contains (1) JASPAR[26] motifs for that TF, which are the "gold standard" motifs generated by biomedical researchers, (2) the positive TFBS class-optimized sequence for each architecture (for the given TF of interest), (3) the positive TFBS test sequence of interest, where the JASPAR motifs in the test sequences are highlighted using a pink box, (4) the saliency map from each DNN model on the test sequence, and (5) forward and backward temporal output scores from the recurrent architectures on the test sequence. In the saliency maps, the more red a position is, the more influential it is for the prediction. In the temporal outputs, blue indicates a negative TFBS prediction while red indicates positive. The saliency map and temporal output visualizations are on the same positive test sequence (as shown twice). The numbers next to the model names in the saliency map section indicate the score outputs of that DNN model on the specified test sequence.

Saliency Maps (middle section of dashboard). By visual inspection, we can see from the saliency maps that CNNs tend to focus on short contiguous subsequences when predicting positive bindings. In other words, CNNs clearly model "motifs" that are the most influential for prediction. The saliency maps of RNNs tend to be spread out more across the entire sequence, indicating that they focus on all nucleotides together, and infer relationships among them. The CNN-RNNs have strong saliency map values around motifs, but we can also see that there are other nucleotides further away from the motifs that are influential for the model's prediction. For example, the CNN-RNN model is 99% confident in its GATA1 TFBS prediction, but the prediction is also influenced by nucleotides outside the motif. In the MAFK saliency maps, we can see that the CNN-RNN and RNN focus on a very wide range of nucleotides to make their predictions, and the RNN doesn't even focus on the known JASPAR motif to make its high confidence prediction.

Table 4. JASPAR motif matches against DeMo Dashboard and baseline motif finding methods using Tomtom

	Saliency Map (out of 500)	Conv. Activations[3,4] (out of 500)	Temporal Output (out of 500)	Class Optimization (out of 57)
CNN	243.9	173.4	N/A	19
RNN	138.6	N/A	53.5	11
CNN-RNN	168.1	74.2	113.2	13

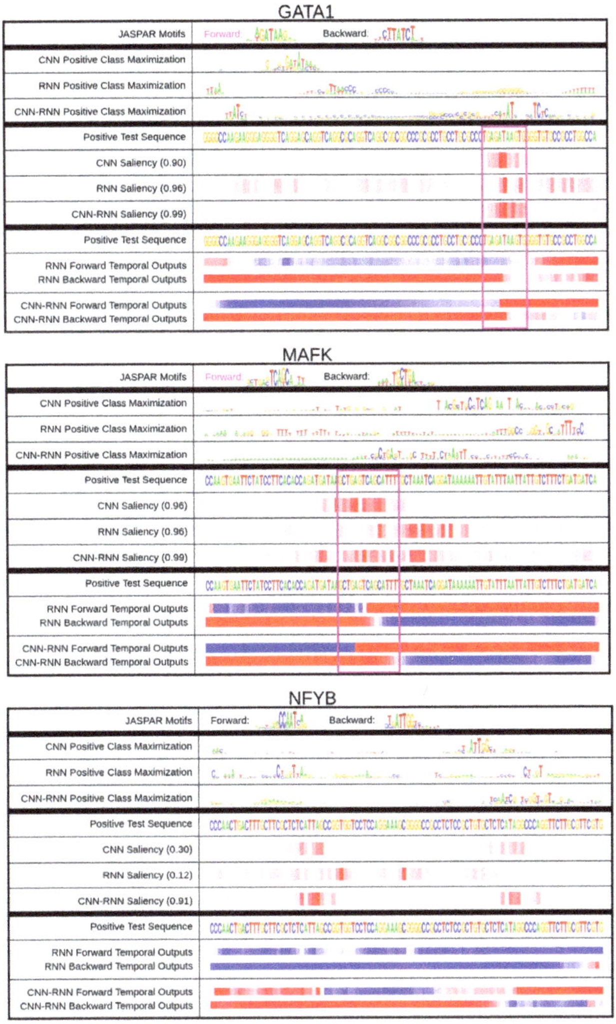

Fig. 2. **DeMo Dashboard**. Dashboard examples for GATA1, MAFK, and NFYB positive TFBS Sequences. The top section of the dashboard contains the Class Optimization (which does not pertain to a specific test sequence, but rather the class in general). The middle section contains the Saliency Maps for a specific positive test sequence, and the bottom section contains the temporal Output Scores for the same positive test sequence used in the saliency map. The very top contains known JASPAR motifs, which are highlighted by pink boxes in the test sequences if they contain motifs.

Temporal Output Scores (bottom section of dashboard). For most of the sequences that we tested, the positions that trigger the model to switch from a negative TFBS prediction to positive are near the JASPAR motifs. We did not observe clear differences between the forward and backward temporal output patterns.

In certain cases, it's interesting to look at the temporal output scores and saliency maps together. An important case study from our examples is the NFYB example, where the CNN and RNN perform poorly, but the CNN-RNN makes the correct prediction. We observe that the CNN-RNN is able to switch its classification from negative to positive, while the RNN never does. To understand why this may have happened, we can see from the saliency maps that the CNN-RNN focuses on two distinct regions, one of which is where it flips its classification from negative to positive. However, the RNN doesn't focus on either of the same areas, and may be the reason why it's never able to classify it as a positive sequence. The fact that the CNN is not able to classify it as a positive sequence, but focuses on the same regions as the CNN-RNN (from the saliency map), may indicate that it is the temporal dependencies between these regions which influence the binding. In addition, the fact that there is no clear JASPAR motif in this sequence may show that the traditional motif approach is not always the best way to model TFBSs.

Class Optimization (top section of dashboard). Class optimization on the CNN model generates concise representations which often resemble the known motifs for that particular TF. For the recurrent models, the TFBS positive optimizations are less clear, though some aspects stand out (like "AT" followed by "TC" in the GATA1 TF for the CNN-RNN). We notice that for certain DNN models, their class optimized sequences optimize the reverse complement motif (e.g. NFYB CNN optimization). The class optimizations can be useful for getting a general idea of what triggers a positive TFBS for a certain TF.

Automatic Motif Extraction from Dashboard. In order to evaluate each DNN's capability to automatically extract motifs, we compare the found motifs of each method (introduced in section 3.4) to the corresponding JASPAR motif, for the TF of interest. We do the comparison using the Tomtom[27] tool, which searches a query motif against a given motif database (and their reverse complements), and returns significant matches ranked by p-value indicating motif-motif similarity. Table 4 summarizes the motif matching results comparing visualization-derived motifs against known motifs in the JASPAR database. We are limited to a comparison of 57 out of our 108 TF datasets by the TFs which JASPAR has motifs for. We compare four visualization approaches: Saliency Map, Convolution Activation,[3,4] Temporal Output Scores and Class Optimizations. The first three techniques are sequence specific, therefore we report the average number of motif matches out of 500 positive sequences (then averaged across 57 TF datasets). The last technique is for a particular TFBS positive class.

We can see from Table 4 that across multiple visualization techniques, the CNN finds motifs the best, followed by the CNN-RNN and the RNN. However, since CNNs perform worse than CNN-RNNs by AUC scores, we hypothesize that this demonstrates that it is also important to model sequential interactions among motifs. In the CNN-RNN combination, CNN acts like a "motif finder" and the RNN finds dependencies among motifs. This analysis shows that visualizing the DNN classifications can lead to a better understanding of DNNs for TFBSs.

5. Conclusions and Future Work

Deep neural networks (DNNs) have shown to be the most accurate models for TFBS classification. However, DNN models are hard to interpret, and thus their adaptation in practice is slow. In this work, we propose the Deep Motif (DeMo) Dashboard to explore three different DNN architectures on TFBS prediction, and introduce three visualization methods to shed light on how these models work. Although our visualization methods still require a human practitioner to examine the dashboard, it is a start to understand these models and we hope that this work will invoke further studies on visualizing and understanding DNN based genomic sequences analysis. Furthermore, DNN models have recently shown to provide excellent results for epigenomic analysis.[8] We plan to extend our DeMo Dashboard to related applications.

References

1. A. Krizhevsky, I. Sutskever and G. E. Hinton, Imagenet classification with deep convolutional neural networks, in *Advances in neural information processing systems*, 2012.
2. I. Sutskever, O. Vinyals and Q. V. Le, Sequence to sequence learning with neural networks, in *Advances in neural information processing systems*, 2014.
3. B. Alipanahi, A. Delong, M. T. Weirauch and B. J. Frey, Predicting the sequence specificities of dna-and rna-binding proteins by deep learning *Nature biotechnology* (Nature Publishing Group, 2015).
4. D. Quang and X. Xie, Danq: a hybrid convolutional and recurrent deep neural network for quantifying the function of dna sequences *bioRxiv* (Cold Spring Harbor Labs Journals, 2015).
5. J. Zhou and O. G. Troyanskaya, Predicting effects of noncoding variants with deep learning-based sequence model *Nature methods* **12** (Nature Publishing Group, 2015).
6. D. R. Kelley, J. Snoek and J. L. Rinn, Basset: Learning the regulatory code of the accessible genome with deep convolutional neural networks *Genome research* (Cold Spring Harbor Lab, 2016).
7. J. Lanchantin, R. Singh, Z. Lin and Y. Qi, Deep motif: Visualizing genomic sequence classifications *ICLR Workshops* 2016.
8. R. Singh, J. Lanchantin, G. Robins and Y. Qi, *Bioinformatics* **32**, i639 (2016).
9. Dashboard definiton `http://www.dictionary.com/browse/dashboard`, Accessed: 2016-07-20.
10. G. D. Stormo, Dna binding sites: representation and discovery *Bioinformatics* **16** (Oxford Univ Press, 2000).
11. E. P. Consortium *et al.*, An integrated encyclopedia of dna elements in the human genome *Nature* **489** (Nature Publishing Group, 2012).
12. P. B. Horton and M. Kanehisa, An assessment of neural network and statistical approaches for prediction of e. coli promoter sites *Nucleic Acids Research* **20** (Oxford Univ Press, 1992).
13. M. Ghandi, D. Lee, M. Mohammad-Noori and M. A. Beer, Enhanced regulatory sequence prediction using gapped k-mer features2014.
14. M. Setty and C. S. Leslie, Seqgl identifies context-dependent binding signals in genome-wide regulatory element maps2015.
15. D. Kingma and J. Ba, Adam: A method for stochastic optimization *arXiv preprint arXiv:1412.6980* 2014.
16. N. Srivastava, G. Hinton, A. Krizhevsky, I. Sutskever and R. Salakhutdinov, Dropout: A simple way to prevent neural networks from overfitting *The Journal of Machine Learning Research* **15**2014.
17. S. Hochreiter and J. Schmidhuber, Long short-term memory *Neural computation* **9** (MIT Press, 1997).
18. K. Simonyan, A. Vedaldi and A. Zisserman, Deep inside convolutional networks: Visualising image classification models and saliency maps *arXiv preprint arXiv:1312.6034* 2013.
19. D. Baehrens, T. Schroeter, S. Harmeling, M. Kawanabe, K. Hansen and K.-R. MÃžller, How to explain individual classification decisions *Journal of Machine Learning Research* **11**2010.
20. M. D. Zeiler and R. Fergus, Visualizing and understanding convolutional networks, in *Computer Vision–ECCV 2014*, (Springer, 2014) pp. 818–833.
21. A. Mahendran and A. Vedaldi, Visualizing deep convolutional neural networks using natural pre-images *International Journal of Computer Vision* (Springer.
22. S. Bach, A. Binder, G. Montavon, F. Klauschen, K.-R. Müller and W. Samek, On pixel-wise explanations for non-linear classifier decisions by layer-wise relevance propagation *PloS one* **10**2015.
23. A. Karpathy, J. Johnson and F.-F. Li, Visualizing and understanding recurrent networks *arXiv preprint arXiv:1506.02078* 2015.
24. J. Li, X. Chen, E. Hovy and D. Jurafsky, Visualizing and understanding neural models in nlp *arXiv preprint arXiv:1506.01066* 2015.
25. P. Machanick and T. L. Bailey, Meme-chip: motif analysis of large dna datasets *Bioinformatics* **27** (Oxford Univ Press, 2011).
26. A. Mathelier, O. Fornes, D. J. Arenillas, C.-y. Chen, G. Denay, J. Lee, W. Shi, C. Shyr, G. Tan, R. Worsley-Hunt *et al.*, Jaspar 2016: a major expansion and update of the open-access database of transcription factor binding profiles *Nucleic acids research* (Oxford Univ Press, 2015).
27. S. Gupta, J. A. Stamatoyannopoulos, T. L. Bailey and W. S. Noble, Quantifying similarity between motifs *Genome biology* **8** (BioMed Central Ltd, 2007).

META-ANALYSIS OF CONTINUOUS PHENOTYPES IDENTIFIES A GENE SIGNATURE THAT CORRELATES WITH COPD DISEASE STATUS

MADELEINE SCOTT*
Stanford University School of Medicine, Stanford University
Stanford, CA 94305, USA

FRANCESCO VALLANIA*
Stanford Institute for Immunity, Transplantation, and Infection, Stanford University
Stanford, CA 94305, USA

PURVESH KHATRI
Stanford Institute for Immunity, Transplantation, and Infection, Stanford University
Division of Biomedical Informatics Research, Department of Medicine, Stanford University
Stanford, CA 94305, USA
Email: pkhatri@stanford.edu

**authors contributed equally to this work*

The utility of multi-cohort two-class meta-analysis to identify robust differentially expressed gene signatures has been well established. However, many biomedical applications, such as gene signatures of disease progression, require one-class analysis. Here we describe an R package, MetaCorrelator, that can identify a reproducible transcriptional signature that is correlated with a continuous disease phenotype across multiple datasets. We successfully applied this framework to extract a pattern of gene expression that can predict lung function in patients with chronic obstructive pulmonary disease (COPD) in both peripheral blood mononuclear cells (PBMCs) and tissue. Our results point to a disregulation in the oxidation state of the lungs of patients with COPD, as well as underscore the classically recognized inflammatory state that underlies this disease.

1. Introduction

Chronic obstructive pulmonary disease (COPD) is a progressive, debilitating lung disease that affects one in 20 people across the globe.[1] It is characterized by declining lung function, as measured by Forced Expiratory Volume (FEV) or Global Initiative for Chronic Obstructive Lung Disease (GOLD) stage.[2,3] FEV is the amount of air that a COPD patient can expel in one second, and decreases as the disease progresses. GOLD scoring is the result of a global effort to reach an agreement on spirometric thresholds for COPD diagnosis and is considered the gold standard of COPD severity.[2,3] An increasing GOLD stage reflects declining lung function, where GOLD stage of 0 represents at-risk patients, while a stage of 4 identifies patients with predicted FEV <30%.[3] The rate of COPD progression varies widely from patient to patient, and there are no current treatment options that effectively halt the disease.[4] There is an urgent, critical unmet need to identify pathways that are robustly and reproducibly associated with COPD severity in order to identify novel targets for therapy.

We have previously described a multi-cohort analysis framework for integrated analysis of heterogeneous datasets, and repeatedly demonstrated its successful application across diverse set of diseases including organ transplant, cancer, and infectious diseases for identifying diagnostic, prognostic, and therapeutic signatures.[5–10] At its core, our multi-cohort analysis framework uses random effects inverse variance meta-analysis to identify differentially expressed genes between two groups of samples (e.g., cases vs controls). However, despite its demonstrated utility, its application is limited to two-class comparisons. One of the drawbacks of this framework is that it does not take into account the stage of disease of the patients.[11,12] Further, many biomedical applications, such as those looking to identify signatures of disease progression, require one-class analysis. Such analyses are indispensible for identifying higher risk patients for more personalized care, and to discover pathways associated with disease progression,[12] which in turn could improve our understanding of the disease.

We have implemented an R package, MetaCorrelator, that addresses this challenge and extends the utility of our multi-cohort analysis framework to analyze continuous phenotypes across multiple datasets (Figure 1). MetaCorrelator follows principles of our framework to identify robust signatures for continuous phenotypes. It provides flexibility to use with different continuous phenotypes and widely heterogenous data.

2. Methods

2.1. *Integration of correlation coefficients across independent datasets*

MetaCorrelator starts by computing a correlation coefficient between a designated continuous phenotype and every gene measured in a given discovery dataset. The correlation coefficients can be computed as Pearson's r, Spearman's ρ, or Kendall's τ. Because Spearman's ρ is defined as the Pearson's r calculated on the ranks,[13] it can used directly as r for the rest of the analysis. Kendall's τ need to be converted into r[14] according to

$$r = \sin(\pi * 0.5 * \tau) \tag{1}$$

Then, each correlation coefficient r is converted into a Fisher's Z effect size, defined as:

$$Z = \frac{1}{2}\ln\frac{1+r}{1-r} \tag{2}$$

with variance, V_z, and standard error, SE_z, defined as

$$V_z = \frac{1}{n-3} \tag{3}$$

and

$$SE_z = \sqrt{V_z} \tag{4}$$

where n is the total number of samples used for correlation. Next, we combine Fisher's z for every gene across all discovery datasets into a summary effect size using a random-effects inverse variance model,[15] which assumes that the true effect sizes across each study are not identical but rather sampled from a distribution of true effects. The summary effect size is calculated as

$$Z_s = \frac{\sum_i^n W_i Z_i}{\sum_i^n W_i} \tag{5}$$

and the corresponding summary standard error was computed as

$$SE_s = \sqrt{\frac{1}{\sum_i^n W_i}} \tag{6}$$

where z_i is the Fisher's Z for a given dataset i and W_i is a weight defined as

$$W_i = \frac{1}{V_i + T^2} \tag{7}$$

where V_i is the variance of the Fisher's Z effect size for a given gene within dataset i and T^2 indicates the in-between-dataset variation. Finally, every gene is assigned a p-value calculated using a two-tailed test defined as

$$p = 2[1 - 2(\phi(|\frac{Z_s}{SE_s}|)] \tag{8}$$

The p-value is then corrected for multiple hypothesis testing using Benjamini-Hochberg.

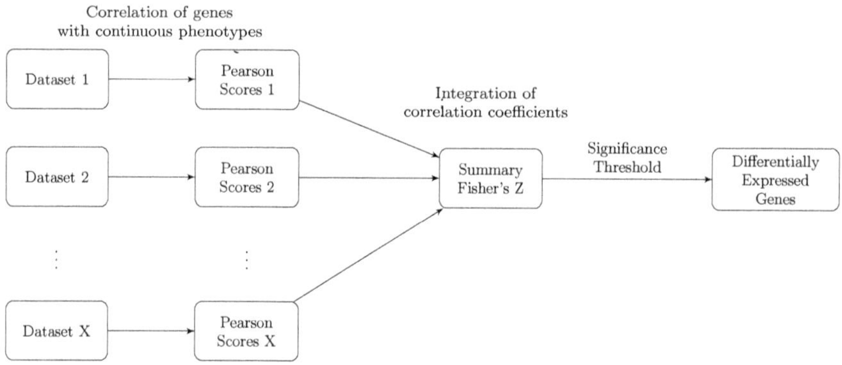

Fig. 1. Schematic Overview of MetaCorrelator. Each dataset is correlated with a continuous phenotype to compute correlation coefficients (example: Pearson's correlation coefficients). These coefficents are then combined into a summary Fisher's Z effect size. A significance threshold is determined to produce the final list of differentially expressed genes.

2.2. *Datasets*

We identified publicly available gene expression datasets from the NCBI GEO that provided lung function in COPD patients using GOLD stage or FEV. In total, we identified six datasets with 642 samples. All six had expression data that was pre-normalized. We used three datasets for discovery and three as validation (Tables 1 and 2). The datasets were highly heterogeneous; five were from lung biopsies and two from PBMCs, spanning collection over seven years and three countries. All probes were matched to Gene IDs based on the platform information available on GEO. Three of the datasets did not have a control group as all of the samples originated from patients with COPD.

2.3. *Selection and validation of COPD signature*

We used FDR $< 5\%$ to identify genes significantly correlated with COPD severity as defined by GOLD stage or FEV. We performed Gene Ontology enrichment analysis using iPathwayGuide (http://www.advaitabio.com). All other statistical analyses were performed using the statistical programming language R.

2.4. *Availability*

Source code is available at http://khatrilab.stanford.edu/metacorrelator. The Khatri lab will provide full results upon request.

Table 1. Datasets Used in Discovery of Lung Function Signature

GEO ID	Tissue	Phenotype	Cases
GSE47460	Lung Biopsy	GOLD Stage and FEV	75
GSE69818	Lung Biopsy	GOLD Stage	70
GSE76705	PBMCs	FEV	229
3 Datasets	2 Tissues		324

Table 2. Datasets Used to Validate Lung Function Signature

GEO ID	Tissue	Phenotype	Cases
GSE42057	PBMCs	FEV	136
GSE38974	Lung Biopsy	GOLD Stage	32
GSE11906	Lung Biopsy	GOLD Stage	150
3 Datasets	2 Tissues		318

3. Results

3.1. *Functional Analysis of Differentially Expressed Genes Identified by MetaCorrelator*

We identified six independent datasets of 692 lung biopsies or PBMC samples from COPD patients that also provided either GOLD stage or FEV for each patient. The samples included in these datasets came from patients across all stages of COPD and covered all the lobes in the lung. We selected three datasets composed of 374 PBMC samples or lung biopsies as discovery datasets (Table 1), and the rest as validation datasets (Table 2). We choose three discovery datasets such that they increased heterogeneity in the discovery. Two datasets were from lung tissue, and had annotation describing the GOLD stage of the samples. Of the two lung tissue datasets, GSE698181 had COPD patients with and without emphysema. Although GSE47460 had both GOLD stage and FEV annotation, only GOLD stage was used for discovery of the gene signature.

MetaCorrelator identified 108 genes (FDR $< 25\%$) that are consistently correlated with COPD severity as measured by GOLD stage or FEV in the three discovery datasets. We performed Gene Ontology enrichment analysis (Figure 2) to explore the functions of the identified genes. Our enrichment analysis highlighted the role of oxidative stress in COPD progression. We identified the disulfide oxidoreductase activity pathway as a highly significant in COPD progression. This is consistent with previous literature that has identified oxidative stress as a sign of COPD progression.[16]

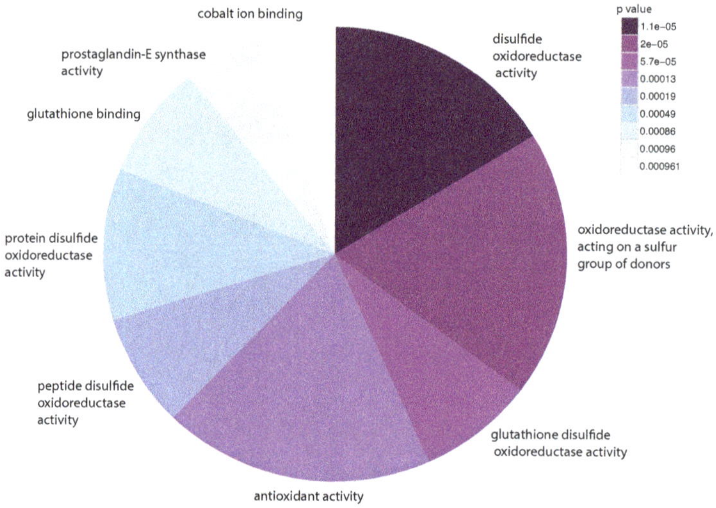

Fig. 2. Functional categories enriched in genes identified by MetaCorrelator

3.2. *Identification and Analysis of a 25-Gene Signature Correlates with COPD Progression*

It is difficult to translate 108-gene signature into a clinical practice. Therefore, to reduce the number of genes, we increased the stringency of our selection criteria by reducing the FDR to 5e-5. We identified 25 genes (7 over-expressed, 18 under-expressed) that are significantly correlated with the COPD severity in the discovery datasets. Enrichment analysis using Gene Ontology of the 25 genes identified matrix remodeling and inflammation as pathways associated with the progression of COPD. Specifically, a subset of the 25 genes, including UDP-Glucuronate Decarboxylase 1 (*UXS1*) and Tetraspanin 13 (*TSPAN13*) are underexpressed genes known to be involved in ECM production and cell adhesion. Importantly, a double tetraspanin knockout mouse (Tetraspanin 28 and 29) has been shown to develop a COPD-like phenotype, underscoring the importance of this protein family to healthy lung function.[18] Inflammatory mediators such as *TLR2* and *FKBP5* are over-expressed in the gene signature, reflecting the inflammatory state of COPD. Previous literature has shown that *TLR2* is over-expressed on Lung CD8+ and CD4+ T-cells as well as CD8+NK T-cells, demonstrating that our results reflect validated biology.[19] Two differentially expressed genes, *TNIK* and *PTPRK* are involved in both ECM and inflammation. Taken together, our results align with previously published literature, which describes COPD as a disregulation of the immune system and subsequent breakdown of the ECM.[20-22]

Next, we defined Lung Function Score (LFS) for a sample as the geometric mean of the expression value of the 25 genes as previously described.[5-7] We observed strong significant correlation between our signature score and FEV in GSE76705 (r = -0.50; p-value = 5.81e-14) (Figure 3). In datasets where GOLD staging was available, we observed a significant score increase in concordance with increasing GOLD stage (JT test; GSE47460: p-value = 9.4e-10; GSE69818: p-value = 5e-4). Interestingly, although only the GOLD stages in GSE47460 were used in the discovery, the LFS strongly correlated with FEV score in GSE47460 (r = -0.57; p-value = 6.23e-4).

3.3. *Validation of the Gene Signature in Three Independent Cohorts*

We validated the 25-gene LFS in three independent cohorts of 318 lung biopsy and PBMC samples from COPD patients (Table 2, Figure 4). Across all three validation cohorts, the LFS was significantly correlated with lung function in the COPD patients (summary effect size = 0.46, p = 3.98e-3). In individual datasets, we observed a significant negative correlation between FEV and LFS in GSE42057 (r = -0.41; p-value = 5.03e-7) and a positive significant correlation with GOLD stages in our remaining two independent cohorts (GSE38974; p-value = 5.539e-6; GSE11906; p-value = 0.02514).

3.4. *Differential Expression in Current vs Never Smokers*

To explore the broader implications of our results, we examined whether any of our 25 identified genes were also significantly expressed in smokers compared to healthy controls. We downloaded seven publicly available datasets from NCBI GEO for a total of 200 samples from smokers and 158 from never smokers (GSE11952, GSE17913, GSE19667, GSE3320, GSE5056, GSE5057, GSE5059). Using the 25-gene LFS derived from MetaCorrelator, two genes, *TSPAN13* and *NR3C2*, were found to be differentially expressed in smokers compared to non-smokers with p value < 0.01. The tetraspanin family has been shown to be critical to normal lung function, and *NR3C2* has been implicated in lung morphogenesis.[23] These results demonstrate the flexibilty of MetaCorrelator to highlight patterns of biological relevance in conjunction with two-class analysis.

4. Discussion

Availability of large amounts of heterogeneous molecular data has necessitated the development of new frameworks to identify patterns and extract new information from these data. We have repeatedly shown the effectiveness of our multi-cohort analysis framework for diagnostic and therapeutic applications across a broad spectrum of human diseases.[5-10] However, this framework is limited to analysis of case-control experiments, and is not suitable for analysis of one-class quantitative phenotypes. Here, we extend our previously established framework to include analysis of gene expression with quantitative quantitative.

Correlation analysis has been a powerful tool for decades, but at this time there does not exist a single framework that can take a collection of datasets and different quantitative

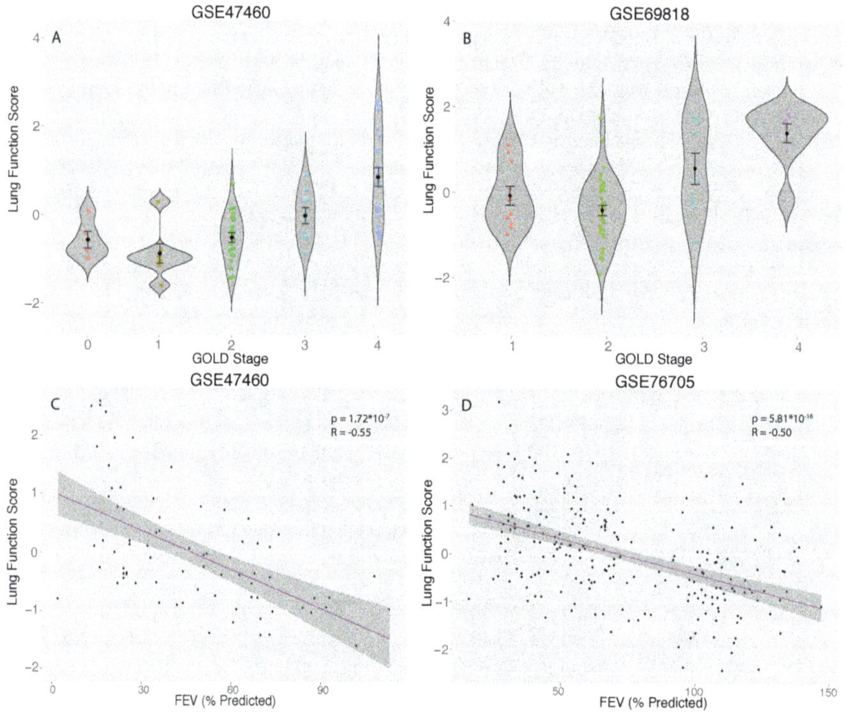

Fig. 3. Lung Function Scores in training cohorts: (A) Violin plots of Lung Function Scores (LFS) in patients distinguished by progressively increasing GOLD stage from GSE47460. (B) Same as (A) but for dataset GSE69818. (C) Correlation plot between LFS and FEV scores in individual patients from GSE47460. (D) Same as (C) but for dataset GSE76705.

phenotypes as input and produce a correlated gene expression signature. There are currently available packages in R, such as metacor, that can compute Fisher's Z values from correlation coefficients; however, MetaCorrelator is uniquely positioned to take multiple datasets as input and correlate gene expression with heterogeneous phenotypes. This is especially relevant in the realm of human disease; methods that are able to integrate different but related organ function phenotypes, such as FEV and GOLD stage, would allow for more powerful analysis that could identify new markers for disease progression.

Our method enables the identification of a gene signature across tissues, thus highlighting the globally relevant differentially expressed genes. By integrating PBMC and lung tissue data, we were able to distill out a gene signature that represents the global differential gene expression of COPD progression. These results emphasize the advantage of integrating multiple tissues. The genes in our signature suggest the importance of inflammation (*TLR2*, *FKBP5*)

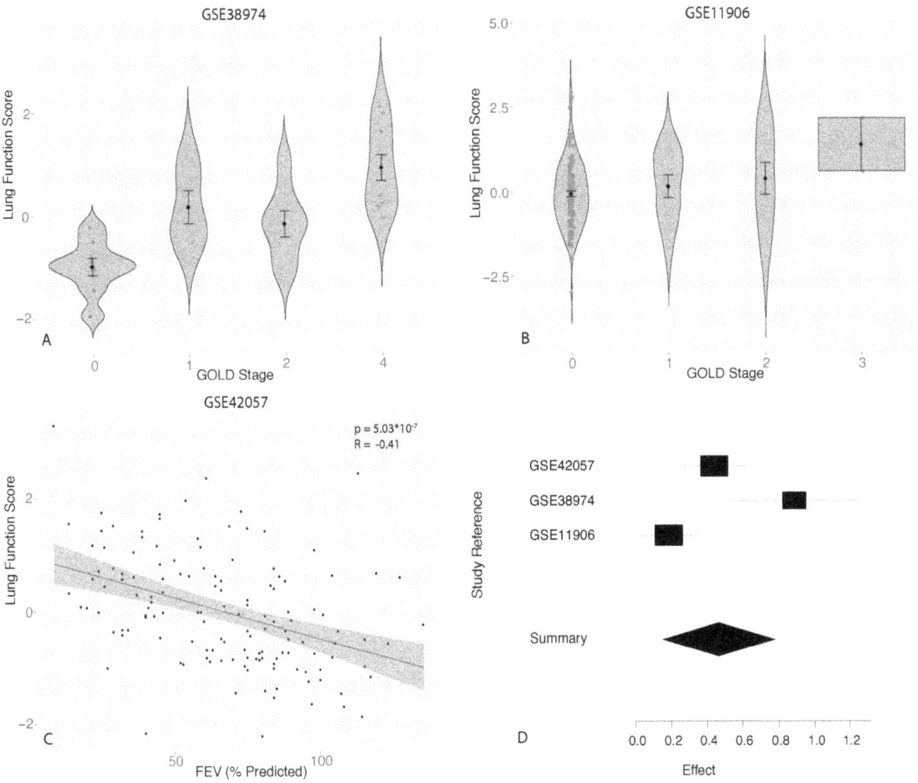

Fig. 4. Lung Function Scores in validation cohorts: (A) Violin plots of Lung Function Scores (LFS) in patients distinguished by progressively increasing GOLD stage from GSE38974. (B) Same as (A) but for dataset GSE11906. (C) Correlation plot between LFS and FEV scores in individual patients from GSE42057. (D) Forest plot representing Fisher's Z values for each of the validation datasets. Squares indicate individual dataset Fisher's Z, with square-size proportional to sample size and horizontal lines indicating individual standard errors (GSE42507 was reverted in sign because of the inverse relationship between GOLD score and FEV). Rhombus indicates summary Fisher's Z with width corresponding to summary standard error.

as well as cell adhesion (*TSPAN13*, *UXS1*), which demonstrates that our framework is able to recapitulate known biology. By integrating the MetaCorrelator framework with established two-class analysis, we can select genes of particular interest. For instance, after identifying differentially expressed genes between smokers and non-smokers, we could further focus on two genes that MetaCorrelator had identified as correlating with COPD progression. MetaCorrelator can be used to correlate any continuous disease phenotype with disease progression. For example, one could identify a gene signature that correlates with prostate specific antigen, a marker of prostate cancer progression. Alternatively, one could correlate a gene signature with ejection fraction of the heart. In summary, MetaCorrelator provides a framework that can correlate whole genome transcriptome across multiple independent datasets with a quantitative phenotype, which in turn can be further explored in case-control studies using the multi-cohort analysis framework.

5. Conclusion

In this study we developed a meta-analysis framework that can integrate multiple gene expression datasets to identify gene signatures that correlate with quantitative phenotypes. Importantly, this method uses the inherent heterogeneity present in multiple cohorts to identify consistently correlated genes and is applicable to datasets that have a single class of sample. Our method can be used in conjunction with other methods that separate samples by class, for example, in order to further differentiate a single group of patients. We applied our method to COPD patients and extracted a 25-gene signature that correlated with lung function in three datasets (two in tissue, one in PBMCs). We then successfully validated our gene signature on three independent datasets. We demonstrated the ability to identify a robust signature with heterogeneous data and phenotypes by correlating the tissue datasets with increasing GOLD stage, and the PBMC dataset with decreasing FEV. Our results suggest an increasing immune response in later stage COPD patients, which has been noted by others, as well as point to a under-appreciated role in sulfur-related oxidative stress. In summary, MetaCorrelator provides a powerful framework to extract a gene signature that is linked to disease progression.

References

1. Halbert, R. J., et al. *European Respiratory Journal* **28**, 523 (2006)
2. Miravitlles, Marc, et al. *Thorax* **64**, 863 (2009).
3. Pauwels, Romain A., et al. *American journal of respiratory and critical care medicine* bf 163, 1256 (2012).
4. Rabe, Klaus F., et al. *American journal of respiratory and critical care medicine* **176**, 532 (2007)
5. Khatri, Purvesh and Roedder, Silke and Kimura, Naoyuki and De Vusser, Katrien and Morgan, Alexander A and Gong, Yongquan and Fischbein, Michael P and Robbins, Robert C and Naesens, Maarten and Butte, Atul J and Sarwal MM. *J. Exp. Med.* **210**, 2205 (2013)
6. M. Andres-Terre, H. M. McGuire, Y. Pouliot, E. Bongen, T. E. Sweeney, C. M. Tato and P. Kha- tri, Immunity **43**, 1199 (December 2015).
7. Timothy E. Sweeney, Aaditya Shidham, Hector R. Wong, and Purvesh Khatri. *Science Translational Medicine* **7**, 287 (2015)
8. Sweeney, Timothy E., Hector R. Wong, and Purvesh Khatri. *Science Translational Medicine* **8** , 346 (2016)
9. R. Chen, P. Khatri, P. K. Mazur, M. Polin, Y. Zheng, D. Vaka, C. D. Hoang, J. Shrager, Y. Xu, S. Vicent, A. J. Butte and E. A. Sweet-Cordero, *Cancer Research* **74**, 2892 (May 2014).
10. T. E. Sweeney, L. Braviak, C. M. Tato and P. Khatri, *The Lancet Respiratory Medicine* **4**, 213 (2016).
11. Walker, Esteban, Adrian V. Hernandez, and Michael W. Kattan. *Cleveland Clinic Journal of Medicine* **75**, 431 (2008)
12. Nordmanna, Alain J., Benjamin Kasendaa, and M. Briel. *Swiss Med Wkly* **142**, w13518 (2012)
13. Myers, Jerome L.; Well, Arnold D. *Research Design and Statistical Analysis* (Routledge, New York, 2003)
14. David A. Walker *JMASM* **2** 525 (2003)
15. M. Borenstein, L.V. Hedges, J.P.T Higgins, and H.R. Rothstein. *Introduction to Meta-Analysis* (Wiley, UK, 2011)
16. Rahman I, Kinnula VL. *Expert Review of Clinical Pharmacology* **5**, 293 (2012).
17. Avrum Spira, Jennifer Beane, Victor Pinto-Plata, Aran Kadar, Gang Liu, Vishal Shah, Bartolome Celli, and Jerome S. Brody. *American Journal of Respiratory Cell and Molecular Biology*, **31**, 601 (2004)
18. Jin, Yingji, et al. *American Thoracic Society* **42**, 633 (2014)
19. Freeman, Christine M., et al. *Respiratory research* **14** (2013)
20. Chung, K. F., and I. M. Adcock. *European Respiratory Journal* **31**, 1334 (2008)
21. Oudijk, EJ D., et al. *Thorax* **60**, 538 (2005)
22. Zandvoort, Andre, et al. *Respiratory research* **9**, 10.1186/1465-9921-9-83 (2008)

23. Duga, Balazs, et al. *Molecular cytogenetics* **7**, 10.1186/1755-8166-7-36 (2014)

PREDICTIVE MODELING OF HOSPITAL READMISSION RATES USING ELECTRONIC MEDICAL RECORD-WIDE MACHINE LEARNING: A CASE-STUDY USING MOUNT SINAI HEART FAILURE COHORT

KHADER SHAMEER[1,2], KIPP W JOHNSON[1,2], ALEXANDRE YAHI [7], RICCARDO MIOTTO[1,2], LI LI[1,2], DORAN RICKS[3], JEBAKUMAR JEBAKARAN[4], PATRICIA KOVATCH[1,4], PARTHO P. SENGUPTA[5], ANNETINE GELIJNS[8], ALAN MOSKOVITZ[8], BRUCE DARROW[5], DAVID L REICH[6], ANDREW KASARSKIS[1], NICHOLAS P. TATONETTI[7], SEAN PINNEY[5] AND JOEL T DUDLEY[1,2,8*]

1. Department of Genetics and Genomics, Icahn Institute of Genomics and Multiscale Biology 2. Institute of Next Generation Healthcare, Mount Sinai Health System 3. Decision Support, Mount Sinai Health System 4. Mount Sinai Data Warehouse, Icahn Institute of Genomics and Multiscale Biology 5. Zena and Michael A. Wiener Cardiovascular Institute, Icahn School of Medicine at Mount Sinai 6. Department of Anesthesiology, Icahn School of Medicine at Mount Sinai 7. Departments of Biomedical Informatics, Systems Biology and Medicine, Columbia University Medical Center, New York 8. Population Health Science and Policy, Mount Sinai Health System, New York, NY

** Corresponding Author, Email: joel.dudley@mssm.edu*

Reduction of preventable hospital readmissions that result from chronic or acute conditions like stroke, heart failure, myocardial infarction and pneumonia remains a significant challenge for improving the outcomes and decreasing the cost of healthcare delivery in the United States. Patient readmission rates are relatively high for conditions like heart failure (HF) despite the implementation of high-quality healthcare delivery operation guidelines created by regulatory authorities. Multiple predictive models are currently available to evaluate potential 30-day readmission rates of patients. Most of these models are hypothesis driven and repetitively assess the predictive abilities of the same set of biomarkers as predictive features. In this manuscript, we discuss our attempt to develop a data-driven, electronic-medical record-wide (EMR-wide) feature selection approach and subsequent machine learning to predict readmission probabilities. We have assessed a large repertoire of variables from electronic medical records of heart failure patients in a single center. The cohort included 1,068 patients with 178 patients were readmitted within a 30-day interval (16.66% readmission rate). A total of 4,205 variables were extracted from EMR including diagnosis codes (n=1,763), medications (n=1,028), laboratory measurements (n=846), surgical procedures (n=564) and vital signs (n=4). We designed a multistep modeling strategy using the Naïve Bayes algorithm. In the first step, we created individual models to classify the cases (readmitted) and controls (non-readmitted). In the second step, features contributing to predictive risk from independent models were combined into a composite model using a correlation-based feature selection (CFS) method. All models were trained and tested using a 5-fold cross-validation method, with 70% of the cohort used for training and the remaining 30% for testing. Compared to existing predictive models for HF readmission rates (AUCs in the range of 0.6-0.7), results from our EMR-wide predictive model (AUC=0.78; Accuracy=83.19%) and phenome-wide feature selection strategies are encouraging and reveal the utility of such data-driven machine learning. Fine tuning of the model, replication using multi-center cohorts and prospective clinical trial to evaluate the clinical utility would help the adoption of the model as a clinical decision system for evaluating readmission status.

1. Introduction

1.1. *Hospital readmission rates – a bottleneck in delivering high value-high volume precision healthcare*

Precision healthcare aims to ensure every patient receive optimal care throughout the onset, maintenance or recovery phases of a disease. Close coordination between different players in the health system is required to integrate and deliver high-quality care. Patients, providers and the care management team play a pivotal role in delivering low-cost, high value and high volume care for patients with diverse healthcare requirements. Improving the quality of healthcare delivery is a challenging task for providers and an important priority for regulatory agencies. As an attempt to reduce healthcare cost, lower healthcare disparities and increase overall quality of care, healthcare regulatory agencies including Centers for Medicaid and Medicare Services (CMS, https://www.cms.gov/) have proposed the Hospital Readmission Reduction Program (HRRP; See: https://www.cms.gov/medicare/medicare-fee-for-service-payment/acuteinpatientpps/readmissions-reduction-program.html). Depending on the performance of a given provider (or hospital) with respect to the regional, state and federal performance rankings, penalties are levied on healthcare providers. In response, in order to reduce readmissions providers have used commercial or in-house readmission assessment tools to predict 30-day readmission rates, but the overall readmission rates still remain high in various provider sites. In 2015, 2,592 U. S hospitals out of 5,627 registered hospitals in the country received penalties from the CMS (http://khn.org/news/half-of-nations-hospitals-fail-again-to-escape-medicares-readmission-penalties/) for not effectively tackling readmission rates. Despite decades of research, interventions, operational improvements and systems engineering methods, readmission remains a major challenge for patients, providers and payers alike.

1.2. *Readmission rate assessment directive by CMS*

The CMS (https://www.medicare.gov/hospitalcompare/Data/30-day-measures.html) directive on unplanned readmission grades the results of five diseases, two surgical procedures and a quantitative estimate of hospital-wide readmission rates. The conditions that CMS evaluates for readmission rates include three specific cardiovascular diseases (heart attack, heart failure, and stroke), one respiratory disease (chronic obstructive pulmonary disease) and an infectious disease (pneumonia). The hospital-wide readmission rates assess the readmission status of patients admitted to internal medicine, surgery/gynecology, pulmonary, cardiovascular, and neurology services. Further, the 30-day mortality measures determine death rates associated these services. Implementing data-driven methods that consider all available clinical variables in a hypothesis-free approach could identify new features driving clinical outcomes. Such an approach could also provide insights into mechanistic or operational factors that could improve clinical outcomes [1-4]. Heart failure is one of the first core measures by The Joint Commission to assess hospital quality initiatives as part of National Hospital Inpatient Quality Measures. Achieving the lowest readmission rates possible is thus critical to provide high-quality care and improve quality assessments (See: https://www.jointcommission.org/core_measure_sets.aspx).

1.3. *Improving quality of healthcare delivery and outcomes using EMR-wide phenomic data*

Implementation of precision phenotyping algorithms and development of prescriptive prediction models models using phenomic data could aid in the discovery of new knowledge from biomedical and healthcare big data generated in the hospital setting[5,6]. Mining of phenomic big data enables the identification of new or unknown features or combinatorial features driving clinical outcomes. Electronic medical records (EMR) provide access to clinical phenome data and enable better understanding of various clinical phenotypes and the associated outcomes in a systematic manner. Design, development, and deployment of predictive and prescriptive models using EMR-based methods could help to accelerate stratification of patients at risk for improved care. Deploying validated predictive patterns in a clinical setting could improve the quality of healthcare delivery and may have a positive impact on patient outcomes. Phenomics[7] is a relatively new omics term used to define collectively the measurement of phenotypic characteristics of biological entities that include the physical and biochemical traits of organisms including humans. Human phenomics can benefit by leveraging EMRs as a longitudinal data source for the collection of clinical and health traits. While the data currently available within EMR for building a complete picture of a human phenomic state is limited, it is rapidly improving with the integration of genomic data, sensor data and other non-clinical data elements[3,4]. Phenome-wide association studies (PheWAS) studies aim to understand the role of a genetic variant identified from genome-wide association studies (GWAS) in increasing or decreasing the likelihood of observing other diseases in a case-control cohort. PheWAS studies are now revealing the molecular architecture of the pleiotropic nature of genetic variants in mediating multiple diseases[1,8].

1.4. *Predictive modeling of readmission rates in heart failure and need for improvement*

Heart failure is a heterogeneous condition characterized by progressive inability of the heart to supply sufficient blood to the organs of the body. HF is associated with high degree of morbidity and mortality, and 50% of patients with HF die within five years of diagnosis. Heart failure accounts for 43% of Medicare spending even though this patient population only makes up 14% of all Medicare beneficiaries. Heart failure is the top cause of readmission for the Medicare fee-for-service patient population and costs approximately 38 billion dollars annually. Several attempts have reported on the utility, accuracy and actionability of predictive models to model and predict potential readmission associated with heart failure hospitalization. Previously reported models have been built using clinical variables and covariates such as age, sex, race, socioeconomic factors, body mass index, laboratory measures, biomarkers (e.g. B-type natriuretic peptide levels), comorbidities (e.g. neurological disorders, type II diabetes mellitus, etc.), behavioral factors, functional phenotyping of cardiovascular systems (e.g. left ventricular ejection fraction), discharge follow-ups and medications [9-12]. Some models have used billing and procedural codes extracted from EMR or other hospital administration databases. Continuous hemodynamic monitoring devices have also been used to predict readmission rates [13-15]. The predictive power of such HF readmission models remains weak, with Area Under Curve (AUC) values generally in the range of 0.6-0.7. Such models provide only modest utility for predicting which patients may return to the hospital for readmission. There is an immediate need for tools that may be used at the bedside or as part of discharge disposition planning to assess and minimize risk for readmission. Studies led by Hosseinzadeh et.al[16] leverage claims data to predict all-cause readmissions, and Duggal et.al[17] used EMR-derived clinical and administrative data to predict readmission in the setting of a diabetes cohort. To the best of our knowledge, our study is one of the first attempts to use phenome-wide data to identify novel factors driving readmissions related to congestive heart failure and develop EMR-wide prediction models with orthogonal validation to predict the readmission event.

2. Methods

The Mount Sinai Institutional Review Board approved the study. An author (JJ) act as the honest data broker to ensure PHI and HIPAA adherence during the data management, analytics and machine learning. Data scientists and research scientists in the project received a deidentified database from the Mount Sinai Data Warehouse. All analyses were performed using the deidentified data.

2.1. *Mount Sinai Heart cohort and characteristics of heart failure cohort*

The study cohort consists of a database of 1,068 individuals admitted to Mount Sinai Heart service during the year 2014. The principal diagnosis of heart failure using the CMS directive was used to compile HF patients. Each patient readmitted to any service of Mount Sinai within 30-days after the discharge of an HF primary encounter is defined as a "case". The remainder of patients who did not return to the hospital within 30-days were defined as "controls". Patients admitted to other locations of Mount Sinai Health System or other hospitals within New York city/state or other states in country were not captured. An author (DR) manually phenotyped the cohort and classified the patients as part of a quality control initiative at Mount Sinai Hospital. As an exploratory study with low case rate, no patient exclusion criteria were applied to the dataset.

2.2. *Clinical data analytics and EMR-wide machine learning*

Data was stored in a MySQL database indexed using a unique hexadecimal identifier associated with the data for the visit about HF. Only data about the primary encounter (admission with HF as primary diagnosis) is employed in the analysis. All figures were generated using Wizard for Mac (http://www.wizardmac.com/) and Weka [18-21]. A Naïve Bayes model is used for machine learning. Exploratory data analyses were performed using Elasticsearch and Kibana (https://github.com/elastic/kibana). All models were independently created using 70% of the dataset for training and 30% of the dataset for testing. Bayesian models were created using features unique to each data element and feature selection was performed using correlation based feature subset selection across two classes. Orthogonal validation of machine

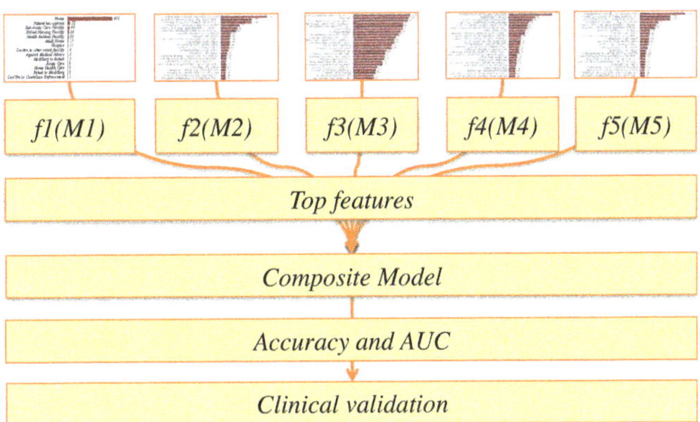

Figure 1: EMR-wide machine learning architecture and predictive modeling strategy to find drivers of readmission rates

learning models was performed with logistic regression. Principal component analyses to understand the variability of features were performed using the Python-based scikit-learn package (http://scikit-learn.org/) and visualized using matplotlib (http://matplotlib.org/). Testing accuracies were estimated using the 5-fold cross validation approach. We define the classification task as a binary classification problem, where RA="Readmitted" patient and NonRA="Not readmitted patient". Weka provides a suite of state-of-the-art machine learning algorithms using a programmatic interface in Java. We used the native Naïve Bayesian classifier in Weka without modification in this exploratory analysis. The algorithm was selected as a rational choice based on prior studies on modeling of readmission prediction[16] Feature ranking and selection[22,23] was performed using a correlation-based feature selection (CFS) method. CFS is a widely used feature selection strategy that aims to find subset of features with significant discriminatory power to perform the classification but which are uncorrelated in feature space. Feature selection is implemented using the "CfsSubsetEval" method in Weka (http://weka.sourceforge.net/doc.dev/weka/attributeSelection/CfsSubsetEval.html). Orthogonal class-specific statistical significance was estimated using Kolmogorov-Smirnov test (distribution estimates), t-test (differences across class-labels), Z-score or Mann-Whitney (median estimates) depending on the data type tested (lab-test, medication, procedure etc.) across the groups (RA and NonRA). An overview of the study design is provided in **Figure 1.**

3. Results

3.1. *Cohort characteristics:*

EMR-wide data mining provides a deep view of various data elements in the cohort (**Figure 2**). A total of 4,205 variables were extracted from EMR. The data from EMR was categorized into five data modalities as diagnosis codes (ICD-9 codes and IMO-codes), procedures (ICD-9, SNOMED-CT and CPT-codes), medications and vital signs. For each patient, the patient encounter specific data is extracted from the EMR. A patient specific filter is used to extract data unique to

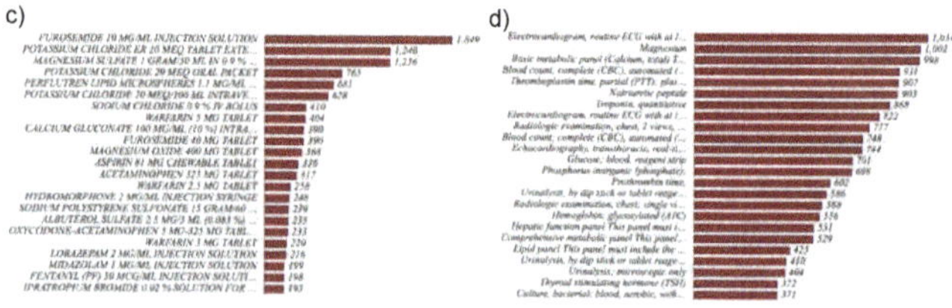

Figure 2: Summary of the study cohort a) case-control ratio: cases are indicated as "1" and controls as "0". Frequency charts of b) diagnoses c) medications and d) procedures.

the visit; the data from the most recent visit of the patients with multiple admissions is incorporated.

Phenomic data extracted from EMR:

1. Diagnoses codes using ICD-9 (n=1,763): ICD-9 codes (http://www.cdc.gov/nchs/icd/icd9.htm) were extracted from Mount Sinai Data Warehouse. The codes were mapped to ICD-9 or IMO codes (https://www.e-imo.com/problemit-terminology-1); all codes were unified to ICD-9 and normalized using UMLS as the bridge (https://www.nlm.nih.gov/research/umls/mapping_projects/icd9cm_to_snomedct.html).
2. Medications (n=1,028): Medications prescribed during the hospitalization were compiled using Epic and extracted from Mount Sinai Data warehouse. Medication name, dosage, route of administration was obtained. All medication data was normalized using RxNorm (https://www.nlm.nih.gov/research/umls/rxnorm/).
3. Laboratory measurements (n=846): Laboratory measures captured in the EMR were compiled; the raw values of the tests without normalization have been used as a matrix of observations with patients as rows and individual tests as columns.
4. Procedures (n=564): Procedures encoded using SNOMED-CT or ICD-9-CM procedures were used.
5. Vital signs (n=4): Pulse, respiration rate, systolic blood pressure, heartbeats and temperature were compiled from bedside monitor logs captured in a MySQL database. Vitals were often captured using multiple monitors and approaches. For example temperature was captured at the bedside as axillary temperature, temperature measured via catheter, oral temperature, rectal temperature, or tympanic temperature.

3.2. *EMR-wide feature selection and predictive modeling using five different data modalities*

The machine learning strategy utilized for our study is outlined in Figure 1. To address the trade-offs in dealing with a broad range of features using a small number of samples and missing data, we first generated distinct models using different data elements and relevant features were selected. Features were also compared using orthogonal metrics including logistic regression and PCA to understand the variable space and their inherent relationships. Finally, a composite model for performing predictions is generated using features selected from the individual models. As a real-world machine-learning task, we had a small subset of cases (16.7%) compared to the controls (83.3%). We used a random subset of age and sex matched controls to control the bias introduced by imbalanced datasets. We first generated five different NB predictors using individual data elements. Medications were the most predictive with an accuracy of 81% and AUC of 0.615. Procedure codes encoded as binary variable fared poorly with AUCs of <0.50 (ICD-9 procedures) and 0.553 (CPT codes). We did not generate an independent model for feature selection using the four vital signs after accounting for the small number of features. Laboratory values also showed lower AUC (0.535). Exploration of the data using principal component analyses also revealed that procedures had low variance compared to medications. From a healthcare delivery standpoint, this is insightful, as most of the patients have undergone the same type of procedures in the cardiac units. However the medication profiles of patients may vary due to individualized disease comorbidities, side effect profiles, age, and gender. Details of individual models and features identified using feature selection method (See Table 1). Detailed analyses of medications could provide better insights into features driving readmissions (Johnson & Shameer *et.al*; *manuscript in preparation*)

3.3. *Feature reduction and model refinement*

Due to the low percentage of the cases in the cohort under investigation, a high-dimension feature array is prone to overfitting in machine learning of binary classification tasks. To address this, we

Data-element	Type	Encoding	Accuracy	AUC	Features
Diagnosis	ICD-9 Diagnosis	Binary	70.3297%	0.605	34/1763
Procedures	ICD-9-Procedure	Binary	77.907%	<0.50	4/273
Procedure	CPT-codes	Binary	72.9858%	0.553	8/564
Medications	Medication name and dosage	Binary	81.9048%	0.615	26/1028
Labs	Non-descriptive lab measurements	Continuous	73.9336%	0.535	29/846
Composite model	Combined features	Hybrid	83.9000%	0.780	105

Table 1: Summary of different Bayesian predictors and features compiled using CFS method

have used a feature reduction approach. Features were tested to assess predictive value using a classifier based method and regression models. Feature selection approach and an orthogonal validation approach provide insights into a subset of highly predictive variables associated with readmitted subset of patients. The AUCs of regression models were 0.5685, 0.6471, 0.7596 and 0.795 (ICD-9 and CPT) for vitals, diagnoses codes, medications, and procedures respectively (See Figure 4 and 5). The final composite model is developed using 105 features with an AUC=0.78 and cross-validation testing accuracy of 83.19%.

A brief summary of features significant in feature selection method and the orthogonal validation approach is provided below (also see Figure 5):

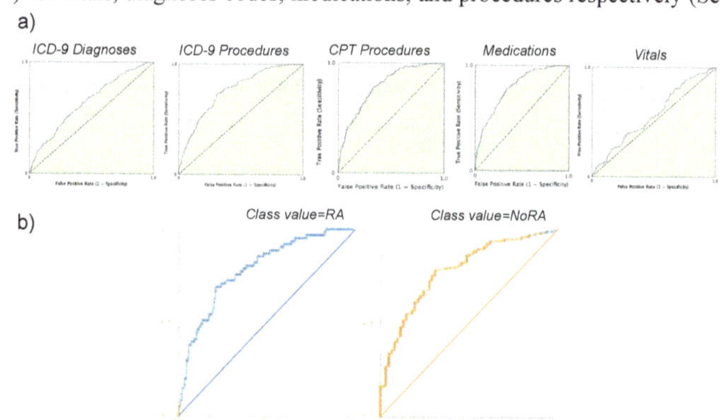

Figure 3: ROC curves a) logistic regression models and b) composite model with 105 features

a) Procedures: out of 12 procedures, codes for invasive procedures including fine needle aspirations with imaging guidance, intravenous catheterization, routine culture and cell count were significant procedures. As procedures were counted as individual events, the subset of readmitted patients has higher frequency of these procedures compared to patients not readmitted. Repetitive tests for culture and cell count could also indicate potential infection or other complications. *b) Medications:* amongst the 1,028 medications, our analyses indicate 28 medications as features with discriminatory power. Three medications (carvedilol 25 mg tablet, ethacrynic acid IVPB and isosorbide dinitrate 30 mg tablet) were validated using logistic regression approach. However, we noted that only 2.7% of the cohort received carvedilol 25 mg, and all of them were part of the readmission subset. Previous work has potentially indicated that increasing in carvedilol dosage may lead to better a outcome on readmission rate[24]. *c) Diagnosis:* chronic conditions like type 1 diabetes (ICD-9 code 250.01), osteoarthritis; manifestations of cancer (ICD-9 code 233); neurological or psychiatric conditions (mood disorders, hallucinations, sleep disturbances cocaine abuse); cardiovascular structural conditions like rheumatic mitral insufficiency and gastrointestinal conditions such as enteritis were conditions significantly associated with readmission rates. Onco-cardiology assessment of patients may also help in reducing the readmission rates in high-risk patients. Assessment of cardiovascular patients for psychosocial aspects and careful evaluation of individual comorbidities could help to reduce the readmission rates and adherence to the medications [25-28]. *d) Laboratory values:* laboratory values were least predictive in the individual modeling stage. During the orthogonal validation step, creatinine kinase, glucose-fluid, fluid triglycerides and lymphocytes were significant. Optimal glycemic control is a key factor in determining positive outcomes in heart failure patients, especially in those with diabetes mellitus [29]. We noted that features identified using our feature selection method are concordant with earlier findings. For example, we have identified glucose-fluid and type-1 diabetes as predictive factors. We have also identified psychiatric illness, a known factor that influences readmission rates in the setting of complex diseases.

a)

Lab tests	*P*
FLUIDTRIGLYCERIDES	0.007
CK(CPK)	0.01
TOTALPROTEINUR24HR	0.016
U-PROTEIN(CONC.)	0.027
LYMPHOCYTE-PER	0.027
HCGTOTALQUANT.	0.03
GLUCOSEFLUID	0.032
IGGQUANT	0.041
HEMOGLOBINPLASMA	0.048
PROTEIN/CREA.RATIOUR	0.048
ABSNEUTROPHILCOUNT	0.051
5804-PSATOTAL	0.071
1412-U-PROTEIN(RANDOM)	0.102
346-APTT-SL	0.112
UREANITROGEN(POCT)	0.14
1201-OSMOLALITYURINE	0.158
PLATELET	0.18
T4TOTAL	0.246
GLUCOSE(POCT)	0.266
HEMATOCRIT-VEN(ISTAT-MPOCT)	0.336
GLUCOSE(MPOCT)	0.344
NEUTROPHIL-PER	0.387
VANCOMYCINTROUGH	0.63
GAMMAGT	0.743
PROGRAF(FK-506)RANDOM	0.771
NEUTROPHIL	0.809

b)

Procedures	*P*
Cell Count, Body Fluid	0.001
Control of epistaxis by cauterization	0.001
Diabetes mellitus without mention of complication, type I [juvenile type], not stated as uncontrolled	0.001
Rheumatic mitral insufficiency	0.001
Carcinoma in situ of breast and genitourinary system	0.002
Cocaine abuse, continuous	0.002
Culture, Urine, Routine	0.002
Fine needle aspiration with imaging guidance	0.002
Hallucinations	0.002
Insertion of a non-tunneled peripherally inserted central venous catheter, without subcutaneous port	0.002
Osteoarthrosis, unspecified whether generalized or localized, shoulder region	0.002
Other and unspecified episodic mood disorder	0.002
Other Cystoscopy	0.002
Personal history of diseases of blood and blood-forming organs	0.002
Regional enteritis of small intestine with large intestine	0.002
Sleep disturbances	0.002
Pulse	0.321

c)

Medications	*P*
ETHACRYNIC_ACID_IVPB	0.002
ISOSORBIDE_DINITRATE_30_MG_TAB	0.002
CARVEDILOL_25_MG_TABLET	0.015

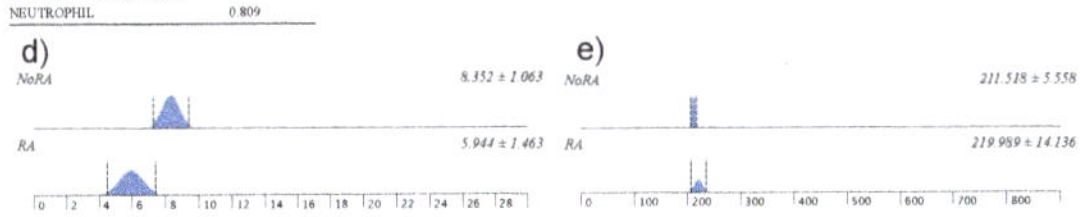

Figure 4: Orthogonal validation of discriminating features a) laboratory tests b) procedures and diagnoses c) medications d) absolute neutrophil count (P=0.051) e) platelet count (P=0.180)

3.4. *Comparison with current heart failure readmission models*

In this work we use EMR-wide feature selection and machine learning to discover novel features and develop new predictors to predict readmission rates. One of the first predictive modeling of hospital readmissions using healthcare data from Quebec, Canada by Hosseinzadeh et.al[16] showed that Naïve Bayes models (0.65) performed better than Random Forest models (0.64). Using a diabetes cohort from a hospital in India, Duggal et.al[17] showed that Naïve Bayes (0.67) showed higher readmission associated savings compared to logistic regression (0.67), Random Forests (0.68), Adaboost (0.67) and Neural Networks (0.62). Futoma et.al[30] showed that Random Forests (0.68) and deep learning using neural networks (0.67) have similar accuracy rate with >1 million patients and > 3 million admission. However, Penalized Logistic Regression had similar accuracy rates as we have shown in our orthogonal validation methods. Compared to existing predictive models for HF readmission rates (AUCs in the range of 0.6-0.7), results from our EMR-wide predictive model (AUC=0.78; Accuracy=83.19%) and phenome-wide feature selection strategies are encouraging and reveal the utility of such data-driven, EMR-wide machine learning.

4. Discussion

Readmission rate is a quality assessment metric routinely used to infer the quality of life index of patient population and the quality of healthcare delivery. Irrespective of the advances in biomedical and healthcare research practices, hospital quality control offices still use traditional readmission risk algorithms and predefined sets of variables to infer the probability patient readmission. However, predictive modeling using big data sourced from different facets of healthcare operations could provide clues to improve the quality of healthcare delivery. Combining predictive analytics with preventive measures would also engage patients, physicians, and payers to participate proactively in improving the health and wellness. Recently we have combined EMR data and genomic data to cluster patients into subtypes with specific genetic variants, disease comorbidities, and medications in a diabetes cohort. Application of deep learning[31,32] in healthcare also shows promise for performing EMR-wide analytics using approaches like Deep Patient[33]. In a recent study, we have created temporal models of disease trajectories that could potentially reveal how the population could cluster into subgroups based on age, gender, self-reported ancestry and comorbidities[34]. Further, we have shown that cognitive machine learning can be utilized for precise phenotyping of high volume echocardiography datasets[35]. We have also applied machine learning to understand various features driving patient satisfaction[36]. Our collective experience in large-scale, automated mining of EMR data suggests that such approaches are useful for both discovery research and the identification of actionable clinical parameters driving diseases or outcomes.

5. Limitations of the current study

In this study, we use all codes without further comprehension; for example, coding systems other than ICD-9 provide an easy way to combine disease. Such an approach could also lead to compiling of similar conditions and hence may not reveal true predictors. For example, we have identified enteritis as a potential diagnosis with readmission. This term would be summarized under gastroenterological conditions. Grouping medication by class or category may also reduce the feature space at the cost of feature resolution. We attempt to capture the best characteristic elements from the real-world data set and hence no data imputation or normalization has been used in our study. The feature selection method may also influence the composition of the models; a systematic assessment of various feature selection algorithms could further enhance the robustness of the model. Healthcare datasets are highly sparse, for example, all patients are not being tested using same laboratory tests except for a few generic tests. Hence, several features may have sparse representations. Even though we had access to EMR-linked genomic data (See BioMe: http://icahn.mssm.edu/research/ipm/programs/biome-biobank), genomic data was not used in this study. Due to a small number of cases; a dramatic increase in feature space would lead to overfitting and high error rates during predictive modeling. We hope to utilize genomic information in a revised version of the model with a larger case dataset. In the current study, we used data from one year of healthcare operations from a single tertiary care healthcare institution. The model should be tested using data from multiple sites and several data-years. Designing of harmonized phenotyping algorithms and data dictionaries leveraging various health information exchanges could help to gather a large number of samples and scale the study using large cohort.

6. Conclusions and Future Directions

A data-driven predictive model is developed to predict readmission rates in heart failure patients. Cases and controls were compiled based on 30-day readmission evidence to the same location. Compared to the existing repertoire of predictive models to assess readmission, our model shows better accuracy using one year of readmission data from a single site. However, the model needs to be updated and calibrated using multiple years of datasets from different sites across the nation. Feature selection provides insights into several novel factors that could help to delineate readmission rates associated with HF. Implementing data-driven methods that EMR-wide variables in a hypothesis-free approach could help us to find new features underlying clinical outcomes. Designing predictive and prescriptive models would help to accelerate stratification of patients at risk for improved care. Such findings and predictive assessments have significant implications for the quality of healthcare delivery and impact on patient outcomes. We envisage that our finding will improve the attempts to develop EMR-wide and scalable phenomics based predictive modeling to find critical events relevant to healthcare delivery and patient outcomes.

7. Acknowledgments

The authors would like to thank the members of the Mount Sinai Health System—Hospital Big Data initiative. This work was supported by a grant from the National Institutes of Health, National Center for Advancing Translational Sciences (NCATS), Clinical and Translational Science Awards (UL1TR001433-01) to KS and JTD.

References

1 Shameer, K. *et al.* A genome- and phenome-wide association study to identify genetic variants influencing platelet count and volume and their pleiotropic effects. *Hum Genet* **133**, 95-109, doi:10.1007/s00439-013-1355-7 (2014).

2 Glicksberg, B. S. *et al.* An integrative pipeline for multi-modal discovery of disease relationships. *Pac Symp Biocomput*, 407-418 (2015).
3 Badgeley, M. A. *et al.* EHDViz: clinical dashboard development using open-source technologies. *BMJ Open* **6**, e010579, doi:10.1136/bmjopen-2015-010579 (2016).
4 Shameer, K. *et al.* Translational bioinformatics in the era of real-time biomedical, health care and wellness data streams. *Brief Bioinform*, doi:10.1093/bib/bbv118 (2016).
5 Hamad, R., Modrek, S., Kubo, J., Goldstein, B. A. & Cullen, M. R. Using "big data" to capture overall health status: properties and predictive value of a claims-based health risk score. *PloS one* **10**, e0126054, doi:10.1371/journal.pone.0126054 (2015).
6 Roski, J., Bo-Linn, G. W. & Andrews, T. A. Creating value in health care through big data: opportunities and policy implications. *Health affairs* **33**, 1115-1122, doi:10.1377/hlthaff.2014.0147 (2014).
7 Houle, D., Govindaraju, D. R. & Omholt, S. Phenomics: the next challenge. *Nat Rev Genet* **11**, 855-866, doi:10.1038/nrg2897 (2010).
8 Karasik, D. How pleiotropic genetics of the musculoskeletal system can inform genomics and phenomics of aging. *Age (Dordr)* **33**, 49-62, doi:10.1007/s11357-010-9159-3 (2011).
9 Thavendiranathan, P. *et al.* Prediction of 30-day heart failure-specific readmission risk by echocardiographic parameters. *Am J Cardiol* **113**, 335-341, doi:10.1016/j.amjcard.2013.09.025 (2014).
10 Padhukasahasram, B., Reddy, C. K., Li, Y. & Lanfear, D. E. Joint impact of clinical and behavioral variables on the risk of unplanned readmission and death after a heart failure hospitalization. *PloS one* **10**, e0129553, doi:10.1371/journal.pone.0129553 (2015).
11 Kansagara, D. *et al.* Risk prediction models for hospital readmission: a systematic review. *JAMA : the journal of the American Medical Association* **306**, 1688-1698, doi:10.1001/jama.2011.1515 (2011).
12 Inouye, S. *et al.* Predicting readmission of heart failure patients using automated follow-up calls. *BMC medical informatics and decision making* **15**, 22, doi:10.1186/s12911-015-0144-8 (2015).
13 Adib-Hajbaghery, M., Maghaminejad, F. & Abbasi, A. The role of continuous care in reducing readmission for patients with heart failure. *J Caring Sci* **2**, 255-267, doi:10.5681/jcs.2013.031 (2013).
14 Bourge, R. C. *et al.* Randomized controlled trial of an implantable continuous hemodynamic monitor in patients with advanced heart failure: the COMPASS-HF study. *J Am Coll Cardiol* **51**, 1073-1079, doi:10.1016/j.jacc.2007.10.061 (2008).
15 Whellan, D. J. *et al.* Development of a method to risk stratify patients with heart failure for 30-day readmission using implantable device diagnostics. *Am J Cardiol* **111**, 79-84, doi:10.1016/j.amjcard.2012.08.050 (2013).
16 Hosseinzadeh, A., Izadi, M., Verma, A., Precup, D. & Buckeridge, D. in *Proceedings of the Twenty-Seventh AAAI Conference on Artificial Intelligence* 1532-1538 (AAAI Press, Bellevue, Washington, 2013).
17 Duggal, R., Shukla, S., Chandra, S., Shukla, B. & Khatri, S. K. Predictive risk modelling for early hospital readmission of patients with diabetes in India. *International Journal of Diabetes in Developing Countries*, 1-10, doi:10.1007/s13410-016-0511-8 (2016).
18 Gewehr, J. E., Szugat, M. & Zimmer, R. BioWeka--extending the Weka framework for bioinformatics. *Bioinformatics* **23**, 651-653, doi:10.1093/bioinformatics/btl671 (2007).

19 Hall, M. *et al.* The WEKA Data Mining Software: An Update. *SIGKDD Explor Newsl* **11**, 10-18 (2009).

20 Pyka, M., Balz, A., Jansen, A., Krug, A. & Hullermeier, E. A WEKA interface for fMRI data. *Neuroinformatics* **10**, 409-413, doi:10.1007/s12021-012-9144-3 (2012).

21 Smith, T. C. & Frank, E. Introducing Machine Learning Concepts with WEKA. *Methods Mol Biol* **1418**, 353-378, doi:10.1007/978-1-4939-3578-9_17 (2016).

22 Guyon, I., Andr, #233 & Elisseeff. An introduction to variable and feature selection. *J. Mach. Learn. Res.* **3**, 1157-1182 (2003).

23 Hastie, T., Tibshirani, R. & Friedman, J. *The Elements of Statistical Learning.* (Springer New York Inc., 2001).

24 Doughty, R. N. & White, H. D. Carvedilol: use in chronic heart failure. *Expert Rev Cardiovasc Ther* **5**, 21-31, doi:10.1586/14779072.5.1.21 (2007).

25 Richardson, L. G. Psychosocial issues in patients with congestive heart failure. *Prog Cardiovasc Nurs* **18**, 19-27 (2003).

26 MacMahon, K. M. & Lip, G. Y. Psychological factors in heart failure: a review of the literature. *Arch Intern Med* **162**, 509-516 (2002).

27 Schweitzer, R. D., Head, K. & Dwyer, J. W. Psychological factors and treatment adherence behavior in patients with chronic heart failure. *J Cardiovasc Nurs* **22**, 76-83 (2007).

28 Ramasamy, R. *et al.* Psychological and social factors that correlate with dyspnea in heart failure. *Psychosomatics* **47**, 430-434, doi:10.1176/appi.psy.47.5.430 (2006).

29 Iribarren, C. *et al.* Glycemic control and heart failure among adult patients with diabetes. *Circulation* **103**, 2668-2673 (2001).

30 Futoma, J., Morris, J. & Lucas, J. A comparison of models for predicting early hospital readmissions. *J Biomed Inform* **56**, 229-238, doi:10.1016/j.jbi.2015.05.016 (2015).

31 LeCun, Y., Bengio, Y. & Hinton, G. Deep learning. *Nature* **521**, 436-444, doi:10.1038/nature14539 (2015).

32 Schmidhuber, J. Deep learning in neural networks: an overview. *Neural Netw* **61**, 85-117, doi:10.1016/j.neunet.2014.09.003 (2015).

33 Miotto, R., Li, L., Kidd, B. A. & Dudley, J. T. Deep Patient: An Unsupervised Representation to Predict the Future of Patients from the Electronic Health Records. *Sci Rep* **6**, 26094, doi:10.1038/srep26094 (2016).

34 Benjamin S. Glicksberg, L. L., Marcus A. Badgeley, Khader Shameer, Roman Kosoy, Noam D. Beckmann, Nam Pho, Jörg Hakenberg, Meng Ma, Kristin L. Ayers, Gabriel E. Hoffman, Shuyu Dan Li, Eric E. Schadt, Chirag J. Patel, Rong Chen, and Joel T. Dudley. Comparative Analyses of Population-scale Phenomic Data in Electronic Medical Records Reveal Race-specific Disease Networks. *Bioinformatics* ISCB Special Issue, doi:10.1093/bioinformatics/btw282 (2016).

35 Sengupta, P. P. *et al.* Cognitive Machine Learning Algorithm for Cardiac Imaging: A Pilot Study for Differentiating Constrictive Pericarditis From Restrictive Cardiomyopathy. *Circ Cardiovasc Imaging* **9**, doi:10.1161/CIRCIMAGING.115.004330 (2016).

36 Li, L., Lee, N. J., Glicksberg, B. S., Radbill, B. D. & Dudley, J. T. Data-Driven Identification of Risk Factors of Patient Satisfaction at a Large Urban Academic Medical Center. *PLoS One* **11**, e0156076, doi:10.1371/journal.pone.0156076 (2016).

LEARNING PARSIMONIOUS ENSEMBLES FOR UNBALANCED COMPUTATIONAL GENOMICS PROBLEMS

ANA STANESCU and GAURAV PANDEY*

Icahn Institute for Genomics and Multiscale Biology and Department of Genetics and Genomic Sciences, Icahn School of Medicine at Mount Sinai New York, NY, USA

**E-mail: gaurav.pandey@mssm.edu*

Prediction problems in biomedical sciences are generally quite difficult, partially due to incomplete knowledge of how the phenomenon of interest is influenced by the variables and measurements used for prediction, as well as a lack of consensus regarding the ideal predictor(s) for specific problems. In these situations, a powerful approach to improving prediction performance is to construct ensembles that combine the outputs of many individual base predictors, which have been successful for many biomedical prediction tasks. Moreover, selecting a *parsimonious* ensemble can be of even greater value for biomedical sciences, where it is not only important to learn an accurate predictor, but also to interpret what novel knowledge it can provide about the target problem. Ensemble selection is a promising approach for this task because of its ability to select a collectively predictive subset, often a relatively small one, of all input base predictors. One of the most well-known algorithms for ensemble selection, CES (Caruana *et al.*'s Ensemble Selection), generally performs well in practice, but faces several challenges due to the difficulty of choosing the right values of its various parameters. Since the choices made for these parameters are usually ad-hoc, good performance of CES is difficult to guarantee for a variety of problems or datasets. To address these challenges with CES and other such algorithms, we propose a novel heterogeneous ensemble selection approach based on the paradigm of reinforcement learning (RL), which offers a more systematic and mathematically sound methodology for exploring the many possible combinations of base predictors that can be selected into an ensemble. We develop three RL-based strategies for constructing ensembles and analyze their results on two unbalanced computational genomics problems, namely the prediction of protein function and splice sites in eukaryotic genomes. We show that the resultant ensembles are indeed substantially more parsimonious as compared to the full set of base predictors, yet still offer almost the same classification power, especially for larger datasets. The RL ensembles also yield a better combination of parsimony and predictive performance as compared to CES.

Keywords: Heterogeneous ensembles; Ensemble selection; Reinforcement learning; Computational genomics.

1. Introduction

Prediction problems in biomedical sciences, such as protein function prediction,[1,2] drug target discovery,[3] and classification of genomic elements[4] are generally quite difficult. This is due in part to incomplete knowledge of how the phenomenon of interest is influenced by the variables and measurements used for prediction, as well as a lack of consensus regarding the ideal predictor(s) for specific problems. From a data perspective, the frequent presence of extreme class imbalance, missing values, heterogeneous data sources of different scales, overlapping feature distributions, and measurement noise further complicate prediction.

In scenarios like these, a powerful approach to improving prediction performance is to construct ensemble predictors that combine the output of many individual base predictors.[5,6] These ensembles have been very successful in producing accurate predictions for many biomedical prediction tasks.[7–13] The success of these methods is attributed to their ability to reinforce accurate predictions as well as correct errors across many diverse base predictors.[14] Diversity among the base predictors is key to ensemble performance: If there is complete consensus (no diversity), the ensemble does not provide any advantage over any of the base predictors. Similarly, an ensemble lacking any consensus (highest diversity) is unlikely to produce confident predictions. Successful ensemble methods strike a balance between diversity and accuracy.[15,16] Popular methods like boosting[17] and random forest[18] generate this diversity by sampling from or assigning weights to training examples. However, they generally utilize a single type of base predictor, such as decision trees, to build the ensemble. Such *homogeneous* ensembles may not be the best choice for problems in biomedical sciences, where the ideal base prediction method is often unclear due to incomplete knowledge and/or data issues.

A more potent approach in this scenario is to build ensembles from the predictions of a wide variety of *heterogeneous* base predictors. Two commonly used heterogeneous ensemble methods are *stacking*,[19,20] and *ensemble selection*.[21,22] Recently, we showed that these methods are more effective than homogeneous ensembles and individual classification methods for complex prediction problems in genomics.[23,24] Other studies have produced similar results.[8,25]

Ensemble selection is an especially promising approach, not only for improving prediction performance, but also because of its ability to select a collectively predictive subset, often a relatively small one, of all input base predictors. This ability to select a *parsimonious* ensemble can be very valuable for biomedical sciences, where it is not only important to learn an accurate predictor, but also to interpret what novel knowledge it can provide about the target problem. For instance, in predicting protein function,[1] it is critical to identify the biological features or principles on the basis of which accurate predictions of protein function are made.[2] It would be easier to reverse engineer a smaller (more parsimonious) ensemble to identify such features or principles than a much larger one, such as all the base predictors taken together. This goal motivated us to develop better algorithms for ensemble selection.

The most well-known algorithm for ensemble selection, which we will refer to as CES (Caruana *et al.*'s Ensemble Selection),[21,22] iteratively grows an ensemble by adding base predictors that produce a gain in prediction performance by (indirectly) enhancing the diversity of the ensemble. Although CES generally performs well in practice, it faces several challenges due to the difficulty of choosing the right values of its various parameters (for details, refer to Section 2.2). For instance, it is unclear how many base predictors should comprise the final ensemble of CES, how many (one or more) should be added in each iteration of the algorithm, and what the right termination condition should be. Since the choices made for these parameters are usually ad-hoc, good performance of CES is difficult to guarantee for a variety of problems or datasets.

To address these challenges with CES and other such algorithms, we propose a novel heterogeneous ensemble selection approach based on the well-established paradigm of reinforcement learning (RL).[26] We demonstrate how RL offers a more systematic and mathematically sound methodology for exploring the many possible combinations of base predictors that can be selected into an ensemble. We test our proposed approaches, and several baselines, on two important computational genomics problem, namely the prediction of protein function and splice sites in eukaryotic genomes. We focus on these unbalanced problems as effective individual (base) predictive models are difficult to learn for them, thus offering a suitable use case for testing ensemble predictors. Our results show that the RL ensembles are indeed substantially more parsimonious with respect to the full set of base predictors, but still offer almost the same predictive power, especially for larger datasets. The RL ensembles also yield a better combination of parsimony and predictive performance as compared to CES. We expect our approaches to be effective for other biomedical problems as well as aid in interpretability, although the latter is a challenging and often subjective task for complex problems. Thus, interpretation of the ensembles we discover is outside the scope of this work.

2. Preliminary Materials and Methods

2.1. *Problem definitions and datasets*

We assess the predictive ability of various ensemble (selection) techniques, such as CES and our RL-based ones, on several protein function (PF) and splice site (SS) prediction datasets.

Protein Function Prediction: Gene expression data are commonly used for predicting protein function, as the simultaneous measurement of gene expression across the entire genome enables effective inference of functional relationships and annotations.[1,2] Thus, for the PFP assessment, we use the gene expression compendium of Hughes *et al.*[27] to predict the functions of roughly 4,000 *S. cerevisiae* genes. Among these genes, the three most abundant functional labels (GO terms) from the list of most biologically interesting and actionable Gene Ontology Biological Process terms compiled by Myers *et al.*[28] are used in our evaluation. These labels are GO:0051252 (regulation of RNA metabolic process), GO:0006366 (transcription from RNA polymerase II promoter) and GO:0016192 (vesicle-mediated transport). We refer to these prediction problems as PF1, PF2, and PF3 respectively (details in Table 1).

Prediction of Splicing Sites: RNA splicing is a naturally occurring phenomenon that contributes to protein diversity. Generally, when creating mature RNA from DNA, introns are removed (or spliced out) from the

Table 1. Details of protein function (PF) and splice site (SS) datasets, including the number of features, number of examples in the minority (positive) and majority (negative) classes, and total number of examples.

Problem	Protein Function Datasets (PF) PF1	PF2 (*S. cerevisiae*)	PF3	Splice Site Datasets (SS) *D. melanogaster*	*C. elegans*	*P. pacificus*	*C. remanei*	*A. thaliana*
#Features	300	300	300	141	141	141	141	141
#Positives	382	344	327	1,598	997	1,596	1,600	1,600
#Negatives	3,597	3,635	3,652	158,150	99,003	156,326	157,542	158,377
#Total	3,979	3,979	3,979	159,748	100,000	157,922	159,142	159,977

gene sequence and exons are retained (or transcribed). Splice sites are conserved nucleotide dimers found at the interfaces between exons and introns. In general, splice sites are *canonical*, as acceptor splice sites are signaled by the occurrence of the consensus dimer "AG" at the 3′ end of the intron, while donor splice sites are characterized by the consensus dimer "GT", situated at the 5′ end of the intron. Such dimers occur frequently throughout most eukaryotic genomes but their presence alone is not sufficient to declare a splice site. Correctly identifying splice sites is an essential step towards genome annotation, and a difficult problem due to the highly unbalanced ratio of bona fide splice sites to decoy dimers.[29] In this work, we focus on identifying acceptor splice sites. In machine learning terms, the problem is formulated as a binary classification of DNA sequences (141-nucleotide-long windows around "AG" dimers, with the dimer situated at position 61) as true acceptor splice sites and decoy dimers. Thus, we assess the ability of ensemble learning to address this important problem on five datasets of acceptor splice sites from five organisms: *D. melanogaster, C. elegans, P. pacificus, C. remanei*, and *A. thaliana*, published by Schweikert *et al.*[4] and Rätsch *et al.*[30]

Note that in some of our experimental evaluations, we investigate the PF and SS datasets results separately due to the substantially different numbers of examples in these datasets.

2.2. *Ensemble selection with CES*

Caruana *et al.*'s Ensemble Selection (CES)[21,22] is arguably the most well-known ensemble selection method. CES begins with an ensemble consisting of the best n individual base predictors ($n = 1$ in our implementation), and iteratively adds new predictors that maximize the performance of the resultant ensemble on a validation set according to a chosen measure. In each iteration, a pool of m of the candidate base predictors are selected randomly without replacement (m = *#base predictors* in our implementation[22]), and the performance resulting from the addition of each individual candidate to the current ensemble is evaluated. The candidate resulting in the best ensemble performance is selected, the ensemble is updated with it, and the process repeats until a maximum ensemble size (set to the *total #base predictors*[22] in our implementation) is reached. We also varied the values of the pool size (m) and maximum size to test the sensitivity of CES to these parameters.

2.3. *Reinforcement Learning (RL)*

The RL machine learning paradigm[26] allows decision-making software agents to learn the ideal behavior within a specific observable environment such that their performance at the specified task is maximized. In order to learn its behavior, an agent requires feedback for its actions, given by the environment in the form of reinforcement signals. Learning what the best action an agent can take given the current state is achieved through trial-and-error interactions with the environment. One of the most basic applications of RL is teaching a novice robot how to traverse a room from one end to another with various obstacles being placed at random locations in the room; for more complex robotics tasks refer to Kober *et al.*[31]

Generally, RL problems are formulated in terms of Markov Decision Processes (MDPs),[32] a commonly adopted framework for modeling environments and sequential decision making. The environment is made up of a finite set of states S, and the actions come from a discrete set A of actions allowed in a given state. The agent's job is to investigate the environment by taking actions and observing the rewards. The Markov property affirms that the current state contains enough information to make a decision about the next action. The goal of RL is to repeat the action-observation process that results in the agent learning a good/optimal strategy, called "policy", for collecting rewards, and completing the task(s) at hand.

Since there is no prior knowledge about any rewards, nor any transition probabilities from any states, (in other words, there is no model available), the agent has to actively "sample" the MDP. Hence the need for *exploration*: the agent tries different actions, often previously untried ones, and assesses their outcome. However, in order to gain sufficient cumulative reward, the agent must also *exploit* its current knowledge about actions already tested that have proved to be beneficial. This exploration/exploitation balance is critical, yet difficult to determine, and represents a fundamental dilemma in RL scenarios. This balance is usually enforced by the classic ϵ-greedy strategy:[33] with probability ϵ, the agent takes a randomly selected action (exploration), and with probability $1 - \epsilon$, the agent chooses the action with the highest estimated payoff (exploitation).

2.4. *Q-Learning*

Following the ϵ-greedy exploration mechanism described above, the agent is able to gather enough information about the environment and create its own model, more specifically, to estimate the quality of each state-action combinations; this mapping is known as the Q-value function. The agent takes an action a_t in a state s_t, ends up in state s_{t+1}, and receives a reward r_t; subsequently, the Q-value associated with action a_t in state s_t is updated. One of the most popular ways for estimating Q-value functions in a model-free framework is the Q-learning algorithm.[34] Under specific conditions/assumptions, Q-learning is able to find an optimal policy (π) regardless of the model the agent adopts, *i.e.*, which action a_t it takes in state state s_t, provided it tries all actions of all states infinitely many times. The policy, which is the agent's learned way of behaving in the environment, is estimated by continuously updating the action-value function $Q(s_t, a_t)$, as described in Equation 1. Here, the learning rate α controls how quickly the learning occurs, and the discount factor γ controls how important future rewards are.

$$Q(s_t, a_t) \leftarrow Q(s_t, a_t) + \alpha \cdot \Big(r_{t+1} + \gamma \cdot \max_a Q(s_{t+1}, a) - Q(s_t, a_t) \Big) \tag{1}$$

3. Proposed Approach

Although CES (Section 2.2) represents an intuitive and straight-forward approach to ensemble selection, several of its inputs/parameters are very difficult to specify beforehand, which is very likely to impact its performance adversely. For instance, the best values of parameters m and n are difficult to specify a priori for a given dataset/problem. Similarly, the termination criteria, *i.e.*, when the CES ensemble should stop growing so as to avoid overfitting, are also often unclear, and are often set in an ad-hoc manner. Due to these factors, CES is only able to perform an ad-hoc sub-optimal search of the possible ensemble space, and not an optimized exhaustive one. We address these challenges of CES and make ensemble selection more rigorous and exhaustive by leveraging concepts from RL (Section 2.3). In particular, we take advantage of the Q-learning algorithm (Section 2.4), which is proven to converge to the optimal solution/policy.[34,35]

To formulate the ensemble selection problem as an RL task, it is necessary to define the following key components of the model. The agent is our proposed ensemble selection algorithm. We define the environment as a deterministic one, in the sense that taking the same action in the same state on two different occasions cannot result in different next states and/or different reinforcement values. More specifically, the environment in which our agent operates consists of all possible subsets of the n base predictors, each serving as a possible ensemble, thus consisting of 2^n states. The environment includes the empty set, which is considered the start state in our implementation. An example environment generated by five base predictors is shown in Fig. 1. It can be viewed as a lattice, and the arrows represent the actions the agent is allowed to take at each state.

The agent investigates the environment by moving from one state (one ensemble) to another in search of better rewards. The reward $R(s_t, a_t, s_{t+1})$ received for the transition from the state s_t to the new state s_{t+1} by executing the action a_t is calculated based on the predictive performance of the ensemble(s) involved in the action. In our experiments, performance is assessed on a validation set separate from the training set, as explained in Section 4. The agent begins its learning at the start state (which corresponds to the empty set) without any prior knowledge about the subsequent states or any of the rewards. Then, the agent moves downward through the lattice from one state to another, until it reaches the final state, *i.e.*, the full ensemble

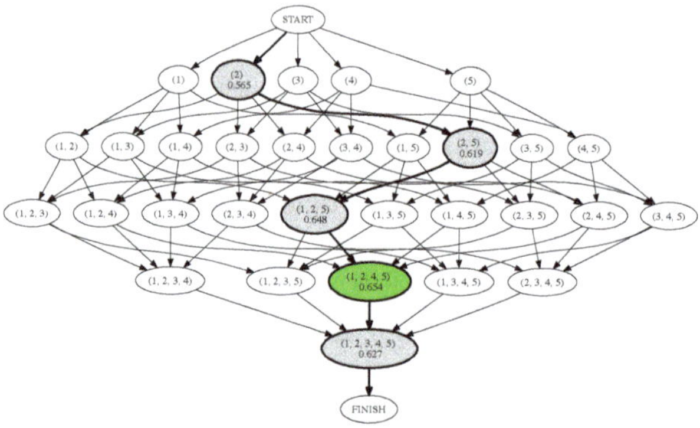

Fig. 1. Example of an environment constructed from five base predictors for the *P. pacificus* splice site dataset. The numbers in the nodes represent the predictive performance of the corresponding ensemble in terms of F-max on the validation set. The path highlighted in grey represents the policy found at the end of the learning process using the RL_greedy strategy ($\epsilon = 0.01$ and ten consecutive episodes yield the same ensemble). The state with the highest reward (*i.e.*, the highest predictive performance) along the path is (1, 2, 4, 5) (green), which is returned by the RL_greedy as the resulting ensemble for the illustrated example. The nodes (2, 4, 5) and (1, 2) represent the ensembles returned by the RL_pessimistic and RL_backtrack strategies respectively. Figure created with Graphviz.[36]

containing all base predictors. Each action is equivalent to adding exactly one other base predictor to the ensemble. In our model, the agent cannot "jump" from a state of k base predictors to a state of $k + 1$ base predictors that is not directly connected in the lattice, such as from (2, 3) to (1, 3, 4). This is necessary to control the complexity of the state-action space. A "traversal" of the lattice from the start state to the finish state is referred to as a "learning episode". We propose Q-learning-based algorithms that employ three different search strategies formulated as different ways of constructing learning episodes and computing rewards. These strategies essentially represent various models of the environment and the interactions between the agent and the environment. The goal, *i.e.*, traversing the lattice and ultimately reaching the full ensemble state, remains the same. These strategies are explained below.

3.1. *Greedy strategy (RL_greedy)*

Our first strategy emulates a "greedy" agent, whose goal is to reach the full ensemble quickly. As a consequence of this plan, the agent rapidly accumulates new base predictors and greedily constructs the ensemble. With each step taken from state s_t to s_{t+1} in the environment, the agent receives state s_{t+1}'s performance as a reward, *i.e.*, $R(s_t, a_t, s_{t+1}) = f(s_{t+1})$, and updates its Q-table as per Equation 1. Each learning episode has a fixed number of steps determined by the depth of the lattice. Fig. 1 shows an example of the path, as well as the resultant ensemble, that RL_greedy finds from a sample lattice constructed from five base predictors.

3.2. *Pessimistic strategy (RL_pessimistic)*

The second strategy resembles a "pessimistic" agent that resets itself to the start state as soon as a less than desirable state is visited, without finishing the traversal of the lattice, and starts a new learning episode. Thus, the lengths of the learning episodes may vary, depending on how soon the agent encounters a "depreciated" state. This strategy is based on the hypothesis that such a depreciated state (negative reward) might indicate a path of overfit and/or under-performing ensembles. For this strategy, the reward is calculated as the difference in performance from s_t to s_{t+1}, *i.e.*, $R(s_t, a_t, s_{t+1}) = f(s_{t+1}) - f(s_t)$.

3.3. *Backtrack strategy (RL_backtrack)*

The last strategy, RL_backtrack, uses the same reward function as RL_greedy. However, unlike the latter strategy, which aims to reach the full ensemble state quickly, the agent "backtracks" (goes back) to the immediately previous state as soon as a decrease in performance is encountered, and resumes its learning from there. A feature of this strategy is that the learning episodes can have a variable numbers of steps, as the agent might wander inside the lattice until it finds an acceptable path ending in the full ensemble.

For all the strategies, the policy derived from the learned action-value function yields a sequence/path of increasingly larger ensembles. We choose the ensemble on this path with the highest performance as the selected ensemble to be evaluated on the test set.

4. Experimental Setup

In all our RL-based experiments, the parameters of the Q-learning algorithm described in Section 2.4, namely α and γ are set to 0.1 and 0.9 respectively, which are commonly used values.[26] The exploration/exploitation trade-off is controlled by the ϵ probability discussed in Section 2.3. A higher probability indicates more exploration. In our work, we experiment with $\epsilon \in \{0.01, 0.1, 0.25, 0.5\}$ for the PF datasets and with $\epsilon \in \{0.01, 0.1, 0.25\}$ for the larger SS ones. The iterative nature of Q-learning requires the initialization of its parameters (values in the Q-table). We initialize the Q-table as a *zero* matrix, and update the values as states and rewards are observed by the agent, as guided by the search strategies defined above.

Although convergence of the Q-learning algorithm and the final policy "optimality" are theoretically guaranteed,[34,35] and achieved when the agent visits all of the environment's states infinitely often, we adopt more practical termination criteria for our search strategies in our experiments. For RL_greedy, the stopping point is reached when the policy induced by the agent produces the same result (*i.e.*, the same ensemble with the highest performance within the currently selected policy) for ten consecutive episodes. In contrast, RL_pessimistic and RL_backtrack are susceptible to longer running times, and even oscillations within the lattice and subsequently of the Q-values learned. For this reason, we set the termination criterion for these strategies as when the agent has taken a fixed number of steps in the environment, specifically 0.5 million. Note that these practical assumptions make it difficult for us to assess the theoretical optimality of our RL algorithms and their results.

In order to assess the relative performance of our RL-based ensemble selection strategies, we also employ several baselines. First, we consider the best base predictor (BP) of each initial set of base classifiers, which is the classifier with the highest classification prediction performance on the validation set. At the opposite end of the spectrum, the full ensemble (FE), consisting of all initial base classifiers, will be the largest ensemble, and our second baseline. Finally, the third baseline is the ensemble produced by CES, implemented as described in Section 2.2.

The general workflow used for the experimental evaluation of the above approaches is shown in Fig. 2. We use 5-fold cross validation to estimate the performance of all the models. All base predictors are learned on the training set (60% of the original data described in Table 1), which is balanced using undersampling of the majority class. The validation set (20% of the data) is used for calculating the rewards of the nodes in the RL environment, and to estimate the performance of the candidate base predictors in the CES approach. The test set (comprising the remaining 20% of the data) is used to assess the overall performance of all studied algorithms. An experiment for an algorithm being tested consists of the collection of these performance scores over all five rounds of this cross-validation.

All performance evaluation, whether internal (on the validation set) or external (on the test set) is conducted using F-max, which is the maximum value of the F-measure across all the values of precision and recall at many thresholds applied to the prediction scores generated by the base classifiers and the resultant ensembles. F-max is appropriate given the highly skewed class distributions of the datasets used in our study, and has been shown to be reliable for performance evaluation in a recent large-scale assessment of protein function prediction.[2] Other metrics that are sensitive to unbalanced problems, *e.g.*, area under the Precision-Recall curve, can also be used.

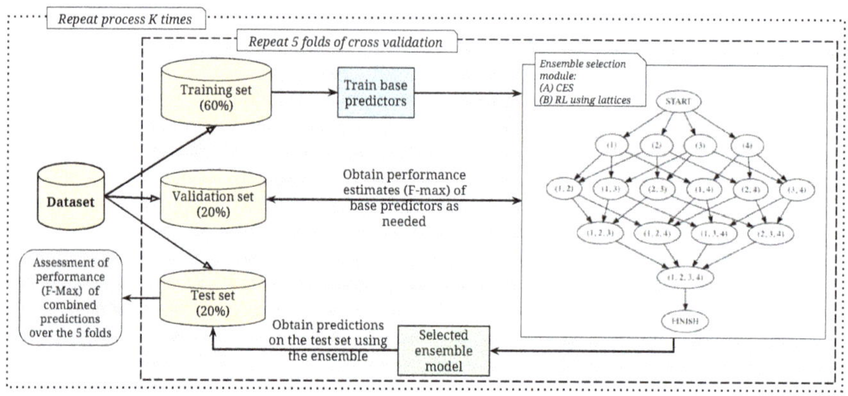

Fig. 2. Visual description of the workflow used to in our experiments. $K = 10$ for the PF datasets and $K = 5$ for the SS datasets.

The ensembles selected by the various approaches we tested (BP, RL, CES) are created by combining the probabilities produced by the constituent base predictors using a weighted average. Here, the importance (weight) of each base predictor is proportional to its predictive performance (measured in terms of the F-max score) on the validation set for all the approaches. We also considered options such as unweighted mean and median but observed that the performance of the ensembles was suffering because of the worst performing individual base predictors among them. In order to efficiently obtain the reward of each state or the performance of the ensemble being considered, we aggregate the predictions using a cumulative moving average.

We train 18 diverse base classification algorithms from Weka,[37] including Naïve Bayes, Multilayer Perceptron, SVM with a polynomial kernel, AdaBoost, Logistic Regression, and Random Forest. Each training set is resampled – to balance the classes – with replacement 10 times, resulting in 180 base predictors. Each ensemble selection algorithm is presented with a pool of base predictors (classification models). We start all our experiments with ten base predictors and increase this set gradually, with steps of ten randomly selected base predictors for each experiment, until we reach the entire set of 180 base predictors. This setup is designed to address the question of how the ensemble selection methods behave with an increasingly larger set of initial base predictors to select from. The performance of the ensembles resulting from of all these methods was evaluated across all these sizes, resulting in curves depicting the dependence of F-max on the number of initial base classifiers. To account for variation, each set of experiments was repeated ten times for the protein function datasets and five times for the splice site datasets (due to time constraints and the much larger size of the SS datasets). We used the area under these curves, denoted auESC (area under Ensemble Selection Curve), as a global evaluation measure for the various algorithms in our study, as it provides a global assessment of ensemble performance across a variety of base predictors. The area is normalized by its maximum possible value, which is the total number of base predictors (the maximum possible value of F-max is one); thus, the maximum value of auESC is one. However, this metric does not follow the same characteristics as auROC, such as random predictors/ensembles producing a score of 0.5. Therefore, auESC is mostly intended for comparative analyses between algorithms running on the same datasets (as done in our experiments), rather than for assessing the absolute performance of these algorithms.

5. Results

In this section, we will investigate various dimensions of classification performance and parsimony of ensembles constructed for the protein function (PF) and splice site (SS) prediction problems.

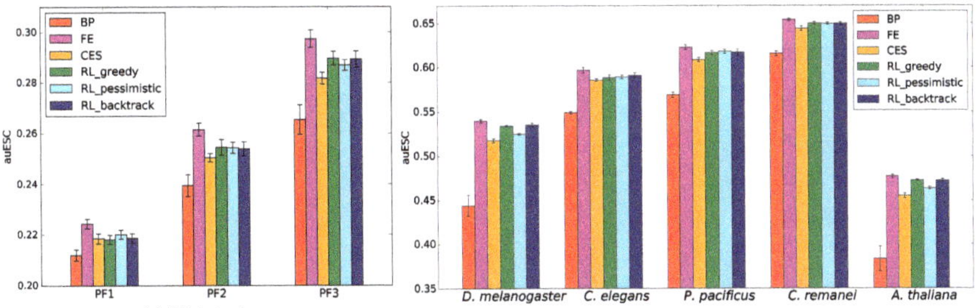

(a) PF datasets (b) SS datasets

Fig. 3. Overall performance (as the auESC score described in Section 4) of all tested algorithms, evaluated across 18 sizes of initial base predictors (from ten to 180, in steps of ten), for (a) PF and (b) SS datasets. The standard errors are calculated over ten repetitions of each experiment for the PF datasets and five repetitions for the SS datasets. The first three bars of each dataset belong to the baselines considered, namely BP (best single base predictor), FE (full ensemble), and the CES algorithm. The last three bars of each dataset represent the RL-based approaches, namely RL_greedy (ten consecutive episodes yielding the same ensemble), RL_pessimistic, and RL_backtrack (both with 0.5 million training steps). For all RL-based approaches, the exploration/exploitation probability $\epsilon = 0.01$.

5.1. *Overall performance of ensemble selection methods*

We first compared the overall classification performance of all the tested ensemble selection algorithms on all the datasets considered. These results are presented in the form of auESC scores in Fig. 3. We also tested the statistical significance of the comparisons of these algorithms in terms of these scores using the Friedman-Nemenyi tests.[38] Several trends can be observed from these figures.

- The best single base predictor (BP) is consistently outperformed by the full ensemble (FE $p = 6.805 \times 10^{-20}$ for the PF datasets and $p = 2.59 \times 10^{-15}$ for the SS datasets) and the RL-based approaches ($p_{RL\_greedy} = 3.29 \times 10^{-4}$, $p_{RL\_pessimistic} = 1.33 \times 10^{-6}$, $p_{RL\_backtrack} = 7.87 \times 10^{-5}$ for the PF datasets and $p_{RL\_greedy} = 5.27 \times 10^{-9}$, $p_{RL\_pessimistic} = 1.105 \times 10^{-5}$, $p_{RL\_backtrack} = 2.78 \times 10^{-8}$ for the SS datasets), thus validating the benefit of ensembles for these problems. In contrast, CES does not show statistically significant improvement over BP ($p > 0.05$ for both types of datasets).
- The biggest ensemble consisting of all the base predictors together (FE) achieves the highest performance across all tested datasets ($p < 0.05$ for pairwise comparisons with all other approaches.)
- For the smaller PF datasets, CES and the RL-based approaches are comparable ($p > 0.05$ for all pairwise comparisons). For the much larger SS datasets, the RL-based RL_greedy and RL_backtrack approaches perform significantly better than CES ($p_{RL\_greedy} = 0.01$, $p_{RL\_backtrack} = 0.025$), while RL_pessimistic does not ($p > 0.05$).
- Among the RL-based approaches, RL_greedy produces the best overall performance in terms of auESC, but the performance of all these approaches is statistically comparable ($p > 0.05$ for all pairwise comparisons for both PF and SS datasets.)

5.2. *Detailed examination of ensemble characteristics over various ensemble sizes*

The analysis based on auESC values shows the overall comparison of the classification performance of the approaches. However, it is critical to examine in detail the dynamics of these ensembles as the number of initial base predictors increases, and as a result, the sizes of the ensembles learnt. Due to lack of space, we only show two representative sets of curves in Fig. 4, one for the PF3 dataset and another for the splice site dataset from *A. thaliana*, showing the variation of ensemble F-max as the number of base predictors increases. These datasets were selected as representatives, as the ensembles showed the most benefit for them over individual base predictors (Fig. 3).

Fig. 4 shows that the performance of several ensemble approaches improves as the number of initial base

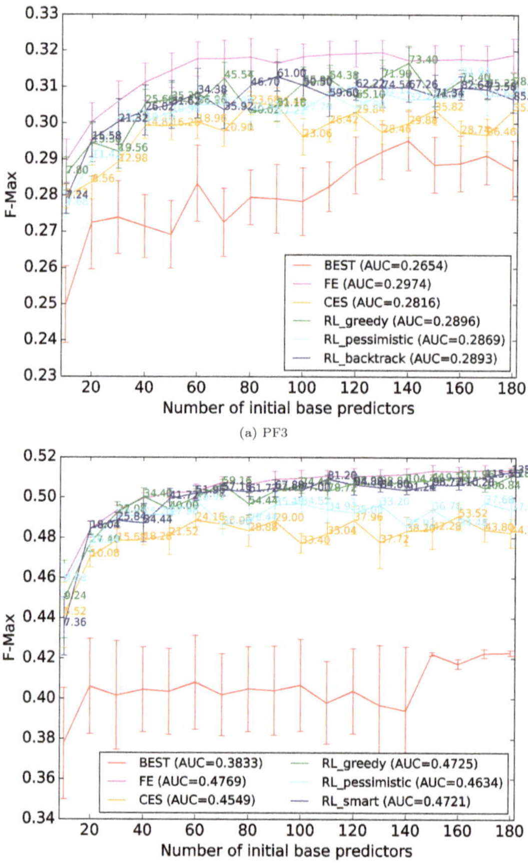

(a) PF3

(b) Splice site dataset from *A. thaliana*

Fig. 4. Performance in terms of F-max of all ensembles as the number of initial base predictors increases from ten to 180 in steps of ten. Each curve corresponds to averages over ten repetitions of each experiment for the (a) PF3 dataset and five repetitions for the (b) *A. thaliana* splice site dataset, along with standard error bars. The curves are annotated with average sizes of the ensembles learnt by the various algorithms at the points shown.

predictors increases, indicating that they are better able to utilize the information learnt. In particular, the curves for the RL approaches track much closer to the FE ones as compared to the CES and BP curves. These observations are stronger for the larger SS datasets (Fig. 4(b)) compared to the smaller PF ones (Fig. 4(a)). Furthermore, the substantial error bars on the CES curves indicate the method's ad-hoc nature. These results show that our RL-based approaches are better able to utilize the information in larger datasets than CES to produce more accurate and stable predictions.

As emphasized earlier, a desirable characteristic of any ensemble selection method is the ability to discover/construct a parsimonious ensemble. To assess this ability of the ensemble selection algorithms we tested,

Table 2. Statistics for the curves shown for PF3 in Fig. 4(a). Column auESC shows the overall performance of the algorithm over all sizes of initial base predictors sets. The ratios of the size and performance of the produced ensembles to those of the best performing approach FE are shown at representative initial base predictors set sizes of 60, 120, and 180.

	auESC	size_ratio@60	size_ratio@120	size_ratio@180	perf_ratio@60	perf_ratio@120	perf_ratio@180
BP	0.2654	0.0167	0.0083	0.0056	0.8919	0.9037	0.9005
FE	0.2974	1	1	1	1	1	1
CES	0.2816	0.31	0.24	0.19	0.9454	0.9493	0.9523
RL_greedy	0.2896	0.62	0.54	0.43	0.9615	0.9621	0.9746
RL_pessimistic	0.2869	0.37	0.24	0.19	0.9601	0.9531	0.9656
RL_backtrack	0.2893	0.57	0.52	0.48	0.9702	0.9714	0.9619

Table 3. Statistics for the curves shown for *A. thaliana* in Fig. 4(b). Column auESC shows the overall performance of the algorithm over all sizes of initial base predictors sets. The ratios of the size and performance of the produced ensembles to those of the best performing approach FE are shown at representative initial base predictors set sizes of 60, 120, and 180.

	auESC	size_ratio@60	size_ratio@120	size_ratio@180	perf_ratio@60	perf_ratio@120	perf_ratio@180
BP	0.3833	0.0167	0.0083	0.0056	0.8118	0.7912	0.8237
FE	0.4769	1	1	1	1	1	1
CES	0.4549	0.40	0.31	0.24	0.9710	0.9577	0.9379
RL_greedy	0.4725	0.50	0.50	0.51	0.9946	0.9927	0.9945
RL_pessimistic	0.4634	0.48	0.29	0.21	0.9919	0.9649	0.9623
RL_backtrack	0.4721	0.87	0.79	0.75	0.9983	0.9919	0.9985

we show in Tables 2 and 3 important statistics of the curves shown in Fig. 4. Specifically, we compute ratios of the selected ensembles' performance and sizes to those of the best performing and largest ensemble, FE. For brevity, we only show statistics at the representative initial base predictors set sizes of 60, 120, and 180.

From the results shown in Tables 2 and 3, it can be seen that CES discovers the smallest ensembles, but these appear not to have enough predictive information, resulting in lower classification performance than FE, especially in the case of the *A. thaliana* SS dataset. The RL-based approaches produce relatively larger ensembles due to their ability to search a much larger portion of the space of ensembles. Due to this same ability, they are able to select ensembles that produce classification performance nearly identical to FE, as shown by the performance ratios. Furthermore, the parsimony ability of these ensembles is enhanced as the number of initial base predictors increases, without significant variation in classification performance. This again validates the parsimonious yet accurate characteristic of the RL ensembles. Finally, our "pessimistic" strategy achieves the best parsimony-performance balance, possibly because of earlier resets of the learning episodes that force the agent to evaluate the upper levels of the lattice, where the smaller ensembles reside. We recommend analyzing summary statistics like the above to identify the ensembles representing the best parsimony-performance balance obtained from various ensemble selection methods. From another perspective, these ensembles might be identified as the point(s) at which the curve(s) like the ones in Fig. 4 start plateauing.

5.3. *Dependence of ensemble selection algorithms' behavior on parameters*

The exploration probability ϵ is critical to the RL-based approaches as it controls the exploration/exploitation management, and consequently how much of the ensemble space is visited. To assess the effect of this parameter on the RL ensembles, we evaluated all three search strategies by executing them with $\epsilon \in \{0.01, 0.1, 0.25, 0.5\}$ for the PF datasets, and with $\epsilon \in \{0.01, 0.1, 0.25\}$ for the SS datasets. We then conducted ANOVA with the F-test to assess the effect of the ϵ values on both ensemble classification and size over the whole performance curves of the type shown in Fig. 4.

For the RL ensemble results from the PF3 dataset (Fig. 4(a)), both in terms of classification performance ($p = 0.26$) and ensemble size ($p = 0.75$), the RL_greedy approach produces similar results for different ϵ values. The RL_pessimistic and RL_backtrack strategies, however, produce statistically variable results with different epsilons, both in terms of classification performance ($p_{RL\_pessimistic} = 7.3 \times 10^{-10}$, $p_{RL\_backtrack} = 0.01$) and ensemble size ($p_{RL\_pessimistic} = 2.7 \times 10^{-13}$, $p_{RL\_backtrack} = 3.85 \times 10^{-4}$). The same trends are observed for PF1 and PF2. Further analysis of RL_pessimistic and RL_backtrack shows that as ϵ increases, the resultant ensemble performance drops, suggesting that too much exploration of the ensemble space may lead to overfitting.

The same analysis of the SS datasets also yielded similar results, with the exception that RL_backtrack did not show a dependence on the value of ϵ for any of the datasets, both in terms of classification performance (*e.g.*, $p = 0.87$ for *A. thaliana*) and ensemble size (*e.g.*, $p = 0.47$ for *A. thaliana*). Given the above observations, as well as the speed of execution, we show results only for $\epsilon = 0.01$ in the previous subsections.

Finally, as mentioned in Section 2.2, CES was run using the recommended values for its parameters.[21,22] To test the impact of these parameters, which are generally difficult to set a priori, we vary their values as the pool size of candidate base predictors $m \in \{N/2, N\}$ and the maximum ensemble size $\in \{N/4, N/2, 3N/4, N\}$ (N = total number of base predictors), and conducted a similar ANOVA analysis as above. The pool size did not have a significant impact on the performance, or the size of the ensembles ($p > 0.05$ for both PF3 and *A. thaliana*). However, the performance is significantly sensitive to the parameter controlling the maximum ensemble size ($p = 3.8 \times 10^{-3}$ for PF3 and $p = 2.8 \times 10^{-4}$ for *A. thaliana*), which also influences the size of the resulting ensemble slightly ($p = 0.08$ for PF3 and $p = 0.01$ for *A. thaliana*). Considering the above results with these, it can be inferred that CES is indeed more sensitive to its key parameters than the RL-based approaches, especially the best performing RL_greedy.

6. Discussion

Ensemble selection offers a powerful approach to addressing biomedical prediction problems, as well as gaining novel knowledge by the analysis of the selected ensembles. This paper presents a framework for selecting ensembles of classifiers using elements of reinforcement learning (RL), which offers a systematic and mathematically sound methodology for exploring the many possible combinations of base predictors that can be selected into an ensemble. This is in contrast to existing ensemble selection algorithms like CES, which often make ad-hoc decisions during ensemble learning and thus cannot offer performance guarantees. Several RL-based methods were implemented in our framework to search the space of possible ensembles as exhaustively as needed.

We tested our methods on two computational genomics problems, namely protein function prediction and splice site detection. We observed that our proposed RL-based methods are able to capture predictive performance close to the full ensembles with a much smaller number of base predictors. This observation is especially strong for the larger splice site datasets, an outcome worth investigating for other biomedical problems, such as cancer phenotype prediction. We also observed that many of the RL parameters, such as the exploration/exploitation probability (ϵ), did not have a significant impact on the downstream performance or sizes of the selected ensembles. It is necessary to examine the intrinsics of RL approaches, such as the effect of ϵ, on a variety of datasets to assess their strengths and weaknesses more comprehensively. Even more fundamentally, the RL approaches we tested can be reformulated, such as by revising the constituent reward functions, to yield better and/or more insightful ensembles. A particular avenue of interest here is to analyze the diversity of the base predictors selected into ensembles by the RL approaches in greater depth, and how that contributes to ensemble performance. Finally, there is also a need to investigate how the performance of our RL algorithms relates to the optimization capabilities and performance guarantees offered by Q-learning.

The RL algorithms studied allow for a larger portion of the space of possible ensembles to be examined more systematically, but this also causes computational requirements to grow substantially, especially as ϵ, the exploration probability, and the number of initial base predictors, grow. For instance, with our basic implementation of these algorithms, which are available as open-source software from `https://github.com/GauravPandeyLab/`, the time (on a 2.3 GHz processor) for one *A. thaliana* RL_greedy experiment varies from approximately one second with ten base predictors to approximately 3 minutes with 180 base predictors for $\epsilon = 0.01$. The same kind of executions took approximately 20 minutes on average for 180 base predictors with $\epsilon = 0.5$. For RL_pessimistic and RL_backtrack, the recorded times for an experiment with 180 base predictors and $\epsilon = 0.01$ are substantially higher, approximately 5.5 and 9.25 hours respectively, due to their more extensive exploration and resets in the search process. The memory requirements also vary similarly with the number of initial base predictors and value of ϵ. To make such executions more computationally feasible, especially for larger datasets, there is a need for developing more parallelized/optimized implementations of these algorithms. We will release such implementations in the future, and invite the community to participate in this effort.

To conclude, our overall effort was towards constructing accurate yet parsimonious (smaller) ensembles, which may in turn be more interpretable, for difficult problems. We acknowledge that interpretation can be a challenging and often subjective task, depending on the target problem, which is why it was out of the scope of our paper, and needs a deeper investigation. Although we didn't explicitly consider this, the interpretability of the base predictors itself would have a major impact on the interpretability of their resultant ensemble. Indeed, this interpretability criterion can be explicitly considered during the ensemble selection/search process to address this challenge more directly.

7. Acknowledgements

This work was partially supported by NIH grant # R01-GM114434 and an IBM faculty award to GP. We thank the Icahn Institute for Genomics and Multiscale Biology and the Minerva supercomputing team for their financial and technical support. We also thank Om P. Pandey and Gustavo Stolovitzky for their technical advice, and the anonymous reviewers for their comments and suggestions.

References

1. G. Pandey, V. Kumar and M. Steinbach, Computational Approaches for Protein Function Prediction: A Survey, Tech. Rep. 06-028, University of Minnesota (2006).
2. P. Radivojac, W. T. Clark, T. R. Oron et al., Nature methods **10**, 221 (2013).
3. M. Kuhn, M. Campillos, P. González, L. J. Jensen and P. Bork, FEBS letters **582**, 1283 (2008).
4. G. Schweikert, G. Rätsch, C. Widmer and B. Schölkopf, Adv. in Neural Info. Processing Systems **22**, 1433 (2009).
5. L. Rokach, Artificial Intelligence Review **33**, 1 (2009).
6. G. Seni and J. F. Elder, Synthesis Lectures on Data Mining and Knowledge Discovery **2**, 1 (2010).
7. P. Yang, Y. H. Yang, B. B. Zhou and A. Y. Zomaya, Current Bioinformatics **5**, 296 (2010).
8. A. Altmann, M. Rosen-Zvi, M. Prosperi et al., PLoS ONE **3**, p. e3470 (2008).
9. A. Khan, A. Majid and T.-S. Choi, Amino Acids **38**, 347 (2010).
10. G. Pandey, B. Zhang, A. N. Chang et al., PLoS Computational Biology **6**, p. e1000928 (2010).
11. G. Yu, H. Rangwala, C. Domeniconi, G. Zhang and Z. Yu, Trans. on Comp. Biol. and Bioinfo. **10**, 1045 (2013).
12. Y. Guan, C. Myers, D. Hess et al., Genome Biology **9**, p. S3 (2008).
13. M. M. Ward, S. Pajevic, J. Dreyfuss and J. D. Malley, Arthritis Care & Research **55**, 74 (2006).
14. K. Tumer and J. Ghosh, Connection Science **8**, 385 (1996).
15. L. I. Kuncheva and C. J. Whitaker, Machine Learning **51**, 181 (2003).
16. T. G. Dietterich, Machine Learning **40**, 139 (2000).
17. R. E. Schapire and Y. Freund, Boosting: Foundations and Algorithms (MIT Press, 2012).
18. L. Breiman, Machine learning **45**, 5 (2001).
19. C. J. Merz, Machine Learning **36**, 33 (1999).
20. D. H. Wolpert, Neural Networks **5**, 241 (1992).
21. R. Caruana, A. Niculescu-Mizil, G. Crew and A. Ksikes, Intl. Conference on Machine Learning **21**, p. 18 (2004).
22. R. Caruana, A. Munson and A. Niculescu-Mizil, International Conference on Data Mining **6**, 828 (2006).
23. S. Whalen and G. Pandey, International Conference on Data Mining **13**, 807 (2013).
24. S. Whalen, O. P. Pandey and G. Pandey, Methods **93**, 92 (2016).
25. A. Niculescu-Mizil, C. Perlich, G. Swirszcz, V. Sindhwani and Y. Liu, J. of Mach. Learn. Research **7**, 23 (2009).
26. R. S. Sutton and A. G. Barto, Intro to Reinforcement Learning, 1st edn. (MIT Press, Cambridge, MA, USA, 1998).
27. T. R. Hughes, M. J. Marton, A. R. Jones et al., Cell **102**, 109 (2000).
28. C. L. Myers, D. R. Barrett, M. A. Hibbs, C. Huttenhower and O. G. Troyanskaya, BMC Genomics **7** (2006).
29. M. B. Shapiro and P. Senapathy, Nucleic acids research **15**, 7155 (1987).
30. G. Rätsch, S. Sonnenburg, J. Srinivasan et al., PLoS Comput Biol **3**, p. e20 (2007).
31. J. Kober, J. A. Bagnell and J. Peters, International Journal of Robotics Research **32**, 1238 (2013).
32. M. L. Puterman, Markov Decision Processes: Discrete Stochastic Dynamic Programming (1994).
33. C. J. C. H. Watkins, Learning from delayed rewards, PhD thesis, University of Cambridge England1989.
34. C. J. C. H. Watkins and P. Dayan, Machine Learning **8**, 279 (May 1992).
35. D. P. Bertsekas and J. N. Tsitsiklis, Neuro-Dynamic Programming, 1st edn. (Athena Scientific, 1996).
36. E. R. Gansner and S. C. North, SOFTWARE - PRACTICE AND EXPERIENCE **30**, 1203 (2000).
37. M. Hall, E. Frank, G. Holmes et al., ACM SIGKDD Explorations Newsletter **11**, 10 (2009).
38. J. Demšar, Journal of Machine Learning Research **7**, 1 (2006).

METHODS FOR CLUSTERING TIME SERIES DATA ACQUIRED FROM MOBILE HEALTH APPS

NICOLE TIGNOR[1], PEI WANG[1], NICHOLAS GENES[1,2], LINDA ROGERS[3], STEVEN G. HERSHMAN[4], ERICK R. SCOTT[1], MICOL ZWEIG[1], YU-FENG YVONNE CHAN[1,2], ERIC E. SCHADT[1]

[1]*Department of Genetics and Genomic Sciences*
Icahn School of Medicine at Mount Sinai, New York, NY, USA

[2]*Department of Emergency Medicine*
Icahn School of Medicine at Mount Sinai, New York, NY, USA

[3]*Department of Medicine, Pulmonary, Critical Care and Sleep Medicine*
Icahn School of Medicine at Mount Sinai, New York, NY, USA

[4]*LifeMap Solutions, Inc, New York, NY, USA*

Email: pei.wang@mssm.edu, eric.schadt@mssm.edu

In our recent Asthma Mobile Health Study (AMHS), thousands of asthma patients across the country contributed medical data through the iPhone Asthma Health App on a daily basis for an extended period of time. The collected data included daily self-reported asthma symptoms, symptom triggers, and real time geographic location information. The AMHS is just one of many studies occurring in the context of now many thousands of mobile health apps aimed at improving wellness and better managing chronic disease conditions, leveraging the passive and active collection of data from mobile, handheld smart devices. The ability to identify patient groups or patterns of symptoms that might predict adverse outcomes such as asthma exacerbations or hospitalizations from these types of large, prospectively collected data sets, would be of significant general interest. However, conventional clustering methods cannot be applied to these types of longitudinally collected data, especially survey data actively collected from app users, given heterogeneous patterns of missing values due to: 1) varying survey response rates among different users, 2) varying survey response rates over time of each user, and 3) non-overlapping periods of enrollment among different users. To handle such complicated missing data structure, we proposed a probability imputation model to infer missing data. We also employed a consensus clustering strategy in tandem with the multiple imputation procedure. Through simulation studies under a range of scenarios reflecting real data conditions, we identified favorable performance of the proposed method over other strategies that impute the missing value through low-rank matrix completion. When applying the proposed new method to study asthma triggers and symptoms collected as part of the AMHS, we identified several patient groups with distinct phenotype patterns. Further validation of the methods described in this paper might be used to identify clinically important patterns in large data sets with complicated missing data structure, improving the ability to use such data sets to identify at-risk populations for potential intervention.

1. Introduction

Handheld mobile devices such as the smartphone are increasingly being utilized by app developers to help users better manage their health and chronic disease conditions. These devices and the mobile health apps that run on them have the potential to provide critical, longitudinal components to an individual's health record. In fact companies such as Apple have greatly facilitated this through their HealthKit, ResearchKit, CareKit, and Homekit platforms, which enable acquisition of very high frequency data over long periods of time, thus providing far more detailed phenotypic user profiles than could ever be reasonably generated in a typical clinical or research setting.

Recently, benefiting from advances in mobile health technologies, we successfully conducted the Asthma Mobile Health Study using an iPhone app.[1] Asthma is a common, highly variable and heterogeneous disease, and it has therefore been difficult to characterize patient disease subtypes precisely enough to inform an optimal individualized treatment plan. Less than half of the 25 million people in the United States with asthma have optimal asthma control, significantly contributing to $56 billion in direct and indirect health care costs annually.[2-3] In order to improve outcomes and reduce costs on a population level, it will be important to acquire large data sets to develop individualized models capable of identifying patients at highest risk to better target resources and tailor therapies. Prior efforts at identifying subgroups of asthma patients have been made based on demographics, lung function tests, biopsy results and blood testing, response to therapy,[4] and recently, genetics.[5-6] Our Asthma Health App, however, for the first time, enables one to collect rich time series data on asthma patients' activities on a daily basis. This opens up the possibility to identify at-risk subgroups of patients based on high-resolution time-course symptom data. The ability to identify clinically relevant patterns of disease could potentially allow targeting of resources to at risk patients to improve disease control.

Participants in the Asthma Mobile Health Study (AMHS) were asked to complete daily surveys to record symptoms and presumed triggers for the duration of the study. Taking the *day symptom outcome* as an example, the collected data of one user is a vector of 0's and 1's indicating whether the user experienced any asthma symptom on each day (1 indicating yes and 0 no symptom experienced). Once collected, the day symptom outcome records of all users can be presented as a 0/1 matrix, which can be used to explore whether subgroups of asthma patients with distinct symptom patterns exist. However, one particular challenge with this type of survey results data is that they contain substantial missing values. While most users may respond to daily survey questions or choose to actively input data on their condition when appropriate, for any given subset of days for which data are being collected, the response rates will be highly varied among different users. Further, for formal studies such as AMHS, users enroll in the study on a rolling basis, such that many non-overlapping periods of enrollment among different users must be accounted for. Lastly, even for the same user, survey response rates often varied over time. Users may be more likely to respond on days when they experience disease symptoms, which further complicates analysis of the data.

A crucial step in handling missing data is to characterize the nature of missing-ness. If the probability of missing data does not depend on the missing values, the missing-data mechanism is referred to as missing-at-random; if so, the mechanism is referred to as not-missing-at-random or non-ignorable. When the proportion of missing values in a data set is large and the missing mechanism is not at random, it is not appropriate to ignore the missing mechanism and perform standard statistical analyses based on the observed values.[7-8] In our AMHS data, since the probability of a user responding to the survey on a particular day depends on the user's asthma symptom on that day, the missing mechanism is non-ignorable. Therefore, in this work, we propose a probability model to characterize the missing mechanism underlying such data and implement a consensus clustering algorithm incorporating multiple imputations. We compare our proposed method with other imputation strategies based on low rank matrix completion procedures.[9] Through extensive simulation studies, we demonstrate the advantage of the probability model based imputation under a range of scenarios reflecting the characteristics of our time series data. While our method is applied to AMHS study and simulated data, the approach can be applied to any time series data in which the missing data mechanism is non-ignorable.

2. Method

Our primary aim was to develop a method that would cluster users in AMHS based on their self-reported day symptom outcome time-series data to identify subgroups of app users with distinct symptom patterns. Given the substantial amount of missing data and that the missing data mechanism is non-ignorable, existing methods were not sufficient for this purpose.

2.1. *A probability based imputation model*

Denote the *day symptom outcome* data matrix as $X_{N\times T} = [\![x_{it}]\!]$, where $i = 1,\cdots,N$, is the index of users and $t = 1,\cdots,T$, is the index of days. Note, since asthma symptoms are often affected by environmental and seasonal changes, we align the profiles of different users according to actual dates instead of arbitrary days in the study. Each x_{it} takes on a value of 1 or 0, depending on whether the ith user reported an asthma symptom on the tth day or not, respectively; x_{it} is set to NA if the ith user did not enroll in the study or did not respond to the daily survey on the tth day.

We further introduce two binary data matrices: $S_{N\times T} = [\![s_{it}]\!]$ to indicate whether users responded to the AMHS survey on each day; and $D_{N\times T} = [\![d_{it}]\!]$ to represent the underlying complete day symptom outcome data. Given these matrices, the observed data $X_{N\times T}$ satisfies: $x_{it} = d_{it}$, if $s_{it} = 1$; and $x_{it} = NA$ if $s_{it} = 0$. If $D_{N\times T}$ was available, existing methods could be employed to cluster users based on this data matrix. However, since we only observe $X_{N\times T}$ and a substantial proportion of $X_{N\times T}$ is NA, we need to impute these missing values first before we can attempt clustering.

The key step for the imputation is to estimate the probability that a given user on a given day had a symptom event that should have been recorded, given the user did not respond to the survey on that day $P(d_{it} = 1|s_{it} = 0)$. In light of the 6-month milestone survey, which is administered to each app user 6 months after the enrollment date in our AMHS, 12% of users indicated that they were more likely to respond to the survey on days when they experienced asthma symptom(s). Given this, we assume that there exists an α_i (≥ 1) for each user, such that $P(s_{it} = 1|d_{it} = 1) = \alpha_i P(s_{it} = 1|d_{it} = 0) = \alpha_i r_{it}^0$, where $r_{it}^0 = P(s_{it} = 1|d_{it} = 0)$. We treat each α_i as a random variable, which takes the value of 1 with probability 0.88, and 2 with probability 0.12, in accordance with feedbacks from AMHS. The choice of 2 is based on the median level of possible range of α_i ($1<\alpha_i<3$) that ensures realistic scenarios given the observed distribution of user response rates. Sensitivity analysis on choices of α_i is shown in Section 3.

We further denote $\bar{p}_{it} = P(d_{it} = 1)$, $p_{it} = P(d_{it} = 1|s_{it} = 1)$, and $r_{it} = P(s_{it} = 1)$. Thus we have that $r_{it} = P(s_{it} = 1|d_{it} = 1)P(d_{it} = 1) + P(s_{it} = 1|d_{it} = 0)P(d_{it} = 0) = \alpha_i r_{it}^0 \bar{p}_{it} + r_{it}^0(1-\bar{p}_{it})$; $p_{it} = \frac{P(s_{it}=1|d_{it}=1)P(d_{it}=1)}{P(s_{it}=1|d_{it}=1)P(d_{it}=1)+P(s_{it}=1|d_{it}=0)P(d_{it}=0)} = \frac{\alpha_i \bar{p}_{it}}{\alpha_i \bar{p}_{it}+(1-\bar{p}_{it})}$. it follows that

$$\bar{p}_{it} = \frac{p_{it}}{\alpha_i(1-p_{it})+p_{it}}, \quad \text{and} \quad r_{it}^0 = r_{it}\frac{\alpha_i(1-p_{it})+p_{it}}{\alpha_i}. \tag{1}$$

And we have that

$$P(d_{it} = 1|s_{it} = 0) = \frac{P(s_{it}=0|d_{it}=1)P(d_{it}=1)}{P(d_{it}=0|s_{it}=1)P(s_{it}=1)+P(d_{it}=0|s_{it}=0)P(s_{it}=0)}$$

$$= \frac{(1-\alpha_i r_{it}^0)\bar{p}_{it}}{(1-\alpha_i r_{it}^0)\bar{p}_{it}+(1-r_{it}^0)(1-\bar{p}_{it})} = \frac{(1-\alpha_i r_{it}^0)p_{it}}{(1-\alpha_i r_{it}^0)p_{it}+\alpha_i(1-r_{it}^0)(1-p_{it})}. \tag{2}$$

We then propose to estimate p_{it} and r_{it} based on the observed data in a time window around the tth day such that

$$\hat{p}_{it} = \frac{\sum_{|t'-t|<\delta} I(s_{it'}=1,\ x_{it'}=1)}{\sum_{|t'-t|<\delta} I(s_{it'}=1)}, \text{ and } \hat{r}_{it} = \frac{\sum_{|t'-t|<\delta} I(s_{it'}=1)}{\sum_{|t'-t|<\delta} 1}, \tag{3}$$

where $I(\cdot)$ is the indicator function, and δ defines the size of the time window. If we plug equation (3) into equations (1) and (2), then we can obtain an estimate of $P(d_{it} = 1|s_{it} = 0)$. In the simulation and real data analysis below, we set δ to be 30 days. This choice resulted from a tradeoff between the robustness to estimate empirical response/symptom rates and sensitivity to capture changes within a short time period.

2.2. *Multiple imputation and consensus clustering*

The probability model in section 2.1 provides a convenient framework for integrating the multiple-imputation procedure[8] and the consensus clustering procedure.[10] Specifically, in the bth imputation run, we first simulate a vector of $\{\alpha_i^b\}_i$. Then to impute an unobserved d_{it}, we calculate $\widehat{P^b}(d_{it} = 1|s_{it} = 0)$ based on α_i^b, and randomly sample a value from a Bernoulli distribution with success probability of $\widehat{P^b}(d_{it} = 1|s_{it} = 0)$. We denote the final imputed complete matrix as $D_{N\times T}{}^b = [\![d_{it}{}^b]\!]$.

Naively, we could perform clustering analysis based on $D_{N\times T}{}^b$. However, when we compare the day symptom profiles of two users, it makes more sense to define distance based on their symptom frequencies over a time window instead of based on events on individual days. For example, suppose there are two users: one has symptoms on Monday, Wednesday and Friday in a given week, while the other has symptoms on Tuesday, Thursday and Saturday in the same week. If we considered the 0/1 vectors of daily symptom events of these two users for this week, they would be extremely different. However, if we consider the symptom frequency over the week, these two users actually show a similar pattern. Therefore, we propose to calculate the frequency profile of each user by performing a running average of the symptom profile:

$f_{it}^b = 1/(2h-1)\sum_{|t'-t|<h} s_{it'}^b$. Then, we can derive clusters of users by performing K-means clustering based on the frequency matrix $F_{N\times T}^b = [\![f_{it}^b]\!]$. We can record the clustering result with an adjacency matrix $(({A_{ij}}^b))_{N\times N}$, where ${A_{ij}}^b = 1$ if the ith user and the jth user are assigned to the same cluster; and ${A_{ij}}^b = 0$ otherwise. We repeat the above imputation-cluster process B times. This gives us B adjacency matrices $\{(({A_{ij}}^b))_{N\times N}\}_b$ corresponding to B sets of clustering results. Intuitively, a large value for A_{ij} suggests a high similarity between the ith and j^{th} user. We can define an average adjacency matrix, $\overline{A_{ij}} = 1/B\sum_b {A_{ij}}^b$, over all adjacency matrices, and then perform the final cluster assignment via another round of K-mean clustering based on the $((\overline{A_{ij}}))$ matrix. We refer to the above procedure as the probability based imputation with consensus clustering (PIC) method. For the special case of $h = 1$, clustering is performed on the imputed day symptom matrix ${D_{N\times T}}^b$. We refer to this special case as the PIC.s method.

One variation on the PIC method worth exploring is to first perform Principal Component Analysis (PCA) on the ${D_{N\times T}}^b = [\![{d_{it}}^b]\!]$ matrix, and then select the loading matrix of the leading L principle components to further perform the clustering analysis. We denote this variation of the PIC procedure as PIC.PC.

3. Simulation Studies

In this section, we investigate the performance of the proposed methods through simulation studies under a range of scenarios reflecting real data conditions.

3.1. *Methods to compare*

In addition to the three methods defined above, PIC, PIC.s and PIC.PC, we also consider performing the probability imputation without taking into account the non-random missing pattern (i.e. set $\alpha_i = 1$). We denote this strategy as "PIC($\alpha_i = 1$)". We also include a few low-rank (LR) matrix completion based approaches for comparison. LR matrix completion has been recently demonstrated to be extremely powerful in recovering large scale matrices[9]. Specifically, we employ the R package *softImpute*,[11] which uses convex relaxation techniques to provide regularized low-rank solutions for large-scale matrix completion problems. We considered three strategies to apply the LR matrix completion (referred to as "LR" in below): (1) we directly apply LR on the raw data matrix ($X_{N\times T}$); (2) for each user, we first imputed the missing data based on the probability model of PIC for days within his/her enrollment period, and then apply LR to impute the missing data on days outside the enrollment period; and (3) similar to (2) except that we further derive the frequency matrix following the imputations. Here, enrollment period of one user is defined as the period from the first to the last instance of non-missing observation based on the empirical day symptom data. In all three strategies, after data imputation, consensus clustering is performed in the same way as for PIC. We denote these three strategies as LR, PIC.S.LR, and PIC.LR, respectively.

3.2. *Simulation settings*

To mimic the data from AMHS, in our simulations (see section 4), we set N=334, T=136, and the total number of clusters to be 3. In addition, we assumed 3 roughly equal-sized clusters (n_1=111,

n_2=111, and n_3=112), so the accuracy of clustering result could be more intuitively assessed. We then generated multiple sets of frequency curves representing a variety of hypothetical symptom frequency profiles (i.e. $\{P\ (d_{it} = 1)\}_t$) (see Fig. 1). We assume the samples belonging to the same cluster share the same underlying symptom frequency profile. To generate time-series data for each sample, we simulated symptom events of the tth day by Bernoulli sampling of 0/1 based on the tth point of the corresponding frequency curve. To simulate non-overlapping enrollment periods, we sampled from the empirically observed enrollment period distribution from the AMHS data.

To further generate non-ignorable missing-ness, we used information from the milestone survey results in AMHS. In this survey, users are asked to provide their reasons for not responding to the daily survey during the study period. Based on users who provided milestone survey responses before April 4, 2016, 12% indicated that they tended to skip the daily surveys on days in which they had no symptoms. Thus, in the simulated data we sampled from a Bernoulli distributed random variable I to identify whether a user was among those whose response depended on symptom state, with p (I = 1) = 0.12. For samples assigned to I = 1, we introduced a parameter Δ to modify the rate of missing data depending on symptom state such that $(s_{it} = 1|d_{it} = 1) = P(s_{it} = 1) + \Delta$ and $P(s_{it} = 1|d_{it} = 0) = P(s_{it} = 1) - \Delta$. For other samples assigned to I = 0, we imposed uniformity over time such that $P(s_{it} = 1|d_{it} = 1) = P(s_{it} = 1|d_{it} = 0) = P(s_{it} = 1)$. For each user, $r_{it} = P(s_{it} = 1)$ was set to be a constant r_i, which is either a pre-determined value or is sampled from an empirical distribution of missing rates calculated from the AMHS data. We then used $P(s_{it} = 1|d_{it} = 1)$ and $P(s_{it} = 1|d_{it} = 0)$ to generate missing data within the enrollment period.

We considered various simulation settings to evaluate how the performances of the different methods were affected by various factors including: (1) the shapes of the frequency profiles, (2) the overall missing percentages, (3) the severity level of the non-random missing, (4) alternative scenarios for setting α_i, and (5) we evaluated the power to detect association between a generic simulated covariate and the inferred cluster assignments derived from the application of each method on simulated data. In the following, we varied one factor at a time, where unless specified, the default setting is to use the frequency profile set labeled b in Figure 1, r_i = 0.4 for all users, Δ = 0.3, and α_i is 1 with probability 0.88 or 2 with probability 0.12. For all settings, the window size h used to derive the frequency profiles is simply set to be a fix value of 15, as we observed that the performance of all strategies are not sensitive to the different choices of h (data not shown).

1. We considered 8 different sets of symptom frequency profiles as illustrated in Figure 1.
2. We considered 4 different ways for setting r_i, where for (1)-(3), r_i = 0.2, 0.4, or 0.6; and for (4) r_i is sampled from an empirical distribution of missing rates calculated from the AMHS data.
3. We varied the value of Δ, where Δ = 0.1, 0.3, or 0.5.
4. We considered 3 alternative scenarios for setting α_i, where for (a1)-(a3): α_i is 1 with probability 0.88 and is 1.5, 2, or 2.5 with probability 0.12.
5. We simulated a binary covariate based on true cluster assignments, where the probability of taking a value of 1 was set to 10% across all clusters (p1), or was set, depending on cluster

assignment, to: (p2) 10%, 15% or 20%, or (p3) 10%, 20% or 30%. For these 3 scenarios, we evaluated the power to detect association between the simulated covariate and the predicted cluster assignments using a p-value cutoff of 0.05 based on Fisher's Exact Test.

3.3. *Simulation results*

For each simulation scenario, we applied each of the strategies in section 3.1 to derive predicted cluster assignments from simulated data sets. True and predicted cluster assignments were compared using the adjusted Rand index.[12] Based on the results from simulation Setting 1, strategies PIC and PICs perform well across a range of symptom profile scenarios (Fig. 1).

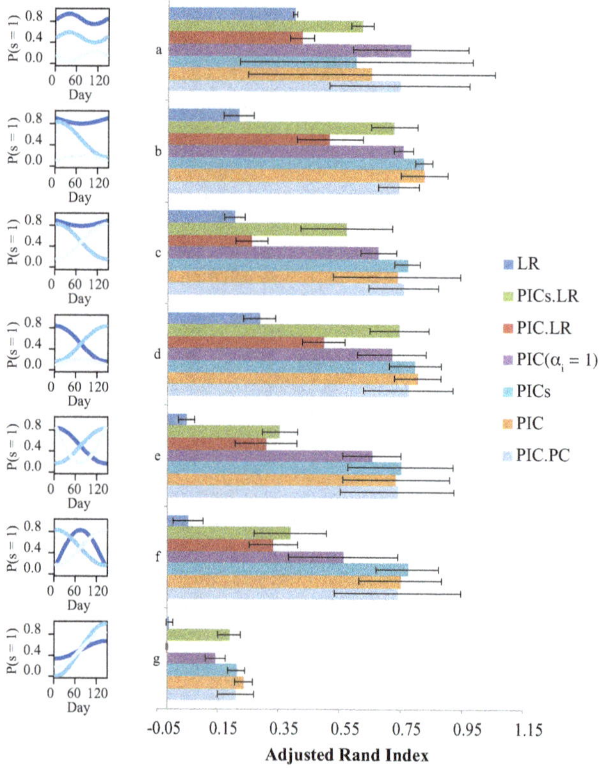

Figure 1. Simulation results for Setting 1, where we consider 8 different sets of symptom frequency profiles while fixing $\Delta = 0.3$, and using $r_i = 0.4$ for all users. Symptom profiles (a-g, left) are defined for sets of 3 clusters, where each cluster is color-coded from highest (dark) to lowest (light blue) overall mean symptom rate. Average adjusted rand indices and their standard deviations across 50 simulations with 100 iterations of imputation each are shown for all strategies.

The strategies involving low rank matrix completion display more variability across symptom profiles, particularly the LR strategy which shows a clear decrease in performance as simulation scenarios become more difficult. The accuracy of all methods tend to decrease with the overall missing rate of the data (Fig. 2A). LR is particularly worse in cases where the overall non-response rate (r_i) or the severity level of non-random missing (Δ) is high (Fig. 2B). We also observe disadvantages of PIC($\alpha_i = 1$) compared to PIC under these same circumstances, due to the lack of treatment of non-ignorable missing (Fig. 2A and Fig. 2B). Most strategies show comparable performance across different α_i scenarios, with the exception of LR, which shows enhanced performance when α_i is set to a2 (Fig. 2C). In the end, Fig. 2D suggests that PIC achieves better power to detect association between covariates and predicted clusters than other clustering strategies when the strength of association is simulated to be more moderate.

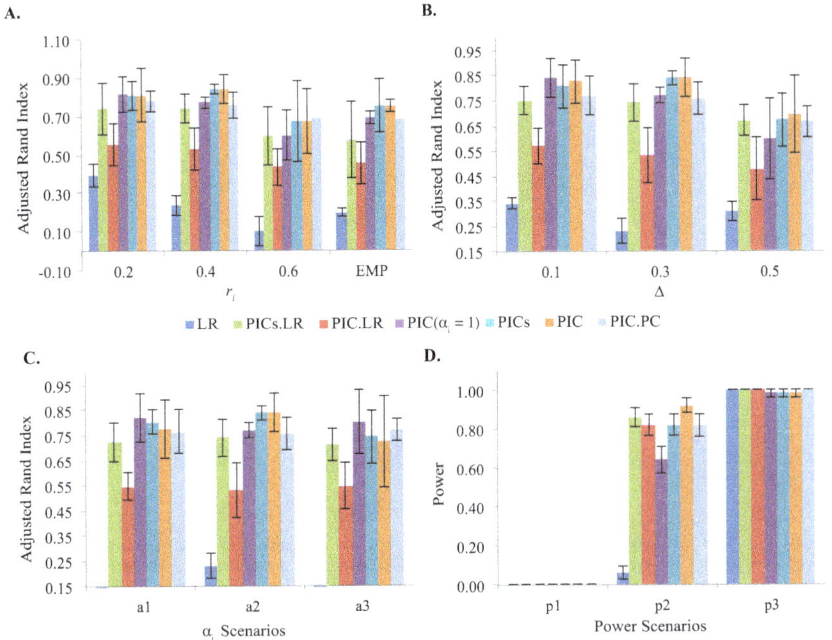

Figure 2. Simulation results based on 50 data simulations with 100 multiple imputations each. A. Results for setting 2, where we consider several values of r_i, including: 0.2, 0.4, and 0.6, where r_i is a constant for all users; and EMP, where r_i is sampled from the empirical distribution of missing rates calculated from AMHS data. B. Results for setting 3, considering several values for Δ. C. Results for setting 4, where we consider different scenarios for assigning values to the random variable α_i, where the maximum fold-difference in $P(s_{it} = 1|d_{it} = 1)/P(s_{it} = 1|d_{it} = 0)$ varies from 1.5 (a1) to 2 (a2) to 2.5 (a3). D. Power analysis based on 3 scenarios of simulated covariate data varying from null (p1) to strongest association (p3) with true cluster assignments.

4. Analysis of the AMHS data using PIC

Clustering analysis was performed for several data types, including daily symptoms and daily self-reports of asthma triggers on air quality, heat, and pollen. Study participants were first clustered into subtypes using daily symptom data collected by the AMHS. To further characterize these subtypes, we tested for associations between predicted cluster assignments and clinical variables (age of diagnosis, GINA control level, smoking status, and weight), demographic variables (gender, income, and ethnicity), as well as self-reported trigger data collected by our app (pollen, heat, and air quality). Tests of association were performed using Fisher's exact tests, where we filtered out categories with fewer than 10 individuals where applicable. Supplemental Table 1 summarizes these results (http://icahndigitalhealth.org/wp-content/uploads/2016/08/Clustering-Supplemental-Data.pdf).

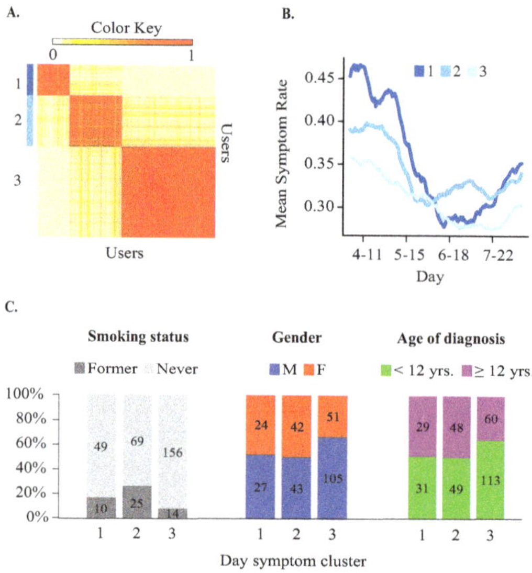

Figure 3. A. Heatmap is based on the adjacency matrix derived from consensus clustering of daily asthma symptoms for 334 users over 136 days using strategy PIC based on 100 iterations of imputation. Three distinct clusters ($n_1 = 60$, $n_2 = 98$, and $n_3 = 176$) are identified by color and enumeration (1-dark, 2-medium, and 3-light blue) where pairs of users most frequently found in the same cluster are found in the red regions along the diagonal. B. Mean curves for the clusters are based on the average of smoothed-imputed data on asthma symptoms. Each curve shows the mean symptom rate for users belonging to each cluster. Clusters are color-coded from dark to light blue by the overall mean symptom rate for each cluster. C. Day symptom clusters are significantly associated with smoking status ($p = 0.0005$), gender ($p = 0.02$), and age of diagnosis ($p = 0.03$) based on Fisher's exact test with simulated p-values based on 2,000 replicates. Barplots show the percentage distribution for each category within each day symptom cluster.

To conduct clustering analysis, we considered daily survey data collected by the app over the 6-month period from March 9, 2015 through August 9, 2015. We restricted our analysis to nonsmokers, defined as either having never smoked or having smoked less than 10 packs per year, without congestive heart failure or lung diseases other than asthma. We further required that each user have at least 50 survey responses over the entire 6-month period. These filterings led to 334 users in total. To ensure adequate overlap among enrollment periods across these users for comparing among methods in our simulation studies, we restricted our analysis to the 136-day period from April 2, 2015 (early spring) to August 8, 2015 (late summer), which corresponds to 136 days in total.

Based on daily survey data from 334 users over a period of 136 days, the average number of surveys provided per user was 70 (SD = 25), with an average per user enrollment period of 109 days (SD = 25). The average within enrollment missing rate was .4 (SD = 0.2). Clustering on the daily asthma symptom data was performed using the PIC strategy. After running the PIC method separately using different cluster numbers ranging from 2 to 5, we determined that users were well grouped into 3 clusters based on visual comparison of heatmaps derived from the adjacency matrices produced during the consensus clustering step of each run (Fig. 3A). Mean curves based on the average symptom rate for the users belonging to each of these clusters is shown in Figure 3B based on the average of the smoothed imputed data across 100 iterations of imputation, where curves are color-coded from dark to light blue to identify clusters with high, middle, and low symptom rates based on averaging across days.

We first sought to characterize our derived day symptom subtypes by comparing them with clinical and demographic variables. We found a significant association between asthma symptoms and smoking status (Fisher's exact test: $p = 5e\text{-}4$; $n = 333$), gender (Fisher's exact test: $p = 0.02$; $n = 292$), and age of diagnosis (Fisher's exact test: $p = 0.03$; $n = 330$). To study the relationship between asthma subtypes and environmental triggers, we used a similar approach to cluster self-reported data on daily asthma triggers collected by the AMHS. In the daily survey, participants were asked to self-report on symptom triggers on a given day. Specifically, users were able to choose from a list of 22 known asthma triggers, including allergens such as pollen, pet dander, and weather conditions. We chose to focus our analysis on air quality, heat, and pollen trigger data based on results from previous validation efforts comparing trigger data with more objective measures (PM2.5, max daily temperature, and pollen counts) using publicly available datasets[1].

Triggers were coded as 0/1 depending on whether a user cited a given trigger on a given day. Although we know that missing data in symptom reports were not random, we have little basis for attributing non-reported symptoms to one trigger over another with greater probability. Therefore, in conducting missing data imputation for trigger data, we used PIC($\alpha_i = 1$). Heatmaps resulting from the application of this method are shown in Supplemental Figure 1A-C (http://icahndigitalhealth.org/wp-content/uploads/2016/08/Clustering-Supplemental-Data.pdf).

Based on these groupings, self-reported asthma triggers were associated with the day symptom cluster groupings. Specifically, with Fisher's exact test, we found highly significant associations between day symptom clusters and clusters derived from self-reported data on pollen ($p = 5e\text{-}4$; $N = 333$), heat ($p = 5e\text{-}4$; $N = 333$), and air quality ($p = 0.02$; $N = 333$) triggers. As expected, we found a significant association between heat and US climate regions[13] broken down by northern and southern regions (Supplemental Table 2), with users belonging to cluster H1, who reported peak heat trigger complaints in late July, more frequently located in the northern US climate regions (72%) ($p = 0.01$; $N = 288$). We found that asthma trigger clusters differentiated by asthma subtype such that users who complain most frequently of pollen and heat are most frequently found in day symptom cluster 1, corresponding to the group with the highest average day symptom levels (Fig. 4A-B). By contrast, individuals frequently citing air quality as their asthma trigger are more frequently found in cluster 3, corresponding to the lowest overall day symptom rate.

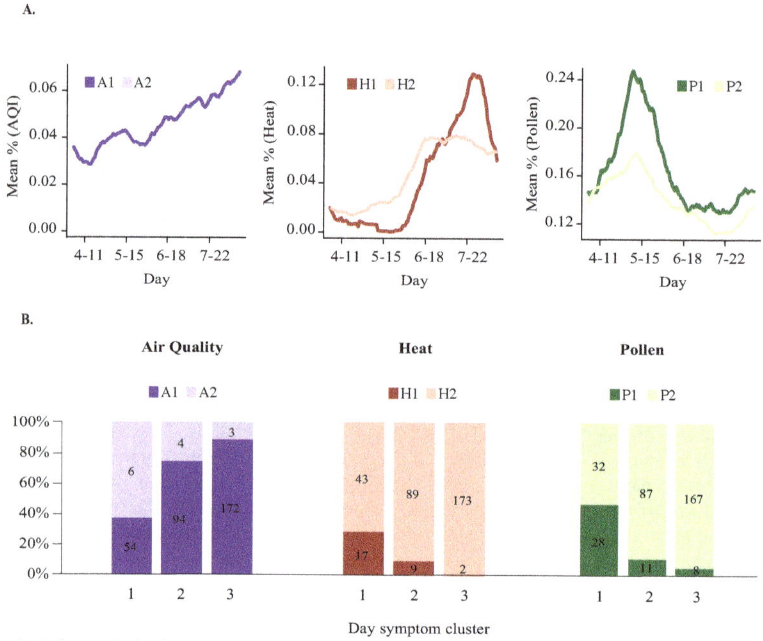

Figure 4. A. Curves depict the mean percentage of users reporting air quality, heat, and pollen for each cluster derived from the application of PIC($\alpha_i = 1$) using 100 multiple imputations. Clusters are color-coded from dark (high) to light (low) according to the overall mean percentage for each cluster averaged across days. B. Day symptom clusters are significantly associated with trigger clusters for air quality (p = 0.02, N = 333), heat (p = 5e-4, N = 333), and pollen (p = 5e-4, N = 333), based on Fisher's exact test with simulated p-values based on 2,000 replicates.

5. Discussion

Here we have considered the problem of clustering time series data collected from mobile health apps in which there is a high proportion of missing data for which the missing data mechanism is at least partially known. For such cases, regular clustering methods cannot be applied directly. To bridge this gap, in this paper, we developed an integrated PIC strategy to both impute the missing data using a probabilistic model and then clustered samples to identify subgroups with distinct patterns. The advantage of our PIC approach over other strategies based on low-rank matrix completion is demonstrated through extensive simulation studies.

When applying PIC on the AMHS data, we identified a unique subgroup of patients who have relatively high symptom rates and are more sensitive to distinct environmental factors with seasonal changes, such as heat and pollen. Furthermore, we noted relatively lower reported symptom rates associated with air quality, which may be attributed to the multi-factorial, reduced variability, and less well defined nature of this asthma trigger. With further validation, the ability to identify unique disease patterns in data sets with non-random missing data could be extremely useful in the conduct of environmental epidemiologic research as it could be used to track and identify novel environmental risk factors linked to worsening asthma. Moreover, it could enable us to identify at risk populations in large data sets and design targeted interventions to apply to reduce risk and improve outcomes. The ability to monitor asthma symptoms longitudinally by mobile technology, and identify specific subgroups of patients who have destabilization of asthma control based on specific triggers creates the opportunity to intervene early therapeutically. For example, if high heat or high pollen conditions are identified using personalized reports available by mobile technology, personalized alerts regarding presence of triggers would allow patients to seek medical advice and potentially adjust therapy in order to avoid the need for urgent care. R code implementing PIC (probability based imputation and consensus clustering) can be found here: http://icahndigitalhealth.org/wp-content/uploads/2016/10/PIC.R.

6. References

1. Chan, Y.-F.Y., et al., *The Asthma Mobile Health Study, a Large Scale Clinical Study Using ResearchKit.* Nature Biotechnology, submitted., 2016.
2. *Asthma-Data, Statistics, and Surveillance: Center for Disease Control and Prevention* 2015.
3. *GINA guidelines: Global Initiative for Asthma.* 2016.
4. Gauthier, M., A. Ray, and S.E. Wenzel, *Evolving Concepts of Asthma.* American Journal of Respiratory and Critical Care Medicine, 2015. **192**(6): p. 660-668.
5. Kaminsky, D.A., *Systems biology approach for subtyping asthma; where do we stand now?* Current opinion in pulmonary medicine, 2014. **20**(1): p. 17-22.
6. Chung, K.F., *Defining phenotypes in asthma: a step towards personalized medicine.* Drugs, 2014. **74**(7): p. 719-728.
7. Rubin, D., *Inference and missing data.* Biometrika 63 (3), 581-592, 1976.
8. Rubin Donald, B., *Multiple imputation for nonresponse in surveys.* 1987, New York: Wiley.
9. EJ Candès, B.R., *Exact matrix completion via convex optimization.* Foundations of Computational Mathematics 9 (6), 717-772.
10. Filkov, V. and S. Skiena, *Integrating microarray data by consensus clustering.* International Journal on Artificial Intelligence Tools, 2004. **13**(04): p. 863-880.
11. Hastie, T. and R. Mazumder, *softImpute: Matrix Completion via Iterative Soft-Thresholded SVD.* R package version, 2015. **1**.
12. Hubert, L. and P. Arabie, *Comparing partitions.* Journal of classification, 1985. **2**(1): p. 193-218.
13. Karl, T. and W.J. Koss, *Regional and national monthly, seasonal, and annual temperature weighted by area, 1895-1983.* 1984: National Climatic Data Center.

A NEW RELEVANCE ESTIMATOR FOR THE COMPILATION AND VISUALIZATION OF DISEASE PATTERNS AND POTENTIAL DRUG TARGETS

MODEST VON KORFF
Research Information Management, Actelion Pharmaceuticals Ltd., Gewerbestrasse 16
Allschwil, 4123, Switzerland
Email: modest.korff@actelion.com

TOBIAS FINK
Research Information Management, Actelion Pharmaceuticals Ltd., Gewerbestrasse 16
Allschwil, 4123, Switzerland
Email: tobias.fink@actelion.com

THOMAS SANDER
Research Information Management, Actelion Pharmaceuticals Ltd., Gewerbestrasse 16
Allschwil, 4123, Switzerland
Email: thomas.sander@actelion.com

A new computational method is presented to extract disease patterns from heterogeneous and text-based data. For this study, 22 million PubMed records were mined for co-occurrences of gene name synonyms and disease MeSH terms. The resulting publication counts were transferred into a matrix $\mathbf{M}_{data}$. In this matrix, a disease was represented by a row and a gene by a column. Each field in the matrix represented the publication count for a co-occurring disease–gene pair. A second matrix with identical dimensions $\mathbf{M}_{relevance}$ was derived from $\mathbf{M}_{data}$. To create $\mathbf{M}_{relevance}$ the values from $\mathbf{M}_{data}$ were normalized. The normalized values were multiplied by the column-wise calculated Gini coefficient. This multiplication resulted in a relevance estimator for every gene in relation to a disease. From $\mathbf{M}_{relevance}$ the similarities between all row vectors were calculated. The resulting similarity matrix $\mathbf{S}_{relevance}$ related 5,000 diseases by the relevance estimators calculated for 15,000 genes. Three diseases were analyzed in detail for the validation of the disease patterns and the relevant genes. Cytoscape was used to visualize and to analyze $\mathbf{M}_{relevance}$ and $\mathbf{S}_{relevance}$ together with the genes and diseases. Summarizing the results, it can be stated that the relevance estimator introduced here was able to detect valid disease patterns and to identify genes that encoded key proteins and potential targets for drug discovery projects.

I. Introduction

Many diseases coexist in a biological context with other diseases [1]. Patients often suffer from more than one disease. Furthermore, it is well known that some diseases cause secondary diseases. A well-established example for a disease with many co-occurring and second-order diseases is diabetes mellitus [2]. In addition, the context of a disease is of crucial importance in drug discovery. The ultimate goal of any new project in drug discovery is to treat or cure the disease in question. Realistically, however, for truly innovative drug discovery projects, in which the selected targets have been only recently identified, only sparse knowledge is available about the relationship between the chosen target and the potential disease. Knowledge about the context of the disease in which the target is involved may help to decide which constellation of diseases to take into account. Later in the drug discovery process, coexisting diseases should be considered for toxicity and DMPK studies. Drugs that are used for the treatment of pre-existing conditions may influence the metabolism of the patient and may result in an interaction with the drug being tested.

Whereas earlier research approaches for studying coexisting diseases were mainly based on phenotypic observations, recent technical advances have paved the way for using genomic and proteomic data. In this context, the "Online Mendelian Inheritance in Man"(OMIM) database was first published as a book and later as an electronic database [3]. This database related genotypes to phenotypes and inspired several research groups to develop computational tools to derive disease–disease associations. One of the first disease–disease association tools was reported by Goh et al. [4]. They developed a human disease network, connected by genes that were associated with two diseases. An enrichment of disease candidate genes via text mining of OMIM descriptions was implemented by van Driel et al. [5]. MeSH (Medical Subject Headings) annotations of MEDLINE articles were analyzed by Liu et al. to extract genetic and environmental factors associated with certain diseases [6]. An overview of the current research in disease associations was recently published by Sun et al. [7]. Hidalgo et al. derived the "Phenotypic Disease Network"(PDN) from 32 million patient records and received about six million co-morbidity relations [8]. This disease association network contains the co-morbidity data for more than 10,000 ICD9-encoded diseases and is one of the largest known so far. Taken together, a review of the existing literature suggests that multiple approaches exist to derive disease associations. However, MEDLINE represents the largest source of data, but it has not been used exhaustively for deriving disease–disease associations so far.

The approach presented here—the Disease–Disease Relevance Miner DDRelevanceMiner—annotated all records in MEDLINE with gene names and MeSH terms. Disease–disease associations were derived by comparing gene name based word vectors. These word vectors are histograms which are extracted from the records in PubMed by mining for co-occurrences of gene names and disease terms. This technique is known as second order co-occurrence and has been used by Schütze for word sense discrimination [9]. He exploited the fact that similar words are often accompanied by groups of identical words. Second-order co-occurrence has the advantage, as it allows calculating the similarity between two words that do not co-occur frequently, but that co-occur with the same neighboring words. Our method differs from Schütze's approach in two ways. The DDRelevanceMiner creates a word vector for a disease term from the complete text corpus and not from a single record. Word vectors, which represent a single text record, are often normalized and then weighted by the inverse document frequency. Weighing by the inverse document

frequency is not feasible for DDRelevanceMiner, because each word vector contains counts from all text records. To overcome this problem we introduced the relevance estimator.
In the next section, the calculation of the relevance estimator is explained in detail. Additionally, the data are described which were used to feed the algorithm and the diseases which were chosen for a thorough test of the results. The presentation and a discussion of the results follow at the end of the manuscript.

II. Methods

A detailed description of the algorithms used by the precursor of DDRelevanceMiner—DDMiner—has been recently published [10]. Here, we describe the new algorithm which significantly improved the results of DDMiner. The new algorithm calculates a relevance estimator for a gene in dependency to a disease. How can one assess the merits of the relevance estimators? We assumed that ranking genes by their relevance estimators should help identifying potential drug targets. We also assumed that calculating similarities between diseases based on the relevance estimators should group these diseases in a meaningful way. Finally, if the relevance estimators are applied to a well-studied disease, it should be possible to prove the importance of the top-ranked genes and the disease patterns by literature. The relevance estimator is loosely related to the scoring scheme that was recently published by Mørk et al. [11].

A. Description of DDRelevanceMiner

DDRelevanceMiner used for analysis all available gene names from a table provided by the HUGO Gene Nomenclature Committee (HGNC) [12]. Gene name synonyms were taken from the HUGO table and other public available sources [13;14]. Every synonym was checked for ambiguity. Every synonym that passed the check was used to form a query for PubMed. A successful query retrieved a number of PubMed records. Index, title, and abstract of the record were searched for disease MeSH terms. All found disease MeSH terms were labeled with the approved gene symbol that was linked to the PubMed record. A detailed description of the search algorithm for gene and disease terms has been previously described in [10].
Querying PubMed with all gene name synonyms and parsing all retrieved records with all disease MeSH terms resulted in the central data matrix $\mathbf{M}_{data}$. A row in the matrix stood for a disease MeSH term and a column – for an approved gene symbol. Each field in the matrix, indicated by a row and a column, contained an integer number indicating how often a disease MeSH term occurred together with a gene. A row m from $\mathbf{M}_{data}$ shows which genes were studied together with the disease m. Vice versa, a column n shows which diseases were reported together with gene n.
However, the pure count of disease–gene co-occurrences is only of limited benefit. Genes with a high frequency of occurrence in the medical literature are often studied in relation to many diseases. But pharmaceutical research is mostly interested in genes that are specific for the disease of interest. Most interesting are genes that are specifically mentioned together with a disease of interest and not together with other genes. A gene with a high number of occurrences with one disease and no mentioning together with other diseases could be assumed to have a high relevance for the disease. Column n was extracted from $\mathbf{M}_{data}$ to calculate the relevance estimator $r_{m,n}$ for a disease with index m and a gene with index n. From column n, only fields with a publication count >0 were considered. For all fields in the column, their rank fraction $f_{m,n}$ was calculated. The rank ρ of a disease m for gene n is the position of the disease after sorting column n according to the number of publications. Diseases with identical publication counts were assigned the same rank.

Consequently, the number of ranks can be smaller than the number of diseases that were mentioned together with gene n. The rank fraction equaled one minus the rank divided by the total number of ranks $f_{m,n}= 1-\rho/\theta$, with θ for the total number of ranks. A fraction of publications $p_{m,n}$ was calculated by dividing the number of publications for disease m found in column n by the sum of all publications for gene n. The fraction of publications for disease m was weighted with the relative rank by $w_{m,n}= p_{m,n} f_{m,n}$. Finally, the relevance estimator $r_{m,n}$ was calculated by multiplying $w_{m,n}$ by the Gini coefficient g_n. The Gini coefficient describes the statistical dispersion for a group of values [15]. A Gini coefficient close to one indicates that all values except one in the group are zero. A Gini coefficient of zero indicates that all values in the group are equal. A relevance estimator $r_{m,n}$ of one is obtained if all three factors in the equation $r_{m,n}= p_{m,n} f_{m,n} g_n$ are equal to one. This means that all publications for gene n refer only to disease m. The calculation of the relevance estimator was done for all fields in the matrix $\mathbf{M}_{data}$. As result, a new matrix $\mathbf{M}_{relevance}$ was obtained, with the same dimensions as the input matrix. This matrix contained the relevance estimators; they covered a range between zero and one. This normalization enabled the meaningful comparison of matrix rows. To reduce the risk of rounding errors and to cut off the influence of very small values, the relevance estimators were multiplied by a factor of 1,000 and converted into integer numbers. Each two $\mathbf{M}_{relevance}$ matrix rows were compared by calculating the generalized Jaccard similarity coefficient. Comparison of two matrix rows gave a similarity value between two diseases. The resulting similarity matrix $\mathbf{S}_{relevance}$ contained the similarity between all diseases.

B. Data

1) Genes and disease MeSH terms

A total of 39,410 approved gene and protein symbols were retrieved from the HUGO table. At least one disease MeSH term was found for 15,203 approved gene symbols. This number defined the number of columns in $\mathbf{M}_{data}$ and $\mathbf{M}_{relevance}$. A number of 5,256 unique MeSH descriptors defined the number of rows in $\mathbf{M}_{data}$ and $\mathbf{M}_{relevance}$.

2) Example diseases to assess the quality of the relevance estimators

Three example diseases, type 2 diabetes mellitus (T2DM), melanoma, and vitiligo were chosen for a detailed analysis of the disease–gene and the disease–disease associations. For each of the three diseases, five genes with the highest relevance estimators—the top genes—and the equal number of genes with the highest publication counts were analyzed. Additionally, for each of the three example diseases, ten most similar diseases were evaluated. Based on the working hypotheses described at the beginning of the Methods section, the following success criteria were defined. The usage of the relevance estimator can be regarded as a success if the top five genes include disease relevant genes. These genes should not be related to numerous other diseases. A relation of an example disease to a similar disease was regarded as valid if there was evidence found in literature.

T2DM is a subtype of diabetes and a complex metabolic disease. It is a complex disease because it involves environmental factors and multiple genes [16;17]. Despite the fact that many anti-diabetic drugs are on the market, the need for new anti-diabetic drugs is still high [18]. Obesity is a dominant risk factor for T2DM, whereas hypertension is one of the main co-occurring diseases. Therefore, we expected to see at least these two disease MeSH terms among the results of the similarity analysis. A number of genetic-driven studies were done for T2DM. Specific genes were found that increase the risk for this disease. Will the relevance estimator be able to identify some of these genes?

Melanoma is a malignant neoplasm (cancer) of the skin and the leading cause of death due to skin disease [19]. It is considered a highly immunogenic tumor [20]. Consequently, we expected to see association between melanoma and the genes that are tumor related but also with the genes that have relevance for the immune response.

Vitiligo is a disease where parts of the skin lose their pigment. Vitiligo is a frequently occurring disease, seen in 0.2%–2% of the population. However, its cause is still unknown [21]. A reason to choose vitiligo as an example disease was the relatively small number of related publications, compared to T2DM and melanoma. Another reason was the existing link between melanoma and vitiligo [22]. Will the disease similarity analysis based on the relevance estimator be able to find the link between these two diseases?

III. RESULTS

After querying PubMed with the synonyms from 39,410 approved gene and protein symbols, 2.7 million unique PubMed records were retrieved. Each of these records contained at least one disease MeSH term together with an unambiguous gene name synonym. The number of rows in $\mathbf{M}_{data}$ and $\mathbf{M}_{relevance}$ were defined by 5,256 unique disease MeSH terms found in the above publications, and the number of columns – by the 15,203 approved symbols. The similarity matrix $\mathbf{S}_{relevance}$ was calculated as described above in Methods. All results described in the following paragraphs were taken from $\mathbf{M}_{data}$, $\mathbf{M}_{relevance}$ and $\mathbf{S}_{relevance}$. The results for all genes and diseases are freely accessible at http://gene2disease.org.

A. *Results for the three example diseases*

Table 1 summarizes the disease–gene associations for the three example diseases. The table contains the disease name, the number of publications for the disease, the total number of genes found in the publications, and the top ten approved gene symbols. In the fourth column, the approved gene symbols are sorted by the relevance estimator from $\mathbf{M}_{relevance}$. In the fifth column, gene symbols are sorted by their publication counts from $\mathbf{M}_{data}$.

For every disease, the ten most similar diseases were extracted from $\mathbf{S}_{relevance}$. Cytoscape was used to visualize the results [23]. A Cytoscape network was created for every example disease (Figs. 1–3). A Cytoscape sketch for a disease contains the results from the corresponding row of Table 1 and from the corresponding table with the similar diseases. Every gene in a sketch is connected to the example disease. A disease is never directly connected to another disease and a gene is never connected directly to another gene. The example disease is marked by white text on the label. Other diseases are depicted as rectangles with black text on the label. The background color of the disease label corresponds to the similarity with the example disease. Similar diseases have a more intense background color than less similar ones. Genes are represented by ellipsoids or diamond shapes. The width of each shape corresponds to the number of connected diseases. A diamond shape symbolizes a gene that is in the top-five list of the genes sorted by relevance in Table 1. An ellipsoid indicates a gene that has similar relevance for two or more diseases, including an example disease, and is among the top five genes for one of the similar diseases.

1) *Type 2 diabetes mellitus*

DDRelevanceMiner found about 31,000 publication records relevant to T2DM which collectively mentioned synonyms for 2,273 genes (Table 1). All five genes with the highest relevance estimators have demonstrated relevance to T2DM. Namely, TCF7L2 is a diabetes susceptibility gene of substantial importance [24], and a potential target for anti-diabetic drugs [25]. GCK (glucokinase) plays an important role in the carbohydrate metabolism and is an attractive target for new drugs [18], whereas GCK mutations are known to cause diabetes. CAPN10, SLC5A2, and SLC30A8 have high relevance for T2DM and are potential drug targets.

Ranking of the same set of genes by publication count differed completely from that by relevance estimators. Four out of the five genes with the highest publication counts, GCG (glucagon), INS (insulin), HBA1 (hemoglobin) and DIANPH, had low relevance estimators. This means that they were mentioned together with many other diseases, i.e., are not specific for T2DM. Low relevance of INS to T2DM was expected, because T2DM is a non–insulin-dependent disease [26]. In contrast, the fifth gene, DPP4 (dipeptidyl peptidase-4), had both a high publication count and a relatively high relevance estimator and is, indeed, a proven drug target for treating T2DM [27].

Table 1. Three selected example diseases and top relevant genes.

Disease	Publications	Gene count	Approved gene symbol (relevance estimator, publications) Top five genes sorted by relevance	publication count
Type 2 diabetes mellitus	31,024	2,273	TCF7L2 (0.140, 386) GCK (0.138, 767) CAPN10 (0.132, 106) SLC5A2 (0.123, 210) SLC30A8 (0.120, 99)	GCG (0.053, 3761) INS (0.036, 3465) HBA1 (0.090, 2374) DIANPH (0.040, 2352) DPP4 (0.113, 1937)
Melanoma	27,000	3,271	MAGEA11(0.236, 14) TYR (0.209, 2357) PMEL (0.206, 23) MIA (0.202, 138) DCT (0.196, 253)	TYR (0.209, 2357) IL2 (0.018, 2191) IFNA1 (0.010, 2007) IFNG (0.010, 1867) IFNA2 (0.010, 1261)
Vitiligo	1,608	380	PCBD1 (0.033, 3) DCT (0.017, 28) PMEL (0.016, 3) LRR1 (0.015, 4) TYR (0.013, 170)	TYR (0.013, 170) CAT (0.0004, 70) IFNG (0, 52) IL2 (0, 49) IFNA1 (0, 46)

Table 2 lists the disease MeSH terms most similar to T2DM based on $\mathbf{S}_{\text{relevance}}$. All ten MeSH terms are known as major co-occurring diseases with diabetes, symptoms of diabetes, or, in case of obesity and body weight, known risk factors for the disease [28]. The size of the common gene set linking each disease with T2DM ranged from 72 to 517 genes. The 'top five genes' column of Table 2 shows genes whose relevance estimators calculated for the disease/MeSH term were most similar to those calculated for T2DM. For this reason, the genes and their order is different from Table 1 showing genes most relevant to T2DM only.

In Table 2, the disease MeSH terms most similar to T2DM are listed. The similarity values were taken from $\mathbf{S}_{\text{relevance}}$. All ten MeSH terms are major co-occurring diseases with diabetes, symptoms of diabetes, or, in case of obesity and body weight, known risk factors for the disease [28]. The size of the common gene set linking each disease with T2DM ranges from 72 to 517 genes. The top five genes are the genes with the relevance estimators that are most similar to those calculated for T2DM. For this reason, there is a difference in the sorting order by the relevance estimators compared with Table 1, where the top genes were those most specific for one disease. Genes that connect more than one disease with T2DM are, e.g., SLC2A4, GCK, PLTP, and APOB. These genes may be regarded as having a key role in the disease pattern. The importance of these key genes is visualized in Fig. 1 where the width of the gene nodes corresponds to the number of connections.

Table 2. Diseases most similar to type 2 diabetes mellitus.

Disease/MeSH term	Similarity	Size of the common gene set	Top five genes
Insulin Resistance	0.81	484	SLC2A4, IRS1, INSR, CAPN10, IRS2
Hyperglycemia	0.79	253	SLC5A2, GCK, PDX1, GCGR, HBA1
Obesity	0.72	517	ADIPOQ, GIP, ADRB3, FTO, GLP1R
Body Weight	0.70	286	GCK, IAPP, SLC2A4, GLP1R, UCP1
Hyperinsulinism	0.66	145	ABCC8, KCNJ11, GCK, SLC2A4, INSR
Hypoglycemia	0.61	72	SLC5A2, DPP4, GLP1R, ABCC8, HBA1
Atherosclerosis	0.58	243	PLTP, PON2, CETP, APOB, APOA1
Dyslipidemias	0.57	122	CETP, APOB, PPARA, PLTP, HMGCR
Hyperlipidemias	0.56	122	VLDLR, APOB, LPL, LIPC, PLTP
Hypertension	0.53	281	DIANPH, HBA1, ACE, ADRB3, REN

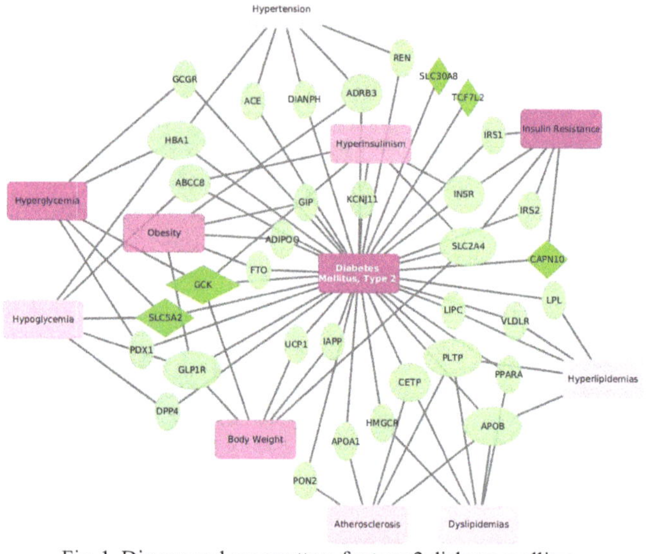

Fig. 1. Disease and gene pattern for type 2 diabetes mellitus

Analysis of gene-disease pattern for T2DM showed that SLC5A2 gene connects T2DM with two out of ten diseases with the highest similarity to T2DM (Fig. 1), which makes SLC5A2 an interesting drug target [29]. CAPN10 connects T2DM with insulin resistance, which is one of its known pre-conditions [30]. Other genes that connect T2DM with more than one disease (e.g., SLC2A4, GCK, PLTP, APOB), may be regarded as having a key role in the gene-disease pattern. The importance of these key genes is highlighted in Fig. 1 by the width of the gene nodes that is proportional to the number of disease connections.

2) Melanoma

Melanoma is a malignant cancer of the skin. The development of cancer includes many genes for cell growth and proliferation. Tyrosinase (TYR) is the gene with the highest publication count, and the top second rank according to the relevance estimator calculated for melanoma. Tyrosinase plays a central role in the process of skin pigmentation. Next four genes with high publication counts, interleukin 2 (IL2) and three interferons, are important for the immune response against cancer but are not specific for melanoma. The lack of specificity is the reason why these genes have low relevance estimators. MAGEA11 is the gene with the highest relevance estimator in the complete examination. Indeed, it is a melanoma antigen. Other genes with high relevance estimators, PMEL, MIA, and DCT, are highly specific for melanoma and are in the focus of ongoing research [31] [32] [33].

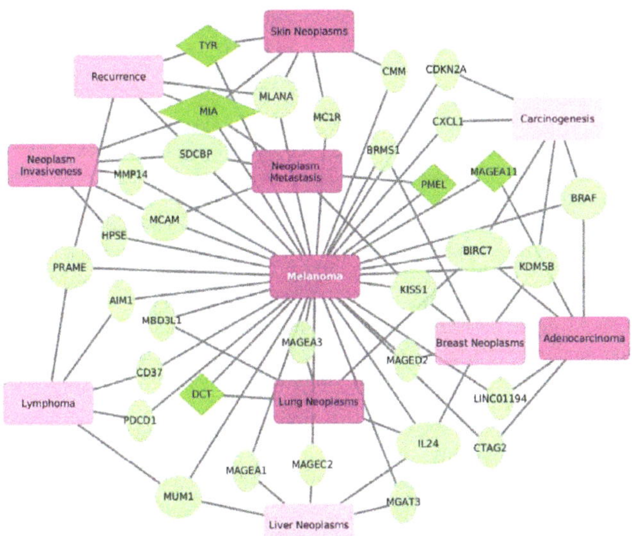

Fig. 2. Disease and gene pattern for melanoma.

The list of similar diseases in Table 3 is determined by diseases generally related to skin cancer and other cancer types. As already mentioned, the number of genes involved in cancer is higher than for other diseases. In the similarity list with the top five genes MIA the only gene that is represented four times. MIA encodes the melanoma-derived growth regulatory protein. The combination of high relevance estimator and the connection of three melanoma-related MeSH terms suggest that MIA is a strong candidate for a drug discovery project.

Table 3. Diseases most similar to melanoma.

Disease/MeSH term	Similarity	Size of the common gene set	Top five genes
Neoplasm Metastasis	0.65	1028	SDCBP, MIA, MCAM, KISS1, PMEL
Lung Neoplasms	0.65	622	MAGEA3, IL24, DCT, BIRC7, MBD3L1
Skin Neoplasms	0.65	366	MC1R, MIA, CMM, MLANA, TYR
Adenocarcinoma	0.63	728	LINC01194, BIRC7, MAGEA11, CTAG2, BRAF
Neoplasm Invasiveness	0.63	508	SDCBP, MIA, MCAM, HPSE, MMP14
Breast Neoplasms	0.60	788	BRMS1, IL24, KDM5B, MAGED2, KISS1
Recurrence	0.59	541	PRAME, MIA, MLANA, SDCBP, TYR
Lymphoma	0.59	581	CD37, PRAME, MUM1, AIM1, PDCD1
Liver Neoplasms	0.58	548	MAGEC2, MGAT3, MUM1, MAGEA1, IL24
Carcinogenesis	0.58	954	KDM5B, CDKN2A, CXCL1, BIRC7, BRAF

3) Vitiligo

Much less is known about vitiligo than for the other two example diseases. The absence of any gene with a high relevance estimator for vitiligo indicates a comparative lack of research.

Table 4 shows the most similar diseases to vitiligo. Coinciding with the low number of publication counts is the small size of the common gene sets. Nevertheless, some genes show multiple connections in the disease–gene network shown in Fig. 3. Tyrosinase has the most connections by linking nine diseases. Dopachrome tautomerase (DCT) connects seven diseases and is one of the most relevant genes for vitiligo. Also seven diseases are connected by the MITF gene encoding melanogenesis associated transcription factor, but this gene is not part of the top relevance genes. PMEL and TYRP1 genes connect six and five diseases, respectively. Fig. 3 shows that all four genes with the most connections (TYR, DCT, MITF, PMEL) relate vitiligo to the same three disease MeSH terms within the skin cancer complex: hypopigmentation, hyperpigmentation and skin neoplasms.

Table 4. Diseases most similar to vitiligo.

Disease/MeSH term	Similarity	Size of the common gene set	Top five genes
Hyperpigmentation	0.72	7	TYR, DCT, MITF, PMEL, TYRP1
Melanosis	0.69	5	TYR, ASIP, LGI3, MITF, MC1R
Melanoma, Experimental	0.63	8	DCT, TYR, PMEL, MITF, TYRP1
Microphthalmos	0.51	10	DCT, TYR, MITF, PMEL, ASIP
Hypopigmentation	0.49	3	TYR, DCT, MITF

Disease/MeSH term	Similarity	Size of the common gene set	Top five genes
Melanoma, Amelanotic	0.49	3	TYR, DCT, MLANA
Epilepsy, Partial, Sensory	0.48	1	LGI3
Skin Neoplasms	0.43	10	TYR, DCT, PMEL, STX17, MITF
Arthritis, Juvenile	0.42	3	PTPN22, NLRP1, PTPN2
Albinism, Oculocutaneous	0.42	5	TYR, PMEL, TYRP1, GCHFR, MC1R

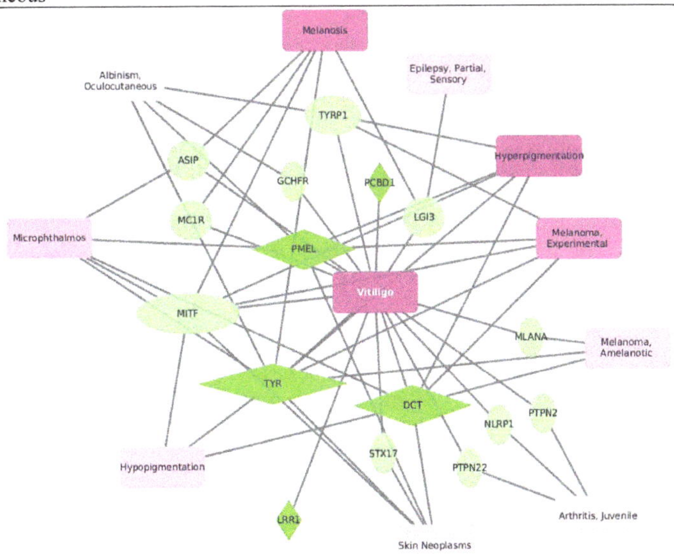

Fig. 3. Disease and gene pattern for vitiligo.

IV. Discussion

The majority of scientific publications contain information as heterogeneous data. And scientific publications are the main source for medical information. The NIH collected abstract records for millions of scientific publications in the PubMed database. It was our goal to extract and to visualize meaningful disease–gene and disease–disease patterns from this plethora of unstructured information. For this purpose, we introduced the relevance estimator described in this study. Three example diseases were chosen to examine this new figure of merit. For each disease, the relation to the most relevant genes was confirmed by literature. Furthermore, ten disease MeSH terms with disease–gene relationships most similar to one of the example diseases were identified and

analyzed. Cytoscape was used to visualize the most relevant genes, the similar diseases and the genes which connect the diseases with the example disease.
The most relevant genes for T2DM and melanoma were found to be highly specific for each disease. The ability of the relevance estimator to link MeSH terms with highly disease-specific genes that may only affect small patient groups makes it interesting for personalized medicine. The gene with the highest relevance estimator, TCF7L2, has been identified as a potential anti-diabetic drug target [25]. Obesity and hypertension were among the top ten disease MeSH terms with highest similarity to T2DM. This finding was pre-defined as a success criterion for the use of the relevance estimator.
For melanoma, a different disease–gene relationship pattern was obtained than for T2DM. All top genes are connecting at least one additional disease MeSH term with melanoma. Immune system relevant genes were listed as top genes by publication counts. However, these immune regulatory genes had low relevance estimators for melanoma. An example is interleukin 2, the protein product of the IL2 gene. This protein is used as a drug in the treatment of melanoma and is known to cause adverse side effects [34]. No genes with a high relevance estimator were found for vitiligo. Here, the combination of low publication counts and low relevance estimators emphasized that vitiligo is a disease with unknown genetic causes. Regardless of the low relevance score, three of the top five genes for vitiligo connected vitiligo to other skin-related disease MeSH terms. Thus, the earlier mentioned link between vitiligo and melanoma was confirmed using relevance estimators.
The evidence provided by the relevance estimators can be summarized as follows:

1. A high relevance estimator together with a low publication count indicates potential drug targets.
2. A low relevance estimator together with a high publication count indicate non-disease-specific genes.
3. A high relevance estimator and a high publication count mark a well-studied gene that is highly specific for the related disease.
4. Genes with low relevance estimators for a certain disease and high connectivity between multiple disease MeSH terms are likely to encode key proteins in a biochemical or signaling pathway.

Concluding, the relevance estimator is a valuable tool to extract disease-gene relation patterns from very large and heterogeneous data sets. Yet, the nature and importance of these patterns can only be evaluated by a scientist.

References

1 M. E. Charlson, P. Pompei, K. L. Ales, and C. R. MacKenzie, *J Chronic Dis* **40**, (1987).
2 K. G. Alberti, and P. Z. Zimmet, *Diabet Med* **15**, (1998).
3 A. Hamosh, A. F. Scott, J. Amberger, C. Bocchini, D. Valle, and V. A. McKusick, *Nucleic Acids Res* **30**, (2002).
4 K. I. Goh, M. E. Cusick, D. Valle, B. Childs, M. Vidal, and A. L. Barabasi, *Proc Natl Acad Sci USA* **104**, (2007).
5 M. A. van Driel, and H. G. Brunner, *Hum Genomics* **2**, (2006).
6 Y. I. Liu, P. H. Wise, and A. J. Butte, *BMC Bioinformatics* **10 Suppl 2**, (2009).
7 K. Sun, J. P. Goncalves, C. Larminie, and N. Przulj, *BMC Bioinformatics* **15**, (2014).
8 C. A. Hidalgo, N. Blumm, A. L. Barabasi, and N. A. Christakis, *PLoS Comput Biol* **5**, (2009).

9 H. Schütze, *Computational linguistics* **24**, (1998).
10 M. Von Korff, B. Deffarges, and T. Sander (2015). *In* "Computational Intelligence, 2015 IEEE Symposium Series on", p. 314. IEEE.
11 S. Mork, S. Pletscher-Frankild, A. Palleja Caro, J. Gorodkin, and L. J. Jensen, *Bioinformatics* **30**, (2014).
12 http://www.genenames.org
13 D. Maglott, J. Ostell, K. D. Pruitt, and T. Tatusova, *Nucleic Acids Res* **33**, (2005).
14 http://www.ncbi.nlm.nih.gov/gene
15 C. Gini, *Colorado College Publication, General Series* **208**, (1936).
16 L. Chen, D. J. Magliano, and P. Z. Zimmet, *Nat Rev Endocrinol* **8**, (2012).
17 R. Sladek, G. Rocheleau, J. Rung, C. Dina, L. Shen, D. Serre, P. Boutin, D. Vincent, A. Belisle, S. Hadjadj, B. Balkau, B. Heude, G. Charpentier, T. J. Hudson, A. Montpetit, A. V. Pshezhetsky, M. Prentki, B. I. Posner, D. J. Balding, D. Meyre, C. Polychronakos, and P. Froguel, *Nature* **445**, (2007).
18 P. Gaitonde, P. Garhyan, C. Link, J. Y. Chien, M. N. Trame, and S. Schmidt, *Clin Pharmacokinet* **55**, (2016).
19 M. A. Papadakis, S. J. McPhee, and M. W. Rabow (2015). *In* "Current Medical Diagnosis & Treatment 2015", p. 101. McGraw-Hill Education.
20 T. H. Nguyen, *Clin Dermatol* **22**, (2004).
21 A. Alkhateeb, P. R. Fain, A. Thody, D. C. Bennett, and R. A. Spritz, *Pigment Cell Res* **16**, (2003).
22 K. U. Schallreuter, C. Levenig, and J. Berger, *Dermatologica* **183**, (1991).
23 P. Shannon, A. Markiel, O. Ozier, N. S. Baliga, J. T. Wang, D. Ramage, N. Amin, B. Schwikowski, and T. Ideker, *Genome Res* **13**, (2003).
24 C. J. Groves, E. Zeggini, J. Minton, T. M. Frayling, M. N. Weedon, N. W. Rayner, G. A. Hitman, M. Walker, S. Wiltshire, A. T. Hattersley, and M. I. McCarthy, *Diabetes* **55**, (2006).
25 M. Ridderstrale, and L. Groop, *Mol Cell Endocrinol* **297**, (2009).
26 C. N. Hales, *Br Med Bull* **53**, (1997).
27 D. J. Drucker, and M. A. Nauck, *Lancet* **368**, (2006).
28 S. E. Kahn, R. L. Hull, and K. M. Utzschneider, *Nature* **444**, (2006).
29 A. Cesar-Razquin, B. Snijder, T. Frappier-Brinton, R. Isserlin, G. Gyimesi, X. Bai, R. A. Reithmeier, D. Hepworth, M. A. Hediger, A. M. Edwards, and G. Superti-Furga, *Cell* **162**, (2015).
30 L. J. Baier, P. A. Permana, X. Yang, R. E. Pratley, R. L. Hanson, G. Q. Shen, D. Mott, W. C. Knowler, N. J. Cox, Y. Horikawa, N. Oda, G. I. Bell, and C. Bogardus, *J Clin Invest* **106**, (2000).
31 F. Shi, Z. Xu, H. Chen, X. Wang, J. Cui, P. Zhang, and X. Xie, *Monoclon Antib Immunodiagn Immunother* **33**, (2014).
32 K. T. Yip, X. Y. Zhong, N. Seibel, S. Putz, J. Autzen, R. Gasper, E. Hofmann, J. Scherkenbeck, and R. Stoll, *Sci Rep* **6**, (2016).
33 S. A. Ainger, X. L. Yong, S. S. Wong, D. Skalamera, B. Gabrielli, J. H. Leonard, and R. A. Sturm, *Exp Dermatol* **23**, (2014).
34 C. Ma, and A. W. Armstrong, *J Dermatolog Treat* **25**, (2014).

NETWORK MAP OF ADVERSE HEALTH EFFECTS AMONG VICTIMS OF INTIMATE PARTNER VIOLENCE

KATHLEEN WHITING

Neuroscience Program, Uniformed Services University, 4301 Jones Bridge Rd, Bethesda, Maryland 20814, USA
Email: kathleen.whiting@usuhs.edu

LARRY Y. LIU

Center of Proteomics and Bioinformatics, Case Western Reserve University, 10900 Euclid Ave, Cleveland, Ohio 44106, USA
Email: lyl14@case.edu

MEHMET KOYUTÜRK

Department of Electrical Engineering & Computer Science, Case Western Reserve University, 10900 Euclid Ave, Cleveland, Ohio 44106, USA
Email: mxk331@case.edu

GÜNNUR KARAKURT

Department of Psychiatry, Case Western Reserve University, 10900 Euclid Ave, Cleveland, Ohio 44106, USA
Email: gkk6@case.edu

Intimate partner violence (IPV) is a serious problem with devastating health consequences. Screening procedures may overlook relationships between IPV and negative health effects. To identify IPV-associated women's health issues, we mined national, aggregated de-identified electronic health record data and compared female health issues of domestic abuse (DA) versus non-DA records, identifying terms significantly more frequent for the DA group. After coding these terms into 28 broad categories, we developed a network map to determine strength of relationships between categories in the context of DA, finding that acute conditions are strongly connected to cardiovascular, gastrointestinal, gynecological, and neurological conditions among victims.

1. Introduction

Domestic abuse is a rampant problem across the globe, contributing to severe economic, health related, and societal costs. The consequences of intimate partner violence (IPV) are devastating and systemic. In 2010, the National Intimate Partner and Sexual Violence Survey found that approximately 30% of women experience physical violence from an intimate partner during their lifetime, with 25% experiencing severe physical violence such as being slammed, hit, or beaten.[1] Although victims of IPV are not exclusively female, women are more likely than men to be the victim and sustain serious physical injury.[1,2]

IPV has been shown to cause numerous adverse health effects, ranging from minor injuries to serious disability and death.[2-4] Physical assault (including sexual violence) is associated with psychological distress such as anxiety, depression, and suicidal ideation,[5] sexually transmitted infections including HIV,[3,6] gynecological problems like pelvic inflammatory disease,[7] and unintended pregnancy and complications relating to the mother's and newborn's health.[3,8-10] Researchers have identified many long-term effects of IPV, finding evidence that victims of violence are more prone than the general populous to suffer from mental health and substance abuse disorders, gastrointestinal problems, chronic pain and physical ailments, and various neurological symptoms.[5,11] This information has not prompted further research into how professionals can prevent and treat the IPV epidemic. Holistic approaches incorporating comprehensive treatment of both physical and emotional ailments have received little attention.

Self-report data indicates that female victims of violence have poorer overall health than female victims of non-violent crimes (and women in general), presenting troubling physical symptoms like tachycardia, tension headaches, menstruation related issues, stomach problems, or skin disorders.[12] Intimate partner violence often involves episodes of physical and sexual violence. No doubt this is a contributing factor to poor victim health, and likely increases the use of healthcare services by victims of violence. For example, sexual assault victims are more likely to seek physical and mental health care within the first six months of the attack, with services increasing 15-24% during the first twelve months alone. Naturally, the corollary of this is a higher cost of health care and treatment for victims than non-victims. Emergency room records indicate un-witnessed episodes of head, neck, and facial injuries are significant markers of IPV,[13] and traumatic brain injury may be more prevalent in this population than previously suspected.[14] Unfortunately, physicians often overlook or misattribute problems associated with violence, which can result in prolonging victims' pain and wasting patient and provider resources. Proper screening and treatment of IPV is critical to ensure that victims of violence receive the necessary care and support for recovery.

While the effects of IPV are known to be serious and diverse, knowledge of specific health effects and their relation to IPV is still limited. In this study, we utilized electronic health records (EHR) to identify frequently occurring symptoms among IPV victims. Our approach is motivated by the notion that EHR data provide valuable information from health care providers that may not be obtained through self-report data. Furthermore, both self-report data and physicians' records are difficult to obtain in large amounts due to topic sensitivity. For these reasons, investigators struggle to compile available symptom data into comprehensive and systematic reviews. The consequences of violence on human health are elusive and complex, and therefore utilization of large-scale data can be useful in identifying correlates that are overlooked by other research. Here, we take a first step toward utilizing EHR data to characterize adverse health effects co-occurring with IPV, identifying statistical associations between IPV and other symptoms and determining the strength of these relationships. It is important to note that our analysis does not target any symptom in particular; rather we mine the entire EHR data (1999 through our original data query point 5/8/14) and test the association of all reported symptoms to identify those statistically significant.

In a previous study,[15] we accessed and analyzed national EHR data through the *Explorys* platform (Explorys Inc., an IBM company), specifically utilizing the "Explorys Enterprise Performance Management (EPM): Explore" web application to identify diseases which seem to be more prevalent among victims of IPV than the general US population. *Explorys* is comprised of EHR, EMR, insurance claims, and billing data sources. A variety of national data sources contribute data to the platform, including affiliated providers, electronic medical systems, health care plans, and care settings. Over twenty major integrated healthcare systems provide data to *Explorys*, bringing together patient information from across America. Over 300,000 providers participate, gathering more than 315 billion clinical, operational, and financial data elements from approximately 50 million unique patients. Data is pooled from clinical EMRs, healthcare system outgoing bills, and adjudicated payer claims. Researchers from a wide range of disciplines use this compiled data to identify patterns and trends in diseases, treatments, and outcomes.[16]

We hoped that our analysis of the data we obtained through the *Explorys* platform would help us differentiate between those health problems which result directly from acute violence and violence related physical injuries, and those health problems which are chronic or persistent and result from multiple non-violent causes. After identifying the diseases occurring significantly more frequently among victims of IPV, we categorized the diseases into 28 broad categories, and found that IPV is predominantly associated with four types of health problems: acute; chronic; gynecological; and mental/behavior health. The results further supported our suspicions that IPV is a systematic problem with multifaceted interactions across a wide range of health issues.

To develop a better understanding of how IPV is related to negative health effects, it is potentially useful to determine the interactions and relationships between symptom categories. Analyzing these relationships may help us discover what physiological systems are more closely associated with experiencing severe consequences of IPV, and could lead to future research into the effects of IPV on the body. For this study, we decided to perform a data-driven analysis and network mapping of the same significantly occurring diseases identified previously to reveal how these terms interact. We chose network mapping because it analyzes the structural relationships and patterns within a network of 'nodes', providing a visual representation of the strength of these relationships. In this case, the nodes are each symptom category, and the connections or 'edges' between these categories indicate how frequently those given categories appear together in our coded symptoms. Through this analysis we specifically hope to identify the strength of connections between different disease categories, in an effort to investigate how these associations may be related to each other. Through this analysis we can explore the many ways IPV affects the overall health of victims.

2. Methods

2.1. *Identification of Terms Prevalent among Domestic Abuse Victims*

The flow chart for the methodology implemented in this study is shown in Figure 1. A complete and detailed description of the data acquisition performed for this study can be found in a previously published manuscript.[15] Here, we provide a brief summary.

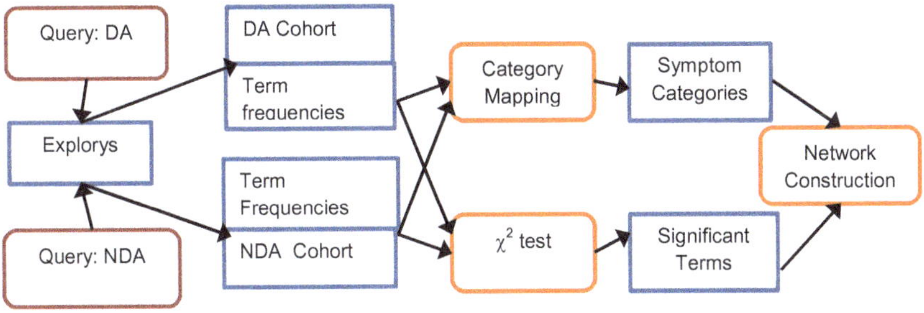

Fig 1. Flowchart for data acquisition, querying, statistical analysis, and network construction. DA: Domestic Abuse. NDA: Non-Domestic Abuse

We obtained data from a population of almost 15 million patients who were adult females aged 18-65 seen in multiple different healthcare systems across the United States with unique EHR from 1999 to the present (5/8/14: original query date). These data were normalized and classified using common ontologies, searchable through the HIPAA-enabled, de-identified "Explorys Enterprise Performance Management: Explore" web application. Using SNOMED clinical terms built into *Explorys*, our search query identified 5870 records (DA cohort) of IPV victims (these records were retrieved by searching for the finding 'domestic abuse', a code option utilized by health professionals for EHR), compared to 14,315,140 records (NDA cohort) of patients who did not have any indication of IPV victimization in their EHR. Racial and age distribution for the DA and NDA cohorts are shown in Figure 2.

It is important to note that in order to protect patient privacy, data is accessible only as frequencies across the cohorts defined by these queries. Similarly, demographic information is available as summaries. For this reason, sophisticated data mining techniques such as association rule mining are not directly applicable. Here, we base our analysis on the comparison of frequencies. Of note, African Americans make up a greater proportion of the DA group than NDA, and while NDA records are relatively evenly distributed across age groups, DA records show higher relative frequencies for ages 25-44.

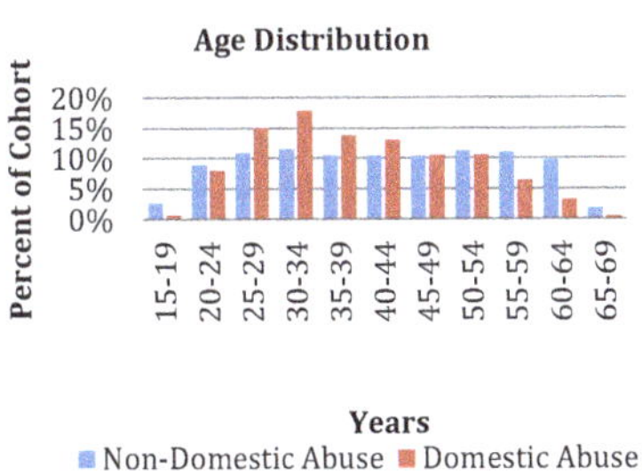

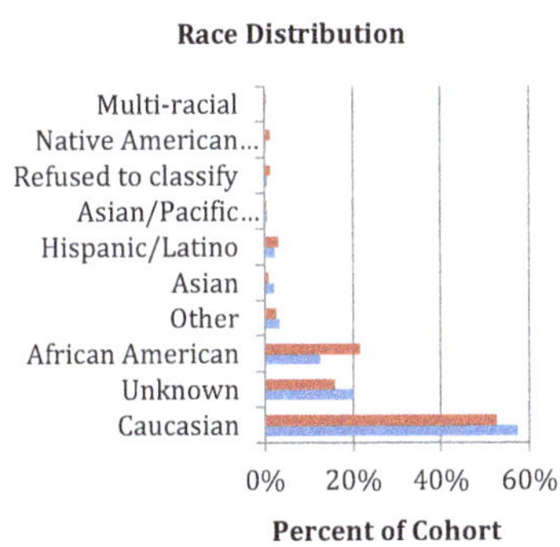

Fig. 2. Demographic summary details

Within the IPV identified records, we found 3458 symptom/diagnosis terms possibly associated with domestic abuse (i.e. the aggregated records contained 3458 medically coded symptoms, diagnoses, and findings – also referred to as "terms" in the following discussion). With a view to identifying conditions that are prevalent among victims of domestic abuse, we compared the frequency of each term in the DA cohort with its frequency in the NDA cohort. For each term, we used χ^2-test to assess the significance of its frequency in the DA cohort with respect to its frequency in the NDA cohort. To adjust for multiple hypothesis testing, we used Bonferroni correction (3458 tests were performed). This analysis suggested that 2430 of the identified terms were significantly more prevalent among IPV victims ($p < 0.05$). Two independent researchers used medical dictionaries to manually code these symptoms into broader, more general categories with high inter-rater reliability, four main classes emerged: chronic symptoms and disorders, acute injuries, mental and behavioral issues, and gynecological problems.

2.2. *Network Construction*

To provide a compact and visually comprehensive view of the diagnosis terms that were significantly more frequent in patients with the finding of "domestic abuse" (DA group), we created a network of diagnosis categories.

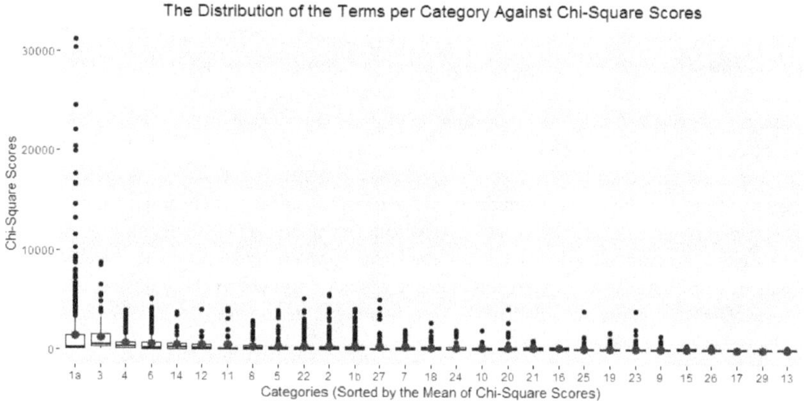

Fig. 3. Distribution of χ^2-statistics among terms assigned to each category. The distributions are shown by box plots, the mean of each distribution is shown by a red. The categories are sorted according to the mean χ^2 statistic

We first categorized the diagnosis terms by assigning each term to 28 specific categories. In this classification, assignment of a term to more than one category was allowed. Subsequently, we selected the 2429 terms that were significantly more frequent in the DA group (p< 0.05 based on the Chi-Square Test). We then counted the frequency of each category among these 2429 terms. The distribution of χ^2-statistics for the terms assigned to each category is shown in Figure 3.

	Disease Classification Category	Percent (%)
1b	Acute Condition	34.1%
1a	Acute Injury	23.1%
2	Chronic	16.2%
6	Disorders	15.6%
19	Cardiovascular	8.6%
8	Pregnancy Related	7.5%
7	Gynecological	7.5%
22	Musculoskeletal	6.2%
4	Mental Health	5.9%
18	Gastrointestinal	5.6%
9	Allergy	5.0%
3	Substance Abuse	4.8%
5	Other	4.6%
20	Nervous system	4.5%
27	Skin related (not burns)	4.0%
21	Respiratory	3.9%
23	Eyes, Ears, Nose & Throat	3.8%
24	Excretory	3.1%
14	Personal History	2.2%
11	Congenital/Hereditary	1.6%
25	Endocrine	1.6%
13	Neoplasm	1.3%
26	Immune System	1.3%
12	Nutrition	1.2%
10	Procedure	0.8%
16	Neuropathy	0.8%
15	Family History	0.7%
17	Diabetes	0.6%

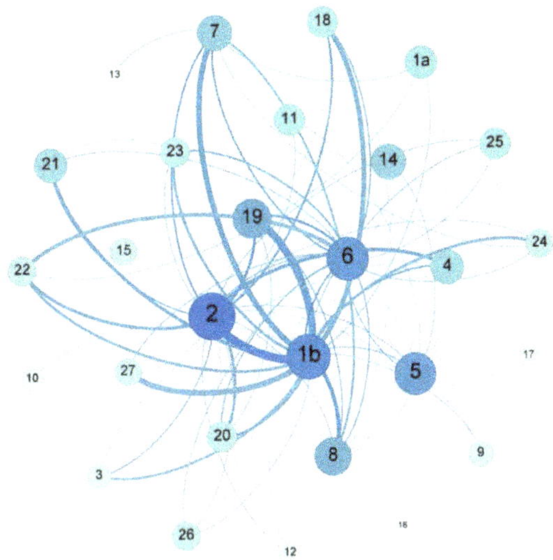

Fig. 4. Network Map of the 28 categories made up of 2429 symptom terms found to be significantly more prevalent among victims of IPV than the general population. Larger nodes indicate higher coding frequency of the category by itself, while thicker edges reveal the strength of relationship between two nodes by signifying the frequency of any two categories being coded together on a single symptom term. Darker nodes appear more frequently among the pairs than lighter nodes. Percentages of total diagnoses found to be significantly frequent in patients with a DA finding for each individual symptom category also shown.

To assess the co-occurrence of each pair of categories in the DA group, we counted the number of terms among these 2429 terms that are assigned both categories. Note that this co-occurrence frequency is not to be confused with co-morbidity of diagnoses; this number rather reflects the likelihood that a diagnosis significantly more frequent in patients with DA will belong to both categories. In total, 208 pairs of categories were assigned together to at least one of the 2429 terms.

After counting frequencies of categories and pairs of categories among the terms that are significantly frequent in the DA group, we visualized the frequencies as a network using GePhi 0.9.0 Beta (Fig. 4). In the figure, the size of each node represents the frequency of the respective category among the terms that are significantly frequent in the DA group. The thickness of each edge represents the number of terms that are assigned both of the respective categories.

3. Results

The results of the data-driven analysis and network mapping are illustrated in Figure 4. The Acute Conditions (1b) and Chronic (2) categories appear to be the most significant in our network map, showing the greatest frequency of occurrence among the coded terms. Chronic (2) exhibits strong connections to Acute Conditions (1b), and fair connections to Mental Health (4), Cardiovascular (19), Nervous System (20), and Musculoskeletal (22). It should be noted that the strong connection between chronic and acute conditions is due to ambiguity of the symptom terms, resulting in the possibility of a term being coded as both chronic and acute because it could be either, depending on the patient's situation. Acute Conditions (1b) has strong connections to Chronic (2) and Cardiovascular (19), with more moderate connections to Gynecological (7), Gastrointestinal (18), Skin Related (27), and Pregnancy Related (8). It also has fair connections to Substance Abuse (3), Nervous System (20), Respiratory (21), Musculoskeletal (22), Eyes, Ears, Nose & Throat (23), and Excretory (24).

Disorders (6) shows some significance in coded frequency, with a moderate connection to the Musculoskeletal (22) category. The Cardiovascular (19), Gynecological (7), Pregnancy Related (8), and Gastrointestinal (18) nodes may also be somewhat significant, but their primary connection is with Acute Conditions (1b), which is itself a significant node in the network overall (Cardiovascular is also fairly connected to Chronic (2), another independently significant node).

Interestingly, although the Acute Injures (1a) node indicates relatively significant frequency of coding, this category is mostly isolated, demonstrating only weak connections.

4. Discussion

4.1. *General*

The results of our analysis and network mapping are fairly consistent with current knowledge of IPV. We found that chronic and acute conditions as well as acute injuries were frequently coded to the symptoms that are more prevalent among victims of IPV. We also found that the categories showing the most individual frequency (chronic and acute conditions) shared strong connections with physiological systems that have been shown to be impaired at a higher rate among IPV victims, including gynecological and pregnancy related issues, as well as gastrointestinal, cardiovascular, and neurological symptoms.[17-19] It is not surprising that 'chronic' and 'acute conditions' had the most significant frequency of coding, since most ailments that require medical attention are either established diseases or emergency issues. 'Acute conditions' shows many strong connections, because most acute conditions would be coded with whatever body system(s) they were related to. 'Chronic' is similarly highly connected, for the same reason. These two categories show a strong connection to each other because the symptoms were often ambiguous, and coded as both chronic and acute because there was not enough information in the symptom term to differentiate it between the categories.

Nodes that represent 'gynecological' and 'pregnancy related' symptoms appear to be fairly significant, which is expected when considering the nature of IPV. Physical and sexual abuse from IPV can be very damaging to the body,[17,20] resulting in trauma, infection, and the contraction of sexually transmitted infection.[6] Victims are at greater risk of experiencing sexual coercion from an intimate partner as well as birth control sabotage, and often fear talking to their partner about pregnancy prevention.[6] Studies have also shown increased risk to mothers' and newborns' health when IPV is experienced during pregnancy, such as miscarriage and low birth weight.[8,10] Pregnancy itself can also be a risk factor for IPV.[21]

We were surprised to see that the node that represents 'mental health' related symptoms was not a significant "hub" in the network. The node itself is significant, i.e., it appears moderately prevalent in the general frequency count, but lacks strong connections. This independence is explained by the fact that most symptoms labeled as 'mental health' would be unlikely to fall into other categories except for 'chronic', since many mental health issues happen to be chronic in nature but generally would not directly interact with other physiological systems.

Stress may be an important factor in the patterns we identified in the network map and analysis. There is a broad research literature describing the interactions between IPV and stress,[17,22,23] as well as the effects of stress on the body.[24-26] IPV causes an increase in cortisol, one of the body's stress hormones, which in turn might cause detrimental impacts on the victim's immune system. This can manifest in a variety of ways, but often affects gastrointestinal and circulatory system functioning. Stress resulting from IPV may also seriously increase the risk of a negative event during and following pregnancy.[27,28] Interestingly, the nodes that represent these categories showed significant frequency in the network of IPV victim health symptoms. It is difficult to verify if stress is the underlying factor in the higher significance and connection of these categories, but it may be possible in future studies to incorporate cortisol-level measurements in the search query. Many of the significant categories in our network map showed strong connections to 'chronic' and 'acute conditions', which only further demonstrates the severity of the negative health effects associated with IPV.

We cannot accurately assess whether this network map reveals subtle patterns or cycles, because our data is a compilation of records that incorporate data without the dimension of time. If we could find a way to apply temporal filters, we might be able to identify a progression of health events common among victims of IPV that would further illustrate exactly how IPV leads to negative health over the victim's lifetime. This will enable health care practices and professionals to more accurately identify, assess, and treat IPV and related illnesses, ultimately utilizing knowledge of how IPV impacts health to improve treatment decision making processes. Analyzing this data will help explore how EHR can be utilized for research. Our findings may demonstrate that it is possible to improve standard screening procedures and treatment plans for victims of violence as well as patients in other circumstances, simply by examining electronic health records. The significant correlations that can be found through this method provide valuable information for both clinical and research applications.

4.2. *Limitations*

Although studies have shown that approximately 1 in every 4 women will experience IPV at some point during their lifetime,[1] the data collected by *Explorys* does not reflect this observation. This is likely due to a variety of factors, and may be most influenced by the underreporting and inadequate screening of IPV victims. Although our query returned NDA records from all 50 states, as well as Puerto Rico, Guam, and APO/FPO (military bases), it returned DA records from only a dozen location categories. The relative distribution of records across these location categories for both DA and NDA cohorts are shown in Figure 5. It would be interesting to examine the laws and regulations of the states that returned EHR for the DA group. It is possible that local regulations influence the likelihood of domestic abuse being screened and recorded by medical professionals.

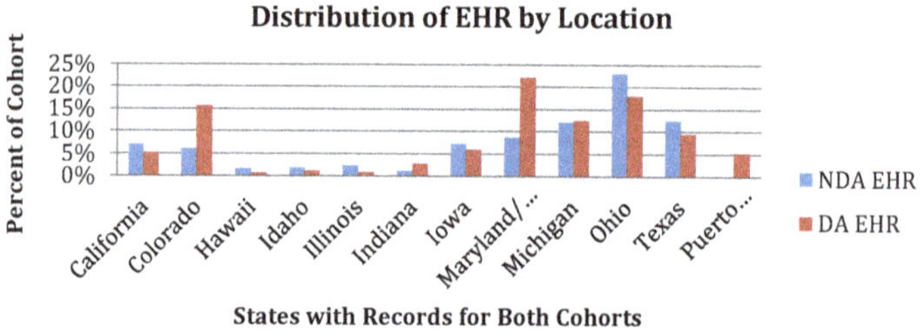

Fig. 5.. Distribution of location categories for the domestic abuse (DA) and non-domestic abuse (NDA) cohorts.

It is interesting to note that the distribution of EHR for the NDA cohort does not correspond with the distribution of the US population as measured by the 2010 US Census Bureau (comparison not shown here, but NDA EHR showed disproportionately high distribution in Texas, Ohio, Michigan, Maryland/DC, Iowa, Idaho, Hawaii, and Colorado compared to relative census population distributions). This difference may reflect the sampling of medical facilities providing data to *Explorys*. These trends are worth investigating to better understand how inherent confounds of the data may influence query results. Many of the records pertaining to this subset of the population may not have been captured by our queries, and thus potentially were not included in our network map and data analysis. This means that it is very likely that records in our NDA group actually belong in the DA group, which would seem to muddy the analysis. However, we feel that due to this underreporting, our DA group likely represents the most severe cases of domestic abuse, and thus highlights the most obvious symptom connections. Further analysis could help us tease out more subtle victim symptom characteristics. This information could then be used to develop targeted queries in the future that may help us to identify high-risk IPV victims through related EHR diagnoses, even if domestic abuse is not listed as a finding. While we were able to extract some demographic characteristics of the cohorts (Fig. 2), the nature of the data does not permit us to match demographics to specific records. However, as Fig. 5 illustrates, there are serious gaps in EHR data for the DA group, and it is not possible to identify confounds resulting from demographics with such limited data.

Domestic abuse is not always noted in a patient's medical records, either because the patient does not reveal that she is a victim, or because medical professionals overlook recording this detail. Research has demonstrated that primary care IPV screening is inadequate and needs attention.[29] Domestic abuse is not considered a 'diagnosis' or 'condition', but is rather described in the *Explorys* database as a 'finding'. This is a reflection of how the medical community labels IPV, and explains why notating this 'finding' may not be a priority when updating a patient's medical records. Since this study most likely captured records with the most severe cases of IPV, where evidence of the abuse was obvious, the current network map may not reflect the more subtle connections associated with less severe instances of IPV. However, the results of our network map and analysis demonstrate how extensive the consequences of IPV may be on victims' health, and further illustrate the vital importance of thoroughly screening for IPV and accurately noting patient findings by members of the medical community. It may be possible to utilize this data to develop more accurate screening procedures in the future.

Utilizing EHR data is challenging, and the currently available query systems leave gaps in data quality. Data is noisy, and we lack the ability to control for a myriad of confounds. Current records allow for the possibility of patients being counted more than once when determining frequencies, and longitudinal data is missing. However, due to the necessity to maintain the strictest levels of privacy, we cannot track specific (though still de-identified) patient records to see how health changes over time. We are exploring how to utilize other tools in *Explorys* and other query and analysis techniques to capture longitudinal data. Examining the changes in symptom presentation in relation to the first appearance of DA on a patient's medical record is a key step to understanding the etiology and health consequences of violence victimization more fully. This could also be an instrumental step in implementing effective risk assessments. However, even at this point, the knowledge gained from the analysis of EHR data can still lead to vast improvements in health care and policy development, and improved queries may improve the quality of data. The techniques demonstrated in this study have implications not only for the care of intimate partner violence victims, but for the health of the entire population as a whole.

5. Conclusion

EHR data is a vital resource in advancing the knowledge of health care professionals. By analyzing the data we can create networks that show how different symptom and disease categories are related to each other, revealing associations which may indicate deeper root causes for deteriorating health. In this study, we were able to examine what health factors are associated with IPV, and how these factors interact. This gives us a more complete and compelling picture of the negative health effects of IPV. With further research it may be possible to develop improved methods and diagnostic tools for successful intervention and treatment, improving victims' quality of life throughout their lives.

Analysis of EHR data gives health providers the information to improve quality of service, especially for victims of IPV. However, it is so important for screening procedures to improve, so that victims are accurately identified and given appropriate medical care. Our network mapping and data analysis demonstrate only a fraction of the far-reaching health consequences of IPV, which cannot be ignored from a medical perspective. We know that many of the victims of IPV were not represented in our DA data set, because they haven't been identified as such in their medical records. It is absolutely imperative that we push to improve screening so that these devastating health effects can be mitigated and prevented. The data from our analysis may help with future research into how we can better identify victims who hesitate to come forward by identifying the tell-tale signs and relationships of their symptoms and conditions. It is clear that mining EHR will reveal many associations between previously independent conditions. Our future research will replicate these analysis techniques with independent datasets to confirm the efficacy of these methods. Doctors and health care providers can use this information to improve the prescription of effective treatment preventions, and identify trends across populations. If we can use this information to develop more effective screening tools and treatments, we will drastically increase the quality of life and healthcare experienced by victims of IPV, and through this the wellbeing of society as a whole.

6. Acknowledgements

This publication was made possible in part by the Clinical and Translational Science Collaborative of Cleveland, NIH/NCRR CTSA KL2TR000440 from the National Center for Advancing Translational Sciences (NCATS) component of the National Institutes of Health and NIH roadmap for Medical Research. Its contents are solely the responsibility of the authors and do not necessarily represent the official views of the NIH.

References

1. M. C. Black, K. C. Basile, M. J. Breiding, S. G. Smith, M. L. Walters and M. T. Merrick, National intimate partner and sexual violence survey 2010 summary report and sexual violence survey (2011).
2. American College of Obstetricians and Gynecologists. *Obstet Gynecol,* 119, 412 (2012).
3. J. Corrigan, M. Wolfe, W. Mysiv, R. Jackson and J. Bogner, *Am J Obstet Gynecol,* 188, S71 (2003).
4. E. Lawrence, A. Oringo and R. Brock, *Partner Abuse,* 3 (2012).
5. G. Karakurt, D. Smith and J. Whiting, *J Fam Violence,* 29 (2014).
6. G. Wingood, R. DiClemente, D. McCree, K. Harrington and S. Davies, *Pediatr,* 107, E72 (2001).
7. E. Letourneau, M. Holmes and J. Chasedunn-Roark, *Women's Health Issues,* 9, 115 (1999).
8. D. El Kady, W. Gilbert, G. Xing and L. Smith, *Obstet Gynecol,* 105, 357 (2005).
9. J. Hathaway, L. Mucci, J. Silverman, D. Brooks, R. Mathews and C. Pavlos, *Am J Prev Med,* 19, 302 (2000).

A. Huth-Bocks, A. Levendosky and G. Bogat, *Violence Vict,* 17, 169 (2002).

10. H. Nelson, C. Bougatsos and I. Blazina, *Ann Int Med, 156,* 796 (2012).

11. J. Campbell, *The Lancet, 359,* 1331 (2002).

12. V. Wu, H. Huff and M. Bhandari, *Trauma Violence Abuse, 11,* 71 (2010).

13. L. Kwako, N. Glass, J. Campbell, K. Melvin, T. Barr and J. Gill, *Trauma Violence Abuse, 12,* 115(2011).

14. G. Karakurt, V. Patel, K. Whiting and M. Koyuturk, *J Fam Violence,* in print, (2016).

15. D.C. Kaelber, W. Foster, J. Gilder, T. E. Love and A. K. Jain, *J Am Med Inf Assoc, 19,* 965 (2012).

16. J. Campbell, A. S. Jones, J. Dienemann, J. Kub, J. Schollenberger, P. O'Campo, A. C. Gielen and C. Wynne, *Arch Int Med, 162,* 1157 (2002).

17. D. J. Sheridan and K. R. Nash, *Trauma Violence Abuse, 8,* 281 (2007).

18. S. Sprague, K. Madden, S. Dosanjh, K. Godin, J. C. Goslings, E. H. Schemitsch and M. Bhandari, *BMC Musculo Disorders, 14* (2013).

19. M. L. Paras, M. H. Murad, L. P. Chen, E. N. Goranson, A. L. Sattler, K. M. Colbenson, M. B. Elamin, R. J. Seime, L. J. Prokop and A. Zirakzadeh, JAMA, 302, 550 (2009).

20. T. L. Taillieu and D. A. Brownridge, *Aggress Violent Behav, 15,* 14 (2010).

21. M. E. Feinberg, D. E. Jones, D. A. Granger and D. Bontempo, *Aggress Behav, 37,* 492 (2011).

22. S. S. Inslicht, C. R. Marmar, T. C. Neylan, T. J. Metzler, S. L. Hart, C. Otte, S. E. McCaslin, G. L. Larkin, K. B. Hyman and A. Baum, *Psychoneuroendocrinology, 31,* 825 (2006).

23. M. Moreno-Smith, S. K. Lutgendorf and A. K. Sood, *Future Oncol, 6,* 1863 (2010).

24. M. Roest, E. J. Martens, P. de Jonge and J. Denollet, *J Am Coll Cardiol, 56,* 38 (2010).

25. J. Shen, Y. E. Avivi, J. F. Todaro, A. Spiro, J. P. Laurenceau, K. D. Ward and R. Niaura, *J Am Coll Cardiol, 51,* 113 (2008).

26. A. Taylor, N. B. Guterman, S. J. Lee and P. J. Rathouz, *Am J Pub Health, 99,* 175 (2009).

27. W. P. Witt, E. R. Cheng, L. E. Wisk, K. Litzelman, D. Chatterjee, K. Mandell and F. Wakeel, *Am J Pub Health, 104,* S81 (2014).

28. P. Tavrow, B. E. Bloom and M. H. Withers, *Violence Against Women,* in print (2016).

DISCOVERY OF FUNCTIONAL AND DISEASE PATHWAYS BY COMMUNITY DETECTION IN PROTEIN-PROTEIN INTERACTION NETWORKS

STEPHEN J. WILSON

Department of Biochemistry and Molecular Biology, Baylor College of Medicine, One Baylor Plaza
Houston, Texas 77030, USA
Email: sw5@bcm.edu

ANGELA D. WILKINS

Department of Molecular and Human Genetics, Baylor College of Medicine, One Baylor Plaza
Houston, Texas 77030, USA
Email: aw11@bcm.edu

CHIH-HSU LIN

Graduate Program in Structural and Computational Biology and Molecular Biophysics, Baylor College of Medicine, One Baylor Plaza
Houston, Texas 77030, USA
Email: Chih-Hsu.Lin@bcm.edu

RHONALD C. LUA

Department of Molecular and Human Genetics, Baylor College of Medicine, One Baylor Plaza
Houston, Texas 77030, USA
Email: lua@bcm.edu

OLIVIER LICHTARGE

Departments of Molecular and Human Genetics, Structural and Computational Biology and Molecular Biophysics, Biochemistry and Molecular Biology, and Pharmacology, Baylor College of Medicine, One Baylor Plaza
Houston, Texas 77030, USA
Email: lichtarg@bcm.edu

Advances in cellular, molecular, and disease biology depend on the comprehensive characterization of gene interactions and pathways. Traditionally, these pathways are curated manually, limiting their efficient annotation and, potentially, reinforcing field-specific bias. Here, in order to test objective and automated identification of functionally cooperative genes, we compared a novel algorithm with three established methods to search for communities within gene interaction networks. Communities identified by the novel approach and by one of the established method overlapped significantly ($q < 0.1$) with control pathways. With respect to disease, these communities were biased to genes with pathogenic variants in ClinVar ($p << 0.01$), and often genes from the same community were co-expressed, including in breast cancers. The interesting subset of novel communities, defined by poor overlap to control pathways also contained co-expressed genes, consistent with a possible functional role. This work shows that community detection based on topological features of networks suggests new, biologically meaningful groupings of genes that, in turn, point to health and disease relevant hypotheses.

1. Introduction

How genes and proteins interact with each other is the basis of molecular biology and disease pathogenesis[1,2]. These functional interactions, which biologists place into pathways, have been characterized through hypothesis-driven experiments and then manually defined in the past[3,4]. This is necessarily knowledge intensive and painstaking, and it stands in sharp contrast to the massive amount of new gene interaction data from high-throughput experiments. Continued reliance on manual recognition of pathways may limit the overall capacity to characterize gene behavior, and potentially focus on already well-known sets of gene interactions. With at least 100,000 interactome hubs in humans, the number of potential interactions to annotate are in the billions[5]. Yet, the current estimate of interactions from the broadly used and expertly curated STRING database[6] that focus solely on proteins are in the millions. This large discrepancy suggests many unrecognized, or "dark," associations and pathways are simply missing.

In order to take a data-driven approach to annotate and detect novel biological pathways, clusters in biological networks were defined based on topological features to isolate functional and disease pathways[5,7]. One topological feature that has been extensively applied in social network analysis[8-10], but has not yet seen widespread use in biology, is community structure[11,12].

Communities are groups of nodes (i.e. proteins) that are more connected to each other than to anything else in a network[8,13]. Often these groups of nodes correspond to a common process, purpose, or function[5,9]. Therefore, it is reasonable to hypothesize that determining communities on biological networks may shed new light on groupings of genes with common biological function or features. Past efforts[13,14] were useful but did not comprehensively test various algorithms in functional and disease contexts. Given appropriate algorithms, community detection has the potential to automatically expand biological pathways, determine novel pathways, and perhaps even predict gene-disease associations.

This study sought to detect communities on a protein-protein interaction network and to evaluate their number and size against existing references. Several methods can evaluate performance in terms of the number and size of the overlap between communities and known control pathways. Moreover, beyond reference pathways, disease data can directly demonstrate the applicability of communities to formulate new and clinically relevant biomedical hypotheses.

2. Results

2.1. *Determining putative biological pathways*

In order to automatically determine putative biological pathways, several possible community detection methods exist. Clauset-Newman-Moore (CNM)[8] and Louvain[10] are well-established and extensively applied algorithms with more than 3000 citations each. BIGCLAM[15] is a more recent alternative that searches for densely overlapping, hierarchically nested communities in an orthogonal approach. Each of these approaches was tested on a STRING protein-protein interaction network[16], limited to high-quality direct biological associations. The communities that were obtained could then be compared to gold standard set of curated biological pathways, such as Reactome[17] and Canonical pathways from the GSEA tool[18], and, for disease pathways, DisGeNET[19].

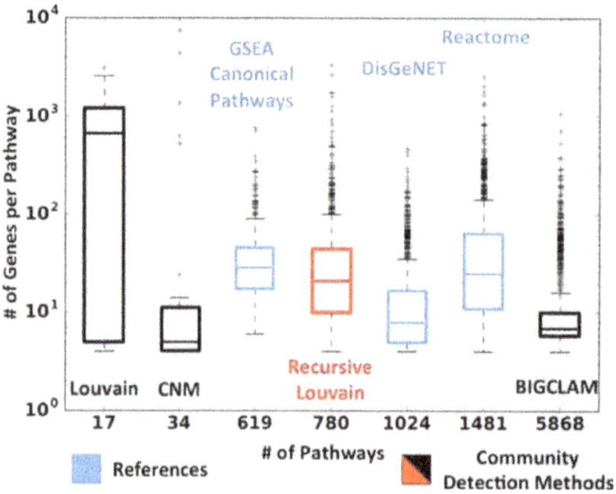

Figure 1: Community Algorithms Detect Variable Numbers and Sizes of Pathways. Recursive Louvain (Red) is a novel community detection method that detects a similar number of gene groups to the references with approximately the same number of genes per gene group given STRING 9.1 protein-protein interaction network.

A first assessment of performance was the granularity of the communities. That is, we compared the number of gene groups and the number of genes in each group in order to determine whether the communities resemble the references. CNM and Louvain community detection found an order of magnitude fewer groupings than the smallest reference set, and BIGCLAM detects five times more groups than the largest reference set (Figure 1). This is not surprising given that the methods were designed for social network analysis. Combined with different numbers of genes per group, these algorithms appear to poorly represent the reference pathways as defined by biologists.

To address this problem, we then introduced a novel community detection algorithm we called, Recursive Louvain (RL). RL applies the Louvain algorithm but iterates on the resulting communities so as to break them down stepwise into smaller and smaller groups until reaching a majority of right-sized communities, an idea that was in fact discussed in the original Louvain community detection paper[10]. In this way, RL generated communities that matched more closely the size of the control pathways (Figure 1, red).

2.2. *Assessing the biological relevance of communities*

Next, when comparing communities to the reference sets, careful consideration of what constitutes a pathway was necessary. First, we removed overly small reference pathways and communities (size ≤ 3 genes) to better focus on significant gene groupings. Additionally, pathways often share many genes, and in the extreme, they can share all but one gene. To avoid the over-counting of a pathway, or community, any with more than 90% of genes in common were combined. Finally, four different metrics were selected to gauge success. Jaccard similarity measures the similarity of a community to a reference by looking at the size of the intersection relative to the union of the genes; the modified Jaccard metric does not punish a community for being larger than the reference; the hypergeometric test measures the likelihood of getting an overlap between a reference and a community given all genes in a given community set; and the F_1 score measures the ability to recover an overlap (see methods for the mathematical details).

To test if communities represent biological information from functional and disease pathways, we compared each community to each reference pathway. This comparison was accomplished with the hypergeometric test, which allows a statistical probability and Benjamini-Hochberg False Discovery Rate (FDR) correction[20]. This correction is essential to account for multiple testing. Encouragingly, many communities were significantly enriched (q-value ≤ 0.1) for a functional pathway (Reactome and Canonical Pathways), a disease pathway (DisGeNET), or a mixture of the two. Depending on the method, between 7-24% of communities were not enriched for any known pathway or disease and were regarded as novel. The exact breakdown of the community classification is shown in Figure 2A, and the majority of communities in BIGCLAM and RL are statistically overlapped with a function pathway and often with a disease pathway. Indeed, RL has the smallest fraction (7%) of novel communities, suggesting a higher positive predictive rate for the references. We noted that the number of genes in each community group generally increases from novel to mixed (Figure 2B). This could have a number of implications, including an observational annotation bias or a biological basis. These data show that community detection methods recover many commonly known functional and disease pathways but also discover new gene associations that possibly suggest novel pathways.

In order to assess the robustness of community detection we tested four metrics of

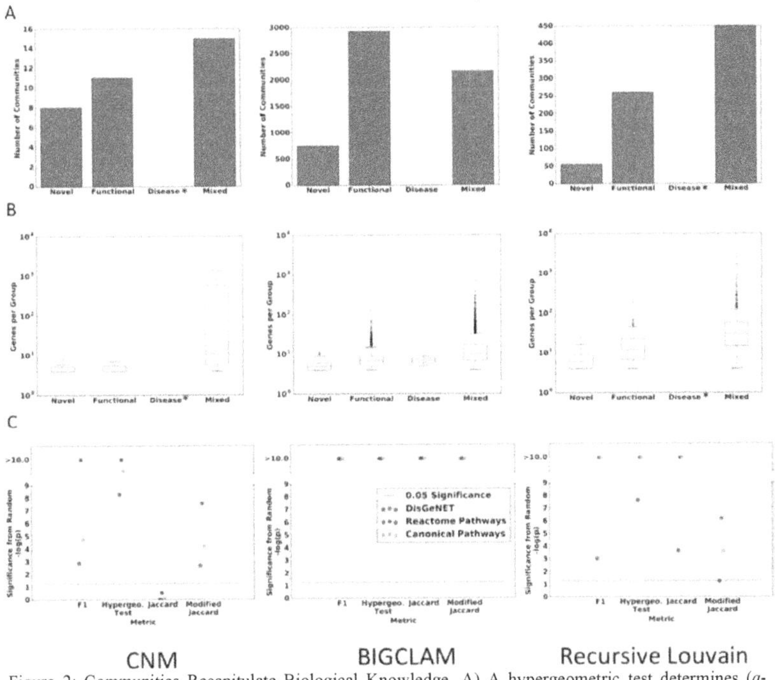

Figure 2: Communities Recapitulate Biological Knowledge. A) A hypergeometric test determines (q-value ≤ 0.1) whether a community is overlapped with a reference, and while many communities were overlapped with a disease or functional pathway, few or none (denoted by an *) were exclusively overlapped with only disease pathways. B) The number of genes in each group generally increases from a novel community to a community enriched for disease and functional pathways. C) All methods were non-randomly associated with every reference according to some metric, and many with p-values smaller than 10^{-10}.

community overlap. Three random controls were generated to match the number and size of a set of communities and then scored against the references. Only the top score of a community or random against all pathways in a reference was kept. The distribution of random scores was then compared against the distribution of community scores using a Kolmogorov-Smirnov test. Due to poor performance and a lack of data (only 17 total communities), Louvain community detection (Supp. Figure 1-2) was assessed, but will not be shown because RL finds more total communities with overlap. As seen in Figure 2C, all three remaining methods were non-random by some metric; however, BIGCLAM and RL were significant on more metrics than CNM. In particular, BIGCLAM and RL appear highly significant in overlap with functional and disease pathways. RL has a higher percentage of communities that are enriched for both functional and disease pathways (Figure 2A), and this may suggest RL is better at recapitulating disease pathways. BIGCLAM has many more communities than the other methods (Figure 1A); this means we are more confident that BIGCLAM is performing different from random because we have more examples of overlap with the references. In contrast, Louvain community detection only found 17 communities, offering fewer opportunities to overlap with the references, and when we break those communities down further with RL, there is now more overlap with the references. These data show that BIGCLAM and RL recapitulate biological knowledge, while CNM appears to be less reliable.

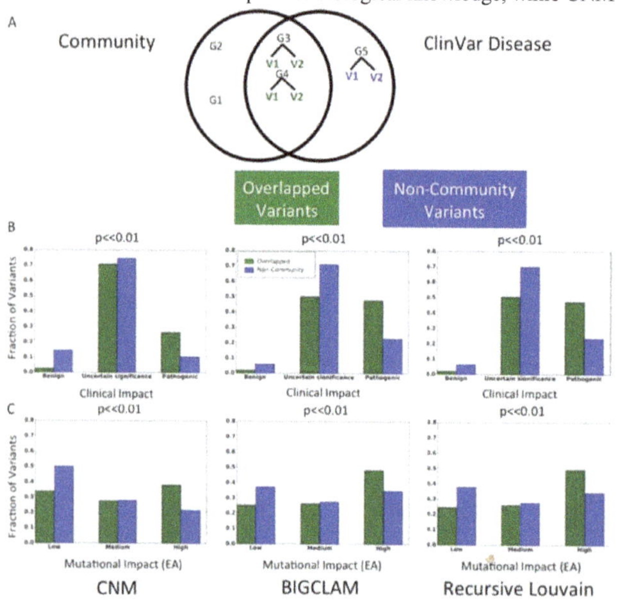

Figure 3: The Overlap between Communities and Diseases is Biased to Highly Pathogenic and Impactful Mutations. A) ClinVar groups variants into diseases. When a community and a disease from ClinVar both share genes, those genes possess a high B) clinical impact and C) mutation impact (p<<0.01) when compared to genes that are not found in communities. This implies that the communities are enriched towards variants that are pathogenic. Overlaps were only taken if the overlap was non-random (q<0.05).

2.3. *Clinical and disease relevance of communities*

A central question is whether these communities have real-world significance with respect to disease mechanisms. In order to address this question, communities were tested for overlap with diseases in the genetics database ClinVar. We hypothesized that communities represent units of biological function, and, if so, disrupting a gene that is part of a community would be more pathogenic than disrupting one that is outside of a community. Indeed, we find that mutations of disease genes that belong to communities have greater impact on the clinical phenotypes and on overall protein fitness

(Figure 3A). This was tested using the extensive annotations ClinVar[21] provides on the clinical impact of disease mutations. Specifically, for every community detection method, genes that fell inside communities showed are biased towards pathogenic variants (Chi-Square p-value << 0.01, Figure 3B). As an orthogonal control to test for bias in the impact of variations on disease genes, the Evolutionary Action (EA) provides an independent assessment of the deleterious impact of a mutation on protein fitness[22]. The same statistically significant trend emerges (Figure 3C, Chi-Square $p << 0.01$). These data show that mutations tend to have greater clinical and evolutionary deleterious impact if they affect genes that are part of communities.

To demonstrate a specific application of communities to disease pathways, we compared communities from BIGCLAM and RL, which outperformed CNM, against two diseases. These two diseases, Zellweger Syndrome (ZS) and Bardet-Biedl Syndrome (BBS) were both statistically associated with communities ($p << 0.01$). To associate diseases to communities, we used the disease-gene association information from two sources: (1) DisGeNet, a disease-gene association database integrating several public data resources and literature, and as shown in Figure 3, (2) ClinVar, a database providing the expert-asserted associations between genetic variants of genes and diseases. These disease-gene associations were used to calculate the statistical overlap between a disease and a community according to a hypergeometric distribution test of the overlap of genes, the unique genes of each, and all human genes. We hypothesized that when a community is statistically associated with a disease, any genes unique to the community are promising novel disease candidates. This hypothesis extends from a guilt-by-association assumption that has been successful in multiple systems[23,24]. As shown in Figure 4, when communities from multiple algorithms are compared to diseases, the overlaps possess high predictive power. For example, ZS is a peroxisomal biogenesis disorder characterized by severe hypotonia, epileptic seizures, and craniofacial abnormalities[25]. Because peroxisomal biogenesis depends highly on protein-protein interactions (PPIs), community detection on a PPI network reliably predicts and expands the disease definition. Indeed, using both community algorithms recovers all genes from DisGeNET and thirty additional genes are predicted. Of these thirty genes, fourteen are already annotated in the literature as being associated or causative of ZS. Two genes predicted to be associated with ZS are *ABCD1* and *ABCD2*, which are not known to be associated with ZS but transport very-long-chain fatty acids (*VLCFA*) across the peroxisome membrane

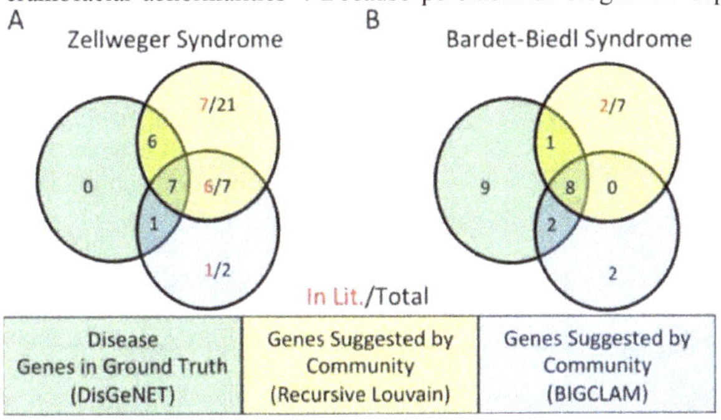

Figure 4: Communities Discover Novel Disease-Gene Associations. A) For Zellweger Syndrome, all known disease associated genes in DisGeNET are recovered between a community from Recursive Louvain and BIGCLAM. Fourteen out of thirty predictions possess some form of evidence in the literature. B) Bardet-Biedl Syndrome (BBS) possesses significant overlap with the ground truth but does not find all known genes. However, there is a high concordance of overlap between the methods and the ground truth, with 2/7 of the Recursive Louvain predictions with literature evidence.

and cause adrenoleukodystrophy, a related peroxisomal disorder[26].

Another example is BBS, a rare ciliopathy that affects multiple body systems, where over half of the known genes were recovered and nine genes were predicted. Of these nine genes, two genes are already known in the literature to be associated with BBS. BBS is characterized by obesity, polydactyly, hypogonadism, intellectual disability, and renal abnormalities[27]. The gene *FOPNL* is suggested by community analysis but possesses no literature evidence. Despite this, *FOPNL* is well known to be associated with the biogenesis of cilia and BBS causative mutations upset cilial function. Furthermore, *FOPNL* interacts with *PCM1*, a known BBS gene that is also suggested by community analysis[28]. For BBS, there is a lack of overlap between community predictions, which points to the fact that each method is dependent on different features and therefore provides unique insight. These data demonstrate that communities can be useful in predicting and expanding sets of genes related to diseases that depend on protein interactions.

We determined if novel communities that lacked overlap with functional and disease pathways are biologically relevant by analyzing the co-expression of community genes in breast invasive carcinoma (BRCA). BRCA was chosen as a test case because it has a large number of patients with whom to power a co-expression study, though other cancers will be investigated in the future. If the genes in a community are co-expressed together more than randomly selected genes within tumor tissue RNA sequencing data, then that community represents a biologically relevant disease module. To validate our co-expression analysis, we examined four Reactome pathways, which are related to breast cancer pathways (PI3K/AKT activation, Signaling to RAS, PI3K/AKT Signaling in Cancer, and Constitutive Signaling by AKT1 E17K in Cancer) and found they are significantly co-expressed/regulated in breast diseased tissues ($q<0.05$). For both BIGCLAM and RL, at least 30% of the communities were co-expressed more than random with a q-value < 0.1 (FDR corrected by Benjamini-Hochberg), and over 52 % of novel RL communities were co-expressed non-randomly (Figure 5). Moreover, CNM preformed weaker than RL and BIGCLAM in comparisons to references, but with co-expression, CNM showed no signal, suggesting that it may be a poor approach for biological analysis. Overall, the figure shows that all classes of communities, including novel communities, have co-expression in BRCA.

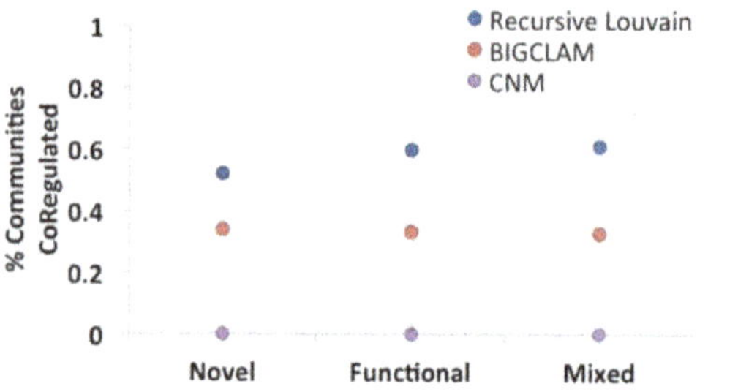

Figure 5: Communities are Significantly Perturbed in Cancer. In order to investigate whether novel communities had biological relevance, novel communities were investigated in the context of Breast Cancer (BRCA) co-expression data from TCGA. According to this analysis, the genes in novel communities are co-expressed to the same degree as communities with statistical overlap to functional and mixed pathways. This suggests that novel pathways represent a promising source of relevant biological knowledge.

As an application, one novel community (no. 657) detected by RL showed significant co-expression in BRCA ($q = 0.00588$) and has 14 gene members. Five members (*GPNTG, ECHS1, NACA, ABHD14B, NKX6-1*) were found to significantly coexist in the same subcellular location, extracellular vesicular exosome (GO:0070062; $q = 0.01456$; see Method). Furthermore, four members (*BTF3, GNPTG, CPEB2*, and *BICC1*) were found to be potentially co-regulated by the same transcription factor, *DACH1*, in a triple-negative breast cancer cell line, MDA-MB-231, ($q = 0.0077$ in ChIP-Seq enrichment analysis; see Method). Although it has been shown that *DACH1* expression level can predict BRCA survival[29] and play roles in breast cancer metastasis[30], *DACH1* currently has no pathway annotation in KEGG and Reactome databases. Therefore, this community might be the pathway related to *DACH1*. These results showed that novel communities could be related not only by expression but also by subcellular location and transcription factors. These data show the potential of communities to expand our knowledge of biology and disease.

3. Discussion

Determining the relationships between genes is essential for molecular biology and medicine. These relationships often cluster together into functional and disease pathways, and the characterization of these pathways is necessary to improve disease classification, patient stratification and, ideally, personalized treatment[5]. Here, we investigated the automated discovery of pathways by comparing several community detection algorithms against known functional and disease pathways and leading us to a novel application of the well-known Louvain algorithm, which we call Recursive Louvain (RL).

First, the communities detected by both BIGCLAM and RL were associated non-randomly with all the control, reference pathways. This strongly supports the biological relevance of these communities. Second, these communities also show a bias towards genes that experience pathogenic and high-impact variants in ClinVar. And third, regardless of the enrichment to a particular reference, these communities are often statistically co-expressed in breast cancer, including those that are new, in the sense that they are not enriched for any known functional or disease pathway. Therefore, these novel communities of genes may point to currently unrecognized biological pathways. Finally, in at least several cases, communities appear to predict genes associated with diseases with high predictive power. In the case of Zellweger Syndrome, six out of seven of the highest confidence predictions were already found in the literature although they were missing from the reference. The data from these approaches therefore consistently show that communities are biologically relevant.

The breadth of information in the input network limits community analysis. With only direct protein-protein interaction information, protein associations via indirect biological mechanisms such as transcription regulation can be missed. Eventually, the addition of transcriptional, post-translational, and epigenetic associations should help better characterize biological processes and extend the ability of community detection to recognize a wider variety of pathways. This is important as we note that, so far, many diseases and pathways are not enriched for communities. Beyond the breadth of information, community detection is also limited by its quality. Low-confidence, spurious associations between proteins surely lead to incorrect memberships of proteins in pathways. Furthermore, the pathways found represent global averages of associations. The future addition of context-specific transcriptional networks, such as from ChIP-seq data in ENCODE[31], should help find context-specific communities relevant to individual tissues or disease states. Despite these limitations, this work reveals the potential of topological network analysis in the identification and expansion of biologically meaningful pathways and shows that diverse results can be achieved through careful algorithm choice.

4. Methods:

4.1. ***Collection of reference sets:*** Reactome was downloaded from http://www.reactome.org, and was filtered for all disease pathways by trimming the disease section of the hierarchy as well as filtering out any pathway with the following words: disorder, hiv, defect, cancer, mutant, host, disease, influenza, toxin, viral, carcinoma, deletions, deficiency, variant, or virus. Canonical Pathways from the GSEA tool were downloaded from http://software.broadinstitute.org/gsea/downloads.jsp. Both KEGG and Reactome pathways are included in the Canonical pathways. All Reactome pathways in this dataset were filtered out to eliminate redundancy and then KEGG pathways related to diseases were filtered out to eliminate overlap with disease pathways from DisGeNET. DisGeNET was downloaded from http://www.disgenet.org/web/DisGeNET/menu/downloads as the curated dataset.

4.2. ***Community detection:*** Louvain community detection was calculated with the python community detection, which can be downloaded at http://perso.crans.org/aynaud/communities/. This base module then was used to create RL. RL runs Louvain, then takes each community larger than ten genes and makes it a subgraph of the original network, then calls Louvain community detection again. It does this iteratively until all communities have been broke down to ten genes or less or a gene has been seen in more than three communities. CNM and BIGCLAM communities were detected using implementations in the SNAP software package[32]. All community detection algorithms were applied onto STRING 9.1 experimental network[16].

4.3. ***Comparison to reference sets:*** All groups of genes were filtered to exclude pathways that contained three or fewer genes. This eliminated pathways that could easily be randomly recapitulated and therefore skew results. Pathways often overlap with each other, with minor differences between them. To prevent over counting from this, pathways that are too similar were collapsed together. Given a set of reference gene groups $R_i \in R$ and a set of community gene groups $C_i \in C$, all gene groupings were collapsed if the Jaccard Similarity > 0.9, where:

$$Jaccard\ Similarity, J(C_i, R_i) = \frac{|C_i \cap R_i|}{|C_i \cup R_i|} \tag{1}$$

To collapse two gene groups, the union of the genes was taken. In addition to the Jaccard Similarity, we then adopted three mathematical measures to evaluate the community detection algorithms outputs against the references, including a Modified Jaccard Similarity, a Hypergeometric Distribution test, and a F_1 score.

$$Modified\ Jaccard\ Similarity,\ J_m(C_i, R_i) = \frac{|C_i \cap R_i|}{|R_i|} \tag{2}$$

$$Hypergeometric\ Test, P(X \geq |C_i \cap R_i|) = 1 - \sum_{j=0}^{|C_i \cap R_i|-1} \frac{\binom{|R_i|}{j}\binom{M-|R_i|}{|C_i|-j}}{\binom{M}{|C_i|}} \tag{3}$$

$$F_1\ Score = \frac{1}{2}(F_R + F_C) \tag{4}$$

$$F_R or\ F_C = 2 \cdot \frac{Precision \cdot Recall}{Precision + Recall} \quad (5)$$

$$Precision = \frac{TP}{TP+FP}\ ,\ Recall = \frac{TP}{TP+FN} \quad (6)$$

Where M=Number of genes in the original network and $F_R or\ F_C$ are the F_1 scores from the perspective of the reference or the community, respectively. TP represents the number of true positives; FP represents the number of false positives; FN represents the number of false negatives.

4.4 ***Comparison to ClinVar:*** In order to compare to ClinVar, we binned variants by clinical impact and Evolutionary Action, and the difference between each group of genes was assessed by a Chi-Square analysis. Only groups of genes from significant overlaps ($q < 0.1$ by hypergeometric analysis) between diseases and communities were assessed.

4.5. ***Generation and evaluation of random controls:*** Random controls were generated for each community set. For each community, a set of randomly generated genes were chosen from the protein interaction network such that the number of genes was identical to the number in the community. This was done three times in order to get a set of random communities that was then compared to the reference sets. The distribution of the random scores was compared against the distribution of the community scores using a Kolmogorov-Smirnov test. Each distribution was built with only the top score for a community or random against all pathways in a reference.

4.6. ***Co-expression analysis in tumor tissues using RNA-seq data:*** To determine if genes in a community have co-expression, RNA sequencing data version 2 of 1104 breast cancer tumor samples were downloaded from The Cancer Genome Atlas (TCGA) database (https://tcga-data.nci.nih.gov/tcga/dataAccessMatrix.htm) dated January, 2015. RNA-Seq by Expectation-Maximization (RSEM) normalized read counts (https://wiki.nci.nih.gov/display/TCGA/RNASeq) were used to represent mRNA expression level. The pairwise Spearman's rank correlation coefficients between the expression levels of pairs of genes in a community were computed. The distribution of absolute values of correlation coefficients was compared to the coefficient distribution of a random gene set, which is three times the size of a community, using a Kolmogorov-Smirnov test. All p-values were adjusted by Benjamini-Hochberg FDR correction[20]. A community was defined as co-expressed if the adjusted p-value is less than 0.1.

4.7. ***Gene Ontology and ChIP-Seq enrichment analysis:*** To understand the subcellular localization and potential upstream transcription factors of genes in a novel community, we analyzed the enrichment of Gene Ontology (GO) Cellular Component 2015 and ChIP Enrichment Analysis (ChEA) 2015 using Enrichr[33] (adjusted p-value < 0.1).

4.8. ***Computation:*** All calculations were done on an Ubuntu OS with 64 GB RAM and 4th Gen. Intel Core i7 3.7 GHz processor or equivalent machine.

4.9. ***Supplemental data*****:** Supplemental data can be seen at:
http://mammoth.bcm.tmc.edu/SupplementalPSB2016Data.pdf

5. Acknowledgements

The authors would like to acknowledge the kind support of Christie Buchovecky, Daniel Konecki, Teng-Kui Hsu, and Panos Katsonis for their discussions and general feedback of the work. Additionally, the authors would like to acknowledge funding by NLM training fellowship (Grant No. T15 LM007093) for SJW, as well as funding from DARPA (N66001-14-1-4027), National Science Foundation (NSF DBI-1356569, NSF DBI-0851393), and National Institutes of Health (NIH-GM079656, NIH-GM066099).

References

1. Pawson, T. & Linding, R. Network medicine. *FEBS Lett* **582**, 1266-1270, doi:10.1016/j.febslet.2008.02.011 (2008).
2. AlQuraishi, M., Koytiger, G., Jenney, A., MacBeath, G. & Sorger, P. K. A multiscale statistical mechanical framework integrates biophysical and genomic data to assemble cancer networks. *Nat Genet* **46**, 1363-1371, doi:10.1038/ng.3138 (2014).
3. Fabregat, A. *et al.* The Reactome pathway Knowledgebase. *Nucleic Acids Res* **44**, D481-487, doi:10.1093/nar/gkv1351 (2016).
4. Kanehisa, M., Sato, Y., Kawashima, M., Furumichi, M. & Tanabe, M. KEGG as a reference resource for gene and protein annotation. *Nucleic Acids Res* **44**, D457-462, doi:10.1093/nar/gkv1070 (2016).
5. Barabási, A.-L., Gulbahce, N. & Loscalzo, J. Network medicine: a network-based approach to human disease. *Nature Reviews Genetics* **12**, 56-68, doi:10.1038/nrg2918 (2011).
6. Szklarczyk, D. *et al.* STRING v10: protein-protein interaction networks, integrated over the tree of life. *Nucleic Acids Res* **43**, D447-452, doi:10.1093/nar/gku1003 (2015).
7. Sharan, R., Ulitsky, I. & Shamir, R. Network-based prediction of protein function. *Mol Syst Biol* **3**, 88, doi:10.1038/msb4100129 (2007).
8. Clauset, A., Newman, M. E. & Moore, C. Finding community structure in very large networks. *Physical review E* **70**, 066111 (2004).
9. Yang, J. & Leskovec, J. Defining and evaluating network communities based on ground-truth. *Knowledge and Information Systems* **42**, 181-213 (2015).
10. Blondel, V. D., Guillaume, J.-L., Lambiotte, R. & Lefebvre, E. Fast unfolding of communities in large networks. *Journal of Statistical Mechanics: Theory and Experiment* **2008**, P10008 (2008).
11. Ahn, Y. Y., Bagrow, J. P. & Lehmann, S. Link communities reveal multiscale complexity in networks. *Nature* **466**, 761-764, doi:10.1038/nature09182 (2010).
12. Palla, G., Derenyi, I., Farkas, I. & Vicsek, T. Uncovering the overlapping community structure of complex networks in nature and society. *Nature* **435**, 814-818, doi:10.1038/nature03607 (2005).
13. Taylor, I. W. *et al.* Dynamic modularity in protein interaction networks predicts breast cancer outcome. *Nat Biotechnol* **27**, 199-204, doi:10.1038/nbt.1522 (2009).
14. Sah, P., Singh, L. O., Clauset, A. & Bansal, S. Exploring community structure in biological networks with random graphs. *BMC Bioinformatics* **15**, 220, doi:10.1186/1471-2105-15-220 (2014).
15. Yang, J. & Leskovec, J. in *Proceedings of the sixth ACM international conference on Web search and data mining.* 587-596 (ACM).

16. Franceschini, A. *et al.* STRING v9.1: protein-protein interaction networks, with increased coverage and integration. *Nucleic Acids Res* **41**, D808-815, doi:10.1093/nar/gks1094 (2013).
17. Croft, D. *et al.* The Reactome pathway knowledgebase. *Nucleic Acids Res* **42**, D472-477, doi:10.1093/nar/gkt1102 (2014).
18. Subramanian, A. *et al.* Gene set enrichment analysis: a knowledge-based approach for interpreting genome-wide expression profiles. *Proceedings of the National Academy of Sciences of the United States of America* **102**, 15545-15550, doi:10.1073/pnas.0506580102 (2005).
19. Pinero, J. *et al.* DisGeNET: a discovery platform for the dynamical exploration of human diseases and their genes. *Database : the journal of biological databases and curation* **2015**, bav028, doi:10.1093/database/bav028 (2015).
20. Benjamini, Y. & Hochberg, Y. Controlling the false discovery rate: a practical and powerful approach to multiple testing. *Journal of the Royal Statistical Society. Series B (Methodological)*, 289-300 (1995).
21. Landrum, M. J. *et al.* ClinVar: public archive of interpretations of clinically relevant variants. *Nucleic Acids Res* **44**, D862-868, doi:10.1093/nar/gkv1222 (2016).
22. Katsonis, P. & Lichtarge, O. A formal perturbation equation between genotype and phenotype determines the Evolutionary Action of protein-coding variations on fitness. *Genome research* **24**, 2050-2058, doi:10.1101/gr.176214.114 (2014).
23. Lee, I. *et al.* A single gene network accurately predicts phenotypic effects of gene perturbation in Caenorhabditis elegans. *Nat Genet* **40**, 181-188, doi:10.1038/ng.2007.70 (2008).
24. McGary, K. L., Lee, I. & Marcotte, E. M. Broad network-based predictability of Saccharomyces cerevisiae gene loss-of-function phenotypes. *Genome Biol* **8**, R258, doi:10.1186/gb-2007-8-12-r258 (2007).
25. Klouwer, F. C. *et al.* Zellweger spectrum disorders: clinical overview and management approach. *Orphanet J Rare Dis* **10**, 151, doi:10.1186/s13023-015-0368-9 (2015).
26. Burtman, E. & Regelmann, M. O. Endocrine Dysfunction in X-Linked Adrenoleukodystrophy. *Endocrinol Metab Clin North Am* **45**, 295-309, doi:10.1016/j.ecl.2016.01.003 (2016).
27. Khan, S. A. *et al.* Genetics of human Bardet-Biedl syndrome, an updates. *Clin Genet* **90**, 3-15, doi:10.1111/cge.12737 (2016).
28. Sedjai, F. *et al.* Control of ciliogenesis by FOR20, a novel centrosome and pericentriolar satellite protein. *J Cell Sci* **123**, 2391-2401, doi:10.1242/jcs.065045 (2010).
29. Wu, K. *et al.* DACH1 is a cell fate determination factor that inhibits cyclin D1 and breast tumor growth. *Mol Cell Biol* **26**, 7116-7129, doi:10.1128/MCB.00268-06 (2006).
30. Zhao, F. *et al.* DACH1 inhibits SNAI1-mediated epithelial-mesenchymal transition and represses breast carcinoma metastasis. *Oncogenesis* **4**, e143, doi:10.1038/oncsis.2015.3 (2015).
31. Gerstein, M. B. *et al.* Architecture of the human regulatory network derived from ENCODE data. *Nature* **489**, 91-100, doi:10.1038/nature11245 (2012).
32. Leskovec, J. & Sosič, R. SNAP: A General-Purpose Network Analysis and Graph-Mining Library. *ACM Transactions on Intelligent Systems and Technology (TIST)* **8**, 1 (2016).
33. Kuleshov, M. V. *et al.* Enrichr: a comprehensive gene set enrichment analysis web server 2016 update. *Nucleic Acids Res* **44**, W90-97, doi:10.1093/nar/gkw377 (2016).

PRECISION MEDICINE:
DATA AND DISCOVERY FOR IMPROVED HEALTH AND THERAPY

ALEXANDER A. MORGAN

Stanford University School of Medicine
Stanford, CA 94305 USA
Email: alexmo@stanford.edu

DANA C. CRAWFORD

Epidemiology and Biostatistics, Institute for Computational Biology
Case Western Reserve University
Cleveland, OH, 44106 USA
Email: dana.crawford@case.edu

JOSH C. DENNY

Vanderbilt University Medical Center
Nashville, TN 37203 USA
Email: josh.denny@vanderbilt.edu

SEAN D. MOONEY

University of Washington
Seattle, WA 98105 USA
Email: sdmooney@uw.edu

BRUCE J. ARONOW

Center for Computational Medicine
Cincinnati Children's Hospital Medical Center and the University of Cincinnati
Cincinnati, OH 45229 USA
Email: bruce.aronow@cchmc.org

STEVEN E. BRENNER

University of California
Berkeley, CA 94720-3012 USA
Email: brenner@compbio.berkeley.edu

The major goal of precision medicine is to improve human health. A feature that unites much research in the field is the use of large datasets such as genomic data and electronic health records. Research in this field includes examination of variation in the core bases of DNA and their methylation status, through variations in metabolic and signaling molecules, all the way up to broader systems level changes in physiology and disease presentation. Intermediate goals include understanding the individual drivers of disease that differentiate the cause of disease in each individual. To match this development of approaches to physical and activity-based measurements, computational approaches to using these new streams of data to better understand improve human health are being rapidly developed by the thriving biomedical informatics research community. This session of the 2017 Pacific Symposium of Biocomputing presents some of the latest advances in the capture, analysis and use of diverse biomedical data in precision medicine.

1. Introduction

The major goal of precision medicine is to improve human health. The researchers presenting work in the 2017 PSB conference session on precision medicine represent a wide range of approaches this challenge. The work ranges from examination of variation in the core bases of DNA and their methylation status, through variations in metabolic and signaling molecules, all the way up to broader systems level changes in physiology and disease presentation. Recent advances in areas as diverse as microfluidics, solid phase chemistry, optics, wireless communication, battery technology, and social networking are supporting the collection and analysis of a whole host of highly multiplex biomedical measurements in increasingly fine temporal resolution of sampling. Whether it is understanding the individual drivers of disease that differentiate the cause of disease in each individual, to the creation of customized drug dosing algorithms, the researchers in this session are advancing data-driven medicine from applying to populations down to individuals.

One common thread that unites much of this work is the value of large datasets combining a wide range of features that encompass causal factors, state measures, and differential outcomes. Whether using a large patient registry focused on specific phenotypes and pathologies (such as autism or cancer) or broad spectrum electronic medical record systems, the linking of data collected as part of healthcare delivery combined with molecular and genomic features has provided an invaluable resource to help create data-precision models for disease understanding and improving care.[1-3] Without these data resources, most of the work in this session would essentially be impossible. Although those who make maximal use of these large datasets have been criticized for taking undue advantage of the labor of others,[4] it is clear that making these large datasets available to biomedical informatics researchers is enabling new methodological developments and new insights to advance clinical care. One recently reported study on the genetics of hypertension used samples from over 300,000 people;[5] at this scale, data should be considered a resource of global importance to health and wellbeing, not part of the academic fiefdom of a single researcher. Newborn screening extends this to its largest scale, addressing every member of a population (e.g., nearly 500,000 per year in California) without bias.[6]

The extensive work reported in this session reflects the diversity of activity in precision medicine and the enthusiasm in the field. However, this enthusiasm must be tempered with healthy caution and skepticism. Last year, the PSB session on precision medicine[7] was accompanied by another session focused on aspects and challenges in reproducibility[8] in research, and this continues to be a challenge in our efforts to develop an individualized understanding of physiology and disease as each person is in effect a sample size of one. This is a challenge across science, and much of the research across the psychological sciences has recently been criticized for its poor level of reproducibility.[9] In parallel to the methodological challenges in reproducibility, there continues be healthy skepticism and cautious evaluation of the continuously evolving techniques and approaches to collecting samples, measuring their properties, and evaluating their biomedical significance in isolation or in combination with other data and properties. A recent evaluation of a direct-to-consumer lab testing technology[10] revealed that any claim of technological advance without appropriate controls, comparisons, and supporting evidence must be examined in open formats by external parties before launching its widespread use in clinical care.

The burden of very careful experimental design and reproducibility does not mean that the field of precision medicine is advancing slowly. For example, recent work has shown that it is possible to develop predictive, customized models of blood glucose level in response to different forms of dietary intake, a huge advance in precision, personalized nutrition.[11] Further research will demonstrate whether these models are stable over time.

The recent CAGI (Critical Assessment of Genome Interpretation)* evaluations have shown the power of the Common Task Framework to allow researchers to compare techniques that make predictions of phenotype from genotype, a key element of precision medicine. However, one of the trade-offs is that improved prediction accuracy often comes at the cost of human interpretability. For example, in the most recent CAGI of 2016, an approach using deep neural networks to predict psychiatric disease status from exome data performed better than other approaches that used far more interpretable models and those that integrated far more human knowledge. Unfortunately, the maturity of performance of these techniques of machine learning currently exceeds the maturity of the tools the to help interpret their predictions, limiting our ability to correct the apparent biases in our human understanding. Consequently, the significance and application of these findings are unclear. Much work has been done in fields like natural language processing and image processing to help visualize and unpack complex predictive AI models;[12] however, successful approaches and visualizations to fully support this increased understanding of many of the currently "black box" models of genomics and precision medicine are continuing to be developed.[13, 14] The many pieces of work presented in this

* https://genomeinterpretation.org

session use a range of visualizations and evaluation metrics, but this continues to be an active area of endeavor in need of new advances.

Concomitant with advances in predictive and analytic approaches, informatics, and machine learning techniques are learning how to perform goal directed tasks, often at better than human levels of performance.[15] It is hoped that we can go beyond simple tasks like playing complex games to guidance of the steps and actions in the delivery of healthcare; however, as noted this will require healthy and active skepticism along with insight into the models developed.

2. Podium presentations

When Hippocrates espoused the idea that physicians should be literate and keep records of patient care and outcomes[16], it was so that these records might be used to improve the understanding of disease and help future patients. It is therefore not a new idea that medical records might be a powerful source of data for advancing biomedicine; the widespread use of electronic medical records systems has allowed several researchers in this session to use these data to deepen our understandings of disease and possible new methods of precision treatment. In particular, **CR Bauer and colleagues** investigate the relationship between genetic variation and 29 common laboratory values. Importantly, they go beyond simply viewing each of the laboratory values as simple quantitative traits, but look at the relationships between those quantitative traits and start to examine compositional quantitative traits derived from those measurements. Although it is common to think about the multiplicity of possible hypotheses derived from examining many genetic variants, little effort is typically spent examining the multiple hypotheses that can be derived from how we partition and divide phenotypes. In the closely related work of **SS Verma and colleagues**, the focus is on the genetic drivers of the variability of common laboratory measurements. Going beyond the conventional central tendency of the laboratory values, they examine genetic associations with heteroscedasticity. This shift in focus from average value or pure prediction accuracy, toward models of higher moments and a focus on understanding what drives dispersion is another theme that runs through several of the papers in this session.

Laboratory values are part of assigning diagnoses, and **MK Beck and colleagues** mine records from 6,923,707 Danish patients to examine issues around the temporal ordering of diagnoses. They focus on the conditions of diabetes and sleep apnea, which often co-occur, but their presence can be hidden from the sufferer for years, and identification of one can lead to ascertainment bias of the other. When mining clinical records, researchers have access to when a disease was diagnosed but little data as to why, which may be impacted by a range of externalities, including differing access to care, but Beck and colleagues investigate patterns of age trajectory and of subsequent disease diagnoses, and data driven methods to stratify patients into subgroups. Moving up from laboratory measurements and diagnoses to directly guiding clinical decision making, but still using data derived from records of clinical care, **LK Wiley and colleagues** evaluate models that determine dosing of a medication with a narrow therapeutic window (warfarin) based on genetic variations in admixed populations, particularly those with African ancestry.

In addition to large datasets of mixed-type patient records, disease registries around specific diseases are a powerful data resource for precision medicine informatics. Three pieces of work in this session focus on techniques for identifying how variants in groups of genes may work together to contribute to phenotype, and much of this work relies on disease specific registry data. **GR Venkataraman and colleagues** use data from an autism patient registry to examine the way *de novo* mutations diffusely spread across sets of genes of shared function and how they may contribute to disease risk. The challenge of polygenic phenotypes is also the focus of the work of **D He and L Parida**, who presently work on disentangling epistasis underlying quantitative traits. **J Gallion and colleagues** have also been working on examining genetic variations in families of genes, in this case families of kinases in cancer, highlighting the shared disease association of variations in homologous locations across genes in a particular family.

Digging in more deeply into cancer, particularly using the data provided by The Cancer Genome Atlas,[17] **JA Thompson and CJ Marsit** present their work combining methylation with gene expression data to predict cancer survival; mixing heterogeneous data with colinearities being a hallmark of much of the cutting edge of precision medicine work. **G Speyer and colleagues** turn focus to drug response in cancer cells. Their work investigates the way expression dependency graphs vary between responsive and non-responsive cells; continuing the theme of mining differences in dispersion, here spread around network connectivity and likelihood, between subgroups. They also make the results of their work available online as a searchable resource.[†]

3. Posters with published papers

The work presented in our poster session with published papers represents a broad range of interests by research groups, with some fairly technical work delving in deeply to new methods of analysis of biomedical data. **A Beck and colleagues** present an approach for using genome uncertainty to modify thresh-holding for tests of Hardy-Weinberg equilibrium; highly relevant to some of the most basic

[†] http://biocomputing.tgen.org/software/EDDY/CTRP/home.html

analysis done in population genomics and often serving as a filter for all the analysis downstream of the variant calling.

The entire *in silico* metabolic modeling of the most simple of single cells is now a reality,[18] and techniques of temporal molecular metabolic flux analysis are advancing dramatically.[19] **A Schultz and colleagues** are working to identify cancer specific metabolic signatures, and we may one day have patient and cancer-specific cellular metabolic models as tools for precision medicine.

As noted, one of the themes of this session has been on the investigation of measures of dispersion as a key biological measures, and **PF Kuan and colleagues** are examining DNA methylation, with methylDMV, a tool that compares not only measures of central tendency but also heteroscedasticity, as a way to highlighting issues like sample bias vs. biological signal.

There is a substantial amount of work in this session delving into methods to better identify cancer subgroups, both as a tool to more precision in individualized prognostic models, but perhaps more importantly to find features that unite these groups that might lead to precisely targeted therapies in those subgroups, or at least provide increased clarity on which existing therapies are likely to more or less efficacious. Cancer driver mutation identification is the focus of the work by **M Ma and colleagues. A Durmaz and colleagues** present work on subgraph analysis with a focus on grouping via dysregulated pathways. **H Kabbat and colleagues** use a competitive endogenous RNA based method combining DNA copy number variation, mRNA expression, and microRNA levels.

The interest in pathway analysis and uniting mRNA with microRNA is united in the work of **D Diaz and colleagues**, which focuses on just that topic. Finally, **T Kamp and colleagues** present work on the value of moving to a more Boolean view of gene expression when doing gene set enrichment analysis in improving analytical output.

4. Acknowledgments

We would especially like to thank the many reviewers who contributed their considerable expertise and precious time to help the many paper submitters refine the presentation of their work. We would also like to thank the PSB 2017 chairs and Tiffany Murray of Stanford University for their efforts in organizing the meeting. This work is supported in part by NIH grants U19 HD077627, R01 AI105776, U41 HG007346, and by a Research agreement with Tata Consultancy Services.

5. References

1. Bush WS, Oetjens MT, Crawford DC. Unravelling the human genome-phenome relationship using phenome-wide association studies. *Nat Rev Genet.* 2016;17(3):129-45.

2. Denny JC, Bastarache L, Roden DM. Phenome-Wide Association Studies as a Tool to Advance Precision Medicine. *Annual Review of Genomics and Human Genetics.* 2016;17(1):353-73.

3. Roden DM, Denny JC. Integrating electronic health record genotype and phenotype datasets to transform patient care. *Clinical Pharmacology & Therapeutics.* 2016;99(3):298-305.

4. Longo DL, Drazen JM. Data Sharing. *New England Journal of Medicine.* 2016;374(3):276-7.

5. Ehret GB, Ferreira T, Chasman DI, Jackson AU, Schmidt EM, Johnson T, et al. The genetics of blood pressure regulation and its target organs from association studies in 342,415 individuals. *Nat Genet.* 2016;48(10):1171-84.

6. Brenner SE, Kingsmore S, Mooney SD, Nussbaum R, Puck J. Use of genome data in newborns as a starting point for life-long precision medicine. *Pac Symp Biocomput.* 2016;21:568-75.

7. Morgan AA, Mooney SD, Aronow BJ, Brenner SE. Precision medicine: data and discovery for improved health and therapy. *Pac Symp Biocomput.* 2016;21:243-8.

8. Manrai AK, Wang BL, Patel CJ, Kohane IS. Reproducible and shareable quantifications of pathogenicity. *Pac Symp Biocomput.* 2016;21:231-42.

9. Open Science Collaboration. Estimating the reproducibility of psychological science. *Science.* 2015;349(6251).

10. Kidd BA, Hoffman G, Zimmerman N, Li L, Morgan JW, Glowe PK, et al. Evaluation of direct-to-consumer low-volume lab tests in healthy adults. *The Journal of Clinical Investigation.*126(7):2773.

11. Zeevi D, Korem T, Zmora N, Israeli D, Rothschild D, Weinberger A, et al. Personalized Nutrition by Prediction of Glycemic Responses. *Cell.* 2015;163(5):1079-94.

12. Ribeiro MT, Singh S, Guestrin C. "Why Should I Trust You?" Explaining the Predictions of Any Classifier. *Proceedings of the 22nd ACM SIGKDD International Conference on Knolwedge Discovery and Data Mining.* 2016:1135-44.

13. Thorvaldsdóttir H, Robinson JT, Mesirov JP. Integrative Genomics Viewer (IGV): high-performance genomics data visualization and exploration. *Briefings in Bioinformatics.* 2013;14(2):178-92.

14. Gehlenborg N, O'Donoghue SI, Baliga NS, Goesmann A, Hibbs MA, Kitano H, et al. Visualization of omics data for systems biology. *Nat Methods.* 2010;7(3):S56-S68.

15. Silver D, Huang A, Maddison CJ, Guez A, Sifre L, van den Driessche G, et al. Mastering the game of Go with deep neural networks and tree search. *Nature*. 2016;529(7587):484-9.

16. Kassell L. Casebooks in early modern England: medicine, astrology, and written records. *Bull Hist Med*. 2014;88(4):595-625.

17. The Cancer Genome Atlas Research N, Weinstein JN, Collisson EA, Mills GB, Shaw KRM, Ozenberger BA, et al. The Cancer Genome Atlas Pan-Cancer analysis project. *Nat Genet*. 2013;45(10):1113-20.

18. Karr Jonathan R, Sanghvi Jayodita C, Macklin Derek N, Gutschow Miriam V, Jacobs Jared M, Bolival Jr B, et al. A Whole-Cell Computational Model Predicts Phenotype from Genotype. *Cell*. 2012;150(2):389-401.

19. Birch EW, Udell M, Covert MW. Incorporation of flexible objectives and time-linked simulation with flux balance analysis. *Journal of Theoretical Biology*. 2014;345:12-21.

OPENING THE DOOR TO THE LARGE SCALE USE OF CLINICAL LAB MEASURES FOR ASSOCIATION TESTING: EXPLORING DIFFERENT METHODS FOR DEFINING PHENOTYPES

CHRISTOPHER R. BAUER[1], DANIEL LAVAGE[1], JOHN SNYDER[1], JOSEPH LEADER[1], J. MATTHEW MAHONEY[2], SARAH A. PENDERGRASS[1]

Biomedical & Translational Informatics, Geisinger Health System
100 N. Academy Ave. Danville, PA 17821, USA
Email: cbauer@geisinger.com

Department of Neurological Sciences, University of Vermont College of Medicine
149 Beaumont Ave. Burlington, VT 05405, USA

The past decade has seen exponential growth in the numbers of sequenced and genotyped individuals and a corresponding increase in our ability of collect and catalogue phenotypic data for use in the clinic. We now face the challenge of integrating these diverse data in new ways new that can provide useful diagnostics and precise medical interventions for individual patients. One of the first steps in this process is to accurately map the phenotypic consequences of the genetic variation in human populations. The most common approach for this is the genome wide association study (GWAS). While this technique is relatively simple to implement for a given phenotype, *the choice of how to define a phenotype is critical.* It is becoming increasingly common for each individual in a GWAS cohort to have a large profile of quantitative measures. The standard approach is to test for associations with one measure at a time; however, there are many justifiable ways to define a set of phenotypes, and the genetic associations that are revealed will vary based on these definitions. Some phenotypes may only show a significant genetic association signal when considered together, such as through principle components analysis (PCA). Combining correlated measures may increase the power to detect association by reducing the noise present in individual variables and reduce the multiple hypothesis testing burden. Here we show that PCA and k-means clustering are two complimentary methods for identifying novel genotype-phenotype relationships within a set of quantitative human traits derived from the Geisinger Health System electronic health record (EHR). Using a diverse set of approaches for defining phenotype may yield more insights into the genetic architecture of complex traits and the findings presented here highlight a clear need for further investigation into other methods for defining the most relevant phenotypes in a set of variables. As the data of EHR continue to grow, addressing these issues will become increasingly important in our efforts to use genomic data effectively in medicine.

1. Introduction

In the past decade, genome wide association studies (GWAS) have revealed more than ten thousand associations between genetic loci and traits [1]. As GWAS continue to grow in number, sample size, and range of phenotypes, they offer an opportunity to begin to untangle the complex network underlying phenotypic variation. One challenge in this pursuit stems from an asymmetry in the genotype-phenotype map. While the range of genetic variation in humans is fairly well characterized and a given genome can be sequenced to arbitrary depth, there is no obvious way to measure a physiologically complete phenome or even outline how to divide it into separate units [2]. Even subtle choices in how a phenotype is defined can affect which loci associate with it [3,

4]. There is a growing need to analyze these choices and their effects if we wish to build a genotype-phenotype map that captures the relationships most relevant to biology and the clinic.

The first human GWAS defined phenotypes based on clinical case control status [5, 6, 7]. Binary phenotypes such as these are a natural choice if our ultimate goal is to predict disease risk, but diseases are typically diagnosed based on a number of underlying quantitative variables and expert opinions. For example, dozens of loci have been implicated in the risk of multiple sclerosis [8]. However, this condition is heterogeneous in its presentation and is diagnosed based on an accumulation of symptoms, quantitative measures, and subjective categorization, only after ruling out other conditions [9]. There are also subtypes of multiple sclerosis as well as other distinct but related demyelinating syndromes [10, 11]. This complexity makes it exceedingly difficult to understand how each of the associated gene variants might be contributing to the disease.

Recently we have begun to see association studies conducted in cohorts that have been given batteries of quantitative assays [12, 13] and comprehensive electronic health record (EHR) data is being used to construct phenotypic profiles. The availability of these large sets of traits has lead to an approach known as the phenome wide association study (PheWAS) where each variant is tested for associations with a range of phenotypes [14, 15]. Recent applications of PheWAS have revealed many novel genotype-phenotype associations and the potential for a large degree of pleiotropy within disease related traits [14, 16, 17]. Variants that associate with multiple traits could be indicative of genetic modules that underlie multiple diseases but in some cases they may simply represent partially redundant measures that correspond to a single disease state.

Given a profile of quantitative traits, multivariate techniques such as principal component analysis allow us to combine related variables into a set of statistically independent measures. Combining different raw measurements into new metrics can identify new associations that may provide important insights into the biology of complex traits and may provide better predictors of disease risk [18, 19]. Consider for example, four GWAS for height, weight, and body mass index (BMI), and type II diabetes. Even though BMI is simply a function of height and weight, the results of these three associations tests will not identify exactly the same sets of loci. Likewise, many variants associate with both BMI and type II diabetes, but a large part of this overlap stems from BMI being a risk factor for type II diabetes [20]. Metabolomic studies have also demonstrated that some gene variants show much stronger relationships with the the ratios of metabolites than they do with the absolute abundances of either molecule [18].

While an EHR can contain thousands of types of data, such as clinical laboratory measures, similar variables may be collected or reported in different ways. Logical observation identifiers names and codes (LOINC) provides unique numerical identifiers to distinguish relevant differences between laboratory measures [21]. Most analyses that have been conducted to date have involved laborious data harmonization procedures to ensure that grouped lab results measure the same quantity in the the same way [22]. With the large numbers and types of measures in the EHR, it is often not feasible to carefully harmonize each and every phenotype. Thus, it is important to explore approaches that will allow for high throughput use of multiple phenotypes.

Here, we have mined the Geisinger Health System EHR for quantitative measures to produce a high dimensional phenotypic profile for a large population of genotyped patients in the MyCode® Community Health Initiative. Using these data, we outline and compare three general strategies for identifying loci that associate with one or more components of this phenomic profile: PheWAS, PCA, and cluster PCA. Our results show that each of these methods can detect associations that are missed by the others and that the significance of a given association can vary by many orders of magnitude based on how a phenotype is defined. These findings set the stage for further use of EHR data in gene associations studies and highlight important considerations as we attempt to improve the predictive power of medical genomics and clinical phenotyping.

2. Methods

2.1. *Genetic Data*

All of the data described in this paper come from a cohort of patients in the MyCode Community Health Initiative at the Geisinger Health System. Each patient was genotyped for 659,010 SNPs with minor allele frequency greater than 1% using Illumina OMNI Express Exome chips. We excluded any SNPs that had call rates < 99%, sample call rates < 99%, as well as 113 SNPs that show large differences between batches. We restricted our analysis to individuals with greater than 99% likelihood of European ancestry, as defined by quadratic discriminant analysis using the first four principal components of ancestry based on the 1000 genomes project.

2.2. *Phenotypic Data*

For 38,269 patients in the Geisinger Health System that met these criteria, we extracted age, sex, BMI and the median values for the following 29 outpatient laboratory measures as defined by LOINC codes (Table 1). Most of the lab measures showed large deviations from normality at the population level, so we first performed Box-Cox transformations on each variable. Each variable was also centered and scaled by subtracting the mean value and dividing by the standard deviation.

2.3. *Imputation*

Within the set of lab data that we analyzed, 7.1% of patient-lab pairs had no results available. Nearly a third of the missing data came from ~6000, mostly young, individuals that lacked lipid measurements (Figure S1). We used predictive mean matching to impute all missing values. Imputation was performed in R, using the MICE package. Due to multicollinearity, within a subset of the 29 variables, we excluded 11 pairs of variables with correlation coefficients greater than 0.5 as predictors of each other. Aside from this restriction, each variable was modeled as a linear function of all other variable, include age, sex, and BMI. We performed 5 separate imputations, selecting among the 5 closest cases, over 120 iterations. Nearly all chains exhibited convergence with 20 iterations. In the majority of cases, the distribution of imputed values was indistinguishable from the original distribution (Figure S2).

Table 1. Definitions of the LOINC codes extracted from electronic health records.

LOINC	Description
718-7	Hemoglobin [Mass/volume] in Blood
4544-3	Hematocrit [Volume Fraction] of Blood by Automated count
787-2	Erythrocyte mean corpuscular volume [Entitic volume] by Automated count
786-4	Erythrocyte mean corpuscular hemoglobin concentration [Mass/volume] by Automated count
785-6	Erythrocyte mean corpuscular hemoglobin [Entitic mass] by Automated count
6690-2	Leukocytes [#/volume] in Blood by Automated count
789-8	Erythrocytes [#/volume] in Blood by Automated count
788-0	Erythrocyte distribution width [Ratio] by Automated count
32623-1	Platelet mean volume [Entitic volume] in Blood by Automated count
777-3	Platelets [#/volume] in Blood by Automated count
2345-7	Glucose [Mass/volume] in Serum or Plasma
2160-0	Creatinine [Mass/volume] in Serum or Plasma
2823-3	Potassium [Moles/volume] in Serum or Plasma
3094-0	Urea nitrogen [Mass/volume] in Serum or Plasma
2951-2	Sodium [Moles/volume] in Serum or Plasma
2075-0	Chloride [Moles/volume] in Serum or Plasma
2028-9	Carbon dioxide, total [Moles/volume] in Serum or Plasma
17861-6	Calcium [Mass/volume] in Serum or Plasma
1743-4	Alanine aminotransferase [Enzymatic activity/volume] in Serum or Plasma by With P-5'-P
30239-8	Aspartate aminotransferase [Enzymatic activity/volume] in Serum or Plasma by With P-5'-P
1975-2	Bilirubin.total [Mass/volume] in Serum or Plasma
2885-2	Protein [Mass/volume] in Serum or Plasma
10466-1	Anion gap 3 in Serum or Plasma
751-8	Neutrophils [#/volume] in Blood by Automated count
2093-3	Cholesterol [Mass/volume] in Serum or Plasma
2571-8	Triglyceride [Mass/volume] in Serum or Plasma
2085-9	Cholesterol in HDL [Mass/volume] in Serum or Plasma
13457-7	Cholesterol in LDL [Mass/volume] in Serum or Plasma by calculation
2965-2	Specific gravity of Urine

2.4. *Principal Component Analysis*

For each imputed dataset, we performed principal component analysis (PCA) in R, using the *prcomp* function. The PCA results were nearly identical within each imputed dataset. The average angle between all ordered pairs of Eigenvectors for the first 19 components was 4.9° and the only angles greater than 20° were caused by an alternation in the order of components 20 and 21 in some of the analyses (Figure S3). Given the minimal differences between the imputed data sets, we chose the first imputed data set to use in all downstream analyses.

2.5. *K-means clustering*

Using K-means clustering, we divided our 29 variables into 7 clusters based on their pairwise absolute correlations (Figure 1). The distance between two LOINC codes was defined as $1\text{-}R^2$. Clustering was performed in R using the *kmeans* function with 200 random starting clusters. Since sum of squares measures did not indicate an optimal number of clusters, we choose the maximum number of clusters where all clusters contained at least 3 phenotypes.

2.6. *GWAS*

We first performed associations between all 29 phenotypes individually (Figure S4). We also performed associations with 29 principle component scores (Figure S5). Finally, we performed associations with scores of the principal components within each cluster (Figure S6). All association tests were performed using PLATO 2.0 (https://ritchielab.psu.edu/plato). In each case, we modelled the principal component score as an additive function of allele count with age, bmi, sex, and the first four principal components of ancestry included in the model as covariates.

3. Results

Our phenotypic dataset comprised 29 outpatient clinical lab measures extracted from Geisinger Health System EHR. In order to ensure compatibility with other datasets, we choose to include only measures that complied with the LOINC standard of medical laboratory observations [23]. For each of the 29 clinical lab measures, we performed a separate GWAS in PLATO. Using these measures, we identified 6361 statistically significant associations (FDR < 0.01). Every lab measure had multiple SNPs associated with it, ranging from 12 SNPs for chloride concentration in blood to 783 SNPs for the number of leukocytes per unit of blood (Figure 1). Of these associations, 31% involved a SNP that was linked to more than one lab measure.

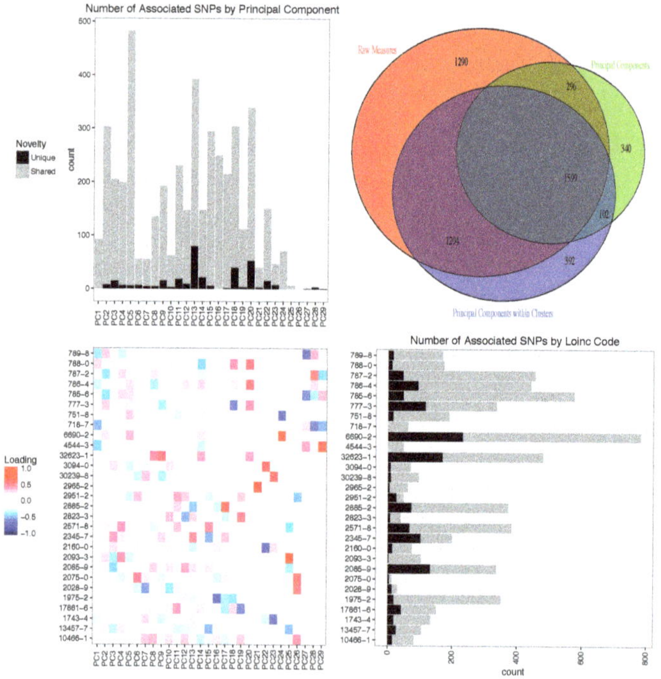

Figure 1. Associations detected with LOINC measures, PCA, and cluster PCA. The Venn diagram in the upper right panel shows the number of unique and shared SNPs that were associated with a phenotypic measure as defined by each of the three methods. The upper left panel shows the number of SNPs that associated with each principal component. The lower right panel shows how the associations were distributed across the LOINC measures. Gray bars represent the total number of SNPs while black shows the number that are unique to that measure. The bottom left panel shows how each of the phenotypes defined by LOINC codes loads onto each of the principal components.

Given that several groups of the lab results had very strong correlations and nearly all showed at least modest correlations with a few other variables (Figure 2) we hypothesized that statistical power might be improved by combining highly correlated measures. To test this, we performed principal component analysis on the combined set of all 29 lab measures. A plot of the cumulative variance explained by each additional component was smooth and increased gradually indicating that even the highest components might be measuring physiologically meaningful traits (Figure S7).

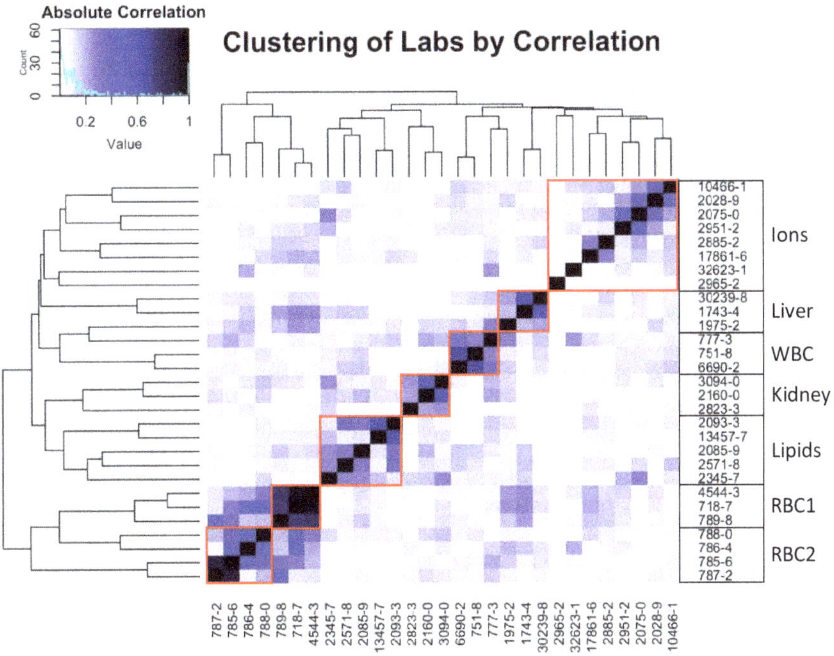

Figure 2.
Clustering of LOINC measures into related groups. The heat map indicates the absolute value of the correlation coefficient between all pairs of LOINC codes. Each cluster, as defined by k-means, is indicated by a red bounding box. The names on right column indicated the functional categories that describe each cluster.

We next performed GWAS for all 29 principal components, just as we did for the original measures (Figure S5). This analysis resulted in 4536 significant associations (FDR = 0.01). We expected to see a reduction in the total number of significant associations as one principal component could capture variation from multiple raw measures. Surprisingly, 48% of these associations involved a SNP that was linked to multiple components. Figure 1 shows a Venn diagram comparing the number of unique and overlapping associations across the various approaches for phenotypes used in this paper. Although 2494 of the SNPs that associated with one or more of the LOINC measures did not show a significant association with any of the principal components we did discover 442 new associations using these scores. 1895 SNPs associated with both a raw measure as well as a principle component (PC).

PC5 had the largest number of significant associations, 482, followed by PC13 with 392 and PC20 with 339. There was no clear pattern in how the significant associations were distributed among the first 24 components, although there were practically no associations with PC25-29 (Figure 1). In PCA, the first few components often capture a large percentage of the variation so it was interesting to see so many SNPs associating with higher components while PC1 only had 92 associated SNPs. Further analyses provide some insights. First, if we include age and BMI in the set of variables prior to PCA, we find that these variables load most strongly onto components 1. This makes sense given that age and BMI contribute to many physiological measures, especially among disease relevant traits. However, since these are both covariates in the GWAS regression model, it would be troubling to see many SNPs associated with PC1.

In PCA, the loadings indicate the magnitudes and directions that the original measures contribute to each component. Analyzing the loadings of some noteworthy components provided some additional clues to the causes of this behavior. PC2 was dominated by a few measures of blood cells: namely, the volume of erythrocytes moving in the opposite direction of cholesterol and the numbers of erythrocytes, white blood cells (WBCs), and platelets (Figure 1). PC5 was similar with WBC counts moving in the opposite directions of platelet volume and cholesterol. These associations may reveal overlap in genetic networks that regulate lipids and and the immune system. A number of studies have previously identified relationships between WBC counts and carotid plaque thickness, body fat percentage, and lipid profiles [24, 25, 26].

While PC2 and PC5 were linked to many loci, these were predominantly the same loci that were linked to one or more of the original measures. More relevant than the total number of associations detected is the number of associations that were unique to a principal component and not detectable using any of the original measures. PC13, PC18, and PC20 were responsible for the majority of these novel associations. PC13 measures a complex relationship among our measures in which serum levels of potassium and glucose vary inversely with total protein and creatinine. This is interesting because potassium and creatinine are highly correlated at the population level and both are diagnostic of kidney function. 89% percent of the associations with PC13 also map to the HLA locus suggesting a relationship between the adaptive immune system and these blood measures. PC18 and PC20 both measure relationships among erythrocyte distribution width, hemoglobin, and platelet measures (Figure 1).

Overall, the principal component approach detected fewer total significant associations than the LOINC measures, but a few components did allow us to identify novel associations. The principal components that proved most useful in this regard seemed to load primarily off of 2-6 measures (Figure 1) and those measures tended to be closely related. Components that were dominated by a single measure or had large number of weak loadings did not yield many novel results. These observations suggested a third approach. If we first divided the original 29 measures into small groups of related traits before performing PCA, we might restrict our range of phenotypes to space that corresponds better to the ways that gene variants actually impact phenotype.

Using K-means clustering, we divided our 29 variables in 7 clusters based on their pairwise absolute correlations. The choice of the number of clusters was somewhat arbitrary as the sums of square both within and between clusters never reached obvious plateaus. The choice of 7 clusters resulted in each group containing 3-8 measures, which corresponded well to our desired range, and it also broke them into groups that made intuitive sense (Figure 2). For example, all of the white blood cell counts formed a single cluster, and all of the lipid measures clustered together with serum glucose. We then performed PCA within each of these clusters and used these principal component scores to run a third GWAS with the same parameters as the previous two (Figure S6).

The genetic variants that associated with the scores of the cluster principal components had much larger overlap with original measures, sharing 2803 SNPs, but it also revealed 392 new SNPs that did not associate with either the original measures or the principal components of the entire data set (Figure 1). The distribution of these new associations varied greatly among each cluster (Figure S8). Within the ions cluster, the majority of the SNPs showed stronger associations with one of original measures than they did with any principal component (Figure S14). Within the three phenotypes that compose the liver cluster (1743-4: alanine aminotransferase, 30239-8: aspartate aminotransferase, and 1975-2: bilirubin), the associations detected for all three principal components correlated almost perfectly with those of one of the original measures (Figure S15). However, within the red blood cell cluster 1 (718-7: hemoglobin, 4544-3: hematocrit, and 789-8: erythrocytes), nearly all of the alleles tested showed their strongest association with one of the principal components (Figure 3).

Within each cluster, the middle components were the most likely to have novel associations. In general, PC1 had associations that were very similar in their significance levels to those found with the original measures. With each successive PC, the p-values would usually become more significant with respect to the LOINC measures, but less significant in absolute terms due to the reduction in total variance with each PC. In the red blood cell 1 (RBC1) cluster, nearly all of the novel significant associations occur with PC2 (Figure 3). A high score in this component corresponds to a low count of erythrocytes per unit volume of blood, but a high hematocrit score, and hemoglobin concentration. Since none of these associations were not found using erythrocyte mean corpuscular volume (787-2) as the phenotype of interest, it seems that there are a large number of gene variants linked to the concentration of hemoglobin within erythrocytes. A Manhattan plot shows that these new associations come from many distinct loci (Figure 4).

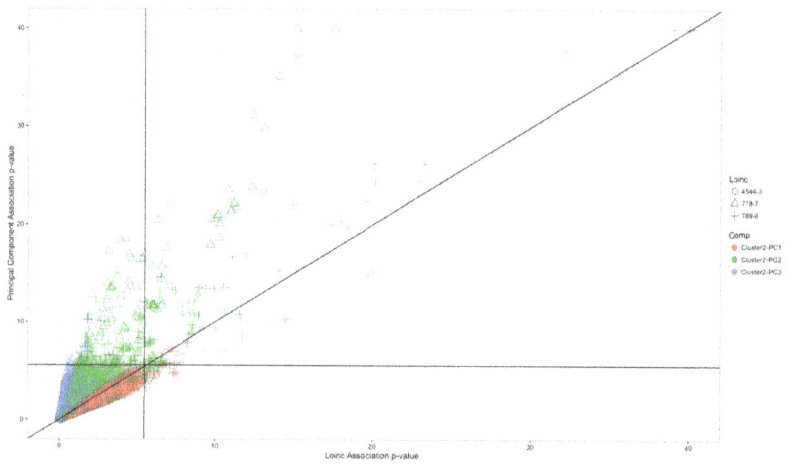

Figure 3.
Comparison of p-values for associations with the principal components and LOINC measures that compose the red blood cell cluster 1. Each point in the scatter plot represent one SNP. Both axes are scaled to the negative log base ten of the p-values. The y-axis indicates the lowest p-value that a given SNP had with any of the principal components. The components are coded by the color or the point. The x-axis indicates the lowest p-value that a given SNP had with any of the LOINC measures. The LOINC measures are coded by the shape of the point.

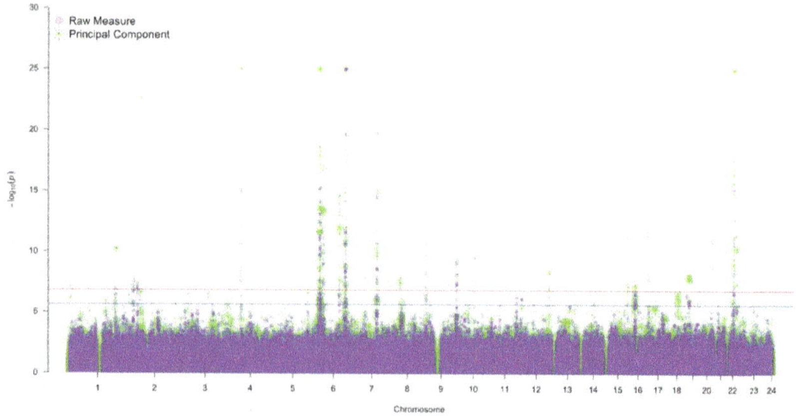

Figure 4.
Manhattan plot of the associations detected for the RBC1 cluster. The x-axis indicates the chromosomal coordinate and the y-axis shows the negative log base ten of the association p-value. Associations with any of principal components and LOINC measures are displayed in green and purple, respectively. The red line indicates a false discovery rate of 0.001 and the blue line indicates a false discovery rate of 0.01.

4. Discussion

Our results demonstrate that the choice of how to define a phenotype can have a large impact on our ability to detect relationships with genetic loci. Given a set of quantitative trait measures, we have outlined three different strategies for defining phenotypes prior to association testing. The standard method is to simply test against whatever phenotypic measures are in hand, without any additional considerations. While some measures of phenotype may be arbitrary or based purely on convenience, this may still be the most reasonable choice in many situations. In this particular case, the original phenotypes come from clinical lab tests that are prescribed because they have proven to be useful diagnostics and we find the greatest number of significant associations using these measures alone.

In spite of this generality, our results also indicate that many genotype-phenotype connections are not apparent when phenotypes are considered individually. Using two different methods based on principal component analysis, we have increased the number of significant associations that we could detect by 19%. Given the extremely large number of hypotheses that are tested in a single GWAS experiment, the p-value threshold for significance must be correspondingly low. Most segregating alleles have relatively small impacts on any given phenotype and we are unlikely to detect a significant association unless the phenotype of interest aligns very well with the effect of the variant. The majority of true positive results will inevitably fail to reach the significance threshold.

Principal component analysis provides one strategy for overcoming some of these obstacles. When performed on the entire dataset, it has the ability to capture relationships between diverse phenotypic measures. In this case, the components with the largest variance did not provide much new information. This is likely because these components capture the covariance of large numbers of measures that relate to the biggest sources of phenotypic variation in populations, such as age. There should not be genetic determinants of age, except in extreme cases, and even if there were, it is common practice to control for the effects of age in regressions.

The greatest utility of this method comes from the middle order components that capture more complex relationships. In our analyses, PC13 was related to a complex interaction between serum concentrations of potassium, creatinine, glucose, and total protein. It is not yet clear how this relates to human physiology, but the fact that 79 SNPs, distributed widely across the HLA locus, associate more strongly with this principal component than they do with any of the measures that contribute to it suggests some underlying mechanistic connection between this combination of variables and the function of the immune system. Perhaps phenotypic profiles such as this will also prove to be useful indicators of disease risk or progression.

This approach also allows us to observe effects that are orthogonal to the primary axis of variation. For example, creatinine and urea levels are both indicative of kidney function and they are very highly correlated at the population level. However, urea is a byproduct of all protein metabolism while creatinine is produced only by muscles so it is reasonable to assume that various genes could influence these traits independently. Indeed, principal component 22 corresponds to an inverse relationship between these two variables and several variants associate with the ratio of creatinine to urea in the blood without a detectable relationship to either variable in isolation.

One of the weaknesses of using principal component scores from a large dataset is that the eigenvectors correspond to the maximal variance within a set of measures which may not have any relationship to how traits are influenced at the gene regulatory level. A SNP that might correspond to elevated total cholesterol is unlikely to affect every other trait that correlates with cholesterol in a population. It can also become difficult to extract meaning from a principal component that is influenced by many, potentially disparate measures. If we hope to translate research findings into clinically relevant information, it can be useful to limit our search space to a number of dimensions that a human can understand. In order to strike balance between exploring the full range of complex interactions in biology and maintaining the ability to interpret our results, we also investigated a third approach that involved clustering our data based on the correlation structure of the variables prior to performing PCA.

While this did not improve our power in all cases, several groups of related measures yielded many more genetic associations, and at least a few new associations were discovered within each cluster. In particular, assays of blood cells and kidney function seem to benefit the most from this technique. The first 3 components of the RBC2 cluster collectively associate with 134 SNPs that do not show significant associations with any other measure that we tested. These components each measure different ways that the variance in erythrocyte size relates to hemoglobin concentration and mean erythrocyte volume. It is interesting to note that PC3 from this cluster had the most unique associations and is related to PC20 from the global PCA, which also identified new SNPs. Within the kidney cluster, PC3 measures the difference between urea and creatinine levels and associates with 41 unique variants. Again, this is related to PC22 from the global analysis. The fact that both clustered and global PCA identify associations with complex interactions between multiple blood cell and kidney function measurements indicates that the genetic regulation of these traits is not captured well by any single measure. It will be interesting to test if these same interactions are linked to the prevalence or prognosis for any disease states.

It is likely that numerous other combinations of the underlying measures would yield even more connections between gene variants and phenotypes but there is no way to exhaustively explore them. As the number of phenotypic measures that we can collect for a GWAS cohort continues to grow, it will be increasingly important to develop better strategies for specifying exactly which measures to choose test for associations. Further investigation into this topic will be critical to gaining insight into gene function and has deep implications for how we think about concepts such as pleiotropy.

Acknowledgements

We would like to acknowledge Marylyn Ritchie, Marta Byrska-Bishop, Anna Basile, Anurag Verma, and H. Lester Kirchner for helpful discussions.

References

1. D. Welter et al., *Nucleic Acids Res.* **1;42**, 1001 (2014).
2. M. Samuels, *Curr Genomics.* **11(7)**, 482 (2010).
3. W. Bush and J. Moore, *PLoS Comput Biol.* **8(12)** (2012).
4. E. Stergiakouli et al., *Obesity.* **10**, 2252 (2014).
5. R. Klein et al., *Science.* **308**, 385 (2005).
6. A. Dewan et al., *Science.* **314**, 989 (2006).
7. The Wellcome Trust Case Control Consortium, *Nature.* **447**, 661 (2007).
8. International Multiple Sclerosis Genetics Consortium et al., *Nat Genet.* **45(11)**, 1353 (2013).
9. S. Katz, *Curr Opin Neurol.* **28(3)**, 193 (2015).
10. T. Avsar et al., *PLoS One.* **5;10(5)**, e0122045 (2015).
11. D. Karussis, *J Autoimmun.* **48-49**, 134 (2014).
12. Y. Kamatani et al., *Nat Genet.* **42(3)**, 210 (2010).
13. K. Suhre et al., *Nat Genet.* **43(6)**, 565 (2011).
14. J. Denny et al., *Nat Biotechnol.* **31**, 1102 (2013).
15. S. Pendergrass et al., *Hum Hered.* **79(3-4)**, 111 (2015).
16. S. Pendergrass et al., *PLoS Genet.* **9(1)**, e1003087 (2013).
17. M. Hall et al., *PLoS Genet.* **4;10(12)**, e1004678 (2014).
18. C. Geiger et al., *PLoS Genet.* **4(11)**, e1000282 (2008).
19. E. Stergiakouli et al., *Obesity.* **22(10)**, 2252 (2014).
20. P. Visscher et al., *Am J Hum Genet.* **13;90(1)**, 7 (2012).
21. A. Forrey et al., *Clin Chem.* **42(1)**, 81 (1996).
22. S Bennett et al., *Genet Epidemiol.* **35(3)**, 159 (2011).
23. J. Deckard et al., *J Am Med Inform Assoc.* **22(3)**, 621 (2015).
24. S. Mitchell et al., *Stroke.* **32**, 842 (2001).
25. M. Farhangi et al., *J Health Popul Nutr.* **31(1)**, 58 (2013).
26. L. Ferreira et al., *Rev. Bras. Hematol. Hemoter.* **35**, 3 (2013).

A POWERFUL METHOD FOR INCLUDING GENOTYPE UNCERTAINTY IN TESTS OF HARDY-WEINBERG EQUILIBRIUM

ANDREW BECK

Department of Biostatistics, University of Michigan
Ann Arbor, MI, 48109 USA
Email: beckandy@umich.edu

ALEXANDER LUEDTKE

Department of Biostatistics, University of California- Berkeley
Berkeley, CA, 94720 USA
Email: aluedtke@berkeley.edu

KELI LIU

Department of Statistics, Harvard University
Cambridge, MI, 02138 USA
Email: kliu@college.harvard.edu

NATHAN TINTLE

Department of Mathematics, Statistics, and Computer Science, Dordt College
Sioux Center, IA 51250, USA
Email: Nathan.Tintle@dordt.edu

The use of posterior probabilities to summarize genotype uncertainty is pervasive across genotype, sequencing and imputation platforms. Prior work in many contexts has shown the utility of incorporating genotype uncertainty (posterior probabilities) in downstream statistical tests. Typical approaches to incorporating genotype uncertainty when testing Hardy-Weinberg equilibrium tend to lack calibration in the type I error rate, especially as genotype uncertainty increases. We propose a new approach in the spirit of genomic control that properly calibrates the type I error rate, while yielding improved power to detect deviations from Hardy-Weinberg Equilibrium. We demonstrate the improved performance of our method on both simulated and real genotypes.

1. Introduction

With recent advances in high-throughput gene sequencing technologies, it is now possible to obtain measurements on millions of single nucleotide variants (SNVs) throughout the human genome. Large scale genetic data sets, whether from microarray, sequencing or imputation, contain genotype uncertainty which, if unaccounted for in downstream analyses, can significantly decrease power to detect disease-variant associations [1,2] if the uncertainty is not associated with the phenotype, or affect the corresponding type I error rate [3,4] if the uncertainty is associated with the phenotype. To minimize the impact of genotype uncertainty, a standard pre-processing step in most studies is to remove markers that are not in Hardy-Weinberg Equilibrium (HWE), since genotyping errors due to factors like DNA contamination and allelic dropout can cause deviation from HWE [5,6].

The standard approach to testing HWE uses a χ^2_{GOF} test whereby observed genotype frequencies at a variant site are used to obtain maximum likelihood estimates (MLEs) of the minor allele frequency (MAF; f) at the site. A one degree of freedom χ^2_{GOF} statistic is then computed to test the null hypothesis that the observed genotype frequencies follow HWE, namely $(1-f)^2, 2f(1-f)$ and f^2 for the major homozygote, heterozygote and minor homozygote, respectively. While this version of the test is the most straightforward and widely used, alternatives exist including methods for testing HWE in datasets with excess correlation between subject genotypes [7,8], missing genotypes [9] and those that account for covariates [10].

Recently, another alternative HWE testing approach was proposed, $\chi^2_{Posterior}$ [6], which extends the standard χ^2_{GOF} approach to allow for the incorporation of genotype uncertainty. The method has widespread application since for all common genotyping technologies (SNP microarray technology [11], imputation [12] and next-generation sequencing technology [13,14]), probabilistic genotypes are obtained as part of the standard genotype calling pipeline. Such probabilistic genotypes typically take the form of a vector of three posterior probabilities for each individual at each variant site, representing the posterior probability that the individual is actually each of the three possible genotypes. While standard analysis techniques typically "call" genotypes by summarizing the posterior probability by a single discrete genotype (e.g., mode posterior probability), researchers are increasingly using alternative approaches. For example, researchers may use of the entire vector of posterior probabilities or they may use the expected genotype (dosage) [15]. The simulation results of Zheng et al. [15], which were recently made rigorous [16], demonstrate substantial power loss from the use of the modal genotype in many realistic situations and approximately equivalent power from use of the dosage or the entire vector of posterior probabilities in case-control tests of genetic association. These results underscore the importance of considering HWE testing methods, which incorporate genotype uncertainty via the underlying posterior probabilities.

The traditional χ^2_{GOF} makes the key assumption that genotype counts are non-negative integers at each variant site, an assumption that is violated with the inclusion of probabilistic calls. A recently proposed alternative approach, $\chi^2_{Posterior}$, allows for the incorporation of probabilistic genotypes. However, $\chi^2_{Posterior}$ has been shown to be overly conservative (empirical type I error rate less than nominal) as uncertainty at the variant site increases [6]. In this manuscript, we explore reasons for the conservative nature of $\chi^2_{Posterior}$ and propose an alternative approach to HWE testing which incorporates genotype uncertainty while maintaining the type I error rate at nominal levels. We then evaluate the type I error and power of the new approach across a variety of realistic HWE and non-HWE settings to identify powerful and robust HWE tests for probabilistic genotypes. Finally, we implement the new method on a real data set illustrating its improved ability to maintain the type I error rate, while improving power to detect variants not in HWE.

2. Methods

2.1. Notation

To facilitate the presentation and evaluation of existing and novel approaches to testing for HWE while incorporating genotype uncertainty, we start by defining some basic notation we will use throughout the manuscript. Genotypes for a given individual i can be represented as a vector of three posterior probabilities, $\alpha_i \triangleq (\alpha_{i0}, \alpha_{i1}, \alpha_{i2})$, where α_{ik} is the posterior probability that individual i has k minor alleles, $k = 0,1,2$ at a variant site of interest. The vector of posterior probabilities, α_i, suggests that the true minor allele count for individual i, denoted $x_i \in 0,1,2$, can be modeled as being a single random draw from a multinomial distribution with probabilities indicated by α_i. We assume that α_i is available for each individual.

2.2. Existing approaches to incorporating genotype uncertainty

The most straightforward and widely used approach to manage genotype uncertainty is to summarize the vector of posterior probabilities α_i with the modal genotype, namely, $m_i \triangleq \arg\max_k(\alpha_i)$ in place of the individuals true genotype. When the modal genotype is used as the true genotype, a standard χ^2 goodness of fit test can be used to test for HWE (χ^2_{Mode}). However, when using such a method we expect an increase in the type I error rate and/or decrease in power due to the introduction of genotype errors caused by ignoring the genotype uncertainty represented in the posterior probabilities vector [2,6]. For example, if $\alpha_{i0} = 0.95$ (the mode), we "call" the individual as having no rare alleles and, thus, there is a 5% chance we are incorrect.

A recently proposed test for HWE, $\chi^2_{Posterior}$, utilizes the entire vector of posterior probabilities [6]. This method starts by computing three, non-discrete, genotype counts based on α_i: $A_0^* = \sum_{i=1}^N \alpha_{i0}$, $A_1^* = \sum_{i=1}^N \alpha_{i1}$, and $A_2^* = \sum_{i=1}^N \alpha_{i2}$, where N is the total sample size and we use A^* to represent genotype counts computed by summing the posterior probabilities across the sample. This approach applies a standard χ^2 goodness of fit test as follows

$$\chi^2_{Posterior} = \chi^2_{GOF}(A^*) = N\left[\frac{\left|\frac{A_0^*}{N}-(1-\hat{f})^2\right|-c/N}{(1-\hat{f})^2}+\frac{\left|\frac{A_1^*}{N}-2(1-\hat{f})\hat{f}\right|-c/N}{2(1-\hat{f})\hat{f}}+\frac{\left|\frac{A_2^*}{N}-(\hat{f})^2\right|-c/N}{(\hat{f})^2}\right] \tag{1}$$

where c is a continuity correction, e.g. 0.5 [17], and where the maximum likelihood estimate (MLE) of the minor allele frequency (MAF), $\hat{f}$, at the site is estimated as $\frac{A_1^* + 2A_2^*}{2N}$. The test uses as its null hypothesis that the variant site is in HWE, and as the alternative hypothesis that the variant site is not in HWE. This approach uses a central χ^2 distribution with a single degree of freedom as the null distribution for $\chi^2_{Posterior}$.

2.3. Direct likelihood approach

As shown via simulation in prior work [6], and confirmed in our simulations (see *Results*), the $\chi^2_{Posterior}$ test has an overly conservative type I error rate, which becomes more pronounced as genotype uncertainty increases. We now argue that the reason for this overly conservative type I error rate is due to a change in the covariance structure of the genotypes when using probabilistic genotypes (α_i). In particular, the $\chi^2_{Posterior}$ test assumes that each individual genotype occurs according to a multinomial distribution. However, this is no longer the case when observed genotype counts are obtained by summing over the posterior probability vectors [18]. Thus, the covariance structure assumed by the $\chi^2_{Posterior}$ test is not true in practice when using probabilistic genotypes. In situations where the alternative covariance structure due to probabilistic genotypes can be explicitly modeled or otherwise controlled for, likelihood based approaches to testing with uncertain genotypes are possible [15,18]. However, that is not the case for HWE testing, as we explain in the following paragraph.

In particular, in order to develop a likelihood ratio test you must have an explicit expression for the likelihood function of the population genotype frequencies, $G_0, G_1, and\ G_2$. Here the likelihood function can be written as $L(G_0, G_1, G_2; \alpha_1, \dots, \alpha_N) = P(\alpha_1, \dots, \alpha_N | G_0, G_1, G_2) = P(\alpha_1, \dots, \alpha_N | g_1, \dots, g_N, G_0, G_1, G_2) P(g_1, \dots, g_N | G_0, G_1, G_2)$, where g_i indicates the true genotype of individual *i*. Thus, you must have knowledge of the true uncertainty mechanism, $P(\alpha_1, \dots, \alpha_N | g_1, \dots, g_N, G_0, G_1, G_2)$ in order to develop a likelihood ratio test based on the posterior probabilities alone. Because explicit knowledge of the true uncertainty mechanism is unlikely, a likelihood approach to HWE testing using $\alpha_1, \dots, \alpha_N$ will not be possible without making unwarranted assumptions.

2.4. Alternative approach

Because of the overly conservative nature of existing approaches and the limitations we describe above when deriving an explicit likelihood approach, we present an alternative strategy: a post-hoc empirical correction in the spirit of genomic-control. Genomic control [19] is a widely-utilized post-hoc correction factor in genome-wide association studies. When systematic inflation of SNP-association statistics occurs in the data, which can occur due to population stratification or differential genotyping errors, dividing the distribution of observed chi-squared statistics by the median observed chi-squared statistic properly controls the empirical type I error rate. Essentially, this approach assumes that when testing thousands of variant sites for association with the phenotype, the vast majority of sites will not be associated with the phenotype. Thus, the observed distribution of test statistics, aside from the extreme upper-tail, can, in essence, be used as its own null distribution.

To extend the notion of genomic control to HWE testing, we argue that in most real testing situations, the majority of variant sites in a sample of many thousands of variants will be in HWE. Thus, we propose computing $\chi^2_{Posterior,j}$ from A^* as shown above for all variants of interest,

j=1,…,m, where m is large. Then the measure of inflation/deflation in the null distribution of test statistics is computed as $\hat{\lambda} = \frac{median(\chi^2_{GOF,1}, \chi^2_{GOF,2}, \ldots, \chi^2_{GOF,m})}{median(\chi^2_1)}$, where $median(\chi^2_1) = 0.455$ [19]. The genomic control-like test statistic for HWE is then computed as $\chi^2_{GC,j} = \frac{\chi^2_{GOF,j}}{\hat{\lambda}}$ for all j=1,…,m. We consider four different versions of χ^2_{GC}: $\chi^2_{GC,overall}$, $\chi^2_{GC,MAF}$, χ^2_{GC,r^2} and χ^2_{GC,MAF,r^2}, where $\hat{\lambda}$ is computed on different subsets of the data. Overall indicates that $\hat{\lambda}$ is computed across all m SNPs in the set. MAF indicates that $\hat{\lambda}$ is computed separately by MAF group (0.05-0.10, 0.1-0.2, 0.2-0.3, 0.3-0.4, 0.4-0.5). r^2 indicates that $\hat{\lambda}$ is computed separately by r^2 group (0-0.5, 0.5-0.75, 0.75-0.85, 0.85-0.95 and 0.95-1), where r^2 is a measure of genotype uncertainty- see next section for details. And, MAF, r^2 computes $\hat{\lambda}$ in groups defined by both MAF and r^2 (25 separate groups).

2.5. Simulation

We simulated genotype data in order to explore the performance of our proposed new approach under a wide variety of situations. We simulated approximately 850,000 SNPs where HWE was maintained (HWE SNPs). To ensure that the characteristics of this simulation reflected both a realistic allele frequency distribution as well as genotype uncertainty, we randomly sampled (f, r^2) pairs with replacement from a large dataset of genotypes from the FUSION study [20] that were imputed using MaCH [12]. For each (f, r^2) pair, we then simulated the 'real' genotypes of 10,000 individuals according to the specified allele frequency, f, assuming the population was in Hardy-Weinberg Equilibrium (HWE) $((1-f)^2, 2f(1-f), f^2)$. To model genotype uncertainty at the appropriate level, r^2, we drew from one of the following Dirichlet distributions conditional on the true genotype [16].

If $g_i = 2$ then $\alpha_i = (\alpha_{i0}, \alpha_{i1}, \alpha_{i2}) \sim Dirichlet(aq^2, 2aq(1-q), a(1-q)^2)$
If $g_i = 1$ then $\alpha_i = (\alpha_{i0}, \alpha_{i1}, \alpha_{i2}) \sim Dirichlet(aq(1-q), a(1-q)^2 + aq^2, aq(1-q))$
If $g_i = 0$ then $\alpha_i = (\alpha_{i0}, \alpha_{i1}, \alpha_{i2}) \sim Dirichlet(a(1-q)^2, 2aq(1-q), aq^2)$

for $a > 0$ and $0 < q < 1$, where a and q are chosen to yield a desired r^2 value. This model is chosen to simulate symmetric noise in posterior probabilities while maintaining HWE. Further details are available in *Appendix #1* and elsewhere [16]. In short, parameter q is the "average" amount of error. For example, if q=0.05 (5% noise/error level in posterior probabilities), then for the major homozygote, $g_i = 2$, $E(\alpha_i) = (\alpha_{i0}, \alpha_{i1}, \alpha_{i2})$ = (0.9025, 0.0.95, 0.0025) and, likewise, if there is no noise/error (q=0), then $\alpha_i = (0,0,1)$. Parameter a is the variation in the error from person to person. For example, as a increases, then $Var(\alpha_i)$ also increases, and so for very small values of a (e.g., a=0.01), there is virtually no variation in the values of α_i from person to person.

We also simulated three sets, each with approximately 75,000 SNPs, that were not in HWE (non-HWE SNPs). To do this we randomly sampled two SNPs (i and j) that were in HWE from the set of 850,000 SNPs described above, keeping track of the difference in the allele frequencies of the two SNPs, $d_{i,j}=f_i-f_j$. We then randomly sampled n(1-k) individuals from SNP i and nk individuals from SNP j, combining the individuals into a single sample of n individuals. We used values of k=0.1, 0.3 and 0.5, and continued to use a total sample size of 10,000. Thus, the resulting sample is not in HWE because the observed genotype frequencies were generated from two subpopulations with different allele frequencies.

The three resulting sets of 75,000 simulated SNP genotypes were analyzed using (a) a standard HWE test on the simulated 'real' genotypes (χ^2_{True}), (b) chi-squared on the modal genotype χ^2_{Mode}, (c) the approach utilizing posterior probabilities ($\chi^2_{Posterior}$) and (d) four different GC-like approaches (χ^2_{GC}; see previous section for details). For the purposes of the GC-like approach we combined random subsets of 25,000 non-HWE SNPs with the 850,000 HWE SNPs and applied the adjustment, keeping the total proportion of non-HWE SNPs in the set below 3%.

Type I error rates were computed on the 850,000 HWE SNPs as the proportion of SNPs that were detected to be 'not in HWE' at a particular significance level and for a particular combination of MAF and r^2 levels. Power was computed as the fraction of non-HWE SNPs with a p-value less than the significance level in 300 separate groups created by values of k (0.1, 0.2, 0.5), difference in MAF between the two SNPs being mixed together (0.1, 0.1-0.2, 0.2-0.3 or >0.3), observed MAF of the combined variant (0.05-0.10, 0.10-0.20, 0.20-0.30, 0.30-0.40 and 0.40-0.50) and observed r^2 of the combined variant (0-0.50, 0.50-0.75, 0.75-0.85, 0.85-0.95 and 0.95-1.0). We examined significance levels of 0.01, 1×10^{-3}, and 1×10^{-5}. We computed power and type I error rates across a variety of subsets of the variants including minor allele frequency, genotype uncertainty (r^2), and deviation from HWE.

2.6. Real data analysis - FUSION

As a proof of concept, we ran χ^2_{Mode}, $\chi^2_{Posterior}$ and χ^2_{GC,MAF,r^2} on 29,361 SNPs imputed with MaCH from chromosome 21 of the FUSION study (n=2456) [20]. We also created 2,377 new variants based on the 29,361 imputed variants, which were out of Hardy-Weinberg equilibrium. These 2,377 new variants were created by first randomly selecting two variants with differences in minor allele frequency of between 0.1 and 0.2 and r-squared values between 0.75 and 0.85. A new variant is created by randomly selecting 10% of the genotypes from one of the variants and 90% from the other. All three Hardy-Weinberg equilibrium tests (χ^2_{Mode}, $\chi^2_{Posterior}$ and χ^2_{GC,MAF,r^2}) were also applied to the 2,377 new non-HWE variants as well. We used a significance level of 1×10^{-5} on the 29,361 real and 2,377 new FUSION variants.

Table 1. Overall type I error rates

Method	Significance level		
	0.01	**0.001**	**1x10^{-5}**
$\chi^2_{Posterior}$	0.0067	0.00057	3.5×10^{-6}
χ^2_{Mode}	0.0134	0.00166	2.6×10^{-5}
$\chi^2_{GC,overall}$	0.0112	0.00127	2.2×10^{-5}
$\chi^2_{GC,MAF}$	0.0112	0.00128	2.3×10^{-5}
χ^2_{GC,r^2}	0.0104	0.0011	1.3×10^{-5}
χ^2_{GC,MAF,r^2}	0.0101	0.00105	1.2×10^{-5}
χ^2_{True}	0.0099	0.00097	8.1×10^{-6}

3. Results

3.1. Type I error simulation

Table 1 gives the overall type I error rates at three different significance levels for each of the six methods applied to posterior probabilities on SNPs in HWE, along with the significance level when using the true genotypes. As expected, use of the true genotypes yields type I error rates at the significance level. Overall, $\chi^2_{Posterior}$ yielded the most conservative type I error rates, while χ^2_{Mode} yielded anti-conservative type I error rates. The χ^2_{GC} corrected approaches tended to yield approximately correct type I error rates, with the version which adjusts statistics both within *MAF* and r^2 (χ^2_{GC,MAF,r^2}) bins providing the best Type I error control. A logistic regression model predicting the type I error rate $\chi^2_{Posterior}$ test across all

850,000 SNPs indicates that both MAF and r^2, as well as an interaction term between MAF and r^2, are significant predictors of the type I error rate, which further supports the necessity to use both bins for both MAF and r^2 when correcting statistics as is done by χ^2_{GC,MAF,r^2}.

The patterns observed in Table 1 remain true across all MAF and r^2 subgroups as shown in Supplemental Table 1. In particular we also see that $\chi^2_{Posterior}$ is the most conservative for less well imputed SNPs, though even well imputed SNPs are treated anti-conservatively by $\chi^2_{Posterior}$ ($8.5x10^{-3}$ for r^2>0.95). In contrast, χ^2_{Mode} is the most anti-conservative for less well imputed SNPs, with some inflation of the type I error rate for moderately well imputed SNPs (e.g., 0.85<r^2<0.95). χ^2_{Mode} only controls the type I error rate for extremely well imputed SNPs (r^2>0.95). χ^2_{GC,MAF,r^2} controls the Type I error rate across MAF and r^2 strata. While Supplemental Table 1 only shows results for a significance level of 0.01, patterns remain the same across other more stringent significance levels (e.g., 0.001, $1x10^{-5}$, detailed results not shown). Figure 1 illustrates the anti-conservative performance of χ^2_{Mode}, the conservative performance of $\chi^2_{Posterior}$ and good control of the type I error rate by χ^2_{GC,MAF,r^2}

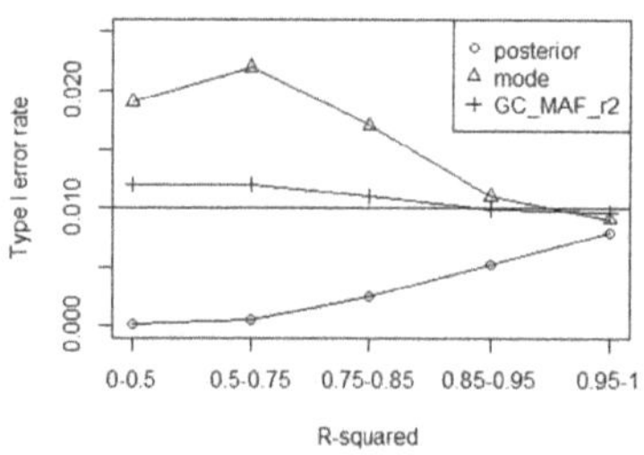

Figure 1. Type I error rate for three different HWE testing methods across different uncertainty levels. Type I error rate is shown across different r^2 settings for three different HWE testing approaches at the 1% significance level. SNPs in the low minor allele frequency range are depicted (MAF between 0.05 and 0.1)

Power simulation

To understand the power of the different approaches for HWE testing, we considered 300 combinations of average minor allele frequency across SNPs i and j, observed r^2, difference in minor allele frequency and k (proportion of individuals from SNP i; where $1-k$ is the proportion of individuals are from SNP j) across 225,000 SNPs which are a mixture of two different allele frequencies. One-hundred twenty-two of the settings yielded 100% power when using all methods, and another 40 combinations yielded no SNPs, and so these 162 settings are eliminated from further consideration. Due to the fact that χ^2_{Mode} has an inflated Type I error rate, we do not consider it in the following comparative analysis of the power of the different methods. Across these 162 settings the median number of SNPs per group was 490 (Min=5; Q1=182; Q3=1708; Max=4353), with only three settings having less than 20 SNPs.

Across the 138 remaining combinations of settings, χ^2_{GC,MAF,r^2} had higher power than $\chi^2_{Posterior}$ 122 times, by an average of 0.038 (SD=0.039). Across the 16 times that $\chi^2_{Posterior}$ yielded higher power than χ^2_{GC,MAF,r^2}, the average power gain was only 0.0029 (SD=0.0024). Table 2 illustrates a subset of 138 simulation settings, illustrating that χ^2_{GC,MAF,r^2} consistently yields higher power than $\chi^2_{Posterior}$ for all but the most certain SNPs, when performance is comparable. Largest gains in power were for the least certain

Table 2. Power[1] by MAF and r^2

MAF	r^2	Number of variants	$\chi^2_{Posterior}$	χ^2_{GC,MAF,r^2}	χ^2_{True}
0.05-0.1	0-0.50	122	0.91	0.98	1
	0.5-0.75	265	0.82	0.94	0.98
	0.75-0.85	166	0.79	0.84	0.98
	0.85-0.95	480	0.86	0.87	0.99
	0.95-1.0	695	0.85	0.85	0.99
0.1-0.2	0-0.50	123	0.67	0.84	0.86
	0.5-0.75	382	0.65	0.78	0.85
	0.75-0.85	441	0.66	0.76	0.82
	0.85-0.95	1411	0.62	0.65	0.82
	0.95-1.0	2561	0.62	0.61	0.81
0.2-0.3	0-0.50	152	0.53	0.66	0.7
	0.5-0.75	365	0.52	0.58	0.7
	0.75-0.85	489	0.56	0.67	0.74
	0.85-0.95	2029	0.53	0.57	0.72
	0.95-1.0	4217	0.52	0.51	0.7
0.3-0.4	0-0.50	81	0.43	0.51	0.57
	0.5-0.75	209	0.39	0.46	0.52
	0.75-0.85	277	0.4	0.52	0.58
	0.85-0.95	1324	0.36	0.41	0.54
	0.95-1.0	3321	0.38	0.38	0.53
MAF>0.4	0-0.50	25	0.32	0.32	0.44
	0.5-0.75	87	0.29	0.36	0.55
	0.75-0.85	160	0.33	0.43	0.48
	0.85-0.95	629	0.37	0.41	0.51
	0.95-1.0	1649	0.38	0.38	0.52

1. At the 1% significance level and when the observed SNP is a mix of two subgroups of individuals with a difference of between 0.10 and 0.20 in minor allele frequency between the two subgroups, and 10% of the individual are from one subgroup and 90% from the other (k=0.1).

SNPs, with overall higher power for all methods with lower MAF. Figure 2 illustrates this relative gain in power. Supplementary Table 1 gives the full power results for all 300 settings.

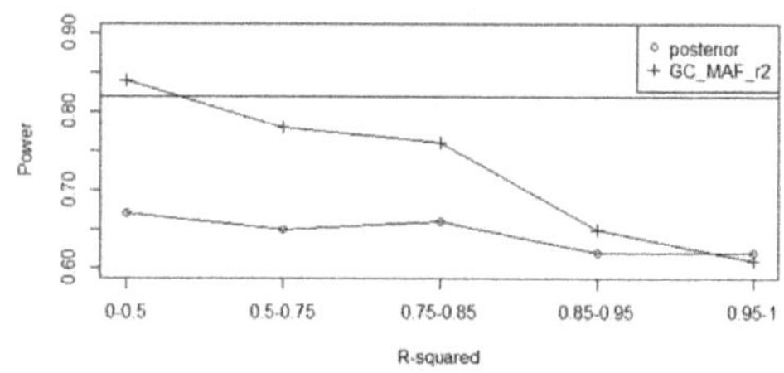

Figure 2 Power for two different approaches to HWE testing across different uncertainty levels
Power is illustrated across different r^2 settings for two different HWE testing approaches at the 1% significance level, with a horizontal line at the power of a test using the real genotypes. Power for SNPs with MAF between 0.1 and 0.2 are depicted, when the observed SNP is a mix of two subgroups of individuals where the difference in MAF between the two subgroups is between 0.1 and 0.2, and the 10% of the individuals are from one subgroup and 90% from the other.

Real data example

When applying the three HWE testing methods to the 29,361 imputed FUSION SNPs, 237 variants were determined to be out of HWE by χ^2_{Mode}, none by $\chi^2_{Posterior}$ and two by χ^2_{GC,MAF,r^2} at a significance level of $1\text{x}10^{-5}$. While true HWE status for these variants is unknown, these results suggest an inflated type I error rate for the χ^2_{Mode} test. When we applied the $\chi^2_{Posterior}$ and χ^2_{GC,MAF,r^2} tests to the 2,377 non-HWE variants, the power was always higher for the χ^2_{GC,MAF,r^2} test (see Table 3).

Table 3. Power to detect pseudo variants not in Hardy-Weinberg Equilibrium from the FUSION study

Observed MAF	Number of variants	$\chi^2_{Posterior}$	χ^2_{GC,MAF,r^2}
0.05-0.10	375	5.3%	8.0%
0.10-0.20	731	28.6%	29.7%
0.20-0.3	412	28.9%	30.8%
0.3-0.4	385	26.8%	32.2%
0.4-0.5	374	2.9%	5.9%
Overall	2277	20.3%	22.8%

4. Discussion

We have proposed a new way to incorporate posterior probabilities in tests of HWE that provides a well-calibrated and more powerful way to incorporate genotype uncertainty. While it is common to use the modal posterior genotype, this approach inflates the type I error rate by failing to incorporate genotype uncertainty---treating uncertain genotypes as if they are error-free. Furthermore, another recent approach which explicitly incorporates posterior probabilities yields an overly conservative test (deflated type I error rate), due to an overestimation of the covariance of the posterior probability genotypes. Our approach applies a post-hoc correction to adjust the test statistic, yielding a calibrated type I error rate and improved power.

The proposed approach is approximately the same as other approaches when genotype uncertainty is low, but shows increasing benefit as genotype uncertainty increases. This result is in line with the fact that the genotype covariance estimates are increasingly biased when using $\chi^2_{Posterior}$ as genotype uncertainty increases. While it is common practice to simply drop markers with very high genotype uncertainty from analyses we've demonstrated that this may not be necessary when using our approach. Furthermore, even if practitioners wish to drop markers with high genotype uncertainty (e.g., r^2<0.5), we've demonstrated that our approach to HWE testing still outperforms other HWE testing procedures for markers with modest genotype uncertainty (0.5<r^2<0.95). Importantly, recent work has shown that simply screening for HWE using r^2 is not sufficient, and that HWE testing is still necessary [21].

While the proposed approach performs well relative to the existing approaches by applying a post-hoc correction, a more explicit approach may also be possible. Preliminary exploration of such methods by our group has taken two separate paths to date. First, we considered multiple imputation by creating many, equally likely, versions of each individual's genotype according to the vector of calibrated posterior genotype probabilities and then computing the standard chi-squared GOF test on each multiply-imputed dataset. Methods for computing significance from a set of multiply-imputed datasets are standard [22–24], but may not be well-calibrated [25]. A lack of calibration was our experience for this application (detailed results not shown). A second approach is a Bayesian approach using the posterior probabilities for each individual's genotype explicitly. Evaluation of this method across a wide-range of simulation settings showed performance comparable to the $\chi^2_{Posterior}$ method and, thus, not as good as χ^2_{GC,MAF,r^2} in many cases (detailed results not shown).

We now make some important notes and comments on limitations of the χ^2_{GC,MAF,r^2}approach. While not considered here, the authors of the $\chi^2_{Posterior}$ approach also considered an exact test for small sample sizes. Future work is needed to evaluate the performance of the post-hoc correction strategy for small sample size situations (e.g., rare variants), though, in principle, there is no reason to believe that an approach in this same spirit is likely to perform well. A key assumption of χ^2_{GC,MAF,r^2} is that a relative small proportion of all markers overall will not be in HWE. In rare cases where a very large proportion of markers are out of HWE, the χ^2_{GC,MAF,r^2} approach may, in fact, be overly conservative by applying a correction factor based on markers not in HWE. However, these cases should be rare as a substantial portion of the markers in the correction set would need to be out of HWE in order to impact the median observed statistic and, hence, the lambda, in a practically significant way. However, since χ^2_{GC,MAF,r^2} computes a separate adjustment for many different MAF, r^2 'bins,' an aggregation of markers not in HWE in any bin could impact results. Finally, the size and quantity of MAF, r^2 bins selected in this study showed good performance, but may need adjustment in practice based on the MAF distribution, r^2 (or other uncertainty metric) distribution and number of variants. Care should be taken to ensure all bins have sufficient markers (generally recommended to be at least 100, but less may be fine) and examination of $\hat{\lambda}$ values within each bin is recommended. Future work may wish to explore the potential for a robust, continuous correction strategy.

Supplemental Files

All supplemental and appendix files are available online at the following URL: http://homepages.dordt.edu/ntintle/hwe.zip

Acknowledgments

This work was funded by the National Human Genome Research Institute (R15HG006915). We acknowledge the use of the Hope College parallel computing cluster for assistance in data analysis. We are grateful for the generosity of the FUSION investigators for access to imputed genotypes on which we based our simulations and demonstrated a proof-of-concept of our proposed approach. The FUSION study is funded by the National Institute of Diabetes and Digestive and Kidney Diseases (U01DK062370).

References

1. Powers S, Gopalakrishnan S, Tintle N. Assessing the impact of non-differential genotyping errors on rare variant tests of association. Hum Hered. 2011;72: 153–60.

2. Gordon D, Finch SJ. Factors affecting statistical power in the detection of genetic association. J Clin Investig. 2005;115.

3. Mayer-Jochimsen M, Fast S, Tintle NL. Assessing the impact of differential genotyping errors on rare variant tests of association. PLoS One. 2013;8: e56626.

4. Moskvina V, Craddock N, Holmans P, Owen MJ, O'Donovan MC. Effects of differential genotyping error rate on the type I error probability of case-control studies. Hum Hered. 2006;61: 55–64.

5. Wang J, Shete S. Testing Hardy-Weinberg proportions in a frequency-matched case-control genetic association study. PLoS One. 2011;6: e27642.

6. Shriner D. Approximate and exact tests of Hardy-Weinberg equilibrium using uncertain genotypes. Genet Epidemiol. 2011;35: 632–7.

7. Li Y. A comparison of tests for Hardy-Weinberg Equilibrium in national genetic household surveys. BMC Genet. 2013;14: 14.

8. She D, Zhang H, Li Z. Testing Hardy-Weinberg equilibrium using family data from complex surveys. Ann Hum Genet. 2009;73: 449–55.

9. Graffelman J, Nelson S, Gogarten SM, Weir BS. Exact Inference for Hardy-Weinberg Proportions with Missing Genotypes: Single and Multiple Imputation. G3 Genes|Genomes|Genetics. 2015;5: g3.115.022111.

10. Schaid DJ, Sinnwell JP, Jenkins GD. Regression Modeling of Allele Frequencies and Testing Hardy Weinberg Equilibrium. Hum Hered. 2013;74: 71–82.

11. Hong H, Su Z, Ge W, Shi L, Perkins R, Fang H, et al. Evaluating variations of genotype calling: a potential source of spurious associations in genome-wide association studies. J Genet. 2010;89: 55–64.

12. Li Y, Willer CJ, Ding J, Scheet P, Abecasis GR. MaCH: Using sequence and genotype data to esimate haplotypes and unobserved genotypes. Genet Epidemiol. 2010;34: 816–834.

13. Li H, Ruan J, Durbin R. Mapping short DNA sequencing reads and calling variants using mapping quality scores. Genome Res. 2008;18: 1851–1858.

14. Nielson R, Paul JS, Albrechtsen A, Song YS. Genotype and SNP calling from next-generation sequencing data. Nat Rev Genet. 2011;12: 443–451.

15. Zheng J, Li Y, Abecasis GR, Scheet P. A comparison of approaches to account for uncertainty in analysis of imputed genotypes. Genet Epidemiol. 2011;35: 102–10.

16. Liu K, Luedtke A, Tintle NL. optimal methods for using posterior probabilities in association testing. Hum Hered. 2013;75: 2–11.

17. Yates F. Contingency table involving small numbers and the X2 test. Suppl to J Roayl Stat Soc. 1934;1: 217–235.

18. Tintle N, Gordon D, McMahon F, Finch SJ. Using Duplicate Genotyped Data in Genetic Analyses : Testing Association and Estimating Error Rates. Stat Appl Genet Mol Biol. 2007;6.

19. Devlin B, Roeder K. Genomic Control for Association Studies. Biometrics. 1999;55: 997–1004.

20. Scott L, Mohlke K, Bonnycastle L, Willer C, Li Y, Duren W, et al. A genome-wide association study of type 2 diabetes in Finns detects multiple susceptibility variants. Science (80-). 2007;316: 1341–5.

21. Shriner D. Impact of Hardy-Weinberg disequilibrium on post-imputation quality control. Hum Genet. 2013;132: 1073–5.

22. Li K-H, Meng X-L, Raghunathan TE, Rubin DB. Signifiacnce levels from repeated p-values with multiply imputed data. Stat Sin. 1991;1: 65–92.

23. Meng X-L, Rubin DB. Performing likelihood ratio tests with multiply-imputed data sets. Biometrika. 1992;79: 103–111.

24. Harel O, Zhou X-H. Multiple imputation: review of theory, implementation and software. Stat Med. 2007;26: 3057–3077.

25. Licht C. New methods for generating significance levels from multiply-imputed data Ph.D. Dissertation. University of Bamberg, Germany. 2010.

TEMPORAL ORDER OF DISEASE PAIRS AFFECTS SUBSEQUENT DISEASE TRAJECTORIES: THE CASE OF DIABETES AND SLEEP APNEA

METTE K. BECK
Novo Nordisk Foundation Center for Protein Research,
University of Copenhagen, Blegdamsvej 3B
Copenhagen, DK-2200, Denmark
Email: mette.beck@cpr.ku.dk

DAVID WESTERGAARD
Novo Nordisk Foundation Center for Protein Research,
University of Copenhagen, Blegdamsvej 3B
Copenhagen, DK-2200, Denmark
Email: david.westergaard@cpr.ku.dk

ANDERS BOECK JENSEN
Novo Nordisk Foundation Center for Protein Research,
University of Copenhagen, Blegdamsvej 3B
Copenhagen, DK-2200, Denmark
Email: anders.b.jensen@cpr.ku.dk

LEIF GROOP
Lund University Diabetes Centre, Department of Clinical Sciences,
Jan Waldenströms gata 35, SE-205 02 Malmö, Sweden
Email: Leif.Groop@med.lu.se

SØREN BRUNAK
Novo Nordisk Foundation Center for Protein Research,
University of Copenhagen, Blegdamsvej 3B
Copenhagen, DK-2200, Denmark
Email: soren.brunak@cpr.ku.dk

Most studies of disease etiologies focus on one disease only and not the full spectrum of multimorbidities that many patients have. Some disease pairs have shared causal origins, others represent common follow-on diseases, while yet other co-occurring diseases may manifest themselves in random order of appearance. We discuss these different types of disease co-occurrences, and use the two diseases "sleep apnea" and "diabetes" to showcase the approach which otherwise can be applied to any disease pair. We benefit from seven million electronic medical records covering the entire population of Denmark for more than 20 years. Sleep apnea is the most common sleep-related breathing disorder and it has previously been shown to be bidirectionally linked to diabetes, meaning that each disease increases the risk of acquiring the other. We confirm that there is no significant temporal relationship, as approximately half of patients with both diseases are diagnosed with diabetes first. However, we also show that patients diagnosed with diabetes before sleep apnea have a higher disease burden compared to patients diagnosed with sleep apnea before diabetes. The study clearly demonstrates that it is not only the diagnoses in the patient's disease history that are important, but also the specific order in which these diagnosis are given that matters in terms of outcome. We suggest that this should be considered for patient stratification.

1. Introduction

Much epidemiological research has focused on simple associations between two diseases. Temporal approaches have been suggested to uncover both causal and genetic links among statistically associated diseases [1-4]. Many recent studies have analyzed more complicated relations between several diseases and have found bidirectional relationships, where one disease increases the risk or severity of the other or vice versa [1–4]. This type of relationship is mostly found for pairs of common diseases such as depression, cardiovascular diseases and diabetes [2,4]. In one example Mezuk et al. reported a 15% increased risk of depression in patients with type 2 diabetes (T2D), but 60% increased risk of developing type 2 diabetes in patients with depression [5]. Since then several papers have confirmed this particular bidirectional observation [6,7]. Similarly, diabetes has been bidirectionally linked with both periodontitis and sleep apnea [1,8,9].

Until now there has not been general studies investigating the effect of the temporal order in which bidirectionally linked diseases are diagnosed, and how the order affects the further disease progression and the general health status of the patients. In this study we highlight the importance of the temporal order using the bidirectionally linked disease pair: diabetes and sleep apnea. Subsequently we generalize this method to a disease-spectrum wide approach for T2D patients.

Sleep apnea is the most common sleep-related breathing disorder, affecting up to 10% of middle-aged women and up to 20% of middle-aged men in high-income and Asian countries [10–12]. It is traditionally stratified into obstructive sleep apnea and central sleep apnea, where obstructive sleep apnea is the most prevalent subgroup that accounts for up to 85% of sleep apnea patients [13–15]. Furthermore, sleep apnea can occur in both children and adults, although these are treated as two different diseases [16–19]. Untreated, sleep apnea increases the risk for cardiovascular, metabolic, and neurocognitive complications and it is therefore a prototypical example of a disease involved in comorbidities [20,21]. Specifically, it is associated with T2D [1,9,22,23].

Although obesity is a predictor of both obstructive sleep apnea and T2D, the bidirectional link between these diseases appears to be independent of weight [1,9,20]. T2D contributes to sleep apnea, by causing neuromyopathy, which impairs reflexes of the upper airway [9,20]. Sleep apnea contributes to the development of T2D by increased activation of the sympathetic nervous system leading to increased insulin resistance [22,24,25]. It has even been suggested that successful treatment of sleep apnea may reduce the risk of T2D, although this is still controversial [9].

To investigate the effect of the order of the diagnoses we combined the Danish National Patient Registry (NPR), which covers all hospital encounters, both public and private, in Denmark from 1994 to 2015, a patient population of nearly seven million individuals with prescription data from the Danish diabetes registry. NPR records diseases using the International Classifications of Diseases, 10th revision (ICD-10), which organizes diseases hierarchically.

Using this unbiased, country-wide data set we describe the comorbidity map of sleep apnea patients in a data driven manner, and show that the diagnostic order of sleep apnea and T2D is close to 50:50. Interestingly, while the order overall appears to be random we show that the order is associated with significantly different frequencies of comorbid diseases, implying two distinct patient groups.

T2D is a chronic disease with a high risk of many servere complications, including cardiovascular, neurological and infectious complications [5,6,26–30]. Consequently, we generalized our approach to all diseases appearing together with T2D. We showed that the disease burden was dependent on the diagnosis order for twelve T2D comorbidities, of which ten show an increase in comorbidities if T2D was diagnosed first.

2. Materials and methods

In this retrospective cohort study we investigated the association between sleep apnea and T2D. We used the NPR, covering all hospital encounters in Denmark from 1994 to 2015, from where we could include 6,923,849 Danish subjects. Specifically, this registry contained 218,750 T2D patients and 95,853 sleep apnea patients.

To define T2D patients we combined the NPR with the Danish Diabetes registry, which contains medical prescription data. We defined T2D patients, as patients diagnosed at least two times with NIDDM but not with IDDM, if oral hypoglycemic agents were prescribed at least two times and they were diagnosed with NIDDM, or if oral hypoglycemic agents and insulin were prescribed at least two times and they were diagnosed with NIDDM and/or IDDM.

Adult sleep apnea patients were defined as patients first diagnosed with sleep apnea at the age of 16 years or older.

2.1. *Comorbidity calculations*

We tested for significant associations between all level three diagnoses in the ICD-10 terminology. The relative risk of a particular disease was calculated using the Cochran–Mantel–Haenszel method, where each bin corresponds to patients of a particular gender and born in a particular decade. We included patients born from 1900 until 2015, giving rise to up to 24 bins per test. We used the Cochran–Mantel–Haenszel test to identify the p-value and accepted results with a Benjamini-Hochberg corrected p-value of 0.05 or below. This method was used both for time-independent and time-dependent analyses.

2.2. *From temporal diagnosis pairs through disease trajectories to disease network*

The method for identifying the trajectories has been described previously in detail [33]. The method consists of three steps: First temporal directed pairs of co-morbid diseases were tested to identify pairs where one disease is associated with an increase in the occurrences of the other. In the second step, the pairs found are tested for directionality (one disease primarily occurring before the other) using a binomial test. Third, the pairs with significant temporality were combined into disease trajectories of three consecutive diseases. Trajectories were only included if at least 100 sleep apnea patients followed them.

2.3. *Difference in mean number of comorbidities*

The difference in mean number of comorbidities was modeled by a Poisson regression using the covariates: years between the two diagnoses, which disease was diagnosed first, age and gender. All four covariates significantly contributed to the model. This Poisson regression was subsequently used to predict the number of comorbidities for all patients to avoid age and/or gender bias. The difference in mean predicted number of comorbidities was tested using student's t-test, stratified by the order of the diagnoses. This was done twenty times, requiring a minimum from zero years up to nineteen years in between the two diagnoses.

2.4. *Diabetes comorbidities selection criteria*

We tested if any level three ICD-10 diagnoses were significantly correlated with T2D using the method for comorbidity calculations. For the diagnosis with a significant association and a relative risk above one, we used a binomial test to ensure lack of directionality. We required the 95% confidence interval to be within 45% - 55% (making the diagnostic order close to 50:50). Lastly, we required a minimum of 1,000 T2D patients to have the disease. For the remaining diseases we performed the method described in "Difference in mean number of comorbidities". We required ten time points to be significant.

3. Results

In the Danish population of 6,923,707 patients, we found 117,913 patients diagnosed with sleep disorders (G47), of these 95,853 patients were diagnosed with sleep apnea (G47.3). The age distribution at which these patients were first diagnosed with sleep apnea is shown in Figure 1A. It has two clearly distinct peaks, the first at age three, and the second major peak just after 50 years, supporting that this diagnosis could cover two distinct disease progression patterns. We computed the relative risk (RR) for both the adult onset of sleep apnea (aged 16 or above) and the childhood onset, compared to all other level three ICD-10 diagnoses. Even though both groups of patients are diagnosed with the same diagnosis, their repertoire of comorbidities is very different (Figure 1B), in part due to the difference in age. We therefore excluded childhood onset of sleep apnea, and investigated sleep apnea in the adult population further. Of the 95,853 sleep apnea patients 90,157 were diagnosed in adult patients, 75% of these were males.

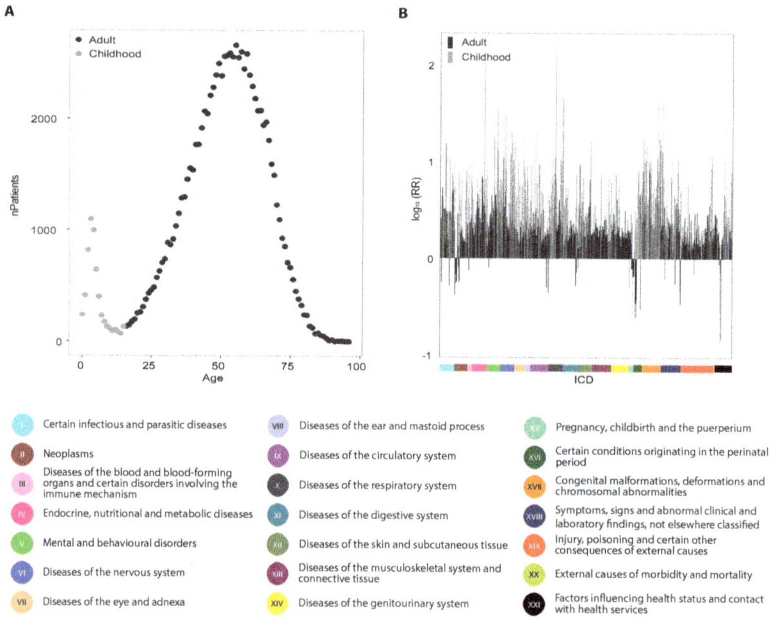

Fig. 3. The increased comorbidity burden for patients diagnosed with T2D before sleep apnea. (A) Distribution of years between T2D and sleep apnea for patients diagnosed with T2D first (pink) and for patients diagnosed with sleep apnea first (blue). (B) The excess number of comorbidities for patients diagnosed with T2D first compared to those diagnosed with sleep apnea first (black line) with the 95% confidence interval (grey area). The x-axis indicates the minimum number of whole years between diagnoses (e.g. 0 years means more than one day but less than a year). The dots indicate the number of patients having minimum x years between the two diagnoses.

3.1. *Temporal disease network reveals no direct connection between diabetes and adult sleep apnea*

We identified all diseases that co-occurred more often than we would expect from their individual frequencies in the patients with adult sleep apnea. For each such disease pair, we testedif one of the diseases occurred significantly more often before the other. This led to the identification of a pool of significant, directed disease-pairs (see Methods). These pairs were combined into linear, temporal disease trajectories of which we found 103 where 100 sleep apnea

Fig. 2. Temporal disease network based on sleep apnea patients. The network was constructed from 103 sleep apnea trajectories and illustrates the number of patients taking a particular step in the disease network (width of arrow). The nodes are colored based on their ICD-10 chapter relationships. The names are written next to their node in the network or mentioned in the legend in alphabetical order.

patients followed three consecutive steps of diseases. Subsequently, the 103 linear trajectories found in the adult sleep apnea patient group were combined into a temporal disease network providing a concerted overview of the comorbidity spectrum (Figure 2). As expected, obesity, a known risk factor for sleep apnea, appears as a statistically significant component in this overview network (present in 20 of the 103 temporal trajectories as either starting or midpoint). Several cardio-vascular complications are also prominent in the network. Additionally, both insulin-dependent diabetes mellitus (IDDM) and non-insulin-dependent diabetes mellitus (NIDDM) are part of the disease network along with several diabetes complications. There is no direct path connecting diabetes and sleep disorders in the disease network, due to the lack of temporality between these diagnoses.

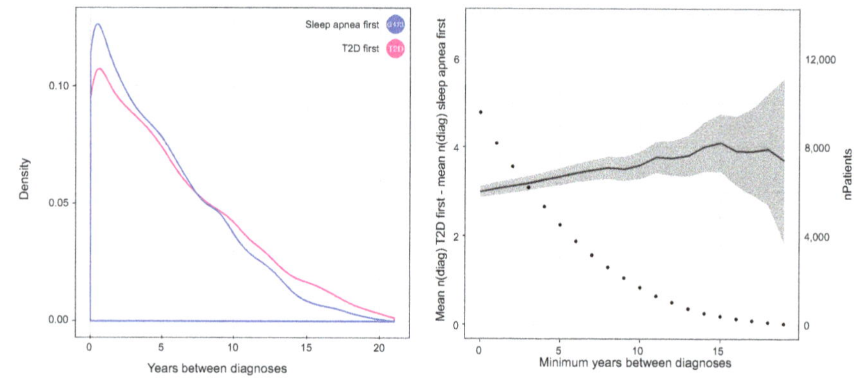

Fig. 3. The increased comorbidity burden for patients diagnosed with T2D before sleep apnea. (A) Distribution of years between T2D and sleep apnea for patients diagnosed with T2D first (pink) and for patients diagnosed with sleep apnea first (blue). (B) The excess number of comorbidities for patients diagnosed with T2D first compared to those diagnosed with sleep apnea first (black line) with the 95% confidence interval (grey area). The x-axis indicates the minimum number of whole years between diagnoses (e.g. 0 years means more than one day but less than a year). The dots indicate the number of patients having minimum x years between the two diagnoses.

3.2. *Diabetes before sleep apnea is associated with an increased amount of comorbidities*

To further investigate the temporal association between sleep apnea and T2D, we defined T2D patients based on the method presented by Lind at al [29,31], using a combination of prescribed drugs and disease codes (see Methods). We found that 11,054 T2D patients have been diagnosed with sleep apnea. A total of 6,061 patients (54,8%) were diagnosed with T2D before sleep apnea, and 4,752 patients were diagnosed with sleep apnea before T2D. In addition, 241 patients were diagnosed with T2D and sleep apnea on the same day. These 241 patients are disregarded in this study, since there is no reliable way to determine which disease came first. Consequently, even though sleep apnea was significantly associated with T2D (RR = 2.87, p < 2.3E-308), NIDDM and sleep apnea does not appear as a temporal pair, due to the lack of a significant temporal order in which these diseases are diagnosed.

To investigate if the patients diagnosed with T2D before adult sleep apnea and patients diagnosed with adult sleep apnea before T2D are two distinct patient groups, we examined the RR for all level three ICD-10 diagnoses for patients with adult sleep apnea and T2D. Those first diagnosed with T2D had on average 3.0 (95% CI: 2.9-3.1) comorbidities more than those diagnosed with adult sleep apnea first. We interpret this as an indicator that the patients first diagnosed with T2D have, on average, a higher disease burden.

The time of diagnosis can be imprecise since neither sleep apnea nor T2D are acute diseases. Consequently, it could be arbitrary which disease was diagnosed first. For some patients the two diagnoses are acquired relatively close to each other, but for many patients there are several years or even decades between the diagnoses (Figure 3A).

We tested if there was a significant difference in the number of diagnoses between these two groups using a Poisson regression model. Covariates include years between the two diagnoses, which disease was diagnosed first, age and gender. We used the fitted model to calculate a point estimate of the number of comorbidities for each patient, given the minimum number of years between sleep apnea and T2D (Figure 3B). The overall difference was 3.0 comorbidities, with patients first diagnosed with T2D being most sick. This difference increases as the number of years between T2D and sleep apnea increases (Figure 3B). Collectively, this clearly illustrates a difference in the general health status of these patients groups.

3.3. *Diabetes before other diseases tends to increase the comorbidity burden*

We applied the same method to investigate if other diabetes comorbidities showed a different comorbidity burden depending on the diagnosis order. We found seventeen diseases positively associated with T2D, and where the diagnostic order for each disease and T2D was close to 50:50 (see Methods). To remove rare disorders we required a minimum of 1,000 T2D patients to have the diagnosis, reducing the number down to sixteen diagnoses of interest. Lastly, we performed an analysis calculating the difference in mean number of comorbidities for patients diagnosed with T2D first compared to patients diagnosed with the other particular diagnosis first. This resulted in twelve diagnoses with a minimum of ten significant time points (Figure 4). Ten out of the twelve diagnoses were associated with a higher comorbidity burden if they were diagnosed with T2D before the other diagnosis, with the two exceptions: Migraine and "Poisoning by psychotropic drugs, not elsewhere classified".

4. Discussion

In this study we examined the complex issue of temporal directionality of disease co-occurrences and used temporal disease trajectories to present a model for stratification of patient groups according to longitudinal patterns.

Using one example analyzed in detail we illustrated the complexities and rediscovered that age of sleep apnea diagnoses follow a bimodal distribution, illustrating two distinct diseases: childhood sleep apnea and adult sleep apnea – a distinction well known in the literature [11,16–18,32]. By investigating the detailed time-ordered relationships between sleep apnea and T2D we confirmed that sleep apnea in the adult population is significantly associated with T2D in the time-dependent analysis. Surprisingly, there was no direct edge between any of the diabetes diagnoses in our temporal disease network, showing that there was no directionality of the T2D and adult sleep apnea diagnoses, in fact we showed that 4,752 patients acquire adult sleep apnea before T2D, 6,061 acquire T2D first while 241 patients acquired the diagnoses on the same day.

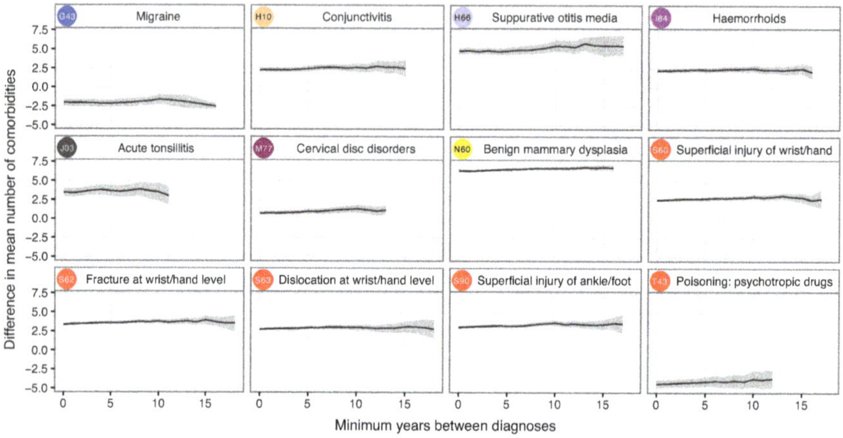

Fig. 4. Comorbidity burden levels as function of time span between diagnoses. Together the panels show that the change in comorbidity burden depends on the disease order. Each disease is indicated by an ICD-10 code colored according to the ICD-10 chapter followed by the name of the disease. The excess number of comorbidities for patients diagnosed with T2D first compared to those diagnosed with the other particular diagnosis (black line) with the 95% confidence interval (grey area). The x-axis indicates the minimum number of whole years between diagnoses.

Interestingly, we found that this order significantly influenced the amount of comorbidities acquired, indicating that patients diagnosed with diabetes before adult sleep apnea have a worse general health status than patients first diagnosed with adult sleep apnea. This is, to our knowledge, the first time this temporal effect of sleep apnea and T2D has been described. To further illustrate the importance of the order, we showed that the difference in the quantity of comorbidities slightly increased with increased time between the diagnoses. Based on these observations we suggest that there is a synergetic effect of T2D and adult sleep apnea, which is dependent on the order of the diagnoses.

We further underlined the importance of order of diagnoses by applying this method to all T2D comorbidities. This resulted in twelve diagnoses with a significant different number of comorbidities depending on the diagnosis order.

Precision medicine attempts to subdivide patients into groups that will benefit from tailor-made treatment. We show in this paper that disease progression patterns can be highly complex even in cases where disease co-occurrence orders appear to be random. The identification of genomic biomarkers could most likely to a higher degree benefit from taking this type of stratification into account in contrast to current models that mostly are based on the case/control paradigm where diseases are investigated individually.

5. Acknowledgements

We would like to acknowledge funding from the Novo Nordisk Foundation (grant agreement NNF14CC0001) as well as the H2020 project MedBioInformatics.

References

1. Aurora, R. N. & Punjabi, N. M. *Lancet Respir. Med.* **1,** 329–338 (2013).
2. Golden, S. H. *et al. JAMA* **299,** 2751–2759 (2008).
3. Hesdorffer, D. C. *et al. Ann. Neurol.* **72,** 184–191 (2012).
4. Lippi, G., Montagnana, M., Favaloro, E. J. & Franchini, M. *Seminars in Thrombosis and Hemostasis* **35,** 325–336 (2009).
5. Mezuk, B., Eaton, W. W., Albrecht, S. & Golden, S. H. *Diabetes Care* **31,** 2383–2390 (2008).
6. Pan, A. *et al. Arch. Intern. Med.* **170,** 1884–91 (2010).
7. Pan, A. *et al. Diabetes Care* **35,** 1171–1180 (2012).
8. Lalla, E. & Papapanou, P. N. *Nat. Rev. Endocrinol.* **7,** 738–48 (2011).
9. Rajan, P. & Greenberg, H. *Nat. Sci. Sleep* **7,** 113–25 (2015).
10. Peppard, P. E. *et al. Am. J. Epidemiol.* **177,** 1006–1014 (2013).
11. Sharma, S. K. & Ahluwalia, G. *Indian J. of Med. Res.* **131,** 171–175 (2010).
12. Ip, M. S. M. *et al. Chest* **119,** 62–69 (2001).
13. Javaheri, S. *Clinics in Chest Medicine* **31,** 235–248 (2010).
14. Morgenthaler, T. I., Kagramanov, V., Hanak, V. & Decker, P. A. *Sleep* **29,** 1203–1209 (2006).
15. Khan, M. T. & Franco, R. A. *Sleep Disord.* 798487 (2014).
16. Tan, H.-L., Gozal, D. & Kheirandish-Gozal, L. *Nat. Sci. Sleep* **5,** 109–23 (2013).
17. Marcus, C. L. *et al. Pediatrics* **130,** 576–84 (2012).
18. Marcus, C. L. *et al. N. Engl. J. Med.* **368,** 2366–76 (2013).
19. Bixler, E. O. *et al. Sleep* **32,** 731–6 (2009).
20. Malhotra, A. & White, D. P. *The Lancet* **360,** 237–245 (2002).
21. Parati, G. *et al. J. Hypertens.* **30,** 633–46 (2012).
22. Cappuccio, F. P., D'Elia, L., Strazzullo, P. & Miller, M. A. *Diabetes Care* **33,** 414–420 (2010).
23. Malhotra, A. *et al. Am. J. Respir. Crit. Care Med.* **166,** 1388–1395 (2002).
24. Chervin, R. D. *Chest* **118,** 372–379 (2000).
25. Barceló, A. *et al. Thorax* **63,** 946–50 (2008).
26. Fowler, M. J. *Clin. Diabetes* **29,** 116–122 (2011).
27. Alves, C., Casqueiro, J. & Casqueiro, J. *Indian J. Endocrinol. Metab.* **16,** 27 (2012).
28. DeFronzo, R. A. *et al. Nat. Rev. Dis. Prim.* **1,** 15019 (2015).
29. Lind, M. *et al. Diabetologia* **55,** 2946–2953 (2012).
30. Bertoni, A. G., Saydah, S. & Brancati, F. L. *Diabetes Care* **24,** 1044–1049 (2001).
31. Lind, M. *et al. N. Engl. J. Med.* **371,** 1972–1982 (2014).
32. Jordan, A. S., McSharry, D. G. & Malhotra, A. *Lancet* **383,** 736–47 (2014).
33. Jensen, A. B. *et al. Nat. Commun.* **5,** 4022 (2014).

MICRORNA-AUGMENTED PATHWAYS (mirAP) AND THEIR APPLICATIONS TO PATHWAY ANALYSIS AND DISEASE SUBTYPING

DIANA DIAZ[1], MICHELE DONATO[3], TIN NGUYEN[1], SORIN DRAGHICI[1,2]

[1]*Department of Computer Science, Wayne State University, Detroit, MI 48202, U.S.A.*

[2]*Department of Obstetrics and Gynecology, Wayne State University, Detroit, MI 48202, U.S.A.*

[3]*Institute for Immunity, Transplantation and Infection, Stanford University Medical Center, Stanford, CA 94305, U.S.A.*

E-mail: sorin@wayne.edu

MicroRNAs play important roles in the development of many complex diseases. Because of their importance, the analysis of signaling pathways including miRNA interactions holds the potential for unveiling the mechanisms underlying such diseases. However, current signaling pathway databases are limited to interactions between genes and ignore miRNAs. Here, we use the information on miRNA targets to build a database of miRNA-augmented pathways (mirAP), and we show its application in the contexts of integrative pathway analysis and disease subtyping. Our miRNA-mRNA integrative pathway analysis pipeline incorporates a topology-aware approach that we previously implemented. Our integrative disease subtyping pipeline takes into account survival data, gene and miRNA expression, and knowledge of the interactions among genes. We demonstrate the advantages of our approach by analyzing nine sample-matched datasets that provide both miRNA and mRNA expression. We show that integrating miRNAs into pathway analysis results in greater statistical power, and provides a more comprehensive view of the underlying phenomena. We also compare our disease subtyping method with the state-of-the-art integrative analysis by analyzing a colorectal cancer database from TCGA. The colorectal cancer subtypes identified by our approach are significantly different in terms of their survival expectation. These miRNA-augmented pathways offer a more comprehensive view and a deeper understanding of biological pathways. A better understanding of the molecular processes associated with patients' survival can help to a better prognosis and an appropriate treatment for each subtype.

1. Introduction

The identification of biological processes underlying conditions is crucial for disease prognosis and treatment programs. As gene signaling pathways are capable of representing complex interactions between genes, pathway databases have become essential for several gene expression analyses. Signaling pathway databases are remarkably important because they allow researchers to analyze high-throughput data in a functional context, reducing complexity and increasing the explanatory power. However, there are other molecules that play important roles in gene regulation, such as microRNAs, which are not included into current pathway databases. MicroRNAs (miRNAs) are small RNA molecules capable of suppressing protein production by binding to gene transcripts. In fact, more than 30% of the protein-coding genes in humans are miRNA-regulated. Additionally, miRNAs have been shown to play an important role in diagnosis and prognosis for different types of diseases[1].

The integration of miRNA into signaling pathways have multiple applications, such as pathway analysis and disease subtyping. Pathway analysis techniques and methods aim to analyze high-throughput data with the goal of identifying pathways that are significantly impacted by a given condition. The typical input of pathway analysis includes gene expression data from two different phenotypes (e.g., condition vs. control) and a set of signaling pathways. Although current pathway analysis methods using gene expression (mRNA) have achieved excellent results[2-4], mRNA expression alone is unable to capture the complete picture of biological processes, as other entities also play important roles. Relevant work has been done to elucidate the important interplay between miRNAs and biological pathways[5-9]. The state-of-the-art approach for miRNA-mRNA pathway analysis is microGraphite[8] which uses an empirical gene set approach. microGraphite's main goal is the identification of signal transduction paths correlated with the condition under study[10].

A second crucial process in the understanding of complex diseases is disease subtyping. Identifying clinically meaningful subtypes in complex diseases is crucial for improving prognosis, treatment, and precision medicine[11]. A typical input of disease subtyping consists of various clinical variables and gene expression data from patients affected by a particular disease. The expected output consists of well-identified groups of patients that highly correlate with one or more variables, such as observed survival (e.g., long-term vs. short-term survival patients). Disease subtyping is typically expressed as a clustering problem with the goal of partition patients in groups based on their genetic similarities with the additional complexity that the number of clusters is unknown. Several methods for disease subtyping using gene expression data have been developed[11-15]. Integrative analysis using clinical data, multi-'omics' data, and prior biological knowledge can leverage current disease subtyping methods.

In this paper, we present a tool for integrating miRNA into signaling pathways (mirIntegrator), a publicly available miRNA-augmented pathway database (mirAP), and we show the applications of such augmentation to pathway analysis and disease subtyping. We have used mirIntegrator previously as a part of our orthogonal meta-analysis approach[16].

Our pathway analysis pipeline uses mirAP and Impact Analysis[3,4], a topology-aware pathway analysis method previously developed by our group. To demonstrate the advantage of our method, we analyze 9 datasets studying 7 different diseases with mRNA and miRNA expression. We show that the proposed approach is able to identify the pathways that describe the underlying diseases as significant. The p-values and rankings of these pathways are significantly smaller than those obtained without data integration as well as when using microGraphite[8].

Our disease subtyping pipeline uses miRNA and mRNA expression data, available clinical variables, and prior biological knowledge. This method includes a feature selection approach based on mirAP to reduce the effective dimensionality of the unsupervised clustering problem. We analyze colorectal cancer miRNA, gene expression data, and clinical records downloaded from the Cancer Genome Atlas (TCGA) with our pipeline and SNF[15], a recently proposed integrative disease subtyping method. The colorectal cancer-relevant pathways and subgroups identified with our approach are significantly different in terms of their survival expectation, outperforming the approach that does not use miRNA, and providing information on biological mechanisms relevant to the difference in survival.

2. Methods

In this section, we propose an algorithm for integrating miRNA into signaling pathways. We also describe two pipelines using miRNA-augmented pathways (mirAP). The first pipeline is for pathway analysis (PA) and the second one is for disease subtyping (DS). The scenarios for these analyses are different. PA is used in biological studies comparing genetic samples from two different phenotypes (e.g., disease vs. control samples), and DS is used in studies with samples of patients undergoing the same disease for which the clinical subtypes are unknown. Our PA pipeline is able to integrate miRNA and mRNA expression data and identify pathways that are related to the disease under study. Our DS pipeline is able of incorporate biological pathways to partition patients into groups with very different survival patterns.

2.1. *Pathway augmentation*

This method augments the graphical representation of original signaling pathways with interactions between miRNAs and their target genes. The input of this method includes a set of signaling pathways and known miRNA-mRNA interactions (Fig. 1a,b). The output is a set of augmented pathways that consists of the original genes, the miRNAs that target those genes and their interactions. Let $P = (V, E)$ denote the graphical representation of the original gene-gene pathway, and $T : M \to V$ a function that identifies the target genes of miRNAs in M. An edge $e \in E$ can be represented as a 3-tuple $e = (g_1, g_2, interaction)$. We augment the nodes and edges of the original pathway as follows:

$$\bar{V} = V \cup \{m \in M | T(m) \cap V \neq \varnothing\}$$
$$\bar{E} = E \cup \{(m, g, inhibition) | m \in V \cap M \wedge g \in T(m)\}$$

We implemented this algorithm in R and published it as the Bioconductor package named mirIntegrator (`http://bit.ly/mirIntegrator`). mirIntegrator is flexible and allows users to integrate user-specific pathway databases with user-specific miRNA-mRNA target databases. Additionally, it generates graphical representations of the augmented pathways (see Fig. 5). We integrated pathways from Kyoto Encyclopedia of Genes and Genomes[17] (KEGG) (version 73) with miRNA targets from miRTarBase[18] (version 4.5) to generate mirAP, a database of miRNA-augmented pathways (`http://www.cs.wayne.edu/dmd/mirAP`).

2.2. *Integrative pathway analysis*

Our pathway analysis pipeline consists of two main steps. In the first step, we augment the signaling pathways with interactions between miRNAs and their targets. Once this is done, the data integration problem is mapped to the original pathway analysis problem for which existing methods can be applied. The difference is that here both miRNA and mRNA expression can be taken into consideration. In the second step, we apply any pathway analysis that uses fold change and p-value as input, e.g., Over-representation analysis[19] (ORA) and Impact Analysis[3,4]. ORA and Impact Analysis are well-known methods developed by our group to identify signaling pathways that are impacted by the effects of diseases. Fig. 1 displays the overall pipeline of our approach.

Impact Analysis[3,4] is a widely used topology-aware method that combines two types of evidence: i) the over-representation (ORA) of differentially expressed (DE) genes in a pathway[19], and ii) the perturbation (PERT) of such a pathway, as measured by propagating expression changes through the pathway topology. These two types of evidence are captured by two independent p-values[4]: p_{ORA} and p_{PERT}. These p-values are combined using Fisher's method to obtain a global p-value per pathway. Each global p-value represents the probability of having the observed number of DE genes, as well as the observed amount of impact just by chance (i.e. when the null hypothesis is true)[4]. To calculate p_{ORA} on mirAP, we assumed that the number of DE entities (genes and miRNAs) on the given pathway follows a hypergeometric distribution. The following information is needed to compute p_{ORA}: i) the total number of measured entities, ii) the number of entities belonging to the given augmented pathway, iii) the total number of DE entities, and iv) the number of DE entities in the given augmented pathway. To calculate p_{PERT} on mirAP, we perform a bootstrap procedure using the following input: i) the log-fold change of DE entities, and ii) the given augmented pathway.

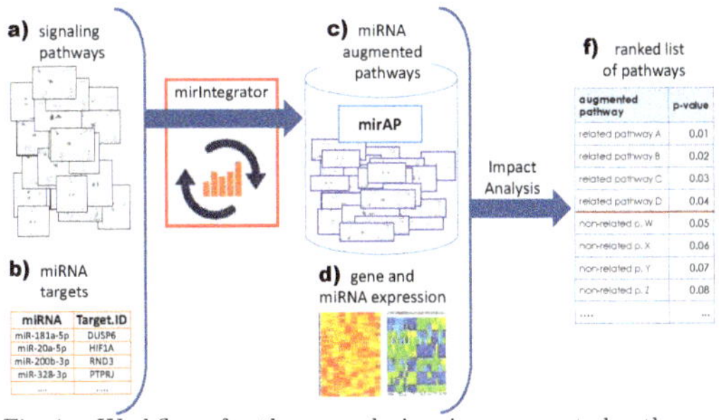

Fig. 1. Workflow of pathway analysis using augmented pathways.

2.3. *Integrative disease subtyping*

Our disease subtyping pipeline is presented on Fig. 2. The input includes: i) mRNA and miRNA sample-matched expression data, ii) survival records, iii) a database of miRNA-target gene interactions, and iv) a database of signaling pathways (see Fig. 2a). The output is a set of selected pathways (Fig. 2f) yielding to subtypes with significantly distinct survival patterns.

First, we obtain the miRNA-augmented pathways from mirAP (Fig. 2b). Second, we partition the patients using the genes and miRNAs provided by each augmented pathway (Fig. 2c). e.g., let us say that we want to analyze gene and miRNA expression from $\mathcal{N}$ number of patients and we obtained $\mathcal{P}$ number of augmented pathways from mirAP. Taking one pathway at the time, we filter the gene expression data by selecting only genes that belong to the pathway. Similarly, we filter the miRNA expression data by selecting only miRNAs that belong to the pathway. Now, we need to combine the filtered gene expression and miRNA data and then perform clustering on the combined data. So, we use Similarity Network Fusion method[15] (SNF) in conjunction with spectral clustering[20] for this purpose. We repeat this process with each pathway to obtain $\mathcal{P}$ different pathway-based clusterings, one per each pathway.

Third, we perform survival analysis on each of the pathway-based clusterings (Fig. 2d). In order to do this, we compute the log-rank test p-value (Cp) of Cox proportional hazards regression analysis by using the input survival information.This p-value represents how significant the difference between the survival curves is. For instance, a Cox log-rank test p-value close to zero may indicate that these groups have well-differentiated survival patterns. Now the question is whether we could obtain the same clustering just by chance[21]. To answer this question we use the random sampling technique. For example, if the pathway has G number of genes and m number of miRNAs, we randomly select G genes and m miRNAs from the measured values. Then, we partition the patients using this randomly selected set of entities and then compute its Cox p-value (rCp). We repeat this random selection a large number of times (e.g., 2,000 times) to construct an empirical distribution of Cox p-values (Fig. 2d). Next, we compare the observed Cox p-value Cp with the distribution of rCp, calculated from randomly selected genes and miRNAs. We estimate the probability of obtaining this Cp by computing the proportion of resampling p-values less than or equal to the observed Cp (e.g., In Fig. 2d the vertical red line indicates the observed Cp). For each pathway, we estimate this probability in order to quantify how likely it is to observe by chance a Cox p-value less than or equal to the one observed with the actual genes and miRNAs in the pathway.

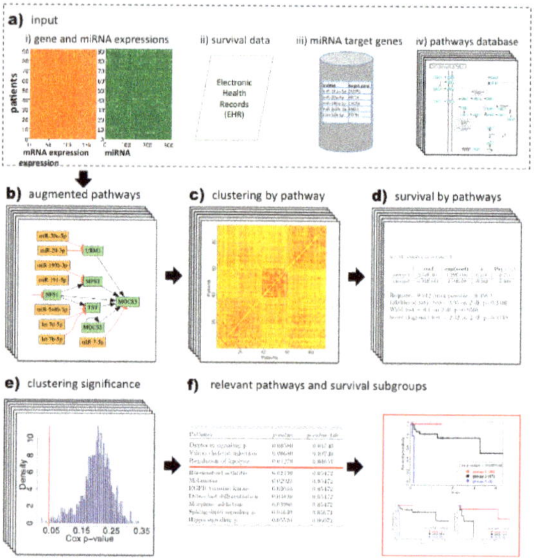

Fig. 2. The proposed pipeline for disease subtyping.

The final step is to select the pathways that are relevant to survival, i.e., pathways yielding to significantly distinct survival curves. To do this, we adjust the p_i p-values for multiple comparisons using False Discovery Rate (FDR). We then rank the pathways by FDR.p-value and select those less than or equal to the significance threshold of 5% as *relevant pathways*. We note that this pipeline can be used in conjunction with other integrative clustering methods.

3. Results

In this section, we present the results of our pathway analysis and disease subtyping pipelines using the miRNA-augmented pathways (mirAP). First, we perform pathway analysis of 9 mRNA/miRNA sample-matched datasets using two different methods (Impact Analysis and ORA) and show that mirAP offers a significant improvement over analyzing mRNA data alone. We also compare the obtained results with the state-of-the-art method (microGraphite)[8]. Second, we perform disease subtyping of a colorectal cancer dataset from TCGA using our subtyping pipeline and compare with the traditional pipeline for subtyping.

3.1. *Validation of our pathway analysis pipeline*

We analyze nine sample-matched datasets from seven different diseases: GSE43592 (multiple sclerosis, 10 controls, 10 cases), GSE35389 (melanoma, 4 controls, 4 cases), GSE35982 (colorectal cancer, 8 controls, 8 cases), GSE26168 (type II diabetes, 8 controls, 9 cases), GSE62699 (alcoholism, 18 controls, 18 cases), GSE35834 (colorectal cancer, 23 controls, 55 cases), GSE43797 (pancreatic cancer, 5 controls, 7 cases), GSE29250 (non-small cell lung cancer, 6 controls, 6 cases), and GSE32688 (pancreatic cancer, 7 controls, 25 cases). For each of these datasets, we used the normalized expression values as found in GEO.[22] The microarray probes were annotated according to their corresponding platform's metadata using GEOquery.[23] Next, we estimated log-fold-change between disease and control groups by fitting to a gene-wise linear model using the R package limma[24]. We use the following two criteria to identify differentially expressed (DE) genes: i) genes with adjusted p-value lower than 5%, and ii) among the genes that satisfy the first criterion, we choose the genes with the highest log-fold change, up to 10% of measured genes. We use the same criteria to identify DE miRNAs.

The nine datasets were selected due to two important reasons. First, these datasets have both mRNA and miRNA measurements for the same set of patients. Second, for each of the underlying diseases, there is a KEGG pathway, henceforth *target pathway*, that was created to describe the underlying mechanisms of the disease. To demonstrate the advantage of the miRNA data integration, we compare the use of the original KEGG pathways with the use of our miRNA augmented pathways (mirAP) by performing two pathway analysis methods that use p-value and fold-change: Impact Analysis (IA)[4] and over-representation analysis (ORA)[19]. The input for IA and ORA using KEGG is mRNA expression data. The input for IA and ORA using mirAP includes both mRNA and miRNA expression data. The output of each method is a list of p-values – one per pathway. These p-values are adjusted for multiple comparisons using False Discovery Rate (FDR)[25].

We also analyze the nine GEO datasets using microGraphite[8] after quantile normalization to compare with our pipeline. The main goal of microGraphite is the identification of signal transduction paths correlated with the condition under study. It is implemented in a four-steps recursive procedure as follows: (i) selection of pathways, (ii) best path identification, (iii) metapathway construction, and (iv) metapathway analysis. Here we only consider the first step of the approach, which is the selection of significant pathways. This selection is based on the significance levels obtained from the test on the mean of the pathways (alpha-mean). The input is the mRNA and miRNA expression data and it does not take in account fold-changes nor differentially expressed entities.

For each dataset, we expect a good method to identify the target pathway as significant, as well as to rank it on top. For instance, in the colorectal cancer dataset which compares colorectal cancer tissue vs. normal, the *Colorectal cancer pathway* must be shown as significant and should be as close to the top of the ranking as possible since this is the pathway that describes the phenomena involved in colorectal cancer. Based on this, we compare the rank and p-value of the target pathway in each disease using the five methods: i) mRNA expression alone using standard KEGG pathways with ORA and ii) IA, iii) mRNA and miRNA expression data using the augmented pathways (mirAP) with iii) ORA and iv) IA, and v) mRNA and miRNA expression data analyzed with microGraphite.

Table 1. Results of target pathway identification using traditional ORA (column 3), traditional IA (col. 4), ORA on mirAP (col. 5), IA on mirAP (col. 6), microGraphite (col. 7)

GEO ID	Target pathway	ORA	IA	ORAmir	IAmir	microGraphite
GSE26168	Type II diabetes mellitus	no	no	no	no	yes
GSE29250	Non-small cell lung cancer	no	no	yes	no	no
GSE35982	Colorectal cancer	no	no	no	no	no
GSE32688	Pancreatic cancer	no	no	yes	yes	no
GSE35389	Melanoma	no	no	yes	yes	no
GSE35834	Colorectal cancer	no	no	yes	yes	no
GSE43592	Amyotrophic lateral scle.	no	no	no	yes	no
GSE43797	Pancreatic cancer	no	no	yes	yes	yes
GSE62699	Alcoholism	no	no	no	yes	no

Table 1 shows the target pathways and their significance for the 9 datasets. The first and second columns display the datasets and their corresponding target pathways while the other five columns indicate whether the target pathways are identified as significant using the five methods: ORA of mRNA expression on KEGG pathways (ORA+KEGG), IA of mRNA expression on KEGG (IA+KEGG), ORA of miRNA and mRNA expression data on mirRNA-augmented pathways (ORA+mirAP), our approach IA of miRNA and mRNA expression on mirAP (IA+mirAP), and miRNA and mRNA expression analysis using microGraphite, respectively. The significance threshold is 5% for FDR p-values. IA and ORA fail to identify any target pathway as significant when using only mRNA whereas our approach (IA+mirAP) correctly identify the target in 6 out of 9 datasets (GSE32688, GSE35389, GSE35834, GSE43592, GSE43797, GSE62699) and ORA+mirAP correctly identify the target pathway as significant in 5 out of 9 datasets (GSE29250, GSE32688, GSE35389, GSE35834, GSE43797). microGraphite correctly identifies the target pathway as significant in only 2 out of 9 datasets (GSE26168, GSE43797). The results demonstrate that our integration of mRNA and miRNA lifts the statistical power for both pathway analysis techniques (ORA and IA) and outperforms microGraphite in target pathway identification.

Fig. 3 shows the p-values and rankings of the target pathways using the five methods. The panel (a) shows the FDR corrected p-values of the target pathways. We compare the lists of p-values using Wilcoxon test. The FDR p-values produced by IA+mirAP are significantly smaller than by IA+KEGG (p=0.007), ORA+KEGG (p=0.005), and microGraphite (p=0.009).

The panel (b) shows the rankings of the target pathways. Again, the rankings produced by IA+mirAP are significantly smaller than those of IA+KEGG (p=0.03 using t-test, and p=0.04 using Wilcoxon test), ORA+KEGG (p=0.03 using t-test and p=0.04 using Wilcoxon test), and microGraphite (p=0.0051 using t-test and p=0.0058 using Wilcoxon test). This confirms that our augmented pathways, mirAP, improve the performance of traditional Impact Analysis and ORA. Also, the results show that the proposed integrative pathway analysis also outperforms microGraphite in terms of both p-values and rankings for target pathway identification.

Furthermore, our pathway database (mirAP) is generated with validated miRNA-mRNA interactions, while microGraphite uses predicted interactions, which increases the number of false positive miRNA-target interactions. Another drawback of microGraphite is it execution time. A typical analysis with microGraphite takes approximately 22 hours while our approach takes only a few minutes. We ran these experiments on a typical desktop workstation with a 2.6 GHz Intel Core i5, 8GB of RAM, on a single thread, and the OS X 10.11 operative system.

3.2. *Validation of our disease subtyping pipeline*

To assess our disease subtyping pipeline we use matched-sample gene and miRNA expression data (level 3 from platforms Agilent G4502A-07 and Illumina GASeq miRNASeq, respectively) of colorectal cancer patients (COAD) downloaded from the Cancer Genome Atlas (TCGA) (`cancergenome.nih.gov`). We selected the largest set of patients with miRNA-mRNA matched samples and available survival records, as were selected in SNF[15]. The number of patients is $M = 92$, the number of genes is $N_g = 17,814$, and the number of miRNAs is $N_m = 705$. We performed unsupervised clustering with the number of clusters set as $k = 3$ according to prior knowledge of the number of subtypes of COAD[15]. We use SNF[15] in conjunction with spectral clustering[20] as integrative clustering method. To perform SNF clustering, we used the SNFtool package with the suggested parameters.

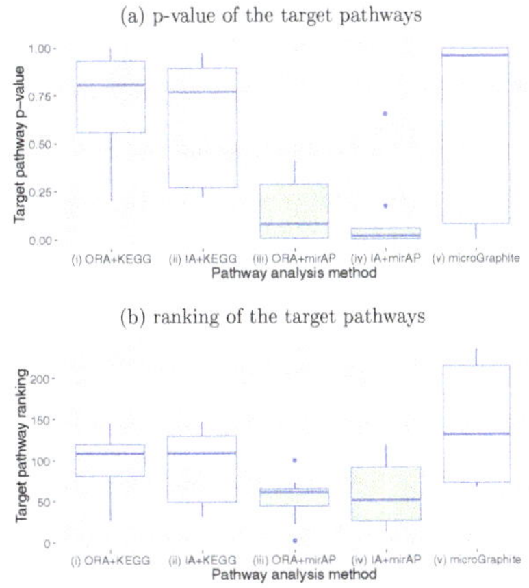

Fig. 3. Corrected p-values and rankings of the target pathways using different methods.

For each miRNA-augmented pathway, our method partitions the patients using the genes and miRNAs in the pathway as clustering features, resulting in a total of 184 clusterings. Then for each pathway-based clustering, we construct the empirical distribution and then estimated the *p-value* of how likely the pathway helps to improve disease subtyping. The *p-values* of the relevant pathways are shown in Table 2. We select the pathways with a FDR-corrected *p-value* ≤ 0.05 as *relevant pathways*. The horizontal red line represents the significance cutoff at 5%. For TCGA-COAD, we identify three relevant pathways: *Oxytocin signaling pathway*, *Vibrio cholerae infection*, and *Regulation of lipolysis in adipocytes*.

We also cluster the 92 patients using SNF with the traditional pipeline, i.e., using all the measured genes and miRNAs. We compare these partitions with those obtained by our pipeline. To assess the correlation between the obtained groups and survival patterns (e.g., long-term vs. short-term survival), we performed survival analysis for all the cases using Kaplan-Meier analysis.

Table 2. List of relevant pathways for colorectal subtyping.

Pathway	*p-value*	*p-value.fdr*
Oxytocin signaling pathway	0.00580	0.0374
Vibrio cholerae infection	0.00680	0.0374
Regulation of lipolysis in adipocytes	0.01270	0.0466
Rheumatoid arthritis	0.02190	0.0547
...	...	...

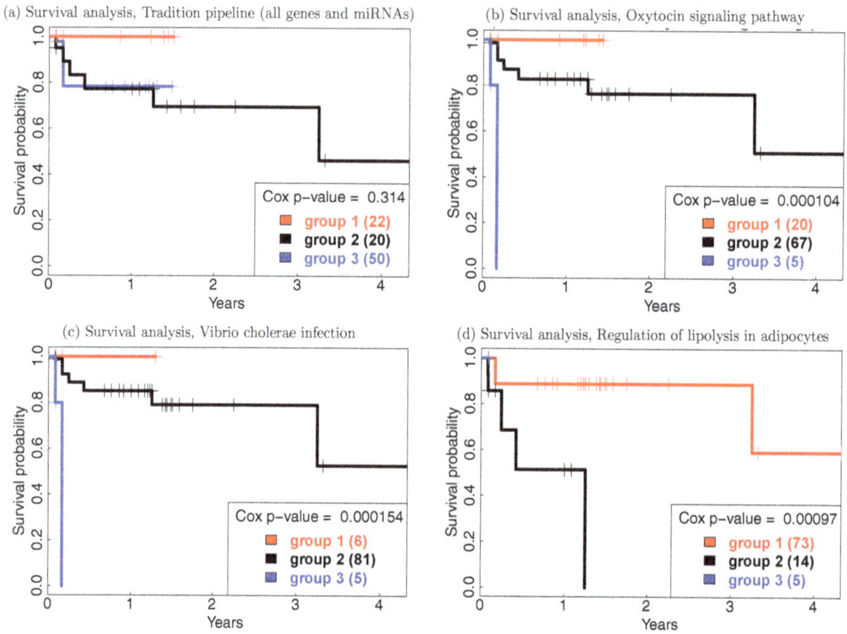

Fig. 4. Kaplan-Meier survival analysis of the obtained COAD subtypes. a) Survival curves using all genes and miRNAs. b), c), and d) Survival curves using relevant pathways.

Fig. 4 shows the Kaplan-Meier plots, each one represents the association of the obtained groups with the observed patient survival. Fig. 4a shows the subtypes obtained with the traditional pipeline using all 17,814 genes and 705 miRNAs. In a Cox proportional hazards regression analysis, we find that there is no statistically significant difference between survival groups obtained with the traditional pipeline (log rank test p-value = 0.314). Fig. 4b, c, and d. shows the resultant clustering on the relevant pathways identified with our approach (Table 2). Clustering based on *Oxytocin signaling pathway* entities gives a log rank test p-value of 0.000104, which indicates a significant difference between the survival curves (Fig. 4b). Similarly, clusterings based on *Vibrio cholerae infection* and *Regulation of lipolysis in adipocytes* augmented pathways indicate significant differences between the survival curves with p-values of $p = 0.000154$ and $p = 0.00097$, respectively (Fig. 4c and d). As we can see, integrative clustering based on relevant mirAP pathways produce subtypes significantly more related to survival data than the traditional subtyping pipeline (approximately 1000 times lower p-values).

Given that our approach requires resampling for computing the pathways' significance (*p-values*), our pipeline is more time consuming than the traditional pipeline. For the computational experiments presented here, we generated 2,000 random clusterings per each pathway. Our pipeline took some hours to subtype the set of patients (approximately 4 hours) while running SNF alone takes only some minutes (less than 3 minutes).

3.2.1. *Biological Significance of relevant Signaling Pathways*

Our pipeline identifies the *Oxytocin signaling pathway* to be related to the survival subtyping of colorectal cancer patients ($p = 0.000104$). Oxytocin (OXT) is a hormone with a well-known effect on uterine smooth muscles and myoepithelial cells. Additionally, it has been shown that oxytocin is expressed along the entire human gastrointestinal (GI) tract, including colon, and it contributes to the control of the GI motility[26]. Moreover, studies have shown that exposure to OXT leads to a significant decrease in cell proliferation for some epithelial cancer cells (e.g., breast and prostate cancer)[27]. In contrast, OXT has a growth-stimulating effect in other types of cancer cells (e.g., small-cell lung cancer, endothelial cancer, and Kaposiâs sarcoma)[28,29]. We think that the evidence of OXT expression on colon and the dual role that OXT has in some cancer cells (as inhibitor and promoter of cancer cells proliferation) may indicate that OXT could also play an important role in differentiating short and long-term survival COAD patients. In addition, OXT is also known to be capable of mitigating symptoms caused by stress, OXT levels increase in acute(short-lived) stress and decrease during chronic stress. Also, it is well-known that chronic stress has an outstanding role in cancer growth and metastasis.[30] From this, we also hypothesize that patients in the short term survival group (Fig. 4b, gr. 3) may have been in a metastatic stage with chronic stress and different OXT expression than patients in the other groups (Fig. 4b,1-2).

Similarly, we identify *Vibrio cholerae infection* pathway as relevant. This pathway describes the colonization of the intestine by Vibrio cholerae bacteria (VC). The main factor involved in this process is Cholera toxin (CTX). Several studies have exhibit relations between gastrointestinal tract bacteria and colon cancer progression. In particular, it has been shown that CTX suppresses carcinogenesis of inflammation-driven sporadic colon cancer[31].

Ultimately, the *Regulation of lipolysis in adipocytes* pathway describes a unique function of white adipose tissue in which triacylglycerols (TAGs) are broken down into fatty acids and glycerol. Fatty acid (FA) pathways play an important role in cancer[32]. In particular, increased gene expression of AGPAT9(PNPLA2), MAGL(MGLL), and HSL(LIPE), FA metabolism regulators, is associated with increased cancer cells proliferation in colorectal cancer[32] (see blue boxes in Fig. 5). By instance, MAGL pharmacological inhibition attenuated aggressiveness of colorectal cancer cells. On the other hand, decreased gene expression of CD36/FAT regulator has been implicated in contributing to colorectal cancer progression, a higher metastasis grade, and low relapse-free survival[33].

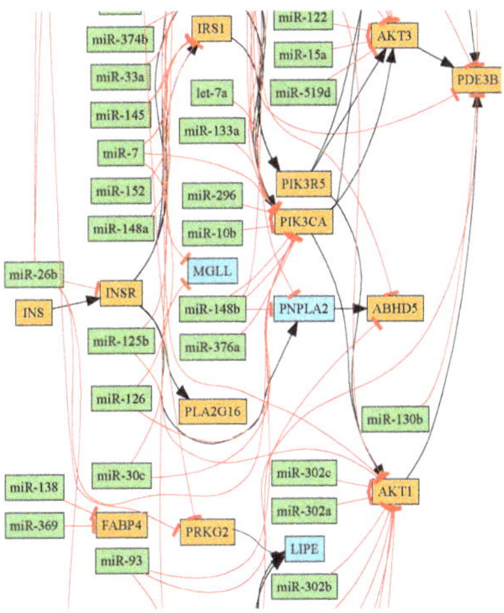

Fig. 5. Portion of the miRNA-augmented *Regulation of lipolysis in adipocytes* pathway.

Fig. 5 shows a portion of the *Regulation of lipolysis in adipocytes* augmented pathway obtained from our database (see the complete pathway at `http://bit.ly/hsa04923`).The green boxes show the protein coding genes while the orange boxes display the miRNAs. The black arrows denote activation and the red bar-headed arrows denote repression.

4. Discussion

In this article, we present a method to augment signaling pathways with miRNA-target interactions. The miRNA-augmented pathways (mirAP) offer a more comprehensive view and a deeper understanding of complex diseases. We also present two pipelines that use mirAP to integrate miRNA and mRNA expression data for the purpose of pathway analysis and disease subtyping. As miRNA expression data are becoming freely accessible, miRNA-mRNA integrative analyses are likely to become a routine.

Our pathway analysis pipeline augments gene-gene signaling pathways with miRNA-target interactions. Then we perform a topology-based pathway analysis that takes into consideration both types of molecular data. We analyze 9 sample-matched datasets that were assayed in independent labs. Our pipeline outperforms traditional methods in identifying target pathways (smaller p-values and rankings of the target pathways). We plan to explore methods for augmenting the pathways using only the process(es) described by each given pathway.

Our disease subtyping pipeline combines gene and miRNA expression data, clinical records, and mirAP. The contribution of our disease subtyping pipeline is two-folds. First, this framework introduces a way to exploit the additional information available in biological databases and integrates clinical data, miRNA and gene expression data for disease subtyping. Second, it identifies pathways associated with survival differentiated subgroups of diseases, which bring us closer to the identification of causal pathways associated with survival. We analyze a colorectal cancer data downloaded from TCGA. Our framework provides pathways relevant to survival patterns and subtypes significantly difference between the survival curves. It greatly improves the former approach with p-values $1,000$ times lower than the former. This pipeline is limited by the availability of datasets containing survival records, miRNA, and mRNA expression matched-samples. We plan to extend this study by investigating more diseases and larger datasets.

Acknowledgments

This work was supported by the National Institutes of Health [R01 DK089167, R42 GM087013]; National Science Foundation [DBI-0965741]; and the Robert J. Sokol Endowment in Systems Biology. Any opinions, findings, conclusions, or recommendations expressed in this material are those of the authors and do not necessarily reflect the views of any of the funding agencies.

References

1. Y. S. Lee and A. Dutta, *Annual Review of Pathology* **4** (2009).
2. P. Khatri, M. Sirota and A. J. Butte, *PLoS Computational Biology* **8**, p. e1002375 (2012).
3. S. Drăghici, P. Khatri, A. L. Tarca, K. Amin, A. Done, C. Voichiţa, C. Georgescu and R. Romero, *Genome Research* **17**, 1537 (2007).
4. A. L. Tarca, S. Drăghici, P. Khatri, S. S. Hassan, P. Mittal, J.-s. Kim, C. J. Kim, J. P. Kusanovic and R. Romero, *Bioinformatics* **25**, 75 (2009).

5. C. Backes, E. Meese, H.-P. Lenhof and A. Keller, *Nucleic Acids Research* **38**, 4476 (July 2010).
6. J. B.-K. Hsu, C.-M. Chiu, S.-D. Hsu, W.-Y. Huang, C.-H. Chien, T.-Y. Lee and H.-D. Huang, *BMC Bioinformatics* **12**, p. 300 (July 2011).
7. I. S. Vlachos, N. Kostoulas, T. Vergoulis, G. Georgakilas, M. Reczko, M. Maragkakis, M. D. Paraskevopoulou, K. Prionidis, T. Dalamagas and A. G. Hatzigeorgiou, *Nucleic Acids Research* **40**, W498 (July 2012).
8. E. Calura, P. Martini, G. Sales, L. Beltrame, G. Chiorino, M. D'Incalci, S. Marchini and C. Romualdi, *Nucleic Acids Research* **42**, p. e96 (2014).
9. S. Nam, M. Li, K. Choi, C. Balch, S. Kim and K. P. Nephew, *Nucleic Acids Research* **37**, W356 (May 2009).
10. P. Martini, G. Sales, M. S. Massa, M. Chiogna and C. Romualdi, *Nucleic Acids Research* **41**, e19 (2013).
11. S. Saria and A. Goldenberg, *IEEE Intelligent Systems* **30**, 70 (2015).
12. T. R. Golub, D. K. Slonim, P. Tamayo, C. Huard, M. Gaasenbeek, J. P. Mesirov, H. Coller, M. L. Loh, J. R. Downing, M. A. Caligiuri, C. D. Bloomfield and E. S. Lander, *Science* **286**, 531 (October 1999).
13. T. Sørlie, R. Tibshirani, J. Parker, T. Hastie, J. S. Marron, A. Nobel, S. Deng, H. Johnsen, R. Pesich, S. Geisler *et al.*, *Proceedings of the National Academy of Sciences* **100**, 8418 (2003).
14. P. Wirapati, C. Sotiriou, S. Kunkel, P. Farmer, S. Pradervand, B. Haibe-Kains, C. Desmedt, M. Ignatiadis, T. Sengstag *et al.*, *Breast Cancer Research* **10**, p. R65 (2008).
15. B. Wang, A. M. Mezlini, F. Demir, M. Fiume, Z. Tu, M. Brudno, B. Haibe-Kains and A. Goldenberg, *Nature Methods* **11**, 333 (2014).
16. T. Nguyen, D. Diaz, R. Tagett and S. Draghici, *Nature Scientific Reports* **6**, p. 29251 (2016).
17. M. Kanehisa and S. Goto, *Nucleic acids research* **28**, 27 (2000).
18. S.-D. Hsu, Y.-T. Tseng, S. Shrestha, Y.-L. Lin, A. Khaleel, C.-H. Chou, C.-F. Chu, H.-Y. Huang, C.-M. Lin, S.-Y. Ho *et al.*, *Nucleic Acids Research* **42**, D78 (January 2014).
19. S. Drăghici, P. Khatri, R. P. Martins, G. C. Ostermeier and S. A. Krawetz, *Genomics* **81**, 98 (2003).
20. U. Von Luxburg, *Statistics and Computing* **17**, 395 (2007).
21. E. Czwan, B. Brors and D. Kipling, *BMC Bioinformatics* **11**, p. 19 (2010).
22. T. Barrett, T. O. Suzek, D. B. Troup, S. E. Wilhite, W. C. Ngau, P. Ledoux, D. Rudnev, A. E. Lash, W. Fujibuchi and R. Edgar, *Nucleic Acids Research* **33**, D562 (2005).
23. S. Davis and P. Meltzer, *Bioinformatics* **14**, 1846 (2007).
24. M. E. Ritchie, B. Phipson, D. Wu, Y. Hu, C. W. Law, W. Shi and G. K. Smyth, *Nucleic Acids Research* **43**, e47 (April 2015).
25. Y. Benjamini and D. Yekutieli, *Annals of Statistics* **29**, 1165 (August 2001).
26. B. Ohlsson, M. Truedsson, P. Djerf and F. Sundler, *Regulatory Peptides* **135**, 7 (July 2006).
27. A. Reversi, V. Rimoldi, T. Marrocco, P. Cassoni, G. Bussolati, M. Parenti and B. Chini, *Journal of Biological Chemistry* **280**, 16311 (April 2005).
28. P. Cassoni, T. Marrocco, S. Deaglio, A. Sapino and G. Bussolati, *Annals of Oncology* **12**, S37 (January 2001).
29. C. Pqueux, B. P. Keegan, M.-T. Hagelstein, V. Geenen, J.-J. Legros and W. G. North, *Endocrine-Related Cancer* **11**, 871 (December 2004).
30. M. Moreno-Smith, S. K. Lutgendorf and A. K. Sood, *Future Oncology* **6**, 1863 (December 2010).
31. M. Doulberis, K. Angelopoulou, E. Kaldrymidou, A. Tsingotjidou, Z. Abas, S. E. Erdman and T. Poutahidis, *Carcinogenesis* **36**, p. bgu325 (December 2014).
32. S. Balaban, L. S. Lee, M. Schreuder and A. J. Hoy, *BioMed Research International* **2015**, p. 274585 (2015).
33. S. M. Rachidi, T. Qin, S. Sun, W. J. Zheng and Z. Li, *PLOS ONE* **8**, p. e57911 (March 2013).

FREQUENT SUBGRAPH MINING OF PERSONALIZED SIGNALING PATHWAY NETWORKS GROUPS PATIENTS WITH FREQUENTLY DYSREGULATED DISEASE PATHWAYS AND PREDICTS PROGNOSIS

ARDA DURMAZ*

Systems Biology and Bioinformatics Graduate Program,
Case Western Reserve University, 10900 Euclid Avenue, Cleveland, Ohio 44106, USA
Email: arda.durmaz@case.edu

TIM A. D. HENDERSON*

Department of Electrical Engineering and Computer Science,
Case Western Reserve University, 10900 Euclid Avenue, Cleveland, Ohio 44106, USA
Email: tadh@case.edu

DOUGLAS BRUBAKER

Department of Biological Engineering,
Massachusetts Institute of Technology, 77 Massachusetts Ave, Cambridge, MA 02139
Email: dkb50@mit.edu

GURKAN BEBEK†

Center for Proteomics and Bioinformatics, Department of Nutrition,
Department of Electrical Engineering and Computer Science,
Case Western Reserve University, 10900 Euclid Avenue, Cleveland, Ohio 44106, USA
Email: gurkan.bebek@case.edu

Motivation: Large scale genomics studies have generated comprehensive molecular characterization of numerous cancer types. Subtypes for many tumor types have been established; however, these classifications are based on molecular characteristics of a small gene sets with limited power to detect dysregulation at the patient level. We hypothesize that frequent graph mining of pathways to gather pathways functionally relevant to tumors can characterize tumor types and provide opportunities for personalized therapies.
Results: In this study we present an integrative omics approach to group patients based on their altered pathway characteristics and show prognostic differences within breast cancer ($p < 9.57E-10$) and glioblastoma multiforme ($p < 0.05$) patients. We were able validate this approach in secondary RNA-Seq datasets with $p < 0.05$ and $p < 0.01$ respectively. We also performed pathway enrichment analysis to further investigate the biological relevance of dysregulated pathways. We compared our approach with network-based classifier algorithms and showed that our unsupervised approach generates more robust and biologically relevant clustering whereas previous approaches failed to report specific functions for similar patient groups or classify patients into prognostic groups.
Conclusions: These results could serve as a means to improve prognosis for future cancer patients, and to provide opportunities for improved treatment options and personalized interventions. The proposed novel graph mining approach is able to integrate PPI networks with gene expression in a biologically sound approach and cluster patients in to clinically distinct groups. We have utilized breast cancer and glioblastoma multiforme datasets from microarray and RNA-Seq platforms and identified disease mechanisms differentiating samples.
Supplementary information: Supplementary methods, figures, tables and code are available at *https://github.com/bebeklab/dysprog.*

*Co-first Author
†Corresponding Author

1. Introduction

Personalized medicine aims to tailor treatment options for patients based on the makeup of their diseases. In the case of cancer, the genetic makeup of tumors is characterized to identify unique tendencies and exploit vulnerabilities of these tumors. However, identifying genomic alterations and molecular signatures that better describe or classify cancer to accomplish this goal has been challenging. Furthermore complex disease phenotypes, such as cancer, cannot be fully explained by individual genes and mutations. Recent studies have explored various approaches to uncover the molecular network signatures of cancers including multivariate linear regression[1] or factor graphs[2] to combine information flow based approaches with copy numbers and DNA methylation data. These techniques identified patient loci with high risk of disease along with genes that are dysregulated for various cancers.[3,4] Gene expression profiles and (in some cases) DNA methylation or metabolomics data have also been used to identify subtypes of the disease.[3-7] However prognostic classification of tumors still requires attention and it is an important step toward identifying most effective approaches in precision medicine.

Glioblastoma multiforme (GBM) is the most common form of malignant brain tumor in adults. GBM is characterized by a median survival of one year and an overall poor prognosis.[8] There have been numerous attempts to classify GBM by differential gene expression to identify clinically and prognostically relevant subtypes.[9,10] Previously methylation status of the *MGMT* promoter is suggested to be associated with tumor response of gliomas to alkylating agents and later associated with increased survival.[11,12] More recently The Cancer Genome Atlas (TCGA) project also provided supporting findings of the methylation status of the *MGMT* promoter as a prognostic marker through analysis of high dimensional data for 206 GBM tumors.[13] Further work utilizing the TCGA data classified GBM by aberrations and gene expression of *EGFR, NF1, and PDGFRA/IDH1* into four subtypes, Classical, Mesenchymal, Neural, and Proneural.[14] These classifications implied strong relationships between subtypes and neural lineages as well as response to aggressive therapy. Though these studies introduced GBM classification, there remained a need to classify dysregulations in tumors more specifically by survivability. While earlier approaches have focused on identifying gene sets,[10,15-18] these had little impact on finding dysregulated pathway segments. For instance, using nearest shrunken centroid classification method,[18,19] or clustering algorithms,[14] gene sets that stratify samples were identified, yet functionally these were not strongly related. Hence, they present little potential for improved treatment opportunities for patients.

Breast Invasive Carcinoma (BRCA) is the most diagnosed cancer among woman consisting of multiple sub classes with distinct clinical outcomes. Previously, 5 subtypes were identified using expression profiles of and later applied to develop predictors by manually selected genes.[6,20,21] Consecutive studies identified differing number of subtypes similar to inital identification. For instance using expression profiles Sotiriou *et al.* identified 6 subtypes further separating luminal-like and basal-like groups.[22,23] Furthermore a comprehensive study integrating multiple omics data to identify unified classification of the breast cancer samples provided strong evidence for 4 subtypes; *Basal, Her-2 enriched, Luminal-A, Luminal-B*.[4] However studies incorporating network or pathway information either used manual selection of pathways or produced limited results. For instance Gatza *et al.* identified 17 subgroups using pathway based classification with mixed intrinsic subtype signatures.[24]

We describe an integrative omics approach based on frequent subgraph mining (FSM) that brings Protein-Protein Interaction (PPI) networks and gene expression data together to infer molecular networks that are dysregulated in patient samples. We tested our approach using gene expression data for both glioblastoma and breast cancer datasets collected with microarray and next generation sequencing (NGS) approaches. The networks inferred from FSM not only stratify patients into clinically-relevant subtypes, but also provides significant prognostic differences. Our results suggest that a network-based stratification of patients is more informative than using gene-level or feature-based data integration. Identifying personalized dysregulated signaling networks will offer effective means to diagnose and treat patients.

2. Methods

The proposed method uses a novel approach to integrate mRNA expression profiles and PPI networks to identify personalized dysregulated signaling pathways. We hypothesize that dysregulated sub-pathways observed in cancer can discriminate between tumors types which lead to different patient outcomes. We utilized publicly available datasets to develop and validate a method to detect altered molecular signatures in canonical pathways. Our classifications better distinguish patient prognosis in biologically relevant terms than previous studies.[14,25,26]

Our approach is to construct personalized networks of PPIs for cancerous tumors based on mRNA expression data. Section 2.1 details the construction of these networks called *dysregulated signaling pathways*. A network is constructed for each of the patients in each of the datasets used in Section 3. Personalized networks are mined using a new algorithm called QSPLOR (queue explorer) to identify a subset of frequently occurring subgraphs with 4 to 8 proteins as detailed in Sections 2.2 and 2.3. Finally, Non-Negative Matrix Factorization is used to cluster the patients via the frequently occurring subgraphs (Section 2.4 and 2.5).

In Section 3 the clusters are shown to separate patients into short-term and long-term survival groups. The methodology presented has the potential to stratify patients based on their molecular signatures, improve delivery of therapies and assist clinicians and researchers alike to better assess patient prognosis.

2.1. *Dysregulated Signaling Pathways*

Dysregulated Signaling Pathways are labeled graphs (Section 2.2) where vertices represent proteins and edges represent dysregulated activation/inhibition interactions. They are constructed from mRNA expression data (Section 3) and known PPI data.[27,28]

Dysregulation is computed by constructing a matrix $\mathbf{P}$, where $\mathbf{P}_{i,a}$ is the standard score of expression level of gene a for patient i. Then an *interaction matrix* $\mathbf{S}$ constructed from $\mathbf{P}$ in Equation 1. In Equation 1 (ab) represents two genes a and b such that the protein encoded by a interacts with the protein encoded by b. The variable i represents a particular patient.

$$\mathbf{S}_{(ab),i} = \sqrt{\mathbf{P}_{i,a}^2 + \mathbf{P}_{i,b}^2} \tag{1}$$

To determine if the relationship between two genes a and b is dysregulated for patient i the *z-score* for each interaction is computed. In Equation 2, $\mu(\mathbf{S}_{(ab),\cdot})$ and $\sigma(\mathbf{S}_{(ab),\cdot})$ respectively refer to the mean and standard deviation of the dysregulation scores for genes a and b.

$$Z(\mathbf{S})_{(ab),i} = \frac{\mathbf{S}_{(ab),i} - \mu(\mathbf{S}_{(ab),\cdot})}{\sigma(\mathbf{S}_{(ab),\cdot})} \tag{2}$$

If $Z(\mathbf{S})_{(ab),i} > c$ then an edge $a \rightarrow b$ is included in the graph for patient i indicating a and b are dysregulated. In Section 3 the constant c, the z-score threshold, was set to 2 to mine for dysregulation.

2.2. *Frequent Subgraph Mining*

Frequent Subgraph Mining (FSM) is a data mining technique which looks for repeated subgraphs in a graph database. As in Inokuchi *et al.*,[29] the database $\mathcal{D}$ is a set of transactions where each "transaction" is the dysregulated signaling pathways for a patient. FSM detects signaling sub-pathways which are dysregulated in multiple patients.

A dysregulated signaling pathway is a directed labeled graph G consisting of a set of vertices V, a set of edges $E = V \times V$, a set of labels L, and a labeling function which maps vertices (or edges) to labels $l : V|E \rightarrow L$. A graph $H = (V_H, E_H, L, l)$ is a subgraph of $G = (V_G, E_G, L, l)$ if $V_H \subseteq V_G$ and $E_H \subseteq E_G$.

A graph H is a subgraph of G ($H \sqsubseteq G$) if there is an injective mapping $m : V_H \rightarrow V_G$ s.t.

(1) All vertices in H map vertices in G with the same label: $\forall\ v \in V_H\ [l(v) = l(m(v))]$
(2) All edges match: $\forall\ (u, v) \in E_H\ [(m(u), m(v)) \in E_G]$
(3) All edge labels match: $\forall\ (u, v) \in E_H\ [l(u, v) = l(m(u), m(v))]$

Such a mapping m is known as an *embedding*. The problem of determining if a graph H is a subgraph of G is called the *subgraph isomorphism problem* and is NP-Complete.[30] The *frequency* of a subgraph H is the number of graphs (transactions) in $\mathcal{D}$ which H *embeds* into.

The subgraph relationship $\cdot \sqsubseteq \cdot$ induces a *partial order* on the subgraphs of the graphs in $\mathcal{D}$. That partial order is referred to as the *subgraph lattice*. If the subgraphs in the lattice are all *connected* it is known as the *connected subgraph lattice*. The connected subgraph lattice of $\mathcal{D}$ can be viewed as a graph $\mathcal{L}_\mathcal{D} = (V_\mathcal{L}, E_\mathcal{L})$. The vertices $V_\mathcal{L}$ are all of the connected subgraphs of G. If u and v are both vertices of $\mathcal{L}_\mathcal{D}$ then there is an edge between u and v if and only if $u \sqsubseteq v$ and v and be constructed from u by adding one edge and at most one vertex. The k *frequent connected subgraph lattice* k-$\mathcal{L}_\mathcal{D}$ contains only those subgraphs of graphs in $\mathcal{D}$ which are present in at least k graphs in the graph database $\mathcal{D}$. The leaf nodes of the k-$\mathcal{L}_\mathcal{D}$ are the *maximal frequent subgraphs*.

The objective of frequent subgraph mining is to discover the vertices of k-$\mathcal{L}_\mathcal{D}$. If a subgraph does have at least k transactions it is embedded in, it is known as a *frequent subgraph*. Since finding a frequent subgraph requires repeated subgraph isomorphism queries the problem complexity of FSM is exponential. The number of steps in frequent subgraph mining is bounded from above by $\mathcal{O}(2^g g^h)$ where g is the size of the graph and h is the size of the largest frequent subgraph. The term 2^g is an upper bound on the number of subgraphs of g. Tighter bounds can be obtained if one has more specific knowledge of the graph. The term g^h is an upper bound on number of steps to check if a graph of size h is a subgraph of g.

We present QSPLOR, a new algorithm to find a subset of frequent subgraphs in Section 2.3. It is used to find frequently dysregulated signaling sub-pathways. QSPLOR uses a fixed

```
1  # param start: frequent single vertex subgraphs
2  # param score: a function to score queue items
3  # param max_size: the max size of the queue
4  # param min_sup: int, amount of support
5  # returns: a generator of frequent subgraphs
6  def qsplor(start, score, min_sup):
7      while not start.empty():
8          queue = [ start.pop() ]
9          while not queue.empty()
10             lattice_node = take(queue, score)
11             kids = lattice_node.extend(min_sup)
12             for ext in kids: add(queue, score, ext, max_size)
13             yield subgraph
14 def add(queue, score, item, max_size):
15     queue.append(item)
16     while len(queue) >= max_size:
17         i = argmin(score(idx, queue) for idx in sample(10, len(queue)))
18         queue.drop(i)
19 def take(queue, score):
20     i = argmax(score(idx, queue) for idx in sample(10, len(queue)))
21     return queue.take(i)
```

Fig. 1. QSPLOR: a new algorithm for mining a subset of frequent subgraphs.

amount of memory and a user defined scoring heuristic to guide the search. The algorithm only reports the maximal frequent subgraphs found for compactness. We report only a subset, and not all of frequently dysregulated signaling pathways because (i) it is much faster to report only some of the frequent subgraphs and (ii) using a greater number of frequent subgraphs does not necessarily lead to a more discriminating clustering of samples in our analysis.

There have been a variety of FSM algorithms developed over the last two decades and there are several recent surveys available.[31,32] In recent years interest in collecting representative subsets of frequent subgraphs has emerged.[33,34] Both studies employ random walks on the frequent connected subgraph lattice to collect a sample of the frequent subgraphs. Finally, Leap Search[35] was proposed to find interesting patterns as defined by an objective function.

2.3. *QSPLOR: Mining a Subset of Frequent Subgraphs*

Figure 1 shows pseudo code for QSPLOR a new algorithm to mine a subset of frequent subgraphs. It proceeds as a graph traversal of k-$\mathcal{L}_\mathcal{D}$ (the k frequent connected subgraph lattice of the graph database). It begins the traversal at each lattice node representing a frequent subgraph containing only one vertex. At each outer step it initializes a queue with one of the starting lattice nodes. Then in each inner step it removes an item of the queue. The `take` function removes one item from a uniform sample of the queue such that a user supplied scoring function is maximized.

On line 11, the lattice node is extended. This involves finding all possible one edge extensions to the subgraph represented by the lattice node. The ones that are frequent are returned by the extend method. After the extensions are found they are added to the queue with the `add` method. If the queue is at the maximal size after the addition, one item from the queue is dropped. The dropped item is from a uniform sample of the queue and minimizes the user supplied score function. After all extensions have been processed the subgraph is output.

The key to our algorithm is the user supplied scoring function which guides the traversal. The simplest scoring function simply returns a uniform random number. This will cause the traversal to be unguided. Complex scoring functions can prioritize certain labels or structures.

The best general scoring functions are those that prioritize *queue diversity* such that traversal is encouraged to explore as much of the lattice as possible. We use a distance function which captures both structural and labeling differences between graphs as the scoring function for this paper. See the supplementary methods for more details on QSPLOR.

2.4. *Non-Negative Matrix Factorization*

Clustering via Non-Negative Matrix Factorization (NMF) is used to partition patients into subgroups. Section 3 shows that the partitions are prognostically discriminative between the patient subgroups. NMF method was first proposed by Lee and Seung[36] with the aim of decomposing images into explanatory basis vectors. NMF has also been used on gene expression data.[37] For a description of our usage of NMF see the supplementary methods.

2.5. *Clustering Metrics*

Use of NMF requires careful evaluation of the results. Since NMF is based on random initialization of the initial stratification we have applied consensus clustering approach. Using R package NMF[38] we have applied method *'nsNMF'* and random seed with 150 runs. To identify best clustering rank k cophenetic correlation coefficient, silhouette values, residual metrics are evaluated. Cophenetic correlation coefficient is first suggested by Brunet *et al.*[37] to quantify the stability of the clusters. It is calculated as the correlation between sample distances obtained from consensus matrix and the cophenetic distances obtained from hierarchical clustering of the consensus matrix. Brunet *et al.* suggested to choose the ranks where cophenetic correlation coefficient starts to decrease. Silhouette is another method for quantifying cluster stability.[39] The values range between -1 and 1. Intuitively the average silhouette value represents how similar each sample is to the cluster the sample belongs to and how distant from neighbor clusters. Clustering with silhouette values > 0.7 are considered strong as patterns. Residual is the error of the NMF method. Since the method produces an approximation of the original matrix, the residuals represent how close the factorization is to the original data. Note that the residuals decrease naturally as the rank of factorization increases since more variables are added to represent the original matrix.

2.6. *Data Sources*

PPI networks were downloaded from Reactome(v56). Reactome is an expert curated publicly available repository which stores multiple types of relations including reactions, indirect and direct complexes.[27,28] Gene expression data was obtained from previously published studies and TCGA using UCSC Cancer Browser.[40] Clinical data is obtained from both TCGA and corresponding publications (See Figure 2).

3. Results

3.1. *Breast Cancer (Microarray)*

Curtis *et al.*[41] used genomic variations to identify novel subgroups in breast cancer and validated on a sample of 995 patients. Using the same discovery dataset we were able to identify 5 groups with significant differences in survival. QSPLOR mined 145 sub-pathways, with 4-8 proteins each, dysregulated in at least 25 patients.

Fig. 2. Summary of Data including sample and network numbers, median days and interquartile range, sample count of alive and dead event status. In this study both microarray (MA) and RNA-Seq data for breast cancer (BRCA) (MA: [41] and RNA-Seq:[4]) and late stage brain tumors (GBM) (MA:[14] and RNA-Seq:[42]) was utilized.

DataSet	Patients	Sub-Pathways	Median Days	Alive/Dead
BRCA MA	995	145	1449	645/350
BRCA RNA-Seq	200	200	1230	685/106
GBM MA	197	553	375	22/175
GBM RNA-Seq	163	548	335	50/113

Consensus clustering and utilization of clustering metrics identified 5 patient groups. The clustering results are similar to clustering of patient samples reported in Curtis *et al.*[41] Identified clusters 1 and 2 matched with clusters 10 and 5 respectively in Curtis *et al.* study as shown in Figure 3b. Furthermore given clusters also match with Basal and Luminal B intrinsic subtypes with further stratification. Compared to previously established subtypes based on the PAM50 classifier, identified clusters are significantly separated in terms of survival(Figure 3a). Enrichment analysis for Reactome pathways in short survivor group revealed pathways that are functionally relevant or predictor of poor survival, i.e. Nonsense-Mediated Decay (NMD),[43] SRP Dependent cotranslational protein targeting to membrane,[44] Selenocysteine synthesis,[45] Signaling by WNT.[46] In contrast, long survivor group was enriched in Neuronal System,[1,45] GABA receptor activation,[47] Signaling by GPCR[48] (See Supplementary Tables S1-S5).

3.2. *Breast Cancer (RNA-Seq)*

To test the proposed method on breast cancer with data from a different platform, we obtained 791 RNA-Seq samples from TCGA with matching clinical data. QSPLOR identified 200 dysregulated subgraphs. Note that the dataset was not filtered based on prior treatment or patient characteristics hence a heterogeneous dataset was utilized in contrast with breast cancer microarray dataset above. The clustering identified 8 clusters based on cophenetic correlation coefficient and silhouette values. However 8 clusters did not result in significant survival differences hence we have utilized 5 clusters to test whether informative groups were obtained with significant survival differences ($p < 0.05$) (Figure 4a). Reactome pathway enrichment for short survivor group resulted in processes related to cellular division; Mitotic Prometaphase, Separation of Sister Chromatids, Activation of ATR in response to replication stress. Furthermore APC/C-mediated degradation of cell cycle proteins and mitotic proteins pathways were significantly dysregulated. Long survivor group was enriched in immune system related processes; MHC class II antigen presentation, TCR signaling, Cytokine signaling.

We have applied the subgraphs found in microarray dataset to RNA-Seq dataset to check cross-platform application of the proposed method. We were able to identify 5 clusters with significant survival differences. The identified clusters 3 and 4 matched previously identified Basal and Her2 subtypes respectively with further stratification (Figure S16). Pathway enrichment for short and long survivor groups resulted in Keratin metabolims, Signaling by Rho GTPases, Signaling by WNT, Gastrin-CREB signaling pathway via PKC and MAPK, Axon guidance for short survivor group and Signaling by GPCR, EGFR, VEGF, FGFR4, Interleukin-2 signaling for long survivor group (See Supplementary Tables S11-S15).

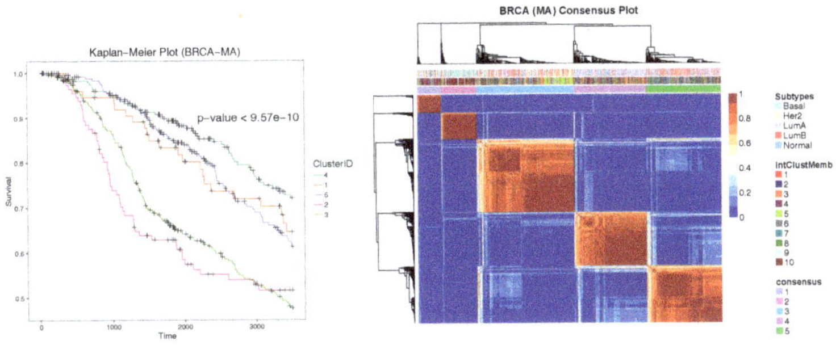

(a) (b)

Fig. 3. Results for breast cancer data analysis used in Curtis *et al.*.[41] **(a)** The Kaplan-Meier plot for 5 groups are shown (Log-rank test $p-value < 9.57E-10$).The x-axis represents days of survival. **(b)** Consensus clustering obtained using NMF is shown. Top bars show novel subtypes clusters, intrinsic subtypes and classification. IntClustMemb shows clusters identified in the Curtis *et al.* study

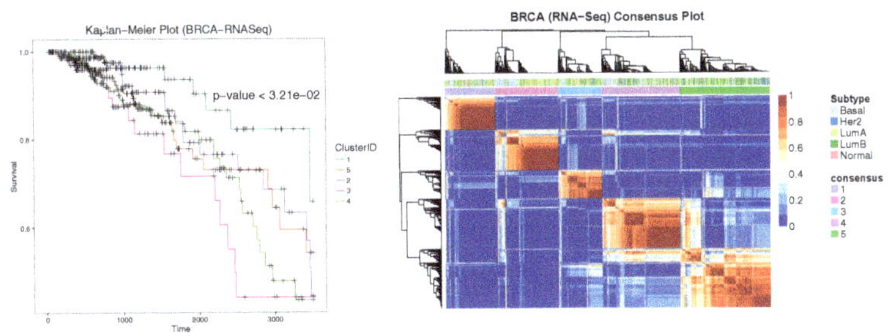

(a) (b)

Fig. 4. **(a)** Kaplan-Meier and consensus clustering results for breast cancer data obtained from TCGA (Log-rank test $p - value < 3.21E - 02$). Survival is represented as days. **(b)** Top bar in figure shows intrinsic subtypes previously defined, lower bar shows our novel pathway based groups.

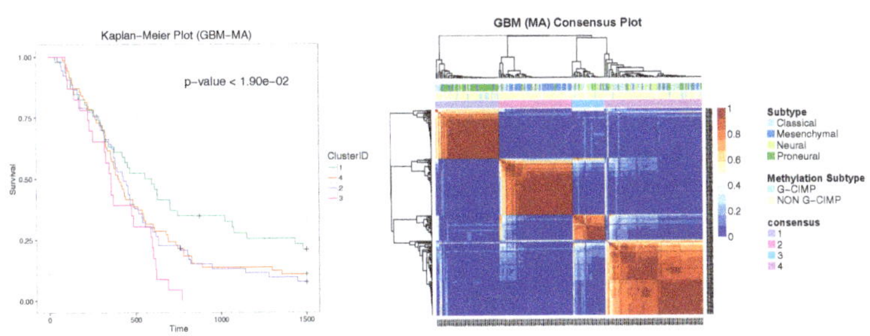

(a) (b)

Fig. 5. (a) Survival and consensus clustering results for glioblastoma multiforme microarray data used in.[14] Survival is represented as days and there is a significant difference (Log-rank test $p - value < 1.9E - 02$). (b) Top bar in consensus clustering shows previous classification of GBM patients.

3.3. *Glioblastoma Multiforme (Microarray)*

Using 11861 genes from GBM microarray dataset[14] our method revealed 4 clusters with statistically significant stratification in survival curves ($p-value < 0.05$). The long survivor group 1 consists mostly of proneural subtypes, which also supports the biological implication of our method. A new stratification is visible in Figure 5b for the short survivor group 3.

To identify biological implications, we conducted over-representation analysis for Reactome pathways. The long survivor group revealed pathways related to extracellular matrix organization and immune system; axon guidance, collagen degradation, TNFSF mediated activation cascade. The short survivor group was enriched in cell cycle related pathways including: replication, strand elongation and repair. Group 2 shows enrichment for trafficking of GPCR signaling, the Glutamate neurotransmitter release cycle, signaling by Wnt, Gastrin-CREB signaling pathway via PKC and MAPK. Group 4 shows enrichment for respiratory electron transport chain, mitochondrial translation and translation related processes. Overall, the analysis suggests new targets to study for GBM therapy (See Supplementary Tables S16-S19).

3.4. *Glioblastoma Multiforme (RNA-Seq)*

Using GBM data from TCGA[42] which included 15739 genes, our method revealed 4 groups with significant survival *(p-value <0.01)* stratification clustered based on 548 identified subgraphs. As in the microarray data analysis, mesenchymal groups were mostly clustered together in group 3 including the classical subtype. Group 4 is comprised of multiple subtypes suggesting a new classification scheme (Figure 6b). Pathway enrichment results may reveal new biomarkers. Short survivor group 3 was enriched in processes related to cell division; Mitotic prometaphase, Separation of Sister Chromatids, G2/M Transition, DNA Replication. In contrast, long survivor group 1 based on 1 year survival is enriched in Assembly of the primary cilium, Cytokine Signaling in Immune System, Gastrin-CREB Signaling pathway via PKC and MAPK, VEGFA-BEGFR2 Pathway and RET signaling. Interestingly Assembly of the primary cilium is found to be associated with GBM tumors[49,50] (See Supplementary Tables S20-S23).

4. Validation

We compared our method against 2 recently published work integrating PPI and pathway information; *Pathifier* and *NCIS*. (Details of the methods are given in supplementary document) Pathifier identified 6 groups with significant differences in survival (Figure S14a). The number of samples in each group does not suggest biologically relevant clustering (n = 6, and the larger clusters are not significant in terms of survival). The separation distances between groups are not robust with cophenetic correlation coefficient 0.61(Figure S14b). NCIS[25] identified 4 previously established subtypes in the GBM microarray dataset in conjunction with a curated PPI network. The network was constructed by the authors from Reactome, NCI-Nature Curated PID, and KEGG. It consists of 11,648 genes, 211,794 interactions matching 7,183 genes in the GBM dataset. The identified subtypes are similar to established subtypes and have significant differences in survival. However, it is not clear how the patients are clus-

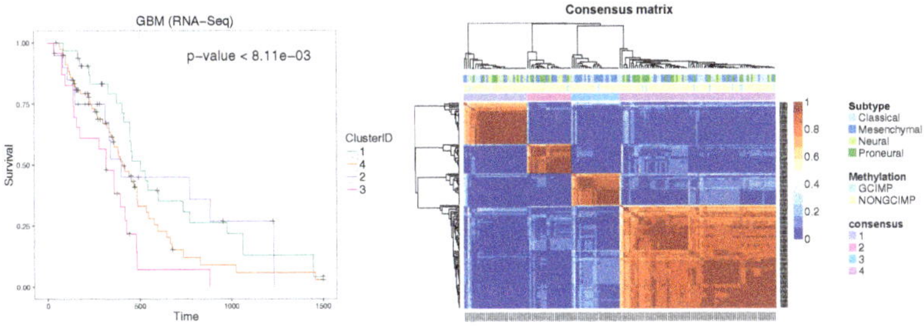

(a) (b)

Fig. 6. (a) Kaplan-Meier and (b) consensus clustering results for glioblastoma multiforme samples obtained from TCGA. The RNA-Seq data set showed significant survival difference (Log-rank $p - value < 8.11E - 03$)

tered since previously identified subtypes do not provide overall significant survival difference (Figure S4). Using the data from NCIS study we have identified 5 clusters (based on the clustering metrics) which show separation of survival curves (Figure S15a). We were able to cluster previously proposed mesenchymal and proneural subtypes with further stratification of mesenchymal group (Figure S15b). Based on the survival analysis, proneural clustered groups show the longest survival curves in agreement with previous findings. These results suggest that the proposed method performed better than the NCIS and Pathifier algorithms in terms of significance of survival stratification and relevance of the identified genes and pathways which can be used as precursor targets for future therapeutic studies.

5. Discussion

The proposed method aims to integrate PPI data with gene expression data using a novel approach. In this study we were able to identify networks that play predictive role in clinical outcome and also networks that crosstalk between the established pathways. A crucial development for improving current prognostic methodologies. The presented method is also more general as it does not require apriori identification of important genes.

Several studies have investigated molecular correlation of prognosis and clinical subclasses in GBM. Earlier studies have identified tumor grade as one of the strong predictors of disease outcome,[51] such as *TP53* mutation and *EGFR* amplifications were claimed to stratify patients into subgroups,[52,53] while a later study contests the validity of this classification.[54] Further studies have identified various gene sets that would separate the patient samples by their molecular characterization,[10,15–18] and some have reported prognostic value of these gene sets. However, most of these have identified different sets of genes, a consensus on the functional delivery has not been reached. These proposed subtype classification methods also identified different sets of patient subtypes, classifications greatly rely on selected patient groups and sample size.

Overall the results suggest possible targets and pathways for cancer progression, mechanisms and survival. Additionally enrichment using long and short survivor groups from RNA-Seq data resulted in similar gene targets. Note that results are 'reversed' for RNA-Seq dataset compared to microarray analyzed samples, however since the stratification is based on dysregulation, the method includes both overexpression or underexpression. Hence genes are categorized as possible markers rather than specific targets for long or short survival.

Our validation of the results we presented here, which reproduced similar survival curves over independent studies, presents great potential for prognostic value for this method. Moreover, finding significant mechanisms that can describe the underlying effects of survival and treatment responses can be easily done within these parameters and provide candidate pathways for therapeutic intervention. While follow up studies are needed to further asses the prognostic value, and possible effect of treatments, analysis that we have conducted provide an initial look of the biological mechanisms underlying in these patient groups with different survival which are also supported by various studies.

Gathering multiple omics datasets to better characterize individuals and associating these with extensive phenotype information has been the hallmark achievement of recent years.[3,4,14,41,42] These datasets have paved the road to improved personalized medicine, promising better disease characterization and diagnosis, identification of patient-specific treatment options and improved monitoring of patients in need. While personalized medicine offers great benefit to individuals, the computational approaches to integrate these multiple omic datasets and statistical methods to leverage the underlying disease and patient traits is still under development. This study tackled this problem of integration network data with transcriptomics data to identify classification scheme for both breast and late stage brain tumors (GBM). Our method can be used to group patients in an unsupervised manner, and have prognostic value. The significant separation of patient samples will allow further studies and utility, since these classifications are based on functionally related frequently altered pathway segments. In the future, we plan to investigate the utility of this method for other cancer types, integrating additional genomic features and investigate its value in improving treatment options.

Acknowledgments

Thank you Leigh Henderson for thoughtful discussions and reading drafts of this paper. This research was partially supported by a Grant from NIH/NCRR CTSA KL2TR000440 to GB.

References

1. Q. Li *et al.*, *Cell* **152**, 633 (Jan 2013).
2. C. J. Vaske *et al.*, *Bioinformatics* **26**, i237 (2010).
3. TCGA, *Nature* **474**, 609 (2011).
4. TCGA, *Nature* **490**, 61 (2012).
5. K. Holm *et al.*, *Breast Cancer Res* **12**, p. R36 (2010).
6. T. Sørlie *et al.*, *PNAS* **98**, 10869 (2001).
7. S. Tardito *et al.*, *Nat Cell Biol* **17**, 1556 (Dec 2015).
8. H. Ohgaki and P. Kleihues, *Acta neuropathologica* **109**, 93 (2005).
9. Y. Liang *et al.*, *PNAS* **102**, 5814 (2005).
10. C. L. Nutt *et al.*, *Cancer research* **63**, 1602 (2003).
11. M. Esteller *et al.*, *New England Journal of Medicine* **343**, 1350 (2000).

12. M. E. Hegi *et al.*, *New England Journal of Medicine* **352**, 997 (2005).
13. TCGA, *Nature* **455**, 1061 (Oct 2008).
14. R. G. Verhaak *et al.*, *Cancer cell* **17**, 98 (2010).
15. H. Colman *et al.*, *Neuro-oncology* , p. nop007 (2009).
16. W. A. Freije *et al.*, *Cancer research* **64**, 6503 (2004).
17. J. M. Nigro *et al.*, *Cancer research* **65**, 1678 (2005).
18. H. S. Phillips *et al.*, *Cancer cell* **9**, 157 (2006).
19. R. Tibshirani *et al.*, *PNAS* **99**, 6567 (2002).
20. C. M. Perou *et al.*, *Nature* **406**, 747 (2000).
21. J. S. Parker *et al.*, *J Clin Oncol* **27**, 1160 (Mar 2009).
22. C. Sotiriou *et al.*, *PNAS* **100**, 10393 (2003).
23. C. Fan *et al.*, *New England Journal of Medicine* **355**, 560 (2006).
24. M. L. Gatza *et al.*, *PNAS* **107**, 6994 (2010).
25. Y. Liu *et al.*, *BMC bioinformatics* **15**, p. 1 (2014).
26. Y. Drier, M. Sheffer and E. Domany, *PNAS* **110**, 6388 (2013).
27. D. Croft *et al.*, *Nucleic acids research* **42**, D472 (2014).
28. M. Milacic *et al.*, *Cancers* **4**, 1180 (2012).
29. A. Inokuchi *et al.*, An Apriori-Based Algorithm for Mining Frequent Substructures from Graph Data, in *Principles of Data Mining and Knowledge Discovery*, jul 2000 pp. 13–23.
30. S. A. Cook, The complexity of theorem-proving procedures, in *ACM Symposium on Theory of Computing*, (ACM Press, New York, New York, USA, 1971).
31. C. Jiang, F. Coenen and M. Zito, *The Knowledge Engineering Review* **28**, 75 (mar 2013).
32. H. Cheng, X. Yan and J. Han, Mining Graph Patterns, in *Frequent Pattern Mining*, (Springer International Publishing, 2014) pp. 307–338.
33. V. Chaoji, M. Al Hasan, S. Salem, J. Besson and M. J. Zaki, *Stat. Anal. Data Min.* **1**, 67 (2008).
34. M. Al Hasan and M. J. Zaki, Output Space Sampling for Graph Patterns, in *Proceedings of VLDB*, (VLDB Endowment, aug 2009).
35. X. Yan, H. Cheng, J. Han and P. S. Yu, Mining Significant Graph Patterns by Leap Search, in *Proceedings of ACM SIGMOD ICMD*, 2008.
36. D. D. Lee and H. S. Seung, *Nature* **401**, 788 (1999).
37. J.-P. Brunet, P. Tamayo, T. R. Golub and J. P. Mesirov, *PNAS* **101**, 4164 (2004).
38. R. Gaujoux and C. Seoighe, *BMC bioinformatics* **11**, p. 1 (2010).
39. P. J. Rousseeuw, *Journal of computational and applied mathematics* **20**, 53 (1987).
40. J. Z. Sanborn *et al.*, *Nucleic acids research* , p. gkq1113 (2010).
41. C. Curtis *et al.*, *Nature* **486**, 346 (Jun 2012).
42. C. W. Brennan *et al.*, *Cell* **155**, 462 (Oct 2013).
43. L. B. Gardner, *Mol Cancer Res* **8**, 295 (Mar 2010).
44. J. Simões, F. M. Amado, R. Vitorino and L. A. Helguero, *Oncoscience* **2**, 487 (2015).
45. R. L. Schmidt and M. Simonović, *Croat Med J* **53**, 535 (Dec 2012).
46. G.-B. Jang *et al.*, *Sci Rep* **5**, p. 12465 (2015).
47. S. Z. Young and A. Bordey, *Physiology (Bethesda)* **24**, 171 (Jun 2009).
48. A. Singh, J. J. Nunes and B. Ateeq, *Eur J Pharmacol* **763**, 178 (Sep 2015).
49. J. J. Moser, M. J. Fritzler and J. B. Rattner, *BMC cancer* **9**, p. 448 (2009).
50. J. J. Moser, M. J. Fritzler and J. B. Rattner, *BMC clinical pathology* **14**, p. 1 (2014).
51. M. D. Prados and V. Levin, Biology and treatment of malignant glioma., in *Semin Oncol*, 2000.
52. A. von Deimling, D. N. Louis and O. D. Wiestler, *Glia* **15**, 328 (1995).
53. K. Watanabe *et al.*, *Brain pathology* **6**, 217 (1996).
54. Y. Okada *et al.*, *Cancer research* **63**, 413 (2003).

HUMAN KINASES DISPLAY MUTATIONAL HOTSPOTS AT COGNATE POSITIONS WITHIN CANCER

JONATHAN GALLION[†]

Structural Computational Biology and Molecular Biophysics, Baylor College of Medicine, One Baylor Plaza Houston, TX, 77030, USA
Email: Jonathan.Gallion@bcm.edu

ANGELA D. WILKINS

Immunology, Lichtarge Laboratory BCM, One Baylor Plaza Houston, TX, 77030, USA
Email: aw11@bcm.edu

OLIVIER LICHTARGE

Structural Computational Biology and Molecular Biophysics, Baylor College of Medicine, One Baylor Plaza Houston, TX, 77030, USA
Email: lichtarge@bcm.edu

The discovery of driver genes is a major pursuit of cancer genomics, usually based on observing the same mutation in different patients. But the heterogeneity of cancer pathways plus the high background mutational frequency of tumor cells often cloud the distinction between less frequent drivers and innocent passenger mutations. Here, to overcome these disadvantages, we grouped together mutations from close kinase paralogs under the hypothesis that cognate mutations may functionally favor cancer cells in similar ways. Indeed, we find that kinase paralogs often bear mutations to the same substituted amino acid at the same aligned positions and with a large predicted Evolutionary Action. Functionally, these high Evolutionary Action, non-random mutations affect known kinase motifs, but strikingly, they do so differently among different kinase types and cancers, consistent with differences in selective pressures. Taken together, these results suggest that cancer pathways may flexibly distribute a dependence on a given functional mutation among multiple close kinase paralogs. The recognition of this "mutational delocalization" of cancer drivers among groups of paralogs is a new phenomena that may help better identify relevant mechanisms and therefore eventually guide personalized therapy.

[†]The authors gratefully acknowledge support from the National Institutes of Health (GM066099 and GM079656), from the National Science Foundation (DBI-1356569), and from DARPA (N66001-15-C-4042)

1. Introduction

A major focus of recent cancer sequencing projects, such as the TCGA, is to identify causal driver mutations responsible for tumorigenesis (*1*) . To this end, many computational tools have been produced to predict the impact of mutations on protein function in order to screen out null function or low impact mutations (*2*). The efforts of these approaches have identified many proteins and mutations driving cancer progression. Unfortunately, the inherent mutational heterogeneity displayed within cancer often limits the statistical power of these methods so as to capture only the most frequent driver mutations in a large cohort of patients (*3*). By contrast, low frequency drivers or smaller patient cohorts suffer from a lack of statistical significance and are therefore easily missed.

While infrequent mutations in a single gene may, at first glance, appear to indicate insignificance in cancer progression, this may be an oversimplification. Driver mutations in cancer may not only target a single gene but rather groups of genes or functional pathways, distributing the mutational burden across many functionally related genes (*4, 5*); while a single gene may lack significance, combining mutations across a regulatory pathway can increase the power of the analysis and identify gene groups driving cancer progression (*3, 6*). Prior studies have taken these groups from databases such as KEGG (*7*), Reactome (*8*), and analyses of gene association networks like STRING (*9*). However, these approaches are not limited to functional or hierarchical pathways but rather could be applied to any group of proteins that share functionality such as, Gene Ontology terms or even groups of protein homologs sharing significant functional overlap.

Further confounding the prediction of cancer drivers, single gene analyses group mutations regardless of their structural location and, therefore, do not account for the functional heterogeneity of these mutations. To account for these difference, an analysis in Colon and Breast Cancers grouped mutations from various genes occurring in homologous protein domains, finding specific domains enriched for high frequency mutations across many individual proteins (*10*). Furthermore, an analysis of disease-related mutations across all human kinases showed that these mutations preferentially localized in specific structural domains, affected certain residues types, and had conserved amino acid substitutions (*11*). These studies show disease-related mutations can preferentially occur at specific structural domains in homologous proteins, such as kinases, and that kinase mutations share conserved patterns of substitution. Here, we expand upon this work and ask whether there are mutational biases in individual positions in the context of cancer.

For the purpose of this study, we focus on human kinases in order to better understand this essential protein family and how it contributes to cancer. There are over 500 human kinases sharing substantial homology in both the kinase structure and the catalytic mechanism (*12*). The kinase family has been further subdivided into 7 classes based on substrate specificity and evolutionary lineage. Kinases are ubiquitous proteins involved in a diverse array of cellular functions; as a result, numerous perturbations in kinase coding regions, translation, and expression lead to disease and cancer progression (*13*). Moreover, after G protein-coupled receptors, kinases are the second most drugged protein family (*11*). While some kinases such as BRAF, EGFR, and PI3-kinase demonstrate a remarkably high mutation rate within cancer (*14, 15*), many kinases are mutated at a much lower frequency making it difficult to access their influence on cancer progression

Here, we hypothesize that some closely related kinases may act as a single functional group from the perspective of a cancer type. That is, mutations at the same (cognate) position across a group of kinases may have a similar functional effect and fulfill the same selective pressure, leading to positional enrichment of impactful mutations within the cancer. To test this possibility, we used kinase alignments and exomic mutations from the TCGA to group all mutations occurring at the same sequence position and then quantified the predicted functional impact using Evolutionary Action (EA). We identified highly conserved, functionally related positions with a significantly increased mutation rate in a pan-cancer and pan-kinase analysis. Additionally, mutational differences are clear between the various kinase subclasses and additional differences across cancer types. This work shows a novel method that moves beyond a single gene approach and which suggests that functionally related homologous proteins may bear driver mutations that substitute for each other to support cancer progression.

2. Methods

2.1. *Evolutionary Trace and Action Analysis*

To identify evolutionarily important residues, we performed Evolutionary Trace (ET) analysis on each of the kinase sub-families as previously described (*16*). ET utilizes changes in genotype and corresponding phenotypic divergences in the phylogenetic tree to score the evolutionary importance of each residue in a protein sequence. In previous work, ET has identified functional sites and their determinants so as to guide mutational engineering in case studies (*17, 18*).

Evolutionary Action (*19*) builds upon ET to predict the impact a mutation has on protein function by multiplying the importance of the position (ET) by the magnitude of the substitution (evolutionary substitution odds). Prediction scores are then normalized for each individual kinase so the range falls between a predicted effect that is null, 0, to one that is most impactful, 100. EA has been repeatedly validated. It was shown to correctly predict mutation impact in multiple systems (e.g. P53, RecA, bacteriophage T4 lysozyme, etc.), it also outcompeted state of the art methods in the past 3 CAGI challenges (Critical Assessment of Genome Interpretation) (19), and in a clinical context, it can stratify patients with head and neck cancer based on their p53 mutational status (28). Using this technique we score each mutations predicted impact.

2.2. *Kinase Alignment, Mutation Acquisition and Mapping*

In order to compare mutations across all human kinases, we aligned separately each of the 7 major subclasses from The Human Kinome project (*20*). These alignments were used as a translation tool, in order to map mutations across human kinases onto canonical protein sequences. Representative crystallized structures were selected for each sub-family to visualize analysis. Representative proteins can be found in the supplement and were manually chosen based on: 1) the availability of a high resolution crystal structure 2) their similarity to other proteins within that class and finally 3) with a focus on longer proteins so as to limit the number of blank alignment positions when mapping other proteins onto the structure.

Mutation data was acquired from the TCGA for 21 major cancer types using the computationally annotated calls. Chromosome positions were converted to protein position using ANNOVAR (*21*) and then were each mapped onto the representative sequence within the alignments. In this way we were able to measure how mutations within kinases distribute throughout the conserved kinase domain.

Unless otherwise stated, all mutation numbering is in relation to the representative structure from TKL kinases (ACTR2B-2QLU). For visualization purposes on the structure, sphere size of each position was scaled based on frequency of high impact mutations (EA>40) according to the equation:

$$\text{Sphere Size}=2*(\text{Frequency}/\text{Maximal Frequency}) \quad (1)$$

Initially this analysis was performed on each of the seven kinase subclasses (358 individual kinases total) using separate alignments and representative structures for each subclass. CK1 kinases were dropped from the analysis due to insufficient mutations. The remaining six individual representative structures were then aligned and merged into a complete pan-cancer analysis.

2.3. *Random Controls*

See Supplement for additional Methods at http://mammoth.bcm.tmc.edu/GallionEtAlPSB/

3. Results

3.1. *Evolutionary Trace Identifies Functionally Important and Divergent Kinase Positions*

In order to gauge the impact of kinase mutations we first sought to identify key functional residues and sites in kinases. This was done using Evolutionary Trace (ET). Figure 1 shows the ET ranks from most to least important (red to blue) mapped onto the structure of ACTR2B, 2QLU (PDB-ID). As expected, functionally essential motifs, such as the magnesium binding DFG motif and the catalytic HRD motif emerge as ET hotspots. ET also suggests functionally relevant residues

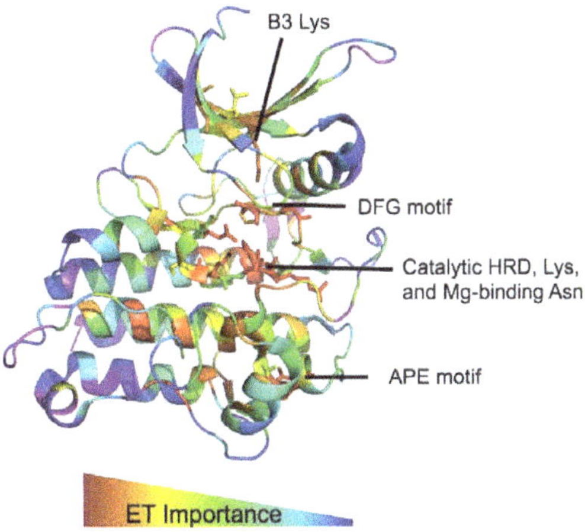

Fig 1: Evolutionary Trace Analysis of ACTR2B (2QLU) identifies evolutionarily important residues corresponding to known motifs.

throughout the substrate pocket and allosteric sites consistent with known protein functionality. Positions predicted to be the least important tend to cluster near the edges of helices, the loop regions, and near solvent exposed positions. Repeating ET analysis on each individual class, we are able to identify positions important to each group. These results confirm that in kinases, ET is able to identify both universally important positions as well as the positions that are evolutionarily divergent among subfamilies correlating to divergent functionalities.

3.2. *Kinase Mutations Demonstrate Non-Random Structural Pattern In TCGA*

To explore structural biases of kinase mutations in cancer, we next conducted a pan-cancer analysis of TCGA data. This analysis grouped mutations occurring at the same sequence position across kinase evolutionary history. This broad pan-cancer analysis identifies 77 residues with a statistically significant mutation rate (p-value<0.01) compared to control (See Supplementary). Then, in order to focus on the subset of impactful mutations and screen out low impact polymorphisms, we repeated the above analysis only using mutations with EA scores greater than 40, and mapped them onto the ET analysis of ACTR2B (Figure 2A). All positions are numbered based on the 2QLU structure unless otherwise specified. For example, the well-known driver mutations from BRAF-V600 (equivalent position V344 in figure) and CHEK2-K373 (R345 in figure) are the most frequently mutated, high impact mutations. Other frequently mutated positions with high impact substitutions occur at known functional residues, such as the glycine-rich region G199, the DFG motif D339, the HRD domain R320 and D321, and a conserved ion-pairing residue R468. Since these mutations involve positions with large ET scores, they are likely to impair protein function. By contrast, and as seen in Figure 2B, there are 54 residues mutated at a lower rate than expected (p-value<0.01). These seldom mutated positions, shown by the small

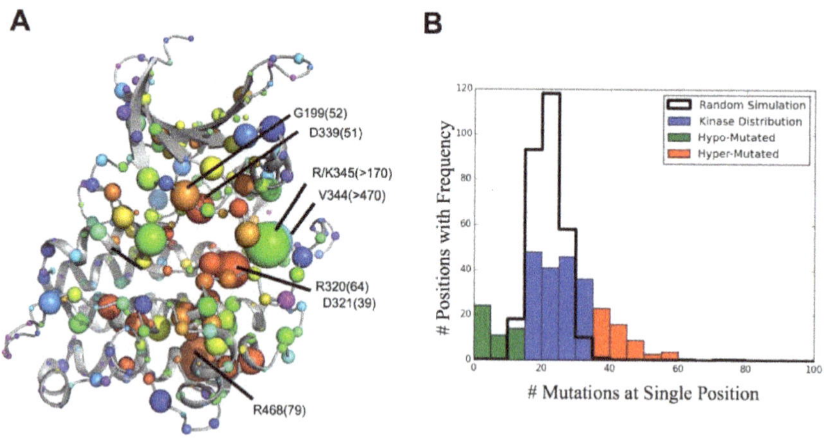

Fig 2: Kinase Mutation Pattern Pancancer (A) Pan-cancer mutations, with an EA>40, mapped onto ACTR2B structure where sphere size=frequency, color=ET importance. (B) Actual mutation frequency significantly varies from Poisson Distribution.

sphere size in Figure 2A fall preferentially in the solvent exposed loop regions of the kinase that are evolutionarily less important according to ET and thus unlikely to have much functional consequence. These data show that kinase mutations in cancer are not evenly distributed throughout the structure. Rather many mutations preferentially fall non-randomly so as to recurrently involve functionally important cognate positions within conserved motifs, where they are likely to be disruptive; conversely, in the loop regions, which are less important, mutations are more rare and involve positions of lesser importance.

3.3. *Frequently Mutated Positions are Enriched for Mutations Predicted to Have a Significant Impact on Protein Function*

To further explore the functional consequences of these mutations, we used EA to predict the functional impact of each mutation on protein function. EA combines the evolutionary importance of the position (ET) with the likelihood of that substitution, based on all evolutionary history, in order to predict the impact of a mutation on protein function. We compared the EA score distribution of frequent positions and infrequent positions (p-value<0.01) to the distribution of all kinase domain mutations from the TCGA using a two-sided t-test (Figure 3A). In agreement with the structural and ET biases, the frequently mutated positions are predicted to have a higher impact on protein function (p-value$=10^{-28}$) while the infrequently mutated positions are biased towards lower impact mutations (p-value$= 10^{-5}$). These data show the frequently mutated positions from the TCGA are further enriched for high impact mutations, while those positions infrequently mutated are predicted to have little functional effect.

3.4. *Frequently Mutated Positions Occur in Many Different Kinases at a Low Individual Frequency*

While these cancer somatic mutations demonstrate site specificity, we next investigated which individual kinases carried these mutations and whether specific proteins drove this pattern. The mutation frequency of each individual kinase is displayed in Figure 3B and is compared against a random simulation in which the same number of mutations were randomly distributed to an equal number of proteins. The random distribution had a mean value of 21.4 mutations per kinase while the experimental distribution, after dropping out the outliers BRAF and CHEK2 (550 and 160 mutations, respectively), had a mean of 19.5. We note that the mutation rate in individual kinases is more variable than expected. Overall the distribution is leftward shifted compared to control with a select number of proteins hypermutated: 29% of kinases were mutated at a decreased frequency (p-value<0.05) while only 14% of kinases were significantly hypermutated (p-value<0.05). Of the hypermutated kinases, nine were mutated at an exceptionally high rate (>50 mutations/protein); many of these however, represent known, high frequency driver mutations occurring at the same location in the same kinase (e.g. BRAF, CHEK2, and EGFR). These data show that within cancer cells, certain kinases experience a remarkably increased mutation rate while the majority of the remaining kinases are hypomutated, typically with fewer that 20 SNVs across a pan-cancer analysis.

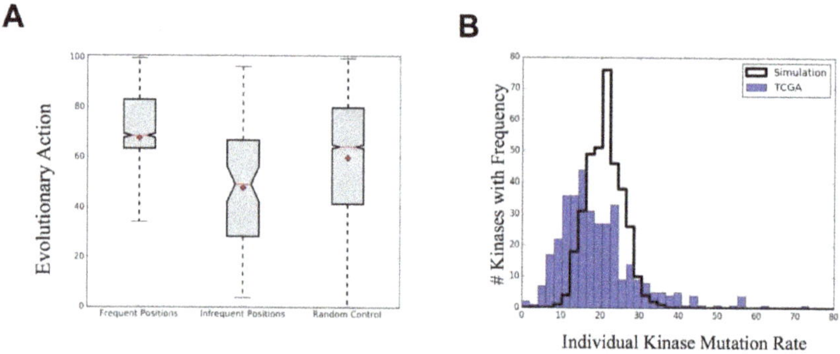

Fig 3: (A) High frequency mutations are significantly biased towards high impact mutations (Pvalue=$5*10^{-28}$) while low frequency mutations are biased towards low predicted impact (Pvalue $2.5*10^{-05}$). Mean=diamond Median=red line Whisk=2STD (B) Observed kinase mutation rate compared to computer simulation of random mutations. BRAF and CHEK2 (550 and 160 mutations, respectively) not shown on plot.

However, while this analysis recapitulates known drivers such as L858R within EGFR, it further identifies mutations at a single residue that individually occur at a low frequency but, taken as a whole, occur at a high frequency. For instance, Table 1 displays a random selection of mutations occurring at the Asp residue of the HRD domain (*p*-value=$2x10^{-4}$). While each individual mutation has a conserved amino acid transition, individual proteins are mutated infrequently with a median value of 1 and a maximal value of 5 mutations (occurring within MAP2K7). Of the original 54 positions with a *p*-value<0.01 only 6 are at least partially driven by a single protein (1 protein with >20% of the mutations), while all remaining positions were significant only through this combination. These data show that while individual mutations may occur at low frequency, they frequently occur at homologous structural positions with the same native residue and amino acid substitution. Furthermore this pattern is distributed across many individual kinases without a single driver protein.

Table 1. Random sample of mutations occurring at catalytic Asp residue from the HRD domain.

Protein	Substitution	EA Score	Kinase Class	Cancer Type
PRKCI	D378N	64.36	AGC	PAAD
PRKG2	D576Y	98.63	AGC	READ
CHEK1	D130Y	98.73	CAMK	LUSC
STK17B	D158N	74.06	CAMK	SKCM
CLK4	D286N	63.19	CMGC	COAD
MAPK4	D149G	90.95	CMGC	STAD
MAP2K3	D190N	71.77	STE	SKCM
MAP2K7	D243N	61.93	STE	COAD
MAP2K7	D243N	61.93	STE	PAAD
PAK7	D568N	75.15	STE	SKCM
EPHA3	D746N	43.15	TK	SKCM
FES	D683E	67.14	TK	BRCA
ROR2	D615N	74.82	TK	SKCM
MAP3K7	D156Y	96.77	TKL	LIHC

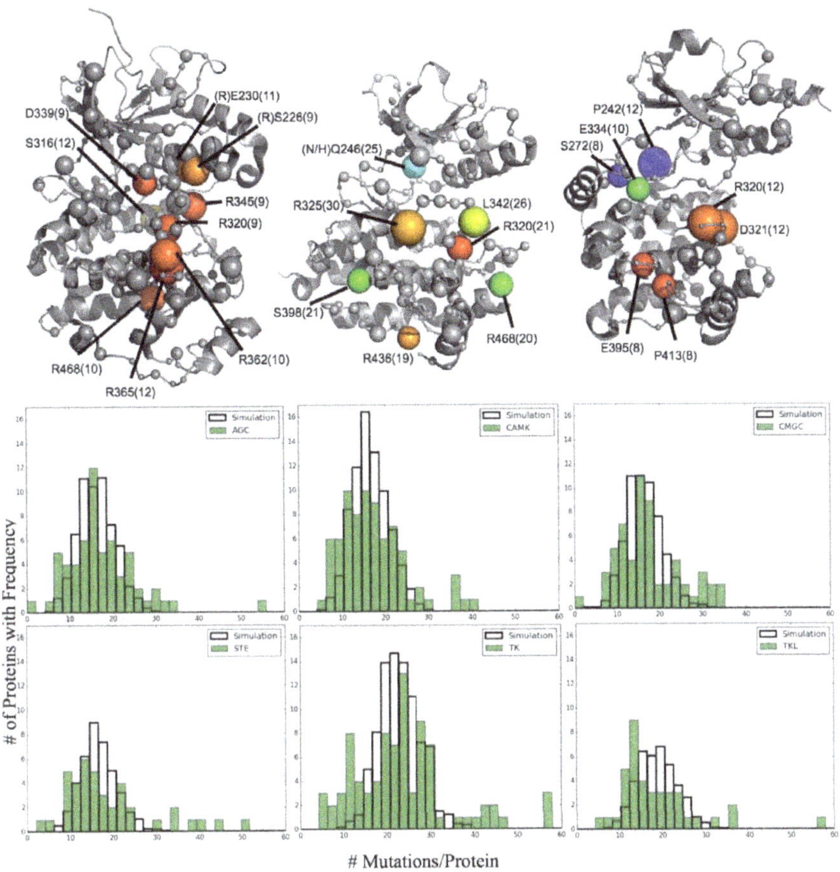

Fig 4: Individual kinase subclasses are frequently mutated at distinct positions (Left to Right: CMGC, TK, STE kinases). Sphere Size=Frequency, Color= ET importance from high to low (red to blue) for each representative kinase: ERK1, EphA5, PAK1 (respectively). All labels are based on ACTR2B numbering. (B) The protein mutation rate for each kinase class was compared against a simulated random distribution specific to the total number of mutations and proteins in each class. BRAF (TKL) and CHEK2 (CAMK) are not shown on their respective figures

3.5. *Individual Kinase Classes Show Unique Mutational Patterns*

Individual subclasses of kinases display marked functional and structural differences corresponding to their target specialization (12). To test if our conclusions held true despite these differences, we repeated the above analysis for each kinase class. As an example, three of these classes are displayed in Figure 4A. While the general location of these residues tend to stay near the catalytic site, the frequently mutated positions from each class vary. Nine residues in CMGC kinases form a statistically significant cluster (z-score=4.58) roughly localized around and occurring within the HRD domain. Seven residues in TK kinases are more broadly distributed throughout the structure with the three most frequent near the HRD domain. Finally, STE kinases seem to show two distinct areas of mutation, the HRD region and the ATP-binding hinge region. In all three cases, similar to the pan-cancer analysis, the most frequent positions tend to occur at evolutionarily important residues in functional motifs with high impact mutations. In addition to these differences, significant positions from the pan-kinase analysis are still significant in multiple classes (e.g. R320 (HRD motif) and R468 (Ion pair)). These data indicate that within cancer, certain positions are preferentially enriched in select kinase subclasses while other positions demonstrate broad enrichment across many or all kinase types.

We further note differences in the mutation frequency of proteins from each of the kinase classes (Figure 4B). In each class, some proteins are mutated at a significantly higher rate than expected. Proteins from the AGC kinase class are normally distributed with an exaggerated variance compared to random simulation, indicating that mutations within this class are fairly distributed to many proteins. Likewise, the mutation rate in CMGC and TK kinases is even more varied but still follow a roughly normal distribution centered around the expected mean. The distribution from CAMK, STE, and TKL kinases match the pan-kinase analysis with a leftward shifted distribution displaying many hypomutated proteins and several hypermutated proteins. As 43% of all mutations within TKL kinases occur in BRAF, we have removed these mutations from this analysis. However, creating a random distribution for this class without first removing this outlier shifts the random distribution right (mean=30). This data shows that, in addition to structural differences, the individual kinase classes are mutated at different rates, with some classes having broadly distributed mutations to many individual proteins while other classes are primarily mutated in a select few proteins.

3.6. *Kinases Further Demonstrate Cancer Type Specific Mutational Patterns*

Variances between kinase subtypes led us to next speculate that certain protein positions could have varying functional importance to specific cancer types as well. The above analysis was repeated, now grouping all kinases together and instead performing a cancer-specific analysis for 7 cancer types within the TCGA [Breast invasive carcinoma (BRCA), Bladder Urothelial Carcinoma (BLCA), Colon adenocarcinoma (COAD), Head and Neck squamous cell carcinoma (HNSC), Lung adenocarcinoma (LUAD), Skin Cutaneous Melanoma (SKCM), and Stomach adenocarcinoma (STAD)]. The most frequently mutated position for all but BRCA and STAD was 345 and 346, driven by the high frequency driver mutations BRAF-V600 and CHEK2-K373 (respectively); these mutations were then removed from this analysis in order to search for novel other positions. Figure 5 shows a selection of positions that were significantly mutated within specific cancer types. Interestingly, the analyses from LUAD and STAD resulted in clusters of mutations within the kinase domain. Some positions were significant in two cancer types, such as

L325 in LUAD and BRCA. In agreement with the pan-cancer analysis, R468 was frequently mutated in many cancer types including STAD and COAD. These data indicate that individual cancer types are enriched for varying structural positions across many individual kinases.

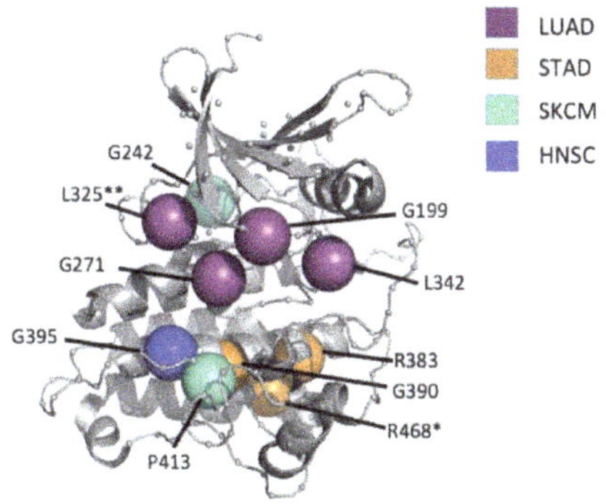

Fig 5: Cancer types demonstrate some specificity towards certain mutation positions. *Occurs in STAD and COAD **Occurs in both LUAD and BRCA

4. Discussion

In order to better predict driver mutations within cancer, computational methods have been extended from gene-by-gene analyses to consider instead groupings of mutations in functional pathways or subnetworks (*3, 6, 22*). In this manner, driver proteins mutated at a low frequency due to the heterogeneity within cancer that are missed by a single gene analysis can still be identified despite their low individual frequency. Being able to predict these diverse infrequent drivers of cancer helps move medicine closer to personalized diagnoses and care. Here, as an alternate way to group genes, we explored protein homology rather than curated hierarchical pathways and gene interactions. Strikingly, we find that among kinases, mutations are structurally biased to functional motifs and evolutionarily important residues.

Mutations providing a benefit to cancer cells become clonally enriched, as that cell proliferates more efficiently than others in the tumor population (*5*). From the pan-kinase analysis, we identified positions frequently mutated across many individual kinases. While the known high frequency driver genes were captured in this analysis, an additional 39 positions were mutated at a low frequency in any given kinase but were significantly mutated across the kinase family. These high frequency positions were preferentially biased for high impact mutations, strongly suggesting a significant effect on protein function. In contrast, the infrequently mutated positions all occurred at evolutionarily unimportant loop regions with a bias towards low impact mutations. These data indicate that enrichment is correlated to functional impact. Presumably, the high-impact mutations across many kinases provide a functional benefit within the cancer cell and are therefore enriched, whereas low-impact mutations, providing little benefit to the cancer cell, are lost from the population resulting in a low mutation rate at those positions.

Previous work in kinases has demonstrated that identical mutations in two different kinases can result in the same phenotype (*23, 24*). For instance, mutations conferring resistance to kinase inhibitors in EGFR occur at the same position as drug resistance mutations in BCR-ABL, PDGFRA and KIT (*25, 26*). A systematic study of mutation locations built upon these observations and demonstrated the existence of 'domain hotspots': frequently mutated regions in many proteins leading to the same functional consequence (*22*). In the context of this analysis of exomic mutations from TCGA, these frequently mutated positions, across many different kinases, with the exact same substitution strongly suggests a conserved functional mechanism driving enrichment: the same mutation in two different kinases likely producing a similar benefit in cancer.

The kinase catalytic mechanism itself is highly conserved across all kinases and is orchestrated by groups of functional motifs; these same positions in all kinases are responsible for the same functions (*12*). These motifs are themselves frequently mutated at a high frequency within cancer; in fact, the majority of the frequently mutated positions occur at or nearby conserved motifs. These positions are often studied in the context of kinases enabling us to speculate on their functional consequences within cancer. For instance, the catalytic Asp and Arg residues of the HRD domain are both mutated in many diverse kinases and are furthermore mutated to the same types of residues (D→N and R→Q/W/H respectively) in each case. Previous characterization of the D→N mutations within the Drosophila Src64 kinase indicate this mutation is equivalent to a gene knockout (*27*). Cancer cells carrying this SNV would therefore experience a loss of kinase activity within this protein, possibly suggesting a tumor suppressing mechanism. What remains to be determined is how far these characterization studies can be extrapolated to other kinases. Further experimental studies are needed in which the same mutation is characterized in multiple proteins to assess how universal these conclusions are. However, given 1) the initial conserved function of these positions, 2) the significant enrichment of the same substitution across many proteins, and 3) the sizeable predicted consequence of these mutations: it becomes tantalizing to suggest that the same mutation in different kinases may produce the same functional benefit in cancer, regardless of the kinase where it occurs.

This hypothesis is further supported by the kinase subclass and cancer type specific analyses. The individual kinase sub-families are evolved to phosphorylate different types of proteins (12). As a result, they have diverged. While the overall structure is conserved, some positions are specific for given target proteins and therefore differ among kinase classes. Likewise, while some positions are broadly mutated, the class specific analyses demonstrate appreciable differences; some positions are enriched in one class but do not occur in another. These variations between kinase classes likely stem from their functional divergence. Mutations occurring at important positions in one class may be beneficial, while the same position in a different class may not, resulting in differential enrichment. Furthermore, cancer types themselves display heterogeneity among their causal driver mutations (*5*), a heterogeneity reflected within kinase mutations as well. Different cancer types are enriched for different kinase positions, again suggesting that some positions may be preferentially beneficial for one cancer type more so than another, and therefore clonally enriched. When the selection pressure varies, either by differing cancer types or by the different kinase classes, the positions of the enriched mutations also vary. This further suggests that a conserved functional mechanism drives this mutational enrichment across many individual kinases.

Cellular homeostasis and function is often maintained by a complex network of proteins with significant functional overlap and crosstalk between functional homologues. For this reason, a single gene approach to predicting driver mutations in cancer may be overly simplistic, therefore requiring a methodology to combine mutations based on functional similarity. Here, we propose that in addition to curated pathways, mutations can also be grouped across homologous protein families. Within kinases, we have demonstrated that individual proteins are enriched for mutations occurring at cognate positions utilizing the same substitutions. These results suggest that the selection pressure within certain cancers may be specific to the mutation's location and not differentiate between which kinase carries the mutation. Taken together, these data show individual kinases may behave in a functionally redundant manner in cancer and that a combined analysis of their mutations could identify individually infrequent driver mutations, previously missed, that occur frequently across the entire class. The conserved nature of these mutations allows speculation as to their predicted functional effect by extrapolating previous characterization studies, in a single protein, to the other kinases. Finally, while these results are specific to kinases, similar analyses could be broadly applicable across many protein families, thereby shifting focus from a 'protein specific' to a 'paralog-wide, cognate position specific' analysis of cancer driver mutations.

References

1. J. S. Kaminker *et al.*, *Cancer Res* **67**, 465-473 (2007).
2. P. Katsonis *et al.*, *Protein Sci* **23**, 1650-1666 (2014).
3. F. Vandin, P. Clay, E. Upfal, B. J. Raphael, *Pac Symp Biocomput*, 55-66 (2012).
4. B. Vogelstein, K. W. Kinzler. *Nat Med* **10**, 789-799 (2004).
5. D. Hanahan, R. A. Weinberg. *Cell* **144**, 646-674 (2011).
6. P. Jia, Z. Zhao. *PLoS Comput Biol* **10**, e1003460 (2014).
7. M. Kanehisa, S. Goto. *Nucleic Acids Res* **28**, 27-30 (2000).
8. D. Croft *et al. Nucleic Acids Res* **42**, D472-477 (2014).
9. D. Szklarczyk *et al.*, *Nucleic Acids Res* **39**, D561-568 (2011).
10. N. L. Nehrt, T. A. Peterson, D. Park, M. G. Kann. *BMC Genomics* **13 Suppl 4**, S9 (2012).
11. A. Torkamani, N. J. Schork. *Genomics* **90**, 49-58 (2007).
12. J. A. Endicott, M. E. Noble, L. N. Johnson. *Annu Rev Biochem* **81**, 587-613 (2012).
13. C. Greenman *et al. Nature* **446**, 153-158 (2007).
14. H. Davies *et al. Nature* **417**, 949-954 (2002).
15. E. Lengyel, K. Sawada, R. Salgia. *Curr Mol Med* **7**, 77-84 (2007).
16. A. Wilkins, S. Erdin, R. Lua, O. Lichtarge, *Methods Mol Biol* **819**, 29-42 (2012).
17. H. J. Kang, A. D. Wilkins, O. Lichtarge, T. G. Wensel. *J Biol Chem* **290**, 2870-2878 (2015).
18. S. M. Peterson *et al. Proc Natl Acad Sci U S A* **112**, 7097-7102 (2015).
19. P. Katsonis, O. Lichtarge. *Genome Res* **24(12)**, 2050-2058 (2014).
20. G. Manning, D. B. Whyte, R. Martinez, T. Hunter, S. Sudarsanam. *Science* **298**, 1912-1934 (2002).
21. K. Wang, M. Li, H. Hakonarson. *Nucleic Acids Res* **38**, e164 (2010).
22. P. Yue *et al. Hum Mutat* **31**, 264-271 (2010).
23. J. L. Marks *et al. PLoS One* **2**, e426 (2007).
24. H. Davies *et al. Cancer Res* **65**, 7591-7595 (2005).
25. W. Pao *et al. PLoS Med* **2**, e73 (2005).
26. S. Kobayashi *et al. N Engl J Med* **352**, 786-792 (2005).
27. T. C. Strong, G. Kaur, J. H. Thomas. *PLoS One* **6**, e28100 (2011).
28. Neskey *et al.* Cancer Research 75,(7); 1527-36 (2015).

MUSE: A MULTI-LOCUS SAMPLING-BASED EPISTASIS ALGORITHM FOR QUANTITATIVE GENETIC TRAIT PREDICTION

DAN HE and LAXMI PARIDA

IBM T.J Watson Research
Yorktown Heights, NY
E-mail: {dhe, parida}@us.ibm.com

Quantitative genetic trait prediction based on high-density genotyping arrays plays an important role for plant and animal breeding, as well as genetic epidemiology such as complex diseases. The prediction can be very helpful to develop breeding strategies and is crucial to translate the findings in genetics to precision medicine. Epistasis, the phenomena where the SNPs interact with each other, has been studied extensively in Genome Wide Association Studies (GWAS) but received relatively less attention for quantitative genetic trait prediction. As the number of possible interactions is generally extremely large, even pairwise interactions is very challenging. To our knowledge, there is no solid solution yet to utilize epistasis to improve genetic trait prediction. In this work, we studied the multi-locus epistasis problem where the interactions with more than two SNPs are considered. We developed an efficient algorithm MUSE to improve the genetic trait prediction with the help of multi-locus epistasis. MUSE is sampling-based and we proposed a few different sampling strategies. Our experiments on real data showed that MUSE is not only efficient but also effective to improve the genetic trait prediction. MUSE also achieved very significant improvements on a real plant data set as well as a real human data set.

Keywords: Genetic Trait Prediction, Mutual Information, Epistasis, Weighted Maximum Independent Set

1. Introduction

Given its relevance in the fields of plant and animal breeding as well as genetic epidemiology,[1–3] whole genome prediction of complex phenotypic traits using high-density genotyping arrays recently received great attentions. Complex traits prediction and association are crucial to translate the findings in genetics to precision medicine. Given the genotype values encoded as $\{0, 1, 2\}$ of a set of biallelic molecular markers (we use "feature", "marker", "genotype" interchangeably), such as Single Nucleotide Polymorphisms (SNPs), on a collection of plant, animal or human samples, quantitative genetic traits, such as weight, height, fruit size etc. of these samples can be predicted effectively. More accurate genetic trait prediction can help breeding companies to develop more effective breeding strategies.

One of the most popular algorithms for the genetic trait prediction problem is *rrBLUP* (Ridge-Regression BLUP),[1,4] which assumes all the markers contribute to the trait value more or less. The algorithm fits an additive linear regression model where all the markers are invovled. It fits the coefficient computed for each marker, which quantifies the importance of the marker. The rrBLUP method has the benefits of the underlying hypothesis of normal distribution of the trait value and the marker effects (well suited for highly polygenic traits). Its performance is as good as or better than other popular predictive models such as Elastic-Net, Lasso, Ridge Regression,[5,6] Bayes A, Bayes B,[1] Bayes Cπ,[7] and Bayesian Lasso,[8,9] as well as other machine learning methods.

Epistasis is the phenomenon where different markers, or genes, can interact with each other. The problem of epistasis detection has been widely studied in GWAS (Genome Wide Association Studies). Lots of work, mainly greedy strategies,[10–16] have been proposed to detect epistasis effects. These greedy strategies all assume that significant epistasis effects come from only strong marginal effects, or the markers that are highly relevant to the trait. While most existing methods target epistasis detection on GWAS, some recent developments have been achieved on quantitative genetic trait prediction. He et al.[17] proposed a sampling-based method MINED to detect significant pairwise epistasis effects and to improve the genetic trait prediction. He and Parida[18] further proposed a two-stage sampling algorithm SAME to handle multi-locus epistasis effects where the number of markers involved can be greater than two. They showed that the prediction can be significantly improved with the help of epistasis. In the meanwhile SAME has a few advantages over the existing methods: It is highly scalable; It captures epistasis effects from both strong and weak marginal effects. However, SAME still has a few drawbacks: Its sampling strategy is based on random sampling where for all interactions the same number of samplings is conducted; It does not check the redundancy of the sampled interactions thus many sampled interactions might be redundant given the huge sample space; Its interaction values are based on multiplications of the genotype values, which does not distinguish all the possible genotype combinations.

In this work, we studied the multi-locus epistasis problem where the interactions with more than two SNPs are considered. We developed an efficient algorithm MUSE (Multi-locus Sampling-based Epistasis algorithm) to improve the genetic trait prediction with the help of multi-locus epistasis. MUSE conducts bidirectional sampling: It samples k-locus interactions from (k-1)-locus interactions and it decomposes the k-locus interactions into multiple (k-1)-locus interactions for further sampling. The motivation comes from the observation made in[17] that when a (k-1)-locus interaction is involved in a significant k-locus interaction, no matter whether it is a strong marginal effect or not, it is likely to be involved in multiple significant k-locus interaction. The main contribution of this work is a set of sampling strategies, including constraint-based sampling, encoding-based sampling and iterative sampling. More details will be given in the method section. Our experiments showed that MUSE is not only efficient but also effective to improve the genetic trait prediction. We also observed significant improvements on a real plant data set as well as a real human data set over the state-of-the-art methods.

2. Preliminaries

Genetic trait prediction problem is usually represented as the following linear regression model:

$$Y = \beta_0 + \sum_{i=1}^{d} \beta_i X_i + e$$

where Y is the phenotype and X_i is the i-th genotype value, d is the total number of genotypes and β_i is the regression coefficient for the i-th marker, e is the error term which usually follows a normal distribution. We call the above model *single marker model.*

Epistasis is the phenomenon where different markers can interact with each other. With the pairwise epistasis effects, the traditional linear regression model becomes the following non-linear additive model:

$$Y = \beta_0 + \sum_{i=1}^{d} \beta_i X_i + \sum_{i,j}^{d} \alpha_{i,j} X_i X_j + e \tag{1}$$

where $X_i X_j$ is the product of the genotype values of the i-th and j-th marker and it denotes the interaction of the two genotypes.

Multi-locus epistasis model is more complicated as more than two markers are involved in the interactions. When n-markers are involved in the interaction, we call it *n-locus* interaction or *n-way* interaction, which are interchangeable and we call n as the *order* of the interaction. The model is shown as below:

$$Y = \beta_0 + \sum_{i=1}^{d} \beta_i X_i + \sum_{i,j}^{d} \alpha_{i,j} X_i X_j + \cdots + \sum_{i_1, i_2, \ldots, i_n}^{d} \alpha_{i_1, i_2, \ldots, i_n} X_{i_1} X_{i_2} \ldots X_{i_n} + e \tag{2}$$

For example, the regression model involving both 2-locus and 3-locus interactions is:

$$Y = \beta_0 + \sum_{i=1}^{d} \beta_i X_i + \sum_{i,j}^{d} \alpha_{i,j} X_i X_j + \sum_{i,j,k}^{d} \alpha_{i,j,k} X_i X_j X_k + e$$

3. Multi-locus Sampling-based Epistasis Algorithm

In this work, we follow the pipeline of SAME[18] to conduct the bi-directional search. We start sampling in a forward manner from the significant (k-1)-locus interactions to obtain the significant k-locus interactions. Then we search in backwards where we take the significant k-locus interactions to guide what extra (k-1)-locus interactions we should consider to sample. This is based on the observations made in the work of He et al.[17] that if a (k-1)-locus interaction is involved in a significant k-locus interaction, no matter whether this (k-1)-locus interaction is significant or not, it is likely to be involved in multiple significant k-locus interactions.

We first use a queue Q to store the features (can be 1-locus to (k-1)-locus interactions) from which the sampling is conducted. We define *sampling* a t-locus effect as that for the t-locus effect, we randomly sample a set of single markers to be combined with the t-locus effect to obtain (t+1)-locus effects. We define a feature is *significant* if its r^2 (The square of the Pearson's correlation coefficient between the feature vector and the trait vector) to the trait is higher than a threshold s (We will show how to determine the threshold later). We use r^2 here as it is the most popular metric for genetic trait prediction (or genomic selection). We start from significant single markers and store all of them in Q. Then we sample each single marker X to obtain a set of significant 2-locus interactions where the marker X is involved in. If the 2-locus interaction is significant, we store it in Q. Then for the significant 2-locus interaction, we decompose it into two 1-locus effects, or two single markers. One of the markers will be X, the other one is either a strong or weak marginal effect. If the other marker is not in Q yet, we store it in Q so that it will be sampled later on.

We then repeat the sampling process for 2-locus interactions upto (k-1)-locus interactions. When we sample a (k-1)-locus interaction, if we obtain a significant k-locus interaction, we then decompose the k-locus interaction into k (k-1)-locus interactions and store them in Q. For example, given a significant 2-locus interaction AB, we randomly sample one single marker and by chance we obtain a significant 3-locus interaction ABF. Then we decompose it into three 2-locus interactions AB, BF, AF and store them in Q if they have not been stored yet. They will be sampled in a later stage.

3.1. *Significance Threshold*

The significance threshold s is determined dynamically. This is because we only keep the top K most significant features and thus the threshold is set naturally as the r^2 of the top K-th feature. We maintain a sorted list of the features according to their r^2 score (notice we consider both epistasis effects and single marker effects). When we check an interaction, we insert the interaction into the top-K feature set if its r^2 score is better than s and we remove the last feature from the list. If the interaction does not have a higher r^2 score than s, we do not change the list. We then set the threshold s as the r^2 score of the current K-th feature. We keep on updating the threshold as we insert more interactions, while keeping the order of the list according to the r^2 scores. As the threshold becomes higher, it becomes harder for an interaction to be selected.

3.2. *P-value*

As the feature space is extremely large, in order to avoid over-fitting problem, we also computed the p-value of the features. We ignore features with high r^2 score if the p-value of the features are not small enough. Similar to GWAS, where a typical p-value threshold is 5×10^{-8} after Bonferroni corrections for multiple testing, we used very small p-values. We observed that we can not use a fixed p-value. Instead, for larger feature space, we need to use smaller p-values. For example, for a feature space of size $O(10^7)$, we use p-value 5×10^{-6}. For a feature space of size $O(10^{10})$, we use p-value as 5×10^{-8} to 5×10^{-11}. The p-value to be used is determined by a grid search using cross validation.

3.3. *Estimate Interaction Probability*

Another thing to notice is that when we conduct the sampling, we do not sample all the single markers as it would be very time consuming for a large number of markers. We conduct an initial sampling with size f. It is shown in[17] that the scores follow a truncated normal distribution. Then using the f sampled r^2 scores, we can fit the truncated normal distribution to estimate the mean and the standard deviation. Using this distribution, and given the total number of single markers as d, we compute the probability of seeing at least one significant r^2 score out of the $O(d)$ possible interactions, where a score is significant if it is higher than the current significance threshold s. If the probability is higher than a threshold P, we will test the interactions between the marker and all the remaining markers. In order to capture as many epistasis interactions as possible, we generally use a small value for P, say 0.005.

As we can see, the performance of MUSE is heavily dependent on the sampling strategy. In SAME,[18] a simple random sampling is conducted which has been shown to have certain disadvantages. Next we introduce three sampling strategies that could significantly improve the random sampling:

3.4. *Sampling Strategies*

3.4.1. *Constraint-based Sampling*

Significant interaction selection can be considered as a feature selection process if we consider each significant interaction as a feature. A popular feature selection criteria is called MRMR (Maximum Relevance and Minimum Redundancy),[19] where the objective is to select a set of features which are maximumly relevant to the trait but minimally redundant with each other. It is shown[19] that minimizing the redundancy of the selected features leads to better prediction. In our approach, the selection of the top-k most significant interactions is equivalent to maximizing the relevance of the selected interactions to the trait. However, the redundancy of the selected interactions is not taken into consideration yet.

It is observed in[17] that a t-locus interaction might be involved in multiple significant (t+1)-locus interactions. However, these multiple significant (t+1)-locus interactions might be highly redundant with each other, as all of them share the same t-locus interaction. As the size k is fixed for the top-k most significant interactions, including many redundant interactions might not improve the prediction according to the MRMR criterion. An extreme case is that all the top-k most significant interactions are redundant, which is equivalent to using only one interaction for prediction. This will obviously lead to poor performance.

Thus here we add a constraint on the sampling process: we require every t-locus interaction involved in at most N (t+1)-locus interactions. We call N the *overlap threshold.* Therefore, any of the top-k interactions should at most overlap with N other top-k interactions, where overlap means two (t+1)-locus interactions share the same t-locus interaction. We call this sampling *Constraint-based Sampling.*

To solve the constraint-based sampling problem, we construct an *Interaction Graph*, where the nodes are (t+1)-locus interactions, the edges indicate that the two (t+1)-locus interactions share the same t-locus sub-interaction. Each node is associated with a weight, indicating the r^2 of the node to the trait. Notice we build a graph for each t. Once we moved from t to t+1, we build a new graph and delete the old graph. As an example, we can see in Figure 1, the interaction ABC share the sub-interaction AB with the interaction ABD. Thus the number of edges associated with a node indicates the degree of overlaps of the node and we call it *connectivity.* In this example, the node ABC has connectivity as 3, the node ABD has connectivity as 4. If we set the overlap threshold N as 1, we can only select the nodes that is connected to one other node.

The constraint-based sampling problem is then converted to the problem where we would like to select a set K of k nodes such that the total weights of the nodes is maximized and in the meanwhile the constraint is satisfied, namely in the node set K, there is no node with more than N edges connecting to the other nodes in the set. The problem is similar to a Weighted Maximum Independent Set (WMIS) problem. The WMIS problem seeks to select a set of

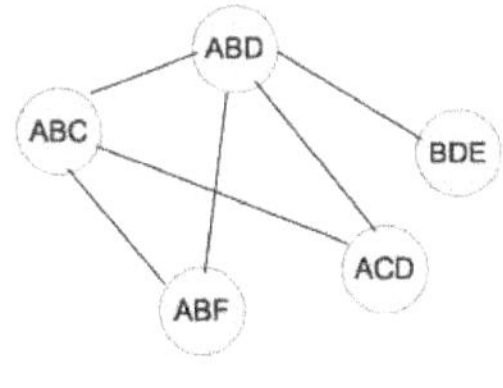

Fig. 1. An example of interaction graph.

nodes from a graph to form an independent set, where all the nodes are not adjacent, such that the sum of the weights on the nodes is maximized. As all the nodes are not adjacent in the independent set, all selected interactions are guaranteed non-overlapping. This is equivalent to allowing the degree of connectivity as 0. In our case, we set the degree of the connectivity of the selected nodes to be no greater than N.

The WMIS problem is NP-complete and what's more, it requires generating the complete interaction graph. However, in our problem, we sample the t-locus interactions one by one. Thus we conducted a greedy algorithm, where we maintain a count for every t-locus interaction. During the samplings, when we sample a t-locus interaction I and find one significant $(t+1)$-locus interaction, we increase the count of I by one. If the count is less than N, we keep on sampling. Otherwise we have two options:

(1) We stop the sampling immediately
(2) We do not stop the sampling, instead we continue the sampling process. However, we maintain only N significant $(t+1)$-locus interactions sampled from I and we call the set S. Once we identify a significant $(t+1)$-locus interaction I', we compare its r^2 score with the r^2 scores of the interactions in S. If its r^2 score is greater than the minimum r^2 score in S, we remove the interaction in S with the minimum r^2 score and replace it with I'.

Obviously, by taking option one, the sampling process can be terminated quickly but it may miss the significant $(t+1)$-locus interactions that might arrive later. By taking option two, we can guarantee that all significant $(t+1)$-locus interactions could be captured. However, we only store N significant $(t+1)$-locus interactions and thus the constraint can be satisfied. By setting N small, we could include more $(t+1)$-locus interactions that have different sub-interactions so that the redundancy of the top-k interactions can be reduced. In MUSE, we choose option two.

3.4.2. *Encoding-based Sampling*

By using the multiplication model and assuming the genotypes are encoded as $\{0, 1, 2\}$, a pairwise epistasis effect contains only 4 different possible values $\{0, 1, 2, 4\}$ (by pairwise multiplication of the values from $\{0, 1, 2\}$) while in reality there are nine different possible combinations of the alleles. It is not clear why a pair of markers with genotypes (0, 1) should have the same interaction value 0 as the pairs with genotypes (0, 2). Thus instead of using the values

$\{0,1,2,4\}$, we could consider using nine different values $\{0, 1, 2, 3, 4, 5, 6, 7, 8\}$ to differentiate the nine different combinations. However, there is no order for the combinations. For example, we can not determine the order of "AA/Bb" and "Aa/BB". Similarly, we can not determine the order of "Aa/bb" and "aa/Bb". Thus we do not have a systematic way to assign the nine different values $\{0, 1, 2, 3, 4, 5, 6, 7, 8\}$ to the nine different combinations.

Therefore, we developed the following encoding formula:

$$encoding = \sum_{i=1}^{n} X_i \times 10^{(n-i)}$$

where n is the number of markers involved, X_i is the encoding of the genotype of the i-th marker, which is one of $\{0,1,2\}$. Thus instead of multiplication, we use the above encodings for the n-way epistasis interactions. For example, for pairwise interactions, assuming the encoding $\{0,1,2\}$ are for "AA, Aa, aa" respectively and the same for "BB, Bb, bb" respectively, we have the following encodings for the nine combinations:

$$AA/BB = 0 \times 10 + 0 = 0, \quad AA/Bb = 0 \times 10 + 1 = 1, \quad AA/bb = 0 \times 10 + 2 = 2$$
$$Aa/BB = 1 \times 10 + 0 = 10, \quad Aa/Bb = 1 \times 10 + 1 = 11, \quad Aa/bb = 1 \times 10 + 2 = 12$$
$$aa/BB = 2 \times 10 + 0 = 20, \quad aa/Bb = 2 \times 10 + 1 = 21, \quad aa/bb = 2 \times 10 + 2 = 22$$

Thus using this encoding, we guarantee that different combinations of epistasis effects have different encodings and we do not need to worry about the assignment of different values to these combinations. Another benefit is that the encoding can be applied to any t-locus interactions in a systematic way. We call this sampling *Encoding-based Sampling.*

3.4.3. *Iterative Sampling*

As we are using sampling to estimate the mean and standard deviation of the normal distribution, it is critical to determine the sample size first. Given an expected error rate, we could estimate the sample size via Equation 3.

$$ME = z\frac{s}{\sqrt{n}} \tag{3}$$

Where ME is the desired margin of error, z is the z-score that depends on the desired confidence level, s is the standard deviation and n is the sample size we want to find. Given the desired margin of error and the confidence level, if we know the standard deviation or we could make a guess on it, we could compute the required sample size n.

However, our problem is much more complicated in that every t-locus interaction has different mean and standard deviation. Therefore it is not appropriate to use an universal sample size and there is no systematic way to estimate the standard deviation for each t-locus interaction.

To address the problem, we propose an iterative sampling method. In iteration one, for every t-locus interaction, we start from a small initial sample size, say, 500, and estimate mean μ_1 and standard deviation δ_1. Then we increase the sample size by 500 for every iteration. In iteration i, we estimate mean μ_i and standard deviation δ_i. If $\frac{abs(\mu_i - \mu_{i-1})}{\mu_i} \leq \epsilon$ and $\frac{abs(\delta_i - \delta_{i-1})}{\delta_i} \leq \epsilon$, where ϵ is a small number such as 0.01, or the number of iterations is greater than a pre-specified number, such as 10, we say that the sampling converges.

Notice that MUSE selects the top-k most significant interactions. After the selection, we combine these interactions with the original set of single markers as a new data set. Regression methods such as rrBLUP are then applied on the new data set to make predictions. Notice k is a user defined parameter. The smaller k is, the more efficient MUSE is. Ideally k could be selected using cross-validation. However, given the extremely large feature space, it is not feasible to try all possible k's. Therefore in our work, we just simply set k as 500, a small number. Our experiments showed that by setting k as 500, we could already achieve significant improvements and yet the program is highly efficient.

4. Experimental Results

We first evaluated MUSE on a plant data set: Maize data set,[2] the Dent and Flint panels, developed for the European CornFed program. We do not consider using simulated data here as the rational for how high order multi-locus interactions contribute to the trait is indeed not clear. As the number of multi-locus interactions is extremely high when the order is high, it is not clear what is a reasonable number of the interactions that contribute to the trait.

The Maize data set indeed consists of 6 sub data sets. The Dent panel were genotyped using a 50k SNP array, which after removing SNPs with high rate of missing markers and high average heterozygosity, yielded 29,094 and 30,027 SNPs respectively. Both of them contain 261 samples and three traits. In all experiments, we perform 10-fold cross-validations and measure the average r^2 between the true and the predicted outputs, where higher r^2 indicates better performance. The parameters are learned from the training data. The baseline method is rrBLUP with single marker model using all markers. For a fair comparison, we use the top-500 most significant interactions (for k-locus interactions where $k \geq 2$) captured by MUSE and we combine them with the original set of single markers as a new data set where rrBLUP is then applied. This will indicate whether the extra information from the interactions benefit the prediction. Notice we mark the performance as "NA" for cases where no significant interaction is captured.

We evaluate the performance of MUSE with the constraint-based sampling (MUSE-C), with the encoding-based sampling (MUSE-E) and with iterative sampling (MUSE-I). We consider only 2-locus scenarios where the p-value p=5×10^{-8}. For the constraint-based sampling, overlap threshold N=5. The baseline method is rrBLUP with single marker model using all markers. As we can see in Table 1, MUSE improves the performance over rrBLUP significantly. As MINED does not use p-values as a criteria to select interactions, its performance is worse than SAME and MUSE. MUSE with the constraint-based sampling (MUSE-C) generally is able to improve the prediction accuracy over SAME, as the constraint-based sampling is able to naturally reduce the redundancy of the sampled interactions, which further leads to improvement on the prediction. MUSE with both the constraint-based sampling and the encoding-based sampling (MUSE-CE) achieve better results except for Flint Trait 3, indicating that both constraint-based sampling and encoding-based sampling are effective in improving the prediction accuracy. For Flint Trait 3, when constraint-based sampling is used, MUSE can not capture any interaction with p-value lower than 5×10^{-8}. However, after we conducted the iterative sampling, MUSE is able to capture interactions with p-value lower than 5×10^{-8} and thus MUSE-CEI achieved the best performance among all the methods. This clearly indicates the power of iterative sampling. In general combining all three sampling strategies gives us the best performance.

Table 1. The r^2 of rrBLUP, MINED, SAME, MUSE on Maize Dent and Flint data sets. We show only 2-locus scenarios where p-value p=5 × 10^{-8}, overlap threshold N=5. For MUSE-C and MUSE-CE, the number of initial sampling is 500. Here for MUSE, "-C" stands for constraint-based sampling, "-E" stands for encoding-based sampling, "-I" stands for iterative sampling.

Trait	rrBLUP	MINED	SAME	MUSE-C	MUSE-CE	MUSE-CEI
Dent Trait 1	0.59	0.59	0.615	0.65	0.65	0.67
Dent Trait 2	0.552	0.552	0.583	0.572	0.59	0.61
Dent Trait 3	0.321	0.356	0.432	0.39	0.486	0.49
Flint Trait 1	0.47	0.476	0.514	0.558	0.576	0.595
Flint Trait 2	0.301	0.316	0.356	0.364	0.419	0.429
Flint Trait 3	0.057	0.096	0.113	NA	NA	0.135

In Table 2, we evaluated 2-locus, 3-locus and 4-locus interactions for MUSE. As we have already shown that MUSE-CEI in general achieves the best performance, we only evaluate the performance of MUSE-CEI. We also varied the overlap thresholds as 5, 20, 50. The running times for MUSE-CEI are 226 sec., 979 sec. and 2056 sec. respectively. As we can see, although the size of the feature space increased exponentially, the running time of MUSE-CEI did not change much, indicating that MUSE-CEI is highly scalable due to its effective sampling process. The baseline method is again rrBLUP with single marker model using all markers.

Overall, we can see that MUSE-CEI achieved very significant improvements over rrBLUP on the single marker model (For Dent data, 21% for trait 1, 22% for trait 2, 59% for trait 3. For Flint Data, 33% for trait 1, 46% for trait 2, 138% for trait 3). We can see that both the p-value and the overlap threshold N are critical to the prediction. The best p-value and N are usually different without clear pattern for different traits and we need to use grid search to find their best values.

By varying the p-values, the prediction performance varies significantly. In general, the p-value should be small enough to achieve the best prediction. However, we do not see a clear pattern on setting the p-values. For different traits, the best p-value could be different. And it is not necessarily the case that using smaller p-value leads to better prediction accuracy. This is because smaller p-values may only produce a small set of statistically significant epistasis effects where larger p-values may produce a larger set of statistically significant epistasis effects. If the size of the set of statistically significant epistasis effects is too small and in the meanwhile they do not have very high r^2 score, they might not be able to improve the prediction performance. In the worst case, we might not be able to identify any significant k-locus interaction given a too small p-value might lead, such as Dent Trait 3 with 3-locus

$p = 5 \times 10^{-12}$ and Flint Trait 3 with 3-locus $p = 5 \times 10^{-12}$ and 4-locus $p = 5 \times 10^{-11}$. As we did not observe a clear pattern between p-values and the prediction performance, grid search with cross-validation should be applied in order to detect the best p-value.

Table 2. The r^2 of rrBLUP and MUSE on Maize Dent and Flint data set. For MUSE, we tested 2-locus, 3-locus and 4-locus interactions with different p-value thresholds. We applied all the sampling strategies. We vary the p-value and the constraint threshold N.

Methods	N=5	N=20	N=50	N=5	N=20	N=50
	Dent Trait 1			**Flint Trait 1**		
rrBLUP	0.59			0.47		
MUSE-CEI 2-locus (p=5×10^{-8})	0.67	0.581	0.58	0.595	0.591	0.568
MUSE-CEI 2-locus (p=5×10^{-10})	0.645	0.655	0.616	**0.626**	0.615	0.586
MUSE-CEI 2-locus (p=5×10^{-11})	0.63	0.693	0.656	0.56	0.583	0.556
MUSE-CEI 3-locus (p=5×10^{-10})	0.538	0.644	0.491	0.578	0.618	0.59
MUSE-CEI 3-locus (p=5×10^{-11})	0.675	**0.714**	0.59	0.617	0.62	0.57
MUSE-CEI 3-locus (p=5×10^{-12})	0.606	0.65	0.673	0.601	0.61	0.581
MUSE-CEI 4-locus (p=5×10^{-11})	0.27	0.384	0.601	0.47	0.488	0.301
	Dent Trait 2			**Flint Trait 2**		
rrBLUP	0.552			0.301		
MUSE-CEI 2-locus (p=5×10^{-8})	0.61	0.552	0.563	0.429	0.412	0.403
MUSE-CEI 2-locus (p=5×10^{-10})	0.663	0.557	0.564	0.413	0.427	0.394
MUSE-CEI 2-locus (p=5×10^{-11})	**0.671**	0.595	0.574	0.417	0.415	0.373
MUSE-CEI 3-locus (p=5×10^{-10})	0.608	0.459	0.459	0.428	**0.439**	0.418
MUSE-CEI 3-locus (p=5×10^{-11})	0.623	0.491	0.491	0.423	0.421	0.402
MUSE-CEI 3-locus (p=5×10^{-12})	0.582	0.625	0.549	0.382	0.395	0.399
MUSE-CEI 4-locus (p=5×10^{-11})	0.3	0.335	0.258	0.37	0.365	0.298
	Dent Trait 3			**Flint Trait 3**		
rrBLUP	0.321			0.057		
MUSE-CEI 2-locus (p=5×10^{-8})	0.49	0.424	0.361	0.135	0.12	0.087
MUSE-CEI 2-locus (p=5×10^{-10})	0.355	0.476	0.466	0.115	0.126	0.103
MUSE-CEI 2-locus (p=5×10^{-11})	0.332	0.397	0.465	0.097	0.067	0.048
MUSE-CEI 3-locus (p=5×10^{-10})	0.482	0.391	0.443	0.089	0.111	0.103
MUSE-CEI 3-locus (p=5×10^{-11})	0.453	0.347	0.398	0.120	**0.136**	0.119
MUSE-CEI 3-locus (p=5×10^{-12})	NA	NA	0.358	NA	NA	0.026
MUSE-CEI 4-locus (p=5×10^{-11})	0.341	**0.511**	0.444	NA	NA	0.046

Another observation is that smaller N in general leads to better performance. This clearly indicates the effects of redundancy: when N is large, we allow more redundant interactions to be selected and thus the performance drops. However, a small N may prevent selecting significant interactions as the pool of interactions to be sampled is dramatically reduced for small N. For example, for Dent Trait 3, p=5×10^{-12}, when N=5 and 20, MUSE can not capture any 3-locus significant interactions. However, when $N = 50$, MUSE could capture some 3-locus significant interactions. Similarly, for Flint Trait 3, 3-locus and 4-locus significant interactions are only captured when $N = 50$.

In summary we observed that although there is no clear pattern for the optimal p-value and overlap threshold N, we see that in general a too large N or a too small p-value lead to poorer performance. Also for higher order interactions, the number of detected significant interactions might be too small to lead improvements.

One more thing to notice is that we do not conduct biological validation on the interactions MUSE selected. This is because we assume all the interactions contribute to the trait more or less. The selected interactions also have lots of peers which have similar r^2 scores. However, we are only able to select a small set of interactions due to efficiency concerns. These interactions are selected by random chance from the pool of interactions with similar r^2 scores. But our experiments illustrated that a small set of interactions is sufficient to improve the genetic trait prediction accuracy dramatically.

Besides plant traits, we also conducted experiments on complex trait for humans. Complex traits prediction and association are crucial to translate the findings in genetics to precision medicine. We studied the data set from the Finland-United States Investigation of NIDDM Genetics (FUSION) study,[20] which is a long-term effort to identify genetic variants that predispose to type 2 diabetes (T2D) or that impact the variability of T2D-related quantitative traits. The dataset has 5000 individuals, 317503 SNPs and 10 traits. For illustration purpose, we show the results on two randomly selected traits (trait 2: HDL-cholesterol, trait 10: Height).

In Table 3, we showed the performance of MUSE on two human complex traits. We can see that in general the predictions are poor, indicating the difficulties of complex trait prediction. However, even on complex traits, we see that by integrating interactions into the predictive model, we can still achieve significant improvements. And by tuning the parameters carefully, MUSE can achieve better performance compared with existing methods. Again, we see that with relatively small N and p-value, MUSE achieved better performance.

5. Conclusion and Future Work

In this work, we studied the multi-locus epistasis problem where the interactions with more than two SNPs are considered. We developed an algorithm MUSE which is very efficient for multi-locus epistasis model. We also showed that the algorithm is very effective in improving the performance of the genetic trait prediction. Three sampling strategies are developed which could improve the overall prediction accuracy. More accurate trait predictions can be very helpful to develop breeding strategies and is crucial to translate the findings in genetics to precision medicine.

References

1. T. Meuwissen, B. Hayes and M. Goddard, *Genetics* **157**, 1819 (2001).
2. R. Rincent, D. Laloë, S. Nicolas, T. Altmann, D. Brunel, P. Revilla, V. M. Rodriguez, J. Moreno-Gonzalez, A. Melchinger, E. Bauer *et al.*, *Genetics* **192**, 715 (2012).
3. M. A. Cleveland, J. M. Hickey and S. Forni, *G3: Genes— Genomes— Genetics* **2**, 429 (2012).
4. J. Whittaker, R. Thompson and M. Denham, *Genet Res* **75**, 249 (2000).

Table 3. The r^2 of rrBLUP, MINED, SAME and MUSE on Finland data set. For MUSE, we tested 2-locus, 3-locus and 4-locus interactions with different p-value thresholds. We applied all the sampling strategies. We vary the p-value and the constraint threshold N.

Methods	N=5	N=20	N=50	N=5	N=20	N=50
	Trait HDL-cholesterol			**Trait Height**		
rrBLUP		0.11			0.03	
MINED		0.15			0.07	
SAME		0.18			0.10	
MUSE-CEI 2-locus ($p=5\times10^{-8}$)	0.14	0.15	0.16	0.04	0.03	0.04
MUSE-CEI 2-locus ($p=5\times10^{-10}$)	0.15	0.17	0.17	0.05	0.07	0.05
MUSE-CEI 2-locus ($p=5\times10^{-11}$)	0.16	0.18	0.19	0.05	0.06	0.06
MUSE-CEI 3-locus ($p=5\times10^{-10}$)	0.16	0.18	0.18	0.07	0.08	0.06
MUSE-CEI 3-locus ($p=5\times10^{-11}$)	0.17	0.2	0.19	0.08	0.11	0.1
MUSE-CEI 3-locus ($p=5\times10^{-12}$)	0.2	**0.22**	0.21	0.1	0.09	0.08
MUSE-CEI 4-locus ($p=5\times10^{-11}$)	0.12	0.15	0.18	0.1	**0.12**	0.11

5. R. Tibshirani, *Journal of the Royal Statistical Society, Series B* **58**, 267 (1994).
6. S. S. Chen, D. L. Donoho, Michael and A. Saunders, *SIAM Journal on Scientific Computing* **20**, 33 (1998).
7. K. Kizilkaya, R. Fernando and D. Garrick, *Journal of animal science* **88**, 544 (2010).
8. A. Legarra, C. Robert-Granié, P. Croiseau, F. Guillaume, S. Fritz *et al.*, *Genetics research* **93**, p. 77 (2011).
9. T. Park and G. Casella, *Journal of the American Statistical Association* **103**, 681 (June 2008).
10. K. A. Pattin, B. C. White, N. Barney, J. Gui, H. H. Nelson, K. T. Kelsey, A. S. Andrew, M. R. Karagas and J. H. Moore, *Genetic epidemiology* **33**, 87 (2009).
11. J. Marchini, P. Donnelly and L. R. Cardon, *Nature genetics* **37**, 413 (2005).
12. N. R. Cook, R. Y. Zee and P. M. Ridker, *Statistics in medicine* **23**, 1439 (2004).
13. C. Yang, Z. He, X. Wan, Q. Yang, H. Xue and W. Yu, *Bioinformatics* **25**, 504 (2009).
14. Y. Zhang and J. S. Liu, *Nature genetics* **39**, 1167 (2007).
15. G. Fang, M. Haznadar, W. Wang, H. Yu, M. Steinbach, T. R. Church, W. S. Oetting, B. Van Ness and V. Kumar, *PloS one* **7**, p. e33531 (2012).
16. X. Zhang, S. Huang, F. Zou and W. Wang, *Bioinformatics* **26**, i217 (2010).
17. D. He, Z. Wang and L. Parada, Mined: An efficient mutual information based epistasis detection method to improve quantitative genetic trait prediction, in *Bioinformatics Research and Applications*, (Springer, 2015) pp. 108–124.
18. D. He and L. Parida, Same: a sampling-based multi-locus epistasis algorithm for quantitative genetic trait prediction, in *Proceedings of the 6th ACM Conference on Bioinformatics, Computational Biology and Health Informatics*, 2015.
19. H. Peng, F. Long and C. Ding, *Pattern Analysis and Machine Intelligence, IEEE Transactions on* **27**, 1226 (2005).
20. E. Zeggini, L. J. Scott, R. Saxena, B. F. Voight, J. L. Marchini, T. Hu, P. I. de Bakker, G. R. Abecasis, P. Almgren, G. Andersen *et al.*, *Nature genetics* **40**, 638 (2008).

ceRNA SEARCH METHOD IDENTIFIED A MET-ACTIVATED SUBGROUP AMONG EGFR DNA AMPLIFIED LUNG ADENOCARCINOMA PATIENTS

HALLA KABAT*

Outreach Program, miRcore, 2929 Plymouth Rd.
Ann Arbor, MI 48105, USA
Email: halla203@gmail.com

LEO TUNKLE*

Outreach Program, miRcore, 2929 Plymouth Rd.
Ann Arbor, MI 48105, USA
Email: leotunkle@gmail.com

INHAN LEE

miRcore, 2929 Plymouth Rd.
Ann Arbor, MI 48105, USA
Email: inhan@mircore.org

Given the diverse molecular pathways involved in tumorigenesis, identifying subgroups among cancer patients is crucial in precision medicine. While most targeted therapies rely on DNA mutation status in tumors, responses to such therapies vary due to the many molecular processes involved in propagating DNA changes to proteins (which constitute the usual drug targets). Though RNA expressions have been extensively used to categorize tumors, identifying clinically important subgroups remains challenging given the difficulty of discerning subgroups within all possible RNA-RNA networks. It is thus essential to incorporate multiple types of data. Recently, RNA was found to regulate other RNA through a common microRNA (miR). These regulating and regulated RNAs are referred to as competing endogenous RNAs (ceRNAs). However, global correlations between mRNA and miR expressions across all samples have not reliably yielded ceRNAs. In this study, we developed a ceRNA-based method to identify subgroups of cancer patients combining DNA copy number variation, mRNA expression, and microRNA (miR) expression data with biological knowledge. Clinical data is used to validate identified subgroups and ceRNAs. Since ceRNAs are causal, ceRNA-based subgroups may present clinical relevance. Using lung adenocarcinoma data from The Cancer Genome Atlas (TCGA) as an example, we focused on EGFR amplification status, since a targeted therapy for EGFR exists. We hypothesized that global correlations between mRNA and miR expressions across all patients would not reveal important subgroups and that clustering of potential ceRNAs might define molecular pathway-relevant subgroups. Using experimentally validated miR-target pairs, we identified EGFR and MET as potential ceRNAs for miR-133b in lung adenocarcinoma. The EGFR-MET up and miR-133b down subgroup showed a higher death rate than the EGFR-MET down and miR-133b up subgroup. Although transactivation between MET and EGFR has been identified previously, our result is the first to propose ceRNA as one of its underlying mechanisms. Furthermore, since MET amplification was seen in the case of resistance to EGFR-targeted therapy, the EGFR-MET up and miR-133b down subgroup may fall into the drug non-response group and thus preclude EGFR target therapy.

* These authors contributed equally to this work.

1. Introduction

Lung cancer accounts for more deaths than any other cancers, with a 5-year survival rate of 10% [1]. Several gene mutations have been shown to play a role in lung adenocarcinoma (LUAD), including KRAS and EGFR [2]. Multiple drugs have been developed to target EGFR proteins and are actively used for those with EGFR mutation cancers. However, some patients do not respond to the targeted therapy and many initially responded patients develop resistance to such drugs. Since diverse molecular pathways are associated with any mutated genes, additional information other than DNA mutation is needed to properly identify which subgroup will benefit from the targeted therapy.

Though mRNA expression data have been used to categorize tumors, correlations across mRNA expression data alone are often difficult to decipher within high dimensional data. Moreover, the most correlated genes or samples often do not provide clinically useful insight. To increase the signals, other types of data such as DNA methylation, microRNA (miR) expression, proteomics, and metabolic data have been incorporated with mRNA expression. In terms of RNA levels, mRNA and miR expression correlations have been heavily mined. However, the lack of known miR targets and excessive false positive target predictions hinder the computational search for significant miR-target gene networks. Worse, since miR effects on most target genes are small in degree, in vitro experimental confirmation is difficult, although the effects may contribute to long term clinical outcomes. Usual mRNA-miR expression analyses calculate correlations among RNAs for all samples.

Studies have recently demonstrated that RNAs can compete with one another for the same regulating miRNAs [3]. One of the earliest of these studies, focused on expression of PTEN, hypothesized that expression levels of "competing endogenous" RNAs (ceRNAs) affected PTEN expression. When siRNAs were used to deplete these RNAs, PTEN expression levels also decreased. Decreased ceRNA levels resulted in fewer miRNAs (which target both the ceRNA and PTEN) being "used-up" in regulation. This frees more of these miRNAs to target PTEN, thereby decreasing its expression. Overall, a decrease in expression of a ceRNA results in a corresponding decrease in PTEN. The same study also demonstrated that an increase in expression of a ceRNA corresponded with an increase in PTEN. This is likely applicable not just to PTEN but to other genes, such as a gene and a similar pseudogene or two genes regulated by the same miRNA. Note that ceRNA by definition entails causality whereas usual mRNA-miR expression results are correlative. RNA expression changes cause other RNA expression changes through miR manipulation.

This RNA-RNA regulation inspired two lines of investigation: biochemical inquiry to identify individual ceRNA pairs [4-6] and bioinformatics research to identify global RNA-RNA networks using RNA expression data along with miR-target predictions [7,8]. Ideal conditions for miRs and ceRNAs have also been explored [9,10]. However, global ceRNA networks are difficult to discern due to imprecise miR target prediction and because, again, the miR effect on one target gene is usually small. Such small degree changes are difficult to identify from multi-layer RNA-RNA regulations of diverse samples though ceRNAs have been associated with diseases and have the potential to uncover disease progression [11].

The Cancer Genome Atlas (TCGA) [12] provides a large amount of various types of data from multiple cancers, enabling new ways of data analysis. For example, LUAD data include the mRNA and miRNA (miR) expressions of 551 patients that could provide insight into multiple biological processes within tumors. This large mass of patient data allows for identification of subgroups based upon very specific traits.

In this study, the concept of ceRNAs was utilized to identify a subgroup related to DNA mutations. We focused on patients with amplified EGFR to identify those who could benefit from EGFR targeted therapy, analyzing multiple datasets including copy number variation (CNV), RNAseq, and miRNAseq from TCGA in order to find the EGFR amplification signature. RNA and miR interactions were then identified using a database of experimentally validated miR-target genes from miRTarBase [13]. Our findings suggest that miR-133b, which targets EGFR, is downregulated due to high mRNA expression for EGFR caused by its DNA amplification, which in turn leads to the upregulation of MET, another gene targeted by miR-133b. In short, EGFR amplification is linked to MET mRNA upregulation through miR-133b, which targets both EGFR and MET in a manner reminiscent of the ceRNA interactions mentioned above. To our knowledge, our research is the first to identify disease subgroups based upon ceRNA interactions, an approach with potential application to other gene mutations or in other types of cancers.

2. Methods

Most research into downstream effects of DNA mutations has focused on protein functions. Here we propose using the ceRNA concept to analyze downstream events of DNA mutation to complement conventional protein-centric biology and to identify RNA-RNA networks (Fig. 1).

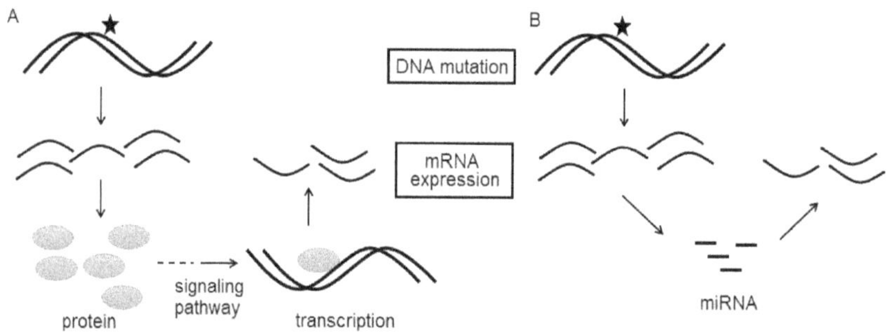

Fig. 1. Underlying biological concepts in mRNA expressions related to DNA mutations. (A) Protein-centric concept. A DNA mutation leads to protein expression changes, resulting in other mRNA changes through signaling pathways. These downstream mRNAs are RNAs of interest. (B) ceRNA concept. If DNA mutation leads to ceRNA upregulation, the "used-up" miRs would fail to regulate the ceRNA pair and thus increase mRNA expression. Similarly, if ceRNA is downregulated, the pairing ceRNA would be downregulated. miR expression data and miR target information are needed to elucidate this process.

2.1. *Overview of data analysis pipeline*

Fig. 2 shows the overall data analysis pipeline to identify subgroups related to a certain DNA amplification [deletion]. Including DNA information may reveal DNA mutation-related ceRNAs, reducing the search space for ceRNA networks. The overall process requires downloading copy number variation (CNV), mRNAseq, miRseq, and clinical data from TCGA and miR-target pairs with strong experimental evidence from miRTarBase.

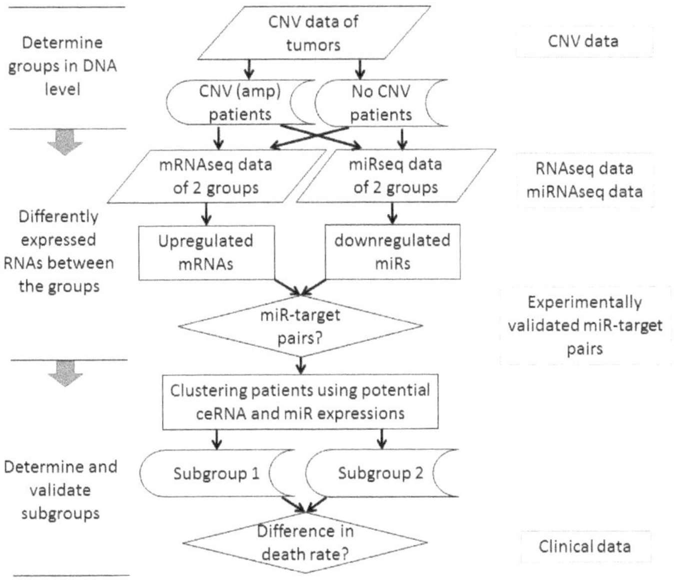

Fig. 2. Overall data analysis process to identify ceRNA-based subgroups. Here an example of amplified CNV genes is shown, with only upregulated mRNAs for clarity.

2.2. *TCGA CNV data analysis*

We downloaded CNV data of all LUAD tumor samples from TCGA and translated chromosome locations to gene-level information using TCGA-Assembler [14]. Overall tumor characteristics were assessed by average CNV values of each gene in chromosome seven (EGFR location) for all tumor samples. Individual tumors' EGFR CNV values were then sorted to determine if the sample number of the EGFR group was adequate. We used DNA copy numbers greater than 3 to define the EGFR amplified group (EGFR amp) and defined the control group as having a copy number between 1.97 and 2.03, yielding a sample size similar to that of EGFR amp. The corresponding log2(CNV/2) for the amp and the control groups is 0.58 and -0.02 to 0.02, respectively.

2.3. *TCGA mRNA and miRNA expression data analysis*

To analyze mRNA expression data, rsem.genes.normalized_results files for RNASeqV2 data of all samples were downloaded using TCGA-Assembler. Data for the EGFR amp and control groups were then extracted. Some patients did not have available rsem-normalized RNAseq data or miRseq data, and were removed from any further analysis. After confirmation of normalization across samples, a student t-test was conducted to compare the amp and the control group data.

To analyze miR expressions, isoform.quantification files for miRNAseq data were downloaded from the TCGA Data Matrix and converted to mature miR values. These individual files were then combined to make a matrix file for all patients. The R code for this function can be found in GitHub (https://github.com/rptashkin/TCGA_miRNASeq_matrix). Upper quartile normalization was applied for student t-test analysis between the amp and control groups, upon which the miRNAs with p-values < 0.05 were separated into up- and downregulated groups.

2.4. *Validated miR target finding*

To see if the miRNAs and genes had potential interactions, data from miRTarBase, a database of miRNA-target interactions, were used. The upregulated genes and downregulated miRNAs were compared to the miR-target pairs with strong experimental evidence to search for any pairs.

2.5. *Subgroup determination and validation*

A heatmap of potential ceRNAs and miRNAs of interest was used to determine the subgroups formed. The patients were clustered using Pearson correlation, and subgroups were determined based on the clustering trees where the mRNA and miRNA expressions of all patients within the trees exhibit negative correlations between miR-targets and positive correlations between ceRNAs. A survival graph was prepared using R and the death rate differences between the groups were tested using student t-test.

3. Results

3.1. *EGFR-amplified patients with lung adenocarcinoma*

The average CNV of genes on chromosome seven from 551 LUAD tumor samples was calculated to assess overall CNV signatures across the entire chromosome (Fig. 2A). One of the two peaks in chr7 corresponds to the EGFR location, confirming the existence of EGFR amplification in these tumor samples strong enough for analysis. To understand the EGFR CNV status of individual patients' tumors, we sorted 551 tumor samples in terms of EGFR CNV values (Fig. 2B). The number of tumors with amplified EGFR copy numbers is much more than that with reduced copy numbers; some tumors showed distinctively amplified EGFR. Using the CNV cutoff value of three, there were a total 50 patients in the EGFR amp group and 56 patients in the control group.

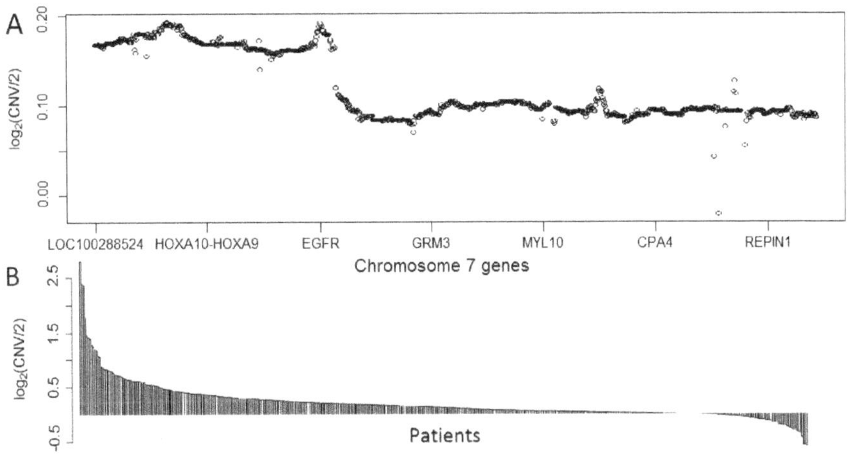

Fig. 3. CNV data for LUAD patients. (A) Average $\log_2$(CNV) values of genes on chromosome 7 for all patients, ordered by chromosome position. (B) EGFR values for the 551 tumor samples.

3.2. *RNA and miRNA expression analysis*

After we downloaded the rsem-normalized data from TCGA, we confirmed the normalization status using box plots. Using the patient lists in the EGFR amp and control groups identified from CNV data, mRNA expression data were extracted and organized for each group. We used isoform.quantification data to obtain mature miR reads for miR expression data analysis. The isoform data were translated to mature miR names and all reads corresponding to the same mature miRs were combined. All EGFR amp and control group patient miR data were merged into a matrix file. Upper quartile normalization was used for miR data and box plots of data before and after normalization were compared to ensure the normalization status. We used only those samples having both mRNA and miR data for further analysis, leaving 42 amp and 35 control patients.

Student t-test was used to identify differently-expressed genes between the two groups of patient samples since the sample number is large. A heatmap of mRNAs with student t-test p-value < 0.0001 (for visual purpose) is shown in Fig. 4A and that of miRs with p-value < 0.05 in Fig. 4B, together with the EGFR amp and control ID labels on top of each heatmap. The unsupervised hierarchical clustering of mRNA expressions identified two large groups: one mostly control and the other mostly amp group. Additionally, the amp group displays a greater number of upregulated genes than does the control group. The mRNA expression of EGFR (p-value of 1.62×10^{-6}), is excluded in this heatmap. The miR clustering also identified two large groups: one with all amp and the other generally with control samples.

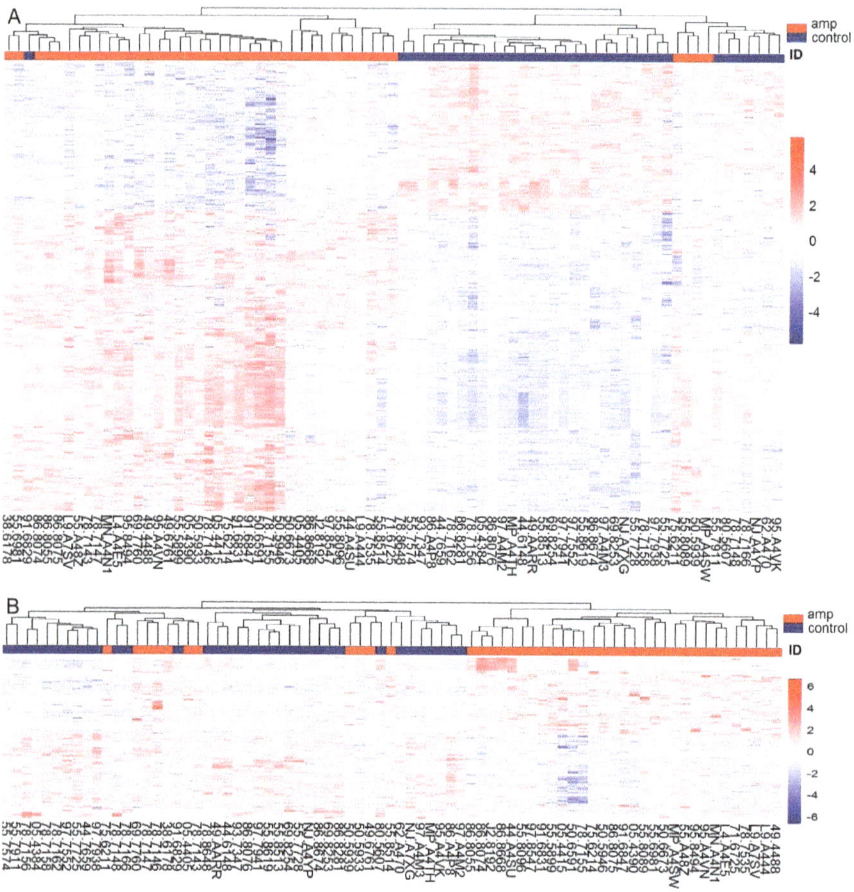

Fig. 4. Hierarchical clustering of mRNA (A) and miRNA expression (B). ID above the heatmap represents the amp group in red and the control in blue. The patient IDs for each group can be found below the heatmap.

3.3. *Identifying miR-target RNA pairs*

We used all mRNAs and miRs with p-values less than 0.05 to find experimentally validated miR-target pairs, since such pairs are still highly limited. To ensure miR-target pair validity, we only used pairs found through strong experimental evidence from miRTarBase. Strong evidence includes validating with a reporter assay, a western blot analysis, or qPCR experiments. Also, since we are looking into direct downstream events of EGFR amplification, only upregulated mRNAs and downregulated miRs in EGFR amp groups were considered.

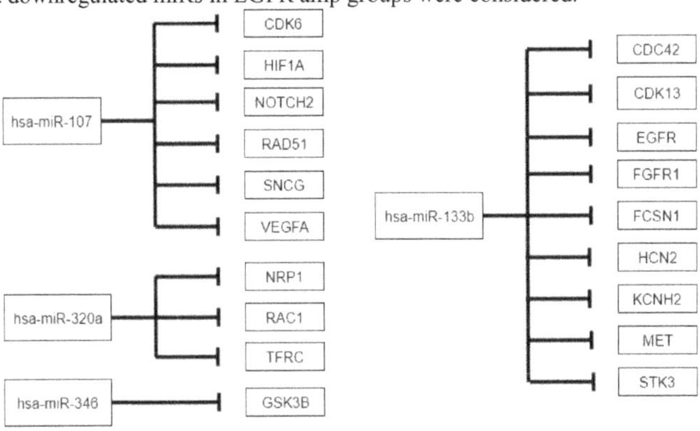

Fig. 5. Validated miRNA-RNA target pairs. The validated target pairs from upregulated mRNAs and downregulated miRNAs with $p < 0.05$.

A total 19 miR-target pairs were identified in the up-mRNAs and down-miR groups, including 4 miRNAs and 19 genes (Fig. 5). One of these pairs included EGFR, a known target of miR-133b. Interestingly, previous studies found miR-133 mediating ceRNAs of mRNA pairs, making miR-133b a good candidate mediator for ceRNAs. Eight other miR-133b targets were found in the upregulated mRNAs, with $p < 0.05$, some possibly functioning as ceRNAs for EGFR through miR-133b in certain patient tumors.

Among them, we decided to focus on MET, given its well-established EGFR and MET crosstalk [15,16], particularly related to drug resistance [17]. The fold changes of EGFR, MET, and miR-133b between EGFR amp and control groups are 6.68, 1.79, and 0.318, respectively; and corresponding p-values for MET and miR-133b are 0.0065 and 0.00085. To exclude other ways of increasing MET mRNA expressions in our dataset, we confirmed that 1) MET copy numbers did not vary in the EGFR amp groups; 2) the expression values of ETS1/2, PAX3, and TCF4, known transcription factors of MET [18], are not upregulated; and 3) ERBB3, known to activate MET [19], is not activated in the EGFR amp groups.

3.4. *Subgroup identification*

To identify patients with potential EGFR-miR-133b-MET interactions, unsupervised hierarchical clustering with only miR-133b, EGFR, and MET were calculated using Pearson correlation distance (Fig. 6A). With a tree cutting of four groups, a subgroup featuring high EGFR-MET and low miR-133g (24 patients) and another subgroup with low EGFR-MET and high miR-133b (24 patients) were identified (boxed in Fig. 6A). Overall Pearson correlation coefficients between

EGFR and MET, EGFR and miR-133b, and MET and miR-133b for all 77 patients are 0.082, -0.030, and 0.082, respectively, unlikely to be identified by global RNA-RNA network analysis of all patients. The correlation coefficients across these 48 patients became 0.22, -0.24, and -0.23, respectively.

To validate these two subgroups, we downloaded patient clinical data from TCGA. As seen in the survival curve (Fig. 6B), these two groups presented different survival rates (student t-test p-value 0.016). Given the known EGFR-MET transactivation, we wondered if subgrouping may also emerge using EGFR and MET expressions alone. We could not see a clear pattern in the clustered heatmap using Pearson correlation distance method, but two clusters showed up using the Euclidean method. The p-value of survival rate differences between these groups was 0.15. Therefore, subgroups identified from EGFR-miR-133b-MET expression data presented stronger clinical implications.

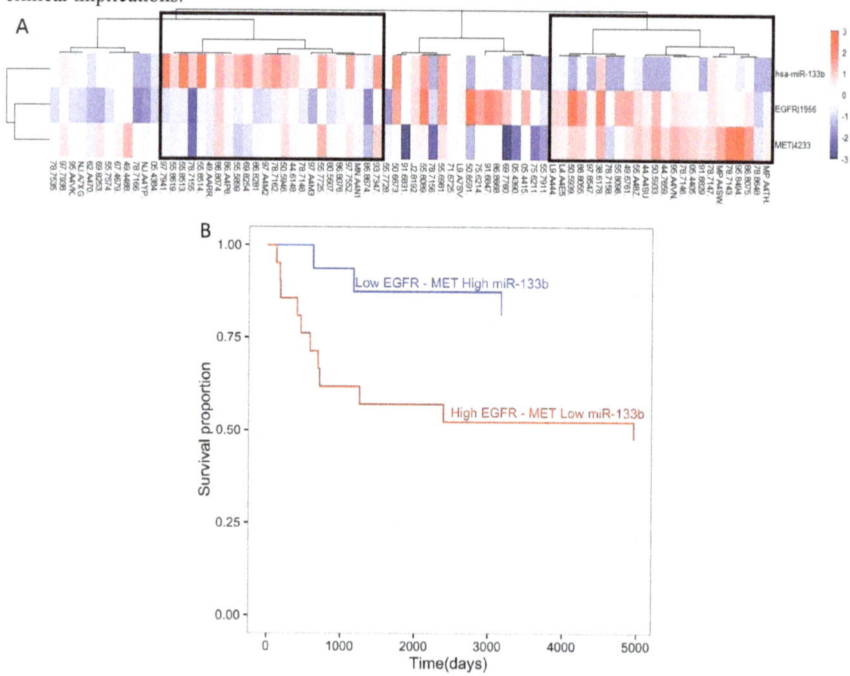

Fig. 6. Subgroup selection and survival curve. (A) Clustered heatmap of EGFR, MET, and miR-133b. The two boxes show the subgroups made through clustering. These subgroups have high miR-133b, low EGFR, and low MET or low miR-133b, high EGFR, and high MET. (B) Survival curves for the subgroups.

4. Discussion

EGFR is one of the more common mutations in lung adenocarcinoma and there exist targeted therapy options for those with this mutation. These currently include drugs such as gefitinib and erlotinib [20]. Though these therapies work well for many patients initially, most patients encounter drug resistance. Of the tumors that develop resistance to these drugs, around 20% have MET amplification [21].

MET, like EGFR, is a growth factor receptor that leads to several signaling cascades including those within the RAS-ERK pathway, which is often targeted by cancer drugs. When functioning normally, MET is essential to such processes as angiogenesis, wound healing, and liver regeneration [22].

Since there is a correlation between MET amplification and drug resistance to an EGFR-targeted therapy, studies have focused on transactivation of EGFR and MET [16-18] though their mechanism has not been cleared elucidated. On the other hand, searching for ceRNA pairs as signature components of DNA level changes, we identified MET as a potential ceRNA for EGFR, suggesting ceRNA as one such mechanism. For a certain subgroup of patients, EGFR and MET were upregulated while their shared regulating miRNA was downregulated. This would fit well with the ceRNA concept, leading to the hypothesis that EGFR CNV amplification "uses up" the regulatory miR-133b, which is then less likely to regulate MET so that EGFR indirectly upregulates MET. Since MET upregulation may be due to MET amplification, we also checked MET CNV values for both the amp and control groups. We found no MET amplification in these groups, confirming that the MET RNA upregulation was not due to DNA amplification.

While we have not biochemically confirmed MET and EGFR to be ceRNAs, EGFR-miR-133b-MET expression clustering could provide subgroups with significantly different survival rates. Since such survival rate difference was not found in groups considering only EGFR-MET expressions, identifying patients with ceRNA function was essential. On the other hand, an EGFR-MET ceRNA pair could have not been found without considering subgroups. Using our method of utilizing multiple-level data consisting of DNA copy number, mRNA expression, and miR expression together with biological information, we may find more clinically relevant potential ceRNA pairs as well as subgroups worthy of pursuit.

Our method can be automated by changing tree distance cutoff values (Pearson correlation distance) in identifying other ceRNAs and related subgroups, which can be validated with survival rates. However, overfitting using survival rate should not be done. Since we started from EGFR CNV-amplified patients, we hypothesized EGFR as the causal mRNA, fit well with ceRNA concept. This kind of biological knowledge is essential to our method.

Acknowledgments

We thank Ryan Ptashkin for the R script to generate a mature miRNA expression matrix file from individual isoform miRNAseq data; Axel Martin for the R script to make mRNA and miRNA matrix files in the same patient order; and Cristina Castillo and Andre Zapico for help in R. This project was extended from a 2015 computational biology summer camp for high school students supported by the University of Michigan WISE (Women in Science and Engineering).

References

1. Survival statistics for lung cancer | Cancer Research (2016) UK.Cancerresearchuk.org.
2. OMIM Entry Search - lung adenocarcinoma. (2016). Omim.org.
3. Cancer Genome Atlas Network. *Nature* **490**, 61–70 (2012).
4. L. Poliseno, *et al. Nature* **465**, 1033–1038 (2010).
5. M. S. Kumar, *et al. Nature* **505**, 212-217 (2013).
6. Y. Tay, *et al. Cell* **147**, 344–357 (2011).
7. F. A. Karreth, *et al. Cell* **147**, 382–395 (2011).
8. P. Sumazin, *et al. Cell* **147**, 370–381 (2011).
9. Y. C. Chiu, T. H. Hsiao, Y. Chen, E. Y. Chuang. *BMC Genomics* **16** Suppl 4, S1 (2015).
10. L. M. Wee, C. F. Flores-Jasso, W. E. Salomon, P. D. Zamore. *Cell* **151**, 1055–1067 (2012).
11. Y. Yuan, *et al. Proc. Natl. Acad. Sci. U.S.A.* **112**, 3158-3163 (2015).
12. Y. Tay, J. Rinn, P. P. Pandolfi. *Nature* **505**, 344-352 (2014).
13. C. Chou, *et al. Nucleic Acids Res*, **44** (D1), D239-D247 (2015).
14. Y. Zhu, P. Qiu, Y. Ji. *Nature Methods* **11**, 599–600 (2014).
15. https://github.com/rptashkin/TCGA_miRNASeq_matrix (2015).
16. A. Guo, *et al. Proc. Natl. Acad. Sci. U.S.A.* **105**, 692-697 (2008).
17. N. Puri, R. Salgia. *J Carcinog.* **7**, 9 (2008).
18. M. Acunzo, *et al. Proc Natl Acad Sci U.S.A.* **110**, 8573-8578 (2013).
19. S. L. Organ, M. S. Tsao. *Ther Adv Med Oncol.* **3**, S7-S19 (2011).
20. J. A. Engelman, *et al. Science* **316,** 1039-1043 (2007).
21. https://clinicaltrials.gov/ct2/show/NCT01024413 (2016).
22. K. Nguyen, S. Kobayashi, D. Costa. *Clinical Lung Cancer*, **10**, 281-289 (2009).
23. http://www.genecards.org/cgi-bin/carddisp.pl?gene=MET (2016).

IMPROVED PERFORMANCE OF GENE SET ANALYSIS ON GENOME-WIDE TRANSCRIPTOMICS DATA WHEN USING GENE ACTIVITY STATE ESTIMATES

THOMAS KAMP

Department of Mathematics, Statistics, and Computer Science, Dordt College
Sioux Center, IA 51250, USA
Email: Thomas.Kamp@dordt.edu

MICAH ADAMS

Department of Mathematics, Statistics, and Computer Science, Dordt College
Sioux Center, IA 51250, USA
Email: Micah.Adams@dordt.edu

CRAIG DISSELKOEN

Department of Mathematics, Statistics, and Computer Science, Dordt College
Sioux Center, IA 51250, USA
Email: Craig.Disselkoen@dordt.edu

NATHAN TINTLE

Department of Mathematics, Statistics, and Computer Science, Dordt College
Sioux Center, IA 51250, USA
Email: Nathan.Tintle@dordt.edu

Gene set analysis methods continue to be a popular and powerful method of evaluating genome-wide transcriptomics data. These approach require *a priori* grouping of genes into biologically meaningful sets, and then conducting downstream analyses at the set (instead of gene) level of analysis. Gene set analysis methods have been shown to yield more powerful statistical conclusions than single-gene analyses due to both reduced multiple testing penalties and potentially larger observed effects due to the aggregation of effects across multiple genes in the set. Traditionally, gene set analysis methods have been applied directly to normalized, log-transformed, transcriptomics data. Recently, efforts have been made to transform transcriptomics data to scales yielding more biologically interpretable results. For example, recently proposed models transform log-transformed transcriptomics data to a confidence metric (ranging between 0 and 100%) that a gene is active (roughly speaking, that the gene product is part of an active cellular mechanism). In this manuscript, we demonstrate, on both real and simulated transcriptomics data, that tests for differential expression between sets of genes using are typically more powerful when using gene activity state estimates as opposed to log-transformed gene expression data. Our analysis suggests further exploration of techniques to transform transcriptomics data to meaningful quantities for improved downstream inference.

1. Introduction

Gene set analysis methods are a popular approach to assessing statistical significance on *a priori*, biologically defined sets of genes, as opposed to on a gene by gene basis [1]. These approaches have now been widely applied to SNP and RNA microarrays, and, more recently, RNA and DNA sequencing. The hope and promise of these methods is a combination of both statistical and biological improvements. Statistically, by analyzing sets of genes, instead of each gene individually, multiple testing penalties can be reduced. Furthermore, by potentially aggregating multiple independent effects (in different genes in the set), the true signal may more easily rise above the 'noise' of other genes in the set. Both reduced multiple testing penalties and aggregated effects have the potential to improve the statistical power of gene set tests. Biologically, by defining gene sets using *a priori* defined sets of genes, there is the increased potential for testing specific and more complex biological hypotheses (e.g., defining a set of genes as all genes in a pathway).

Previously, we discussed application of gene set analysis methods to testing for differential levels of gene expression in a genome-wide transcriptomics setting for bacteria [2]. In particular, we evaluated the performance of novel methods of testing for differential gene expression finding that the novel methods often outperformed, other popular methods, like Fisher's Exact Test (FET) [3]. These novel methods of testing for differential gene expression between two experiments (or bacterial strains) utilize the entire vector of normalized gene expression values for all genes in the set, instead of first defining an arbitrary cutoff (as is the case in FET). By leveraging the entire vector of expression values, instead of suffering from the information loss due to defining an arbitrary cutoff, the methods are generally more powerful than FET.

While gene set analysis typically focus on analyzing 'raw' gene expression data, many current approaches to understanding genome-wide transcriptomics data attempt to further leverage the data by classifying genes into one of two states: *active* (roughly speaking, the gene product is part of an active cellular mechanism) or *inactive* (the cellular mechanism is not active) [4]–[6]. We label this classification a determination of the *gene activity state.* Recently, we published a novel approach, *MultiMM* [7], to address documented deficiencies in many of the current state of the art methods. *MultiMM* is a parametric Bayesian mixture modelling approach which addresses limitations in existing methods as demonstrated through a rigorously grounded statistical framework, better performance than existing methods on simulated and real transcriptomics data, and through improved consistency with well-accepted biological realities and fluxomics data. Full details of, and links to, software for the *MultiMM* method are available elsewhere [7]. Ultimately, the *MultiMM* method yields a confidence estimate, $a_{ij} \in [0,1]$, that gene i is active in condition j. One stated goal of the

MultiMM method is to improve inference in downstream interpretations of gene expression data.

In this manuscript we consider the performance of a variety of gene set analysis methods on both raw gene expression data, as well as on a_{ij} values (confidence estimates that gene i, is active in experiment *j*) in order to determine if a_{ij} values are advantageous for use when conducting gene set analysis.

2. Methods

2.1. Methods of gene set testing

We consider three broad classes of gene set analysis methods [2], [3], [8].

First, we consider the burden test type of gene set testing method, with test statistic defined as:

$$B_m = \left|\sum_{i=1}^{k} e_{ij_1}^{m} - \sum_{i=1}^{k} e_{ij_2}^{m}\right|^{\frac{1}{m}} \quad (1)$$

Where e_{ij} is the expression value of the ith gene measured in the jth condition, m is a positive constant (including infinity), and *k* is the number of genes in the set. As is discussed elsewhere [8], the Burden (B_m) test class of methods of conducting gene set analysis assumes that the effects of the genes within the test will tend to be in the same direction. For example, all genes in the set of interest are either not changing in underlying expression values, or are increasing, but none are decreasing. In the framework of 'activity states' this means that all genes are either moving from inactive to active (across the two experiments being compared) or are in the same state in both experiments. When this assumption is not met, Burden tests tend to be low powered since effects 'cancel out.' As *m* increases, increasing weight is put on the most expressed genes, such that if m=∞, $\sum_{i=1}^{k} e_{ij_1}^{m} = argmax(e_{ij_1})$.

The Variance Components class of test methods was envisioned primarily in response to the fact that Burden tests could not appropriately handle changes in multiple directions within the same set of genes (e.g., some genes move from inactive to active and others from active to inactive when comparing two experiments) [9]. The general form of a Variance Components gene set test statistic, VC_m, is given as:

$$VC_m = \left(\sum_{i=1}^{k} \left|e_{ij_1} - e_{ij_2}\right|^{m}\right)^{\frac{1}{m}}$$

Similar to the behavior for Burden tests, Variance components tests put increasing weight on pairwise differences in expression values as m increases, such that when $m=\infty$, the VC statistic takes the value of the largest observed pairwise difference in expression values.

The third class of tests we considered was Fisher's Exact Test (FET). In this approach, an arbitrary cutoff, *c*, is first chosen, such that if $|e_{ij_1} - e_{ij_2}| > c$, then the gene is coded '1' (changing state; differentially expressed) and otherwise is coded '0' (not changing state; not differentially expressed). The proportion of genes in the set of interest which are deemed to be differentially expressed (>*c*) is compared to the proportion of genes not in the set of interest which are deemed to be differentially expressed using Fisher's Exact test, which uses a hypergeometric distribution to assess statistical significance.

2.2. Implementation of methods of gene set testing

In this manuscript we consider nine different tests, applied to both raw expression data (e_{ij}) and gene activity state estimates (a_{ij}; see next section for details). The nine tests are $B_1, B_2, B_\infty, VC_1, VC_2, VC_\infty, FET(1SD), FET(2SD)$ *and* $FET\ (3SD)$. The test statistic equations for B and VC are given in the previous section, along with a description of the FET approach. For the FET approach, we use 1SD, 2SD and 3SD to denote how determine a cutoff value, *c*. In short, we find the average within gene SD across genes and experiments for which data is available, and then use that value (1SD), 2 times that value (2SD) or 3 times that value (3SD) to determine the cutoffs. For e_{ij} $1SD = 0.75$ and, for a_{ij}, 1SD=0.3. FET determines statistical significance using the hypergeometric distributions. All other tests are evaluated for statistical significance by comparing the observed statistic to a null distribution of 10,000 randomly generated statistics obtained by randomly choosing 10,000 sets of the same size as the gene set being evaluated and finding the fraction of randomly chosen sets with larger statistics than observed (the *p*-value).

2.3. Moving from raw expression values to estimates of gene activity states

The *MultiMM* algorithm takes as input a genome-wide matrix of transcriptomics data ***E*** across numerous experimental conditions, such that the entries in ***E*** are denoted e_{ij} and represent the estimated gene expression of gene *i* in condition *j*. Additionally, if available, *MultiMM* allows for *a priori* identification of sets of genes which are known to be co-regulated such that in the same experimental condition, the co-regulated genes are all active or all inactive. The *MultiMM* algorithm starts by using the Bayesian Information Criterion (*BIC*) to assess the fit of a 1-component (univariate or multivariate) Gaussian mixture distribution (gene is always active or inactive in the set of conditions represented) vs. a 2-component mixture distribution (gene is sometimes active and sometime inactive in the set of conditions represented) using the *R* package *Mclust* [10]. Following Raftery et al. [11] we require the BIC to be at least 12 points better for the 1-component model to be chosen vs. the 2-component model. Second, for all genes estimated to come from a 2-component mixture distribution, a Gaussian mixture model is fit and a Gibbs sampler is used in order to yield estimates of the means and standard deviations of the components of the mixture model, along with an estimate of the proportion of experiments for which the gene is active. In the case of co-regulated sets of genes this mixture model is multivariate, whereas for genes that are not known to be co-regulated with other genes, the mixture model is univariate. Finally, the estimated mixture distribution parameters can be used to yield a confidence estimate, $a_{ij} \in [0,1]$, that gene *i* is active in condition *j*. For genes inferred as being always active or always inactive in the dataset in step one of the algorithm, multiple imputation is used to impute a_{ij} values. Full details of, and links to, software for the *MultiMM* method are available elsewhere [7].

2.4. Simulation of gene expression data

We simulated expression data with 'known' gene activity states (active/inactive). The simulation of expression data was informed by the *E. coli* expression data described later. We first ran the Screening Method described above (BIC with MClust) and dropped all operons (co-regulated gene sets), including single gene operons, for which the two-component model did not yield the highest BIC (n=697 dropped). We then randomly selected 26.3% (=697/2648) of the remaining 1951 operons to be single component in the simulated data, with each of the single component operons having an equal likelihood of being always active or always inactive.

To calculate the mixing parameter, π, used in the simulation for the 1438 two-component operons we chose a random value for π between 0.2 and 0.8. Values for $\vec{\mu}_0, \vec{\mu}_1, \Sigma_0=\Sigma_1$ are all as estimated by the *MultiMM* method computed on the real expression data. To generate simulated expression values, ϵ_{ij}^s, we drew $907(\pi_i)$ random values from a multivariate normal distribution $(\vec{\mu}_{1i}, \Sigma_{1i})$ and $907(1-\pi_i)$ random values from a multivariate normal distribution $(\vec{\mu}_{0i}, \Sigma_{0i})$. Thus, we generated a 907 by 3435 matrix of ϵ_{ij}^s values. Prior analysis has shown this simulated data to have good properties and behave in reasonable ways [7].

2.5. Simulation of gene sets for analysis

We used the simulated gene expression data described above to generate random sets of genes for evaluation of different methods of gene set analysis. We selected random sets of 8, 20 or 40 genes from among genes which were not changing or changing states between the two experiments of interest. In particular, we looked at the following proportions of genes in the set which were not changing state (0, 25, 50, 75 and 100%), and either 0%, 50% or 100% of the genes in the set active in the first experiment. Thus, we explored 45 simulation settings (3 (set size) by 5 (not changing) by 3 (starting state). Of these 45 simulation settings, 9 represent settings for which we can evaluate the empirical type I rate and 36 will be used to evaluate statistical power. Each of the nine test statistics is computed for the set, and then each of the nine statistics is compared to a distribution of the same statistic across 10,000 randomly selected sets of the same size (an approach termed 'gene sampling' which uses a 'competitive null hypothesis'[12]). We considered 1000 randomly selected sets at each of the 45 simulation settings. Full simulation results are available in Supplemental File #1. We also analyzed 574 *a priori* defined operon (co-regulated) sets based on operon definitions for *E. coli* as provided by Microbes Online [13]. Full results are available in Supplemental File #2. Supplemental Files are available at: http://homepages.dordt.edu/ntintle/gsa_supp.zip

2.6. Real data

We also used genome-wide gene expression data from 907 different microarray data sets collected on 4329 *Escherichia coli* genes via the M3D data repository [14]–[16] both to inform simulated data analysis and when considering the actual performance of the methods. Raw data from Affymetrix [17] CEL files were normalized using RMA [18]. Details of data processing are described elsewhere [19], [20].

2.7. Statistical analysis

Empirical power and type I error rate estimates are computed as the proportion of times that the p-value was less than the significance level for a particular test and simulation setting. We considered significance levels of 5%, 0.5% and 0.05%.

Results

Across 36 simulation settings where at least one gene in the set changed activity states, power was consistently better when using gene activity state estimates, than raw expression data (see Table 1 for overall summary). Across the 9 simulation settings where none of the genes in the set changed state (type I error setting), the Type I error rate was generally controlled for all methods (detailed results not shown). Table 1 shows that gains in power can be high across all methods, whereas when power is worse when using activity states, the reduction in power is usually quite minimal (19 to 82 average percentage point increase vs. 0.3 to 2.3 average percentage point decrease).

Table 1. Power improvements comparing raw expression data to gene activity state estimates using a variety of gene set analysis approaches

Gene set analysis approach		Proportion of 36 simulation settings where power is better using a_{ij}	Average (SD) power gain when power is better using a_{ij}[1]	Proportion of 36 simulation settings where power is the same using a_{ij}	Proportion of 36 simulation settings where power is worse using a_{ij}	Average (SD) power loss when power is worse using a_{ij}[2]
Fisher's exact test	Cutoff=3SD	73.1%	24.9% (21.8%)	17.6%	9.3%	0.3% (0.2%)
	Cutoff=2SD	63.9%	28.4% (20.5%)	16.7%	19.4%	0.7% (0.9%)
	Cutoff=1SD	66.7%	25.2% (21.0%	16.7%	16.7%	0.9% (0.8%)
Burden	m=1	48.1%	19.1% (16.4%)	29.6%	22.2%	1.4% (1.1%)
	m=2	46.3%	22.1% (17.7%)	25.0%	28.7%	1.2% (1.3%)
	m=∞	55.6%	39.0% (29.2%)	10.2%	34.3%	2.3% (2.9%)
Variance components	m=1	61.1%	28.9% (20.6%)	19.4%	19.4%	0.8% (0.6%)
	m=2	64.8%	40.4% (28.0%)	10.2%	25.0%	1.0% (1.1%)
	m=∞	100%	82.2% (17.4%)	0	0	-

1. In situations when the power is better using a_{ij} vs. $e_{ij,}$ what is the difference in powe$_r$ estimates between the two different methods. For example, for VC_∞ the difference power between using a_{ij} and e_{ij} averaged 82.2% percentage points, reflecting the fact that VC_∞ is substantially better when using a_{ij}
2. In situations when the power is worse using a_{ij} vs. $e_{ij,}$ what is the difference in powe$_r$ estimates between the two different methods. For example, for B_∞ the difference power between using a_{ij} and e_{ij} averaged 2.3% percentage points, reflecting the fact that B_∞ is not much worse using a_{ij} and e_{ij} in the 34.3% of cases when it is worse

For each of the thirty-six simulation settings used to estimate power, the power was always highest across all 18 methods (nine different test statistics using either e_{ij} or a_{ij}) for a method using gene activity state estimates. This was true for each of the 3 different significant levels. VC_∞ was frequently the most powerful approach (16 out of 36 times for significance level 5%; 26 out of 36 times for significance level 0.5% and 33 times for significance level 0.05%). While other B and VC methods were periodically most powerful, notably, the FET methods were never the most powerful, even when using gene activity state estimates (a_{ij}).

Figure 1 illustrates typical performance of the VC methods as the proportion of genes in the set changes, by highlighting the performance of the methods on sets of size 8. VC_∞ is most robust to lower proportions of genes in the set changing state, while all methods perform well when the proportion of genes in the set changing state is relatively large.

Figure 1. Power of different VC tests as the proportion of genes in the set changing state varies

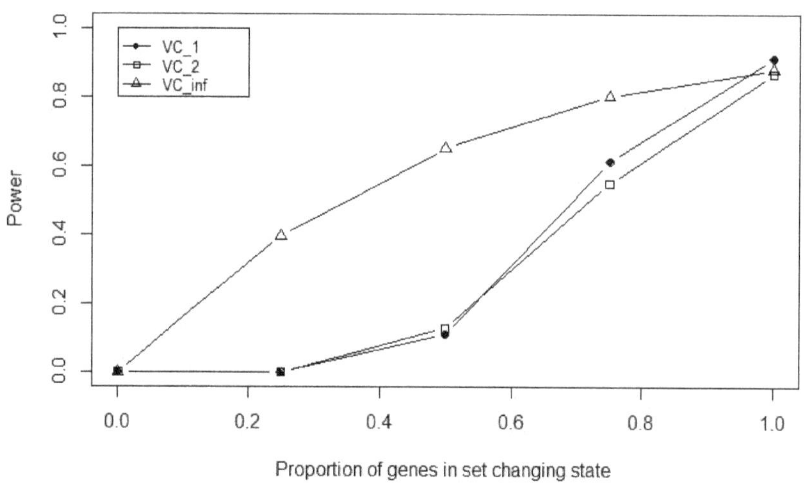

Analysis of the 574 real, operon based sets of genes showed similar performance to the randomly generated gene sets, with even better performance of the activity state informed methods in many cases (detailed results not shown).

Real data example

The L-arabinose (*ara*) operon is a well-studied set of three co-located genes (*araB, araA, araD*) which encode enzymes needed for the catabolism of arabinose in *E. coli* [52]. Across the 907 experiments in our dataset, L-arabinose is present in the media in 227 cases. We randomly selected 1000 pairs of experiments where one experiment had L-arabinose present in the media and one experiment did not. We then computed different gene set analysis test statistics for the L-arabinose operon using both raw expression data and activity state estimates, as compared to 100,000 randomly selected sets of 3 genes. Table 2 illustrates that methods using activity state estimates were always more powerful than methods which were based on raw expression values.

Table 2. Empirical power estimates for detecting significant changes in activity for the L-arabinose operon in *E. coli* when comparing an experiment with L-arabinose present in the media vs. one without

Sig. Level	Method	B_1	B_2	B_∞	VC_1	VC_2	VC_∞
0.05%	Raw expression (e_{ij})	96.6%	98.1%	1.6%	95.7%	52.3%	1.7%
	Activity state estimates (a_{ij})	100%	100%	99.6%	100%	100%	99.6%
0.005%	Raw expression (e_{ij})	85.3%	86.1%	0%	58.0%	3.9%	0%
	Activity state estimates (a_{ij})	99.6%	99.6%	99.6%	99.6%	99.6%	99.6%

4. Discussion

Gene set analysis remains a statistically promising and biological relevant approach to the analysis of genome-wide transcriptomics data. Here we demonstrate that, in line with previous work [2], methods which don't arbitrarily introduce a cutoff and lose information, are generally more powerful than methods that do (e.g., Fisher's exact test). We also demonstrate that using a more statistically grounded metric to quantify gene expression (activity state estimates, a_{ij}) generally leads to more powerful tests than using raw gene expression data (e_{ij}) on simulated data, with promising results also observed on real data in well-understood biological systems.

We note that the VC_∞ method performed particularly well, especially at low significance thresholds. This finding reflects the use of gene-sampling (a competitive null hypothesis). Briefly, when using gene sampling to assess statistical significance, test statistics generated for the gene set of interest, are compared to randomly chosen gene sets. The VC_∞ method performs relatively better as compared to other methods as the significance level decreases because it is focused on the most extreme observed difference in activity state estimates and, thus, is more robust than other methods to small numbers of randomly selected sets of genes with extreme values of the test statistic. This performance was particularly notable in the example with the L-arabinose operon, where the VC_∞ method using activity state estimates (a_{ij}) outperformed its performance on raw expression values (e_{ij}) by nearly 100%. While other test statistics did not show as large of a difference, in all cases the power was higher when using activity state estimates. Thus, when attempting to determine if sets of genes are differentially active in two conditions, inferring gene activity state estimates prior to applying gene set analysis methods will maximize the likelihood of identifying differential activity. In short, use of these methods will maximize our ability to identify sets of genes associated with differential activity between two conditions.

We note numerous opportunities for future work, including (1) the ability to expand these methods to incorporate information from multiple, similar experimental conditions, instead of only comparing two conditions, (2) integrating directionality and/or gene set topology, (3) potential improvements by further leveraging the statistical properties of well-calibrated a_{ij} (the posterior likelihood that gene *i* is active in gene *j*), (4) potential further improvements in power by using non-competitive null hypotheses, which may be possible through statistical quantification of the null distributions of particular methods when using well-calibrated a_{ij}'s and (5) use of this general framework to test for whether a set of genes in a single experiment shows evidence of significant 'activity' (vs. only a change in activity levels between two experiments, as we considered here).

The most notable limitation of our analysis here is the limited application to real data, though initial results are promising and performance on real (operon-based sets) was also quite encouraging. Further work is necessary to ensure transferability of these promising initial findings to additional organisms. For example, to determine if these methods will successfully distinguish sets of differentially active genes between diseased and non-diseased tissue. Furthermore, further work is necessary to explore validation in other well-understood biological systems and as compared to the results of other –omics data (e.g., genome-scale metabolic models; fluxomics, etc.).

Acknowledgments

This work is supported by NSF MCB-1330734. We gratefully acknowledge the use of the Silicon Mechanics grant funded beaker computer cluster on the campus of Dordt College for computations.

References

[1] C. de Leeuw, B. M. Neale, T. Heskes, and D. Posthuma, "The statistical properties of gene-set analysis," *Nat. Rev. Genet.*, vol. 17, pp. 353–364, 2016.

[2] N. L. Tintle, A. A. Best, M. DeJongh, D. Van Bruggen, F. Heffron, S. Porwollik, and R. C. Taylor, "Gene set analyses for interpreting microarray experiments on prokaryotic organisms.," *BMC Bioinformatics*, vol. 9, no. 1, p. 469, 2008.

[3] P. Khatri and S. Drăghici, "Ontological analysis of gene expression data: Current tools, limitations, and open problems," *Bioinformatics*, vol. 21, no. 18, pp. 3587–3595, 2005.

[4] S. Abel, T. Bucher, M. Nicollier, I. Hug, V. Kaever, P. Abel zur Wiesch, and U. Jenal, "Bi-modal Distribution of the Second Messenger c-di-GMP Controls Cell Fate and Asymmetry during the Caulobacter Cell Cycle," *PLoS Genet.*, vol. 9, no. 9, p. e1003744, 2013.

[5] C. A. Gallo, R. L. Cecchini, J. A. Carballido, S. Micheletto, and I. Ponzoni, "Discretization of gene expression data revised," *Brief. Bioinform.*, no. May, pp. 1–13, 2015.

[6] J. E. Ferrell, "Self-perpetuating states in signal transduction: positive feedback, double-negative feedback and bistability," *Curr. Opin. Cell Biol.*, vol. 14, no. 2, pp. 140–148, 2002.

[7] C. Disselkoen, B. Greco, K. Cook, K. Koch, R. Lerebours, C. Viss, J. Cape, E. Held, Y. Ashenafi, K. Fischer, A. Acosta, M. Cunningham, A. A. Best, M. DeJongh, and N. L. Tintle, "A Bayesian framework for the classification of microbial gene activity states," *Front. Microbiol.*, vol. 7, no. 1191, 2016.

[8] K. Liu, S. Fast, M. Zawistowski, and N. L. Tintle, "A geometric framework for evaluating rare variant tests of association," *Genet. Epidemiolgoy*, vol. 37, no. 4, pp. 712–722, 2013.

[9] M. C. Wu, S. Lee, T. Cai, Y. Li, M. Boehnke, and X. Lin, "Rare-variant association testing for sequencing data with the sequence kernel association test.," *Am. J. Hum. Genet.*, vol. 89, no. 1, pp. 82–93, Jul. 2011.

[10] C. Fraley, A. Raftery, L. Scurcca, T. B. Murphy, and M. Fop, "mclust: Normal mixutre modelling for model-based clustering, classification and density estimation," *CRAN*, 2015. [Online]. Available: https://cran.r-project.org/web/packages/mclust/index.html.

[11] A. Raftery, "Bayesian model selection in social research," *Sociol. Methods*, vol. 25, pp. 111–163, 1995.

[12] J. Goeman and P. Buhlmann, "Analyzing gene expression data in terms of gene sets:

methodological issues," *Bioinformatics*, vol. 23, no. 8, pp. 980–987, 2007.

[13] M. N. Price, K. H. Huang, E. J. Alm, and A. P. Arkin, "A novel method for accurate operon predictions in all sequenced prokaryotes.," *Nucleic Acids Res.*, vol. 33, no. 3, pp. 880–92, Jan. 2005.

[14] "Many Microbes Database." [Online]. Available: http://m3d.mssm.edu.

[15] J. J. Faith, M. E. Driscoll, V. a Fusaro, E. J. Cosgrove, B. Hayete, F. S. Juhn, S. J. Schneider, and T. S. Gardner, "Many Microbe Microarrays Database: uniformly normalized Affymetrix compendia with structured experimental metadata.," *Nucleic Acids Res.*, vol. 36, no. Database issue, pp. D866–70, 2008.

[16] J. J. Faith, B. Hayete, J. T. Thaden, I. Mogno, J. Wierzbowski, G. Cottarel, S. Kasif, J. J. Collins, and T. S. Gardner, "Large-scale mapping and validation of Escherichia coli transcriptional regulation from a compendium of expression profiles.," *PLoS Biol.*, vol. 5, no. 1, p. e8, Jan. 2007.

[17] "Affymetrix." [Online]. Available: http://www.affymetrix.com.

[18] T. Irizarry, R. a., Bolstad, Benjamin, Collin, Francois, Cope, Leslie, Hobbs, Bridget, Speed, "Summaries of Affymetrix GeneChip probe level data," *Nucleic Acids Res.*, vol. 31, no. 4, p. 15e–15, Feb. 2003.

[19] S. Powers, M. DeJongh, A. a Best, and N. L. Tintle, "Cautions about the reliability of pairwise gene correlations based on expression data.," *Front. Microbiol.*, vol. 6, no. June, p. 650, Jan. 2015.

[20] N. Tintle, A. Sitarik, B. Boerema, K. Young, A. Best, and M. De Jongh, "Evaluating the consistency of gene sets used in the analysis of bacterial gene expression data.," *BMC Bioinformatics*, vol. 13, no. 1, p. 193, Jan. 2012.

methylDMV: SIMULTANEOUS DETECTION OF DIFFERENTIAL DNA METHYLATION AND VARIABILITY WITH CONFOUNDER ADJUSTMENT

PEI FEN KUAN*, JUNYAN SONG and SHUYAO HE
Department of Applied Mathematics and Statistics,
Stony Brook University,
Stony Brook, NY 11794, USA
**E-mail: peifen.kuan@stonybrook.edu*
http://www.stonybrook.edu/commcms/ams2/

DNA methylation has emerged as promising epigenetic markers for disease diagnosis. Both the differential mean (DM) and differential variability (DV) in methylation have been shown to contribute to transcriptional aberration and disease pathogenesis. The presence of confounding factors in large scale EWAS may affect the methylation values and hamper accurate marker discovery. In this paper, we propose a flexible framework called `methylDMV` which allows for confounding factors adjustment and enables simultaneous characterization and identification of CpGs exhibiting DM only, DV only and both DM and DV. The proposed framework also allows for prioritization and selection of candidate features to be included in the prediction algorithm. We illustrate the utility of `methylDMV` in several TCGA datasets. An R package `methylDMV` implementing our proposed method is available at `http://www.ams.sunysb.edu/~pfkuan/softwares.html#methylDMV`.

Keywords: DNA methylation; Differential variability; Feature selection; Elastic net.

1. Introduction

DNA methylation is an important hallmark of genomic imprinting, transcriptional regulation, X-inactivation and chromosomal stability.[1] The most common DNA methylation process in human involves the addition of a methyl group to the 5-carbon of the cytosine ring. In human, this modification mostly occurs at a CpG site in which a cytosine nucleotide is followed by a guanine nucleotide. Aberrant patterns of DNA methylation have been shown to be a critical mechanism in the development and progression of various diseases, in particular cancer.[2] DNA methylation is one of the most widely studied epigenetics event and has been profiled extensively in large consortiums including the Cancer Genome Altas (TCGA), NIH Roadmap and the Encyclopedia of DNA Elements (ENCODE) projects. These efforts provide research opportunities for secondary analyses of the large datasets to further understand the biology of the disease.

Most of the work in DNA methylation have been focused on identifying DNA methylation markers that exhibit differential average or mean methylation (DM).[3,4] These epigenetic markers have been shown to be promising biomarkers in designing platform for disease diagnosis.[5] Over the last few years, there has been an increasing interest in identifying DNA methylation markers that exhibit differential variability in various diseases, including cancer[6–8] and obesity.[9] These epigenetic variabilities can be attributed to increased plasticity arising from changing environment including varying oxygen tension[10] and is associated with the risk of morphological and neoplastic transformation.[11] These studies opened up new avenues to the study of DNA methylation, which indicated that simultanenous investigation of both differential mean and variability may delineate the complex patterns of epigenetic regulation in pathophysiology and development of diseases.

One of the most widely used DNA methylation platforms is the Illumina Infinium HumanMethylation450 BeadChip which profiles more than 450,000 CpGs genome wide. The latest phase of the Illumina methylation array is the MethylationEPIC BeadChip which covers approximately 850,000 methylation sites including CpG islands, enhancers and regulatory regions identified from the ENCODE project. The methylation value for each CpG is represented as a *beta* (β) value, which is the ratio of methylated probe intensities to the total probe intensities, where $0 \leq \beta \leq 1$; $\beta = 0$ and $\beta = 1$ indicate that the CpG is fully unmethylated and methylated, respectively.

An important aspect of differential methylation analysis is to identify CpGs which exhibit differential mean or variance in large scale hypothesis testing. Statistical tests for detecting CpGs which exhibit differential mean methylation include t-tests, non-parametric Wilcoxon rank sum test or limma[12] based on linear models and empirical Bayes approach. On the other hand, several algorithms have been proposed in recent years to identify CpGs which exhibit differential variability in large scale hypothesis testing. For instance, Teschendorff et al. (2012)[8] proposed a regularized version of the Bartlett's test, Ahn et al. (2013)[13] used a score test from generalized regression model, Phipson et al. (2014)[14] proposed a modification of Levene's test, Wahl et al. (2014)[15] introduced a generalized additive models for location, scale and shape (GAMLSS) framework and Kuan (2014)[16] proposed a general linear model with propensity score method for detecting CpGs with differential variability.

CpGs which exhibit differential mean methylation have been utilized in classification algorithm to define methylation signatures for disease subtypes.[17,18] As the methylation arrays encompass $> 450,0000$ CpGs, a common approach in training the classification algorithm is to pre-select features ranked highly by the univariate differential mean methylation as candidate CpGs in the classification algorithm to improve the stability of the algorithm. Motivated by the biological insights of differential variability in methylation, Teschendorff et al. (2012)[8] proposed a method which selected differential variable CpGs using Bartlett's test for inclusion in the prediction algorithm.

Large scale differential methylation analysis requires proper adjustment for confounders to reduce the biases associated with the identified methylation markers. For instance, age[19,20] and cigarette smoking[21,22] have been shown to be associated with DNA methylation; thus in studies to identify methylation markers for cancer or other disease phenotypes, appropriate adjustment for these factors is necessary. In the analysis of differential mean methylation, this can be achieved via a regression framework where confounders are included as covariates in the model. However, in the analysis of differential variability, potential biases due to confounding variables are usually ignored.[8,14]

This paper aims to develop a unified framework to address the limitation of existing work: (1) incorporates adjustment for confounding variables that potentially affect methylation levels, and allows for simultaneous detection of differential mean (DM) and differential variability (DV) in methylation analysis, (2) systematic selection of CpGs which exhibit differential mean and/or differential variability in the prediction algorithm to improve prediction accuracy and biological interpretation. In Section 2, we describe our proposed approach. This is followed by simulation studies and real data applications in Sections 3 and 4, respectively. The paper concludes with a discussion in Section 5.

2. Methods

2.1. *A framework for simultaneous detection of differential mean (DM) and differential variability (DV)*

Without loss of generality, we describe our proposed framework for detecting differential mean and differential variability between two conditions or groups (e.g., tumor versus normal). A common distribution to model the *beta* values from Illumina methylation arrays is the beta distribution.[23] Since the variance of a beta distribution is a function of the mean, the β values exhibit significant heteroscedasticity.[24] To overcome the heteroscedasticity issue, we consider a variance stabilizing transformation via the logit function to the β values, i.e., $logit(\beta) = \log[\beta/(1-\beta)]$. Let x_{ij} denote the logit transformed methylation value for sample i and CpG j. We first define a deviation measure $r_{ij} = |x_{ij} - \text{wt.med}_i(x_{ij})|$ where $\text{wt.med}_i(x_{ij})$ is the weighted median of CpG j with weights $w_i = 1/2n_{g_i}$, $g_i = 0$ if sample i is a control and $g_i = 1$ if sample i is a case, and n_0 and n_1 are the respective sample sizes.

We recast the model for simultaneous detection of differential mean and differential variable CpGs using a logistic regression model. Let y_i denote the group membership of sample i, where $y_i = 0$ if the sample is a control/normal and $y_i = 1$ if the sample is a case/tumor. y_i is assumed to follow a binomial distribution with $P(y_i = 1) = \pi_i$ and $\log[\pi_i/(1-\pi_i)] = \theta_i$. We consider the four competing models for each CpG:

Model 1: $\theta_i = \beta_0 + \sum_{k=1}^{K} \gamma_k Z_{ik}$ (no DM or DV)
Model 2: $\theta_i = \beta_0 + \beta_m x_{ij} + \sum_{k=1}^{K} \gamma_k Z_{ik}$ (DM only)
Model 3: $\theta_i = \beta_0 + \beta_v r_{ij} + \sum_{k=1}^{K} \gamma_k Z_{ik}$ (DV only)
Model 4: $\theta_i = \beta_0 + \beta_m x_{ij} + \beta_v r_{ij} + \sum_{k=1}^{K} \gamma_k Z_{ik}$ (both DM and DV)

In all models, $\mathbf{Z}_k = (Z_{ik})'$ corresponds to confounding variable k, for instance age, smoking status or alcohol consumption. Model 1 is the baseline model which adjusts for confounding variables and assumes that the phenotype is not associated with differential mean (DM) or differential variability (DV). Model 2 (Model 3) assumes that the phenotype is associated with DM (DV) after adjusting for confounders, whereas Model 4 assumes that the phenotype is associated with both DM and DV for a CpG. To identify CpGs which exhibit DM, one can compare Model 1 to Model 2 using likelihood ratio tests or score tests.[25] On the other hand, Model 3 can be compared to Model 1 to obtain p-values associated with DV for each CpG. The comparison of Model 4 and Model 1 identifies CpGs which exhibit either DM or DV. The vector of p-values from each analysis are adjusted via the false discovery rate (FDR)[26] to account for multiple testings. In addition to large scale hypothesis testing framework to identify DM and DV CpGs, another advantage of our proposed model is that it allows for automatic classification of the CpGs into the four classes (1) no DM or DV, (2) DM only, (3) DV only and (4) both DM and DV. This is carried out via a Bayesian Information Criterion

(BIC) to rank the four models for each CpG, i.e., the CpG is categorized into the class with the smallest BIC score.

2.2. *Candidate feature selection for prediction modeling*

The BIC used for model ranking within each CpG can also be utilized to aid candidate feature selection to improve the stability of the prediction algorithm. The proposed framework provides flexibility to the user for including top ranking features in constructing prediction model. For instance, if the user is interested in a prediction model using CpGs which exhibit the largest discriminative power in terms of both DV and DM after adjustment for confounding variables, then the subset of CpGs which show the lowest BIC scores for Model 4 are selected as candidate features. On the other hand, if the user is interested in a prediction model using only DM CpGs , then the candidate features correspond to the CpGs which identify Model 2 as the best model using BIC scores.

The selected candidate features are used in the prediction algorithm for constructing classification rule discriminating case from control. In this paper, we consider the elastic net algorithm.[27] The objective function of elastic net consists of a loss function + penalty:

$$\min_{\beta} ||\mathbf{y} - \mathbf{X}\beta||^2 + \lambda \left\{\alpha||\beta||_1 + (1-\alpha)||\beta||^2\right\}$$

where $||\beta||_1 = \sum_{j=1}^{p} |\beta_j|$ and $||\beta||^2 = \sum_{j=1}^{p} \beta_j^2$. The parameters λ and α are tuned via cross-validation. Other types of machine learning prediction algorithm can also be used on the selected candidate features, for instance the random forest[28] which is a non-parametric ensemble approach based on a large number of classification trees trained on bootstrap samples.

An R package `methylDMV` implementing our proposed method for testing DM and DV, as well as CpGs ranking by BIC and candidate feature selection is available at `http://www.ams.sunysb.edu/~pfkuan/softwares.html#methylDMV`.

3. Simulation studies

We carried out simulation studies to evaluate the effect of confounders on CpG ranking. Specifically, denote Z_{i1} and Z_{i2} as the two confounders, where $Z_{i1} \sim N(0,1)$ and $Z_{i2} \sim \text{Bernoulli}(0.6)$ for sample i, $i = 1, 2, \ldots, n$. The group indicator y_i was generated from the following model

$$\text{logit}(p_i) = \gamma_0 + \gamma_1 Z_{1i} + \gamma_2 Z_{2i}$$
$$y_i \sim \text{Bernoulli}(p_i)$$

For each CpG j ($j = 1, 2, \ldots, p$), the measurements x_{ij}'s were generated from the Gaussian distribution under the assumption that the beta values have been properly transformed (e.g., logit or arcsine transformation), i.e., $x_{ij} \sim N(\mu_{ij}, \sigma_{ij}^2)$ where

(i) $\mu_{ij} = \mu_0 + \alpha_1 Z_{i1} + \alpha_2 Z_{i2}$ and $\sigma_{ij}^2 = \sigma_0^2$ if CpG j is from Model 1 (no DM or DV)
(ii) $\mu_{ij} = \mu_0 + \alpha_g y_i + \alpha_1 Z_{i1} + \alpha_2 Z_{i2}$ and $\sigma_{ij}^2 = \sigma_0^2$ if CpG j is from Model 2 (DM only)
(iii) $\mu_{ij} = \mu_0 + \alpha_1 Z_{i1} + \alpha_2 Z_{i2}$ and $\sigma_{ij}^2 = \sigma_0^2 + \beta_g y_i$ if CpG j is from Model 3 (DV only)
(iv) $\mu_{ij} = \mu_0 + \alpha_g y_i + \alpha_1 Z_{i1} + \alpha_2 Z_{i2}$ and $\sigma_{ij}^2 = \sigma_0^2 + \beta_g y_i$ if CpG j is from Model 4 (both DM and DV)

The proportion of CpGs from Models 1-4 were drawn from a multinomial distribution with $\pi = (\pi_1, \frac{1-\pi_1}{3}, \frac{1-\pi_1}{3}, \frac{1-\pi_1}{3})$. We set $\gamma_0 = 1, \gamma_1 = 2, \gamma_2 = -2$ to obtain approximately equal number of cases and controls; and $\alpha_g = 1, \beta_g = 1, \mu_0 = 0, \sigma_0^2 = 1$. We varied $\alpha_1 = \alpha_2 = 0, 0.5, 1, 3, 5$ to reflect the different degrees of confounding in the methylation measurements and $\pi_1 = 0.4, 0.6, 0.8$ for the different mixing proportions of DM and DV CpGs. To evaluate the effect of confounders on the phenotype, i.e., case/control, we also considered the case in which the y_i's were not affected by confounders. Under this scenario, $y_i = 0$ for $i = 1, 2, \ldots, n/2$ and $y_i = 1$ for $i = n/2 + 1, \ldots, n$. For each scenario, the simulation was conducted for $n = 200$ samples and $p = 10000$ CpGs over 100 iterations.

We compared the average accuracy of the BIC ranking procedure in classifying the CpGs into Models 1-4 with (BICadj) and without (BICnoadj) adjustment for confounders. We also included comparison to method which performed tests for DM and DV separately. Two sample t-test and Levene's test were used to identify DM and DV CpGs, respectively. CpG j was classified as DM (DV) if the p-value from t-test (Levene's test) adjusted via the Benjamini-Hochberg procedure[26] $\leq$ FDR. We considered FDR 0.05 and 0.1, and referred to this method as SepTest0.05 and SepTest0.1, respectively.

Figure 1 summarizes the average accuracy for the four methods across the different settings. In scenarios where both the phenotype (case/control status) and methylation measurements were affected by confounders (top row of Figure 1 for $\alpha_1 \neq 0$), the methods which did not adjust for confounders exhibited poor accuracy across different mixing proportions π_1. For the case where $\alpha_1 = 0$, i.e., methylation measurements were not affected by confounders, the BICadj method showed a slight decrease in accuracy compared to other methods. Bottom row of Figure 1 displays the results for the scenarios where only the methylation measurements were confounded while the phenotype was not affected by confounders. For these cases, the performance of the methods were comparable for $\alpha_1 \leq 1$. The advantages of adjusting for confounders were apparent for $\alpha_1 = 3, 5$, i.e., strong confounding effect in the methylation measurements even in the absence of confounding in case/control status.

4. Case studies

4.1. *Data preprocessing and normalization*

We illustrated our proposed method, `methylDMV` on three datasets, namely the breast cancer (BRCA), kidney cancer (KIRC) and liver cancer (LIHC) dataset. The breast cancer dataset consisted of 909 samples downloaded from the TCGA data portal and the NCBI gene expression omnibus under accession number GSE67919, whereas the kidney and liver cancer consisted of 475 and 404 samples from the TCGA data portal, respectively. All the samples were profiled using the Illumina Infinium HumanMethylation450 BeadChip.

Preprocessing of the methylation data at the 485,557 CpGs were performed as follows. Probes with detection p-value > 0.05 were set to missing and probes with more than 20% missing were filtered. A beta mixture quantile (BMIQ) normalization[29] was applied to the beta values for correction of bias due to the type I and type II probes. Non-specific, cross-hybridized probes,[30,31] probes overlapping with a SNP and probes mapping to repeat regions were filtered. For KIRC and LIHC, we further filtered for CpGs mapping to chromosomes X

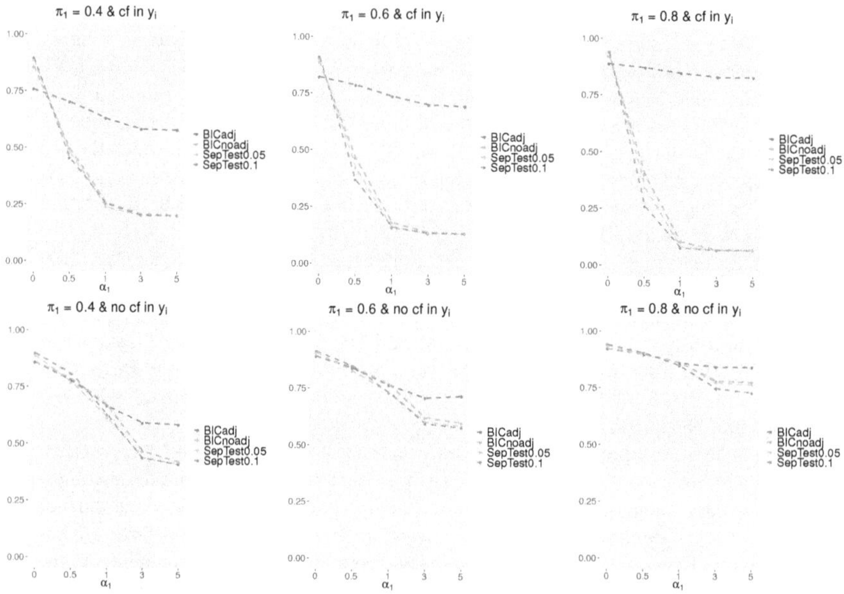

Fig. 1. Average accuracy of CpG classification across α_1's for our proposed BIC ranking with confounding adjustment (BICadj, orange), BIC ranking without confounding adjustment (BICnoadj, green), separate two-sample t-test and Levene's test for DM and DV at FDR 0.05 (SepTest0.05, turquoise) and 0.1 (SepTest0.1, purple). Each panel corresponds to a specific π_1 value and whether the case control status was affected by confounders (top row: $y_i \sim$ Bernoulli(p_i), i.e, affected by confounders; bottom row $y_i = 0, i = 1, \ldots n/2$, $y_i = 1, i = n/2+1, \ldots n$, i.e, not affected by confounders).

and Y. The normalized datasets consisted of 374,680, 365,896 and 365,658 CpGs for BRCA, KIRC and LIHC, respectively. We performed the following pairwise comparisons:

(i) **KIRC (tumor vs normal)**: Models 1-4 were fitted on $n_0 = 156$ normal (control) and $n_1 = 319$ tumor (case), adjusting for age and race.

(ii) **LIHC (tumor vs normal)**: Models 1-4 were fitted on $n_0 = 47$ normal (control) and $n_1 = 357$ tumor (case), adjusting for age and race.

(iii) **BRCA (tumor vs normal)**: Models 1-4 were fitted on $n_0 = 180$ normal (control) and $n_1 = 729$ tumor (case), adjusting for age and race.

(iv) **BRCA (basal vs luminal A)**: Models 1-4 were fitted on $n_0 = 93$ luminal A (control) and $n_1 = 30$ basal (case), adjusting for age and race.

(v) **BRCA (basal vs luminal B)**: Models 1-4 were fitted on $n_0 = 40$ luminal B (control) and $n_1 = 30$ basal (case), adjusting for age and race.

(vi) **BRCA (luminal B vs luminal A)**: Models 1-4 were fitted on $n_0 = 93$ luminal A (control) and $n_1 = 40$ luminal B (case), adjusting for age and race.

4.2. *Feature ranking by BIC scores*

In tumor versus normal comparison within KIRC, LIHC and BRCA datasets, majority of the CpGs were showing either DM or DV or both as shown in Table 1. A large number of CpGs ranked Model 4 (DM and DV) as the best model which indicated that both differential mean and differential variability play important role in distinguishing tumor from normal. In KIRC and BRCA, CpGs showing DM only (Model 2) were enriched in CpG islands, first exons, 200 bp upstream of the transcription start sites (TSS200); whereas CpGs showing DV only (Model 3) were enriched in CpG shores and gene body as shown in Figures 2 and 3. In LIHC, the proportions of DM and DV CpGs mapping to CpG islands were fairly similar, whereas the proportion of DM CpGs mapping to gene body was higher compared to DV CpGs. On the other hand, the subtypes comparison within BRCA identified fewer number of CpGs exhibiting DM or DV. In basal versus luminal A or luminal B comparisons, the proportions of DV CpGs mapping to CpG island and TSS200 were higher than DM CpGs.

Among the lists of DM only CpGs (Model 2) identified by tumor versus normal comparison within KIRC, LIHC and BRCA datasets, 4814 CpGs were in common. On the other hand, there were 1223 and 46885 common CpGs in DV only (Model 3) and both DV and DM (DM&DV) (Model 4) categories, respectively. DAVID (`https://david-d.ncifcrf.gov/home.jsp`) functional annotation enrichment analysis was performed on the genes of mapping to each of the top 1000 common DM only CpGs, DV only CpGs and DM&DV CpGs to identify enriched canonical pathways and biological process ontologies. At FDR ≤ 0.05, enriched canonical pathways for DM only CpGs include Rho GTPase cycle, Rap1 signaling pathway and NRAGE signals death through JNK; whereas DM&DV CpGs identified olfactory transduction and signaling pathway among the top enriched pathways. On the other hand, DM only CpGs, DV only CpGs and DM&DV CpGs identified processes related to GTPase regulation, regulation of transcription from RNA polymerase II promoter and regulation of ion transmembrane transport, respectively.

Table 1. Number of CpGs identified for each model based on BIC scores for the different datasets and comparisons.

Data	Model 1	Model 2	Model 3	Model 4
KIRC: tumor vs normal	18685	94948	44291	207972
LIHC: tumor vs normal	85769	52315	83296	144278
BRCA: tumor vs normal	33735	104575	43880	192490
BRCA: basal vs luminal A	201378	131085	23193	19024
BRCA: basal vs luminal B	198192	124764	31393	20331
BRCA: luminal B vs luminal A	290963	47145	31327	5245

4.3. *Elastic net predictive modeling*

The elastic net algorithm[27] was applied to each dataset for constructing a prediction model differentiating case from control. We randomly split the dataset into 80% training and 20%

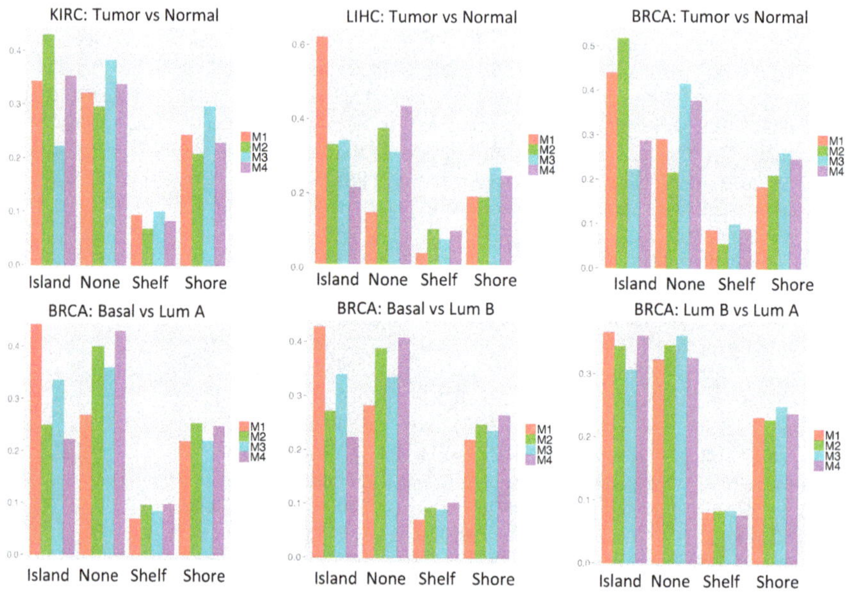

Fig. 2. CpG island, shelf and shore annotation for the proportion of CpGs identified by each model (color code: orange (Model 1), green (Model 2), turquoise (Model 3), purple (Model 4)) for the different datasets and comparisons.

test set. The parameters λ and α were tuned using 10 fold cross-validation on the training set. The random partitioning of data into training and test set was repeated 10 times. We compared the following methods for selecting top 2000 CpGs from the training set to be included as candidate features:

(i) **Set 1**: Logit transformed beta values x_{ij} of the top 2000 CpGs among the CpGs which ranked model 2 as the best model.
(ii) **Set 2**: Absolute deviation measure r_{ij} of the top 2000 CpGs among the CpGs which ranked model 3 as the best model.
(iii) **Set 3**: Both the logit transformed beta values x_{ij} and absolute deviation measure r_{ij} of the top 2000 CpGs among the CpGs which ranked model 4 as the best model.

We evaluated the performance of the prediction algorithm on the test set in terms of area under the receiver operating characteristics curve (AUC), accuracy (Acc)= $\frac{TP+TN}{n_0+n_1}$, sensitivity (Sn)= $\frac{TP}{TP+FN}$, specificity (Sp)= $\frac{TN}{TN+FP}$ and Matthew's correlation coefficient (Mcc)= $\frac{TP\times TN-FP\times FN}{\sqrt{(TP+FP)(TP+FN)(TN+FP)(TN+FN)}}$, averaged over the 10 iterations. The results are presented in Table 2. The prediction model for predicting tumor from normal in KIRC, LIHC

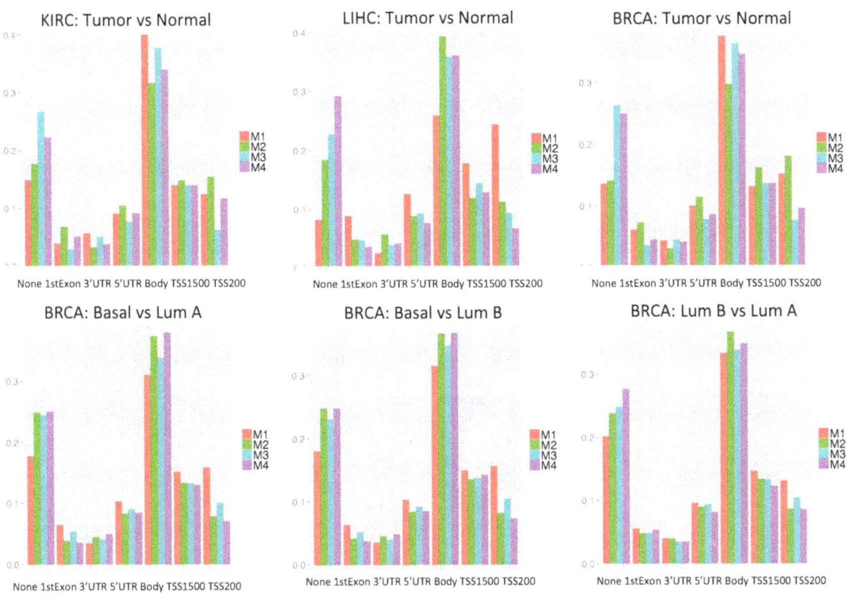

Fig. 3. Gene annotation for the proportion of CpGs identified by each model (color code: orange (Model 1), green (Model 2), turquoise (Model 3), purple (Model 4)) for the different datasets and comparisons.

and BRCA had high accuracy and AUC, and were comparable across the different candidate feature sets. Similar patterns were observed in basal versus luminal A and basal versus luminal B comparisons, indicating that DNA methylation was able to differentiate the more aggressive subtype (basal) from the less aggressive subtypes (luminals A and B) regardless of whether DM or DV CpGs were used. On the other hand, the prediction algorithm for predicting luminal A from luminal B subtypes exhibited lower accuracy compared to the previous comparisons, indicating that it is harder to differentiate these two subtypes based on DNA methylation.

5. Discussion

The promise and power of DNA methylation for therapeutics and diagnostics have been demonstrated in various diseases including cancer. Advancements in biotechnology enable large scale and population based epigenome-wide profiling of DNA methylation for identifying differential mean (DM) and differential variability (DV) CpGs. In these studies, covariates such as demographic and clinical factors may be confounded with both DNA methylation and disease phenotypes. One way to circumvent this problem is via randomization. However, this approach is not always feasible especially in case control studies. Moreover, in DNA

Table 2. Average AUC, Mcc, Accuracy (Acc), Sensitivity (Sn) and Specificity (Sp) for the different datasets and comparisons.

Candidate feature	AUC	Mcc	Acc	Sn	Sp
KIRC: tumor vs normal					
Set 1	1.000	0.998	0.999	0.998	1.000
Set 2	1.000	0.991	0.996	0.994	1.000
Set 3	1.000	0.998	0.999	0.998	1.000
LIHC: tumor vs normal					
Set 1	0.996	0.933	0.986	0.992	0.940
Set 2	0.994	0.913	0.981	0.985	0.950
Set 3	0.997	0.929	0.984	0.987	0.960
BRCA: tumor vs normal					
Set 1	1.000	0.976	0.992	0.997	0.975
Set 2	0.999	0.969	0.990	0.993	0.978
Set 3	1.000	0.976	0.992	0.997	0.972
BRCA: basal vs luminal A					
Set 1	0.996	0.947	0.980	0.950	0.989
Set 2	0.987	0.848	0.944	0.817	0.984
Set 3	0.995	0.947	0.980	0.950	0.989
BRCA: basal vs luminal B					
Set 1	0.998	0.905	0.950	0.950	0.950
Set 2	0.996	0.905	0.950	0.967	0.938
Set 3	0.998	0.889	0.943	0.950	0.938
BRCA: luminal B vs luminal A					
Set 1	0.798	0.339	0.741	0.425	0.874
Set 2	0.720	0.287	0.722	0.413	0.853
Set 3	0.791	0.380	0.767	0.413	0.916

methylation studies using whole blood sample, the different cell types have been shown to be confounded with the measured methylation levels.[32] In such cases, confounding factors need to be properly accounted for to avoid biases in DNA methylation biomarker detection. There are several approaches for DM analysis which allow for confounders adjustment,[33] however to the best of our knowledge existing DV analysis approaches are not tailored for confounders adjustments, except for our earlier work[16] which proposed a DV only analysis in the presence of confounders within large scale hypothesis testings framework. This paper extends our earlier work which allows for simultaneous detection of DM and DV in large scale hypothesis testings framework, and at the same time provides a candidate feature selection mechanism for the prediction algorithm.

We showed that the analysis on KIRC, LIHC and BRCA TCGA datasets identified DM and DV CpGs which mapped to different CpG and gene annotations. For instance, in tumor versus normal comparisons, a larger proportion of DM CpGs mapped to CpG island and TSS200, whereas in basal versus luminal A or B comparisons, a larger proportion of DV CpGs mapped to these regions, suggesting that DM and DV CpGs regulate transcription differently. An R package `methylDMV` implementing this flexible framework is available at `http://www.ams.sunysb.edu/~pfkuan/softwares.html#methylDMV`.

DNA methylation generated from high resolution arrays including Illumina Infinium HumanMethylation450 BeadChip may induce a natural correlation structure among neighboring CpGs. An immediate extension of our current framework is to model the dependence structure and borrow information from nearby CpGs to improve the power of detecting DM and DV CpGs. Two of such approaches are (1) the hidden Markov model and local index of significance method as in Kuan et al. (2012),[34] and (2) the smoothing and bump hunting method as in Jaffe et al (2012),[7] which can possibly be adapted into our current `methylDMV` framework for detecting DM and DV CpGs.

Acknowledgments

The authors thank the reviewers for their constructive comments and suggestions.

References

1. V. Rakyan, T. Down, N. Thorne, P. Flicek, E. Kulesha, S. Graf, E. Tomazou, L. Backdahl, N. Johnson, M. Herberth, K. Howe, D. Jackson, M. Miretti, H. Fiegler, J. Marioni, E. Birney, T. Hubbard, N. Carter, S. Tavare and S. Beck, *Genome Research* **18**, 1518 (2008).
2. M. Esteller, *Annual Review Pharmacological Toxicology* **45**, 629 (2005).
3. R. Irizarry, C. Ladd-Acosta, B. Carvalho, H. Wu, S. Brandenburg, J. Jeddeloh, B. Wen and A. Feinberg, *Genome Research* **18**, 780 (2008).
4. P. Wang, Q. Dong, Z. Chong, P. Kuan, Y. Liu, W. Jeck, W. Jiang, G. S. nd T. Tan, J. Andersen, T. Auman, J. Hoskins, A. Misher, C. Moser, S. Yourstone, J. Kim, K. Cibulskis, S. Getz, H. Hunt, S. Thorgerisson, L. Roberts, D. Ye, K. Guan, Y. Xiong, L. Qin and D. Chiang, *Oncogene* **32**, 3091 (2012).
5. K. Conway, S. Edmiston, Z. Khondker, P. Groben, X. Zhou, H. Chu, P. Kuan, H. Hao, C. Carson, M. Berwick, D. Olilla and N. Thomas, *Pigment Cell and Melanoma Research* **24**, 352 (2011).
6. K. Hansen, W. Timp, H. Bravo, S. Sabunciyan, B. Langmead, O. McDonald, B. Wen, H. Wu, Y. Liu, D. Diep, E. Briem, K. Zhang, R. Irizarry and A. Feinberg, *Nature Genetics* **26**, 768 (2011).
7. A. Jaffe, A. Feinberg, R. Irizarry and J. Leek, *Biostatistics* **13**, 166 (2012).
8. A. Teschendorff and M. Widschwendter, *Bioinformatics* **28**, 1487 (2012).
9. X. Xu, S. Su, V. Barnes, C. Miguel, J. Pollock, D. Ownby, H. Shi, H. Zhu, H. Snieder and X. Wang, *Epigenetics* **8**, 522 (2013).
10. A. Feinberg and R. Irizarry, *Proc. Natl Acad. Sci. USA* **107**, 1757 (2010).
11. A. Teschendorff, A. Jones, H. Fiegl, A. Sargent, J. Zhuang, H. Kitchener and M. Widschwendter, *Genome Medicine* **4**, p. DOI: 10.1186/gm323 (2012).
12. G. Smyth, *Statistical Application in Genetics and Molecular Biology* **3**, p. 3 (2004).
13. S. Ahn and T. Wang, *Pacific Symposium of Biocomputing* , 69 (2013).

14. B. Phipson and A. Oshlack, *Genome Biology* **15**, DOI: 10.1186/s13059 (2014).
15. S. Wahl, N. Fenske, S. Zeilinger, K. Suhre, C. Gieger, M. Waldenberger, H. Grallert and M. Schmidt, *BMC Bioinformatics* **15**, DOI: 10.1186/1471 (2014).
16. P. Kuan, *Statistical Applications in Genetics and Molecular Biology* **13**, 645 (2014).
17. O. Stefansson, S. Moran, A. Gomez, S. Sayols, C. Arribas-Jorba, J. Sandoval, H. Hilmarsdottir, E. Olasfdottir, L. Tryggvadottir, J. Jonasson, J. Eyfjord and M. Esteller, *Molecular Oncology* **9**, 555 (2015).
18. J. Zhuang, M. Widschwendter and A. Teschendorff, *BMC Bioinformatics* **13**, DOI: 10.1186/1471 (2012).
19. S. Horvath, *Genome Biology* **14**, p. R115 (2013).
20. M. Jung and G. Pfeifer, *BMC Biology* **13**, doi: 10.1186/s12915 (2015).
21. M. Dogan, B. Shields, C. Cutrona, L. Gao, F. Gibbons, R. Simons, M. Monick, G. Brody, K. Tan, S. Beach and R. Philibert, *BMC Genomics* **15**, DOI: 10.1186/1471 (2014).
22. K. Lee and Z. Pausova, *Frontiers in Genetics* **4**, p. doi: 10.3389/fgene.2013.00132 (2013).
23. A. Houseman, B. Christensen, R. Yeh, C. Marsit, M. Karagas, M. Wrensch, H. Nelson, J. Wiemels, S. Zheng, J. Wiencke and K. Kelsey, *BMC Bioinformatics* **9**, doi:10.1186/1471 (2008).
24. P. Du, X. Zhang, C. Huang, N. Jafari, W. Kibbe, L. Hou and S. Lin, *BMC Bioinformatics* **11** (2010).
25. C. Rao, *Proceedings of the Cambridge Philosophical Society* **44**, 50 (1948).
26. Y. Benjamini and Y. Hochberg, *Journal of the Royal Statistical Society: Series B (Statistical Methodology)* **57**, 289 (1995).
27. H. Zou and T. Hastie, *Journal of the Royal Statistical Society, Series B* **67**, 301 (2005).
28. L. Breiman, *Journal of Machine Learning* **45**, 5 (2001).
29. A. Teschendorff, F. Marabita, M. Lechner, T. Bartlett, J. Tegner, D. Gomez-Cabrero and S. Beck, *Bioinformatics* **29**, 189 (2013).
30. E. Price, A. Cotton, L. Lam, P. Farre, E. Emberly, C. Brown, W. Robinson and M. Kobor, *Epigenetics and Chromatin* **6** (2013).
31. Y. Chen, M. Lemire, S. Choufani, D. Butcher, D. Grafodatskaya, B. Zanke, S. Gallinger, T. Hudson and R. Weksberg, *Epigenetics* **8**, 203 (2013).
32. A. Houseman, W. Accomando, D. Koestler, B. Christensen, C. Marsit, H. Nelson, J. Wiencke and K. Kelsey, *BMC Bioinformatics* **13**, 189 (2012).
33. M. Ritchie, B. Phipson, D. Wu, Y. Hu, C. Law, W. Shi and G. Smyth, *Nucleic Acids Research* **43**, p. e47 (2015).
34. P. Kuan and D. Chiang, *Biometrics* **68**, 774 (2012).

IDENTIFY CANCER DRIVER GENES THROUGH SHARED MENDELIAN DISEASE PATHOGENIC VARIANTS AND CANCER SOMATIC MUTATIONS

MENG MA[1], CHANGCHANG WANG[2], BENJAMIN S. GLICKSBERG[1], ERIC E. SCHADT[1], SHUYU D. LI[1]*, RONG CHEN[1]*

[1]*Department of Genetics and Genomic Sciences, Icahn School of Medicine at Mount Sinai, 1 Gustave L. Levy Pl. New York City, NY 10029, USA*

[2]*School of Computer Science, Anhui University, Anhui, P.R. China*
**Email: rong.chen@mssm.edu; shuyudan.li@mssm.edu.*

Genomic sequencing studies in the past several years have yielded a large number of cancer somatic mutations. There remains a major challenge in delineating a small fraction of somatic mutations that are oncogenic drivers from a background of predominantly passenger mutations. Although computational tools have been developed to predict the functional impact of mutations, their utility is limited. In this study, we applied an alternative approach to identify potentially novel cancer drivers as those somatic mutations that overlap with known pathogenic mutations in Mendelian diseases. We hypothesize that those shared mutations are more likely to be cancer drivers because they have the established molecular mechanisms to impact protein functions. We first show that the overlap between somatic mutations in COSMIC and pathogenic genetic variants in HGMD is associated with high mutation frequency in cancers and is enriched for known cancer genes. We then attempted to identify putative tumor suppressors based on the number of distinct HGMD/COSMIC overlapping mutations in a given gene, and our results suggest that ion channels, collagens and Marfan syndrome associated genes may represent new classes of tumor suppressors. To elucidate potentially novel oncogenes, we identified those HGMD/COSMIC overlapping mutations that are not only highly recurrent but also mutually exclusive from previously characterized oncogenic mutations in each specific cancer type. Taken together, our study represents a novel approach to discover new cancer genes from the vast amount of cancer genome sequencing data.

1. Introduction

Significant efforts in the past several years in cancer genomic sequencing by individual investigators and large consortium such as The Cancer Genome Atlas (TCGA) and The International Cancer Genome Consortium (ICGC) have uncovered a large number of novel oncogenic drivers. These studies not only advanced our understanding on the genetic basis of tumorigenesis and cancer progression, but also significantly enabled the development of personalized cancer therapeutics [1, 2]. Cancer genome or exome sequencing data have been generated from approximately 25,000 tumor samples covering more than 50 tumor types [3, 4], representing a comprehensive cancer genomic atlas. While data generation has been greatly facilitated by rapid technology development, interpretation of cancer sequence information still remains a major challenge. As most solid tumors harbor a median of 40-80 non-synonymous somatic mutations per tumor, only three to six of them are driver mutations [5]. The most commonly used approach to distinguish a small number of driver mutations from those background passenger mutations is to identify significantly mutated genes in a cohort study [6]. The underlying rationale is if a gene is mutated at significantly greater rate than the background mutation rate, it is more likely to be oncogenic, as the mutations conferring tumor growth advantage are evolutionarily selected during cancer development. To complement this approach, various computational tools have been developed to assess the effects of missense mutations on protein functions [7]. While such an approach has further characterized numerous novel cancer drivers and oncogenic pathways from cancer genomic sequencing data, it requires a large number of samples to uncover those drivers mutated at low population frequency in a given tumor type. This is particularly problematic for those cancers with high background mutation rates such as melanomas and lung cancers. For example, it has been estimated that it would require approximately 4,000 melanoma patient samples to detect cancer genes mutated at 2% frequency, and more than 20,000 samples for genes mutated at 1% with 90% power for 90% of genes [8].

Many human genetic diseases are Mendelian disorders caused by one or more aberrations in the genome. These diseases are often heritable as the disease causing, pathogenic variants are passed on from parents' genome. To date, approximately 180,000 genetic variants in more than 7,000 genes have been identified as pathogenic for more than 4,000 Mendelian diseases [9]. Some of the first established cancer genes with frequent somatic mutations were originally identified from their associations with familial cancer syndromes. The first tumor suppressor RB1 was discovered by studying the familial form of retinoblastoma [10]. The most frequently mutated gene in cancers, p53, was also identified as a tumor suppressor inactivated in Li–Fraumeni syndrome, a rare cancer predisposition hereditary disorder. Other well-known cancer genes harboring high frequency somatic mutations and that are associated with Mendelian diseases include VHL in Von Hippel-Lindau syndrome, MLH1, MSH2, MSH6 in Lynch syndrome, TSC1, TSC2 in Tuberous sclerosis, and ATM in ataxia-telangiectasia [11]. Notably, a recent study has revealed potentially novel cancer-associated genes through analysis of comorbidity between cancers and Mendelian diseases [12].

By definition, germline pathogenic variants impact the functions of key proteins involved in the developmental process and consequently cause heritable diseases. If the same germline pathogenic variants occur as somatic mutations in cancers, these mutations would also alter protein functions and may play a role in tumor initiation and progression, even though the same proteins can have very different functions during development than in adult tissues. Indeed in a recent report, several genes sharing identical mutations in Mendelian diseases and cancers were proposed as novel cancer genes [13]. Based on this underlying hypothesis, we carried out a systematic comparative analysis of the reported pathogenic variants in Mendelian diseases and cancer somatic mutations. There are several repositories for pathogenic variants. A comparison of four of the most comprehensive databases showed that HGMD is currently the largest collection of human disease variants, although each database has its own advantages in terms of the information collected as well as database infrastructure [9]. For cancer somatic mutations, COSMIC is recognized as the most comprehensive resource for somatic mutations in human cancers [14], with more than 1.4 million confirmed somatic mutations identified from 1.1 million tumor samples including genome-wide sequencing data from more than 20,000 tumors. In this study, we first identified overlapping mutations between pathogenic variants in HGMD [15] and cancer somatic mutations from the COSMIC database [14]. Further characterization of these mutations show that the mutation-harboring genes are significantly enriched for known cancer genes, supporting the above described hypothesis. We then examined those genes harboring the shared pathogenic variants and somatic mutations in cancers by applying additional filters such as the number of overlapping HGMD/COSMIC mutations in a given gene or the frequency of overlapping mutations in each tumor type. Moreover, those overlapping mutations with high recurrence in cancers were subjected to mutual exclusivity analysis with known oncogenes in each tumor type in order to identify novel oncogenic drivers. Taken together, our study represents a novel approach to discover new cancer genes from the vast amount of cancer genome sequencing data.

2. Methods

COSMIC V73 was downloaded from sftp-cancer.sanger.ac.uk using GUI client WinSCP under protocol sftp and port 22. HGMD Professional can be accessed from https://www.qiagenbioinformatics.com/products/human-gene-mutation-database/ with an authorized license. 1000 Genome Phase3 was downloaded from ftp://ftp.1000genomes.ebi.ac.uk/vol1/ftp/release/20130502/. ExAC database was downloaded from ftp://ftp.broadinstitute.org/pub/ExAC_release/. All RefSeq Exons were downloaded from UCSC table refGene through UCSC Table Browser (clade: Mammal, genome: Human, assembly: Feb.2009 (GRCH37/hg19), group: Genes and Gene Predictions, track: RefSeq Genes, Table: refGene). Cancer Gene Census dataset was downloaded from http://cancer.sanger.ac.uk/census/.

All the analyses were performed using shell scripts, mysql scripts and R scripts. The mutual exclusivity heat map was generated using gitools (http://www.gitools.org/). The survival analysis was done through cBioPortal (http://www.cbioportal.org/). Several major scripts for database query and statistical analyses are available on github (https://github.com/CosmicHGMD/CancerMendelian).

3. Results

3.1. *Identification of overlapping pathogenic variants in HGMD and somatic mutations in COSMIC*

HGMD includes six classes of variants, and we only included disease-causing mutations (DM and DM?) in our analysis. The DM class variants have been demonstrated in literature to confer the associated clinical phenotype of the affected individuals. The DM? class variants have some degree of uncertainty, but nevertheless have strong evidence supporting their pathogenicity. At the time of this writing, there are a total of 153,593 DM/DM? class variants in HGMD database. 11,523 of these variants are present in the COSMIC database, representing 0.54% of the total mutations in COSMIC (Table 1). When we only include the confirmed somatic mutations in COSMIC, there are 8,582 mutations (0.6%) that overlap with HGMD DM/DM? variants. As the majority of the somatic mutation data in COSMIC are from cancer genomic sequencing studies, some of these mutations are likely false positives, particularly those from early whole genome/exome sequencing when computational methods for calling somatic mutation were less reliable or if the identified somatic mutations were not validated by a different sequencing platform. Therefore, we further restrict COSMIC data to include only those somatic mutations occurred in more than one tumor samples. Although the total number of overlapping mutations with HGMD DM/DM? variants is reduced to 3,470, using this limited but more reliable somatic mutation list, the percentage with respect to the total number of these recurrent mutations (215, 436) in COSMIC increases to 1.6% (Table 1), suggesting Mendelian disease pathogenic variants are over-represented in recurrent somatic mutations in cancers.

Then we randomly selected the same number of genetic variants (153,593) from 1000 genome (exonic region) or the ExAC database as control variant datasets, and performed the same analysis. The analysis of randomly selected, mostly non-pathogenic common genetic variants was repeated 1000 times, and the results indicated that percentages of common non-pathogenic variants overlapping with COSMIC mutations are lower than the HGMD pathogenic variants (Table 1). The statistical significance was assessed based on the distribution of results from 1000 simulations. This finding supports our initial hypothesis that overlapping pathogenic variants in HGMD with cancer somatic mutations could enable identification of novel cancer genes.

Table 1. Enrichment of HGMD pathogenic variants in cancer somatic mutations.

Variant dataset (total number of variants)	**Randomly selected variants**	**Overlap with COSMIC mutations (percentage)**		
		All mutations in COSMIC (2,132,117)	**Somatic mutations in COSMIC (1,425,978)**	**Recurrent somatic mutations in COSMIC (215,436)**
HGMD DM/DM? (153,593)	-	11,523 (0.54%)	8,582 (0.60%)	3,470 (1.6%)
1000 Genome exonic region (2,156,973)	153,593	8,092 (0.38%); $p<0.001$	5,983 (0.42%); $p<0.001$	1,975 (0.92%); $p<0.001$
ExAc (10,450,722)	153,593	6,919 (0.32%); $p<0.001$	4,841 (0.34%); $p<0.001$	1,325 (0.62%); $p<0.001$

Next, we tested if HGMD/COSMIC overlapping mutations are more likely to occur at high frequency in cancers than those somatic mutations non-overlapping with HGMD. We first divided the confirmed COSMIC somatic mutations into two groups. The first group includes those mutations overlapped with HGMD DM/DM? variants and the second group includes the rest of somatic mutations that are only present in the COSMIC database. Then, for a given recurrence frequency cutoff *c*, we computed the percentage of somatic mutations with recurrence frequency (*f*) greater than *c* in group 1 (denoted as $\%G1_{f>c}$) and those in group 2 (denoted as $\%G2_{f>c}$). This is followed by computing the ratio of $\%G1_{f>c}$ over $\%G2_{f>c}$ at various mutation frequencies. As illustrated in Figure 1A, as the recurrence frequency (x-axis) increases, this ratio (y-axis) also increases. For example, the ratio is approximately 25 for recurrence frequency 20, indicating that COSMIC mutations overlapping with HGMD pathogenic variants are 25 fold more likely to occur in more than 20 tumor samples than those not overlapping with HGMD variants. We also directly plotted $\%G1_{f>c}$ and $\%G2_{f>c}$ (Figure 1B), and it clearly shows the HGMD/COSMIC overlapping mutations have higher mutation frequencies than those mutations only in the COSMIC database with mean recurrence in 8.0 and 1.3 tumors respectively (p = 1.5E-5, one-sided t-test). Because the likelihood that a somatic mutation is a cancer driver increases with its mutation frequency in cancers, this result is consistent with the hypothesis that cancer mutations overlapping with germline disease pathogenic variants in HGMD are more likely to be oncogenic. We further examined the presence of known cancer genes in the two groups using cancer gene census annotation [16]. While only 4.3% of the COSMIC somatic mutations do not overlap with HGMD are in the cancer gene census list, there are approximately 10% of the somatic mutations overlapping with HGMD occur in cancer census genes.

To determine if the combination of somatic mutation frequency and the presence of overlap with HGMD pathogenic mutations would facilitate cancer gene discovery, we computed the percentage of somatic mutations mapped to cancer census genes in all COSMIC confirmed somatic mutations or only in those overlapped with HGMD DM/DM? variants. This procedure was then repeated for mutations with increasing frequencies (Figure 2). Two observations were notable from the results. First, a somatic mutation is more likely to be in a cancer gene as its frequency increases, evidenced by increasing percentage of cancer census genes. Second, the probability that the mutation-harboring genes are cancer-related increases if there are overlapping with HGMD variants (Figure 2, red bars vs. blue bars).

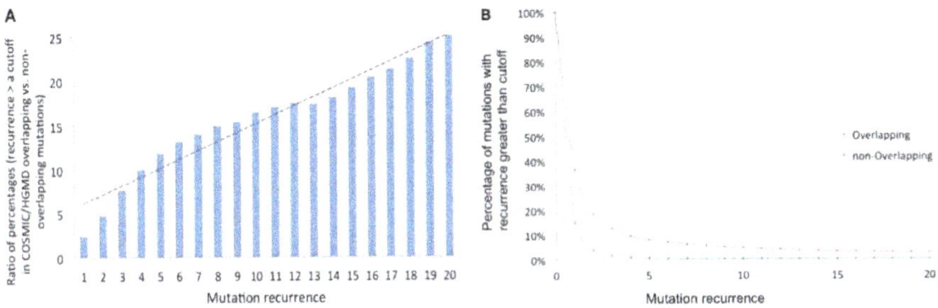

Figure 1. Overlap of HGMD variants with cancer somatic mutations is correlated with high mutation recurrence in cancers.

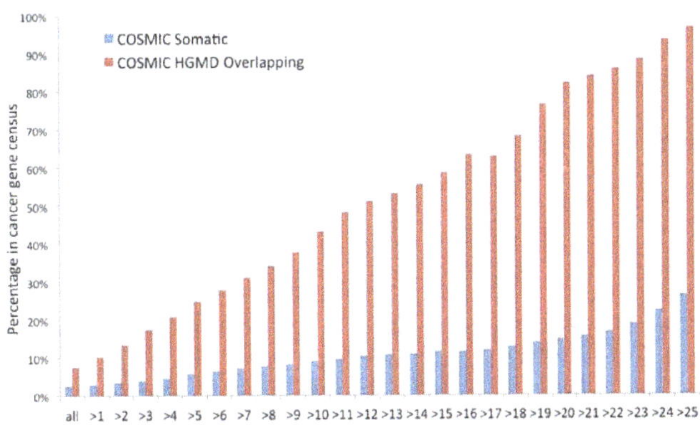

Figure 2. Novel cancer gene discovery through overlapping with HGMD and high mutation recurrence. X-axis represents mutation recurrence in COSMIC.

3.2. *Identification of potential tumor suppressors*

We examined whether the number of distinct overlapping HGMD/COSMIC mutations in a given gene is associated with the probability that the gene is a cancer gene. Figure 3 shows that as the number of distinct overlapping HGMD/COSMIC mutations in a given gene increases, the percentage of genes that belong to cancer gene census increases as well. Of those genes with more than six distinct overlapping mutations, approximately 25% are present in cancer gene census. It is generally recognized that while oncogenes are often mutated recurrently at certain positions (referred as hotspots), tumor suppressors tend to lack such mutational hotspots and are mutated at many positions across the gene sequences. Therefore, we reason that identifying genes with high number of distinct overlapping HGMD/COSMIC mutations would allow us to discover potentially novel tumor suppressors.

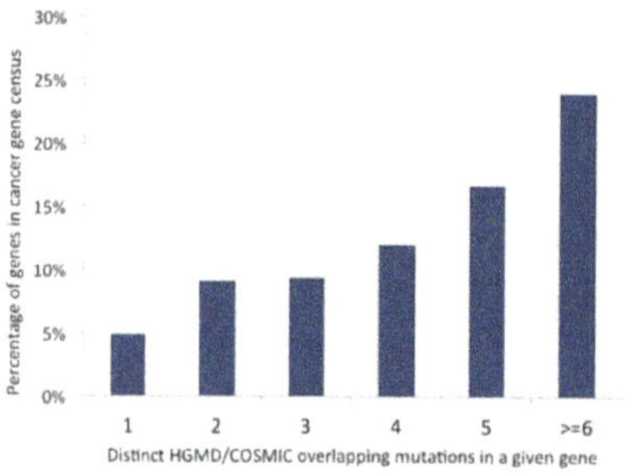

Figure 3. Identification of novel tumor suppressors based on the number of distinct HGMD/COSMIC overlapping mutations.

Accordingly, we ranked all the genes in COSMIC based on the number of distinct somatic mutations that overlap with HGMD DM/DM? variants and provided both cancer gene census annotations as well as oncogene/tumor suppressor classifications according to Vogelstein et al. [5] in Table 2. Almost half (23/48) of the genes with at least 20 overlapping HGMD/COSMIC mutations are in the cancer gene census list and/or annotated as an oncogene or a tumor suppressor, furthering the notion that HGMD pathogenic variant annotation may help distinguish driver oncogenic mutations from the passenger mutations in tumors. As expected, most of those genes with oncogene/tumor suppressor annotations are classified as tumor suppressors (19/21, 90%; Table 2). A literature search has provided support that some of the remaining genes are likely novel tumor suppressors. There are several genes that encode ion channels with many HGMD/COSMIC overlapping mutations, including SCN5A (67 overlapping mutations), SCN1A (52), CFTR (48), RYR1 (36) and RYR2 (30) (Table 2). While ion channels have not been recognized as a major class of cancer related genes, emerging evidence suggest at least some ion channels are involved in promoting malignancy. For example, CFTR, the cystic fibrosis (CF) gene, has been postulated to be a tumor suppressor because loss of CFTR enhanced tumor cell proliferation and epithelial-to-mesenchymal transition, and is associated with poor prognosis in several cancer types [17, 18, 19]. We also observed multiple collagen family genes with significant overlap between HGMD and COSMIC mutations, such as COL3A1 (29 overlapping mutations), COL7A1 (22) (Table 2), COL1A2 (19), COL4A5 (18), COL2A1 (14), COL6A3 (13), COL1A1 (13), and COL4A4 (11) (data not shown). Although collagens are considered as a barrier to suppress angiogenesis since they are key components of extracellular matrix in tumor microenvironment, only recent functional studies have shown a causal relationship between loss of collagens and tumor progression [20]. Our results suggest that collagens may represent another new class of tumor suppressors. Notably, two genes FBN1 and TGFBR2, associated with a genetic disorder of connective tissue known as Marfan syndrome [21, 22], had 43 and 22 HGMD/COSMIC overlapping mutations respectively. Upon further investigation, we found that the two genes are mutated frequently in lung squamous cell carcinomas (SCCs) with a combined mutation frequency 10% in the TCGA cohort [23]. Moreover, FBN1 and TGFBR2 mutations are associated with poor survival. As shown in Figure 4, FBN1 mutation-harboring lung SCCs had poor disease progression free survival (DFS) (Figure 4A), and those patients with TGFBR2 mutations had both poor DFS and overall survival (OS) (Figure 4B, 4E). The combined FBN1 and TFGBR2 mutations are associated with both poor DFS and OS (Figure 4C, 4F).

Table 2. Genes ranked by the number of distinct overlapping HGMD-COSMIC mutations. Only genes with at least 20 overlapping mutations are shown. CGC: cancer gene census. TSG: tumor suppressor gene.

Gene	Mutations	CGC	Oncogene/TSG	Gene	Mutations	CGC	Oncogene/TSG
TP53	198	Yes	TSG	F9	33		
APC	192	Yes	TSG	DMD	33		
VHL	173	Yes	TSG	PKHD1	31		
NF1	148	Yes	TSG	SMAD4	31	Yes	TSG
PTEN	145	Yes	TSG	PTPN11	31	Yes	Oncogene
RB1	91	Yes	TSG	RYR2	30		
SCN5A	67			COL3A1	29		
CDKN2A	66	Yes	TSG	MLH1	29	Yes	TSG
NF2	65	Yes	TSG	BRCA1	29	Yes	TSG
KMT2D	56	Yes		MSH2	28	Yes	TSG
F8	56			VWF	27		
MYH7	54			TSC2	27	Yes	
SCN1A	52			STK11	27	Yes	TSG
USH2A	50			PTCH1	27	Yes	TSG
ATM	50	Yes	TSG	ATP7B	25		
MEN1	49	Yes	TSG	WT1	24	Yes	TSG
CFTR	48			TGFBR2	22		
FBN1	43			PAH	22		
HNF1A	39	Yes	TSG	IRF6	22		
RET	39	Yes	Oncogene	COL7A1	22		
ABCA4	37			CASR	22		
RYR1	36			APOB	22		
BRCA2	36	Yes	TSG	GCK	21		
LDLR	35			MYBPC3	20		

3.3. *Identification of potential oncogenes*

To identify putative oncogenes from the overlapping HGMD/COSMIC mutations, we applied two criteria. First, as most well-known oncogenic, activating mutations are highly recurrent in a specific tumor type, we ranked HGMD/COSMIC overlapping mutations by their mutation frequency. This was done separately for each tumor type in COSMIC. Second, because different oncogenic mutations in a given tumor type are often mutually exclusive, we performed mutual exclusivity analysis to identify those HGMD/COSMIC overlapping mutations that are not only highly recurrent but also mutually exclusive from mutations in known oncogenes based on oncogene classification by Vogelstein et al [5].

To achieve sufficient statistical power in mutual exclusivity analysis, we only analyzed 19 tumor types with at least 200 samples that had whole genome or exome sequencing data in COSMIC and focused on those HGMD/COSMIC overlapping mutations in non-cancer genes (oncogene or tumor suppressor according to Vogelstein et al.) that are mutated in at least 1% of the total samples in a specific tumor type. Interestingly, of the 19 tumor types we analyzed, only endometrium, large intestine, and upper aero-digestive tract (UADT) cancers had such mutations, indicating that while only a very small percentage of COSMIC somatic mutations overlap with HGMD pathogenic variants (Table 1), even fewer are mutated in cancers with high recurrence. Notably, the ACVR1 R206H mutation occurred in 3 endometrium cancer samples, and an additional endometrium tumor harbors the ACVR1 G356D mutation. Mutual exclusivity analysis revealed that 3 of these 4 samples are mutually exclusive from the most frequently mutated oncogene PIK3CA, CTNNB1 and KRAS in this tumor type (p-value = 0.078; Figure 5).

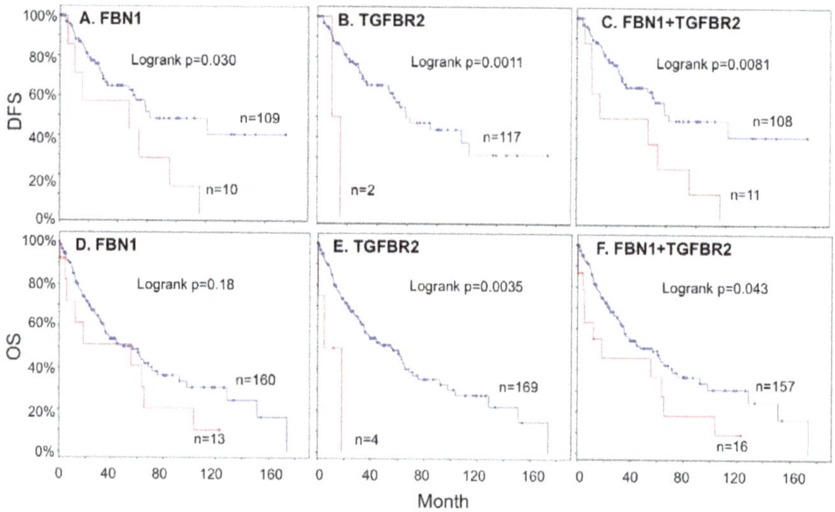

Figure 4. FBN1 and TGFBR2 mutations are associated with poor survival in lung squamous cell carcinomas. Disease free survival (DFS) are shown in panel A-C, and overall survival (OS) are shown in panel D-F. Red curves represent patients harboring somatic mutations for the indicated gene and blue curves represent patients with wild type gene. Sample size in red and blue curves, and logrank p-values in survival analysis are shown in each panel.

Figure 5. Mutual exclusivity of HGMD/COSMIC overlapping ACVR1 mutations from most frequently mutated oncogenes in endometrium cancers. Each column represents a tumor sample. The presence of a mutation in each gene in a given tumor sample is indicated by the blue color.

Since the above approach combining high mutation frequency and mutual exclusivity from known oncogenic drivers in each specific tumor type led to very few candidates as putative oncogenic mutations, we ranked somatic mutations only based on frequency across all cancer types in COSMIC without taking mutual exclusivity into consideration (Table 3). Many oncogenes have multiple mutational hotspots, and therefore for each gene we only show the mutation (at amino acid level) with the highest recurrence. Of genes with the most recurrent amino acid change occurring in at least 15 tumors, 18 had oncogene/tumor suppressor annotations (Table 3). While 50% (9/18) are classified as oncogenes, the presence of many tumor suppressors is not surprising because mutational hotspots (typically dominant negative mutations) are also observed in some tumor suppressors such TP53 [24]. The remaining genes without oncogene/tumor suppressor annotations provide possible candidate oncogenes due to the presence of mutational hotspots. It is noteworthy that there are 3 protein kinases that had a recurrent somatic mutation detected in more than 10 but less than 15 tumor samples: RAF1, p.S257L, 13 tumors; FGFR4, p.G388R, 12 tumors; TYK2, p.V362F, 12 tumors. Although the 3 kinases are not recognized as oncogenes, there are strong evidences that these recurrent mutations are activating and/or oncogenic [25, 26, 27], suggesting RAF1, FGFR4 and TYK2 are likely novel oncogenes.

4. Discussion

Owing to technological advancement and cost reduction, genomic sequencing is a new paradigm in cancer research and personalized cancer therapeutics. A large number of cancer somatic mutations have been described from whole genome/exome sequencing studies. As only a small percentage of somatic mutations are cancer drivers, it is of paramount importance to distinguish those driver mutations from a background of predominantly passenger mutations. Although many computational methods have been developed to predict the functional consequences of mutations [7], it has been indicated that their utility is limited [28]. In this study, we applied an alternative approach to discover cancer drivers from genomic sequencing data. By overlapping cancer somatic mutations and well-defined pathogenic disease-causing germline variants in Mendelian diseases, we identified putative tumor suppressors and oncogenes, which warrant follow-up functional studies. Our analyses suggested that ion channels, collagens and Marfan syndrome-related genes may represent new classes of tumor suppressors. More significantly, mutations in two Marfan syndrome-related genes FBN1 and TGFBR2 are associated with poor prognosis in lung squamous cell carcinomas, providing novel biomarkers with potential clinical relevance in areas of prevention, diagnosis and treatment [29]. Although the previous report by Zhao and Pritchard [13] also interrogated overlapping pathogenic mutations in inherited diseases and cancer somatic mutations, we applied a novel approach to identify candidate tumor suppressors and oncogenes separately based on different criteria. Our approach is particularly useful in identifying the above highlighted putative tumor suppressors.

Table 3. Genes ranked by mutation frequency of the most recurrent HGMD-COSMIC overlapping mutation (at amino acid level) for each gene. Tumor samples with genome-wide sequencing data were used in the analysis. Only genes with the most recurrent amino acid change in at least 15 tumors are shown. TSG: tumor suppressor gene.

Gene	Mutation	Tumors	Oncogene/TSG	Gene	Mutation	Tumors	Oncogene/TSG
KRAS	p.G12D	524	Oncogene	TMEM106B	p.T185S	19	
IDH1	p.R132H	293	Oncogene	TAS2R43	p.H212R	19	
PIK3CA	p.H1047R	274	Oncogene	ROCK2	p.T431N	18	
TP53	p.R175H	226	TSG	PRNP	p.M129V	18	
APC	p.R1450*	66	TSG	GZMB	p.P94A	18	
PTEN	p.R130Q	42	TSG	PON2	p.S311C	17	
CDKN2A	p.R80*	40	TSG	KRT14	p.A94T	17	
CHEK2	p.Y390C	37		HNF1A	p.I27L	17	TSG
SMAD4	p.R361H	31	TSG	FGFR2	p.S252W	17	Oncogene
ABCD1	p.S606P	29		NRAS	p.G13D	16	Oncogene
KMT2C	p.T316S	26		IL1A	p.A114S	16	
OPRD1	p.C27F	25		HLA-DPB1	p.M105V	16	
PRDM9	p.T681S	24		EME1	p.I350T	16	
IDH2	p.R140Q	24	Oncogene	ALK	p.R1275Q	16	Oncogene
ARID1A	p.R1989*	24	TSG	ABCA1	p.R219K	16	
AR	p.Q58L	24	Oncogene	POU5F1B	p.E238Q	15	
UGT2A1	p.R75K	23		LTF	p.K47R	15	
PRSS1	p.K170E	23		IFIH1	p.A946T	15	
UGT1A7	p.N129K	22		HLA-A	p.L180*	15	
USH2A	p.C3416G	21		GRIN3B	p.T577M	15	
TGFB1	p.P10L	20		FGFR3	p.Y373C	15	Oncogene
RAD21L1	p.C90R	20		BRCA2	p.N372H	15	TSG
HRG	p.P204S	20		ATM	p.R337C	15	TSG

From our analyses, we rediscovered genes with cancer predisposing mutations, including TP53, APC, VHL, RB1 and many others (Table 2), which enhanced our confidence in the approach. However, as these genes have been well studied with respect to both germline mutations in familial cancer syndromes and somatic mutations in cancers, our focus lies on those genes with unknown connections between Mendelian diseases and specific cancers associated with the identical mutations. As genes often function differently in development versus in adult tissues, it is critical to further investigate the molecular pathways modulated by those genes in order to understand the mechanisms by which the same mutations can cause Mendelian diseases during development and drive tumor growth in adult tissues. This is best illustrated by an example in our oncogene discovery that revealed 5 HGMD/COSMIC overlapping mutations in ACVR1 gene cumulatively occurred in 19 central nervous system (CNS) cancers (data not shown). While the 5 ACVR1 mutations in germline cause fibrodysplasia ossificans progressiva (FOP), an autosomal dominant disorder of skeletal malformation and disabling heterotopic ossification [30], the same mutations are somatic oncogenic drivers in a subtype of CNS cancers, specifically diffuse intrinsic pontine glioma (DIPG) [31]. Functional studies have demonstrated the ACVR1 mutations in germline activate the canonical bone morphogenic protein (BMP) pathway to promote osteogenic differentiation and endochondral bone formation resulting in FOP, and the same BMP pathway activated by these mutations in astrocyte cells in the brain accelerates cell proliferation ultimately leading to malignancy [32]. Therefore, these seemingly unrelated two diseases involving different tissue and cell types might be connected by the same molecular pathway activated by identical mutations in germline or in somatic cells. As described in the results section, two of these five mutations are also present in endometrium cancers, and they are largely mutual exclusive from the most frequently mutated oncogenes (Figure 5), suggesting that deregulated activation of the BMP pathway in uterus epithelial cells is likely a key oncogenic mechanism in at least some cases of endometrium cancers. Interestingly, the ACVR1 mutations and their potential oncogenic roles in endometrium cancers were also discussed in a recent study [13].

We recognize the limitations in our study. Since the percentage of cancer somatic mutations overlapping with germline pathogenic variants is small (0.6% of somatic mutations, 1.6% of recurrent somatic mutations in COSMIC; Table 1), our approach will not be applicable to the majority of the somatic mutation data from cancer genomic sequencing. Furthermore, identification of putative oncogenes based on high recurrence and mutual exclusivity from known oncogenes yielded few candidates. This is partly due to the fact that very few HGMD/COSMIC overlapping mutations have high recurrence in cancers. In addition, lack of mutual exclusivity with known oncogenes does not necessarily preclude the mutations as cancer drivers. Our goal was only to identify potentially novel cancer genes with high confidence. As more cancer genomic sequencing data become available in COSMIC, our approach will likely lead to the identification of additional putative oncogenes. Another limitation is that most of the candidate cancer genes from our analysis lack apparent functional connection to cancer development. This is somewhat expected due the inherent nature of our approach using Mendelian diseases pathogenic variants to aid novel cancer gene discovery. Accordingly, our study demonstrates a powerful technique for hypothesis generation to identify associations that warrant further experimental validation.

Acknowledgments

We thank Robert Maki from Icahn School of Medicine at Mount Sinai for valuable discussions. This work was supported in part through the computational resources and staff expertise provided by the Department of Scientific Computing at the Icahn School of Medicine at Mount Sinai.

References

1. Chmielecki J, Meyerson M. *Annual review of medicine* **65**, 63-79 (2014).
2. Garraway LA, Lander ES. *Cell* **153**, 17-37 (2013).
3. Hudson TJ, *et al. Nature* **464**, 993-998 (2010).
4. Tomczak K, Czerwinska P, Wiznerowicz M. *Contemporary oncology (Poznan, Poland)* **19**, A68-77 (2015).
5. Vogelstein B, Papadopoulos N, Velculescu VE, Zhou S, Diaz LA, Jr., Kinzler KW. *Science (New York, NY)* **339**, 1546-1558 (2013).
6. Lawrence MS, *et al. Nature* **499**, 214-218 (2013).

7. Gnad F, Baucom A, Mukhyala K, Manning G, Zhang Z. *BMC genomics* **14 Suppl 3**, S7 (2013).
8. Lawrence MS, *et al. Nature* **505**, 495-501 (2014).
9. Peterson TA, Doughty E, Kann MG. *Journal of molecular biology* **425**, 4047-4063 (2013).
10. Friend SH, *et al. Nature* **323**, 643-646 (1986).
11. Nagy R, Sweet K, Eng C. *Oncogene* **23**, 6445-6470 (2004).
12. Melamed RD, Emmett KJ, Madubata C, Rzhetsky A, Rabadan R. *Nature communications* **6**, 7033 (2015).
13. Zhao B, Pritchard JR. *PLoS genetics* **12**, e1006081 (2016).
14. Forbes SA, *et al. Nucleic acids research* **43**, D805-811 (2015).
15. Stenson PD, Mort M, Ball EV, Shaw K, Phillips A, Cooper DN. *Human genetics* **133**, 1-9 (2014).
16. Futreal PA, *et al. Nature reviews Cancer* **4**, 177-183 (2004).
17. Than BL, *et al. Oncogene*, (2016).
18. Xie C, *et al. Oncogene* **32**, 2282-2291, 2291.e2281-2287 (2013).
19. Zhang JT, *et al. Biochimica et biophysica acta* **1833**, 2961-2969 (2013).
20. Martins VL, *et al. Journal of the National Cancer Institute* **108**, (2016).
21. Loeys B, *et al. Human mutation* **24**, 140-146 (2004).
22. Mizuguchi T, *et al. Nature genetics* **36**, 855-860 (2004).
23. The Cancer Genome Atlas. *Nature* **489**, 519-525 (2012).
24. Stracquadanio G, *et al. Nature reviews Cancer* **16**, 251-265 (2016).
25. Imielinski M, *et al. The Journal of clinical investigation* **124**, 1582-1586 (2014).
26. Tomasson MH, *et al. Blood* **111**, 4797-4808 (2008).
27. Ulaganathan VK, Sperl B, Rapp UR, Ullrich A. *Nature* **528**, 570-574 (2015).
28. Miosge LA, *et al. Proceedings of the National Academy of Sciences of the United States of America* **112**, E5189-5198 (2015).
29. Iyengar P, Tsao MS. *Surgical oncology* **11**, 167-179 (2002).
30. Kaplan FS, *et al. Human mutation* **30**, 379-390 (2009).
31. Zadeh G, Aldape K. *Nature genetics* **46**, 421-422 (2014).
32. Taylor KR, Vinci M, Bullock AN, Jones C. *Cancer research* **74**, 4565-4570 (2014).

IDENTIFYING CANCER SPECIFIC METABOLIC SIGNATURES USING CONSTRAINT-BASED MODELS

A. SCHULTZ[1], S. MEHTA[1], C.W. HU[1], F.W. HOFF[2], T.M. HORTON[3], S.M. KORNBLAU[2] and A.A. QUTUB[1*]

[1]*Department of Bioengineering, Rice University, Houston, Texas 77005, U.S.A*
**E-mail: aminaq@rice.edu*

[2]*Department of Leukemia, The University of Texas M.D. Anderson Cancer Center, Houston, Texas 77030, U.S.A*

[3]*Department of Pediatrics, Baylor College of Medicine and Texas Children's Hospital, Houston, Texas 77030, U.S.A*

Cancer metabolism differs remarkably from the metabolism of healthy surrounding tissues, and it is extremely heterogeneous across cancer types. While these metabolic differences provide promising avenues for cancer treatments, much work remains to be done in understanding how metabolism is rewired in malignant tissues. To that end, constraint-based models provide a powerful computational tool for the study of metabolism at the genome scale. To generate meaningful predictions, however, these generalized human models must first be tailored for specific cell or tissue sub-types. Here we first present two improved algorithms for (1) the generation of these context-specific metabolic models based on omics data, and (2) Monte-Carlo sampling of the metabolic model flux space. By applying these methods to generate and analyze context-specific metabolic models of diverse solid cancer cell line data, and primary leukemia pediatric patient biopsies, we demonstrate how the methodology presented in this study can generate insights into the rewiring differences across solid tumors and blood cancers.

Keywords: Genome-scale metabolic reconstructions, constraint-based models, tissue-specific models, Flux Balance Analysis, cancer metabolism.

Introduction

Cancer tissues exhibits significant metabolic differences when compared to their healthy counterparts, such as the *Warburg effect*[1] and *glutamine addiction*.[2] In recent years it has been revealed that these metabolic transformations are largely driven by oncogenes and subdued by tumor suppressor genes.[3,4] This regulation suggests that cancer metabolism plays an important role in tumor progression, as opposed to being a consequence of the tumor microenvironment.[5] These findings have led to a renewed interest in the field of cancer metabolism,[6] with particular interest in exploiting metabolic differences as therapeutic targets.[7] Cancer metabolism, however, is also extremely heterogeneous across cancer types,[8] and treatments targeting metabolic pathways need to be carefully tailored to specific cancer phenotypes. Consequently, a better understanding of the metabolic differences across cancer sub-types, and between healthy and cancerous tissues will greatly assist the development of novel therapeutic strategies.[7,8]

Genome-Scale Models: To help elucidate the metabolic differences between cancer and healthy tissues, computational approaches can be extremely helpful. In particular, genome-scale models (GEMs) have proven extremely useful in studying human metabolism at the genome level,[9,10] with many studies dedicated specifically to cancer metabolism.[11–13] These studies have, for example, identified glycosaminoglycans as a marker for clear cell renal cell carcinoma,[14] identified carnitine palmitoyltransferase 1 as a potential target for hepatocellular carcinoma,[15] and identified MLYCD as a potential target for leukemia and kidney cancer.[16]

GEMs are defined at the core by a *stoichiometric matrix* S, where each row corresponds to a metabolite, each column to a metabolic reaction, and each entry to the stoichiometric coefficient of that particular metabolite in that particular reaction.[17] For any given *stoichiometric matrix*, flux distribution column vectors (v) can be defined where each element v_i gives the metabolic flux (e.g. rate of metabolite conversion) through each reaction i. The matrix multiplication $S \cdot v = m$ then yields a vector m where each element m_j gives the rate of change of concentration of metabolite j given the reaction fluxes defined by v. A steady-state flux distribution is one where $S \cdot v = 0$. A more detailed description of the constraint-based model formulation is available in the *supplemental information*.

Metabolic Model Analysis: Although a wide array of methods have been developed to study GEMs,[18] many of them are dependent on an *objective function*, which is most often assumed to be cellular growth.[19] Mammalian cells, however, do not have a well established objective, and do not seek to optimize biomass production. One prominent unbiased and objective-independent method for GEM analysis, suited for the study of mammalian cells, is Monte-Carlo sampling (MCS). This method finds normally distributed steady-state flux distributions inside the solution space of $S \cdot v = 0$ defined by lower (lb) and upper (ub) reaction bounds, such that $lb_i \leq v_i \leq ub_i$. Valuable insight into the metabolic capabilities of the model in question can be obtained by analyzing how different MCS conditions (e.g. different lower and upper bounds) affect the sampled reaction flux values. This approach has been used, for example, to model the metabolic exchange between *M. tuberculosis* and human macrophages,[20] and between different cell types in the human brain;[21] to study aspirin resistance in platelet cells;[22] and to characterize metabolic differences between healthy and cancerous tissues.[23]

Mammals also have a complex and compartmentalized metabolism, where not every metabolic reaction takes place in all cells of the body. In order to generate predictions specific to different cell types, cancer categories or patients, generalized human GEMs then need to be tailored to specific contexts.[24] We recently introduced the Cost Optimization Reaction Dependency Assessment (*CORDA*) tissue-specific algorithm,[23] which builds tissue-specific metabolic models based on omics data and a generalized human metabolic reconstruction. The algorithm is based on a *dependency assessment* (DA), where reactions associated with little experimental evidence, called negative confidence reactions (NC), are assigned an arbitrarily high cost. This cost is then minimized while enforcing a small flux through medium (MC) or high (HC) confidence reactions (i.e. reactions with medium or considerable experimental evidence) in order to identify which NC reactions are beneficial for MC or HC reactions to carry flux. This DA is then used to build a tissue-specific model including all HC reactions and as many MC reactions as possible, while minimizing the inclusion of NC reactions. For additional details on the original algorithm we refer readers to the original *CORDA* publication.[23]

Need for New Analyses: MCS of large metabolic networks is computationally expensive, and static approaches are only feasible for extremely small networks.[25] For MCS of higher dimensional networks, the Artificially Centered Hit and Run (ACHR) algorithm[26] is most frequently used. Given a set of points, or steady-state flux distributions, inside the solution space, ACHR calculates a center point as the average of all points, then moves each point i randomly along the directional vector defined by the trajectory between the center and another random point j. ACHR sampling of large networks can be extremely time consuming, however, and even small relative increments in computational efficiency can lead to fewer hours of computational time. Although alternatives to ACHR have been proposed, many of these methods are limited by sample distributions that are significantly different than ACHR outputs,[27–29] by their dependence on objective functions,[27] by long computational times,[30] or by lack of validation and parametrization in larger metabolic networks.[31]

Introduction of CORDA2 and mfACHR: Here we present two improved algorithms for the study of human GEMs. We first introduce an improved version of the *CORDA* algorithm to build tissue-specific metabolic models,[23] referred to here as *CORDA2*. *CORDA2* yields tissue models very similar to the ones given by the previous algorithm, but it is considerably faster than *CORDA* computationally. *CORDA2* is also noise-independent, thus providing unique model outputs for any given set of parameters, which facilitates the comparison of metabolic models across different modeling conditions (i.e. different cancer categories). We next introduce a new formulation of the ACHR algorithm,[26] referred to here as the matrix-form ACHR (mfACHR), which performs significantly faster than previous formulations.

Integrating the two new methods, we generate a panel of cell-line specific metabolic models using *CORDA2* and experimental data from the Human Protein Atlas[32] (HPA), and illustrate how flux samples generated using *mfACHR* can provide valuable insights into the metabolic profile of different cancer types, including pediatric leukemia. While we had previously shown that MCS of CORDA models can identify metabolic differences between healthy and cancerous tissues, here we show that this framework can also pinpoint metabolic differences between different cancer categories. The methods presented in this study provide significant advances in the generation and analysis of context-specific metabolic models.

Methods

Cost Optimization Reaction Dependency Assessment 2

In this work we present two modifications to *CORDA*, defining a new version of the algorithm referred to here as *CORDA2*. First, in the original algorithm, reversible reactions were split into forward and backward rates during every DA to ensure cost production regardless of directionality. That is, a reaction '$A \Leftrightarrow B$' was split into '$A \Rightarrow B + cost$' and '$B \Rightarrow A + cost$'. Since thousands of DAs are performed throughout the model building process, this modification was then repeated thousands of times during the algorithm. In *CORDA2*, this modification is performed at the beginning of the algorithm, and forward and backward rates are treated separately throughout the model building process, speeding the computational time. Furthermore, while in *CORDA* the reaction directionality in the tissue-model was imported from the generalized human reconstruction, *CORDA2* assigns directionality based on whether the forward, backward, or both reaction parts are included in the final tissue model.

Second, pathways with similar costs are captured in *CORDA* by adding a small amount of noise to reaction costs during every DA. This noise-driven approach leads to different

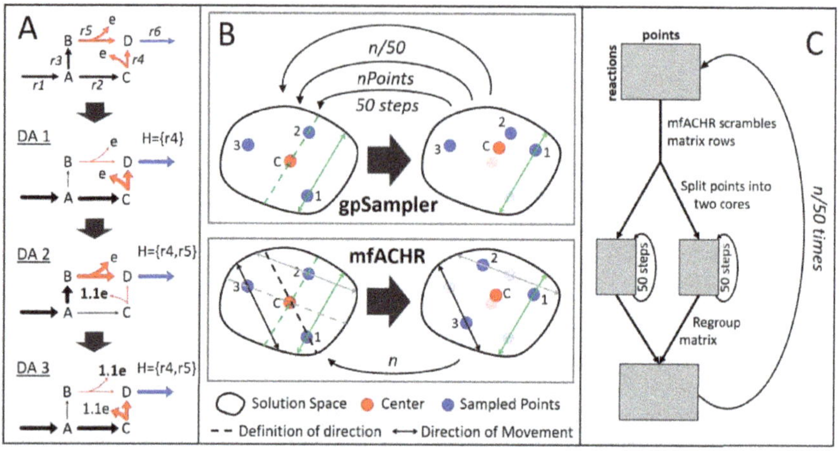

Fig. 1. **Representation of the CORDA2 and mfACHR algorithms.** (A) Identification of undesirable reactions (red) beneficial for the desirable reaction (blue) to carry flux through three DAs. Pathways taken during each DA are highlighted, and H represents the set of undesirable reactions taken up to that point. After an undesirable reaction is used, its cost (e) is increased. The process is repeated until H is unchanged. (B) gpSampler moves one point at a time, 50 steps at a time. The mfACHR algorithm identifies all possible directions of movement at once and moves all points simultaneously. Vectors defining the trajectory of movement, taken as the difference between j and the center point, and the corresponding path of movement of i are color-coded. (C) During parallelization of the MCS process, the matrix of sampled points is divided into 2 cores, which are sampled for 50 steps, then re-combined.

reconstructions after every run of the algorithm, and it is not guaranteed to include every alternative pathway. This approach is also inefficient since the same pathway can be sampled multiple times. In *CORDA2*, only undesirable reactions are assigned an arbitrarily high cost (while in *CORDA* all reactions received a basal cost value). This cost is then minimized during the DA, and the high cost reactions used are saved in a set H. The cost associated with the reactions in H is then increased, and the DA is performed again (**Fig. 1A**). This process is repeated iteratively until H is unchanged. This way, once a pathway is used, its cost is increased and another pathway with similar but now slightly lower cost is identified in the next DA. Additional details of the *CORDA2* formulation, as well as the MATLAB code for its implementation, can be found in the *supplemental information*.

Matrix-Form Artificially Centered Hit and Run

One of the most widely implemented ACHR formalisms is *gpSampler*.[33] *GpSampler* starts by moving a given point 50 steps as described by the ACHR algorithm, then repeats the process for each point being sampled. This whole process is then repeated $\frac{n}{50}$ times for a total of n ACHR steps (**Fig. 1B**). Here we propose a slightly different ACHR formulation, termed matrix-form ACHR (mfACHR). In *mfACHR*, all possible directions of movement are first calculated as the directional vectors defined by each sampled point and the center (dashed lines in **Fig. 1B**). These trajectories are then randomly assigned to each point, and each point is moved randomly along its assigned direction of movement (solid lines in **Fig. 1B**) within the bounds of the solution space. This whole process is repeated a total of n times for a desired number of steps. Both *gpSampler* and *mfACHR* can also be implemented in multiple cores. For that, the points being sampled are first divided into i groups, i being the number of cores used. Each group is then assigned to a core and mixed for 50 steps. All points are then re-combined and the process is repeated $\frac{n}{50}$ times for a total of n steps (**Fig. 1C**).

Cancer Cell Proteomics and Model Generation

Cell line gene and protein expression data were obtained from the HPA[32] in order to build the cell-line specific models. Gene expression data was measured using RNA-seq and protein expression was measured by immunohistochemistry using an extensive library of well validated antibodies. Forty-four models were generated using gene expression data and fifty-two models were generated using the proteomics data. Protein expression was available for 523 (35.0%) gene products, and gene expression data was available for 1,474 (98.7%) of the 1,494 unique genes in the generalized human reconstruction Recon1.[34] All gene and protein expression values were categorized into not detected, low/medium, and high expression in line with threshold values from the HPA, then used to categorize reaction confidence values used in the *CORDA2* algorithm. Following the reconstruction all models were sampled using *mfACHR*. Details of how these models were generated and sampled can be found in the *supplemental information*. For additional details on how the dataset was collected we refer readers to the HPA.[32]

Leukemia Patient Samples: Pediatric leukemia data was obtained from bone marrow biopsies of 95 acute myeloid leukemia (AML), 57 B-cell acute lymphoblastic leukemia (B-ALL), and 16 T-cell acute lymphoblastic leukemia (T-ALL) pediatric patients, and were collected at the Texas Children's Hospital. Protein expression level was measured using reverse phase protein array (RPPA) using 194 strictly validated antibodies.[35] Additional information on the pediatric leukemia data is available in the *supplemental information*.

Results and Discussion

Results of our study demonstrate the robustness of the *CORDA2* and *mfACHR* methods, and their utility in analyzing diverse cell line and primary leukemia cancer metabolism. A summary of the *CORDA2* and *mfACHR* validation is provided below, while a complete description of the algorithm validation and analysis is provided in the *supplemental information*.

CORDA2 Validation

In order to validate the *CORDA2* algorithm, outputs of this formulation were compared to 108 tissue-specific metabolic models generated using *CORDA* and similar model parameters (e.g. same dataset and overlapping algorithm parameters). Overall, at least 99.7% of MC reactions, 88.9% of NC reactions, and 93% of unclassified reactions included in each of the previous 108 models are also included in the *CORDA2* model, showing significant overlap between the output of both algorithms. Furthermore, *CORDA2* was approximately 2.5 times faster than *CORDA* when the later was performed with five DAs for every reaction tested. Although performing fewer DAs in *CORDA* led to computational times comparable to *CORDA2*, the reconstructions returned in that case are not as comprehensive. In the original *CORDA* publication, models reconstructed using one DA were on average 2.3% smaller than models built using multiple DAs. The *CORDA2* algorithm also showed very similar results across multiple metabolic tests when compared to the previous formulation. This analysis shows that *CORDA2* yields models similar to *CORDA* in composition and behavior, while being faster and noise independent.

mfACHR Validation

To assess the performance of *mfACHR* when compared to *gpSampler*, flux distributions and convergence speed of both formulations were compared for three different metabolic models: a red blood cell (RBC) model,[36] a platelet model,[22] and the generalized human reconstruction Recon1.[34] These models have 453, 1,008, and 2,473 active reactions respectively, and were sampled for $3{\cdot}10^4$, $7{\cdot}10^4$, $3{\cdot}10^5$ steps respectively. As an initial step in this validation, MCS outputs of four algorithm formulations (*mfACHR*, *mfACHR* parallel, *gpSampler*, and *gpSampler* parallel) were compared, and all four formulations were shown to converge to similar steady states (*supplemental information*).

Next, convergence speed was assessed by computational time and number of algorithm steps. Convergence based on number of steps was measured as the percentage of reactions at any given point with a Kullback-Leibler divergence (KLD) of sampled flux values below 0.05 of the final distribution. KLD represents the expected logarithmic difference between two probability distributions, and it has been previously used with a similarity threshold of 0.05 to compare sets of sampled flux distributions in metabolic models.[31] The four tested formula-

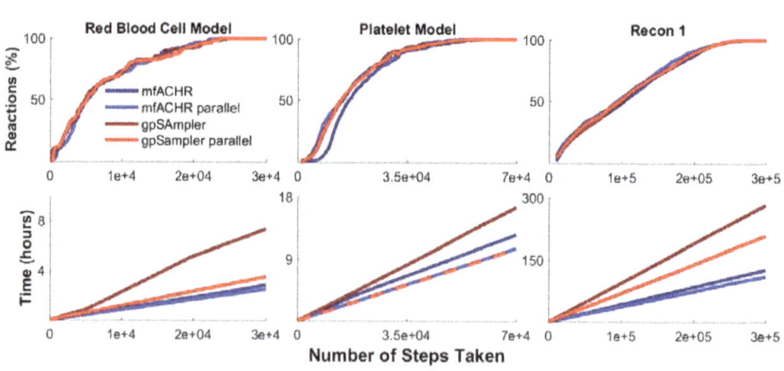

Fig. 2. **Conversion speed of *mfACHR* and *gpSampler*.** (Top) Percentage of reactions in the model with a KLD below 0.05 when compared to the final set of sampled points. (Bottom) Computational running time per number of algorithm steps.

tions showed nearly identical conversion curves when considering the number of steps taken (**Fig. 2**). When considering computational times, *mfACHR* performed significantly better than *gpSampler* when both methods were performed without parallelization. When considering parallelization, *mfACHR* showed very similar computational times in the platelet model, slightly better times in the RBC model, and significantly better times in Recon1. Differences in computational time can be partially attributed to the fact that matrix operations performed by *mfACHR* are automatically parallelized in MATLAB, while the *for* loops performed by *gpSampler* are not. This allows for *mfACHR* to perform significantly faster than *gpSampler* even when the latter is performed with parallelization, and explains the low relative increase in efficiency when explicit parallelization is implemented in *mfACHR*. Overall, *mfACHR* showed consistently faster computational times when compared to *gpSampler*, often in the order of hours, while converging at the same speed in terms of number of algorithm steps.

Cancer Cell Models

Following the validation of both algorithms, a series of cell-line specific models were generated using *CORDA2* and sampled using *mfACHR*, as described in the methods section. Twenty-six of the cancer metabolic models were combined into four tissue categories as presented in the HPA: myeloid, lymphoid, brain, and female reproductive system (FRS) cancer cell lines. These cancer types were chosen since they had the most number of cell lines. We then identified metabolic reactions that have significantly different sampled flux distributions between the four cancer categories (**Fig. 3**). MCS of *CORDA* models previously highlighted metabolic differences between healthy and cancerous tissues.[23] That is, using *CORDA* we correlated high sampled flux values with metabolic pathways known to take place in healthy or cancerous phenotypes. Analogously, in this study we demonstrate that *mfACHR* sampling of *CORDA2* models generated using HPA expression data can also highlight metabolic characteristics between different cancer categories. These characteristics include:

Brain tumors produce high levels of triglyceride: Lipid synthesis is an important factor for cancer survival and progression, and it has been previously suggested as a therapeutic target.[37–40] However, while most cancer types divert fatty-acids predominantly towards the production of phospholipids, not triglycerides,[39,41] glioma cells have been shown to synthesize triacylglycerol at high rates for membrane complex lipids.[42,43] Glioma cells, as well as healthy astrocytes and neurons, can also produce fatty acids from ketone-bodies,[44,45] a metabolic characteristic of brain cells which can further explain the high rate of fatty acid production in glioma cells. In the MCS results presented here, brain tumors present a significantly higher flux through glycerol-3-phosphate acyltransferase (**Fig. 3**) and 1-acylglycerol-3-phosphate O-acyltransferase, enzymes responsible for triacylglycerol synthesis.

Brain and lymphoid tumors have highly active glutamine metabolism: Glutamine plays an essential role in cancer metabolism,[46,47] and different tumors have been shown to utilize glutamine differently.[47] Brain tumors, in particular, have been shown to accumulate glutamine both *in vitro* and *in vivo*.[48,49] Glutamine metabolism has also been shown to play an important role in lymphoid tissues.[50] The role of this pathway in breast cancer, on the other hand, is not well defined, since basal but not luminal breast cancer cells show glutamine-

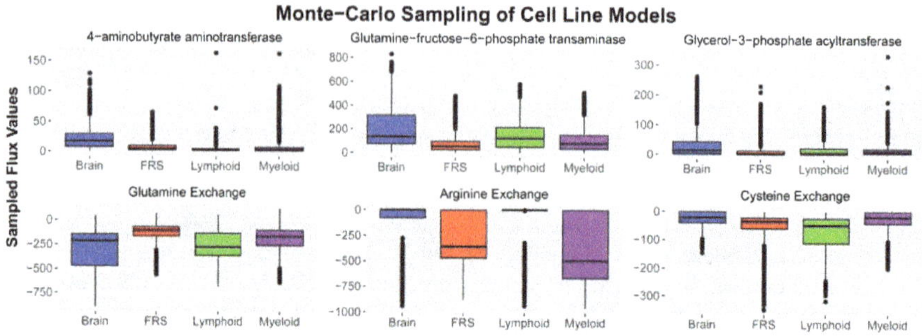

Fig. 3. **MCS results.** Sampled flux values for six different reactions across four model categories. Boxplots represent combined flux values for a particular reaction in all models in that cancer category. For exchange reactions, negative values represent uptake of the particular metabolite, while positive values represent secretion. Colored boxes represent values within the interquartile range (IQR), ranging from the 25th to the 75th percentile. Horizontal line represented the median value (50th percentile), and vertical lines indicate values within 1.5 IQR of the 25th and 75th percentiles. Outliers are represented by dots.

dependence.[51] In the results presented here, brain and lymphoid cell lines show high levels of glutamine uptake, while cell lines of the FRS show relatively low levels (**Fig. 3**).

Lymphoid tissues are cysteine dependent: While cysteine is not considered an essential amino-acid, lymphoid tumors have been shown to contain much lower levels of cystathionase, the last enzyme in the cysteine production pathway, when compared to healthy lymphoid tissues, and are dependent on cysteine for growth.[52] Targeting cysteine transporters has also been shown to selectively target lymphoma cells,[53] and cysteine uptake has been associate with malignant progression in lymphoma cells.[54] In this study, lymphoid models presented much higher levels of cysteine uptake (**Fig. 3**).

Tumors show different levels of arginine dependence: Different types of cancer respond differently to arginine deprivation.[55] A study performed on 26 healthy and cancerous cell lines found that tumor cells are much more sensitive to arginine deprivation than healthy cells.[56] Furthermore, while premyelocytic and lymphoblastic leukaemia cell lines die in about two days of arginine deprivation, cell lines of the FRS died largely in three to four days, and glioma cell lines died in four to five days.[56] Interestingly, levels of arginine dependence presented in the study by Scott et. al.[56] correspond to sampled flux values of arginine uptake in the present study. Myeloid cancers, the most arginine dependent, were predicted to uptake the largest amounts of arginine, followed by models of the FRS, then brain tumors, the least arginine dependent. Acute myeloid leukemia tumors have also been shown to be dependent on arginine for proliferation.[57]

Brain tumors were also predicted to have higher fluxes through the enzyme glutamine-fructose-6-phosphate transaminase (GF6PTA) (**Fig 3**), the rate limiting step in the hexosamines synthesis pathway (HSP), a nutrient sensor pathway.[58,59] When excess nutrients such as glucose and free fatty-acids are available, the HSP prevents cells from uptaking excess amounts from the bloodstream.[60] Furthermore, overweight and obese patients, which have excess amounts of nutrients in the bloodstream, are at an overall increased risk of mortality due to cancer.[61] Interestingly, sampled flux values through the HSP presented here are anti-correlated with the increase in risk of mortality in cancer patients. According to a study of over 57,000 cancer patients, obese patients with brain tumors have a modest increase in mortality compared to non-obese glioma patients, while patients with cancer of the FRS have a high increase in risk, and patients with Non-Hodgkins lymphoma, multiple myeloma, and leukemia have a medium increase.[62] Accordingly, brain tumor models in this study present high GF6PTA flux values, while tumors of the FRS present low fluxes, and lymphoid and myeloid tumors present intermediate values (**Fig. 3**). One possible explanation for this correlation is that higher fluxes through the HSP can prevent cells from uptaking excess amounts of nutrients, which in turn leads to a lower relative increase in malignancy. Further work should help elucidate these observations in context.[63,64]

Sampled flux values also predict a high flux through the enzyme 4-aminobutyrate aminotransferase in brain cancer cells. This result is expected since this enzyme is responsible for GABA production, a pathway highly active in brain tissues. In brain cancer cells, however, this enzyme can help produce acetyl-CoA for energy production, since larger amounts of nutrients are diverted away from glycolysis and into the HSP. A diagram of this proposed mechanism is presented in **Fig 4A**.

Primary pediatric leukemia models: We next analyzed sampled flux values in three different types of leukemia blood sample models (AML, T-ALL, and B-ALL) to clinical pro-

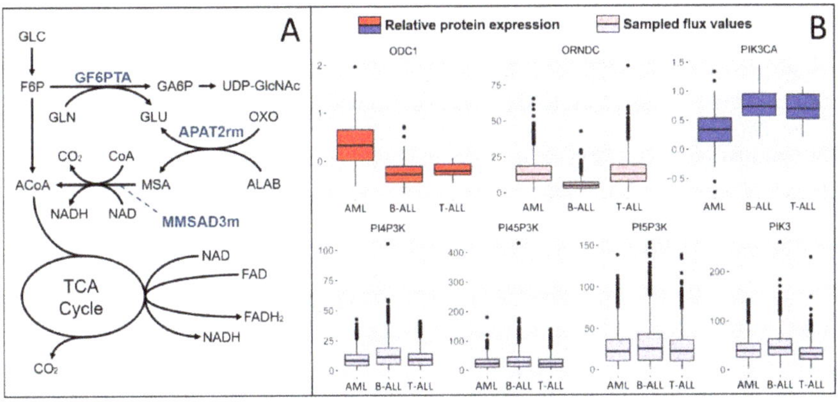

Fig. 4. **Model Predictions.** (A) Pathways with increased activity in brain tumors. Metabolites are glucose (GLC), fructose-6-phosphate (F6P), acetyl-CoA (ACoA), glutamine (GLN), glutamate (GLU), glucosamine-6-phosphate (GA6P), Uridine diphosphate N-acetylglucosamine (UDP-GlcNAc), oxoglutarate (OXO), beta-alanine (ALAB), and malonate semialdehyde (MSA). (B) Relative protein expression and sampled flux values for proteins differentially expressed between AML and ALL pediatric patients. OCD1 participates in the reaction Ornithine Decarboxylase (ORNDC), and PIK3CA participates in reactions PI4P3K, PI45P3K, PI5P3K, and PIK3. All reactions are labeled as in the BiGG database.[65]

teomics data collected from 168 pediatric leukemia patients as described in the methods section. Seven proteins were present both in the leukemia blood sample models and the clinical dataset, of which two were significantly differentially expressed between AML and ALL patients. The relative protein expression of these two proteins, along with the sampled flux values of reactions associated with these proteins, are presented in **Fig 4B**.

Sampled flux values follow trends that correlate with protein expression in both the B-ALL and AML models. That is, while AML patients show significantly higher expression levels of ODC1, the AML model showed significantly higher fluxes through Ornithine Decarboxylase (ORNDC), an ODC1 participating reaction, when compared to the B-ALL model. Likewise, while AML patients showed significantly lower expression of PIK3CA, the AML model also showed significantly lower sampled flux values through the PIK3CA reactions (**Fig 4B**). Sampled flux values between the AML and T-ALL model did not seem to match the differential protein expression, however. One possible explanation for this is the fact that there were considerably fewer T-ALL patients in the clinical dataset, and fewer T-ALL samples were used to generate the proteomics data used in the models building process (2 compared to 3 B-ALL and 4 AML). For instance, in the HPA, T-ALL OCD1 and PIK3CA protein scores are in between B-ALL and AML values, as opposed to much closer to B-ALL values like we see in the pediatric clinical data. This first example application to integrating RPPA leukemia data with metabolic pathway analysis demonstrates how *CORDA2* and *mfACHR* can also be used to analyze clinical data and provide insight into patient-specific metabolic behaviors.

Conclusion

This work illustrates how Monte-Carlo sampling of metabolic models generated using *CORDA2* can generate valuable predictions about context specific cancer metabolism. In applying these new optimized methods to different cancer systems, we show how this work goes beyond the identification of metabolic differences between healthy and cancerous tissues. It identifies differences in metabolism between different cancer types, paving the way to patient-specific metabolic models of cancer. In sum, the *CORDA2* platform elucidates metabolic differences across cancers and provides valuable knowledge of context-specific metabolic behavior that can help guide future directed cancer therapies.

Acknowledgments

This work was funded by NSF grant numbers 1354390 and 1150645, and NIH grant numbers GM106027 and CA164024.

Author Contributions

AS and AAQ developed the computational methods. AS, SM, and AAQ applied and validated the methods. TH and SMK collected the AML and ALL clinical data. FH and CWH processed the clinical data.

Supplemental Information

Supplemental files are available at www.qutublab.org/psb

References

1. O. Warburg *et al.*, *Science* **123**, 309 (1956).
2. H. Eagle, *Science* **122**, 501 (1955).
3. R. J. DeBerardinis, N. Sayed, D. Ditsworth and C. B. Thompson, *Current opinion in genetics & development* **18**, 54 (2008).
4. R. D. Michalek and J. C. Rathmell, *Immunological reviews* **236**, 190 (2010).
5. C. Munoz-Pinedo, N. El Mjiyad and J. Ricci, *Cell death & disease* **3**, p. e248 (2012).
6. R. A. Cairns, I. S. Harris and T. W. Mak, *Nature Reviews Cancer* **11**, 85 (2011).
7. M. G. Vander Heiden, *Nature reviews Drug discovery* **10**, 671 (2011).
8. J. R. Cantor and D. M. Sabatini, *Cancer discovery* **2**, 881 (2012).
9. A. Bordbar and B. O. Palsson, *Journal of internal medicine* **271**, 131 (2012).
10. A. Mardinoglu and J. Nielsen, *Journal of internal medicine* **271**, 142 (2012).
11. K. Yizhak, B. Chaneton, E. Gottlieb and E. Ruppin, *Molecular systems biology* **11**, p. 817 (2015).
12. I. Goldstein, K. Yizhak, S. Madar, N. Goldfinger, E. Ruppin and V. Rotter, *Cancer Metab* **1**, 10 (2013).
13. C. Frezza, L. Zheng, O. Folger, K. N. Rajagopalan, E. D. MacKenzie, L. Jerby, M. Micaroni, B. Chaneton, J. Adam, A. Hedley *et al.*, *Nature* **477**, 225 (2011).
14. F. Gatto, N. Volpi, H. Nilsson, I. Nookaew, M. Maruzzo, A. Roma, M. E. Johansson, U. Stierner, S. Lundstam, U. Basso *et al.*, *Cell reports* **15**, 1822 (2016).
15. R. Agren, A. Mardinoglu, A. Asplund, C. Kampf, M. Uhlen and J. Nielsen, *Molecular systems biology* **10**, p. 721 (2014).
16. K. Yizhak, E. Gaude, S. Le Dévédec, Y. Y. Waldman, G. Y. Stein, B. van de Water, C. Frezza and E. Ruppin, *Elife* **3**, p. e03641 (2014).
17. J. D. Orth, I. Thiele and B. Ø. Palsson, *Nature biotechnology* **28**, 245 (2010).
18. N. E. Lewis, H. Nagarajan and B. O. Palsson, *Nature Reviews Microbiology* **10**, 291 (2012).
19. A. M. Feist and B. O. Palsson, *Current opinion in microbiology* **13**, 344 (2010).
20. A. Bordbar, N. E. Lewis, J. Schellenberger, B. Ø. Palsson and N. Jamshidi, *Molecular systems biology* **6**, p. 422 (2010).
21. N. E. Lewis, G. Schramm, A. Bordbar, J. Schellenberger, M. P. Andersen, J. K. Cheng, N. Patel, A. Yee, R. A. Lewis, R. Eils *et al.*, *Nature biotechnology* **28**, 1279 (2010).
22. A. Thomas, S. Rahmanian, A. Bordbar, B. Ø. Palsson and N. Jamshidi, *Scientific reports* **4** (2014).
23. A. Schultz and A. A. Qutub, *PLoS Comput Biol* **12**, p. e1004808 (2016).
24. S. R. Estévez and Z. Nikoloski, *Front. Plant Sci* **5**, 10 (2014).
25. N. D. Price, J. Schellenberger and B. O. Palsson, *Biophysical journal* **87**, 2172 (2004).
26. D. E. Kaufman and R. L. Smith, *Operations Research* **46**, 84 (1998).
27. N. Chaudhary, K. Tøndel, J. Puchałka, V. A. M. dos Santos and R. Bhatnagar, *Molecular BioSystems* (2016).
28. W. Megchelenbrink, M. Huynen and E. Marchiori, *PloS one* **9**, p. e86587 (2014).
29. S. Bordel, R. Agren and J. Nielsen, *PLoS Comput Biol* **6**, p. e1000859 (2010).
30. P. A. Saa and L. K. Nielsen, *Bioinformatics* , p. btw132 (2016).
31. D. De Martino, M. Mori and V. Parisi, *PloS one* **10**, p. e0122670 (2015).
32. M. Uhlén, L. Fagerberg, B. M. Hallström, C. Lindskog, P. Oksvold, A. Mardinoglu, Å. Sivertsson, C. Kampf, E. Sjöstedt, A. Asplund *et al.*, *Science* **347**, p. 1260419 (2015).
33. J. Schellenberger, R. Que, R. M. Fleming, I. Thiele, J. D. Orth, A. M. Feist, D. C. Zielinski, A. Bordbar, N. E. Lewis, S. Rahmanian *et al.*, *Nature protocols* **6**, 1290 (2011).
34. N. C. Duarte, S. A. Becker, N. Jamshidi, I. Thiele, M. L. Mo, T. D. Vo, R. Srivas and B. Ø. Palsson, *Proceedings of the National Academy of Sciences* **104**, 1777 (2007).

35. T. M. Horton, Y. Qiu, G. Jenkins and S. M. Kornblau, *Blood* **124**, 3784 (2014).
36. A. Bordbar, N. Jamshidi and B. O. Palsson, *BMC systems biology* **5**, p. 110 (2011).
37. T. Mashima, H. Seimiya and T. Tsuruo, *British journal of cancer* **100**, 1369 (2009).
38. J. A. Menendez and R. Lupu, *Nature Reviews Cancer* **7**, 763 (2007).
39. F. P. Kuhajda, *Cancer research* **66**, 5977 (2006).
40. T. Migita, S. Okabe, K. Ikeda, S. Igarashi, S. Sugawara, A. Tomida, R. Taguchi, T. Soga and H. Seimiya, *The American journal of pathology* **182**, 1800 (2013).
41. F. P. Kuhajda, K. Jenner, F. D. Wood, R. A. Hennigar, L. B. Jacobs, J. D. Dick and G. R. Pasternack, *Proceedings of the National Academy of Sciences* **91**, 6379 (1994).
42. H. Cook and M. Spence, *Canadian Journal of Biochemistry and Cell Biology* **63**, 919 (1985).
43. H. W. Cook and M. W. Spence, *Biochimica et Biophysica Acta (BBA)-Lipids and Lipid Metabolism* **918**, 217 (1987).
44. M. S. Patel, J. J. Russell and H. Gershman, *Proceedings of the National Academy of Sciences* **78**, 7214 (1981).
45. L. M. Roeder, S. E. Poduslo and J. T. Tildon, *Journal of neuroscience research* **8**, 671 (1982).
46. R. J. DeBerardinis and T. Cheng, *Oncogene* **29**, 313 (2010).
47. C. T. Hensley, A. T. Wasti and R. J. DeBerardinis, *The Journal of clinical investigation* **123**, 3678 (2013).
48. E. A. Maher, I. Marin-Valencia, R. M. Bachoo, T. Mashimo, J. Raisanen, K. J. Hatanpaa, A. Jindal, F. M. Jeffrey, C. Choi, C. Madden *et al.*, *NMR in biomedicine* **25**, 1234 (2012).
49. I. Marin-Valencia, C. Yang, T. Mashimo, S. Cho, H. Baek, X.-L. Yang, K. N. Rajagopalan, M. Maddie, V. Vemireddy, Z. Zhao *et al.*, *Cell metabolism* **15**, 827 (2012).
50. M. Ardawi and E. Newsholme, Glutamine metabolism in lymphoid tissues, in *Glutamine metabolism in mammalian tissues*, (Springer, 1984) pp. 235–246.
51. H.-N. Kung, J. R. Marks and J.-T. Chi, *PLoS Genet* **7**, p. e1002229 (2011).
52. J. Iglehart, R. M. York, A. P. Modest, H. Lazarus and D. Livingston, *Journal of Biological Chemistry* **252**, 7184 (1977).
53. P. Gout, A. Buckley, C. Simms and N. Bruchovsky, *Leukemia (08876924)* **15** (2001).
54. P. Gout, Y. Kang, D. Buckley, N. Bruchovsky and A. Buckley, *Leukemia* **11**, 1329 (1997).
55. D. N. Wheatley, *Seminars in Cancer Biology* **15**, 247 (2005).
56. L. Scott, J. Lamb, S. Smith and D. Wheatley, *British journal of cancer* **83**, p. 800 (2000).
57. F. Mussai, S. Egan, J. Higginbotham-Jones, T. Perry, A. Beggs, E. Odintsova, J. Loke, G. Pratt, A. Lo, M. Ng *et al.*, *Blood* **125**, 2386 (2015).
58. M. G. Buse, *American Journal of Physiology-Endocrinology And Metabolism* **290**, E1 (2006).
59. M.-J. J. Pouwels, C. J. Tack, P. N. Span, A. J. Olthaar, C. Sweep, F. C. Huvers, J. A. Lutterman and A. R. Hermus, *The Journal of Clinical Endocrinology & Metabolism* **89**, 5132 (2004).
60. L. Wells, K. Vosseller and G. Hart, *Cellular and Molecular Life Sciences CMLS* **60**, 222 (2003).
61. E. E. Calle and R. Kaaks, *Nature Reviews Cancer* **4**, 579 (2004).
62. E. E. Calle, C. Rodriguez, K. Walker-Thurmond and M. J. Thun, *New England Journal of Medicine* **348**, 1625 (2003).
63. T. N. Sergentanis, G. Tsivgoulis, C. Perlepe, I. Ntanasis-Stathopoulos, I.-G. Tzanninis, I. N. Sergentanis and T. Psaltopoulou, *PloS one* **10**, p. e0136974 (2015).
64. T. Niedermaier, G. Behrens, D. Schmid, I. Schlecht, B. Fischer and M. F. Leitzmann, *Neurology* **85**, 1342 (2015).
65. Z. A. King, J. Lu, A. Dräger, P. Miller, S. Federowicz, J. A. Lerman, A. Ebrahim, B. O. Palsson and N. E. Lewis, *Nucleic acids research* **44**, D515 (2016).

DIFFERENTIAL PATHWAY DEPENDENCY DISCOVERY ASSOCIATED WITH DRUG RESPONSE ACROSS CANCER CELL LINES [*]

GIL SPEYER, DIVYA MAHENDRA[†], HAI J. TRAN[†], JEFF KIEFER

The Translational Genomics Research Institute
Phoenix, AZ 85004, U.S.A.
Email: gspeyer@tgen.org, mahendradivya@gmail.com, hjtran@brown.edu, jkiefer@tgen.org

STUART L. SCHREIBER, PAUL A. CLEMONS

Broad Institute of Harvard and MIT
Cambridge MA 02142, U.S.A.
Email: stuart_schreiber@harvard.edu, pclemons@broadinstitute.org

HARSHIL DHRUV, MICHAEL BERENS, SEUNGCHAN KIM

The Translational Genomics Research Institute
Phoenix, AZ 85004, U.S.A.
Email: hdhruv@tgen.org, mberens@tgen.org, skim@tgen.org

The effort to personalize treatment plans for cancer patients involves the identification of drug treatments that can effectively target the disease while minimizing the likelihood of adverse reactions. In this study, the gene-expression profile of 810 cancer cell lines and their response data to 368 small molecules from the Cancer Therapeutics Research Portal (CTRP) are analyzed to identify pathways with significant rewiring between genes, or differential gene dependency, between sensitive and non-sensitive cell lines. Identified pathways and their corresponding differential dependency networks are further analyzed to discover essentiality and specificity mediators of cell line response to drugs/compounds. For analysis we use the previously published method EDDY (Evaluation of Differential DependencY). EDDY first constructs likelihood distributions of gene-dependency networks, aided by known gene-gene interaction, for two given conditions, for example, sensitive cell lines vs. non-sensitive cell lines. These sets of networks yield a divergence value between two distributions of network likelihoods that can be assessed for significance using permutation tests. Resulting differential dependency networks are then further analyzed to identify genes, termed *mediators*, which may play important roles in biological signaling in certain cell lines that are sensitive or non-sensitive to the drugs. Establishing statistical correspondence between compounds and mediators can improve understanding of known gene dependencies associated with drug response while also discovering new dependencies. Millions of compute hours resulted in thousands of these statistical discoveries. EDDY identified 8,811 statistically significant pathways leading to 26,822 compound-pathway-mediator triplets. By incorporating STITCH and STRING databases, we could construct evidence networks for 14,415 compound-pathway-mediator triplets for support. The results of this analysis are presented in a searchable website to aid researchers in studying potential molecular mechanisms underlying cells' drug response as well as in designing experiments for the purpose of personalized treatment regimens.

[*] This work was supported in part by the National Cancer Institute, National Institutes of Health [1U01CA168397] and a grant from Dell, Inc. via its *Legacy of Good* program that seeks to put technology and expertise to work where it can do the most for people and the planet.
[†] D. Mahendra and H. Tran were supported by the Helios Education Foundation through the Helios Scholars at TGen summer internship program in biomedical research at the Translational Genomics Research Institute in Phoenix, AZ.

1. Introduction

The effort to personalize treatment plans for patients involves the identification of drug treatments that can effectively target the disease while minimizing the likelihood of adverse reactions. The advent of high-throughput –omics and drug-screening data has given rise to the development of complex analytical approaches to identify biomarkers and drug-targets) [1]. Considering complex molecular mechanisms underlying complex diseases such as cancer, the discovery of such biomarkers and subtype-specific drug targets must be based on activities of multiple genes rather than individual genes. Gene Set Enrichment Analysis (GSEA) [2] is one popular method of testing for differential expression of gene sets between conditions. As pathways are capable of complex rewiring between conditions, network-based analyses have become increasingly attractive for extraction of biological hypotheses from big data [3]. For example, the approaches to identify individual differential dependencies‡ [4-8] or condition-specific sub-networks from genome-wide dependency networks such as a protein-protein interaction networks have gained much interest [9-11] for the determination of biomarkers and subtype-specific therapeutic vulnerabilities.

Recently, we developed a novel computational method *Evaluation of Differential DependencY* (EDDY) that identifies pathways enriched with differential dependencies and that discovers mediators as potential therapeutic targets. The method has been further improved by incorporating known gene interactions as prior knowledge. The method has been successfully applied to the study of glioblastoma (GBM) [12, 13] and adrenocortical carcinoma (ACC) [14].

In this study, we present results from an integrated analysis of large-scale transcriptomic data of 810 cancer cell lines and large-scale high-throughput screening data of the same cancer cell lines across 368 compounds using EDDY algorithm. The analysis not only identified the pathways enriched with differential dependencies between sensitive and non-sensitive cancer cell lines to each compound, but also discovered mediators as potential novel targets of the compound via graphical analysis of differential dependency networks. Identified compound-pathway-mediator triplets were further queried across known drug-gene database as well as a known gene-gene interaction database to identify corroborating evidence to support newly discovered compound-pathway-mediator triplets. We also developed a searchable website to aid researchers in studying potential molecular mechanisms underlying cells' drug response and in designing experiments for the purpose of personalized treatment regimens, publicly available at http://biocomputing.tgen.org/software/EDDY/CTRP.

2. Methods

2.1. *High-Throughput Drug Screening of Cancer Cell Lines*

The Cancer Cell Line Encyclopedia (CCLE) project is an effort to conduct detailed genetic characterization of a large panel of human cancer cell lines. The CCLE provides public access to DNA copy number, mRNA expression, and mutation data for 1,000 cancer cell lines,

‡ In this manuscript, we use 'dependency' to denote statistical dependencies derived from data such as co-expression, conditional dependencies, and 'interaction' to denote *known* relationships between genes or related molecules.

encompassing 36 different tumor types [15].

The Center for the Science of Therapeutics at Broad Institute performed analysis of sensitivity of CCLE cell lines using ~500 small molecules as perturbagens, and made the data available at the Cancer Therapeutics Response Portal (CTRP; http://www.broadinstitute.org/ctrp/). The "Informer Set" consists of 481 small compounds, including 70 FDA approved drugs, 100 clinical candidates and 311 small-molecule probes. In this study, we used the transcriptomic profile and CTRP drug-response data to identify pathways with condition-specific rewiring of gene dependencies in the context of drug sensitivity [16, 17]. All of these aforementioned processed data is publicly available on the CTD² data portal (https://ctd2.nci.nih.gov/dataPortal/).

2.2. *EDDY: Evaluation of Differential Dependency*

EDDY is a statistical approach that combines pathway-guided and differential dependency analyses in a probabilistic framework [12, 13]. The algorithm queries each pathway (gene set) in a database such as BioCarta (http://cgap.nci.nih.gov/Pathways/BioCarta_Pathways) or REACTOME [18] to test for differential dependencies across the set of genes between two or more conditions, by comparing gene-dependency networks constructed for each condition. In evaluating differential dependency, EDDY uses a network likelihood distribution over multiple networks constructed via resampling for each condition and compares the distributions between the conditions, instead of just using the single, most probable network from each condition. The statistical significance of

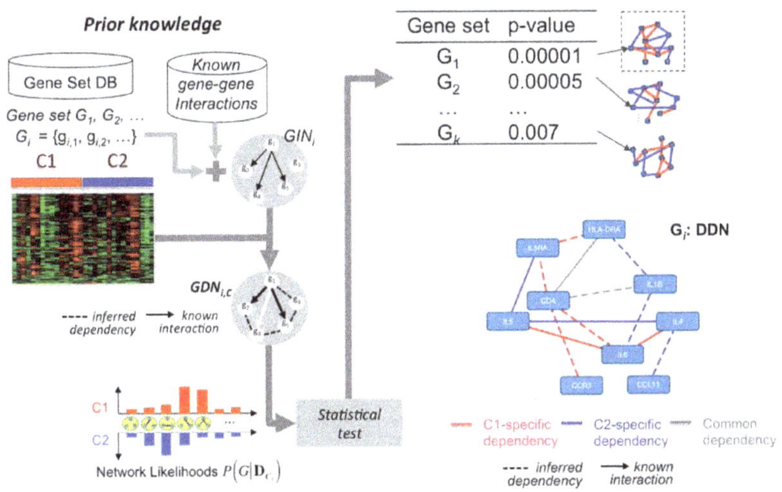

Figure 1. Knowledge-assisted EDDY Workflow. $GDN_{i,C}$ is a gene-dependency network constructed for a gene set G_i, for condition C, aided by gene interaction network GIN_i. A network likelihood distribution over multiple networks is constructed via resampling for each condition and the network score distributions between the conditions are compared. Permutation testing assesses the significance of the divergence between the distributions of scores. Differential dependency networks can then be constructed for statistically significant gene sets.

the divergence is then estimated using asymptotic approximation of Jensen-Shannon divergence based on a beta distribution whose parameters are estimated using a permutation test. Probabilistic and gene-set assisted approaches together contribute to significantly higher sensitivity and specificity of EDDY, compared to other methods, such as GSEA and Gene Set Co-expression Analysis (GSCA) [12].

Incorporation of Prior Knowledge into EDDY: Known interactions from the Pathway Commons 2 (http://www.pathwaycommons.org) database are integrated into EDDY as prior knowledge (Figure 1). This integration has been shown to improve the interpretability of results from EDDY. Prior weight (W_p) is specified to determine the degree of weight that is given to the prior knowledge in evaluating new edges to be included in the proposed dependency structure. Since prior knowledge is not condition-specific, large prior weight could decrease EDDY's sensitivity to detect differential dependency while reducing discovery of false-positive dependencies. For this analysis, a prior weight of $W_p = 0.5$ was used, meaning that any edges with half the support from data were included in the dependency network. The choice was based on extensive analysis of various data sets where $W_p = 0.5$ seemed to give the best compromise between sensitivity and false discovery rate when varying prior weight, as reported in Speyer et. al. [13].

2.3. *Input Data*

Transcriptomic data: BAM files of 935 CCLE cell lines downloaded from the Cancer Genomics Hub (https://cghub.ucsc.edu) were converted to a FASTQ format and transcript quantification was performed using Salmon [19] to obtain quantitative estimate of mRNA expression in TPM (transcripts per million). These mRNA expression values were $\log_2$ transformed and quantized to values -1 (under-expressed), 0 (intermediate), and 1 (over-expressed). For each gene, median average deviation (MAD) was computed and used to determine under-expression (MAD < -1), over-expression (MAD > 1), and intermediate.

Drug sensitivity: The cell lines were grouped into sensitive and non-sensitive classes using the Small-Molecule Cancer Cell Line Sensitivity Profiling CTRP 2.0 2015 Dataset, acquired from CTD^2 (Cancer Target Discovery and Development). CTRP summarizes drug sensitivity between each cell line and drug pair using the area-under-percent-viability-curve (AUC) values [16, 17]. We used the 'extremevalues' R package to identify outliers in AUC values and group the cell lines into sensitive (-1; lower-end outliers), non-sensitive (1; upper-end outliers), and intermediate (0; non-outliers) groups for each compound.

In order to conduct a statistically meaningful analysis using EDDY, only those drugs that had at least 50 samples in each sensitive and non-sensitive class were analyzed. This reduced the number of drugs that could be analyzed to 368 drugs.

2.4. *Identification of Mediators*

For each compound, the results from EDDY analysis (Figure 2) are summarized into 1) a list of pathways enriched with differential dependency of statistical significance, and 2) a differential dependency network (DDN) that captures how gene dependency changes between sensitive and non-sensitive cell lines. We identified those genes that seemed to play a significantly different role (based on statistical dependencies) between cell lines that were sensitive to a drug and cell lines that were non-sensitive, and termed them as *mediators*.

Essentiality mediators: Each DDN is split into condition-specific dependency networks (CDNs) where each CDN is composed of dependencies manifested in each condition. We then compute between-ness centrality for each gene in both CDNs and compute the difference of the betweenness centrality. The genes with the most differential betweenness centrality are termed *essentiality mediators*, as the genes with highest betweenness centrality in gene regulatory network are often interpreted as essential genes [20].

Specificity mediators: We also analyzed how many dependencies for each gene change between the CDN from sensitive cell lines and the CDN from non-sensitive cell lines. Formally, Let $P_C = E_C/(E_C + E_S)$, a proportion of condition-specific edges (E_C) across the overall number of edges ($E_C + E_S$), and E_{C_i} be the number of condition-specific edges and E_{S_i} be number of shared edges, of a gene i. Note $E_C = \sum_i E_{C_i}$ and $E_S = \sum_i E_{S_i}$. We can then compute the probability, $\Pr(k \geq E_{C_i})$, that a gene i can have E_{C_i} or more condition-specific edges by random chance, via binomial probability $B(k, E_{C_i} + E_{S_i}, P_C)$. If this probability, $\Pr(k \geq E_{C_i}) < 0.05$, we termed gene i as *specificity mediator*.

2.5. *Evidence Networks*

However, uncertainty in interpreting these drug-pathway-mediator triplets hinders prioritization of hypotheses or experimental design to explore these potentially valuable results. We address this challenge by constructing evidence networks built with protein and drug interactions from the STRING and STITCH interaction databases. STITCH and STRING are sister knowledge-bases that store scored drug-protein interactions and protein-protein interactions, respectively [21, 22]. As compounds can have multiple names, from commercial and generic labels to chemical formula

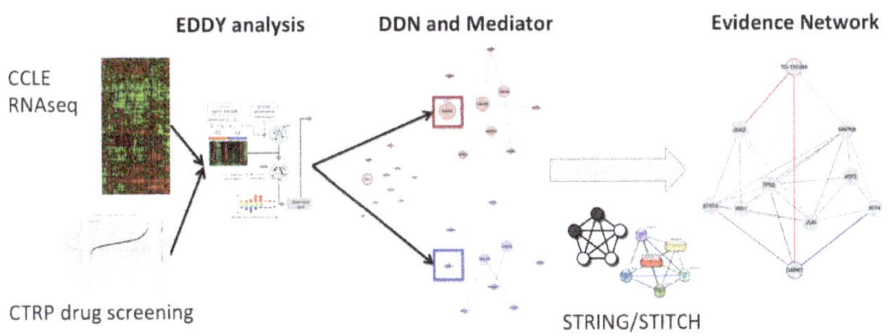

Figure 2. Overall workflow of EDDY analysis of CCLE and CTRP data. EDDY identifies significant pathways from RNA expression and compound-response categorization of cancer cell lines. Graphical analysis of output networks (edge color indicating condition) identifies important genes, termed mediators. Mining knowledge bases yields evidence networks for compound-mediator pairings (edge color here indicating evidence type).

and IUPAC ID, the database employed a unifying InChIKey to maximize comprehensiveness and to avoid false negatives.

Evidence networks were generated using a modified Yen's *K*-shortest paths algorithm [23] with a weight function of W(EDGE) = 1 – EDGE.SCORE, so that edges with higher scores would be preferred over edges with lower scores (all scores are within the interval [0,1] and are based on how compelling the supporting evidence is). To generate the evidence networks, shortest paths were continually found and added to the network until there were no more paths from the drug to the gene or there were at least *N* distinct nodes in the sub-network, where *N* is some arbitrary threshold. *N* was not a strict floor as sometimes the last path added to the sub-network would add two or more distinct nodes pushing the total number of distinct nodes over the threshold. Instead, *N* was used simply as a stopping condition and was chosen in order to prevent generation of evidence networks that would be too overwhelming for users to interpret. Choosing $N = 5$ yielded abundant evidence nets without excessive density. Dijkstra's shortest-path algorithm with a Fibonacci heap was used as the supporting shortest-path algorithm in the modified Yen's *K*-shortest-paths algorithm [24, 25].

3. Results

3.1. *Pathway and Mediator Analysis*

EDDY analysis identified a total of 8,811 statistically significant pathways and 26,822 compound-pathway-mediator triplets. Of these, 534 pathways out of 685 BIOCARTA and REACTOME pathways were identified for at least one compound, and 2,401 genes out of 4,298 unique BIOCARTA and REACTOME genes were identified as mediators for at least one compound. On average each compound identified about 24 pathways and 73 mediators. We found that for 125 compounds, EDDY identified pathways that had the compound's intended target in their DDN, and 29 mediators were identified as intended targets. Only 248 out of the 368 compounds had intended targets that EDDY could potentially identify within the REACTOME and BIOCARTA pathways. Hence, EDDY identified pathways that included the intended target for 125 out of 248 compounds (50.4%). We tabulated (Table 1 & Table 2) the top 10 statistically significant pathways and mediators, respectively, which were identified by the largest number of compounds. We can see that the top two pathways that were statistically significant were ERYTH (erythrocyte

Table 1. Top 10 most commonly identified statistically significant pathways that were statistically significant

Pathway	# Compounds	Database
Erythrocyte differentiation (ERYTH)	78	BIOCARTA
Cells and molecules involved in local acute inflammatory response (LAIR)	61	BIOCARTA
CBL mediated ligand-induced downregulation of EGF receptors (CBL)	55	BIOCARTA
TERMINATION OF O GLYCAN BIOSYNTHESIS	52	REACTOME
SIGNALING BY HIPPO	49	REACTOME
NUCLEOTIDE LIKE PURINERGIC RECEPTORS	48	REACTOME
ZINC TRANSPORTERS	46	REACTOME
GRANULOCYTES	45	BIOCARTA
SYNTHESIS OF SUBSTRATES IN N GLYCAN BIOSYTHESIS	44	REACTOME
PURINE CATABOLISM	43	REACTOME

differentiation pathway) and LAIR (pathway for cells and molecules involved in local acute inflammatory response) from BIOCARTA. The erythrocyte differentiation pathway is the pathway responsible for the formation of red blood cells from the bone marrow. It is expected that this pathway would be altered in hematopoietic cancers and that its alteration would be involved in immune responses. The genes found in this pathway include TGFB2 and cytokines IL1A, IL3, IL6, IL9, and IL11. Cytokines are involved in various immune responses and inflammatory processes. The LAIR pathway includes mechanisms associated with the releases of cytokines IL1A and IL6. The genes IL1A and IL6 are among the top fourteen mediators identified by compounds in EDDY and they are also intended targets for the ERYTH and LAIR pathways. IL1A gene is a cytokine involved in various immune responses, inflammatory processes, and hematopoiesis. This protein is released in response to cell injury. IL6 is also a cytokine that functions in inflammation and maturation of B cells [26]. Indeed, upon further examination of the response data for the compounds differentially dependent for the ERYTH and LAIR pathways, hematopoietic cell lines were on average six times more prevalent in the sensitive versus the non-sensitive groups.

Table 2. The top 10 most commonly identified mediators

Pathway	# Compounds
MAPK1	185
MAPK3	171
GRB2	168
NUP210	158
HRAS	136
NUP37	125
AKT1	120
ORC4	114
MAPK14	114
CDK1	114

The MAPK signaling pathway is an important signaling pathway in cancer studies because it is altered in many different cancer types and regulates processes such as cell proliferation, cell differentiation, and cell death. MAPK1, MAPK3 and MAPK14 are mitogen-activated protein kinases and are members of the MAP kinase family. These genes act in signaling pathways (MAPK signaling, immune response) and various other cellular processes such as proliferation, differentiation, and cell cycle progression. MAPK14 is activated by environmental stresses and cytokines associated to inflammatory responses. MAP kinases play important roles in cascades of cellular responses and lead to direct activation of transcription factors [27].

3.2. *Evidence Network Analysis*

EDDY-CTRP analysis identified 26,822 drug-pathway-mediator triplets. Among these pairs, 19,222 of them consisted of a drug or a gene that is contained within the STRING and STITCH databases. Mining STITCH and STRING for each of 19,222 unique compound-pathway-mediator triplets yielded 14,415 evidence networks (~75%) of a path with 3 or fewer intermediate genes. These evidence networks are integrated into the main EDDY-CTRP portal as searchable tables (Table 3).

We note that 102 evidence networks indeed were direct compound and mediator relations, among which only 34 were intended targets defined in the CTRP data and annotation. This indicates

Table 3. Distribution of the number of intermediate genes in shortest path between drug and mediator pair.

	Direct targets	Indirect targets		
		# of intermediate genes in shortest path		
		1	2	3
# of pairs	102	988	3,410	9,915

STITCH/STRING contain drug-target relations that were not included in the CTRP database, but EDDY-CTRP analysis was able to discover those relations. Most of these evidence networks were for drug-pathway-mediator triplets where mediators were not direct targets of drug but had some known functional association to the drug (based on STITCH/STRING database). Note that known "hub" genes such as TP53 turned out to have high prevalence in the constructed evidence networks. In future development, the algorithm will introduce weighting to counter this bias.

3.3. *Interactive and Searchable Web-Portal for EDDY-CTRP Results*

The web-portal of the CTRP analysis (http://biocomputing.tgen.org/software/EDDY/CTRP) consists of two main views: CTRP compound-centric and mediator-centric. These views provide alternate perspectives on hypothesis-testing data from the EDDY analysis. CTRP compound-centric view (Figure 3) provides pathways enriched with differential dependencies for each of 368 compounds uncovered by EDDY. For each compound, a user can explore each identified pathway, corresponding DDNs, and mediators. Mediator-centric view (Figure 4) lists all compound-pathway-mediator triplets uncovered across all compounds and all identified pathways. For each triplet, a user can also explore evidence networks as well as corresponding DDNs and pathways.

4. Case Studies: Potential Alternative Drug Targets

4.1. *DAPK3 as an Alternative Target for TG-101348*

TG-101348 was developed as a selective inhibitor of JAK2 kinase for the treatment of myeloproliferative disorder [28]. EDDY identified 29 pathways significantly enriched with differential dependency, and 66 mediators. One of the pathways is the EPONFKB pathway, which has JAK2 as an identified mediator, and, examining this DDN, JAK2 has exclusively sensitive-specific edges. We obtained the evidence networks for 59 of 66 mediators, and one of those mediators with evidence network is DAPK3 which is identified as a direct target of TG-101348, based on STITCH database. DAPK3 was identified as a mediator for the "ROLE OF DCC IN REGULATING APOPTOSIS" pathway which has an altered differential dependency

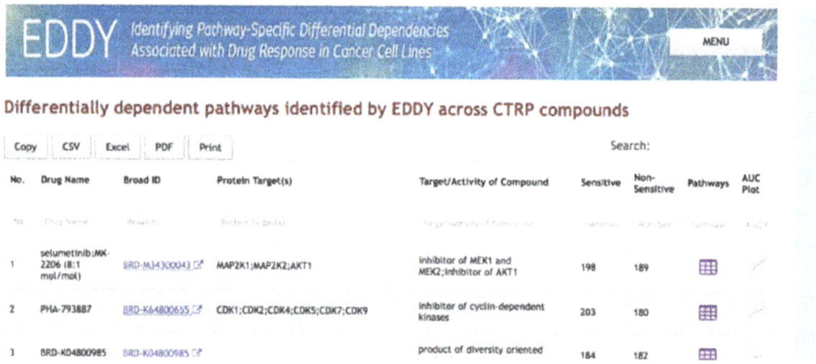

Figure 3. CTRP compound-centric view

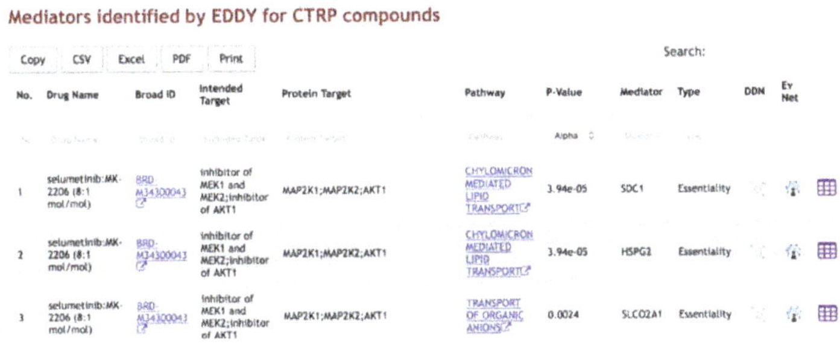

Figure 4. CTRP mediator-centric view

network for TG-101348. The gene product of DAPK3 was a mediator in this pathway due to high change of essentiality (betweenness centrality) between the condition-specific dependency networks TG-101348 sensitive cancer cell lines and non-sensitive cancer cell lines. In TG-101348-sensitive cell lines, DAPK3 is highly connected in the network (Figure 5a), consistent with DAPK3 playing a central role in a functioning apoptotic network. In the non-sensitive cell lines, however, DAPK3 is not connected to the rest of the network (Figure 5b), corroborating the indication that disconnected DAPK3 may confer insensitivity to TG-101348 sensitivity.

The evidence network built for TG-101348 - DAPK3 supports this hypothesis by showing a direct association between TG-101348 and DAPK3, discovered from the STITCH database (Figure 5c). Indeed, the evidence link was from a study that showed TG-101348 can inhibit the kinase activity of DAPK3, indicating that TG-101348 actually does target DAPK3 in addition to JAK2. Additionally, an association between the downstream JAK2 modulator and DAPK3 was revealed suggesting further signaling interactions targeted by TG-101348 [29]. So, while this target was not annotated in CTRP annotation for known targets of TG-101348, EDDY-CTRP

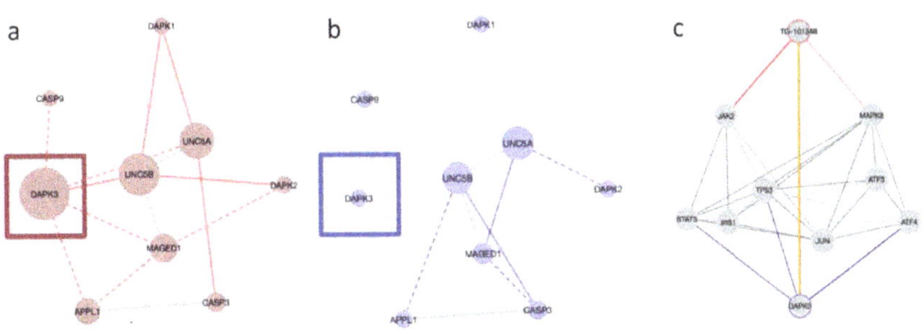

Figure 5. **(a)** Condition-specific dependency network (CDN) for TG-101348-sensitive cell lines. Dashed lines represent statistical dependencies while solid lines known interactions. Size of nodes represents node essentiality. **(b)** CDN for TG-101348-insensitive cell lines. **(c)** Evidence network for the TG-101348 – DAPK3 drug-mediator pair. All edges represent a known association based from the STRING/STITCH databases. Blue edges represent mediator-gene associations, red edges drug-gene associations, and yellow edge a direct drug-mediator association.

analysis was able to detect this relationship. This example illustrates EDDY can discover potentially novel targets of a compound and how the evidence network provides further contextual information regarding the possible mechanisms of how mediators selected in the EDDY analysis function to alter individual drug responses.

4.2. *HIF1A as an Alternative Target for Indisulam*

Indisulam is a carbonic anhydrase IX (CA9) inhibitor [30]. CA9 activity in cancer is associated with an acidic microenvironment that favors tumor cell survival and growth [31]. EDDY identified the HIF pathway as a DDN associated with indisulam response. The HIF pathway is important for cancer-cell survival in hypoxic conditions often seen in tumors [32]. In the non-responsive HIF pathway DDN two genes, HIF1A and JUN exhibit high essentiality compared to the responsive HIF DDN (Figure 6a). HIF1A is a major gene that signals for cell survival in hypoxic conditions [32]. The evidence network for indisulam and HIF1A reveals a direct link between CA9 and HIF1A (Figure 6c). This would not be evident if investigator had only HIF pathway DDN evidence. Inspection of the evidence from STRING shows that HIF1A positively regulates CA9 expression. Cancer cells may be non-responsive to indisulam because HIF1A increases CA9 levels such that the drug is not effective at tested concentration in fully inhibiting CA9. This example shows how the evidence network is able to mechanistically link EDDY DDNs to drug targets and expand understanding of signaling events associated with drug response.

5. Conclusions

While the current CTRP dataset allows the study of the correlations between genetic features with sensitivity to compounds, and while there are previous studies associating genes with compound sensitivity [33], this paper presents an unprecedented identification of pathways with differential dependency networks across a large number of cancer cell lines with drug-screening data. Additionally we have created a web repository to allow clinicians and researchers to view the results of our analysis. The web repository provides an interactive method to view the results for

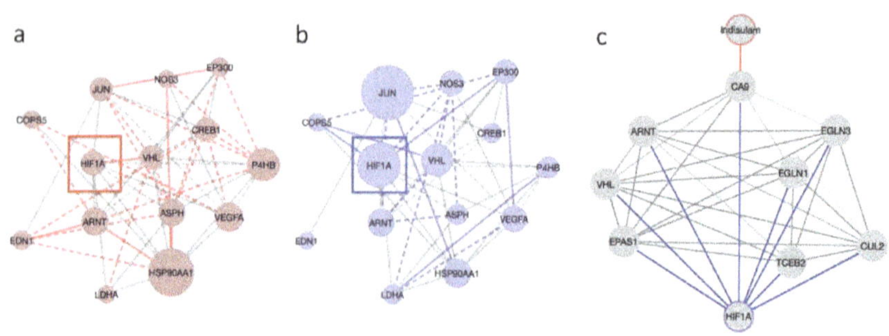

Figure 6. **(a)** Condition-specific dependency network (CDN) for indisulam for drug-sensitive cell lines. **(b)** CDN for indisulam for drug-insensitive cell lines. **(c)** Evidence network for the indisulam – HIF1A drug-mediator pair.

specific drugs. Researchers can query the intended targets, genes, or pathways to identify types of drugs, known targets, and to discover hitherto unknown mediators. We integrated quick unique links to the CTRP database, MSigDB Database, and Gene Cards, for each of the compounds, pathways, and genes. These links allow users to view the analysis and information about the drug, pathway, or gene seamlessly. We also provide links to the interactive DDN and condition-specific CDNs so that users can move around the nodes and edges to better analyze the results. In addition we provide links to generate the Oncoprints for the sensitive and non-sensitive cell lines for each DDN. These links allow the users to look at the mutation data used to generate the DDN.

This resource can be valuable for researchers to explore potential targets of their interest and allow them to look at differential dependencies across a large number of cell lines and compounds. It may aid in studying potential molecular mechanisms underlying cells' response to drug as well as designing experiments for the purpose of personalized treatment regimens.

Computational methods that can efficiently predict the effectiveness of drugs based on the genetic makeup of tumors would provide a major breakthrough towards personalized therapy for cancer patients based on their tumor's molecular markers. To strengthen the validity of our analysis and resource, experimental validation of the pathways identified by EDDY is warranted. We anticipate that this web repository will be a living resource for clinicians and researchers to use for designing experiments and identifying potential personalized treatment regimens.

References

1. Roden, D.M. and A.L. George, Jr., *The genetic basis of variability in drug responses.* Nat Rev Drug Discov, 2002. **1**(1): p. 37-44.
2. Subramanian, A., et al., *Gene set enrichment analysis: a knowledge-based approach for interpreting genome-wide expression profiles.* Proceedings of the National Academy of Sciences of the United States of America, 2005. **102**(43): p. 15545-50.
3. Califano, A., *Rewiring makes the difference.* Mol Syst Biol, 2011. **7**: p. 463.
4. Lai, Y., et al., *A statistical method for identifying differential gene-gene co-expression patterns.* Bioinformatics, 2004. **20**(17): p. 3146-3155.
5. Hu, R., et al., *Detecting intergene correlation changes in microarray analysis: a new approach to gene selection.* BMC Bioinformatics, 2009. **10**(1): p. 20.
6. Mentzen, W., M. Floris, and A. de la Fuente, *Dissecting the dynamics of dysregulation of cellular processes in mouse mammary gland tumor.* BMC Genomics, 2009. **10**(1): p. 601.
7. Zhang, B., et al., *Differential dependency network analysis to identify condition-specific topological changes in biological networks.* Bioinformatics, 2009. **25**(4): p. 526-32.
8. Zhang, B., et al., *DDN: a caBIG(R) analytical tool for differential network analysis.* Bioinformatics, 2011. **27**(7): p. 1036-8.
9. Hwang, T. and T. Park, *Identification of differentially expressed subnetworks based on multivariate ANOVA.* BMC Bioinformatics, 2009. **10**(1): p. 128.
10. Kim, Y., et al., *Principal network analysis: Identification of subnetworks representing major dynamics using gene expression data.* Bioinformatics, 2010.
11. Ma, H., et al., *COSINE: COndition-SpecIfic sub-NEtwork identification using a global optimization method.* Bioinformatics, 2011.
12. Jung, S. and S. Kim, *EDDY: a novel statistical gene set test method to detect differential genetic dependencies.* Nucleic Acids Res, 2014. **42**(7): p. e60.

13. Speyer, G., et al., *Knowledge-Assisted Approach to Identify Pathways with Differential Dependencies.* Pac Symp Biocomput, 2016. **21**: p. 33-44.
14. Zheng, S., et al., *Comprehensive Pan-Genomic Characterization of Adrenocortical Carcinoma.* Cancer Cell, 2016. **29**(5): p. 723-36.
15. Barretina, J., et al., *The Cancer Cell Line Encyclopedia enables predictive modelling of anticancer drug sensitivity.* Nature, 2012. **483**(7391): p. 603-7.
16. Seashore-Ludlow, B., et al., *Harnessing Connectivity in a Large-Scale Small-Molecule Sensitivity Dataset.* Cancer Discov, 2015. **5**(11): p. 1210-23.
17. Rees, M.G., et al., *Correlating chemical sensitivity and basal gene expression reveals mechanism of action.* Nat Chem Biol, 2016. **12**(2): p. 109-16.
18. Fabregat, A., et al., *The Reactome pathway Knowledgebase.* Nucleic Acids Res, 2016. **44**(D1): p. D481-7.
19. Patro, R., G. Duggal, and C. Kingsford, *Accurate, fast, and model-aware transcript expression quantification with Salmon.* biorxiv, 2015.
20. Khuri, S. and S. Wuchty, *Essentiality and centrality in protein interaction networks revisited.* BMC Bioinformatics, 2015. **16**: p. 109.
21. Kuhn, M., et al., *STITCH 4: integration of protein-chemical interactions with user data.* Nucleic Acids Res, 2014. **42**(Database issue): p. D401-7.
22. Szklarczyk, D., et al., *STRING v10: protein-protein interaction networks, integrated over the tree of life.* Nucleic Acids Res, 2015. **43**(Database issue): p. D447-52.
23. Yen, J., *Finding the K Shortest Loopless Paths in a Network* Management Science, 1971. **17**(11): p. 712-716.
24. Dijkstra, E.W., *A note on two problems in connexion with graphs.* Numerische Mathematik, 1959. **1**(4): p. 269-271.
25. Fredman, M.L. and R.E. Tarjan, *Fibonacci heaps and their uses in improved network optimization algorithms.* Journal of the Association for Computing Machinery, 1987. **34**(3): p. 596-615.
26. Seruga, B., et al., *Cytokines and their relationship to the symptoms and outcome of cancer.* Nat Rev Cancer, 2008. **8**(11): p. 887-99.
27. Lei, Y.Y., et al., *Mitogen-activated protein kinase signal transduction in solid tumors.* Asian Pac J Cancer Prev, 2014. **15**(20): p. 8539-48.
28. Wernig, G., et al., *Efficacy of TG101348, a selective JAK2 inhibitor, in treatment of a murine model of JAK2V617F-induced polycythemia vera.* Cancer Cell, 2008. **13**(4): p. 311-20.
29. Sato, N., et al., *Physical and functional interactions between STAT3 and ZIP kinase.* Int Immunol, 2005. **17**(12): p. 1543-52.
30. Supuran, C.T., *Indisulam: an anticancer sulfonamide in clinical development.* Expert Opin Investig Drugs, 2003. **12**(2): p. 283-7.
31. Swietach, P., et al., *New insights into the physiological role of carbonic anhydrase IX in tumour pH regulation.* Oncogene, 2010. **29**(50): p. 6509-21.
32. Masson, N. and P.J. Ratcliffe, *Hypoxia signaling pathways in cancer metabolism: the importance of co-selecting interconnected physiological pathways.* Cancer Metab, 2014. **2**(1): p. 3.
33. Gottlieb, A. and R.B. Altman, *Integrating systems biology sources illuminates drug action.* Clin Pharmacol Ther, 2014. **95**(6): p. 663-9.

A METHYLATION-TO-EXPRESSION FEATURE MODEL FOR GENERATING ACCURATE PROGNOSTIC RISK SCORES AND IDENTIFYING DISEASE TARGETS IN CLEAR CELL KIDNEY CANCER

JEFFREY A. THOMPSON[1] and CARMEN J. MARSIT[2]

[1]*Program in Quantitative Biomedical Science, Geisel Medical School at Dartmouth College, Lebanon, NH 03756, USA*

[2]*Department of Environmental Health, Rollins School of Public Health at Emory Unveristy, Atlanta, GA 30322, USA*

E-mail: carmen.j.marsit@emory.edu

Many researchers now have available multiple high-dimensional molecular and clinical datasets when studying a disease. As we enter this multi-omic era of data analysis, new approaches that combine different levels of data (e.g. at the genomic and epigenomic levels) are required to fully capitalize on this opportunity. In this work, we outline a new approach to multi-omic data integration, which combines molecular and clinical predictors as part of a single analysis to create a prognostic risk score for clear cell renal cell carcinoma. The approach integrates data in multiple ways and yet creates models that are relatively straightforward to interpret and with a high level of performance. Furthermore, the proposed process of data integration captures relationships in the data that represent highly disease-relevant functions.

Keywords: prognostic; survival; cancer; data integration; eQTL; m2eQTL; m2eGene

1. Introduction

The recent abundance of large datasets of diverse molecular features have vastly increased our knowledge of cellular processes disrupted in disease; yet, these datasets, taken individually, have frequently failed to reveal useful biomarkers for complex diseases, such as cancer [1, 2].

Despite the clear utility of individual 'omic' datasets, such as gene expression, DNA methylation, copy number alteration, etc., in better understanding disease etiology and in some cases providing useful prognostic or predictive value [3], it is equally clear that each of these data types can only capture part of the disease signature in a cell. Therefore, interest has been growing in more holistic methods, which integrate data of different types. As of yet, these approaches have met with mixed success. For example, a study of long-term survival in patients with glioblastoma multiforme (an aggressive form of brain cancer), found that joint regression of different types of data did not improve predictive accuracy [4]. Another study across five different cancer types came to a similar conclusion [5]. Nevertheless, a more nuanced approach, based on integrating separate models built from individual datatypes for ovarian cancer outcomes did show a higher predictive accuracy for integration across datatypes [6].

A recent review of data integration approaches classified them as falling into one of two broad categories: multi-stage and meta-dimensional integration [7]. Multi-stage integration techniques are currently the most developed and wide-spread. These involve using separate analyses of multiple types of data, with the results from one data type used to filter, and presumably increase the power of, another. The most commonly used example of multi-stage integration is expression-quantitative trait loci (eQTL) analysis, wherein single nucleotide polymorphisms (SNPs) are associated with changes in gene expression, which in turn are associated with disease [8, 9]. Meta-dimensional techniques consist of integrated models, in which all data are used as part of a joint model or analysis, which might involve joint regression, or integration at the level of individual models [10, 11].

Important prognostic information may in some cases be obscured by noise. However, it is much less likely that noise will obscure that information from different types of data for the same features. For example, it may be that repression of a gene promoter through DNA methylation represents a disease state. Nevertheless, that gene's expression may be altered in healthy individuals through alternative regulation. Therefore, it may not be enough to capture the gene expression data alone. Furthermore, one type of data may capture nascent information of disease progression that is not yet apparent in other data types. In some cases, there may not be one superior type of data for predicting prognosis. Finally, there may be informative interactions between data types that are not possible to assess when using only one type of data. For these reasons, we hypothesize that an appropriate multi-omic data-integrated approach will create superior prognostics to those using only a single data type.

In this work, we developed a data integration approach for combining gene expression data with DNA-methylation to create prognostic models for clear cell renal cell carcinoma (the most common form of kidney cancer). Our data integration approach is a hybrid method combining both multi-stage and meta-dimensional elements but results in a model that is easily interpreted by those familiar with traditional statistical approaches. Furthermore, it is amenable to extremely high dimensional data but runs quickly compared to other methods. We demonstrate the viability of our approach in the context of creating prognostic markers for kidney cancer and compare it to two other methods that have proven successful in this context: random survival forests and penalized Cox regression [12–14].

We chose to integrate DNA methylation and gene expression because they have proven to be prognostically useful data sources for a number of cancers [12] and are highly related. DNA methylation controls tissue specific expression of genes. Therefore, if we can exploit this redundant information, we may be able to create a more informative prognostic model. Furthermore, it has long been suspected that aberrant DNA methylation itself is related to carcinogenesis [15], although only recently has evidence begun to mount for a causative role [4, 16]. Given that DNA methylation tends to be a more stable mark than gene expression [17, 18], in certain cases it may be informative where gene expression is not. In cancer, hypermethylation of gene promoters silences tumor suppressors and other genes throughout the genome [19]. Hypomethylation of other regions is associated with genomic instability [20]. Thus, disruption of DNA methylation patterns may be a potentially relevant etiological factor, which could increase the utility of our approach.

2. Methods

We used M2EFM, and two other approaches, to model overall survival in clear cell renal cell carcinoma. For the main analysis, gene expression and DNA methylation profiles from untreated, resected tumors for patients with clear cell renal cell carcinoma were created by The Cancer Genome Atlas (TCGA) project [21] on the Illumina HiSeq 2000 sequencing and

Illumina Infinium HumanMethylation450 platforms respectively. RNA-seq data normalization was performed by TCGA and normalized data were downloaded from the UCSC Cancer Genomics Browser [22] (Table 1). The RSEM normalized read counts were log_2 transformed by the UCSC, and we left them in that form. DNA methylation data were obtained from the National Cancer Institute's Genomic Data Commons. These were functionally normalized using the `minfi` package [23, 24] for the R statistical environment [25].

A separate smaller dataset of methylation profiles (from the same platform) was also used by our method to identify differentially methylated loci between 46 paired tumor and tumor-adjacent normal clear cell kidney cancer samples obtained through the National Center for Biotechnology Information's Gene Expression Omnibus (GSE61441) [26]. Again, we used functional normalization for these data.

Table 1. Distribution of Samples in TCGA Clear Cell Renal Cell Carcinoma (Clear Cell Kidney Cancer) Data

	RNA-seq (%)	450k (%)	Overlap (%)
Samples w/ overall survival data	525	311	310
Male	341 (64.95)	201 (64.63)	201 (64.84)
Female	184 (35.05)	110 (35.37)	109 (35.16)
Stage I	262 (49.90)	150 (48.23)	150 (48.39)
Stage II	56 (10.67)	30 (9.65)	30 (9.68)
Stage III	126 (24.00)	75 (24.16)	74 (23.87)
Stage IV	81 (15.43)	56 (18.01)	56 (18.06)
Grade 1	12 (2.29)	7 (2.25)	7 (2.26)
Grade 2	228 (43.43)	132 (42.44)	132 (42.58)
Grade 3	202 (38.48)	119 (38.26)	119 (38.39)
Grade 4	75 (14.29)	49 (15.76)	48 (15.48)
Grade X	5 (0.95)	2 (0.64)	2 (0.65)
Missing Grade	3 (0.57)	2 (0.64)	2 (0.65)
Deaths	166 (31.62)	99 (31.83)	98 (31.61)
Mean Age	60.65	61.43	61.48

There was no evidence of significant differences in the distribution of staging or tumor grade for cases in the RNA-seq and DNA-methylation data (χ^2 test, p = 7.77e-01 and p = 9.54e-01 respectively). For all data types, there were 8 cases missing survival data, with 5 having no clinical annotation at all. The remaining 3 were female, had a mean age of 70.33 years, and contained 2 stage I and 1 stage II tumors. Other than the 5 with no clinical annotation, there were no samples missing on clinical predictors, therefore we decided to remove the 8 samples missing outcomes from the analysis.

Beta values were transformed into M-values [27], and we removed probes on the X or Y chromosomes, containing SNPs [28, 29], or with cross-hybridization issues [30]. Finally, probes with values missing for greater than 50% of samples were removed and the remaining values were imputed using the k-nearest neighbors method, with k=10, from the `impute` package [31, 32] for R.

2.1. *M2EFM*

We developed a data-integrated modeling approach we call Methylation-to-Expression Feature Model (M2EFM). The basis of this approach is to find loci that are differentially methylated between matched pathologic and non-pathologic data and to associate those loci with significant differences in gene expression in the disease state. The process is analogous to expression quantitative trait loci (eQTL) analysis, except that instead of associating SNPs with changes in gene expression, we associate differentially methylated loci. The loci are then called m2eQTLs (for methylation-to-expression QTLs) and the genes are called m2eGenes.

The approach consists of five primary steps (summarized in Fig. 1):

(1) **Filtering probes and genes for variability.** Gene expression values were filtered to remove very low variability genes (usually genes with no expression) by removing genes with a median absolute deviation of .05 or less, leaving 16907 genes. Methylation probes were filtered to remove those with a median absolute deviation of less than 0.8 (after transformation to M-values). This left 27700 probes for the kidney cancer data.

(2) **Identifying differentially methylated loci.** Differential methylation was identified using the empirical Bayes method from the `limma` package [33] for R. We used 46 paired tumor and tumor-adjacent normal samples from a separate dataset than used in the rest of the analysis. This initial step was used to identify which loci to focus on. We passed the 500 CpG loci with the lowest adjusted p-values (Benjamini-Hochberg) for differential methylation on to the the next step.

(3) **Identifying methylation-to-expression quantitative trait loci (m2eQTLs).** m2eQTL analysis involves associating methylation levels at the loci identified in the previous step with gene expression levels genome-wide. In terms of an eQTL analysis, the proportion of methylated alleles for a particular loci is equivalent to the genotype at a single nuncleotide polymorphism (SNP), although it is a continuous, rather than discrete value. Identification of m2eQTLs was performed using the `MatrixEQTL` package [34] for R, which builds linear models to test association in a computationally efficient manner. In this way, the M-value of probes in the training data that were found to be differentially methylated in the first step were tested for their association with gene expression patterns in both *cis* and *trans* in a manner analogous to that used in typical eQTL analysis. An m2eQTL was defined to act in *cis* if it was associated with a gene within 10000bp, otherwise it was defined to act in *trans*. The top 150 *cis* and *trans*-m2eQTLs (by effect size) and their associated m2eGenes were passed on to the next step. This number was simply chosen to identify around 200 relevant genes and may not be optimal.

(4) **Building integrated models from m2eQTLs and m2eGenes.** From the previous results we built a joint regression model across both probes and genes involved in the m2eQTLs. Given that these were bound to have collinearity, to prevent overfitting we used Cox regression with Ridge penalty [35]. The linear predictor from the Cox model was used as a molecular risk score for all training samples (see Supplementary File 1, http://dx.doi.org/10.5061/dryad.b1t61).

(5) **Integrating clinical variables.** M2EFM uses a second regression to integrate clinical variables. For this step, we performed an unpenalized Cox regression on the molecular risk

score from the previous step and the values of clinical variables. This allows the hazards in the model to be more interpretable and keeps the clinical covariates from being penalized. In a typical Cox proportional hazards model, there is a rule of thumb that there should be no more than about 10 events in the data per variable in the model. Each training dataset in our data will have about 69 events (depending on the split of the data), meaning the model should have only about 7 variables. Clinical variables used for cancer prognosis vary but can include TNM staging, tumor grade, AJCC stage, patient sex, and age at diagnosis. We tried a few alternative clinical models on the training data only and picked the one with the highest discrimination (measured by concordance index, Table S1). Although the results were close for TNM staging and AJCC stage (the difference was significant at p = 1.04e-05), TNM staging would add 17 variables to the model and AJCC stage only 4, so our final model includes patient age at diagnosis, sex, tumor stage, and risk score. Although this is 8 variables, relaxing the rule to 9 events per variable has been shown to be acceptable [36] and can moreover be judged to some degree from our results.

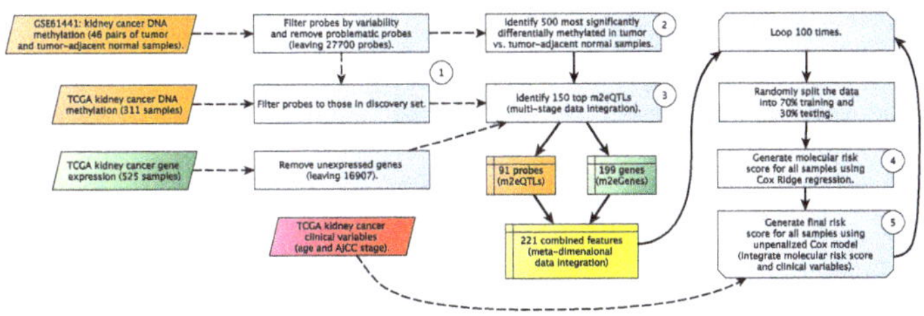

Fig. 1. Workflow for M2EFM analysis of clear cell renal cell carcinoma.

2.2. *Experimental Design*

We built 100 different M2EFM models of overall survival in clear cell kidney cancer for 100 different random splits of the data, using 70% training and 30% testing data sets. This process was repeated for different combinations of data (clinical variables only, gene expression only, methylation only, expression and clinical variables, and methylation and clinical variables).

The results of our approach were compared with two other methods that have previously been shown to successfully integrate molecular and clinical data to generate prognostic markers: penalized Cox regression and random survival forest [12] (although that work did not attempt molecular data integration). We used Cox Ridge regression, rather than LASSO (which was used in [12]), because it generally has better predictive performance. The model was built using the `glmnet` package [37] for R, and the lambda parameter was found using 10-fold cross validation for each split of the data. The random survival forest was built using the `randomForestSRC` package [38] for R. The run time of the random forest prevented cross validation of the parameters, so these were left at the defaults, as in [12]. The performance of the models was evaluated using concordance or C-index, a commonly used measure of discrimination in prognostic models. The C-index is a measure of how likely it is, in any given pair of individuals, that the individual with the higher risk score has the event first.

2.3. *Functional Analysis Approach*

Although it is not a requirement that the genes used in a prognostic model are functionally related to the disease, models built from functional relationships can reveal important insight into why one patient might have a better prognosis than another, which can lead to improved treatment decisions and a higher probability of model validation. Therefore, we performed a functional analysis of the gene set used in our model. The m2eQTL genes were used to perform a gene set network enrichment analysis using the online tool WEB-based GEne SeT AnaLysis Toolkit (WebGestalt) [39] to identify genes in our gene set that were enriched in sub-networks of protein-protein interactions that were, in turn, enriched for biological functions. We also used it to perform enrichment analysis for GO biological process terms. For both of these analyses we required at least 5 genes to overlap the gene module or pathway.

Our goal with this work was to demonstrate a method by which a biomarker can be identified. We do not identify a specific gene and DNA methylation probe set, in part because an independent validation dataset would be required.

3. Results

3.1. *M2EFM Prognostics*

The m2eQTL phase of M2EFM identifies differentially methylated loci that are associated with changes in gene expression throughout the genome. An example is shown in Fig. S1.

For the M2EFM-based risk score, the median C-index over 100 random splits of the data of the score from combined clinical and molecular variables (M2EFM Exp+Meth+Clin) reflects the highest prognostic accuracy of any method or data type used at .792. The median C-index of the risk score from clinical variables alone (M2EFM Clin) was .776 and the median C-index of the risk score from molecular variables alone (M2EFM Exp+Meth) was .702 (Fig. 2). The improvement in C-index for the combined clinical and molecular model over the clinical variables alone was significant at $p = 4.25e\text{-}06$ by two tailed Wilcoxon signed-rank test.

The M2EFM expression without methylation models had only slightly lower accuracy than models built using both data types. For these models, the median C-index for the combined clinical and expression models (M2EFM Exp+Clin) was .791 and for the expression only models (M2EFM Exp) was .703. The improvement in C-index for M2EFM Exp+Clin over the clinical variables alone was significant at $p = 1.50e\text{-}08$.

The M2EFM methylation without expression models were not as accurate as the other M2EFM models. The median C-index for the combined clinical and methylation models (M2EFM Meth+Clin) was .755 and for the methylation only models (M2EFM Meth) was .643. In this case, the clinical variable model had generally stronger C-index values than M2EFM Meth+Clin at $p = 2.068e\text{-}08$.

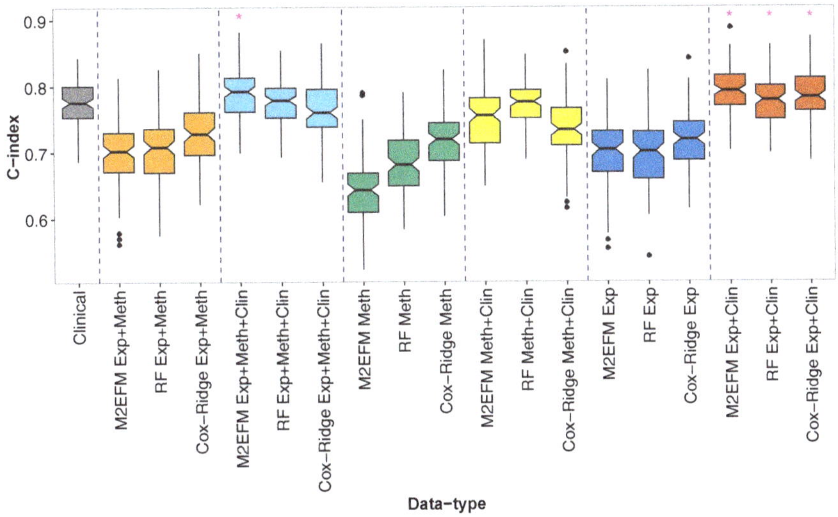

Fig. 2. C-index across 100 random splits into training and testing data of the various approaches. If one method was significantly better than another than the notches in the box plots will not overlap. For convenience, if a method resulted in significantly better results than clinical data alone, it is marked with "*".

3.2. *Random Survival Forest Prognostics*

Random survival forest was not as effective at exploiting the integrated expression and methylation data as our guided M2EFM approach. The median C-index for the combined clinical and molecular features (RF Exp+Meth+Clin) over the same 100 random splits of the data was .776 and the median C-index of the models built from the molecular data alone (RF Exp+Meth) was .696. The addition of the molecular data using random survival forest models was no more discriminatory than the clinical variables alone.

The performance of the expression without methylation random survival forest model was similar to the model with both data types. The median C-index for RF Exp+Clin model was .777, which was very slightly but significantly stronger than the clinical only model ($p = 7.16e\text{-}03$) while the median C-index for the RF Exp model was .694.

The performance of the methylation without expression random survival forest model was slightly worse than when both data types were used. The median C-index for the RF Meth+Clin model was .776, but the RF Meth model was a significant improvement over M2EFM Meth ($p = 8.05e\text{-}13$) and had a median C-index of .682.

3.3. *Cox-Ridge Prognostics*

Cox regression with ridge penalty [40] outperformed M2EFM when it came to the molecular data alone, but its molecular risk score was less independent of the clinical variables, thus its accuracy for the full model was less than that of M2EFM. The median C-index for the combined clinical and molecular features (Cox-Ridge Exp+Meth+Clin) of the same 100 random splits of the data was .760 and the median C-index of the models built from molecular data alone (Cox-Ridge Exp+Meth) was .727 and was improved over M2EFM Exp+Meth (p = 3.10e-05).

The performance of Cox-Ridge Exp+Clin model was slightly worse than the M2EFM Exp+Clin model (p = 9.14e-13) with a median C-index of .782. Again, the performance of the molecular data only model, Cox-Ridge Exp, was somewhat better than M2EFM Exp (p = 2.35e-06) with a median C-index of .718.

Finally, the Cox-Ridge Meth+Clin model did not perform as well as the M2EFM model. It achieved a median C-index of .735, which was significantly worse than the M2EFM Meth+Clin model (p = 6.37e-15). Nevertheless, the Cox-Ridge Meth model, with a median C-index of .705, performed better than the M2EFM Meth model (p = 1.62e-12).

3.4. *Comparison to Yuan et al.*

A direct comparison of our approach to that used in [12] on the same data was not possible, because the data they deposited included only the pre-filtered DNA methylation values, which did not include the same probes we identified in our discovery set. Nevertheless, we attempted to run our method on this subset of probes (which necessarily created different models than those used above). The highest mean C-index of any method listed in [12] on the kidney cancer data as .767 for a model including microRNA and clinical variables. On the same data (normalized by Yuan et al.), we achieved a mean C-index of .775 for the M2EFM Meth+Exp+Clin model and a mean C-index of .773 for the M2EFM Exp+Clin.

3.5. *Functional Analysis*

3.5.1. *Gene Set Network Enrichment*

Next we performed gene set network enrichment analysis using the online tool WebGestalt, requiring a minimum of 5 genes to overlap a gene module. All significant results (after multiple testing correction) are shown in Table 2. The full list of genes found in each pathway is given in Supplementary File 2. This approach revealed enrichment for gene modules associated with immune response, proliferation, and other functions. As an example, a portion of the largest sub-network our model was enriched in (which is enriched for the JAK-STAT Cascade) is shown in Fig. 3 (visualized using Cytoscape [41]). The genes from our gene set are shown in green and are highly connected nodes in the network.

3.5.2. *Biological Process Enrichment*

We further tested the straight enrichment for biological process terms in the Gene Ontology using our gene set (without network enrichment), again requiring a minimum of 5 genes to overlap a pathway. The results in Table 3 show the top 5 most enriched GO terms, with a clear enrichment for immune system related genes. The full list of genes enriched in each pathway is given in Supplementary File 3.

Table 2. Protein Interaction Network Module Enrichment

Pathway	Observed	Expected	Adj. p
T Cell Costimulation	7	.33	2.69e-07
Regulation of Defense Response to Virus by Host	11	1.18	2.69e-07
JAK-STAT Cascade Involved in Growth Hormone Signaling Pathway	34	19.64	3.80e-03
Complement Activation	6	1.82	3.43e-02

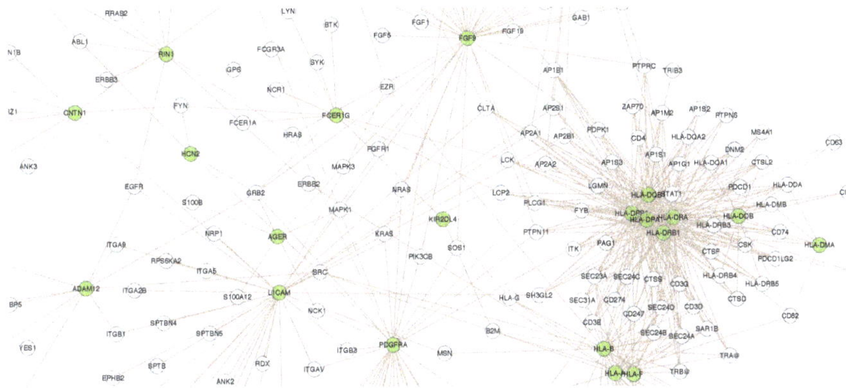

Fig. 3. Portion of the gene module enriched for the JAK-STAT cascade. Genes from our gene set are shown in green.

Table 3. Enriched for GO Biological Process

Pathway	Observed	Expected	Adj. p
Antigen Processing and Presentation of Exogenous Antigen	13	2.04	1.60e-05
Antigen Processing and Presentation of Exogenous Peptide Antigen	13	2.01	1.60e-05
Antigen Processing and Presentation	15	2.6	1.60e-05
Response to Interferon-Gamma	11	1.30	1.60e-05
Cellular Response to Interferon-Gamma	10	1.07	1.60e-05
Exogen	13	2.15	2.09e-05
Interferon-Gamma-Mediated Signaling Pathway	9	.89	2.09e-05
Antigen Processing and Presentation of Peptide Antigen	13	2.23	2.87e-05
Immune Response	32	12.24	2.95e-05

4. Discussion

All of the approaches described show it is possible to attain a meaningful level of prognostic discrimination using a joint regression on both gene expression and DNA-methylation values, if collinearity is properly accounted for. However, our approach, which first identifies dysregulation of DNA methylation in cancer, then associates that dysregulation to differences in gene expression, and finally builds prognostic markers from genes and CpG loci that are associated with this loss of regulation, was able to build models with a higher level of prognostic discrimination than either a random survival forest approach or Cox regression with Ridge penalty as well as a model built from traditional clinical variables. The results for our joint molecular regression with M2EFM were about the same as using expression data alone, leaving it unclear if this form of meta-dimensional integration is helpful on top of the multi-stage integration, which selected the features, thus more work on this part of the approach is needed.

The median C-index of .792 achieved by our M2EFM Exp+Meth+Clin model was the most accurate predictor of overall survival achieved by any approach in this study. This result was achieved through three data integrations, including different types of molecular data, as well as clinical variables. Notably, we showed that M2EFM's combination of a molecular risk score with clinical variables was a significant improvement over the clinical variables alone. Furthermore, our m2eQTL analysis identified 199 genes with high relevance to clear cell kidney cancer, without *a priori* knowledge of those genes' association to the disease. In fact, one of the top results from our gene set network enrichment analysis was for the JAK-STAT Cascade pathway, which is a known factor in kidney cancer progression [42]. That we identified this pathway by associating differentially methylated CpGs with differences in gene expression may suggest a role for dysregulation of methylation in the development of the disease, although caution in this interpretation is warranted, due to the cross-sectional nature of our study. An additional limitation was our lack of an independent dataset containing samples with gene expression and DNA methylation profiles as well as clinical data for validation.

The high enrichment we observed for genes involved in the immune system may indicate the utility of our approach in identifying survival differences based on dysregulation of immune functions. Given that immunotherapy has emerged over the last several years as an important component of kidney cancer treatment [43] and the pressing need for biomarkers that can identify the patients that will benefit from treatment [43], further development of this approach may be warranted in this regard. Another interesting result was our identification of *CA9*, which is currently of interest as a possible serum biomarker for kidney cancer [44], as a potential target for radioimaging [45], and as a potential therapeutic target [46]. Taken together, our results suggest that our approach is able to identify functionally relevant, and not just prognostic, genes. This is promising in terms of eventual validation of our approach.

Most of our results were better than those in a recent study including kidney cancer prognostics [12], but in a couple of cases, either the random forest or the Cox-Ridge approach did not perform as well as the methods in that work. However, they used fewer samples in that study and included inferred cancer subtypes from non-negative matrix factorization (NMF), in addition to gene and probe level measurements. Using only the DNA methylation and gene expression data from that study, which handicapped our method in discovery, M2EFM still showed slightly higher discrimination than any other approach. However, our goal was to develop a method based primarily on feature selection, rather than transformative dimensionality reduction techniques, in order to reduce the complexity of the models. Although interpretability is still limited by our use of Cox Ridge regression in generating the molecular risk score, it is over a limited number of genes that appear to be functionally related, mitigating this issue. It is notable that our m2eQTL-based approach creates models that outperform those using NMF, through a motivated feature selection technique that selects for putative regulatory relationships. We also note that Cox-Ridge in most cases outperformed the Cox-LASSO approach used in [12], and in some subsets of the data performed slightly better than M2EFM for prognostic accuracy. However, this accuracy comes at the cost of interpretability. The Cox-Ridge models contain thousands of genes or probes, telling us little in terms of the function of prognostic genes and creating unwieldy biomarkers in terms of real world use.

5. Conclusions

We developed a new data-integrated approach to modeling cancer prognostics and applied it to clear cell renal cell carcinoma data. M2EFM uses both a multi-stage data integration that links changes in methylation between tumor and normal tissues to levels of gene expression, and a meta-dimensional data integration that combines DNA methylation and gene expression values as part of a joint regression for outcome prediction. M2EFM was shown to identify not only prognostic, but functionally relevant features that may be associated with therapeutic response and that were highly connected in relevant protein-protein interaction networks.

6. Acknowledgements

We would like to acknowledge our funding sources: NIH-NIMH R01MH094609, NIH-NIEHS R01ES022223, and NIH-NIEHS P01 ES022832/EPA RD83544201 and a Hopeman award from the Norris Cotton Cancer Center.

References

[1] M. Huang, A. Shen, J. Ding and M. Geng, *Trends Pharmacol Sci* **35**, 41 (2014).
[2] S. E. Kern, *Cancer Res* **72**, 6097 (2012).
[3] A. S. Coates, E. P. Winer, A. Goldhirsch, R. D. Gelber, M. Gnant, M. Piccart-Gebhart, B. Thürlimann, H.-J. Senn, F. André, J. Baselga *et al.*, *Ann Oncol* **26**, 1533 (2015).
[4] J. Lu, M. C. Cowperthwaite, M. G. Burnett and M. Shpak, *PloS ONE* **11**, p. e0154313 (2016).
[5] L. Xu, L. Fengji, L. Changning, Z. Liangcai, L. Yinghui, L. Yu, C. Shanguang and X. Jianghui, *PloS ONE* **10**, p. e0142433 (2015).
[6] D. Kim, R. Li, S. M. Dudek and M. D. Ritchie, *BioData Min* **6**, p. 23 (2013).
[7] M. D. Ritchie, E. R. Holzinger, R. Li, S. A. Pendergrass and D. Kim, *Nat Rev Genet* **16**, 85 (2015).
[8] A. C. Nica and E. T. Dermitzakis, *Phil Trans R Soc B* **368**, p. 20120362 (2013).
[9] R. Breitling, Y. Li, B. M. Tesson, J. Fu, C. Wu, T. Wiltshire, A. Gerrits, L. V. Bystrykh, G. De Haan, A. I. Su *et al.*, *PLoS Genet* **4**, p. e1000232 (2008).
[10] E. R. Holzinger, S. M. Dudek, A. T. Frase, S. A. Pendergrass and M. D. Ritchie, *Bioinformatics* , p. btt572 (2013).
[11] P. K. Mankoo, R. Shen, N. Schultz, D. A. Levine and C. Sander, *PLoS ONE* **6**, p. e24709 (2011).
[12] Y. Yuan, E. M. Van Allen, L. Omberg, N. Wagle, A. Amin-Mansour, A. Sokolov, L. A. Byers, Y. Xu, K. R. Hess, L. Diao *et al.*, *Nat Biotechnol* **32**, 644 (2014).
[13] G. Ambler, S. Seaman and R. Omar, *Stat Med* **31**, 1150 (2012).
[14] F. R. Datema, A. Moya, P. Krause, T. Bäck, L. Willmes, T. Langeveld, B. de Jong, J. Robert and H. M. Blom, *Head Neck-J Sci Spec* **34**, 50 (2012).

[15] M. Ehrlich *et al.*, *Oncogene* **21**, 5400 (2002).
[16] D.-H. Yu, R. A. Waterland, P. Zhang, D. Schady, M.-H. Chen, Y. Guan, M. Gadkari and L. Shen, *J Clin Invest* **124**, 3708 (2014).
[17] J.-P. Issa, *J Clin Oncol* **30**, 2566 (2012).
[18] P. W. Laird, *Nat Rev Cancer* **3**, 253 (2003).
[19] M. Esteller *et al.*, *Oncogene* **21**, 5427 (2002).
[20] K. L. Sheaffer, E. N. Elliott and K. H. Kaestner, *Cancer Prev Res (Phila)* **9**, 534 (2016).
[21] Cancer Genome Atlas Research Network *et al.*, *Nature* **499**, 43 (2013).
[22] M. S. Cline, B. Craft, T. Swatloski, M. Goldman, S. Ma, D. Haussler and J. Zhu, *Sci Rep* **3** (2013).
[23] J.-P. Fortin, A. Labbe, M. Lemire, B. W. Zanke, T. J. Hudson, E. J. Fertig, C. M. Greenwood and K. D. Hansen, *Genome Biol* **15**, p. 1 (2014).
[24] M. J. Aryee, A. E. Jaffe, H. Corrada-Bravo, C. Ladd-Acosta, A. P. Feinberg, K. D. Hansen and R. A. Irizarry, *Bioinformatics* **30**, 1363 (2014).
[25] R Core Team, *R: A Language and Environment for Statistical Computing.* R Foundation for Statistical Computing, Vienna, Austria, (2016).
[26] J.-H. Wei, A. Haddad, K.-J. Wu, H.-W. Zhao, P. Kapur, Z.-L. Zhang, L.-Y. Zhao, Z.-H. Chen, Y.-Y. Zhou, J.-C. Zhou *et al.*, *Nat Commun* **6** (2015).
[27] P. Du, X. Zhang, C.-C. Huang, N. Jafari, W. A. Kibbe, L. Hou and S. M. Lin, *BMC Bioinformatics* **11**, p. 587 (2010).
[28] 1000 Genomes Project Consortium *et al.*, *Nature* **491**, 56 (2012).
[29] L. Butcher, *Illumina450ProbeVariants.db: Annotation Package combining variant data from 1000 Genomes Project for Illumina HumanMethylation450 Bead Chip probes*, (2013). R package version 1.1.1.
[30] Y.-a. Chen, M. Lemire, S. Choufani, D. T. Butcher, D. Grafodatskaya, B. W. Zanke, S. Gallinger, T. J. Hudson and R. Weksberg, *Epigenetics* **8**, 203 (2013).
[31] O. Troyanskaya, M. Cantor, G. Sherlock, P. Brown, T. Hastie, R. Tibshirani, D. Botstein and R. B. Altman, *Bioinformatics* **17**, 520 (2001).
[32] T. Hastie, R. Tibshirani, G. Sherlock, M. Eisen, P. Brown and D. Botstein, Imputing missing data for gene expression arrays (1999).
[33] M. E. Ritchie, B. Phipson, D. Wu, Y. Hu, C. W. Law, W. Shi and G. K. Smyth, *Nucleic Acids Res* **43**, p. e47 (2015).
[34] A. A. Shabalin, *Bioinformatics* **28**, 1353 (2012).
[35] A. E. Hoerl and R. W. Kennard, *Technometrics* **12**, 55 (1970).
[36] E. Vittinghoff and C. E. McCulloch, *Am J Epidemiol* **165**, 710 (2007).
[37] N. Simon, J. Friedman, T. Hastie and R. Tibshirani, *J Stat Softw* **39**, p. 1 (2011).
[38] H. Ishwaran, U. B. Kogalur, E. H. Blackstone and M. S. Lauer, *Ann Appl Stat* , 841 (2008).
[39] J. Wang, D. Duncan, Z. Shi and B. Zhang, *Nucleic Acids Res* **41**, W77 (2013).
[40] H. Zou and T. Hastie, *J R Stat Soc Series B Stat Methodol* **67**, 301 (2005).
[41] P. Shannon, A. Markiel, O. Ozier, N. S. Baliga, J. T. Wang, D. Ramage, N. Amin, B. Schwikowski and T. Ideker, *Genome Res* **13**, 2498 (2003).
[42] S. Li, S. J. Priceman, H. Xin, W. Zhang, J. Deng, Y. Liu, J. Huang, W. Zhu, M. Chen, W. Hu *et al.*, *PLoS ONE* **8**, p. e81657 (2013).
[43] M. W. Ball, M. E. Allaf and C. G. Drake, *Discov Med* **21**, 305 (2016).
[44] M. Takacova, M. Bartosova, L. Skvarkova, M. Zatovicova, I. Vidlickova, L. Csaderova, M. Barathova, J. Breza, P. Bujdak, J. Pastorek *et al.*, *Oncology Lett* **5**, 191 (2013).
[45] P.-C. Lv, J. Roy, K. S. Putt and P. S. Low, *Mol Pharm* **13**, 1618 (2016).
[46] J. Tostain, G. Li, A. Gentil-Perret and M. Gigante, *Eur J Cancer* **46**, 3141 (2010).

DE NOVO MUTATIONS IN AUTISM IMPLICATE THE SYNAPTIC ELIMINATION NETWORK*

GUHAN RAM VENKATARAMAN

Department of Bioengineering, Stanford University, 318 Campus Drive
Stanford, CA 94305, USA
Email: guhan@stanford.edu

CHLOE O'CONNELL, FUMIKO EGAWA, DORNA KASHEF-HAGHIGHI, DENNIS P. WALL

Department of Pediatrics and Biomedical Data Science, 1265 Welch Road
Stanford, CA 94305, USA
Email: dpwall@stanford.edu

Autism has been shown to have a major genetic risk component; the architecture of documented autism in families has been over and again shown to be passed down for generations. While inherited risk plays an important role in the autistic nature of children, *de novo* (germline) mutations have also been implicated in autism risk. Here we find that autism *de novo* variants verified and published in the literature are Bonferroni-significantly enriched in a gene set implicated in synaptic elimination. Additionally, several of the genes in this synaptic elimination set that were enriched in protein-protein interactions (CACNA1C, SHANK2, SYNGAP1, NLGN3, NRXN1, and PTEN) have been previously confirmed as genes that confer risk for the disorder. The results demonstrate that autism-associated *de novos* are linked to proper synaptic pruning and density, hinting at the etiology of autism and suggesting pathophysiology for downstream correction and treatment.

* This work was supported in part by the Hartwell Foundation's Autism Research and Technology Initiative.

1. Introduction

Autism Spectrum Disorder (ASD) is a neurodevelopmental disorder that impairs social skills, communication, and normal behavior. About 1% of the world's population has ASD, and this number is rapidly rising: the prevalence of autism more than doubled between 2002 and 2012[1]. ASD-linked impairment leads to higher lifespan costs[2] and a significant reduction in ability to procure both postgraduate education and jobs[3].

Both inherited (present in mother or father) and *de novo* (germline) mutations have been shown to contribute to the disease[4]. Although several of each type appear to contribute to ASD risk, there is still not a clear picture or full map of what leads to ASD[5]. Several hypotheses exist for the genetic etiology of autism; one of note is referred to as the "synaptic elimination hypothesis," the exploration of which is the focus of this paper.

Synaptic elimination is a normal neurodevelopmental process, starting in the fifth week of development and continuing throughout life. The process occurs in parallel with synaptic formation, which relies on input from both presynaptic and postsynaptic neurons. Elimination eventually outpaces formation in adolescence and adulthood[6,7]. During the development of the central nervous system, neurons form multiple synapses in excess of functional need. These redundant synapses are later eliminated through various means: 1) loss of signals necessary from either presynaptic or postsynaptic neurons to maintain synaptic stability[8]; 2) apoptosis of synapses; 3) ubiquitination of synaptic proteins for proteosomal degradation[9]; 4) macro-autophagy[7]; or 5) phagocytosis of synapses as a result of opsonization by synaptic elimination-mediating complement factors, microglia, and astrocytes[10-13].

Previous work has shown that faulty synaptic formation and maturation contribute to ASD[6]. However, given that increases in both dendritic spine density and brain weight (both of which are characteristic of autism) can be caused by mutations in genes regulating synaptic elimination, the hypothesis developed that autism could *also* be a disease of abnormal synaptic elimination[8,14-16].

Currently, the major pathway and ontology databases (KEGG, GO, Panther, and Reactome) do not contain any gene sets that pertain to synaptic elimination or synaptic pruning. As part of this study, we endeavored to create a robust and manually curated list of genes contributing to synaptic elimination; our goal was to test the hypothesis that the curated gene set would be enriched for *de novo* mutations (see Supplemental Materials 2 and 3 for list of genes and references used to generate this list, respectively). We hypothesized that increased burden of mutations in synaptic elimination genes would lead to the synaptic pruning abnormalities observed in autism, such as increased dendritic spine density and increased brain weight.

We used the *dnenrich* package[17], a network burden analysis tool, to test for enrichment in the synaptic elimination gene set on a comprehensive set of exomes from family-based trios having one child with autism. The package has been shown to be particularly powerful for identifying *de novo* mutations with small individual association to phenotype, but large effect in combination. We used dnenrich on previously documented autism-associated gene sets and autism-associated *de novos* as a pilot. This was done to verify that the program was suitable for use with autism *de novos* and that our list of *de novos* was large enough to provide sufficient power to detect enrichment of certain gene sets. After doing so, we tested the hypothesis that our list of genes involved in synaptic elimination would have a higher burden of autism *de novos* than would be expected by chance.

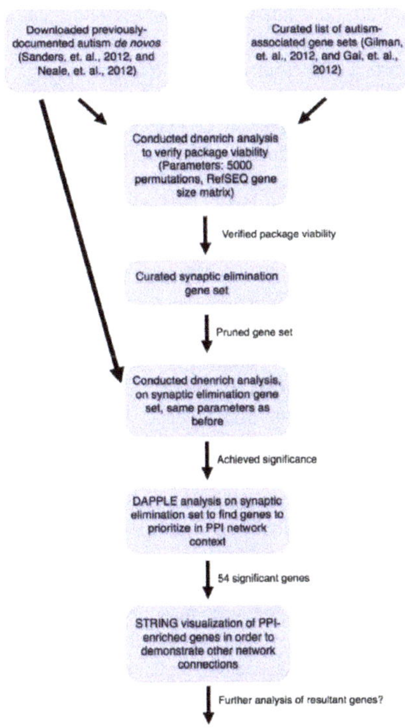

Fig 1. Schematic describing the overall flow of our experiment.

2. Methods

2.1. Autism *de novo* variants

We downloaded genomes of 3982 autism family trios from the Autism Sequencing Consortium (ASC) and the Simons Simplex Collection (SSC)[18-20]. These cohorts have been studied from a single-variant perspective, but have not yet been examined for their potential relationship to the synaptic elimination network.

Focusing on previously published full exome data, we built a comprehensive database of genomic variants to test for enrichment of synaptic elimination[18-21]. Specifically, we selected 189 autism trios and 31 unaffected siblings from SSC and then filtered out samples known to carry large *de novo* CNVs. Whole-exome sequencing was completed for 238 families selected from SSC, 200 of which included an unaffected sibling[23]. 15,480 DNA samples in 16 sample sets were analyzed, integrating *de novo*, inherited, and case-control loss-of-function counts and *de novo* missense variants predicted to be damaging. *De novos* were called using enhancements of previously published methods[18].

The full variant list from this collection, which also includes ASC cohorts, were compared to a larger set of 1,779 other exomes to confirm their putative roles in autism, and all *de novo* events were validated via PCR amplification and Sanger sequencing[22]. Family quads selected from SSC were sequenced with enrichment for higher functioning probands[19,20]. These *de novos* were interpreted using pipeline tools at each respective participating data center.

2.2. *Dnenrich Pilot Study*

Dnenrich simulates peppering the genome with random *de novos* by taking into account tri-nucleotide contexts, gene sizes, sequencing coverage, and functional effects of mutations. After permuting this process for a user-defined number of times, it then calculates one-sided P values, testing whether the observed number of mutations (in each gene set) is greater than the average simulated number of mutations (again, in each gene set).

We assembled 37 candidate gene sets to test their enrichment in our curated list of autism de novo variation. These 37 gene sets included Gene Ontology sets from previous autism network analyses[24,25], as well as genes shown to interact with FMRP (a mutation in which causes Fragile X syndrome, one of the most common causes of autism spectrum disorders)[21,26]. A full list of genes in each set tested for enrichment can be found in Supplemental Materials 1, along with their sizes.

We then performed an extensive process of literature mining and curation, through combined database search, hand-search, and related reference review, to assemble the synaptic elimination gene set. Our Pubmed search (conducted between May and June., 2016) included use of the terms "synapse," "synaptic," "elimination," "pruning," and "gene." We then performed additional hand-searches of *Nature* and *Cell* using the same terms. References of included studies, review articles, and related references were screened for additional relevant studies based on title and abstract review. Our screening criteria for inclusion in the synaptic elimination set was the presence of the following terms: "synaptic elimination," "synaptic pruning," "synaptic stabilization," "synaptic destabilization," and "synaptic plasticity." In all searches we excluded the following terms: "axon scaling," "viral infection," "axon repulsion," "axon retraction," and "neuromuscular junction." Studies pertaining to synaptic formation, maintenance, and/or elimination within the peripheral nervous system were excluded on the basis of arising from separate embryologic origin than the central nervous system. Abstracts and unpublished data were excluded. The synaptic elimination gene set was curated through careful review of 120 selected studies and related reviews yielding 274 genes related to synaptic formation or elimination. Gene function was cross-referenced in ClinVar (accessed July 11, 2016), and 213 genes of interest were selected based on their role in synaptic elimination (see Supplemental Materials 2). After its curation, we tested the synaptic elimination gene set for enrichment for autism *de novos*.

Table 1. The 213 genes in the synaptic elimination gene set.

C1QA	PROS1	TYROBP	CD200	IGFBP4	TLR4	NFKB1
C1QB	CXC3L1	NFKB1	ITGAX	EDNRB	BDNF	NFKB2
C1QC	CX3CR1	CREB	ITGB2	TIMP2	C5	CAMK2G
C3	DAP12	MAPK14	GDNF	COL1Q2	H2-D	NCKAP5L
Mac-2	TREM2	NPTX2	CSF1	FN1	CCL7	NRXN2
CRK	CR-1	NGFR	CNTF	IRF8	CCL2	NRXN3
ELMO1	PGRN	APP	PTGER2	TGRBR2	CDC42	NRTK2
RAC1	CD68	PILRB	C1QBP	CFB/MHCIII	MBP	CRMP1
BAI1	CASP8	CD247	CALR	FCGR1B	CXCL13	CRK
MEGF10	CASP3	B2M	CR2	AIF1	Uba1	PLXA3
GULP1	CASP6	KLRA1	CD33	IL10BR	Mov34	PLXA4
ABCA1	CLU	TAP1	TNFRSF19	NOS1AP	Rpn6	TBR1
TYRO3	HLA-DR	C4	PDGFRA	MASP1	USP2	DPYSL2
AXL	HLA-C	CR-3	LEP	CD46	UFD2A	ADNP
MERTK	HLA-A	CD22	LEPR	CD55	MEF2	SPARC
GAS6	DR6	CD47	IGFBP3	TLR2	MEF2A	DYRK1A
EN2	GDA	TSPAN7	PAK3	CTNNB1	WNT2	FOXP1
MEF2B	REL	NLGN1	SEMA3A	BDNF	CHN2	RCAN1
MEF2C	RELA	NLGN2	SEMA3F	DHCR7	MAPK3	CHD8
MEF2D	RELB	NLGN3	NRP1	FMR1	MAPK1	RAC1
PARK2	SERPINA3	NLGN4	NRP2	AUTS10	TSC1	OPHN1
caspases	CUL3	SHANK1	RhoA	LAMC3	TSC2	FOXP2
hdc	ESCRT-I	SHANK2	ROCK1	MECP2	DOCK1	ARC
MIB1	shrub	SHANK3	OTX1	THBS1	EPHA4	CASK
UBE3A	ESCRT-III	CNTN4	DISC1	THBS2	EPHB3	DLG4
UBE3B	CHMP2B	CNTNAP2	KATNAL2	THBS4	EFNA4	HOMER1
PCDH10	mop	CNTNAP4	NTNG1	MAP2	EFNB3	PTEN
ATG5	Kat60L	CACNA1C	SYNGAP1	KALRN	NCK2/GRB4	
Atg7	IKBKG	SCN1A	Mek-1	KALRN	EB3	
LC3-II	Mical	SCN2A	Mek-2	CDC42	NGFR	
p62	NRXN1	RELN	SPARCL1	PPP1R9B	GRM5	

3. Results

We tested the 37 initial gene sets with dnenrich with the default gene size matrix provided on the dnenrich website (as adjusting for per-trio joint sequencing coverage "[does] not have a noticeable effect on results"[17]). We ran the simulation on the downloaded autism *de novos* for 5000 permutations without weighting any genes. Of the 37 gene sets tested, 10 were significantly enriched for *de novos* after Bonferroni adjustment for 37 hypotheses. These sets are listed in Table

2. Given the enrichment of *de novos* in known autism networks calculated by dnenrich, we felt confident in using both this set of previously-published *de novos* and dnenrich to test the single hypothesis that synaptic elimination genes would have an exceedingly high burden of *de novos*.

Table 2. Bonferroni-significant gene sets enriched for autism *de novos* using dnenrich. Unadjusted p-values were obtained directly from dnenrich; adjusted p-values were Bonferroni-corrected by the number of sets tested.

	p-value		Number of Mutations		Location	
Gene Set Name	Unadjusted	Adjusted	Observed	Expected	Reference to Autism	Source
Developmental Process	1.9996×10^{-4}	8.798×10^{-3}	731	648.659	Gai et. al. (2012)	GO
FMRP	1.9996×10^{-4}	8.798×10^{-3}	412	285.33	Darnell et. al. (2011)	Paper
Learning and/or Memory	1.9996×10^{-4}	8.798×10^{-3}	78	44.1032	Gilman et. al. (2011)	GO
M3	1.9996×10^{-4}	8.798×10^{-3}	206	151.955	Parikshak et. al. (2013)	Paper
Protein modification process	1.9996×10^{-4}	8.798×10^{-3}	577	496.748	Gai et. al. (2012)	GO
Synaptic transmission	1.9996×10^{-4}	8.798×10^{-3}	163	114.367	Gai et. al. (2012)	GO
Axonogenesis	3.9992×10^{-4}	1.7596×10^{-2}	136	93.8364	Gilman et. al. (2011)	GO
Cell-cell signaling	3.9992×10^{-4}	1.7596×10^{-2}	241	188.513	Gai et. al. (2012)	GO
Neuron development	5.9988×10^{-4}	2.6395×10^{-2}	253	207.273	Gilman et. al. (2011)	GO
Axon	9.998×10^{-4}	4.3991×10^{-2}	121	87.635	Gilman et. al. (2011)	GO

Consistent with the synaptic elimination hypothesis, the synaptic elimination set also proved to be significantly enriched for autism *de novo* mutations ($p = 1.9996*10^{-4}$). It exceeded the observed-to-expected mutation ratio of all other significantly enriched gene sets (Figure 2).

Fig 2. Ratio of observed-to-expected mutations per enriched gene set. The dnenrich software calculated expected number of mutations by simulating and averaging the number of *de novo* events in each gene set using information like tri-nucleotide context, gene size, etc. The systematically-generated synaptic elimination set has the highest ratio of observed-to-expected mutations by a significant margin.

To narrow the list of 213 genes in the synaptic elimination set down to a shorter list of genes to prioritize, we used DAPPLE. Developed by Elizabeth Rossin of the Broad Institute, the algorithm marks genes that are ripe for further study[28]. DAPPLE relies on protein-protein interaction databases such as InWeb (populated with hundreds of thousands of known protein-protein interactions). When researchers input a network, the algorithm compares the network it to what would be expected by pure probability by permuting proteins (linked to the inputted genes) many times. It determines if genes in all of the inputted regions could play a role in disease, and tests whether or not the network is more connected than would be expected by chance. For our purposes, this analysis would point to genes of interest within our synaptic elimination network that have higher levels of interconnectivity than expected.

Using DAPPLE (Figure 3) on the synaptic elimination gene set yielded fifty-four genes significantly enriched for protein-protein interactions (PPI), which are listed in Table 3. Six of these fifty-four (CACNA1C, SHANK2, SYNGAP1, NLGN3, NRXN1, and PTEN) have already been confirmed as genes associated with autism risk[27]. Those genes that were enriched were visualized using the STRING database (Figure 4) in order to examine other known (and predicted) gene interactions. Further inquiry into these resultant genes involved in synaptic elimination could elucidate etiology and shed light on related ASD risk.

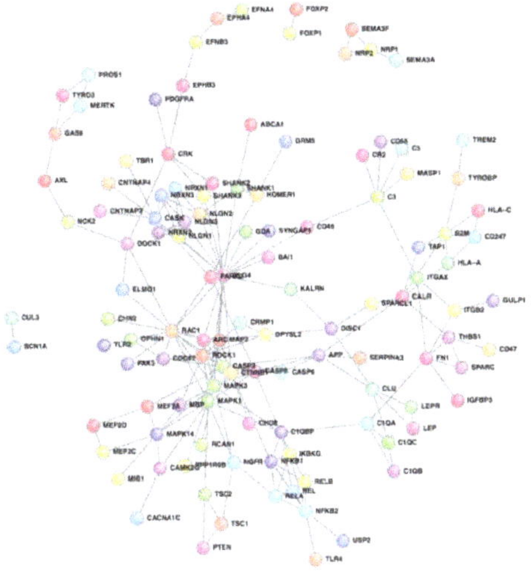

Fig 3. DAPPLE visualization of the synaptic elimination gene set. DAPPLE analyzes the protein-protein interaction network generated by the genes in the set; it marks the genes that are significantly more connected in the network than by chance (PPI-enriched). The nodes represent genes in the network, and the edges represent interactions between proteins downstream of the connected genes. The graphic is arbitrarily colored and is meant to show connectivity only.

Table 3. DAPPLE PPI-Enriched Genes in the synaptic elimination gene set. The table pairs genes with their DAPPLE significances.

Gene Name	*p*-value	Gene Name	*p*-value
DOCK1	0.005985024	CASK	0.001997004
HLA-A	0.045426102	TYRO3	0.005985024
CLU	0.00797604	HOMER1	0.025805363
SHANK1	0.00797604	SEMA3F	0.00797604
CD33	0.005985024	AXL	0.001997004
PARK2	0.003992012	SYNGAP1	0.00996506
ITGAX	0.005985024	CASP3	0.003992012
MERTK	0.045426102	TREM2	0.001997004
ELMO1	0.037601759	MAPK1	0.001997004
OPHN1	0.003992012	CACNA1C	0.035640683
CASP6	0.033677611	SHANK2	0.003992012
MAPK3	0.001997004	SHANK3	0.001997004
NRP1	0.01790118	CALR	0.013937112
NLGN1	0.001997004	NLGN2	0.001997004
GDA	0.011952084	NFKB2	0.013937112
NRXN3	0.001997004	DLG4	0.001997004
C3	0.001997004	TLR2	0.049326298
CTNNB1	0.001997004	ROCK1	0.00996506
CASP8	0.001997004	NRXN1	0.00797604
NGFR	0.001997004	MAP2	0.001997004
REL	0.023832312	EFNB3	0.037601759
ARC	0.001997004	MEF2A	0.01988022
RAC1	0.001997004	B2M	0.011952084
GAS6	0.001997004	NLGN3	0.001997004
NRP2	0.027776419	PTEN	0.00797604
GRM5	0.00996506	THBS1	0.039560839
NRXN2	0.001997004	NFKB1	0.001997004

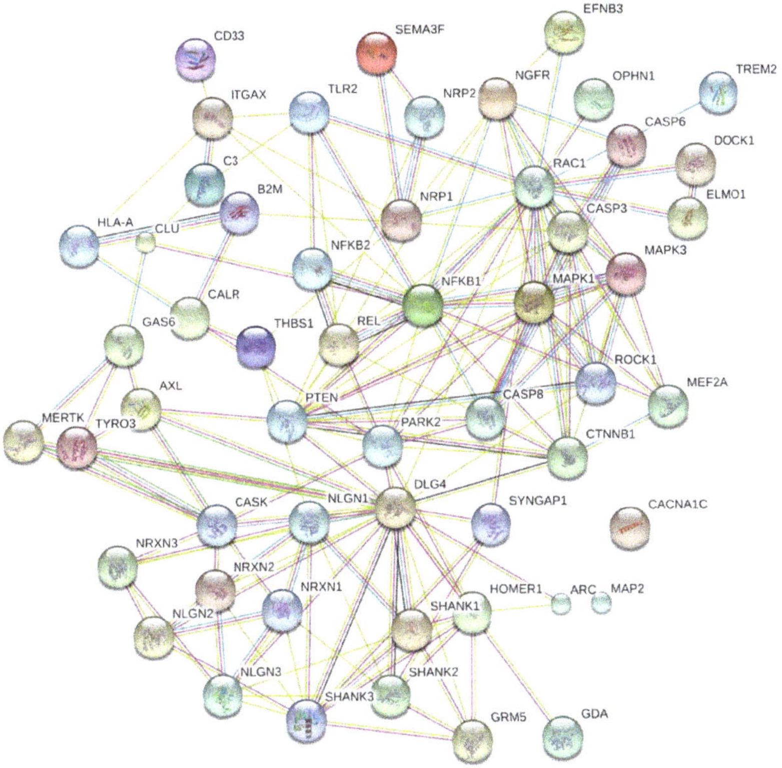

Fig 4. STRING visualization of DAPPLE PPI-enriched genes in the synaptic elimination gene set. Colored nodes represent query proteins and the first shell of interactors. White nodes represent the second shell of interactors. Cyan edges represent known interactions from curated databases; purple edges represent known interactions that are experimentally determined; green edges represent a gene neighborhood predicted interaction; red edges represent a gene fusion predicted interaction; blue edges represent a gene co-occurrence predicted interaction; yellow edges represent textmining; black edges represent co-expression; light blue edges represent protein homology.

4. Discussion

Overall, our results supported the hypothesis that genes involved in synaptic elimination are significantly enriched for autism *de novo* mutations, pointing to deregulation in synaptic elimination as a potential pathogenic mechanism for ASD. Synaptic elimination, as part of the larger synaptic homeostatic mechanism, contributes to higher structural and functional connectivity underlying cognitive functions through the removal of synaptic structures. Several of the genes that were PPI-enriched in the gene set were confirmed autism disease genes, suggesting that the genes central to the synaptic elimination network may play an important role in influencing genetic risk for autism. Given the biological plausibility of this pathway, along with the enrichment for *de novos* in known autism cases, the additional genes in this pathway may serve as candidate genes in the future investigation of the genetic etiology of autism spectrum disorder.

Within the context of the hypothesis that *de novo* mutations contribute to the risk of developing ASD in families with no previous history, previous gene enrichment studies have focused on identification of these *de novo* mutations and their interconnections as a multifaceted network without exploration of specific neurodevelopmental processes[22]. In the present study, we took advantage of a large collection of full exomes from trios with one affected child. This enabled us to explore the role of de novo mutations in synaptic density and pruning, confirming that there is a strong link and supporting the potential value of these de novos for use in increasing precision in early diagnosis/prognosis.

The lack of a validation cohort is a drawback of this study. A new set of *de novos* is currently undergoing quality control procedures; we will attempt to replicate this signal in a much larger collection of families. A consortium that includes our group has amassed over 5000 whole genomes (30x coverage) in multiplex families containing 2 or more children with autism. This is the largest database of its kind and valuable for determining whether the de novo signal seen replicates across siblings and families with varying levels of autism severity. In addition, it may be worthwhile to consider the genes involved in synaptic formation or maintenance in addition to those involved in elimination. Gene sets like the GO Neuron Development or Cell-cell Signaling sets, which showed significant *de novo* mutation enrichment, provide a good starting point for future studies, as neuronal activity and signaling play a definitive role in determining synapse strength and number.

More work is necessary to determine the biological implications of the association between synaptic elimination and autism. For the PPI-enriched genes in the synaptic elimination network, many of which have validated associations with autism, the exact process by which they affect brain development leading to behavioral change is unclear. The true role of these genes in the pathophysiology of autism must be elucidated by future science.

Network analyses like these have successfully been able to identify and validate gene sets that contribute risk to ASD and other neuropsychiatric disorders. The high likelihood that these findings are reproducible in the context of newer, more complete, and more specific datasets bolsters the hope of eventually having a more complete picture of ASD risk factors that impact precision care of this complex disorder. Such a map would be invaluable to both the diagnosis and subsequent treatment of ASD; synaptic elimination may play a key role in that map.

Supplemental Materials

Supplemental Materials 1 – Gene set sizes and gene/gene set mappings – https://drive.google.com/open?id=0B4nOSzAytcrBdlBLVWJWNzFpcGc

Supplemental Materials 2 – synaptic elimination genes and sources – https://drive.google.com/open?id=0B2UCU6mZg1CuSXNKaEJOODNLcmc

Supplemental Materials 3 – synaptic elimination curation references – https://drive.google.com/open?id=0B2UCU6mZg1CudVNiYzJmWGxTWjA

Acknowledgements

This work was supported in part by the Hartwell Foundation's Autism Research and Technology Initiative. It was also conducted with support from a TL1 Clinical Research Training Program of the Stanford Clinical and Translational Science Award to Spectrum (NIH TL1 TR 001084).

References

1 Keen, D. & Ward, S. Autistic spectrum disorder a child population profile. *Autism* **8**, 39-48 (2004).

2 Buescher, A. V., Cidav, Z., Knapp, M. & Mandell, D. S. Costs of autism spectrum disorders in the United Kingdom and the United States. *JAMA pediatrics* **168**, 721-728 (2014).

3 Shattuck, P. T. *et al.* Services for adults with an autism spectrum disorder. *The Canadian Journal of Psychiatry* **57**, 284-291 (2012).

4 Nord, A. S. *et al.* Reduced transcript expression of genes affected by inherited and de novo CNVs in autism. *European Journal of Human Genetics* **19**, 727-731 (2011).

5 Sung, Y. J. *et al.* Genetic investigation of quantitative traits related to autism: use of multivariate polygenic models with ascertainment adjustment. *The American Journal of Human Genetics* **76**, 68-81 (2005).

6 Bourgeron, T. From the genetic architecture to synaptic plasticity in autism spectrum disorder. *Nature Reviews Neuroscience* **16**, 551-563, doi:10.1038/nrn3992 (2015).

7 Tang, G. *et al.* Article Loss of mTOR-Dependent Macroautophagy Causes Autistic-like Synaptic Pruning Deficits. *Neuron* **83**, 1131-1143 (2014).

8 Ebert, D. H. & Greenberg, M. E. Activity-dependent neuronal signalling and autism spectrum disorder. *Nature* **493**, 327-337, doi:10.1038/nature11860.Activity-dependent (2013).

9 Toro, R. *et al.* Key role for gene dosage and synaptic homeostasis in autism spectrum disorders. *Trends in Genetics* **26**, 363-372, doi:10.1016/j.tig.2010.05.007 (2010).

10 Bialas, A. R. & Stevens, B. TGF-Beta Signaling Regulates Neuronal C1q Expression and Developmental Syanptic Refinement. *Nature Neuroscience* **16**, 1773-1782, doi:10.1038/nn.3560.TGF- (2013).

11 Stevens, B. *et al.* The Classical Complement Cascade Mediates CNS Synapse Elimination. *Cell* **131**, 1164-1178, doi:10.1016/j.cell.2007.10.036 (2007).

12 Paolicelli, R. C. & Gross, C. T. Microglia in development: linking brain wiring to brain environment. *Neuron glia biology* **7**, 77-83, doi:10.1017/S1740925X12000105 (2011).

13 Chung, W.-S., Allen, N. J. A. & Eroglu, C. Astrocytes Control Synapse Formation, Function, and Elimination. *Cold Spring Harb Perspect Biol* **7**, doi:10.1530/ERC-14-0411.Persistent (2015).

14 Siegel, A. & Sapru, H. N. *Essential neuroscience.* (Lippincott Williams & Wilkins, 2006).

15 Caglayan, A. O. Genetic causes of syndromic and non-syndromic autism. *Developmental Medicine and Child Neurology* **52**, 130-138, doi:10.1111/j.1469-8749.2009.03523.x (2010).

16 Geschwind, D. H. & Levitt, P. Autism spectrum disorders : developmental disconnection syndromes. *Current Opinion in Neurobiology* **17**, 103-111, doi:10.1016/j.conb.2007.01.009 (2007).

17 Fromer, M. *et al.* De novo mutations in schizophrenia implicate synaptic networks. *Nature* **506**, 179-184 (2014).

18 De Rubeis, S. *et al.* Synaptic, transcriptional and chromatin genes disrupted in autism. *Nature* **515**, 209-215 (2014).

19 Iossifov, I. *et al.* De novo gene disruptions in children on the autistic spectrum. *Neuron* **74**, 285-299 (2012).

20 Iossifov, I. *et al.* The contribution of de novo coding mutations to autism spectrum disorder. *Nature* **515**, 216-221 (2014).

21 Parikshak, N. N. *et al.* Integrative functional genomic analyses implicate specific molecular pathways and circuits in autism. *Cell* **155**, 1008-1021 (2013).

22 O'Roak, B. J. *et al.* Sporadic autism exomes reveal a highly interconnected protein network of de novo mutations. *Nature* **485**, 246-250 (2012).

23 Sanders, S. J. *et al.* De novo mutations revealed by whole-exome sequencing are strongly associated with autism. *Nature* **485**, 237-241 (2012).

24 Gilman, S. R. *et al.* Rare de novo variants associated with autism implicate a large functional network of genes involved in formation and function of synapses. *Neuron* **70**, 898-907 (2011).

25 Gai, X. *et al.* Rare structural variation of synapse and neurotransmission genes in autism. *Molecular Psychiatry* **17**, 402-411, doi:10.1038/mp.2011.10 (2012).

26 Darnell, J. C. *et al.* FMRP Stalls Ribosomal Translocation on mRNAs Linked to Synaptic Function and Autism. *Cell* **146**, 247-261, doi:10.1016/j.cell.2011.06.013 (2011).

27 Sanders, S. J. *et al.* Insights into autism spectrum disorder genomic architecture and biology from 71 risk loci. *Neuron* **87**, 1215-1233 (2015).

28 Rossin, Elizabeth J., et al. "Proteins encoded in genomic regions associated with immune-mediated disease physically interact and suggest underlying biology." *PLoS Genet* 7.1 (2011): e1001273.

IDENTIFYING GENETIC ASSOCIATIONS WITH VARIABILITY IN METABOLIC HEALTH AND BLOOD COUNT LABORATORY VALUES: DIVING INTO THE QUANTITATIVE TRAITS BY LEVERAGING LONGITUDINAL DATA FROM AN EHR*

SHEFALI S. VERMA[1], ANASTASIA M. LUCAS[1], DANIEL R. LAVAGE[1], JOSEPH B. LEADER[1], RAGHU METPALLY[2], SARATHBABU KRISHNAMURTHY[1], FREDERICK DEWEY[3], INGRID BORECKI[3], ALEXANDER LOPEZ[3], JOHN OVERTON[3], JOHN PENN[3], JEFFREY REID[3], SARAH A PENDERGRASS[1], GERDA BREITWIESER[2], MARYLYN D. RITCHIE[1]

Department of Biomedical and Translational Informatics, Geisinger Health System, Danville, PA[1]
Department of Functional and Molecular Genomics, Geisinger Health System, Danville, PA[2]
Regeneron Genetics Center, Tarrytown, NY[3]

A wide range of patient health data is recorded in Electronic Health Records (EHR). This data includes diagnosis, surgical procedures, clinical laboratory measurements, and medication information. Together this information reflects the patient's medical history. Many studies have efficiently used this data from the EHR to find associations that are clinically relevant, either by utilizing International Classification of Diseases, version 9 (ICD-9) codes or laboratory measurements, or by designing phenotype algorithms to extract case and control status with accuracy from the EHR. Here we developed a strategy to utilize longitudinal quantitative trait data from the EHR at Geisinger Health System focusing on outpatient metabolic and complete blood panel data as a starting point. Comprehensive Metabolic Panel (CMP) as well as Complete Blood Counts (CBC) are parts of routine care and provide a comprehensive picture from high level screening of patients' overall health and disease. We randomly split our data into two datasets to allow for discovery and replication. We first conducted a genome-wide association study (GWAS) with median values of 25 different clinical laboratory measurements to identify variants from Human Omni Express Exome beadchip data that are associated with these measurements. We identified 687 variants that associated and replicated with the tested clinical measurements at $p<5x10^{-08}$. Since longitudinal data from the EHR provides a record of a patient's medical history, we utilized this information to further investigate the ICD-9 codes that might be associated with differences in variability of the measurements in the longitudinal dataset. We identified low and high variance patients by looking at changes within their individual longitudinal EHR laboratory results for each of the 25 clinical lab values (thus creating 50 groups – a high variance and a low variance for each lab variable). We then performed a PheWAS analysis with ICD-9 diagnosis codes, separately in the high variance group and the low variance group for each lab variable. We found 717 PheWAS associations that replicated at a p-value less than 0.001. Next, we evaluated the results of this study by comparing the association results between the high and low variance groups. For example, we found 39 SNPs (in multiple genes) associated with ICD-9 250.01 (Type-I diabetes) in patients with high variance of plasma glucose levels, but not in patients with low variance in plasma glucose levels. Another example is the association of 4 SNPs in *UMOD* with chronic kidney disease in patients with high variance for aspartate aminotransferase (discovery p-value: $8.71x10^{-09}$ and replication p-value: $2.03x10^{-06}$). In general, we see a pattern of many more statistically significant associations from patients with high variance in the quantitative lab variables, in comparison with the low variance group across all of the 25 laboratory measurements. This study is one of the first of its kind to utilize quantitative trait variance from longitudinal laboratory data to find associations among genetic variants and clinical phenotypes obtained from an EHR, integrating laboratory values and diagnosis codes to understand the genetic complexities of common diseases.

* This work is supported by funds from Geisinger Health System and the Regeneron Genetics Center. Supplementary material can be found at: http://ritchielab.psu.edu/publications/supplementary-data/psb-2017/CBC-Met-Labs.

1. Introduction

In this era of personalized medicine, emphasis is on preventive care facilitated by integration of a patient's medical and genomic information. De-identified electronic health records (EHR) and bio-repositories represent significant resources of information that have been widely used for association studies in past decade[1]. Electronic health record (EHR) data is primarily designed for clinical care and is represented in both structured (such as ICD-9 codes, medication information, clinical laboratory values) as well as unstructured (physician notes) forms. Many association studies have utilized ICD-9 codes as well as clinical lab variables (structured forms of EHR data) to identify variants associated with EHR-derived phenotypes that might be of clinical relevance[2–4]. The number of association studies using EHR-derived phenotypes (both structured and unstructured data) has been increasing rapidly[5].

The complete blood count (CBC) panel and comprehensive/basic metabolic panel (CMP/BMP) are part of routine medical care for all medical practices. These panels are comprised of tests that help clinical practitioners identify underlying causes for conditions like weakness and fatigue, as well as to identify chronic illnesses (e.g., kidney failure, heart disease). These tests are generally conducted on patients that show some signs of illness, but these routine measurements are conducted from time to time on healthy individuals as well. Thus, utilizing these panels can help us understand overall health of patients by comparing these measurements across all patients in an EHR. These tests are recorded as quantitative variables for which units of measurements can be standardized across multiple clinical practices. ICD-9 codes and clinical measurements go hand in hand for a patient's medical record as a diagnosis code may either initiate the lab test which confirms the code or the code may be entered as a result of the test. Thus, integrating both clinical laboratory measurements and diagnosis codes present powerful approaches for understanding genetic variants that show similar associations with both data types obtained from an EHR[3]. The majority of association studies that use quantitative traits derived from an EHR as phenotypes use either mean/median values[3,6] or most recent measurements[7]. While this approach has been successful, utilizing only mean/median values limits the understanding of these traits by neglecting the variability over time that may be present in an individual patient's clinical history. This can be captured for analysis by using unique longitudinal information from EHR. Longitudinal data provides a better picture of the patient's health by actually pinpointing the time of disease onset, or time in which the quantitative trait became out of the normal range, which is especially important for the diseases that are more heterogeneous in nature and progress over time/age. A strategy such as this has been applied to family-based studies, using a mixed effects model to find associations among candidate genes and longitudinal data[8]. Utilizing the longitudinal data in some way other than considering one value also provides the opportunity to consider not just the average, but also the variability in these traits over time. In this study, our goal was to develop a strategy to embrace the longitudinal data in a population-based dataset, using trait variance, rather than a measure of central tendency approach such as median values, by binning patients in high and low variance groups separately to then test for associations. This strategy allows for the integration of clinical lab measurements as quantitative traits, embracing the variability in the traits, along with ICD-9 code PheWAS associations as well as SNPs.

2. Materials and Methods

2.1 Genotype Data

The MyCode® Community Health Initiative is a research initiative to engage Geisinger Health System patients in research and integrate their clinical EHR data along with genetic information to make discoveries in health and disease[9]. Over 109,000 Geisinger patients have consented to participate in MyCode and approximately 50,000 participants have whole exome sequencing and genome-wide genotype data generated. For this study, we used participants that have been genotyped using the Illumina Human Omni Express plus Exome beadchip. This dataset contains 45,899 samples and ~600K variants after some initial quality control procedures. For this analysis, after sample QC (removing one sample from pairs of highly related samples up to 1st cousins and removing any samples that did not pass a sample call rate filter of 90%), we divided the total dataset into two random sets to perform discovery and replication analyses. We included only European American samples with age >18 years. Our discovery dataset consisted of 17,347 samples and our replication dataset consisted of 17,348 samples (see **Supplementary Table** 1 for demographic information on these samples). We also filtered the variants that did not pass a genotype call rate filter of 99% to keep only high quality SNP data. To test common variants only, we applied a minor allele frequency (MAF) filter of 1%. This resulted in a total of 629,274 variants that were considered for association testing in the discovery dataset and 629,016 variants tested in the replication dataset.

2.2 Phenotype Data

Twenty-five clinical laboratory variables were extracted from EHR outpatient data and checked for consistency of unit measurements. A list of all 25 variables is provided in Table 1, along with information on the panel from which they were obtained. The phenotype data is extracted from the EHR as longitudinal data for all patients across their clinical history. Thus, each sample has multiple entries for each variable. The first step in conducting our GWAS analysis was to obtain median values for all 25 variables across patients' longitudinal data. We wanted to be able to compare the GWAS on median values with the analyses in the high-variance and low-variance groups. We visually inspected the clinical lab variable distributions to determine which variables needed a natural log transformation. We also removed all outliers that were more than 2.5 standard deviations from the mean. While this could lose some very interesting data points, for this pilot analysis, we wanted to be sure to remove gross errors in lab variable coding/data entry. Supplementary Figure 1 and 2 show the distribution of discovery and replication datasets, respectively, after removing outliers and performing natural log transformation wherever necessary. Table 1 lists the name of the variable, how the sample is collected (i.e. Blood or Serum/Plasma), which panel the variable is obtained from (i.e. Complete Blood Count (CBC) or Comprehensive Metabolic Panel (CMP) or Basic Metabolic Panel (BMP)), the total sample size for each phenotype in both discovery and replication datasets, and whether or not the data were transformed.

Table 1. List of 25 clinical laboratory measurements that are used in the analysis.

Clinical Laboratory Measurement	Panel type	Discovery Sample Size	Replication Sample Size	Transformation
ALANINE AMINOTRANSFERASE (ALT) - SERUM/PLASMA	CMP	15527	15393	Yes
ALBUMIN - SERUM/PLASMA	CMP	15519	15439	Yes
ALKALINE PHOSPHATASE - SERUM/PLASMA	CMP	15189	15088	Yes
ANION GAP - SERUM/PLASMA	BMP/CMP	15954	15849	No
ASPARTATE AMINOTRANSFERASE (AST) - SERUM/PLASMA	CMP	15406	15310	Yes
BILIRUBIN - SERUM/PLASMA	CMP	15224	15141	Yes
CALCIUM (CA) - SERUM/PLASMA	BMP/CMP	16164	16098	No
CARBON DIOXIDE (CO2) - SERUM/PLASMA	BMP/CMP	16309	16203	No
CHLORIDE (CL) - SERUM/PLASMA	BMP/CMP	16235	16130	No
CREATININE - SERUM/PLASMA	BMP/CMP	16403	16323	Yes
Erythrocyte Distribution Width (RDW) - BLOOD	CBC	16032	15974	Yes
GLUCOSE - SERUM/PLASMA	BMP	16184	16137	Yes
Hematocrit (HCT) - BLOOD	CBC	16213	16184	No
HEMOGLOBIN - BLOOD	CBC	16234	16186	No
Mean Corpuscular Hemoglobin (MCH) - BLOOD	CBC	16175	16120	No
Mean Corpuscular Hemoglobin Concentration (MCHC) - BLOOD	CBC	16166	16114	No
Mean Corpuscular Volume (MCV) - BLOOD	CBC	16220	16161	No
PLATELET - BLOOD - COUNT	CBC	16122	16099	No
Platelet Mean Volume (MPV) - BLOOD	CBC	16281	16247	No
POTASSIUM (K) - SERUM/PLASMA	BMP/CMP	16255	16165	No
PROTEIN - SERUM/PLASMA	CMP	15002	14932	No
RBC-COUNT-BLOOD	CBC	16187	16142	No
SODIUM (NA) - SERUM/PLASMA	BMP/CMP	16222	16144	No
UREA NITROGEN - SERUM/PLASMA	BMP/CMP	16147	16049	No
WBC-COUNT-BLOOD	CBC	16478	16455	Yes

For the variance based analysis, we first calculated the variance for each sample across their longitudinal clinical data from EMR. For each clinical lab variable, we visually inspected scatterplots of the variance distribution and determined a threshold for discovery and replication datasets separately (**Supplementary Table 2**). Next, samples were divided into high and low variance groups. For the high-variance/low-variance PheWAS analyses, we extracted all ICD-9 codes from the EHR. Participants were defined as cases if they had 3 or more instances of a particular ICD-9 code; less than 3 instances per participant were set to missing; and for no occurrence of an ICD-9 code, participants were designated control status. This resulted in testing a total of 541 ICD-9 codes.

Figure 1. Flow chart describing the analyses for median lab variable linear regression GWAS on 25 clinical labs

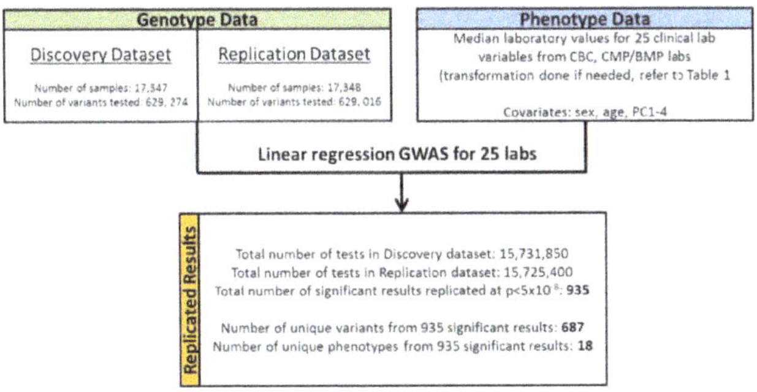

2.3 Analysis Methods

We performed the analysis for this study as a two-step process. First we performed a GWAS on median values for 25 different clinical lab variables (**Figure 1**). Next, we took the SNPs associated with the median trait values and performed an ICD-9 code PheWAS after grouping the participants into high-variance and low-variance groups for each clinical lab variable (Figure 2). Each of these analyses is described in more detail in the following sections.

2.3.1 Genome wide association analysis for 25 median clinical laboratory measurement

We performed a genome-wide association study (GWAS) to identify associations among all variants from the data (after quality control data cleaning) with median lab values for each of the 25 phenotypes. Linear regression analysis was performed using PLATO[10] (http://ritchielab.psu.edu/software/plato-download). All models were adjusted for age, sex and first 4 principal components to control for confounding influences in the analysis. Approximately 15M (~600,000 SNPs and 25 variables) tests were performed for each patient for both discovery and replication datasets. This analysis was repeated for both discovery and replication datasets separately and then we identified p-values for all variant and clinical lab combinations that were below genome-wide significance (p-value $5x10^{-8}$) in both datasets (discovery and replication).

2.3.2 *Variance-based analysis to identify associations with ICD-9 codes*

For all phenotypes from the median lab GWAS that has statistically significant replicating results (18 out of the 25 clinical lab variables, see **Figure 1**), we obtained longitudinal data for each patient across the EHR and calculated the trait variance for each lab variable. Next, for each of the 18 variables, we created scatterplots of the variance to identify samples that can be categorized as high

Figure 2. Flow chart describing the PheWAS analyses for high/low variance based datasets

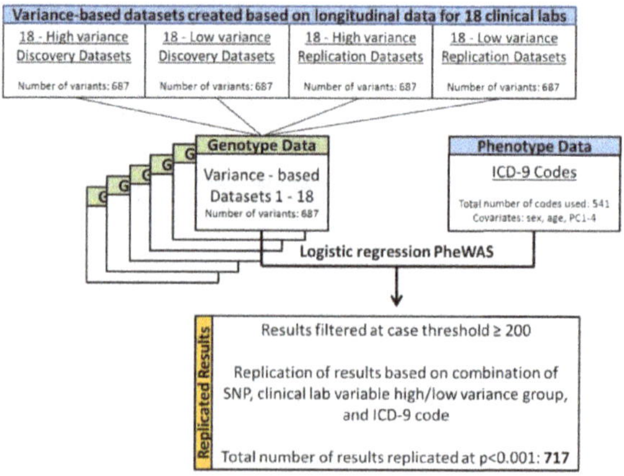

and low variance. Individual scatter plots for all of these variables are shown in **Supplementary Figure 3 and 4** for the discovery and replication datasets. For each variable, we created high variance and low variance groups based on a user-defined threshold to allow for PheWAS analyses separately in groups with high variability or low variability in each of the clinical lab variables. **Supplementary Table 2** lists the thresholds and samples sizes for low and high variance categories in both discovery and replication datasets. Participants below the chosen thresholds (based on looking at individual scatterplots) were categorized as low variance and above threshold were categorized as high variance.

The genotype data was filtered to include only those variants (687 SNPs) that were significantly associated in both the discovery and replication datasets for one or more clinical lab variables in the GWAS of median clinical lab values. Here, we are interested in the following question: Are genetic variants that are associated with a median clinical lab variable, also associated with diagnosis codes in patients with high variability or low variability in that lab variable? In other words, are there diseases that show association with that SNP in patients who are highly variable in their lab values or perhaps have low variability in their lab values? To investigate diagnosis codes that are associated with these variants, we performed logistic regression analysis for ICD-9 codes using PLATO (http://ritchielab.psu.edu/software/plato-download) by adjusting all models by age, sex and first 4 principal components. We only considered ICD-9 codes that had at least 200 or more cases with the code to reduce any false positive associations. Thus, for each sample 371,667 tests were erformed (687 SNPs and 541 ICD-9 codes). Lastly, we report the PheWAS results below a p-value threshold of 0.001 that replicate in low variance and/or high variance categories.

3. Results

Genome-wide association studies for median values from 25 clinical laboratory variables produced 935 SNP-phenotype associations that are present in discovery and replication sets at p-value less than $5x10^{-8}$. Association results below p-value 0.1 are shown in **Figure 3** as Manhattan plots for both discovery and replication datasets. Among the top results are multiple variants in the *UGT1A* gene family associated with serum bilirubin levels, where p-values for both discovery and replication datasets is 3.29×10^{-83}. This association has been identified and extensively reported by candidate gene and genome-wide association studies[11]. Hyperbilirubinemia results from a mutation in the *UGT1A1* gene which causes the non- or slow elimination of bilirubin from the body. We also identified variants in *SLCO1B1* associated with bilirubin levels, as suggested by previous GWAS studies [12–14] (rs4149081, Discovery p-value: $8.18x10^{-31}$ Replication p-value:$3.81x10^{-22}$).
Another association we identified is between missense variant, rs855791, on chromosome 22 in

Figure 3. Manhattan plots for GWAS performed on all 25 clinical lab variables. X-axis represents the chromosome and base pair location of each SNP and Y-axis represent the –log10 of p-value from association analysis. The two colors represent p-value for discovery and replication datasets. Direction of effect (positive or negative) is shown by the direction of arrows. Results at p-value <0.1 are shown in the plot. Black line indicates genome-wide significance (5e-08) threshold.

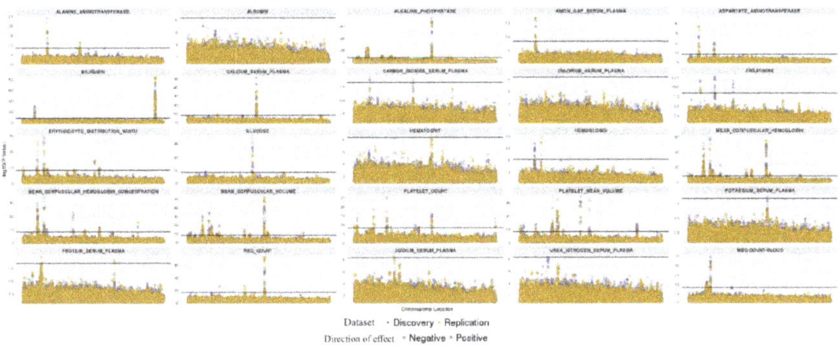

gene *TMPRSS6* (Discovery p-value: $2.04x10^{-60}$ (beta=-0.27); Replication p-value: $1.73x10^{-51}$ (beta=-0.25)). This association was identified by previous GWAS studies with hemoglobin levels as well as hemoglobin concentration[15,16]. It has been suggested that *TMPRSS6* is essential for maintaining iron levels in blood as it is involved in the control of iron homeostasis [16,17]. In addition, our GWAS analyses also identified many more previously reported associations, including variants in the *ABO* gene with alkaline phosphatase[18] (rs505922, discovery p-value: 2.41×10^{-52}, replication p-value:8.48×10^{-65}), the *CASR* gene with calcium levels[19,20] (rs17251221, discovery p-value: 6.55×10^{-44}, replication p-value:2.31×10^{-51}), and the *TCF7L2* gene with glucose levels[21] (rs7903146, discovery p-value: 1.41×10^{-35}, replication p-value:6.23×10^{-24}).

To explore pleiotropic associations among variants where one SNP is associated with multiple phenotypes, we generated a phenogram plot[22] shown in **Figure 4.** This plot shows, for example, multiple associations on chromosome 10 in gene *JMJD1C* to be associated with platelet mean volume as well as alkaline phosphatase (red box on **Figure 4**). Different GWAS studies performed separately on blood and metabolic panels have identified these associations[23,24] and our study serves as confirmation for these associations when both panels are combined together and analysis is run on the same patients. In our analysis, we see opposite directions of effect for both of these associations, i.e. erythrocyte distribution width (discovery beta: -0.004 and replication beta: -0.004) and mean corpuscular hemoglobin (discovery beta: 0.09 and replication beta: 0.12) which confirms the relationship observed in anemic patients, where elevation in RDW and decrease in hemoglobin is observed.

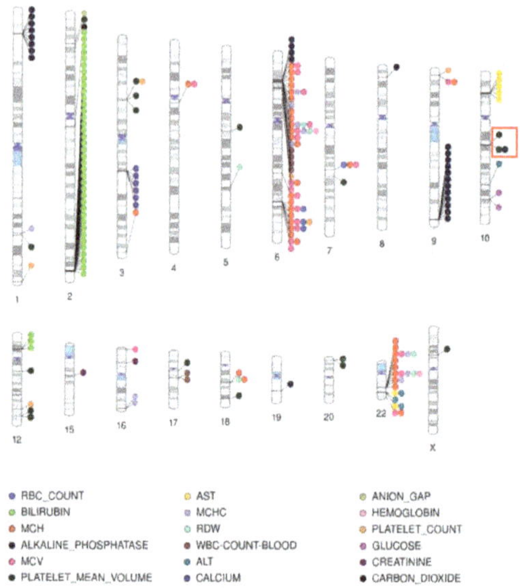

Figure 4. Phenogram plot representing pleiotropic associations. Here each colored circle is a SNP and its location is represented on the chromosome. SNPs are color coded based on the phenotype colors as shown in the legend. SNPs are also pruned to LD threshold of 0.4. Here MCH is Mean Corpuscular Hemoglobin; MCHC is MCH is Mean Corpuscular Hemoglobin Concentration; AST is Aspartate Aminotransferase; RDW is Erythrocyte Distribution Width; ALT is Alanine Aminotransferase.

Among our novel associations are intronic variant rs8095374 in gene *C18orf25* associated with erythrocyte distribution width known as RDW (discovery p-value: 8.79×10^{-10}, and replication p-value: 2.16×10^{-10}) and mean corpuscular hemoglobin (discovery p-value: 3.57×10^{-9}, and replication p-value: 1.84×10^{-13}). Both laboratory measurements are for red blood cells and could be useful in understanding the etiology of anemia.

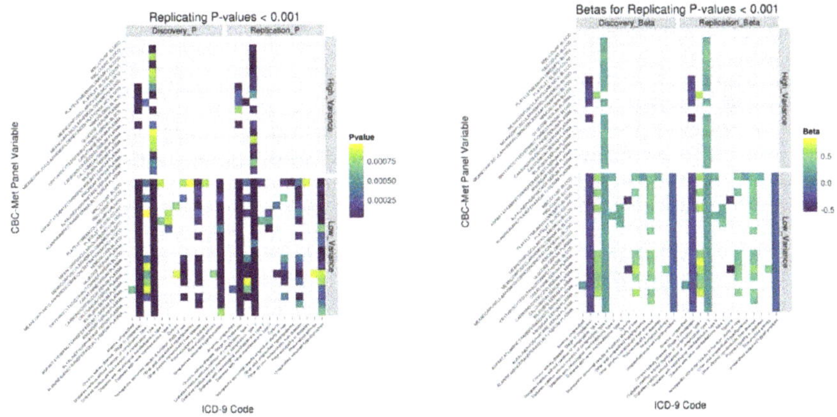

Figure 5. Heat map representing p-values (on left) and beta (on right) from variance based analysis for the combination of a SNP, ICD-9 code and clinical lab measurement in both high and low variance categories. Each point is the replicating SNP with the color gradient showing the range of p-value and beta. The results are only shown for replicating results at p-value<0.001 for both discovery and replication datasets in both high variance and low variance categories. X-axis lists all the ICD-9 codes and Y-axis lists the corresponding clinical lab variable for which replicating association is observed.

Our next approach was to integrate ICD-9 code data along with clinical lab variables to identify variants that we have found to be associated with median values of quantitative traits, and are *also* linked to diagnosis codes in the EHR. To perform this analysis, we wanted to utilize longitudinal data, rather than a measure from a single point in time. Hence, we divided patients into categories of high and low variance as described in *Methods*. Replication was observed based on the combination of SNP, clinical lab variable, ICD-9 code, and variance category (high or low). Replicated results are shown in form of a heat map in **Figure 5**. These heat maps show that in our study, the majority of our replicating associations occur in the low variance category. The primary reason for this is likely due to low sample size in the high variance groups gave us less statistical power to detect associations; although we would like to continue to explore this to determine whether there is a biological explanation for this. In total, this analysis resulted in 717 replicated associations.

We observed 39 SNPs on chromosome 6 that map to multiple genes (*C6orf10, FKBPL, BAT3, BAT2, EGFL2, RDBP, MSH5, TNXB*, C6orf27, *CSNK2B and BAT1*) are associated with Type 1 Diabetes (ICD-9 code 250.01) when the samples with high variance glucose levels were evaluated. These associations were not seen in samples in the low variance glucose category. One of the most interesting associations identified is between four SNPs in the *uromodulin (UMOD)* gene and ICD-9 code 585.3 (Chronic kidney disease) in patients with low variance for aspartate aminotransferase (discovery p-value: 8.71×10^{-9} and replication p-value: 2.03×10^{-6}). It has been observed by previous studies that patients with chronic kidney disease usually have low levels of aminotransferase in serum[25]. This association was not replicated in the high variance aspartate aminotransferase group. Association of variants in the *UMOD* gene with chronic kidney disease, kidney stones, and end stage renal diseases has been previously established[26,27] but an association with aspartate aminotransferase levels has not been identified by previous studies. Next, to integrate both the GWAS results and variance-based grouping PheWAS results, we generated networks of all genome-wide significant results from GWAS analysis and replicated results from variance based PheWAS analysis using Cytoscape[28] as shown in **Figure 6**. We explored the integrated results for SNP-Clinical lab variable- ICD-9 code in order to identify the three-way associations that are indicative of disease diagnosis. This figure shows the three top integrated networks from our analysis where both ICD-9 codes and clinical lab variables are linked via a SNP. One thing to note here is that all these networks resulted from the low variance groups only.

From the network visualization, we determined three variants in gene *TCF7L2* are associated with Type 2 Diabetes (T2D) and glucose levels. This association is expected because these variants have been reported by many previous studies to be associated with T2D[21,29,30]. Similarly, from this network analysis we also observed variants in the *UMOD* gene associated with chronic kidney disease and creatinine levels obtained from serum which has been previously reported by GWAS[26,27,31]. Lastly, a novel network obtained from this analysis is a link between rs3132941 (mapped to gene, *EGFL8*) with WBC count and Type I Diabetes. A high WBC has been observed in a few studies in T1D patients[32,33]. The *EGFL8* gene maps near the MHC region (Major-histocompatibility complex) on chromosome 6 and thus its association with T1D can be easily

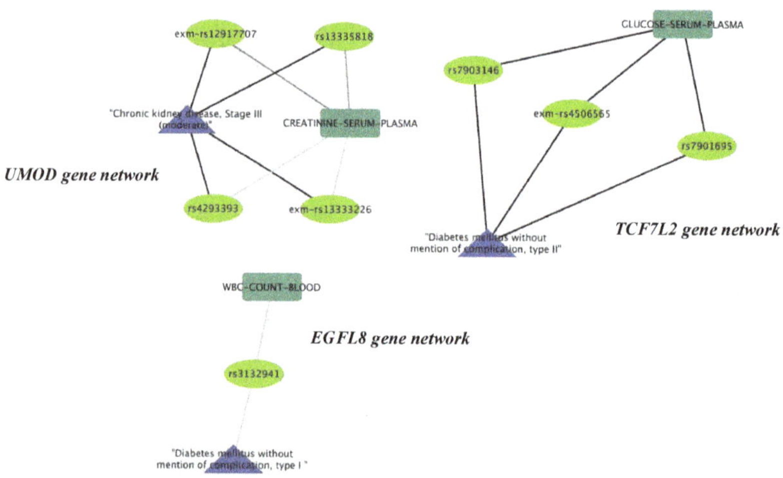

Figure 6. Network visualization generated by Cytoscape using replicated results from both GWAS and variance based analysis. Here, triangles represent ICD-9 code description, rectangles represent clinical lab variable, and ovals represent SNP. Darker edges represent more significant associations.

established[34,35] but its association with WBC has not been found in any previous studies. Our study presents this novel result which warrants further investigation.

4. Discussion

Genome-wide association studies have been tremendously successful in unravelling the etiologies of common complex diseases and the use of EHR in conducting such genome-wide and phenome-wide studies has shown resounding progress. Many researchers are now working on approaches to incorporate longitudinal information from the EHR into these studies. As a proof of concept, in this study we aimed at advancing the use of longitudinal information from laboratory values by looking at the variance for each outpatient clinical lab value rather than just mean/median or most recent value. We first conducted a GWAS for 25 clinical lab median values and then, based on variance, we divided participants into high and low variance groups. Next, we conducted a PheWAS to identify which SNPs are associated with median clinical lab variable *and* ICD-9 codes. This study represents a proof-of concept approach for utilizing trait variance and the longitudinal data as we successfully identified and confirmed many previously known associations. We also described several novel associations observed from our study. Variance, rather than mean/median may better capture the richness of the longitudinal data. In this pilot analysis, we demonstrate that this approach can be used to identify networks which reveal trends of associations among SNPs, laboratory measurements, and diagnosis codes. In the future, we plan to replicate this analysis with a larger sample size and in an independent EHR system. We also plan to use variance as the outcome for an association study in all 50,000 patients from Geisinger MyCode dataset and replicate in an independent dataset. One limitation of our approach here is that the use of longitudinal data in the way shown in this study ignores the fact that in an EHR, the duration of longitudinal information varies from patient to patient. Future approaches should also focus on developing methods which adjust for the duration of longitudinal information. Developing approaches, such as the one described in this manuscript, to explore the longitudinal nature of EHR data will provide greater opportunities for discovery and understanding of the genetic and clinical architecture of common diseases.

5. References

1. Manolio, T. A. Biorepositories--at the bleeding edge. *Int J Epidemiol* **37,** 231–233 (2008).
2. Moore, C. B. *et al.* Phenome-wide Association Study Relating Pretreatment Laboratory Parameters With Human Genetic Variants in AIDS Clinical Trials Group Protocols. *Open Forum Infectious Diseases* **2,** ofu113–ofu113 (2015).
3. Verma, A. *et al.* INTEGRATING CLINICAL LABORATORY MEASURES AND ICD-9 CODE DIAGNOSES IN PHENOME-WIDE ASSOCIATION STUDIES. *Pac Symp Biocomput* **21,** 168–179 (2016).
4. Namjou, B. *et al.* Phenome-wide association study (PheWAS) in EMR-linked pediatric cohorts, genetically links PLCL1 to speech language development and IL5-IL13 to Eosinophilic Esophagitis. *Front Genet* **5,** (2014).
5. Wei, W.-Q. & Denny, J. C. Extracting research-quality phenotypes from electronic health records to support precision medicine. *Genome Med* **7,** (2015).
6. Kullo, I. J., Ding, K., Jouni, H., Smith, C. Y. & Chute, C. G. A Genome-Wide Association Study of Red Blood Cell Traits Using the Electronic Medical Record. *PLoS One* **5,** (2010).
7. Namjou, B. *et al.* EMR-linked GWAS study: investigation of variation landscape of loci for body mass index in children. *Front Genet* **4,** (2013).
8. Luan, J. 'an *et al.* A multilevel linear mixed model of the association between candidate genes and weight and body mass index using the Framingham longitudinal family data. *BMC Proc* **3 Suppl 7,** S115 (2009).
9. Carey, D. J. *et al.* The Geisinger MyCode community health initiative: an electronic health record-linked biobank for precision medicine research. *Genet Med* (2016). doi:10.1038/gim.2015.187

10. Grady, B. J. *et al.* Finding unique filter sets in PLATO: a precursor to efficient interaction analysis in GWAS data. *Pac Symp Biocomput* 315–326 (2010).
11. Lin, J.-P. *et al.* Association between the UGT1A1*28 allele, bilirubin levels, and coronary heart disease in the Framingham Heart Study. *Circulation* **114,** 1476–1481 (2006).
12. de Azevedo, L. A. *et al.* UGT1A1, SLCO1B1, and SLCO1B3 polymorphisms vs. neonatal hyperbilirubinemia: is there an association? *Pediatr Res* **72,** 169–173 (2012).
13. Kang, T.-W. *et al.* Genome-wide association of serum bilirubin levels in Korean population. *Hum. Mol. Genet.* **19,** 3672–3678 (2010).
14. Johnson, A. D. *et al.* Genome-wide association meta-analysis for total serum bilirubin levels. *Hum. Mol. Genet.* **18,** 2700–2710 (2009).
15. Chambers, J. C. *et al.* Genome-wide association study identifies variants in TMPRSS6 associated with hemoglobin levels. *Nat Genet* **41,** 1170–1172 (2009).
16. van der Harst, P. *et al.* Seventy-five genetic loci influencing the human red blood cell. *Nature* **492,** 369–375 (2012).
17. Benyamin, B. *et al.* Common variants in TMPRSS6 are associated with iron status and erythrocyte volume. *Nat Genet* **41,** 1173–1175 (2009).
18. Li, J. *et al.* Genome-wide association study on serum alkaline phosphatase levels in a Chinese population. *BMC Genomics* **14,** 684 (2013).
19. Bonny, O. & Bochud, M. Genetics of calcium homeostasis in humans: continuum between monogenic diseases and continuous phenotypes. *Nephrol. Dial. Transplant.* **29,** iv55–iv62 (2014).
20. Kapur, K. *et al.* Genome-Wide Meta-Analysis for Serum Calcium Identifies Significantly Associated SNPs near the Calcium-Sensing Receptor (CASR) Gene. *PLOS Genet* **6,** e1001035 (2010).
21. Billings, L. K. & Florez, J. C. The genetics of type 2 diabetes: what have we learned from GWAS? *Ann N Y Acad Sci* **1212,** 59–77 (2010).
22. Wolfe, D., Dudek, S., Ritchie, M. D. & Pendergrass, S. A. Visualizing genomic information across chromosomes with PhenoGram. *BioData Min* **6,** 18 (2013).
23. Yuan, X. *et al.* Population-based genome-wide association studies reveal six loci influencing plasma levels of liver enzymes. *Am. J. Hum. Genet.* **83,** 520–528 (2008).
24. Qayyum, R. *et al.* A meta-analysis and genome-wide association study of platelet count and mean platelet volume in african americans. *PLoS Genet.* **8,** e1002491 (2012).
25. Ray, L., Nanda, S. K., Chatterjee, A., Sarangi, R. & Ganguly, S. A comparative study of serum aminotransferases in chronic kidney disease with and without end-stage renal disease: Need for new reference ranges. *Int J Appl Basic Med Res* **5,** 31–35 (2015).
26. Gudbjartsson, D. F. *et al.* Association of Variants at UMOD with Chronic Kidney Disease and Kidney Stones—Role of Age and Comorbid Diseases. *PLOS Genet* **6,** e1001039 (2010).
27. Reznichenko, A. *et al.* UMOD as a susceptibility gene for end-stage renal disease. *BMC Medical Genetics* **13,** 78 (2012).
28. Smoot, M. E., Ono, K., Ruscheinski, J., Wang, P.-L. & Ideker, T. Cytoscape 2.8: new features for data integration and network visualization. *Bioinformatics* **27,** 431–432 (2011).
29. Wellcome Trust Case Control Consortium. Genome-wide association study of 14,000 cases of seven common diseases and 3,000 shared controls. *Nature* **447,** 661–678 (2007).
30. Lyssenko, V. *et al.* Mechanisms by which common variants in the TCF7L2 gene increase risk of type 2 diabetes. *J. Clin. Invest.* **117,** 2155–2163 (2007).
31. Pattaro, C. *et al.* A meta-analysis of genome-wide data from five European isolates reveals an association of COL22A1, SYT1, and GABRR2 with serum creatinine level. *BMC Med. Genet.* **11,** 41 (2010).
32. Xu, W. *et al.* Correlation between Peripheral White Blood Cell Counts and Hyperglycemic Emergencies. *Int J Med Sci* **10,** 758–765 (2013).
33. Twig, G. *et al.* White Blood Cells Count and Incidence of Type 2 Diabetes in Young Men. *Diabetes Care* **36,** 276–282 (2013).
34. Nejentsev, S. *et al.* Localization of type 1 diabetes susceptibility to the MHC class I genes HLA-B and HLA-A. *Nature* **450,** 887–892 (2007).
35. Abraham, R. S., Wen, L., Marietta, E. V. & David, C. S. Type 1 Diabetes-Predisposing MHC Alleles Influence the Selection of Glutamic Acid Decarboxylase (GAD) 65-Specific T Cells in a Transgenic Model. *J Immunol* **166,** 1370–1379 (2001).

STRATEGIES FOR EQUITABLE PHARMACOGENOMIC-GUIDED WARFARIN DOSING AMONG EUROPEAN AND AFRICAN AMERICAN INDIVIDUALS IN A CLINICAL POPULATION

LAURA K. WILEY

Div. of Biomedical Informatics and Personalized Med., University of Colorado, 13001 E. 17th Pl. MS F-563 Aurora, CO 80045, USA

Email: laura.wiley@ucdenver.edu

JACOB P. VANHOUTEN

Dept. of Biomedical Informatics, Vanderbilt University, 2525 West End Ave Ste. 1475 Nashville, TN 37203, USA

Email: jacob.p.vanhouten@vanderbilt.edu

DAVID C. SAMUELS

Dept. of Mol. Physiology & Biophysics, Vanderbilt Genetics Inst., Vanderbilt University, 2215 Garland Ave. Nashville, TN 37232, USA

Email: david.c.samuels@vanderbilt.Edu

MELINDA C. ALDRICH

Dept. of Thoracic Surgery, Div. of Epidemiology, Vanderbilt University Medical Center, 609 Oxford House Nashville, TN 37232, USA

Email: melinda.aldrich@vanderbilt.edu

DAN M. RODEN

Dept. of Medicine, Vanderbilt University, 2215B Garland Ave Nashville, TN 37203, USA

Email: dan.roden@vanderbilt.edu

JOSH F. PETERSON

Dept. of Biomedical Informatics, Dept. of Medicine, Vanderbilt University, 2525 West End Ave Ste.1050 Nashville, TN 37203, USA

Email: josh.peterson@vanderbilt.edu

JOSHUA C. DENNY

Dept. of Biomedical Informatics, Dept. of Medicine, Vanderbilt University, 2525 West End Ave Ste.1475 Nashville, TN 37203, USA

Email: josh.denny@vanderbilt.edu

The blood thinner warfarin has a narrow therapeutic range and high inter- and intra-patient variability in therapeutic doses. Several studies have shown that pharmacogenomic variants help predict stable warfarin dosing. However, retrospective and randomized controlled trials that employ dosing algorithms incorporating pharmacogenomic variants under perform in African Americans. This study sought to determine if: 1) including additional variants associated with warfarin dose in African Americans, 2) predicting within single ancestry groups rather than a combined population, or 3) using percentage African ancestry rather than observed race, would improve warfarin dosing algorithms in African Americans. Using BioVU, the Vanderbilt University Medical Center biobank linked to electronic medical records, we compared 25 modeling strategies to existing algorithms using a cohort of 2,181 warfarin users (1,928 whites, 253 blacks). We found that approaches incorporating additional variants increased model accuracy, but not in clinically significant ways. Race stratification increased model fidelity for African Americans, but the improvement was small and not likely to be clinically significant. Use of percent African ancestry improved model fit in the context of race misclassification.

1. Introduction

Warfarin is a commonly used anticoagulant with a narrow therapeutic index and high rate of significant adverse reactions from both over- and under-dosing.[1] A number of pharmacogenomic variants are associated with stable warfarin dose,[2] and many studies have developed dosing algorithms using these variants.[1,3] Genotype-guided dosing is part of the United States Food and Drug Association (FDA) product label for warfarin.

The two largest randomized controlled trials of pharmacogenomic-guided warfarin dosing, EU-PACT[4] and COAG[5], yielded discordant findings on the clinical utility of incorporating pharmacogenomics into current dosing strategies. The EU-PACT study showed significantly increased percent time in therapeutic range (PTTR) over 12 weeks for the pharmacogenomic group while the COAG trial did not see a significant difference in PTTR over a 4-week time period. One of the reasons highlighted for these inconsistent findings across trials was the higher frequency of African descent individuals in COAG (27%) compared to EU-PACT (0.9%).[6] In COAG, African Americans with genotype-guided dosing spent an average of 8% less time in therapeutic range than the clinical algorithm group. Studies have shown that the *CYP2C9**2/*3 variants used by both COAG and EU-PACT are less frequent among those of African descent.[7] There are also variants important for dosing among individuals of African descent alleles that were unaccounted for in these trials.[7–11] Drozda found that failing to take into account these expanded variants resulted in significantly worse dose predictions among African Americans.[12] Additionally, Limdi found that using a race stratified dosing approach resulted in significantly more dose variation explained in both whites and blacks compared to a race-combined dosing model.[13]

Although much work has been conducted in this area, there remain outstanding questions that need to be answered. For example, because the algorithm proposed by Drozda was developed only in African Americans, its generalizability to individuals of European descent is unknown. Additionally, clinical dosing algorithms using a stratified approach, as advocated by Limdi have not been robustly tested to determine clinical validity. Further, in other clinical predictive models, using percent African ancestry as a more nuanced and biologically accurate measure of race provided better predictive performance than categorical race.[14] This study seeks to expand on previous warfarin dosing algorithm development efforts within Vanderbilt's EMR-linked biobank[15] to account for new variants associated with warfarin dose in African Americans. Additionally, we investigate whether race-stratified models or models using percent African ancestry result in clinically significant improvements (≥0.5-1mg/day) in dose prediction accuracy.

2. Methods

2.1. *Study Population*

Using BioVU, the Vanderbilt University biobank linked to electronic medical records (EMR), we selected all adult patients (≥18 years old) with DNA available who also had warfarin mentioned in the active prescription section of their problem list or a note from one of the hospital's anticoagulation clinics as of July of 2015. We used two approaches to extract stable warfarin dose based on whether the patient's warfarin was managed by an anticoagulation clinic or an individual physician.

We used a previously published and validated algorithm[15] to extract stable warfarin dose from patients with their dose managed by a Vanderbilt anticoagulation clinic or, for a subset of African Americans, where the dose was managed by their primary care provider. This approach identifies stable warfarin dose windows, as summarized in **Figure 1**. A stable dose window is defined as the presence of two or more notes from the anticoagulation clinic (or problem list entries for those managed by a primary-care provider) at least three, but not more than 12, weeks apart. During this time (from 7 days before the first note through the second note) the patient must also have two or more International Normalized Ratio (INR) measurements at least one day apart and all INR measurements in the window must be between 2 and 3. For anticoagulation clinic patients the INR goal range at the time of the stable dose window was required to be between 1.9-3.2. Primary-care managed patients were assumed to have an INR goal range of 2-3 unless otherwise specified (where ranges outside of 1.9-3.2 resulted in exclusion from the study). Warfarin dose was extracted from every anticoagulation clinic note in the window using regular expressions. The first window with identical prescribed warfarin doses throughout the window was selected as the stable warfarin dose. Patients lacking a window with identical warfarin doses throughout the window were manually reviewed to confirm accurate dose extraction. If multiple doses were prescribed during the window, the median dose was used. All primary care managed patient records were manually reviewed to extract warfarin dose and verify INR goal range because problem lists are susceptible to copy/paste redundancies and computational extraction may be invalid.

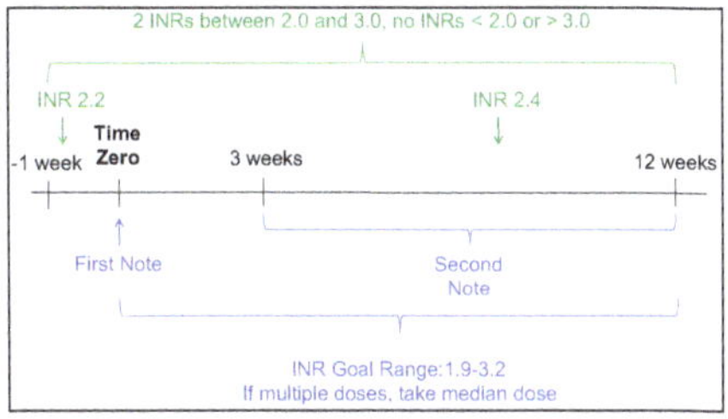

Figure 1. Stable Warfarin Dose Window Algorithm

Clinical covariates influencing stable warfarin dose were extracted with a variety of methods. Concomitant therapies (amiodarone, carbamazepine, phenytoin, and rifampin) listed in the problem list before or during the dose window were manually reviewed to confirm the prescriptions were active during the window. Smoking status was identified combination of natural language processing (NLP) and tobacco use International Classification of Disease version 9 (ICD9) codes,[16,17] followed by manual review to confirm active smoking at the time of the stable dose window. Body surface area,[18] was calculated using the median height and weight across the stable dose window or the closest height and weight measurement available within 3-6 months before or after the window (extracted via manual review). Age was defined as the age at the first anticoagulation clinic note or problem list warfarin entry in the stable dose window. "EMR recorded race" is defined by the care provider, but has shown concordance with genetic ancestry.[19] Indication for warfarin treatment, blood clots (i.e., deep venous thrombosis [DVT] or pulmonary embolism[PE]) or atrial fibrillation, was determined through ICD9 codes.[20,21]

2.2. *Genotyping*

This study genotyped twenty-one single nucleotide polymorphisms (SNPs) that had ever been associated with warfarin dose in European or African-descent populations and recorded in the

Table 1. Overview of Dose Prediction Models Tested

	Genetic Model				
	Clinical Only	**Limited Genetic**	**Expanded Genetic**	**Combined SNP**	**Haplotype**
Clinical Vars. (All)	Age (in decades); Body surface area; Smoking status; Amiodarone; Enzyme inducer				
Race Adj. (one of:)	1) Unadjusted 2) EMR Race 3) %African Ancestry 4) White Only 5) Black Only				
Genetic Variables	-	*VKORC1*-1639	*VKORC1*-1639	*VKORC1*-1639	*VKORC1*-1639
		*CYP2C9**2	*CYP2C9**2	*CYP2C9**2	rs2359612
		*CYP2C9**3	*CYP2C9**3	*CYP2C9**3	rs2884737
			*CYP2C9**5	rs7200749	rs7200749
			*CYP2C9**6	rs9934438	rs8050894
			*CYP2C9**8	rs17886199	rs9934438
			*CYP2C9**11	rs10871454	rs17886199
			rs2359612	rs2108622	rs10871454
			rs2884737	rs11676382	rs2108622
			rs7200749	rs12714145	rs11676382
			rs8050894	rs339097	rs12714145
			rs9934438	rs12777823	rs339097
			rs17886199	*VKORC1* Other[1]	rs12777823
			rs10871454	*CYP2C9* Other[2]	*CYP2C9**1/*2
			rs2108622		*CYP2C9**1/*3
			rs11676382		*CYP2C9**2/*2
			rs12714145		*CYP2C9**2/*3
			rs339097		*CYP2C9**3/*3
			rs12777823		*CYP2C9* Other Het.[3]
					CYP2C9 Other Hom.[4]

[1] If individual carries one or more minor allele at rs2359612 or rs2884737 or rs61162043 or rs8050894 then called 1, else 0; [2] If individual carries one or more minor allele at CYP2C9 *5/*6/*8 or *11 then call 1, else 0; [3] CYP2C9 *1/*11, *1/*5, *1/*6, *1/*8); [4] CYP2C9 *3/*8, *5/*8, *5/*11, *8/*8, *8/*11

Pharmacogenomics Knowledge Base (PharmGKB, www.pharmgkb.org).[22] Three variants (rs9923231, rs1799853, rs1057910) were genotyped using a Taqman assay by the Vanderbilt Technologies for Advanced Genomics (VANTAGE) core. A subset of white subjects had previous genotyping for these variants on the Illumina ADME assay and were not included in the Taqman assay. The remaining 17 variants were genotyped across the entire study population with a Sequenom assay performed by the VANTAGE core. Genotyping data were checked for marker efficiency and samples removed if they were missing one or more genotype calls for the tested variants. Duplicates and HapMap controls were validated.

We used existing genotyping data to calculate percent African ancestry across a subset of the population. Individuals were genotyped on one or more of the following platforms: Illumina Exome Beadchip, Human Omni Express Exome v2, Metabochip and/or OmniQuad. For each platform independently, samples with discrepant genders or sample efficiency <99% were removed. Markers with genotyping efficiencies < 99% or minor allele frequencies<5% were dropped. For the Exome chip, thresholds were set to 97% and 98% for genotyping and sample efficiency respectively as has been done previously to account for low frequency variants.[23] Within each platform, percent African ancestry was calculated using the ADMIXTURE supervised learning method with HapMap Phase III CEU and YRI reference populations.[24] The median estimate was used for individuals genotyped on multiple platforms.

2.3. *Analysis*

We fit and tested 25 different dosing models, combing 5 genetic modeling strategies (including exclusion of genetics altogether) with 5 different methods of race/ancestry adjustment. A summary of the 25 models tested are presented in **Table 1**. For race-stratified models, variants that were monomorphic or non-varying clinical factors were not included. To validate model summaries and prevent overfitting, we bootstrapped 1000 samples with replacement, trained a generalized linear model on each bootstrap, and tested the original dataset against each model. We calculated the mean absolute error (in mg/week) and R^2 for each bootstrap model, then calculated the median and an empiric confidence interval using the 2.5^{th} and 97.5^{th} percentiles of the bootstrap summaries. For all combined race models, we calculated these evaluation criteria across the entire test population and then within each EMR recorded race group separately. Because there are different risks for over- and under-dosing, we also calculated these summary evaluation criteria stratified by low (<21mg/week), medium (21-49mg/week), and high (>49mg/week) stable dose across the entire test

Table 2. Summary of Previous Algorithms Tested for Warfarin Dosing

Algorithm	Clinical Predictors	Genetic Predictors	Notes
Fixed 35mg Weekly Dose	-	-	-
FDA Dosing Table[1]	-	VKORC1-1639 CYP2C9*2 CYP2C9*3	Used mean of dosing range given.
IWPC (International Warfarin Pharmacogenetics Consortium)[2]	Age (in decades) Height Weight Asian African American Amiodarone Enzyme Inducers	VKORC1-1639 CYP2C9*2 CYP2C9*3	-
Ramirez et. al.[3]	Age (in years) Race Sex Body Surface Area Smoking Status DVT/PE Atrial Fibrillation	VKORC1-1639 CYP2C9*2 CYP2C9*3 CYP2C9*6 CYP2C9*8 rs2108622 rs339097	-
Hernandez et. al.[4]	Age (in years) Weight DVT/PE	VKORC1-1639 VKORC1, rs61162043 CYP2C9*2 CYP2C9*3 CYP2C9*5 CYP2C9*8 CYP2C9*11 rs7089580 rs12777823	Performed on subset of population with genotyping for rs61162043. Missing CYP2C9 rs7089580 due to probe failure. Set all patients to reference allele

[1] www.accessdata.fda.gov/drugsatfda_docs/ label/2010/009218s108lbl.pdf; [2] Klein *et. al.* NEJM. 2009; [3] Ramirez *et. al.* Future Medicine. 2010; [4] Hernandez *et. al.* The Pharmacogenomics Journal. 2014.

Table 3. Population Demographics

	Combined (n = 2181)	Whites (n = 1928)	Blacks (n = 253)
Weekly Warfarin Dose, mg/wk (median, sd)	35.0 (±17.6)	35.0 (±17.0)	40.8 (± 19.9)
Age, years (mean, sd)	66 (± 15)	66 (± 15)	60 (± 16)
Female (n, %)	911 (41.8%)	784 (40.7%)	127 (50.2%)
African American (n, %)	253 (11.6%)	-	-
% African Ancestry (median, sd)[1]	0.99 (± 31)	0.65 (± 4.5)	81.6 (± 10.3)
Height, cm (median, sd)	173 (± 13.5)	174 (±13.0)	170 (±16.1)
Weight, kg (median, sd)	89 (± 24.0)	88 (± 23.9)	91 (±24.7)
Body Surface Area, m^2 (median, sd)	2.0 (± 0.29)	2.0 (± 0.29)	2.0 (± 0.30)
Current Smokers (n, %)	209 (9.6%)	168 (8.7%)	41 (16.2%)
Amiodarone (n, %)	229 (10.5%)	202 (10.5%)	27 (10.7%)
Enzyme Inducers (n, %)	20 (0.92%)	15 (0.78%)	5 (1.98%)
Indication			
VTE (n, %)	414 (19.0%)	337 (17.5%)	77 (30.4%)
Atrial Fibrillation (n, %)	1592 (73%)	1443 (75%)	149 (59%)

[1] %-African ancestry available for 987 individuals (808 whites, 179 blacks)

population and then within each EMR recorded race separately. To evaluate the validity of our models and compare to existing algorithms, we also calculated mean absolute error and R^2 for a number of existing algorithms. The algorithms tested are summarized in **Table 2**.

3. Results

A total of 3,498 patients (3188 whites, 310 blacks) had a stable dose window (all features in **Figure 1**, except INR goal range filtering) and were genotyped on the Sequenom platform. Of these, 596 whites had *VKORC1*-1369 and *CYP2C9**2/*3 genotypes from the ADME platform, all other individuals were genotyped via Taqman. 291 individuals were missing one or more genotypes (with exceptions of rs7089580 and rs61162043 due to poor probe performance described below) and were removed from the analysis. Of the remaining 3,207 individuals 2,419 (2,192 whites and 227 blacks) had warfarin managed by the anticoagulation clinic. Filtering this population for INR goal ranges between 1.9-3.2 removed a further 233 individuals (212 whites, 21 blacks). Manual review to confirm stable warfarin dose, height and/or weight was performed for 203 whites and 28 blacks. This review removed 52 whites and 9 blacks for missing warfarin dose, height and/or weight. A total of 56 black individuals had warfarin managed by their primary care provider and were manually reviewed to extract warfarin dose and INR goal range. Combining the anticoagulation clinic and primary care populations yielded a final population of 2,181 individuals (1,928 whites, and 253 blacks).

Population demographics are presented in **Table 3**. Blacks had higher warfarin doses (40.8 vs 35mg/week), were younger (60 vs 66 years), were more likely to be current smokers (16% vs 8%), were more likely to be on anticoagulants due to thromboembolic events (30.4% vs 17.5%), and less likely to be on anticoagulants due to atrial fibrillation (59% vs. 75%) than whites. All other demographics factors were similar between blacks and whites.

One marker, rs7089580, failed genotyping in the Sequenom pool. Genotyping efficiency rates and minor allele frequencies are presented for the remaining 20 variants in **Table 4**. One variant, rs61162043 had lower genotyping efficiency (failed genotyping in 111 whites and 21 blacks) and was excluded from the expanded variants model. However, this variant was included in the *VKORC1* combined variable for the Combined Variant model. A summary of

Table 4. Genotyping Quality Control and Minor Allele Frequencies

Gene	SNP	Minor Allele	Call Rate[a]	Minor Allele Frequency (%) Combined (n=2181)	Whites (n=1928)	Blacks (n=253)
VKORC1	rs9923231	T	99.79[b]	35.1	38.3	10.5
VKORC1	rs2359612	A	100	36.4	38.5	20.8
VKORC1	rs2884737	C	99.96	23.3	25.3	3.8
VKORC1	rs61162043	A	93.82	37.2	35.8	49.6
VKORC1	rs7200749	A	99.96	2.6	0.3	20.2
VKORC1	rs8050894	G	99.92	37.3	38.8	26.5
VKORC1	rs9934438	A	99.96	35.1	38.4	10.5
VKORC1	rs17886199	G	100	0.5	0	4.2
STX4	rs10871454	T	99.96	35.3	38.5	10.9
*CYP2C9*2*	rs1799853	T	99.79[b]	13.5	14.9	2.4
*CYP2C9*3*	rs1057910	C	99.95[b]	6.1	6.7	2.0
*CYP2C9*5*	rs28371686	G	100	0.2	0.05	1.6
*CYP2C9*6*	rs9332131	del	99.79	0.2	0	1.4
*CYP2C9*8*	rs7900194	A	100	0.9	0.03	7.3
*CYP2C9*11*	rs28371685	T	99.96	0.4	0.3	1.4
CYP4F2	rs2108622	T	100	28.2	30.4	11.3
GGCX	rs11676382	G	99.96	8.8	9.7	2.6
GGCX	rs12714145	T	100	42.1	41.6	45.8
CALU	rs339097	G	99.83	1.6	0.2	12.3
CYP2C-cluster	rs12777823	A	99.92	16.1	14.6	28.1

[a]Call rates for completed genotyped population – not the final study population (as valid genotypes required for all but rs61162043); [b]Call rate for Taqman group only. ADME QC according to typical procedures.

the frequency of observed diplotypes for *CYP2C9* is presented **Table 5**. The majority of both racial/ethnic populations had a *1/*1 diplotype. Homozygotes and compound heterozygotes for the *2 and *3 variants (i.e., *2/*2, *3/*3, and *2/*3) were only observed in whites. Homozygotes and compound heterozygotes of the less common *5, *6, *8, and *11 alleles were only observed in blacks.

Within our final study population, 978 individuals (800 whites and 178 blacks) had genome-wide data available. A total of 764 individuals were genotyped on two platforms, 98 had genotyped data from three platforms, and 5 individuals had genotyping on four platforms. Of these individuals, the majority (n=437) had a maximum difference of less than 1% between estimates across platforms. Only 7 individuals had estimates across platforms that differed by more than 5% (maximum range of 9.8%). Three individuals had an EMR-recorded race of white, but had more than 50% African ancestry. The median ancestry estimate was used for all analyses.

A summary of the mean absolute error and percent variation explained (R^2) for all twenty-five fitted models, as well as the performance of existing dosing algorithms are provided in **Table 6**. Comparing all new and existing algorithms, the Expanded Genetic unadjusted, Expanded Genetic

Table 5. CYP2C9 Diplotype Frequencies

CYP2C9 Haplotype	Combined (n = 2181)	Whites (n = 1928)	Blacks (n = 253)
*1/*1	1402 (64.3%)	1222 (63.4%)	180 (71.2%)
*1/*2	357 (16.4%)	345 (18%)	12 (4.8%)
*1/*3	214 (9.8%)	205 (10.6%)	9 (3.6%)
*1/*5	8 (0.4%)	2 (0.1%)	6 (2.4%)
*1/*6	7 (0.3%)	-	7 (2.8%)
*1/*8	28 (1.3%)	1 (0.1%)	27 (10.7%)
*1/*11	15 (0.7%)	11 (0.6%)	4 (1.6%)
*2/*2	100 (4.6%)	100 (5.2%)	-
*2/*3	31 (1.4%)	31 (1.6%)	-
*3/*3	11 (0.5%)	11 (0.6%)	-
*3/*8	1 (<0.1%)	-	1 (0.4%)
*5/*8	1 (<0.1%)	-	1 (0.4%)
*5/*11	1 (<0.1%)	-	1 (0.4%)
*8/*8	3 (0.1%)	-	3 (1.2%)
*8/*11	2 (0.1%)	-	2 (0.8%)

EMR recorded race adjusted, Haplotype unadjusted, and Haplotype EMR recorded race adjusted models had the lowest mean absolute error across the combined population, with the Haplotype models explaining slightly more dose variance (54.4% vs 54.1%). The Expanded Variant model with percent ancestry adjustment had the lowest mean absolute errors in whites, and the Expanded Genetic stratified model had the lowest mean absolute error in blacks.

The algorithm performance with respect to mean error within low, medium, and high weekly dose groups are presented in **Figure 2**. When broken down by dose range 362 individuals (336 white and 26 black) had low warfarin requirements (<21mg/week), 1,313 individuals (1,173 whites and 140 blacks) had moderate warfarin requirements (21-49mg/week), and 486 individuals (402 whites and 84 blacks) had high warfarin requirements (>49mg/week). Within the medium dose requirement group (60% of the study population), dose predictions in whites were less than 5mg/week overestimated, while dose predictions in blacks were ~5mg/week overestimated. For the 17% of individuals with low warfarin dose requirements, mean dosing error was <10mg/week overestimated in whites, and 10-20mg/week overestimated in African Americans. The existing algorithm with the best performance among low-dose requiring African Americans was Ramirez *et. al.* (overestimating warfarin dose by 11.6mg/week). Within the high dose requirement individuals (22%), all races were consistently underestimated by 10-20mg/week.

Table 6. Performance of Predictive Dosing Algorithms

Algorithm	Mean Absolute Error (mg/week) Median (95% Confidence Interval)			Percent Variation Explained (R^2) Median (95% Confidence Interval)		
	Combined	Whites	Blacks	Combined	Whites	Blacks
Existing Algorithms						
Fixed 35 mg/week	13.5	13.2	16.1	-2.3	-1.1	-23.7
US FDA Table mid-range	12.0	11.7	14.7	17.5	18.3	1.1
IWPC	9.5	9.0	13.4	42.7	45.2	20
Ramirez *et. al.*	9.2	8.8	12.9	47.5	49.5	28.7
Hernandez *et. al.*	10.4	9.9	14.2	37.3	39.4	17.2
New Models						
Clinical						
Unadjusted	12.0 (11.9-12.0)	11.7 (11.7-11.8)	14.0 (13.8-14.2)	20.3 (19.9-20.5)	19.6 (19.0-19.9)	12.3 (9.9-14.9)
Race Adjusted	11.9 (11.9-11.9)	11.7 (11.7-11.7)	13.6 (13.4-13.8)	21.5 (21.1-21.6)	19.8 (19.3-20.0)	20.8 (18.9-22.1)
% Ancestry Adjusted	11.5 (11.5-11.7)	10.9 (10.8-11.0)	14.5 (14.3-14.9)	23.8 (22.9-24.2)	20.6 (19.3-21.3)	21.7 (17.9-24.2)
Race Stratified	-	11.7 (11.7-11.7)	13.4 (13.3-13.7)	-	19.8 (19.4-20.0)	21.7 (18.0-23.1)
Limited Genetic						
Unadjusted	9.3 (9.2-9.3)	8.8 (8.8-8.8)	12.9 (12.8-13.1)	51.5 (51.2-51.6)	53.8 (53.4-54.1)	27.8 (25.4-29.7)
Race Adjusted	9.3 (9.2-9.3)	8.8 (8.8-8.8)	12.8 (12.7-13.0)	51.8 (51.5-52.0)	54.0 (53.6-54.2)	30.0 (27.8-31.2)
% Ancestry Adjusted	9.5 (9.4-9.5)	8.5 (8.4-8.6)	13.7 (13.5-14.0)	49.8 (49.0-50.2)	52.8 (51.5-53.4)	32.4 (28.9-34.7)
Race Stratified	-	8.8 (8.8-8.8)	12.7 (12.5-13)	-	54.0 (53.7-54.2)	30.9 (27.1-32.5)
Expanded Genetic						
Unadjusted	**9.0 (9.0-9.1)**	8.6 (8.6-8.7)	12.2 (11.9-12.6)	54.1 (53.6-54.4)	55.4 (54.9-55.7)	37.7 (34.0-40.0)
Race Adjusted	**9.0 (9.0-9.1)**	8.6 (8.6-8.7)	12.2 (11.9-12.6)	54.1 (53.5-54.4)	55.4 (54.9-55.8)	37.5 (33.5-40.1)
% Ancestry Adjusted	9.2 (9.1-9.3)	**8.4 (8.3-8.5)**	12.9 (12.5-13.4)	52.5 (50.8-53.3)	54.2 (52.5-55.1)	39.7 (32.9-43.8)
Race Stratified	-	8.6 (8.6-8.7)	**11.9 (11.6-12.4)**	-	55.7 (55.2-55.9)	39.5 (33.8-42.5)
Combined SNP						
Unadjusted	9.9 (9.9-10.0)	9.5 (9.5-9.6)	12.8 (12.6-13.1)	44.0 (43.5-44.3)	44.7 (44.2-45.0)	30.2 (27.2-32.5)
Race Adjusted	9.9 (9.9-10.0)	9.6 (9.5-9.6)	12.8 (12.6-13.1)	44.0 (43.5-44.3)	44.7 (44.2-45.1)	30.3 (27.0-32.6)
% Ancestry Adjusted	10 (9.9-10.1)	9.2 (9.1-9.3)	13.6 (13.3-14.0)	44.4 (43.3-45.0)	44.8 (43.2-45.8)	34.2 (29.5-37.9)
Race Stratified	-	9.6 (9.5-9.6)	12.5 (12.2-12.9)	-	44.9 (44.3-45.1)	33.9 (29.6-36.4)
Haplotype						
Unadjusted	**9.0 (9.0-9.1)**	8.6 (8.6-8.7)	12.1 (11.8-12.5)	**54.4 (54.0-54.7)**	**55.8 (55.3-56.1)**	38.0 (34.5-40.1)
Race Adjusted	**9.0 (9.0-9.1)**	8.6 (8.6-8.7)	12.1 (11.9-12.5)	**54.4 (53.9-54.7)**	**55.8 (55.3-56.1)**	37.8 (34.3-40.2)
% Ancestry Adjusted	9.2 (9.1-9.3)	**8.4 (8.3-8.5)**	12.7 (12.4-13.3)	53.1 (51.6-53.9)	54.6 (52.3-55.5)	**41.5 (36.1-44.5)**
Race Stratified	-	8.6 (8.5-8.6)	12.0 (11.7-12.5)	-	56.2 (55.7-56.4)	39.6 (34.6-42.1)

Bold and shaded cells indicate the best performing algorithm for each population.

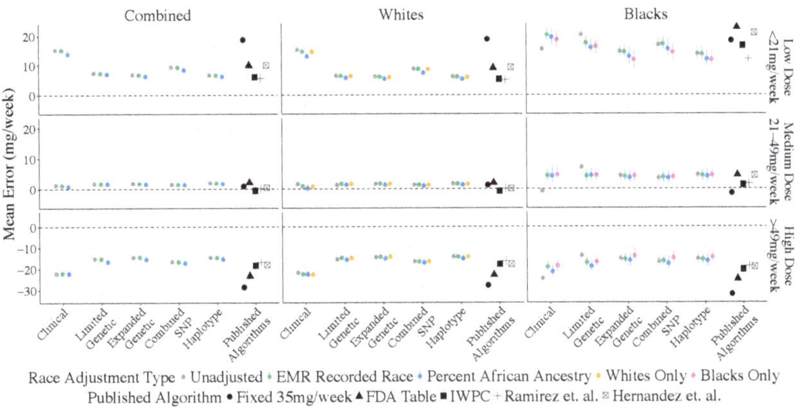

Figure 2. Performance of Dosing Algorithms by Stable Dose Range
This figure shows the algorithm performance (mean error in mg/week) divided by EHR recorded race and the stable dose range, e.g. patient's stable warfarin dose is a low weekly dose (<21mg/week), medium weekly dose (21-49mg/week), or high weekly dose (>49mg/week). Mean errors greater than 0 indicate over dosing, while mean errors less than 0 indicate underdosing.

4. Discussion

The goal of this study was to: 1) account for variants associated with warfarin dose in African Americans, 2) investigate whether race-stratified dosing leads to clinically significant improved dose predictions, 3) investigate whether race adjustment using percent ancestry offers improved prediction accuracy compared to EMR recorded race. The last goal was predicated on a study of lung function predictions (a continuous trait that, like warfarin dose, differs by race) that found improved model fit when they included percent African ancestry.[14] This hypothesis was bolstered by a study among Caribbean Hispanics that found adjusting for admixture improved warfarin dose prediction.[25]

Although this study required that individuals have DNA available in our biobank, because we took a complete cross-section of all individuals with warfarin exposure and DNA, the relative percentage of African Americans in this study (~10%) is consistent the broader Vanderbilt clinical population. As previously observed in the literature,[13] our black study population had a higher incidence of DVT/PE as an indication for anticoagulation. The genetics of our population were consistent with expected allele frequencies from the HapMap populations, with African Americans having allele frequencies lying between the Yoruba in Ibadan, Nigeria (YRI) and African Americans in the Southwest USA (ASW). Ancestry estimates for the black population were as expected with African Americans having approximately 80% African ancestry,[26] and allele frequencies for *CYP2C9**2/*3 and *VKORC1*-1639 were consistent with other studies within the Vanderbilt clinical population (that are not necessarily part of the biobank population).[27] Importantly, *CYP2C9* *2 and *3 homozygotes and compound heterozygotes were only observed in our white population, lending support to the notion that use of only *CYP2C9**2/*3 for warfarin dosing algorithms may be insufficient for African Americans.

Examining algorithm performance over the entire study population, the inclusion of additional variants associated with warfarin dose did increase dosing accuracy (mean absolute error) and percentage of dose variation explained for the combined, white and black populations. In all three populations one of the novel algorithms using SNPs independently (Expanded Genetic) or combined by *CYP2C9* diplotype (Haplotype) outperformed existing algorithms, the Clinical, and the Limited Genetic models. When considering confidence intervals, the Expanded Genetic and Haplotype models performed at similar levels across all populations. This is important for future clinical implementation as algorithms such as MyDrugGenome use *CYP2C9* diplotype. These diplotypes do not always have unambiguous assignments and are subject to change as the number of known genetic variants in a gene rise.[28] Our results suggest that algorithms utilizing unique SNPs can perform at similar levels to those using diplotypes and may be preferable due their more stable identification.

When considering only mean absolute error, stratified dosing models outperformed combined models only in African Americans. Interesting, stratified dosing did not result in improved performance over combined models in whites. This may be due to race misclassification of the three individuals recorded as white in the EMR, but who nevertheless had greater than 50% African ancestry. We chose not to manually change these individuals' race, as this misclassification is a real, generalizable[14] problem in the clinic, and would have an effect on algorithms' accuracy if clinically deployed. Although stratified dosing did improve algorithm performance among African Americans, it did not increase percent of warfarin dose explained by the model as has been seen in other studies.[13]

Correcting for race with percent ancestry yielded interesting results. Within the clinical model, percent ancestry improved model fit (lower mean absolute error, higher R^2) in the combined population, but not when pharmacogenomic markers were added into the model. Interestingly, percent ancestry improved dosing among whites across all models including those with pharmacogenomic markers. It is possible that the race misclassification also affected the algorithms using percent African ancestry. While this misclassification would be an important limitation in clinical implementation, at the current time this is less important because genetic ancestry is typically unavailable in current clinical systems. However, should this information have increased clinical utility in the future, panel testing of ancestry informative markers could enable implementation of these data.

While the algorithms developed in this study outperformed existing algorithms when considering the mean absolute error of prediction, we advocate using **Figure 2** to evaluate algorithm performance for desired implementation. We also caution that to determine the overall "best" algorithm, one must think within the context of clinical implementation of these algorithms. "Best" needs to be defined not just by performance, but also the generalizability and feasibility of implementation. For example, the Ramirez *et. al.* algorithm outperforms all algorithms for blacks with low warfarin doses and performs similarly to the best algorithms across most other race/dose requirement groups. However, the Ramirez *et. al.* algorithm requires the reason for anticoagulation (DVT/PE or atrial fibrillation), information typically computationally unavailable at the time of warfarin initiation. Many settings implementing prospective pharmacogenomic testing rely on automated clinical decision support and active intervention at the time of ordering to tailor the prescription. Although our overall best performing algorithm/s are not clinically significantly improved over the

Ramirez *et. al.* algorithm, they can all be computed with information readily available in a patient's medical record, allowing for immediate calculation of starting warfarin dose at the time of prescription.

In addition to the question of implementation one must also consider that the clinical impact of dose misclassification is not consistent across all dosing groups. Overdosing individuals with low warfarin requirements (warfarin dose <21mg/week) can lead to serious bleeding events, while under-dosing those with high warfarin requirements (doses >49mg/week) can lead to clotting events.[29,30] Although the IWPC algorithm performs similarly to the highest performance algorithms, it is particularly poor at predicting doses of low dose African Americans (~4.5 mg worse than the best performing algorithms). Depending on the frequency of low dose African Americans in the health system (determined with retrospective data), the IWPC algorithm may not be the best option. However, if the health system had a significant Asian population, use of the IWPC algorithm may be preferred because it takes these variables into account even if performance among low dose African Americans is reduced.

An important limitation of this study is that one of the previously tested algorithms, Ramirez *et. al.* was derived on a subset of patients included in this study. Thus it is possible that the high performance of the Ramirez algorithm in our population is inflated and may not be generalizable. The novel algorithms were also likely positively biased given the lack of an external validation set. Further, the results of the Hernandez *et. al.* algorithm were likely negatively biased as two SNPs predicting higher dose in African Americans were not included in this study due to poor genotyping quality. This study was also limited by the small number of African Americans studied. Additionally, since these data are from a single institution the results may not generalize to other populations. Warfarin dose is highly affected by vitamin K intake and the eating habits/cultural norms in the South may not reflect other parts of the US and world. Similarly, since this study only included whites and blacks, it is not clear how well the derived algorithms will perform among other ancestry groups.

In conclusion, expanding the variants in a warfarin dosing model does increase model accuracy, but not in clinically significant ways over existing algorithms in the literature. Similarly, race stratification resulted in the best model fits for African Americans, but the difference is unlikely to be clinically significant. Finally, percent ancestry surprisingly improves model fit – especially in the context of race misspecification in EMR recorded white race. However, the improvement in model fit among the white population is not clinically significant. When determining which dosing model to use, care must be given to selecting a model that not only matches the racial distribution of the population, but is also technically and financially achievable.

Acknowledgements

Thanks to Cara Sutcliffe and Paxton Baker of the VANTAGE core and Sarah Collier of the BioVU team for their help facilitating the genotyping in this project. This project was funded by Vanderbilt University, NCI (K07CA172294), NHLBI (U01HL122904, U19HL065962), NHGRI (U01HG007253, U01HG008672), and NLM (T15LM007450). The data used for this analysis were obtained from BioVU, supported by institutional funding and funding from NCATS (UL1TR000445).

References

1. B.F. Gage, C. Eby, J.A. Johnson, et al., *Clin Pharmacol Ther* **84**, 326 (2008).
2. J.A. Johnson and L.H. Cavallari, *Trends Cardiovasc Med* **25**, 33 (2015).
3. T.E. Klein, R.B. Altman, N. Eriksson, et al., *N Engl J Med* **360**, 753 (2009).
4. S.E. Kimmel, B. French, S.E. Kasner, et al., *N Engl J Med* **369**, 2283 (2013).
5. M. Pirmohamed, G. Burnside, N. Eriksson, et al., *N Engl J Med* **369**, 2294 (2013).
6. S.A. Scott and S.A. Lubitz, *Pharmacogenomics* **15**, 719 (2014).
7. M.A. Perera, L.H. Cavallari, N.A. Limdi, et al., *Lancet* **382**, 790 (2013).
8. M.A. Perera, E. Gamazon, L.H. Cavallari, et al., *Clin. Pharmacol. Ther.* **89**, 408 (2011).
9. Y. Liu, H. Jeong, H. Takahashi, et al., *Clin Pharmacol Ther* **91**, 660 (2012).
10. L.H. Cavallari, M. Perera, M. Wadelius, et al., *Pharmacogenet. Genomics* **22**, 152 (2012).
11. R. Daneshjou, E.R. Gamazon, B. Burkley, et al., *Blood* **124**, 2298 (2014).
12. K. Drozda, S. Wong, S.R. Patel, et al., *Pharmacogenet Genomics* **25**, 73 (2015).
13. N.A. Limdi, T.M. Brown, Q. Yan, et al., *Blood* **126**, 539 (2015).
14. R. Kumar, M.A. Seibold, M.C. Aldrich, et al., *N Engl J Med* **363**, 321 (2010).
15. A.H. Ramirez, Y. Shi, J.S. Schildcrout, et al., *Pharmacogenomics* **13**, 407 (2012).
16. M. Liu, A. Shah, N.B. Peterson, et al., *AMIA Annu Symp Proc* (2012).
17. L.K. Wiley, A. Shah, H. Xu, et al., *J Am Med Inf. Assoc* (2013).
18. D. Du Bois and E.F. Du Bois, *Nutrition* **5**, 303 (1989).
19. L. Dumitrescu, M.D. Ritchie, K. Brown-Gentry, et al., *Genet. Med.* **12**, 648 (2010).
20. E.R. McPeek Hinz, L. Bastarache, and J.C. Denny, *AMIA Annu Symp Proc* **2013**, 975 (2013).
21. S. Khurshid, J. Keaney, P.T. Ellinor, et al., *Am J Cardiol* **117**, 221 (2016).
22. M. Whirl-Carrillo, E.M. McDonagh, J.M. Hebert, et al., *Clin. Pharmacol. Ther.* **92**, 414 (2012).
23. Y. Guo, J. He, S. Zhao, et al., *Nat Protoc* **9**, 2643 (2014).
24. D.H. Alexander, J. Novembre, and K. Lange, *Genome Res* **19**, 1655 (2009).
25. J. Duconge, A.S. Ramos, K. Claudio-campos, et al., *PLoS One* **11**, (2016).
26. F. Zakharia, A. Basu, D. Absher, et al., *Genome Biol* **10**, R141 (2009).
27. S.L. Van Driest, Y. Shi, E.A. Bowton, et al., *Clin Pharmacol Ther* **95**, 423 (2014).
28. J.D. Robarge, L. Li, Z. Desta, et al., *Clin. Pharmacol. Ther.* **82**, 244 (2007).
29. E.M. Hylek, A.S. Go, Y. Chang, et al., *N Engl J Med* **349**, 1019 (2003).
30. E.M. Hylek, C. Evans-Molina, C. Shea, et al., *Circulation* **115**, 2689 (2007).

SINGLE-CELL ANALYSIS AND MODELLING OF CELL POPULATION HETEROGENEITY

NIKOLAY SAMUSIK

Department of Microbiology & Immunology
Stanford Medical School
Stanford 94305 CA, USA
Email: samusik@stanford.edu

NIMA AGHAEEPOUR

Department of Anesthesiology
Stanford Medical School
Stanford 94305 CA, USA
Email: naghaeep@stanford.edu

SEAN BENDALL

Department of Pathology
Stanford Medical School
Stanford 94305 CA, USA
Email: bendall@stanford.edu

Recent technological developments allow gathering single-cell measurements across different domains (genomic, transcriptomics, proteomics, imaging etc). Sophisticated computational algorithms are required in order to harness the power of single-cell data. This session is dedicated to computational methods for single-cell analysis in various biological domains, modelling of population heterogeneity, as well as translational applications of single cell data.

1. Introduction

Inferring the molecular mechanism of cell behavior and linking it to function and dysfunction is one of the ultimate goals of quantitative biology and medicine. Until recently, most measures to classify and characterize cellular behavior have been performed on the 'bulk samples', whereby a large number of cells were physically homogenized and then assayed. Bulk measurements erase the information about the potentially complex heterogeneity of cellular states within the samples. The problem with such approaches becomes obvious from a simple example: whenever researchers observe a difference in average values of a single parameter between samples, it is quite impossible to differentiate between a scenario where there was a homogenous change of a variable in all cells versus a shift in compositional ratios between a differentially expressing populations Besides, the measurements derived from pooled populations of cells lack the specificity to capture outlier cell behavior that might explain cell differentiation and transitions from normal to disease cellular states. The noise, or variance, between the molecular states of different cells -- even among cells assumed to be homogenous – can be correlated with protein expression and function [1] as well as cell morphology and interaction with neighbors[2]. Emergence of cell heterogeneity might be sporadic (e.g., cell-to-cell variation in cell culture[3]), programmed (e.g., cell differentiation[4] or immune receptor recombination[5]), or a result of adaptive evolution and semi-heritable phenotypic plasticity[6].

The ability to quantify molecular events with single cell resolution is intrinsically linked to analytical advances. Unfortunately, many of those variations could not be systematically studied by traditional molecular biology methods, such as PCR, Western Blotting, IP, genome sequencing, microarrays and RNA-seq, because they lack the sensitivity and the throughput that are required for single cell analysis. One notable exception is immunology, which has enormously benefitted from early adoption of the single-cell analysis by flow cytometry and FACS. Flow cytometry has been pivotal to detailed characterization of various immunological processes, such as blood cell development and activation and has enabled systematic mapping of the roles of various immune cell populations in healthy and disease states. Driven by a need to distinguish multiple cell populations, cytometry placed emphasis on multiparametric analysis whereby the cell populations were defined by increasingly complex combinations of protein markers. More recently, the importance of multiparametric analysis has increased with advent of mass cytometry[7]. Many excellent computational tools have been developed for handling cytometry data, including specialized clustering algorithms for automated mapping of cell population[8], machine learning tools that find cell populations that are correlated to clinical outcome[9], data visualization tools that trace cell differentiation trajectories[10,11], a specialized ontology of cell types[12], algorithms for causal inference of signaling networks by leveraging huge training sets of single-cell data [13], data-driven reference maps of immune cell populations[14] and many others.

For many years the single-cell analysis has been associated with flow cytometry and was limited to measuring protein concentrations using tagged antibodies. Recent advances in experimental techniques and automation have greatly expanded the scope of single-cell analysis and introduced completely novel readouts and modalities. Examples include:

1. Genomic sequencing in single cells [15]
2. Single cell RNA-seq [16]
3. Single molecule RNA sequencing in situ [17]
4. Gene expression profiling by flow cytometry [18 19]
5. Histo-cytometry [20]
6. Multiplexed ion beam imaging [21]
7. Mapping of chromatin state in single cells [22]
8. Cell morphology and motility analysis in cell cultures [2]
9. Single cell western blotting[23]

These emerging technologies provide an unprecedented opportunity to capture new biological processes and mechanisms at the single cell level. Given the list of analytical methods with a single cell resolving power now available, a wealth of new information, including: protein abundance, methylation patterns, promoter structure, gene expression, copy number variations, gene function and essentiality, DNA structure, evolutionary plasticity, and selective advantage can now be created for integration. Synthesis and interpretation of various modalities of single cell-level data now depends on novel computational approaches that aim to uncover and model the biological principles behind the cell heterogeneity. Data fusion methods that leverage prior biological knowledge for automated cell type annotation. Most importantly, computational methods are needed to provide a system-level view of the interplay of diverse, fluctuating biological components and identify clinically relevant and actionable modules within the biological system. In this session we feature excellent pieces of original research that broadly cover various aspects of single-cell analysis and modelling of cellular heterogeneity.

2. Session contributions

2.1. *Data normalization and quality control*

Quality control is a cornerstone of quantitative data analysis: rigorous filtering of noisy and spurious signals and correction of systematic variability is lays the solid foundation which ensures that the downstream data analysis captures true biological effects.

Aevermann et al. present a quality control pipeline for single-cell analysis which pioneers the use of objective criteria and machine learning for QC of single-nuclei sequencing data. While many researchers today still rely on subjective assessment of data quality, Aevermann and colleagues designed and trained a classifier that implements a random-forest approach with 79 features per sample to stratify samples into 3 quality classes: 1 pass and 2 types of fails. Analysis of 2272 single-nuclei samples successfully screened out 21% low quality data points. Authors demonstrated that removing the low-quality samples had a marked effect on the quality of the results in the downstream multidimensional manifold embedding analysis.

Fread et al. devised an elegant advance for the quality control and filtering of barcoded mass cytometry (CyTOF) data. They are introducing a concept of per-sample filtering of data following the debarcoding, which allows for proper handling of potentially very significant sample-to-sample variations in barcode intensity. Authors are also pioneering the idea of combining multiple cellular features into semi-artificial filtering parameters and writing them into the FCS files, which gives the human analyst an opportunity to set filtering gates using gating software and adjust the positioning of such gates on as sample-by-sample basis, dynamically monitoring the data quality based on biaxial scatterplots for other parameters. This simple yet elegant improvement dramatically streamlines the process of filtering spurious single-cell events and their publicly available software can be expected to be of a great utility to the CyTOF community.

2.2. *Manifold embedding and tracing with single-cell datasets*

One of the most exciting opportunities in the age of single-cell data is the ability to map the complex processes of cell differentiation by tracing the manifold shapes of single-cell distributions and discovering the local trajectories of cell changes in the marker space. This analysis is complicated by the unpredictable nature of manifolds in the data, high dimensionality of feature space and the instability of the local covariance matrix.

Cordero et al. introduce an approach for linear trajectory tracing in single cell RNA-seq data called SCIMITAR that implements morphing Gaussian model and performs simultaneous estimation of the mean expression levels along the trajectory and the local covariance matrix. The authors introduce a new statistical test to select relevant genes based on correlation of gene expression to the trajectory. They convincingly demonstrate that this test is more sensitive and specific that a conventional group-based comparison, picking up more biologically significant genes than the ANOVA-based statistical test in the original paper[24]. While the SCIMITAR algorithm is currently limited by the assumption of a single curvilinear trajectory, the authors anticipate further extension of this approach that would allow capturing more complex manifolds.

Kim et al. present a new scalable algorithm for fast embedding of multidimensional data based on LargeVis algorithm[25]. Unlike most embedding methods, the algorithm works in linear time, which it very useful given the ever-growing datasets. Authors validate the algorithm on CyTOF data from mouse bone marrow and show that the quality of embedding is superior to the slower tSNE algorithm that is currently popular in the single-cell analysis community.

2.3. *Cross-species alignment of single-cell expression patterns*

Traditionally, comparative cytology and histology relied on qualitative descriptions of tissue architectures and cell functions across different organisms. The availability of single-cell data opens a possibility to quantitatively align differentiation trajectories and cell types between species based on their expression profiles and other quantitative functional features. Such mapping could help us understand better the development and evolution of multicellular organisms and also facilitate the transfer of pre-clinical results from model organisms to human.

Johnsons et al. harness the single-cell RNA-seq data from neural precursors in human and mouse for building the cross-species map of neural cell populations. They take a two-step approach, which starts with defining the list of genes which show concordant expression patterns across major neuronal precursor populations in both species. In the second step, the authors co-cluster neuronal cell distributions of the two species based on the concordant gene subset, thus constructing a cross-species map of cell populations. Despite the lack of a perfect overlap, which is expected due to systematic differences in cell distributions between species, the authors show that the obtained cross-species map can be utilized for transferring the functional annotations of cells subsets between the corresponding population of the two species.

2.4. *Modelling of cell heterogeneity in cancer*

While single-cell readouts provide excellent snapshots of population heterogeneity, creating comprehensive mathematical models of cell interactions, somatic transdifferentiation and clonal evolution is key to attaining detailed understanding of dynamic processes that underpin the population heterogeneity in cancer. By identifying the causal chains of events and iterating through various scenarios, mathematical models of cancer cell populations can yield clinically actionable predictions and assist in optimizing treatment strategies.

Kanigel Winner and Costello present a novel modeling technique to model the treatment regimens for people with metastatic bladder cancer. This form of cancer metastasized to the lung has not been previously modeled and hence is an important and realistic problem since overall survival for this disease has not improved in the past three decades. The authors created a computational model to simulate tumor environment by carefully incorporating quantitative data about cell division rates, in vivo drug concentrations, in vitro IC50 curves for cancer cell lines and vascularization patterns of tumor microenvironment. This model was used to analyze different chemotherapeutic regimens much faster than getting in-vivo data. Authors strikingly demonstrated that the standard-of-care chemotherapeutic regimen that alternates gemcitabine and cisplatin inevitably leads to quick emergence of resistant clones, which goes in line with the abysmal 5-year survival rate (6.8%) for this type of cancer following the aforementioned treatment. Authors also found that any conceivable regimen combining the two drugs will eventually lead to resistance because of randomly surviving cancer cell clones. Key factors that contribute to this resistance is the inhomogeneity of drug distribution in the tissue and the 'dilution effect' whereby rapidly dividing cells effectively drop the drug concentration by splitting it between daughter cells. With further refinement, this model could help design novel therapeutic regimens that would hopefully lead to disease eradication.

3. Acknowledgements

We thank all of the authors who submitted papers for this session and all of the reviewers who contributed their time and expertise. We acknowledge the NIH grant R01GM109836 and the Rachford and Carlota A. Harris Endowed Chair for support. We are grateful to the PSB organizers for their support and especially Tiffany Murray for meeting coordination.

4. References

1. Newman, J. R. S. *et al.* Single-cell proteomic analysis of S. cerevisiae reveals the architecture of biological noise. *Nature* **441,** 840–6 (2006).
2. Battich, N., Stoeger, T. & Pelkmans, L. Control of Transcript Variability in Single Mammalian Cells. *Cell* **163,** 1596–1610 (2015).
3. Spencer, S. L., Gaudet, S., Albeck, J. G., Burke, J. M. & Sorger, P. K. Non-genetic origins of cell-to-cell variability in TRAIL-induced apoptosis. *Nature* **459,** 428–32 (2009).
4. Süel, G. M., Kulkarni, R. P., Dworkin, J., Garcia-Ojalvo, J. & Elowitz, M. B. Tunability and noise dependence in differentiation dynamics. *Science* **315,** 1716–9 (2007).
5. Proudhon, C., Hao, B., Raviram, R., Chaumeil, J. & Skok, J. A. Long-Range Regulation of V(D)J Recombination. *Adv. Immunol.* **128,** 123–82 (2015).
6. Quaranta, V. *et al.* Trait variability of cancer cells quantified by high-content automated microscopy of single cells. *Methods Enzymol.* **467,** 23–57 (2009).
7. Bendall, S. C. *et al.* Single-cell mass cytometry of differential immune and drug responses across a human hematopoietic continuum. *Science* **332,** 687–96 (2011).
8. Aghaeepour, N. *et al.* Critical assessment of automated flow cytometry data analysis techniques. *Nat. Methods* **10,** 228–38 (2013).
9. Bruggner, R. V, Bodenmiller, B., Dill, D. L., Tibshirani, R. J. & Nolan, G. P. Automated identification of stratifying signatures in cellular subpopulations. *Proc. Natl. Acad. Sci. U. S. A.* **111,** E2770-7 (2014).
10. Bendall, S. C. *et al.* Single-Cell Trajectory Detection Uncovers Progression and Regulatory Coordination in Human B Cell Development. *Cell* **157,** 714–725 (2014).
11. Zunder, E. R., Lujan, E., Goltsev, Y., Wernig, M. & Nolan, G. P. A Continuous Molecular Roadmap to iPSC Reprogramming through Progression Analysis of Single-Cell Mass Cytometry. *Cell Stem Cell* **16,** 323–337 (2015).
12. Meehan, T. F. *et al.* Logical Development of the Cell Ontology. *BMC Bioinformatics* **12,** 6 (2011).
13. Sachs, K., Perez, O., Pe'er, D., Lauffenburger, D. A. & Nolan, G. P. Causal protein-signaling networks derived from multiparameter single-cell data. *Science* **308,** 523–9 (2005).
14. Spitzer, M. H. *et al.* An interactive reference framework for modeling a dynamic immune system. *Science (80-.).* **349,** 1259425–1259425 (2015).

15. Navin, N. *et al.* Tumour evolution inferred by single-cell sequencing. *Nature* **472,** 90–4 (2011).

16. Jaitin, D. A. *et al.* Massively parallel single-cell RNA-seq for marker-free decomposition of tissues into cell types. *Science* **343,** 776–9 (2014).

17. Lee, J. H. *et al.* Fluorescent in situ sequencing (FISSEQ) of RNA for gene expression profiling in intact cells and tissues. *Nat. Protoc.* **10,** 442–58 (2015).

18. Lyubimova, A. *et al.* Single-molecule mRNA detection and counting in mammalian tissue. *Nat. Protoc.* **8,** 1743–58 (2013).

19. Frei, A. P. *et al.* Highly multiplexed simultaneous detection of RNAs and proteins in single cells. *Nat. Methods* (2016). doi:10.1038/nmeth.3742

20. Gerner, M. Y., Kastenmuller, W., Ifrim, I., Kabat, J. & Germain, R. N. Histo-cytometry: a method for highly multiplex quantitative tissue imaging analysis applied to dendritic cell subset microanatomy in lymph nodes. *Immunity* **37,** 364–76 (2012).

21. Angelo, M. *et al.* Multiplexed ion beam imaging of human breast tumors. *Nat. Med.* **20,** 436–442 (2014).

22. Buenrostro, J. D. *et al.* Single-cell chromatin accessibility reveals principles of regulatory variation. *Nature* **523,** 486–490 (2015).

23. Hughes, A. J. *et al.* Single-cell western blotting. *Nat. Methods* **11,** 749–55 (2014).

24. Darmanis, S. *et al.* A survey of human brain transcriptome diversity at the single cell level. *Proc. Natl. Acad. Sci. U. S. A.* **112,** 7285–90 (2015).

25. Tang, J., Liu, J., Zhang, M. & Mei, Q. Visualizing Large-scale and High-dimensional Data. *Proc. 25th Int. Conf. World Wide Web* 287–297 (2016). doi:10.1145/2872427.2883041

PRODUCTION OF A PRELIMINARY QUALITY CONTROL PIPELINE FOR SINGLE NUCLEI RNA-SEQ AND ITS APPLICATION IN THE ANALYSIS OF CELL TYPE DIVERSITY OF POST-MORTEM HUMAN BRAIN NEOCORTEX*

BRIAN AEVERMANN[1#], JAMISON MCCORRISON[1#], PRATAP VENEPALLY[1#], REBECCA HODGE[2], TRYGVE BAKKEN[2], JEREMY MILLER[2], MARK NOVOTNY[1], DANNY N. TRAN[1], FRANCISCO DIEZ-FUERTES[1,3], LENA CHRISTIANSEN[4], FAN ZHANG[4], FRANK STEEMERS[4], ROGER S. LASKEN[1], ED LEIN[2], NICHOLAS SCHORK[1], RICHARD H. SCHEUERMANN[1,5,6†]

[1]J. Craig Venter Institute, 4120 Capricorn Lane, La Jolla, CA 92037, USA, [2]Allen Institute for Brain Science, 615 Westlake Avenue North, Seattle, WA 98103, USA, [3]Centro Nacional de Microbiología, Instituto de Salud Carlos III, Madrid, Spain, [4]Illumina, Inc.,5200 Illumina Way, San Diego, CA 02122, USA, [5]Department of Pathology, University of California, San Diego, 9500 Gilman Drive, La Jolla CA 92093, USA, [6]Division of Vaccine Discovery, La Jolla Institute for Allergy and Immunology, 9420 Athena Circle, La Jolla, CA 92037, USA

Next generation sequencing of the RNA content of single cells or single nuclei (sc/nRNA-seq) has become a powerful approach to understand the cellular complexity and diversity of multicellular organisms and environmental ecosystems. However, the fact that the procedure begins with a relatively small amount of starting material, thereby pushing the limits of the laboratory procedures required, dictates that careful approaches for sample quality control (QC) are essential to reduce the impact of technical noise and sample bias in downstream analysis applications. Here we present a preliminary framework for sample level quality control that is based on the collection of a series of quantitative laboratory and data metrics that are used as features for the construction of QC classification models using random forest machine learning approaches. We've applied this initial framework to a dataset comprised of 2272 single nuclei RNA-seq results and determined that ~79% of samples were of high quality. Removal of the poor quality samples from downstream analysis was found to improve the cell type clustering results. In addition, this approach identified quantitative features related to the proportion of unique or duplicate reads and the proportion of reads remaining after quality trimming as useful features for pass/fail classification. The construction and use of classification models for the identification of poor quality samples provides for an objective and scalable approach to sc/nRNA-seq quality control.

---

* This work is supported by the Allen Institute for Brain Science, the JCVI Innovation Fund, and the U.S. National Institutes of Health 1R21AI122100.

# Contributed equally to this work.

† Corresponding author email: rscheuermann@jcvi.org.

1. Introduction

Single cell genomic analysis is poised to revolutionize our understanding of the diversity and complexity of multicellular organisms. One of the key applications of single cell genomics is the determination of transcriptional profiles using next generation sequencing of amplified cDNA synthesized from the RNA content of single cells or single nuclei (sc/nRNA-seq). By avoiding the averaging phenomenon inherent in the analysis of bulk cell populations, sc/nRNA-seq is revealing a level of cell type complexity and dynamics that is unprecedented in comparison with previous technologies.

sc/nRNA-seq has now been applied to explore a wide range of biological questions. It has been used to understand the heterogeneity of somatic mutations acquired in cancer subclones arising from the same progenitor [Patel 2014][Min 2015], providing insights into therapeutic responses and disease progression. sc/nRNA-seq has been used to track cell state transition dynamics during normal tissue differentiation [Nestorowa 2016], cell cycle progression [Scialdone 2015], and *in vitro* trans-differentiation induced using direct reprogramming methodologies [Treutlein 2016]. It has also been used to investigate the dynamics of X chromosome inactivation in preimplantation embryos [Petropoulus 2016], lineage determination during blastocyst development [Blakeley 2015], T cell receptor repertoires in antigen-specific immune responses [Eltahla 2016], T cell progressive cell states [Proserpio 2016], variability in cellular responses to viral infections [Ciuffi 2016], and the similarities between induced pluripotent stem cell-derived neurons and cells from primary tissue and cortical layers [Handel 2016]. And at its most basic level, sc/nRNA-seq is being used to understand the complexity of steady state cell type distributions in normal human tissues [Zeisel 2015][Wang 2016][Lacar 2016][Li 2016], and abnormal tissue disorders [Ramsköld 2012][Glaublomme 2015][Tirosh 2016].

RNA-seq from single *nuclei* (Grindberg, 2013) provides transcriptomes that strongly reflect those obtained from whole cells. Nuclei can be used in place of cells to assess cell type and state, as well as revealing mRNAs and non-coding RNAs that are differentially enriched in the nucleus. The use of nuclei as a starting material also has the advantage of providing individual transcriptomes without the harsh proteolytic treatment required to disperse single cells from intact tissue specimens, which is known to alter gene expression and damage sensitive cell types. snRNA-seq has enabled single neuron studies even from postmortem human brain tissue (Krishnaswami, 2016). Use of nuclei for RNA-seq enabled the first single neuron analysis of immediate early gene expression associated with memory formation in the mouse hippocampus, whereas proteolytic dissociation of neurons yielded artifactual expression in most cells (Lacar, 2016). In this study we use data from single nuclei RNA-seq, however, the QC analysis proposed should be equally applicable to single cell data.

While the promise of sc/nRNA-seq is enormous, the methods used to isolate and specifically amplify the RNA target material in a manner that preserves the molecular structures and abundance levels pushes the limits of these technologies. As a result, the impact of contaminating nucleic acid templates (e.g. chromosomal and other contaminating DNAs, rRNA, mtDNA), technical variability in laboratory reagents and procedures (e.g. variability in the efficiencies of enzymatic reactions, quality of oligonucleotide reagents, plate position effects, reagent stability), biological variability (e.g. eQTL effects) can introduce noise and bias into the resulting sequence read data that can be difficult to control. Thus, the combination of technical noise and intrinsic biological variability makes the detection of and control for technical artifacts challenging. For this reason, the development and implementation of rigorous quality control procedures throughout the entire laboratory and informatics workflow is essential in order to assess, improve and optimize both the wet lab and dry lab component steps in order to obtain optimal transcript expression values for downstream analysis.

Here we describe an approach to quality control (QC) for sc/nRNA-seq assays in which we capture over 70 different quantitative laboratory and data metrics and use these quality metrics to construct QC classification models that can be used to filter out poor quality samples from downstream analysis. We've applied this QC approach in the context of a project to define the cell type complexity of the human brain neocortex in a collaboration involving the Allen Institute for Brain Science, the J. Craig Venter Institute, and Illumina, Inc.

2. Methods and Results

Laboratory and Informatics workflows

Our standard laboratory workflow for single nuclei RNA-seq is summarized in Figure 1 and is based on the detailed protocol described previously [Krishnaswami 2016]. Single nuclei are sorted into 96- or 384-well plates containing 2 μL 0.2% Triton X-100, 2 Units/μL RNase inhibitor, 1:2000000 dilution of ERCC spike-in RNAs (Life Technologies) per well and frozen immediately in an ethanol/dry ice bath. The ERCC external RNA control, consisting of 92 transcripts derived from NIST-certified plasmids that mimic natural eukaryotic mRNAs, is used to measure limits of detection and dynamic ranges, and can also be used to help quantify differential gene expression. Amplified cDNA is prepared using a Smart-Seq2 modification [Ramsköld 2012, Krishnaswami 2016] to our previous method [Grindberg 2013] to improve amplification of transcript 5' ends. cDNA quality is evaluated using Taqman qPCR for selected housekeeping (ACTB), ERCC, and sample-specific genes. Using the single nuclei amplified cDNA, bar coded libraries are prepared and 60 sample pools are used for next generation sequencing using paired end 2 x 150 chemistry

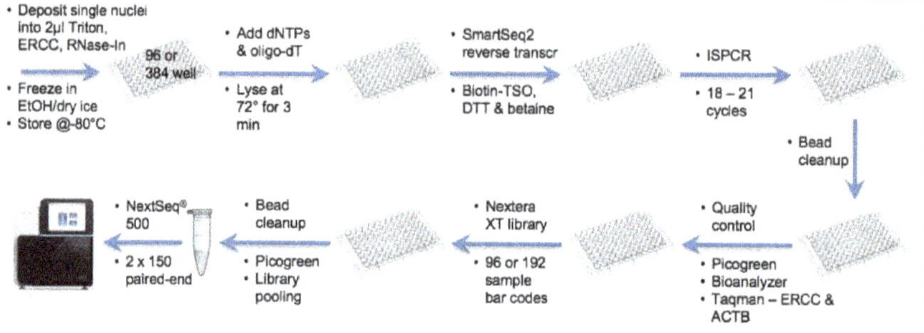

Figure 1. Single cell RNA-seq laboratory workflow – See text and [Krishnaswami 2016] for details. Abbreviations used: External RNA Controls Consortium RNA spike-in control (ERCC), RNase inhibitor (RNase-In), template switching oligonucleotide (TSO), Dithiothreitol (DTT), and beta-actin (ACTB). For Research Use Only. Not for use in diagnostic procedures.

on an Illumina NextSeq® 500 instrument. In each of our pools we also include a small number of positive (diluted, purified human RNA from bulk samples) and negative controls (water only, ERCC only). Sequencing results are quality controlled (QC) as described below, including the use of the laboratory-derived ACTB and ERCC Ct qPCR values, Bioanalyzer length distribution metrics, and picogreen cDNA concentration values.

Our standard operating procedure (SOP) for data processing includes steps for primer and quality trimming, read alignment, transcript assembly, and expression quantification as summarized in Figure 2, and has been described in detail in a recent Nature Protocol publication [Krishnaswami 2016]. After demultiplexing, cDNA, PCR, and library/bar code primer sequences and low quality reads are removed from the primary read-level data using Trimmomatic, producing the input reads for downstream steps. The input reads are fed into several downstream pipelines - RSEM (Bowtie2/EM) for transcript quantification, and TopHat (Bowtie2/Cufflinks), fastQC, MEONCA and SCavenger for quality control metric assessment. MEONCA and SCavenger are in-house developed methods that will be described elsewhere.

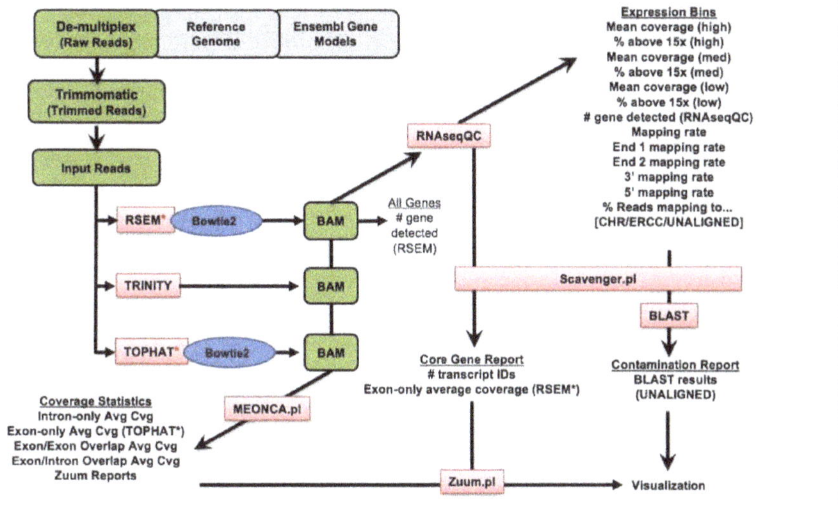

Figure 2. Single cell RNA-seq data processing workflow – Our standard operating procedure is based on the use of the Bowtie2/RSEM combination for sequence alignment, assembly, and transcript quantification. In addition, the ouput of a variety of additional workflows produce quantitative metrics used for assessment of sample and sequence quality. See text for details.

For the data included here, the following software and database versions were used:

- •GENCODE fasta and gtf files (http://www.gencodegenes.org/releases/current.html) Release 21 (GRCh38.p5);
- •FASTX Toolkit (http://hannonlab.cshl.edu/fastx_toolkit/download.html) v0.0.14;
- •fastQC (http://www.bioinformatics.babraham.ac.uk/projects/fastqc/) v0.10.1;
- •Picard toolkit (http://rseqc.sourceforge.net/) v1.137;
- •Trimmomatic (http://www.usadellab.org/cms/?page=trimmomatic) v0.35;
- •Bowtie2 (http://sourceforge.net/projects/bowtie-bio/files/bowtie2/) v2.2.7;
- •SAM tools (http://sourceforge.net/projects/samtools/files/samtools/) v1.3;
- •RSEM: (http://deweylab.biostat.wisc.edu/rsem/) v1.2.28;
- •Tophat (https://ccb.jhu.edu/software/tophat/index.shtml) v2.1.0;
- •Cufflinks (https://cole-trapnell-lab.github.io/cufflinks/) v2.2.1.

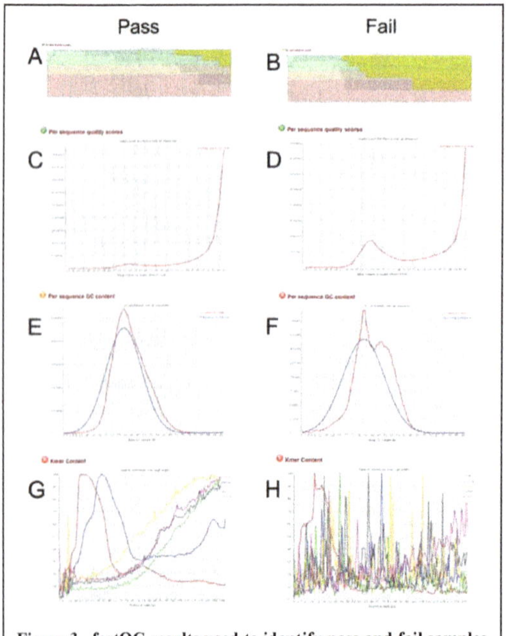

Figure 3. fastQC results used to identify pass and fail samples – Quality statistics produced by fastQC for representative Pass (A, C, E, G) and Fail (B, D, F, H) samples include average Phred score across the length of the read (A, B), average Phred score for the entire read (C, D), GC per read (E, F), and Kmer distribution across the length of the read (G, H).

One of the primary objectives of our informatics pipeline is to identify poor quality samples for possible exclusion, to determine the causes of poor quality for sample preparation process improvement, and to identify marginal quality samples for downstream bioinformatics "normalization". Because the determination of transcriptional profiles at a single cell level pushes the limits of next generation sequencing technologies, the rigorous approach we use for quality control is perhaps the most important aspect of the Single Cell Genomics Lab at JCVI.

Between the laboratory and data processing workflows described above, we collect 79 different quantitative measures that may reflect the quality of the input samples, processing steps, and resulting primary read-level data, which can be used to help address these objectives. Our approach is to use machine learning strategies, specifically random forest approaches, to classify individual sample data as either pass or fail for specific downstream analysis applications. In order to illustrate our approach, we describe the preliminary results from our work to develop a pass/fail classification model for a collaborative project between the JCVI Single Cell Genomics Lab, the Lein Group at the Allen Institute for Brain Science, and Illumina, Inc. to determine the transcriptional profiles for 2272 nuclei isolated from specific neo-cortex regions of post-mortem human brain.

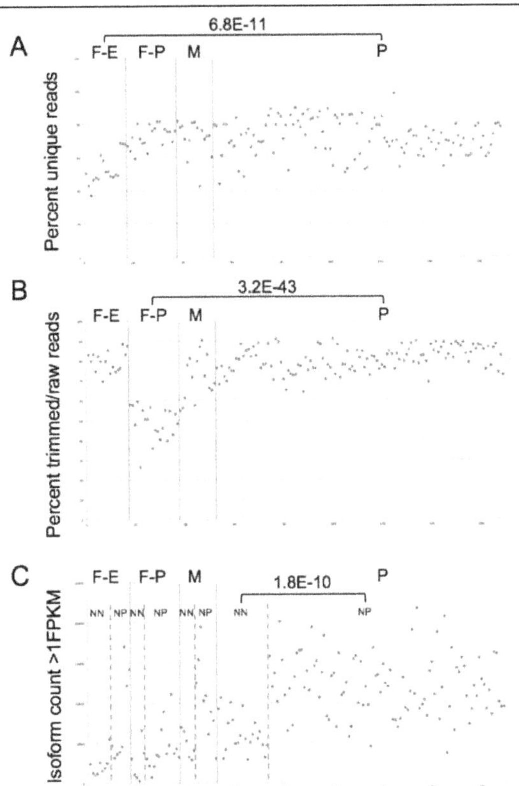

Figure 4. QC metrics in Pass and Fail samples – Single nuclei samples were annotated as Pass (P), Fail-ERCC (F-E), Fail-Phred (F-P), and Marginal (M) based on subjective evaluation of the fastQC results (see text for details). The quantitative levels of three different QC metrics for these four classes of samples are shown. P-values were calculated using a 2-sided student's t-test. NN – NeuN-; NP – NeuN+

Manual evaluation of fastQC results for QC model training

The first step in the development of machine learning classification models is to produce training data for model construction. For our purposes, we used a set of high confidence pass/fail calls for individual samples based on the qualitative assessment of data produced by fastQC, which includes quality Phred scores, GC content, Kmer distributions, and sequence over-representation information, for a random set of selected samples. Examples of these distributions are shown in Figure 3. Pass samples generally exhibit high average quality per read across the entire length of the sequenced fragment (Figure 3A & C). In contrast, Fail samples exhibit a significant number of reads with low mean quality, and quality scores that fall off down the length of the fragments (Figure 3B and D). High quality Pass samples also show an average GC content around 40%, reflecting the GC content of the expressed human transcriptome (Figure 3E). In contrast, some Fail samples show a second peak in the GC content distribution with a mean around 48% GC (Figure 3F); this peak appears to be generated from ERCC reads, which are derived from bacterial genome sequences.

Since we find that some Fail samples show reasonable Phred quality scores but over-representation of ERCC reads and vice versa, we distinguish between Fail samples due to low quality scores (Fail-Phred) and Fail samples due to ERCC over-representation (Fail-ERCC). Finally, Pass samples show a Kmer content distribution in which distinct polyA and polyT peaks can be observed toward the beginning of the read due to the use of oligo-dT priming in 1st strand cDNA synthesis (Figure 3G), whereas Fail sample often show a more random pattern (Figure 3H).

QC metric correlation with QC training data

In order to produce training data for machine learning in the 2272 nuclei study, we selected 196 samples at random, including 169 single nuclei samples and 27 controls (positive and negative), and performed a blinded qualitative evaluation of the fastQC data, producing three classification labels – Pass (152 samples, including all positive controls), Fail-Phred (29 samples), and Fail-ERCC (15 samples) (all negative controls we correctly classified into one of the two Fail categories). Qualitative fastQC evaluation was chosen to produce training data since this approach is independent from the quantitative QC metrics produced by our core data processing workflows described above. A few examples of the correlation between fastQC Pass/Fail calls and the quantitative QC metrics is shown in Figure 4. For Fail-ERCC samples, the "percent unique reads" are significantly lower ($p = 6.8E-11$) than for the Pass samples (Figure 4A), probably due to the fact that with a greater proportion of ERCC reads, more duplicate reads would result. For Fail-Phred samples, the "percent trimmed/raw reads" are significantly lower than for the Pass samples (Figure 4B, $p = 3.2E-43$), presumably due to the fact that Trimmomatic removes reads of poor quality. For Pass samples, the number of transcript isoforms detected tends to be generally higher than the number of transcript isoforms detected in either type of failed sample (Figure 4C). However, we noted that there appeared to be a subset of Pass samples that had relatively low isoform counts, similar to what we observed in the Fail samples. It turns out that during the nuclei isolation step, we stain for the expression of a neuron-specific protein, NeuN, to ensure that we get a selection of both neuronal and non-neuronal cell types for our study. When we compared data for NeuN+ and NeuN- passed samples, we found that the isoform counts were significantly different between the two major cell type categories ($p = 1.8E-10$), with NeuN+ nuclei and NeuN- producing an average of 12,162 and 6,233 transcript isoforms with >1FPKM, respectively.

Machine learning for high throughput QC processing

These quality annotation labels and QC metric values were then used to train the Random Forest algorithm as implemented in KNIME v3.1.2. We generated 100,000 decisions trees that could distinguish the three categories of samples. An example of a high scoring tree is shown in Figure 5 in which "percent trimmed over raw" is used at the first level and is effective at distinguishing Fail-Phred sample from both Pass and Fail-ERCC, and "percent unique reads" is used at the second level to distinguish Pass from Fail-ERCC, as also seen in Figure 4. A summary of the QC features that score high across the entire 100,000 decision tree collection is shown in Figure 6. Using this Random Forest classification model, all 196 samples in the training set were classified correctly with high confidence scores:

- Pass: average confidence = 0.9689; standard deviation = 0.0524
- Fail-Phred (F-P): average confidence = 0.8828; standard deviation = 0.0703
- Fail-ERCC (F-E): average confidence = 0.8286; standard deviation = 0.0959

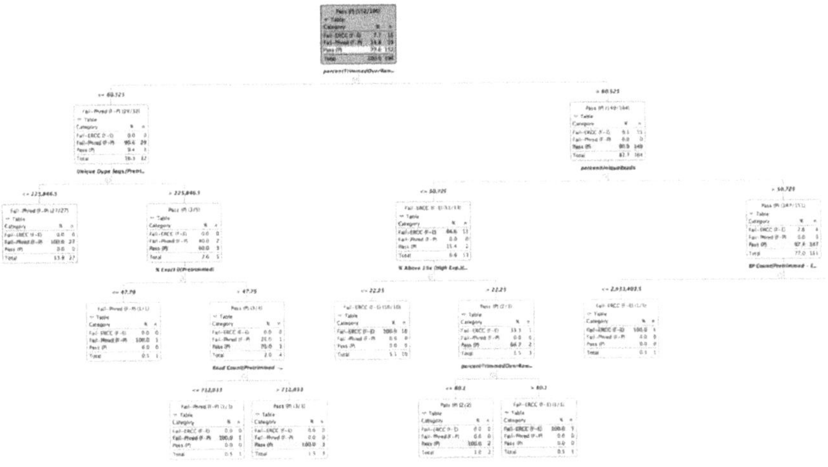

Figure 5. One of the 100,000 decision trees constructed from Random Forest training – The tree shows the different branch point levels, the feature used to segregate the branches at each level, and the segregation of the Pass, Fail-ERCC and Fail-Phred samples at each branch point. The tree was truncated after the first three levels.

To test the classification accuracy of the resulting random forest model, we used an independent test set of 185 single nuclei samples classified using the same fastQC evaluation criteria applied to the training data, with 135 determined to be Pass samples, 29 determined to be Fails and 21 determined to be Marginals. Application of the random forest model to these test Pass and Fail samples resulted in only 8 misclassifications (4.9%), for a classification accuracy of 95%. Marginal samples were split between Pass and Fail classification by the random forest model, with 8 Marginals classified as Pass and 12 classified as Fail.

Using this random forest model applied to the entire dataset, 79% of 2272 single nuclei samples passed quality control. For these samples, the average number of reads after trimming was 16,335,055 (±19,771,224), percent of hg38 mapped read was 33.04 (± 15.50), number of ERCC transcripts detected was 42.43 (± 4.37), and the number of genes detected at a level of >1FPKM was 6794 (± 2131), giving an average coverage of 793 reads per human gene detected. In contrast for Failed-ERCC samples, the average number of reads after trimming was 10,333,560 (±8,589,613), percent of hg38 mapped read was 12.18 (± 13.32), number of ERCC transcripts detected was 42.11 (± 4.73), and the number of genes detected at a level of >1FPKM was 2784 (± 1401), giving an average coverage of 452 reads per human gene detected. For Failed-Phred samples, the average number of reads after trimming was 6,763,387 (±6,167,257), percent of hg38 mapped read was 14.87 (± 12.54), number of ERCC transcripts detected was 39.60 (±12.14), and the number of genes detected at a level of >1FPKM was 2903 (± 1897), giving an average coverage of 346 reads per human gene detected. Removal of these poor quality samples was found to produce tighter cell type clusters in downstream PCA/biSNE analysis (data not shown).

QC Metric	#splits (level 1)	#candidates (level 1)	#splits (level 2)	#candidates (level 2)	#splits (level 3)	#candidates (level 3)	Rank
percentTrimmedOverRawReads	10932	10977	16082	21759	17735	41878	2.16
% ExactDuplicates	7814	10631	6029	21654	5532	41702	1.15
percentUniqueReads	3778	10811	8075	21777	10019	42219	0.96
% ExactDuplicatesAlignedHuRef	6432	10837	4519	21719	3736	41993	0.89
3' Mapping Rate(All Genes)	5420	11068	5164	21734	4984	41857	0.85
IsoformcountsGT1FPKM	4720	10835	4927	21727	5594	42136	0.80
% ExactDuplicatesUnmapped	5743	10707	3859	21738	3104	41890	0.79
ReadCountERCC Aligned	4747	10831	3716	21726	3377	42204	0.69
%InHighExpressionBins	4197	10751	3935	21556	3798	42062	0.66
genecountsGT1FPKM	3164	10860	4716	21605	5758	41733	0.65

Figure 6. QC features most useful in Pass/Fail classification trees – The top ten QC metrics found useful for Pass/Fail sample classification are listed together with the number of trees in which they were used for branching at levels 1, 2, and 3, and the number of times they were considered as candidates at that given level (due to the feature down-sampling used by the Random Forest algorithm. For example, percentTrimmedOverRawReads was considered as a candidate feature in 10977 level 1 branches and was selected as the best feature 10932 times.

Discussion/Conclusion

Many groups using sc/nRNA-seq to identify and quantify cellular diversity in complex tissue samples have recognized the critical importance of quality control procedures to obtain optimal results in downstream data analysis, and have used qualitative and quantitative assessment of single quality metrics for this purpose. These include abnormal expression of housekeeping genes (e.g. ACTB, GAPDH) [Ting 2014, Treutlein 2014], outlier clustering [Zeisel 2015, Jiang 2016], median expression value cutoffs [Pollen 2014], and number of genes detected or read mapping rate [Kumar 2014], each with their advantages and disadvantages. In this paper we have demonstrated the use of a machine learning approach, specifically random forest decision trees with a large combination of wet lab and dry lab quantitative metrics, to objectively perform this QC classification. The advantage of this approach is that not only does it provide for an objective, high-throughput pass-fail classification, but it also identifies those quantitative metrics that are most useful in identifying problematic samples.

In this study, we found that there appear to be at least two classes of failed samples, and that the metrics useful in identifying each are different. Failed samples with a second peak in the %GC content plot apparently due to reads derived from the ERCC spike-in control are identified by metrics like the % of exact duplicates and % of unique reads, presumably due to the fact that a relatively small number of transcripts derived from the ERCC control are responsible for a significant proportion of the total reads obtained from those samples. In contrast, failed samples with relatively poor quality scores (low Phred scores) are identified by metrics like the % of trimmed over raw reads, presumably due to the impact of poor quality data trimming by the Trimmomatic software. While there are some metrics that appear to be effective at identifying both classes of failed samples, e.g. the number of transcript isoforms with FPKM values greater than 1, these do not rank as high as the class-specific metrics in the useful feature list. This suggest that identifying and distinguish different types of failure modes may be useful for building QC classification models using machine learning approaches. And while both the three class prediction model used here and a two class prediction model constructed by grouping both fail categories into one showed perfect classification of the training data, the prediction confidence values for calling pass samples were slightly higher using the three class model.

In addition, we also find that the use of metrics related to the number of genes or transcript isoforms detected for quality control purposes should be approached cautiously since these may vary between different cell types, as we observed between our NeuN+ neurons and our NeuN- glial cells, or between different cellular states (e.g. cell cycle phase or activation state).

Recently, Ilicic et al. reported the use of support vector machine modeling to identify stressed/broken/killed cells, empty capture sites and sites with multiple cells in Fluidigm C1 flow cells using microscopic visualization as the gold standard for model training [Ilicic 2016]. They found seven features that were useful for classification independent of cell type and protocol – cytoplasm and mitochondrially-localized proteins, mtDNA-encoded genes, mapped reads, multi-mapped reads, non-exonic reads, and transcriptome variance. Differences between these and the features reported here could be due to the use of different quality metrics as input, the use of nuclei versus whole cells, or that different sorting platforms give rise to different poor quality modes. In any case, the approach reported here is advantageous because it does not require visual microscopic inspection to produce the gold standard results for model training and therefor can be applied in a high throughput fashion to data from any cell sorting platform. While the random forest model developed here has yet to be applied to a completely independent dataset, the test samples used to assess classification accuracy were derived from separate cDNA synthesis, PCR amplification, and library preparation reactions and sequencing runs. The fact that the model gave a 95% classification accuracy on this semi-independent dataset suggests that the feature included in the model are at least robust to technical batch effects.

In conclusion, the use of both wet lab and dry lab metrics for the production of a QC classification model using random forest machine learning appears to be an effective objective strategy for the quality control of sc/nRNA-seq samples, providing further insights into the data features that are most useful for identifying problematic samples.

Acknowledgements

This work is supported by the Allen Institute for Brain Science, the JCVI Innovation Fund, and the U.S. National Institutes of Health 1R21AI122100.

References

Blakeley P, Fogarty NM, del Valle I, et al. Defining the three cell lineages of the human blastocyst by single-cell RNA-seq. Development. 2015 Sep 15;142(18):3151-65. doi: 10.1242/dev.123547. Epub 2015 Aug 20. Erratum in: Development. 2015 Oct 15;142(20):3613. PubMed PMID: 26293300; PubMed Central PMCID: PMC4582176.

Ciuffi A, Rato S, Telenti A. Single-Cell Genomics for Virology. Viruses. 2016 May 4;8(5). pii: E123. doi: 10.3390/v8050123. Review. PubMed PMID: 27153082; PubMed Central PMCID: PMC4885078.

Eltahla AA, Rizzetto S, Pirozyan MR, et al. Linking the T cell receptor to the single cell transcriptome in antigen-specific human T cells. Immunol Cell Biol. 2016 Jul;94(6):604-11. doi: 10.1038/icb.2016.16. Epub 2016 Feb 10. PubMed PMID: 26860370.

Gaublomme JT, Yosef N, Lee Y, et al. Single-Cell Genomics Unveils Critical Regulators of Th17 Cell Pathogenicity. Cell. 2015 Dec 3;163(6):1400-12. doi: 10.1016/j.cell.2015.11.009. Epub 2015 Nov 19. PubMed PMID: 26607794; PubMed Central PMCID: PMC4671824.

Grindberg RV, Yee-Greenbaum JL, McConnell MJ, et al. RNA-sequencing from single nuclei. Proc Natl Acad Sci U S A. 2013 Dec 3;110(49):19802-7. doi: 10.1073/pnas.1319700110. Epub 2013 Nov 18. PubMed PMID: 24248345; PubMed Central PMCID: PMC3856806.

Handel AE, Chintawar S, Lalic T, et al. Assessing similarity to primary tissue and cortical layer identity in induced pluripotent stem cell-derived cortical neurons through single-cell transcriptomics. Hum Mol Genet. 2016 Mar 1;25(5):989-1000. doi: 10.1093/hmg/ddv637. Epub 2016 Jan 5. PubMed PMID: 26740550; PubMed Central PMCID: PMC4754051.

Ilicic T, Kim JK, Kolodziejczyk AA, et al. Classification of low quality cells from single-cell RNA-seq data. Genome Biol. 2016 Feb 17;17:29. doi: 10.1186/s13059-016-0888-1. PubMed PMID: 26887813; PubMed Central PMCID: PMC4758103.

Jiang P, Thomson JA, Stewart R. Quality control of single-cell RNA-seq by SinQC. Bioinformatics. 2016 Apr 10. pii: btw176. [Epub ahead of print] PubMed PMID: 27153613.

Krishnaswami SR, Grindberg RV, Novotny M, et al. Using single nuclei for RNA-seq to capture the transcriptome of postmortem neurons. Nat Protoc. 2016 Mar;11(3):499-524. doi: 10.1038/nprot.2016.015. Epub 2016 Feb 18. PubMed PMID: 26890679; PubMed Central PMCID: PMC4941947.

Kumar RM, Cahan P, Shalek AK, et al. Deconstructing transcriptional heterogeneity in pluripotent stem cells. Nature. 2014 Dec 4;516(7529):56-61. doi: 10.1038/nature13920. PubMed PMID: 25471879; PubMed Central PMCID: PMC4256722.

Lacar B, Linker SB, Jaeger BN, et al. Nuclear RNA-seq of single neurons reveals molecular signatures of activation. Nat Commun. 2016 Apr 19;7:11022. doi: 10.1038/ncomms11022. PubMed PMID: 27090946; PubMed Central PMCID: PMC4838832.

Li J, Klughammer J, Farlik M, et al. Single-cell transcriptomes reveal characteristic features of human pancreatic islet cell types. EMBO Rep. 2016 Feb;17(2):178-87. doi: 10.15252/embr.201540946. Epub 2015 Dec 21. PubMed PMID: 26691212; PubMed Central PMCID: PMC4784001.

Min JW, Kim WJ, Han JA, et al. Identification of Distinct Tumor Subpopulations in Lung Adenocarcinoma via Single-Cell RNA-seq. PLoS One. 2015 Aug 25;10(8):e0135817. doi: 10.1371/journal.pone.0135817. eCollection 2015. PubMed PMID: 26305796; PubMed Central PMCID: PMC4549254.

Nestorowa S, Hamey FK, Pijuan Sala B, et al. A single cell resolution map of mouse haematopoietic stem and progenitor cell differentiation. Blood. 2016 Jun 30. pii: blood-2016-05-716480. [Epub ahead of print] PubMed PMID: 27365425.

Patel AP, Tirosh I, Trombetta JJ, et al. Single-cell RNA-seq highlights intratumoral heterogeneity in primary glioblastoma. Science. 2014 Jun 20;344(6190):1396-401. doi: 10.1126/science.1254257. Epub 2014 Jun 12. PubMed PMID: 24925914; PubMed Central PMCID: PMC4123637.

Petropoulos S, Edsgärd D, Reinius et al. Single-Cell RNA-Seq Reveals Lineage and X Chromosome Dynamics in Human Preimplantation Embryos. Cell. 2016 May 5;165(4):1012-26. doi: 10.1016/j.cell.2016.03.023. Epub 2016 Apr 7. PubMed PMID: 27062923; PubMed Central PMCID: PMC4868821.

Pollen AA, Nowakowski TJ, Shuga J, et al. Low-coverage single-cell mRNA sequencing reveals cellular heterogeneity and activated signaling pathways in developing cerebral cortex. Nat Biotechnol. 2014 Oct;32(10):1053-8. doi: 10.1038/nbt.2967. Epub 2014 Aug 3. PubMed PMID: 25086649; PubMed Central PMCID: PMC4191988.

Proserpio V, Piccolo A, Haim-Vilmovsky L, et al. Single-cell analysis of CD4+ T-cell differentiation reveals three major cell states and progressive acceleration of proliferation. Genome Biol. 2016 May 12;17(1):103. doi: 10.1186/s13059-016-0957-5. Erratum in: Genome Biol. 2016;17(1):133. PubMed PMID: 27176874; PubMed Central PMCID: PMC4866375.

Ramsköld D, Luo S, Wang YC, et al. Full-length mRNA-Seq from single-cell levels of RNA and individual circulating tumor cells. Nat Biotechnol. 2012 Aug;30(8):777-82. PubMed PMID: 22820318; PubMed Central PMCID: PMC3467340.

Scialdone A, Natarajan KN, Saraiva LR, et al. Computational assignment of cell-cycle stage from single-cell transcriptome data. Methods. 2015 Sep 1;85:54-61. doi: 10.1016/j.ymeth.2015.06.021. Epub 2015 Jul 2. PubMed PMID: 26142758.

Ting DT, Wittner BS, Ligorio M, et al. Single-cell RNA sequencing identifies extracellular matrix gene expression by pancreatic circulating tumor cells. Cell Rep. 2014 Sep 25;8(6):1905-18. doi: 10.1016/j.celrep.2014.08.029. Epub 2014 Sep 18. PubMed PMID: 25242334; PubMed Central PMCID: PMC4230325.

Tirosh I, Izar B, Prakadan SM, et al. Dissecting the multicellular ecosystem of metastatic melanoma by single-cell RNA-seq. Science. 2016 Apr 8;352(6282):189-96. doi: 10.1126/science.aad0501. PubMed PMID: 27124452; PubMed Central PMCID: PMC4944528.

Treutlein B, Brownfield DG, Wu AR, et al. Reconstructing lineage hierarchies of the distal lung epithelium using single-cell RNA-seq. Nature. 2014 May 15;509(7500):371-5. doi: 10.1038/nature13173. Epub 2014 Apr 13. PubMed PMID: 24739965; PubMed Central PMCID: PMC4145853.

Treutlein B, Lee QY, Camp JG, et al. Dissecting direct reprogramming from fibroblast to neuron using single-cell RNA-seq. Nature. 2016 Jun 8;534(7607):391-5. doi: 10.1038/nature18323. PubMed PMID: 27281220; PubMed Central PMCID: PMC4928860.

Wang YJ, Schug J, Won KJ, et al. Single cell transcriptomics of the human endocrine pancreas. Diabetes. 2016 Jun 30. pii: db160405. [Epub ahead of print] PubMed PMID: 27364731.

Zeisel A, Muñoz-Manchado AB, Codeluppi S, et al. Brain structure. Cell types in the mouse cortex and hippocampus revealed by single-cell RNA-seq. Science. 2015 Mar 6;347(6226):1138-42. doi: 10.1126/science.aaa1934. Epub 2015 Feb 19. PubMed PMID: 25700174.

For Research Use Only. Not for use in diagnostic procedures.

TRACING CO-REGULATORY NETWORK DYNAMICS IN NOISY, SINGLE-CELL TRANSCRIPTOME TRAJECTORIES

PABLO CORDERO

JOSHUA M. STUART

UC Santa Cruz Genomics Institute, University of California, Santa Cruz, California, USA

Keywords: single-cell measurements, Gaussian mixtures, transcriptomics, single-cell trajectory reconstruction

The availability of gene expression data at the single cell level makes it possible to probe the molecular underpinnings of complex biological processes such as differentiation and oncogenesis. Promising new methods have emerged for reconstructing a progression 'trajectory' from static single-cell transcriptome measurements. However, it remains unclear how to adequately model the appreciable level of noise in these data to elucidate gene regulatory network rewiring. Here, we present a framework called Single Cell Inference of MorphIng Trajectories and their Associated Regulation (SCIMITAR) that infers progressions from static single-cell transcriptomes by employing a continuous parametrization of Gaussian mixtures in high-dimensional curves. SCIMITAR yields rich models from the data that highlight genes with expression and co-expression patterns that are associated with the inferred progression. Further, SCIMITAR extracts regulatory states from the implicated trajectory-evolving co-expression networks. We benchmark the method on simulated data to show that it yields accurate cell ordering and gene network inferences. Applied to the interpretation of a single-cell human fetal neuron dataset, SCIMITAR finds progression-associated genes in cornerstone neural differentiation pathways missed by standard differential expression tests. Finally, by leveraging the rewiring of gene-gene co-expression relations across the progression, the method reveals the rise and fall of co-regulatory states and trajectory-dependent gene modules. These analyses implicate new transcription factors in neural differentiation including putative co-factors for the multi-functional NFAT pathway.

Introduction

Understanding the dynamics of gene expression progression in a cell population as it traverses a biological process such as differentiation has been an outstanding problem in modern cell biology. These dynamics are characterized not only by the changes in cell-to-cell gene expression levels, but by the rewiring of gene regulatory networks as the cells transform from one transcriptional state to another. Tracking these gene regulatory changes would pinpoint coordination of biological function as gene modules are turned on or off throughout the progression.

Single-cell transcriptomics has given important insights into gene expression dynamics, revealing the stochastic nature of gene expression and characterizing in detail the behavior of small genetic networks.[1–4] In their initial incarnation, these measurements were confined to demanding microscopy protocols that assayed gene expression levels through time of only a handful of genes. In recent years, advances in flow cytometry, microfluidics, and sequencing technologies have enabled the interrogation of up to the whole transcriptome in hundreds to thousands of cells.[5–7] Application of these techniques to biological processes such as development provide snapshots of cell states through time and space.

Many computational methods have emerged to infer trajectories of connected state transitions from the static samplings of single-cell transcriptomes. The goal of these methods is to provide a pseudotemporal ordering of cells in which neighboring cells are similar to each other, capturing an overall biological progression. These approaches have been successfully applied to elucidate complex transcriptional patterns and regulators in myoblast differentiation,[8] B cell development,[9] and haematopoiesis.[10] Nevertheless, cell orderings alone give little insight into the state of gene regulatory networks across time. In addition, while most methods use strategies to tackle biological and technical noise, none account for the dynamic, heteroscedastic nature of the data. Further, only a few take into consideration uncertainties in pseudotime assignments,[11] making error estimates difficult to evaluate.

To address these challenges we propose a strategy, Single Cell Inference of MorphIng Trajectories and their Associated Regulation (SCIMITAR), for inferring gene expression network dynamics throughout biological progression from static, single-cell transcriptomes. SCIMITAR gives a detailed, fully probabilistic description of the expression trajectory that, in contrast with previous methods, explicitly accounts for heteroscedastic noise in the data. In addition, it tracks the changes of gene-gene expression correlations at each point in the progression. The probabilistic nature of SCIMITAR transition models allows for evaluating the shape of the multivariate gene expression distribution as a function of biological progression, which we show can be used to pinpoint co-regulatory cell states.

We benchmarked SCIMITAR's inference capabilities in two scenarios. First, we tested its ability to infer cell ordering and network rewiring from simulated transcriptomic measurements where the underlying cell behavior was known. Second, we asked whether SCIMITAR could yield insights in the developmental trajectory of human fetal neurons by analyzing recently published fetal brain single-cell measurements. A likelihood ratio test designed for SCIMITAR revealed 36 genes that significantly varied throughout the progression but that were missed by standard differential expression between cell groups including genes in cornerstone developmental pathways such as the hypoxia inducible factor 1 α (HIF1α), nuclear factor of activated T cells (NFAT), and androgen receptor (AR) pathways. Further, by tracking SCIMITAR co-expression matrices across pseudotime we were able to detect the evolution of co-regulatory states, gene modules, and genes that gained and lost connectivity throughout the trajectory.

Results

Uncovering the full probability distribution progression underlying static single-cell measurements with SCIMITAR

Recently, there has been an explosion of single-cell transcriptomic data in various biomedical contexts and systems. A projection of the data from three such studies (refs[8,10,12]) in Fig 1A using a locally linear embedding reveals that these datasets are characterized by distinct groups of many cells interspersed with cells that fall along what appear to be isolines between groups. This structure suggests a model that combines distributions for cell population density and evolving cell states with heteroscedastic noise. One such model that could describe these data is a continuous mixture of Gaussian distributions with constraints that allow only for smooth, continuous changes in parameters over the course of the progression. We call such a model a Morphing Gaussian Mixture (MGM, see Methods and Fig 1B). The MGM has a mean function, $\mu : [0,1] \to \mathbb{R}^n$ that threads through the data and is equipped with a covariance matrix function $\Sigma : [0,1] \to \mathbb{R}^{n \times n}$ that defines a Gaussian distribution at each point in the progression, with n being the number of genes. The mean and covariance matrix functions vary continuously throughout the $[0,1]$ interval, defining a probability $P(x|\mu, \Sigma, t)$ for each cell gene expression vector x and pseudo time-point $t \in [0,1]$. To ease inference, these mean and covariance functions can be parametrized with different functional classes, such as polynomials, splines, or Gaussian processes (see Methods). This probabilistic structure maps samples to a smooth curve and allows points to veer away stochastically by modeling the structure of the changing biological and technical noise. $P(x|\mu, \Sigma, t)$ captures the uncertainty of a cell mapping to a particular pseudotime due to the changing covariance nature of the MGM. A key advantage of this approach is that it replaces standard, grouped differential gene expression analysis or differential co-expression analysis with a more sensitive test for potential gene-gene regulatory relationships that change throughout the progression. Details of the MGM model as well as inference of its parameters from data are given in the Methods section.

Benchmarking SCIMITAR in simulated data

To test our strategy, we asked whether SCIMITAR could infer the underlying cell ordering and co-expression networks of simulated data where the ground truth was available. We tested SCIMITAR's cell order inference capabilities in two settings in which noise was added to the system: 1) the noise is *uncorrelated* to the underlying trajectory and 2) the noise is *correlated* with the trajectory. The first setting, adding noise uncorrelated with the trajectory, tests robustness of the method in the presence of genes that are unrelated to the biological progression and that confound ordering inference. The second setting tests how biological and technical noise intrinsic to the system, including gene-gene correlated noise that change over time, affect cell ordering inference.

For the first setting, we simulated data closely following the simulation procedure described in ref.[9] We simulated data in which 3 genes defined the true cell state and 7 genes represented unrelated (uncorrelated) expression programs to the simulated progression. Simulations in this scenario then, 3 dimensions of the data were "signal" while 7 were "noise". To obtain the three-dimensional trajectory, we performed a random walk for 600 steps and sampled a 'cell' from a standardized normal distribution centered at the current point in the walk. We then added seven dimensions of Gaussian noise. We generated several datasets with an increasing noise magnitude (quantified as the standard deviation times the range of the trajectory). We then used SCIMITAR to model these data and obtain the model's optimal cell ordering. We used SCIMITAR with three different functional classes (see Methods): third degree polynomials, cubic splines, and Gaussian Processes with a squared exponential correlation function (GP). We compared SCIMITAR's performance with the cell orderings inferred by two popular methods, Monocle[8] and Wanderlust,[9] and used the Pearson correlation coefficient to compare the approaches (see Fig 2A). The best overall performers were all SCIMITAR models, with Wanderlust coming in close second and Monocle performing slightly worse possibly due to its assumption of linearity in its dimensionality reduction step in agreement with previous studies.[13] All methods were susceptible to the noisy dimensions uncorrelated with the trajectory.

For the second test that adds noise correlated with the trajectory, we simulated a curve, μ_{sim} traversing a 10-dimensional space using 10 randomly-generated quadratic polynomials. The correlated noise was simulated from the evolution of randomly generated Watts-Strogatz networks and an additional set of quadratic polynomials with 6 different settings of signal-to-noise ratios (see Supplemental Methods for a detailed description of this benchmark). We found all methods performed similarly (Fig 2B), suggesting that noise intrinsic to the system, including gene-gene statistical dependencies, equally confounds any cell ordering inference method.

In addition to solving the cell ordering problem, SCIMITAR models track evolving gene-gene correlations. We used the correlated noise simulations to test the accuracy of SCIMITAR's gene network rewiring inference. To this end, we compared the covariance functions inferred by the polynomial, spline, and GP SCIMITAR versions. We measured the concordance of trends between each entry of the predicted matrix functions $\Sigma_{ij}^{pred}(t)$ and the corresponding entry of the simulated values $\Sigma_{ij}^{sim}(t)$ using the Pearson correlation coefficient (see Fig 2C). The spline version of SCIMITAR produced the highest correlation coefficients while all versions were substantially better than randomly-generated covariance matrix functions. Closer examination of the three functional classes revealed that the GP version tended to overfit the data locally, closely following local covariance structure even in regions where a few samples were present while the polynomial version lacked the flexibility to model some complex twists and turns in evolving true covariance structures. The spline version struck a balance between smoothing inferences in intervals of the trajectory with few samples and maintaining flexibility to capture non-linear trends. We therefore chose to use the spline functional class for SCIMITAR models in the remainder of this study.

A differentiation model for human fetal neurons

In a previous study, Darmanis et al. obtained a transcriptomic map of the adult and fetal brain using single-cell RNA-seq measurements.[14] One of the findings of the study was a continuous transition the between fetal replicating and quiescent neurons. We applied SCIMITAR to infer cell ordering and network rewiring of these data to elucidate key regulatory changes across the differentiation process. We downloaded these data from the gene expression omnibus (series identifier GSE67835) and obtained the subset corresponding to all fetal neurons. We focused on all transcription factors that were expressed in at least 10% of the cells, log-transformed the data and controlled for cell-cycle effects using scLVM.[15] We then fit SCIMITAR to the data and visualized the results in a two-dimensional locally linear embedding (see Fig 3A). The visualization suggested a single linear trajectory that traversed the fetal replicating and quiescent neurons which was captured by the SCIMITAR model. To obtain progression associated genes, we used a likelihood ratio test tailored for SCIMITAR models with dynamic noise (see Methods). The test revealed 92 genes with expression that was significantly pseudotemporal-dependent (see Fig 3B). To obtain global insights from these genes, we used hierarchical clustering with the Pearson correlation similarity metric to group them into 5 groups and performed Gene Ontology and KEGG pathway enrichment tests on each group (see color groups in Fig 3B). Early-expressed genes (red and green clusters) were associated with glucocorticoid receptors, heat shock factors, and signal transduction; genes expressed in the middle of the progression (yellow and pink clusters) were enriched with Maf-like proteins and cytokines; and the late-expressed genes (cyan cluster) had apoptosis, neurogenesis, and alternative splicing enrichment. These enrichments correspond to multiple observations in the literature. For example, heat shock factor proteins are well known to be involved in early neurodifferentiation[16] while glucocorticoid receptors and Maf-like proteins are found to be expressed at different stages in hippocampal and developmental neurogenesis, respectively.[17,18] Further, neurodifferentiation has been found to be particularly enriched for alternative splicing events.[19]

We then compared SCIMITAR's progression associated genes to those obtained using an ANOVA differential expression test between cells grouped according to their fetal replicating or quiescent annotations. SCIMITAR uncovered 36 genes missed by ANOVA, most of which were highly expressed in the middle of the progression, a detail that is lost when grouping cells into two groups. These missed genes implicate different pathways whose genes were engaged in progression dynamics. For example, five genes, BHLHE40, SMAD3, SP1, and SMAD4, of the hypoxia inducible factor 1 α (HIF1α) pathway, involved in neural development,[20] were revealed to follow an ordered progression by the SCIMITAR model but missed using grouped ANOVA differential expression (see Fig 3C). SCIMITAR revealed that the progression associated genes of this pathway were mostly active in early stages of differentiation. SCIMITAR also illuminated two other pathways: the Nuclear factor of activated T-cells (NFAT) and the Androgen receptor pathway which is critical for neural stem cell fate commitment[21,22] (see Fig 3C).

We note that SCIMITAR's progression associated genes did not include 7 genes from the ANOVA list, false positives for which the variance was too large or where the statistic was skewed by outliers in an otherwise lowly expressed gene. Nevertheless, three genes that seem to be be differentially expressed by manual inspection (BCL11B, AFF1, and REST) were found by ANOVA but missed by SCIMITAR, presumably due to a small subset of cells driving the change between groups.

Evolving co-expression networks reveal defined co-regulatory states

We then used SCIMITAR's inferred covariance functions to track changes in gene-gene connectivity across the progression. We sampled 100 correlation matrices at regular intervals from the covariance function, restricting the matrices to genes deemed progression associated. We calculated a global distance matrix between networks using Frobenius distance to assess their similarities and plotted the similarity values across pseudotime (see Fig 4A). As expected, the strongest similarities were between networks that were neighbors in pseudotime. However, three network clusters could be appreciated in the matrix, suggesting three different co-regulatory states. We obtained the consensus network of each state by averaging the network members of the cluster. Then, we ranked each gene by comparing their co-expression degree in each state to their co-expression degrees in the other two states using z-scores.

The top 20 genes that gained the most connectivity in each state are listed in Fig 4B. All of the gain-of-connectivity genes include genes that have been established as key players in neurodifferentiation, such as PAX6, DLX1, and NEUROD6 and were enriched with neurodevelopmental and neurogenesis GO terms.

To track highly connected gene modules of each state that significantly changed their connectivity, we obtained gene modules for each co-regulatory state using affinity propagation (with a dampening parameter of 0.5), finding 27 gene modules in total. We annotated these modules by gene set enrichment and ordered them across pseudotime (see Fig 4C). This analysis revealed a coordinated functional response across the trajectory: modules in state 1 were annotated with neural stem cell commitment, immune response, and protein trafficking, while state 2 was enriched with embryonic development, neuron regulation, and pallium development. State 3 had more diverse enrichments, from morphogenesis to membrane organelles, suggesting a stage when cells start taking on mature neuron roles depleted of differentiation potential. Importantly, this analysis pinpointed an NFAT-associated module to be most active in co-regulatory state 2 (see Fig 4D). Most NFAT co-factors involved in neural development are still unknown.[23] The uncovered NFAT-associated module provides putative candidates for this function. The full list of modules and their gene networks can be found in the Supplemental Results (see below).

Discussion

An outstanding goal of systems biology is to understand the principles under which the gene regulatory circuitry of a cell changes during a biological process. Single-cell transcriptomes offer a fast way to obtain transcriptome-wide snapshots of these processes. When properly analyzed, these data can be used to recover the principal trends of the biological progression, but current methods do not model the dynamic gene-to-gene correlations in expression that are the hallmarks of the underlying regulatory circuitry. Here, we presented SCIMITAR, a strategy that leverages morphing Gaussian mixtures to track biological progression and model the rewiring of these gene networks from static transcriptomes. SCIMITAR models account for heteroscedastic noise and increase the statistical power to detect progression-associated genes when compared to traditional differential expression tests. Further, the models allow for detecting modes in co-expression structure in the trajectory: defined co-regulatory states that represent potential metastable and transitionary cell states. We note that Gaussian mixtures with non-diagonal covariance matrices suffer from the curse of dimensionality, which we have tried to control for by using shrinkage estimators. Exploring the robustness of other types of regularized estimators such as the graphical LASSO would be a logical next step to improve confidence in the inferred morphing mixture models.

SCIMITAR is part of a recent wave of probabilistic methods for cellular trajectory reconstruction from single-cell measurements.[11,24] These types of models present several advantages, such as assigning uncertainty estimates of cell orderings and providing a natural way for mapping new samples to a trained model — a necessary task for building queryable trajectory maps with multiple progressions. Although SCIMITAR as presented cannot model branched cellular trajectories such as those corresponding to multiple cell fate decisions, the framework can be readily extended by replacing the single-curve parametrization of the mixtures with a branching structure, which deserves further investigation.

Methods

Morphing Gaussian Mixtures: correlated gene progression modeling with no dimensionality reduction

Single-cell transcriptomic measurements are high-dimensional, with the number of variables measured typically ranging from a few markers (generally no less than 48) to the full transcriptome that can be upwards around 30000 transcripts. However, not every gene or transcript is relevant to the biological system of interest and most are not expressed at all. Further, due to the underlying gene regulatory networks, the expression patterns of many genes are correlated and the strength of this correlation changes throughout the progression as the regulatory system changes from one cell state to the next. These biological constraints put the data in some low-dimensional manifold, a property that is used in various ways by cell ordering algorithms to justify reducing the dimensionality of the dataset to a manageable number of dimensions. Monocle, for example, reduces the data's dimensionality to 2 dimensions using independent component analysis and performs its calculations on a lower dimensional manifold. While the procedure captures general aspects of the trajectory, 2 dimensions is generally not enough to capture all of the relevant variability of the data and the reduction leads to loss of information that can impact trajectory reconstruction (see e.g. our benchmarks in the Results sections and other benchmarks in[13,24]). Other methods, such as Wanderlust, reduce the dimensionality in a more principled way through nearest-neighbor calculations but forego capturing the changes in gene-gene expression correlations over time. To address both of these shortcomings, we introduce a model that retains the dimensionality of the dataset and tracks gene-gene correlations throughout the trajectory. To this end, we extended Gaussian graphical models to accommodate time-dependent changes in the mean and covariances of the model with time being a latent variable.

Gaussian graphical models are one of the dominant frameworks for analyzing gene expression data, where the data is assumed to follow a multivariate Gaussian distribution defined by a mean vector and a covariance matrix. Modeling the data becomes more challenging in the presence of population structure where several different populations, each with its own distribution, are intermixed. Gaussian mixture models, which posit that the data comes from a finite combination of multivariate Gaussians, have been used successfully in this scenario.[25] In static single-cell expression from a group of cells continuously undergoing a biological process, such as differentiation, the boundaries between populations are blurred and the data is best described as a continuous transformation between the first and last states. We model this transformation by assuming that the data comes from a *continuous* Gaussian mixture, parametrized by timepoints within the progression (the so-called pseudotime), which are unknown. Let X be the data, p the number of genes, $\mu : [0,1] \to \mathbb{R}^p, \Sigma : [0,1] \to \mathbb{R}^{p\times p}$ the mean and covariance functions of the evolving populations that are time dependent, and γ a probability distribution on the $[0,1]$ interval representing cell population density at each pseudo time-point. Then the probability of the data given the model $M = \{\mu, \Sigma, \gamma\}$ can be written as:

$$P(X|M) = \int_0^1 \gamma(t) P(X|\mu(t), \Sigma(t)) \tag{1}$$

Here, t stands for the pseudotime in the progression. This model, which we name the morphing Gaussian mixture model (MGM), differs from other mixture models in that we require the mean and covariance structures to be described through continuous functions and generalize other related models such as principal curves by inferring local covariance structure in addition to the mean curve. The changing covariance structure allows the model to both keep the dimensionality of the dataset and track co-expression changes throughout the progression.

To fit the model to the data, we use a maximum likelihood approach. As previously defined, the parameters in the MGM model are difficult to infer, since optimization of the likelihood function requires searching the space of all continuous functions. Additionally, the positive-definite requirement on $\Sigma(t)$ makes fitting the matrix function difficult. Therefore, we recast the problem of fitting $\Sigma(t)$ into fitting its pseudotime-dependant Cholesky decompositions: $\Sigma(t) = C(t)^T C(t), \forall t$ and impose a functional form to the $\mu(t)$ and $C(t)$ functions. We consider three different functional classes: polynomials, Gaussian processes with squared exponential correlation models, and cubic, De Boor smoothing splines, a special case of Gaussian processes.

To fit the parameters of the model, we employ coordinate ascent. In the first step, we are given a fixed set values for M and we calculate, for each sample x, the optimal pseudotime t_{opt} in the $[0, 1]$ interval for which $P(x|\mu(t_{opt}), \Sigma(t_{opt}))$ is maximized. In the second step, given optimal pseudotime values, we calculate the cell density γ by fitting kernel density estimator to the assigned pseudo time-points. Finally, in the third step, given density weights γ and pseudotime assignments, we find the μ and Σ functions that best fit the data. To achieve this, we approximate $\mu(t)$ and $C(t)$ locally by obtaining optimal values at the pseudo time-points $0, 0.1, 0.2, ..., 1.0$, inferring the local mean and covariance using each data point weighted by their probabilities as given by γ, and leveraging these values to fit functions from the desired functional class (e.g. a polynomial, spline, or Gaussian process). Because we may have considerably less samples than genes, we use the Ledoit-Wolf-type estimator in the R `corpcor` package to fit the covariance at each pseudo time-point. We repeat this procedure until convergence, as evaluated by the Pearson correlation coefficient of current and past pseudotimes, with stopping criterion $r > 0.9$. As initial values for pseudotime assignments to our optimization routine, we use a de-noised one-dimensional locally linear embedding.[26]

Visualization of the data and SCIMITAR models

To visualize the data and models, we use 2-dimensional locally-linear embeddings, with number of neighbors set to 80% of the number of samples. We plot SCIMITAR means by sampling 100 equidistant points across the mean function and projecting to the embedding. To obtain a projection of the SCIMITAR model's probability density function, we obtain 1000 samples from the model, evenly spaced across pseudotimes in the $[0, 1]$ interval, project to the embedding, and plot a 2-dimensional kernel density estimator of the 1000 points.

A progression association statistical test

To obtain genes whose expression is progression-dependent, we use a likelihood ratio test to compare the SCIMITAR model of each gene's progression and the null hypothesis where the expression of the gene is 'flat-lined', i.e. does not track with the model's path. Specifically, we calculate the statistic:

$$LR = log(L_{null}(\hat{\mu}, \hat{\sigma})) - log(L_{scim}(\mu, \Sigma)) \tag{2}$$

Where L_{scim}, L_{null} are the likelihood functions of the SCIMITAR and null models, respectively, with the null distribution defined as a normal distribution centered at the empirical mean $\hat{\mu}$ and standard deviation $\hat{\sigma}$ of all the data representing the case where the data is independent of the progression. To assess whether the null hypothesis should be rejected, we obtain the distribution of LR under the null hypothesis using parametric bootstrapping with 1000 samples and compare the resulting ratios to the LR of the data. We use the Benjamini-Hochberg procedure to correct for multiple comparisons, setting an FDR cutoff of 5%.

Acknowledgments

We thank members of the Stuart lab for feedback on the methodology and Rocio Soto Astorga for graphics support. PC and JMS are supported by a grant from the California Institute of Regenerative Medicine working under the auspices of the Stem Cell Genome Center of Excellence. JMS was also supported by NIGMS grant 5R01GM109031.

Method availability and supplementary material

SCIMITAR code and documentation are freely available at `https://github.com/dimenwarper/scimitar`. Supplementary methods and results can be found at `https://github.com/dimenwarper/scimitar/wiki`.

References

1. M. B. Elowitz, A. J. Levine, E. D. Siggia and P. S. Swain, *Science* **297**, 1183 (16 August 2002).
2. J. M. Raser and E. K. O'Shea, *Science* **304**, 1811 (18 June 2004).
3. H. Maamar, A. Raj and D. Dubnau, *Science* **317**, 526 (27 July 2007).
4. J. Paulsson, *Nature* **427**, 415 (29 January 2004).
5. F. Tang, C. Barbacioru, E. Nordman, B. Li, N. Xu, V. I. Bashkirov, K. Lao and M. A. Surani, *Nat. Protoc.* **5**, 516 (March 2010).
6. S. C. Bendall, E. F. Simonds, P. Qiu, E.-A. D. Amir, P. O. Krutzik, R. Finck, R. V. Bruggner, R. Melamed, A. Trejo, O. I. Ornatsky, R. S. Balderas, S. K. Plevritis, K. Sachs, D. Pe'er, S. D. Tanner and G. P. Nolan, *Science* **332**, 687 (6 May 2011).
7. T. Hashimshony, F. Wagner, N. Sher and I. Yanai, *Cell Rep.* **2**, 666 (27 September 2012).
8. C. Trapnell, D. Cacchiarelli, J. Grimsby, P. Pokharel, S. Li, M. Morse, N. J. Lennon, K. J. Livak, T. S. Mikkelsen and J. L. Rinn, *Nat. Biotechnol.* **32**, 381 (April 2014).
9. S. C. Bendall, K. L. Davis, E.-A. D. Amir, M. D. Tadmor, E. F. Simonds, T. J. Chen, D. K. Shenfeld, G. P. Nolan and D. Pe'er, *Cell* **157**, 714 (24 April 2014).

10. V. Moignard, S. Woodhouse, L. Haghverdi, A. J. Lilly, Y. Tanaka, A. C. Wilkinson, F. Buettner, I. C. Macaulay, W. Jawaid, E. Diamanti, S.-I. Nishikawa, N. Piterman, V. Kouskoff, F. J. Theis, J. Fisher and B. Göttgens, *Nat. Biotechnol.* **33**, 269 (March 2015).
11. K. Campbell and C. Yau, *bioRxiv* (5 April 2016).
12. G. Guo, S. Luc, E. Marco, T.-W. Lin, C. Peng, M. A. Kerenyi, S. Beyaz, W. Kim, J. Xu, P. P. Das, T. Neff, K. Zou, G.-C. Yuan and S. H. Orkin, *Cell Stem Cell* **13**, 492 (3 October 2013).
13. J. D. Welch, A. J. Hartemink and J. F. Prins, *Genome Biol.* **17**, p. 106 (23 May 2016).
14. S. Darmanis, S. A. Sloan, Y. Zhang, M. Enge, C. Caneda, L. M. Shuer, M. G. Hayden Gephart, B. A. Barres and S. R. Quake, *Proc. Natl. Acad. Sci. U. S. A.* **112**, 7285 (9 June 2015).
15. A. McDavid, G. Finak and R. Gottardo, *Nat. Biotechnol.* **34**, 591 (9 June 2016).
16. M. T. Loones, Y. Chang and M. Morange, *Cell Stress Chaperones* **5**, 291 (October 2000).
17. C. Mirescu and E. Gould, *Hippocampus* **16**, 233 (2006).
18. H. Motohashi, J. A. Shavit, K. Igarashi, M. Yamamoto and J. D. Engel, *Nucleic Acids Res.* **25**, 2953 (1 August 1997).
19. E. V. Makeyev, J. Zhang, M. A. Carrasco and T. Maniatis, *Mol. Cell* **27**, 435 (3 August 2007).
20. Y. Zhao, M. Matsuo-Takasaki, I. Tsuboi, K. Kimura, G. T. Salazar, T. Yamashita and O. Ohneda, *Stem Cells Dev.* **23**, 2143 (15 September 2014).
21. M. Moreno, V. Fernández, J. M. Monllau, V. Borrell, C. Lerin and N. de la Iglesia, *Stem Cell Reports* **5**, 157 (11 August 2015).
22. L. A. M. Galea, M. D. Spritzer, J. M. Barker and J. L. Pawluski, *Hippocampus* **16**, 225 (2006).
23. T. Nguyen and S. Di Giovanni, *Int. J. Dev. Neurosci.* **26**, 141 (April 2008).
24. K. Campbell and C. Yau, *bioRxiv* (23 June 2016).
25. H. H. Chang, M. Hemberg, M. Barahona, D. E. Ingber and S. Huang, *Nature* **453**, 544 (22 May 2008).
26. H. Chen, G. Jiang and K. Yoshihira, Robust nonlinear dimensionality reduction for manifold learning, in *18th International Conference on Pattern Recognition (ICPR'06)*, (ieeexplore.ieee.org, 2006).

Figures

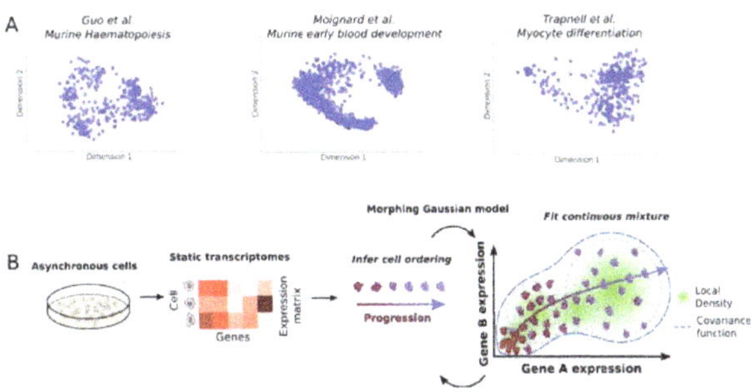

Fig. 1. A. Survey of three different single-cell transcriptomic studies. From left to right: murine haematopoiesis by Guo et al., early blood development by Moignard et al., and myocyte differentiation by Trapnell et al. B. Overview of the SCIMITAR method. Trajectory modeling with dynamic and correlated noise of static transcriptomes of asynchronous cells is achieved by iterating through optimal cell ordering and inference of a continuous set of Gaussian distributions in a morphing mixture of Gaussian models (see Methods in text).

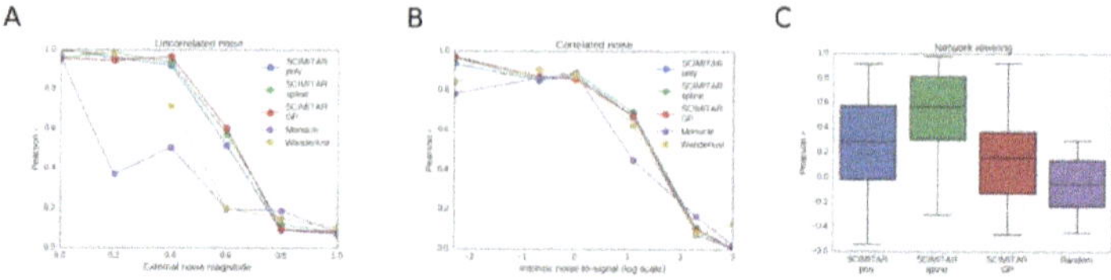

Fig. 2. SCIMITAR *in silico* benchmark. A. Cell ordering results for three functional classes of SCIMITAR (a third degree polynomial, a cubic spline, and Gaussian processes with squared exponential correlation model) and two state-of-the-art methods Monocle and Wanderlust in a setting with noise uncorrelated to the trajectory. B. Cell ordering results for noise correlated with the trajectory. C. Evaluation results of network rewiring across biological progression for SCIMITAR's three functional classes and random covariance functions.

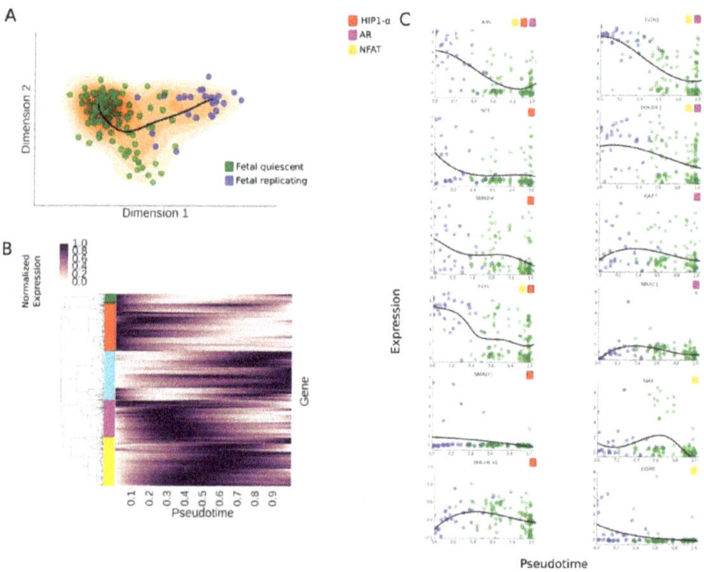

Fig. 3. A. SCIMITAR model for fetal neuron differentiation, projected to a 2-dimensional locally linear embedding. The data is plotted as circles in blue (fetal replicating neurons) and green (fetal quiescent nuerons) while the SCIMITAR model's mean is plotted in black and its projected PDF is plotted in orange. B. Normalized SCIMITAR model means for genes that were deemed progression associated across the progression, clustered into five different clusters using expression correlation throughout psuedotime. C. Expression levels of several genes from three central neurodifferentiation pathways: the HIF1α, NFAT, and Androgen Receptor (AR) pathways that were pinpointed by SCIMITAR associated progression tests.

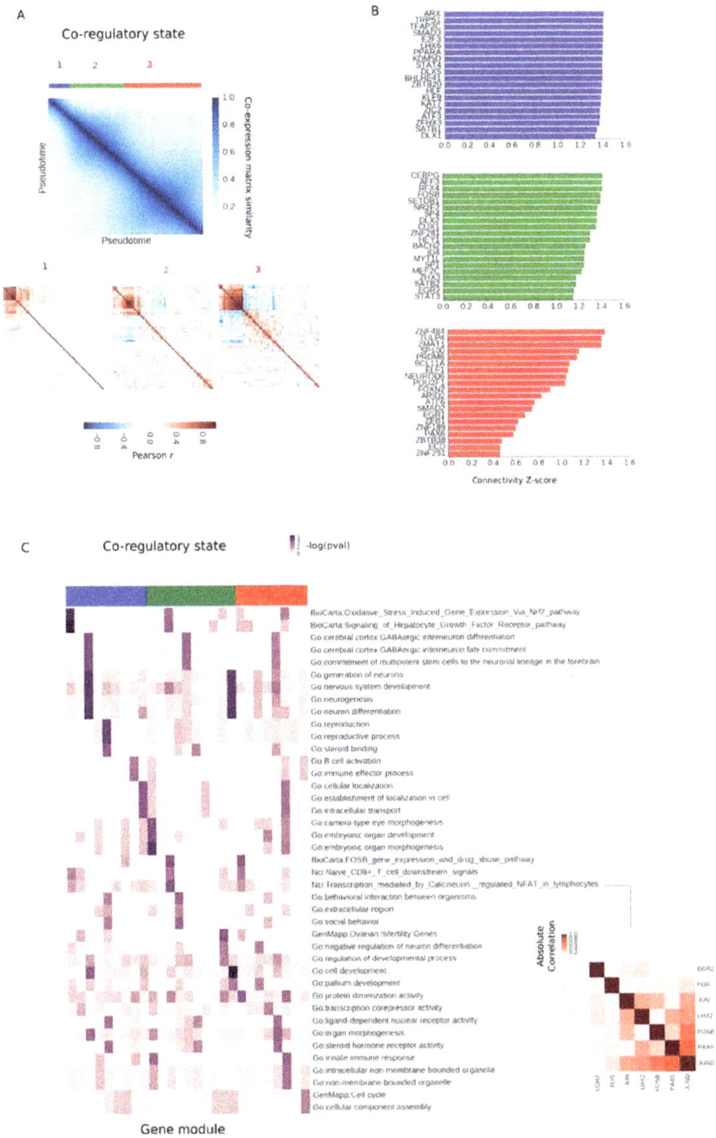

Fig. 4. A. Similarity matrix between co-expression matrices fitted in the SCIMITAR fetal neuron differentiation model across pseudotime. Three different co-regulatory states can be appreciated in the matrix, marked in blue, green, and red. B. Top 20 genes with the most gain-of-connectivity in each co-regulatory state alongside their log co-expression degree. C. Evolution of annotated modules. Each column is a module and each row is a gene annotation — enrichments are shown as $-log(p - value)$ in the heatmap. Column colors denote co-regulatory state. An NFAT-associated module of state 2 is highlighted in the red matrix

AN UPDATED DEBARCODING TOOL FOR MASS CYTOMETRY WITH CELL TYPE-SPECIFIC AND CELL SAMPLE-SPECIFIC STRINGENCY ADJUSTMENT

KRISTEN I. FREAD
Department of Biomedical Engineering, University of Virginia,
Charlottesville, VA 22903, USA
Email: kif5qw@virginia.edu

WILLIAM D. STRICKLAND
Department of Biomedical Sciences, University of Virginia,
Charlottesville, VA 22903, USA
Email: wds2df@virginia.edu

GARRY P. NOLAN
Department of Microbiology and Immunology, Stanford University,
Stanford, California 94305, USA
Email: gnolan@stanford.edu

ELI R. ZUNDER
Department of Biomedical Engineering, University of Virginia
Charlottesville, VA 22903, USA
Email: ezunder@virginia.edu

Pooled sample analysis by mass cytometry barcoding carries many advantages: reduced antibody consumption, increased sample throughput, removal of cell doublets, reduction of cross-contamination by sample carryover, and the elimination of tube-to-tube-variability in antibody staining. A single-cell debarcoding algorithm was previously developed to improve the accuracy and yield of sample deconvolution, but this method was limited to using fixed parameters for debarcoding stringency filtering, which could introduce cell-specific or sample-specific bias to cell yield in scenarios where barcode staining intensity and variance are not uniform across the pooled samples. To address this issue, we have updated the algorithm to output debarcoding parameters for every cell in the sample-assigned FCS files, which allows for visualization and analysis of these parameters via flow cytometry analysis software. This strategy can be used to detect cell type-specific and sample-specific effects on the underlying cell data that arise during the debarcoding process. An additional benefit to this strategy is the decoupling of barcode stringency filtering from the debarcoding and sample assignment process. This is accomplished by removing the stringency filters during sample assignment, and then filtering after the fact with 1- and 2-dimensional gating on the debarcoding parameters which are output with the FCS files. These data exploration strategies serve as an important quality check for barcoded mass cytometry datasets, and allow cell type and sample-specific stringency adjustment that can remove bias in cell yield introduced during the debarcoding process.

* This work is supported in part by the University of Virginia Department of Biomedical Engineering, CIRM Basic Biology II Grant RB2-01592, and NIH F32 GM093508-01.

1. Introduction

1.1. *Sample multiplexing for flow cytometry and mass cytometry with cell barcoding*

Sample multiplexing, also referred to as pooled sample analysis, is a general approach that has been applied to several biological assays, including ELISA immunoassay[1], next-generation DNA sequencing[2,3], fluorescence-based flow cytometry[4], and mass cytometry[5–7]. In this approach, individual samples are labeled with unique identifiers, and then pooled together for processing and measurement. These unique identifiers can be thought of as sample-specific barcodes. After processing and measurement, the pooled sample dataset is deconvolved using these barcodes to recover individual sample data for further analysis (Fig. 1A).

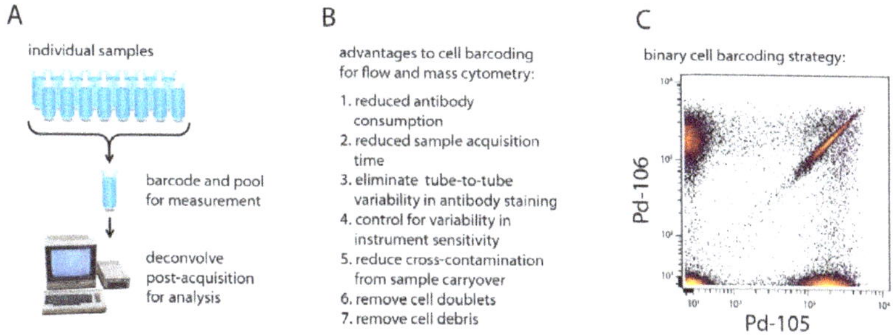

Figure 1. Mass cytometry barcoding overview. (A) General strategy for pooled sample analysis. (B) Flow and mass cytometry-specific advantages to cell barcoding for pooled sample analysis. (C) Binary cell barcoding strategy for flow and mass cytometry, in which every cell is labeled either positively or negatively on barcode-dedicated channels.

The obvious advantages gained by sample multiplexing are a) reducing the time and resources required to analyze multiple samples, and b) improving the comparability between samples, because they are processed identically after pooling. Major advantages specific to flow cytometry and mass cytometry include reduced antibody consumption, increased sample acquisition rate, and the elimination of tube-to-tube variability in antibody staining conditions (Fig. 1B).

Sample multiplexing for fluorescence-based flow cytometry is performed with cell-reactive dyes that bind irreversibly to accessible nucleophiles on the cell[4]. These accessible nucleophiles include free thiols present on cysteine residues, and free amines present on lysine residues and at the N-terminus of proteins. While not strictly required, cell permeabilization greatly improves cell barcoding performance by increasing the number of accessible nucleophiles available on each cell. Multiple levels of fluorophore labeling can be achieved – previous studies have demonstrated 96-sample multiplexing with only 3 dedicated fluorescence channels: Alexa Fluor 700 (4 staining levels), Pacific Blue (4 staining levels), and Alexa Fluor 488 (6 staining levels)[4]. This multi-level staining approach allows for a high level of multiplexing with limited measurement channels, but relies on uniform levels of dye reactivity between all cell types and samples.

If there is considerable variability in labeling reagent uptake between cell types or sample types, a simpler binary cell barcoding approach can be applied to improve the fidelity of cell sample assignment at the deconvolution step. Because each cell sample is labeled either positively or negatively on each barcode-dedicated channel, the two populations are better separated with less potential for overlap (Fig. 1C). This approach is favored for mass cytometry cell barcoding, because the lanthanide and palladium-based barcode reagents react rapidly with cells even at 4°C[5,7], making the labeling reaction effectively stoichiometric and therefore more sensitive to variability between the samples in cell number, cell type/size, the presence of cellular debris, and residual bovine serum albumin (BSA) from the wash buffer. Using a binary barcode scheme requires more barcode-dedicated measurement channels than multi-level labeling, but allows for greater sample assignment fidelity during deconvolution while still permitting over 40 molecular measurements per cell with a staining panel made up of lanthanide-based mass cytometry antibodies, I127-IdU to mark S-phase cells[8], and cisplatin as a viability stain[9].

1.2. *Doublet-filtering cell barcode scheme*

Cell doublets (as well as triplets, quadruplets, and higher-order cell clusters) pose a significant challenge for single-cell analysis. When analyzing or performing fluorescence-activated cell sorting (FACS) on cell samples with known and well-defined cell types, such as whole blood or primary blood mononuclear cells (PBMCs), cell doublets are for the most part an annoyance that can be gated out using cell surface markers and light scatter properties. In certain defined settings, the study of cell doublets by flow cytometry has even proved to be illuminating with respect to cell adhesion and cell-cell interactions[10]. However, during exploratory analysis of uncharacterized cell samples and cell types, cell doublets are especially problematic, because they may be falsely interpreted as a novel cell type that shares the molecular characteristics of its two component cells.

Fluorescence-based flow cytometry has forward scatter (FSC) and side scatter (SSC) parameters that can be used to identify and remove cell doublets by two-dimensional gating[11]. Mass cytometry does not have a comparable measurement parameter, but a binary barcode scheme has been developed that can identify and remove cell doublets as well as higher-order clusters[7]. Instead of using every possible binary combination, this doublet-filtering barcode scheme uses a limited subset of binary combinations, such that any doublet combination will result in an "illegal" combination that is recognized as a doublet and removed from the dataset. A binary barcode scheme with n dedicated measurement channels will provide 2^n unique barcode combinations, but the doublet filtering binary barcode scheme only uses n-choose-k combinations, where $k = n/2$. 6 palladium isotopes are often used for cell barcoding because they are incompatible with the DTPA-based polymer used to label antibodies with lanthanide metals[12]. Instead of multiplexing 64 samples with all binary combinations (2^6) of the palladium isotopes, the doublet-filtering scheme only allows 20-sample multiplexing (6-choose-3) with palladium-based barcoding reagents (Fig. 2A). Because each barcode combination in this scheme is positive for exactly 3 palladium isotopes (Fig. 2B), any cell that is positive for 4 or more palladium isotopes will be identified as a cell doublet and removed from the dataset (Fig. 2C).

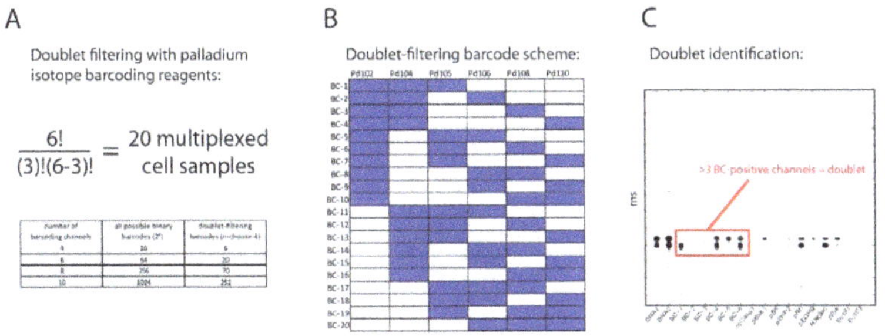

Figure 2. Doublet filtering barcode scheme. (A) Sample multiplexing with exhaustive (2^n) and doublet-filtering (n-choose-k) barcode schemes. (B) Palladium isotope combinations for doublet-filtering barcode scheme. (C) Doublet identification by "illegal" barcode combination viewed in the mass trace scanning window of the mass cytometer.

This doublet-filtering scheme has become part of the standard mass cytometry workflow for many laboratories, and was incorporated into the third-generation Helios™ CyTOF® mass cytometer. Each user should consider the benefits of each approach for their experiment, because in some cases increased sample multiplexing could be more valuable than doublet removal. However, the recent description of ruthenium and osmium-based cell barcoding reagents suggests that high-level multiplexing with simultaneous doublet-filtering is now within reach[13], without having to give up any of the traditional mass cytometry measurement channels such as the lanthanide series metals.

1.3. *Sample deconvolution by sequential gating and Boolean gating strategies*

After pooled sample analysis, sample-specific barcodes are used to recover individual sample data for analysis. Different approaches have been applied to this deconvolution step, including cell type-specific gating followed by sequential 2-D barcode gating[4] or Boolean 1-D barcode gating[5]. Two drawbacks from these gating approaches are 1) time-consuming manual gating, and 2) the potential for cell loss or sample mis-assignment. In situations where the separation between barcoded populations is not large enough to be separable (Fig. 3A), the researcher must decide whether to throw out cells that reside in this intermediate space (Fig. 3B), or to split the populations and accept that some cells may be incorrectly assigned (Fig. 3C). In barcoded samples there is very often at least a small number of cells present in this intermediate zone that cannot be assigned to a specific sample by this debarcoding method. Usually this population is minor as shown in Figure 1C, but results like Figure 3A can also occur, particularly if the cell number in one or more samples is not estimated accurately resulting in uneven barcode labeling between samples. For this 1-D or 2-D gating strategy, boundaries can be drawn algorithmically using distribution shape and percentile cut-points, but the exact placement will depend on how the competing desires for cell yield vs. sample assignment accuracy.

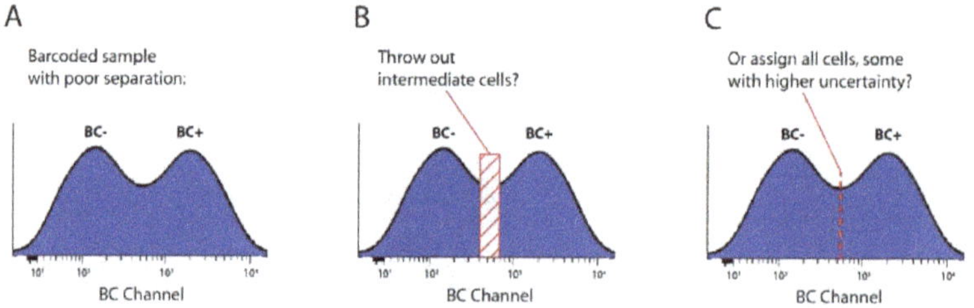

Figure 3. Traditional gating method on poorly-separated barcode sample. (A) Overlapping positive and negative barcode populations. (B) Intermediate cells can be thrown out to increase barcode deconvolution stringency. (C) Intermediate cells can be assigned to increase barcode deconvolution yield.

1.4. Sample deconvolution by single-cell debarcoding algorithm

In order to recover as many cells as possible in an automated and unbiased manner, a novel method for barcode deconvolution was previously developed, termed single-cell debarcoding[7]. This method is designed to perform especially well with the problematic "intermediate zone" cells. Instead of population-based gating, it looks at each cell individually, and asks "which sample barcode does this cell most closely resemble?" Sample assignment and the level of confidence associated with it is calculated by the separation distance between normalized positive and negative barcode channel measurements (Fig. 4A). The choice of separation distance used for this calculation depends on the binary barcode scheme being used. For exhaustive non-doublet-filtering barcode schemes, the largest separation distance is identified. For doublet-filtering barcode schemes, the distance between the top $n/2$ and bottom $n/2$ normalized barcode intensities is used, whether or not this is the largest separation distance present. If the separation distance is large, there is high confidence that the barcode sample assignment is correct. If the separation distance is small, there is low confidence that the barcode sample assignment is correct and these cells may be discarded depending on the deconvolution stringency desired.

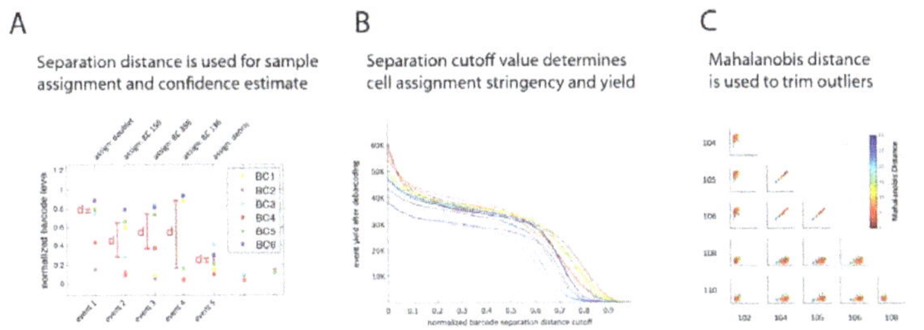

Figure 4. Single-cell debarcoding algorithm. (A) After normalization of the individual barode channel intensities, separation distances (indicated by a red line and the letter "d") are calculated for every cell. In this example, a 6-channel doublet-filtering barcode scheme was used. Therefore, event 1 does not receive a sample assignment because it appears to be a doublet with 4 positive barcode channels and a small separation distance between the top 3 and bottom 3 barcode intensities. Event 5 has low normalized intensities for all 6 barcode measurement channels, and therefore appears to be "debris." (B) The relationship between separation distance cutoff and debarcoder cell yield. Each colored line represents one of the 20 samples in a 6-metal, doublet-filtering, pooled sample dataset. Cell yield decreases with increasing separation distance cutoff stringency, but plateaus somewhat between 0.1 and 0.6. (C) Mahalanobis plots of every barcode-by-barcode biaxial plot for a single assigned cell sample. Every cell is colored by mahalanobis distance, from low (0-red) to high (30-blue).

The single-cell debarcoding software tool was released as a MATLAB standalone executable (https://github.com/nolanlab/single-cell-debarcoder)[7] that does not require a MATLAB installation (http://www.mathworks.com/products/compiler/). This software tool performs debarcoding and sample assignment in a semi-automated manner, presenting the user with visualizations that aid in the choice of two key debarcoding parameters: the separation distance cutoff which affects sample assignment stringency and cell yield (Fig. 4B), and the mahalanobis distance cutoff which is used to trim outliers (Fig. 4C). Standard practice for the single-cell debarcoder is to choose a separation cutoff distance that is as stringent as possible without severe cell loss, such as approximately 0.5 in Figure 4B. Most separation distance plots follow a similar trend, with a plateau in the center flanked by steep declines in the 0-0.1 range (debris and cell doublets) and approaching 1 (all cells will eventually fail the stringency test). Mahalanobis plots are more variable, depending on the mix of cell types in each sample. There is no specific rule or recommendation for setting the mahalanobis distance cutoff, but the default setting of 30 is a good starting point for 6-metal/20-sample palladium-barcoded samples. After the user selects values for the separation distance cutoff and mahalanobis distance cutoff parameters, the single-cell debarcoder tool outputs every deconvolved cell sample as an FCS file.

1.5 Limitations and drawbacks to using fixed-value debarcoding cutoff parameters

Applying the same parameter cutoffs to each sample while debarcoding as previously described[7] is not optimal, because each sample was barcode-stained individually and will therefore vary in barcode staining intensity and population-level variance. If all samples are similar (in terms of cell type, cell number, cellular debris, and residual BSA concentration) and the cell barcoding protocol is performed precisely, then barcode staining will be fairly uniform across every sample. Frequently this is not the case however, resulting in considerable variability of barcode staining between cell samples and large differences in sample behavior with respect to the debarcoding parameters, especially the normalized barcode separation distance cutoff (Fig. 5A).

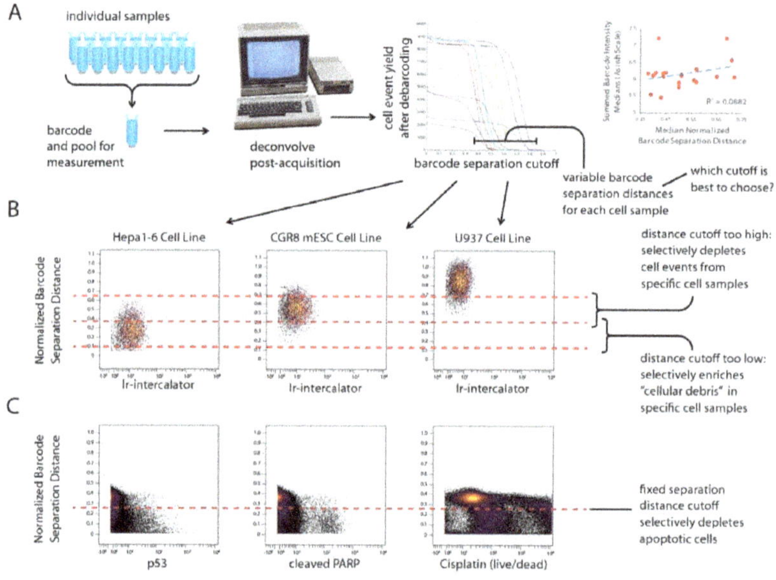

Figure 5. Cell barcode variability and its consequences. (A) Cell samples representing 20 different cell types were barcoded by the 6-palladium doublet-filtering method and then pooled for analysis. The amount of barcode reagent added to each sample was adjusted according to cell number in each sample to normalize barcode staining intensity. Variability in barcode separation distance was observed between the samples, but is only weakly correlated to barcode staining intensity, as measured by the 6-metal summed barcode intensity medians for each sample. (B) Three of the twenty barcoded cell samples, which show highly variable barcode separation distance levels, precluding a single optimal cutoff value. (C) Apoptotic cells with elevated levels of p53, cleaved-PARP, and cisplatin labeling have reduced barcode separation distance, and could be unintentionally discarded from analysis with a typical debarcoding workflow.

In this scenario, no single cutoff value for barcode separation distance is optimal for every sample, forcing the researcher to choose between depleting cells of interest in some samples, or enriching for cellular debris in other samples (Fig. 5B). In addition to cross-sample differences, different cell types can be depleted or enriched within a single sample due to differences in barcode staining behavior based on cell size, cell identity, or cell state (Fig. 5C). These sample-specific and cell type-specific effects are usually minimal, but have the potential to introduce bias into the analysis and conclusions drawn from barcoded mass cytometry experiments. Therefore, each barcoded dataset should be investigated to detect the extent of these effects, and correct for them if necessary.

2. Methods

2.1 Output single-cell debarcoding parameters with each FCS file for visualization and analysis

With the previously released debarcoding tool[7], investigating the possibility for barcode-related enrichment or depletion of specific samples and cell types required laborious and time-consuming back and forth between rounds of debarcoding and FCS file analysis. Side-by-side comparison of FCS files debarcoded with iterative values for the debarcoding parameters was necessary to detect cell type or sample-specific effects. To obviate the need for this slow and inefficient analysis, we have updated the debarcoding software tool to output the debarcoding parameter values for each cell as additional data columns in the FCS file. This update allows for visualization of the barcode parameters, and analysis of how they interact with the other measured parameters and cell types of interest. The MATLAB source code for the updated software tool as well as pre-compiled executable files that do not require MATLAB installation are available to download at https://github.com/zunderlab/single-cell-debarcoder.

2.2 Post-assignment application of debarcode stringency filter and outlier trimming

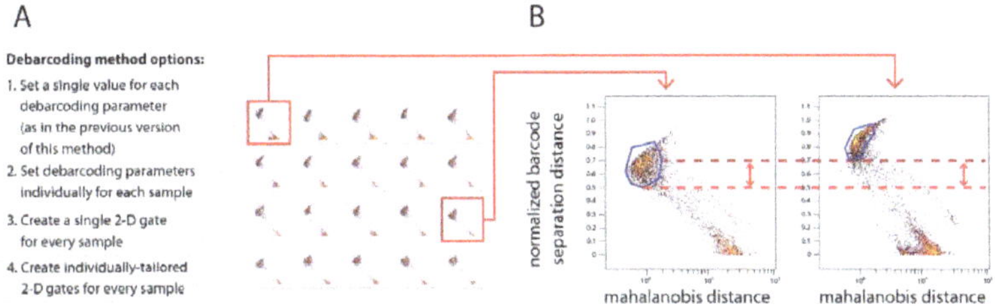

Figure 6. Sample-specific stringency adjustment by individual gating on debarcode parameters. (A) FCS output of the debarcoding parameters allows different strategies for stringency filtering. The option for individually-tailored 2-D gating on normalized barcode separation distance and mahalanobis distance is presented. (B) Two cell samples from Fig. 6A are highlighted to illustrate the population-level differences in barcode parameters between samples.

In addition to visualization and analysis, outputting the debarcode parameters in the FCS file has another practical benefit: stringency filters can be turned off during the debarcoding step and applied after the fact instead. This gives the user flexibility in their choice of stringency filtering: they may apply fixed parameters as in the previous version of this method, or perform sample-specific two-dimensional gating on the debarcode parameters (Fig. 6A). Whichever method is chosen, the user is given the tools to explore these parameters and their relationship to other cell measurements, which will aid in the choice of filtering strategy and its implementation. Some users may prefer fixed parameter stringency filtering because it is simpler and less time consuming, but users with complex, variable samples should consider individually-tailored stringency filtering, which requires more time to implement but helps prevent the introduction of sample-specific biases (Fig. 6B).

3. Results

3.1. Precision Debarcode Stringency Filtering

The newly updated single-cell debarcoding software tool functions identically to the previous version, but with two additions: 1) values for the normalized barcode separation distance and mahalanobis distance are output for every cell, and 2) default parameters for debarcoding are set as "barcode separation threshold = 0" and "mahalanobis distance threshold = inf" (Fig. 7A). These default parameters ensure that every cell is assigned to a sample for FCS output and can be filtered after the fact. This differs from the fixed-parameter filtering which took place at the debarcoding step in the previous software version, resulting in an additional FCS output for unassigned cell events. Outputting the entire dataset (Fig. 7B) with this new method allows for precision stringency filtering by gating on the debarcode parameters (Fig. 7C). This gating will typically be performed using flow cytometry/FCS analysis software, and can be done iteratively and in combination with more fundamental cell type and dataset-specific analyses.

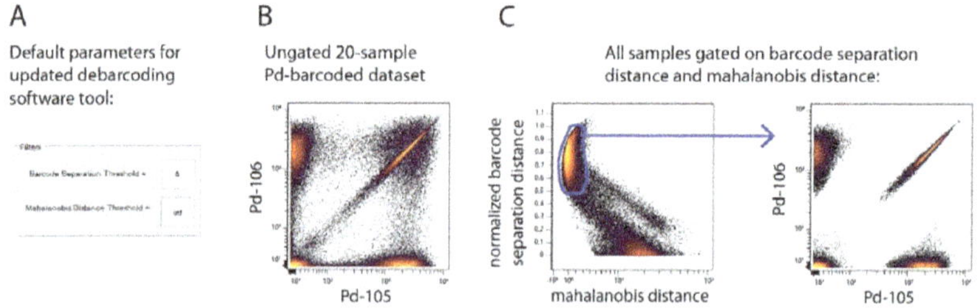

Figure 7. Debarcode stringency gating overview. (A) 20-sample Pd-based doublet-filtering barcode sample, ungated. (B) Barcode stringency trimming by 2-D gating on the normalized barcode separation distance vs. mahalanobis distance.

3.2 Identification and Reduction of Debarcoding-induced Sample Bias

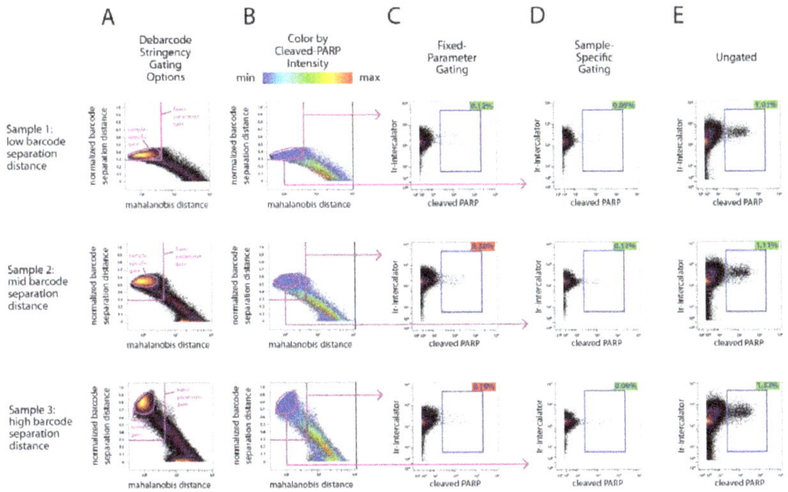

Figure 8. Sample-specific debarcode stringency gating reduces unbalanced enrichment of cleaved-PARP-positive cells. (A) Two options for stringency gating are displayed in magenta: fixed-parameter and sample-specific. (B) Cleaved-PARP intensity color scale applied to the plot from Figure 8A reveals that fewer cells with elevated cleaved-PARP levels fall within the fixed-parameter gate in sample 1 compared to samples 2 and 3. (C-E) The percentages of cleaved-PARP-positive cells present in the bulk-gated, individually-gated, and ungated populations.

Sample-specific debarcode gating on the normalized separation distance and mahalanobis parameters provides the greatest advantage over the previously used fixed-parameter debarcoding method when there is variability in the debarcode parameters between samples (Fig. 8A), which can lead to uneven distribution of specific cell types across the debarcoded samples. Cells with elevated cleaved-PARP levels are associated with lower separation distance and higher mahalanobis distance (Fig. 8B). This leads to disproportionate enrichment for cleaved-PARP cells in some samples when using fixed-parameter debarcode filtering (Fig. 8C), but is ameliorated by sample-specific gating (Fig. 8D), which more closely matches the ungated sample ratios (Fig. 8E).

4. Discussion

This updated method for single-cell mass cytometry debarcoding allows for visualization and analysis of the debarcoding parameters, and how they specifically relate to every other cell measurement. This can be used to detect any cell type-specific or sample-specific effect of the debarcoding process on the underlying cell data of interest. The source code and Win/Mac executable software are available to download from https://github.com/zunderlab/single-cell-debarcoder. We recommend that this analysis be performed on every debarcoded dataset as a data quality check, particularly when mixed cell types and sample types are barcoded together. In addition to data quality verification, the output debarcoding parameters in every assigned FCS file can be used to guide sample-specific stringency filtering that can be performed after the fact rather than during the debarcoding process. This allows multiple stringency levels to be tested rapidly using flow cytometry/FCS analysis software, where multiple iterations of 1-D or 2-D gating can be used while monitoring the effect on cell type-specific and sample-specific cell yield as well as overall data quality. One limitation to this method is that stringency filtering is not automated, and currently relies on hand-drawn gates. While this method is optimally used to reduce cell yield and enrichment bias between cell samples and cell types that vary in barcode staining behavior, sample-specific or cell type-specific manual gating has the potential to introduce bias. As with any other hand-drawn gating analysis, the barcode gating strategy should always be presented in addition to further analysis in order to mitigate this potential for user-introduced bias. In the future, stringency filtering could be automated with sequential, percentile-based gating steps; or more complex computational methods.

5. Acknowledgments

This work was supported by the University of Virginia Department of Biomedical Engineering, CIRM RB2-01592, and NIH F32 GM093508-01.

6. References

1. Fulton et al. *Clinical Chemistry* **43,** 1749–1756 (1997).
2. Meyer et al. *Nucl. Acids Res.* **35,** e97 (2007).
3. Parameswaran *et al. Nucl. Acids Res.* **35,** e130 (2007).
4. Krutzik, P. O. & Nolan, G. P. *Nature Methods* **3,** 361–368 (2006).
5. Bodenmiller *et al. Nature Biotechnology* (2012). doi:10.1038/nbt.2317
6. Behbehani *et al. Cytometry* n/a-n/a (2014). doi:10.1002/cyto.a.22573
7. Zunder *et al. Nat. Protocols* **10,** 316–333 (2015).
8. Behbehani *et al. Cytometry Part A* **81A,** 552–566 (2012).
9. Fienberg *et al. Cytometry Part A* **81A,** 467–475 (2012).
10. Snippert *et al. Cell* **143,** 134–144 (2010).
11. Hoffman, R. A. in *Current Protocols in Cytometry* (John Wiley & Sons, Inc., 2001).
12. Majonis *et al. Biomacromolecules* **12,** 3997–4010 (2011).
13. Catena *et al. Cytometry* **89,** 491–497 (2016).

MAPPING NEURONAL CELL TYPES USING INTEGRATIVE MULTI-SPECIES MODELING OF HUMAN AND MOUSE SINGLE CELL RNA SEQUENCING*

TRAVIS JOHNSON MS,
Dept. Biomedical Informatics, Ohio State University,
250 Lincoln Tower, 1800 Cannon Dr. Columbus, Ohio, 43210
Travis.Johnson@osumc.edu

ZACHARY ABRAMS PhD,
Dept. Biomedical Informatics, Ohio State University,
250 Lincoln Tower, 1800 Cannon Dr. Columbus, Ohio, 43210
Zachary.Abrams@osumc.edu

YAN ZHANG PhD,
Dept. Biomedical Informatics, Ohio State University,
250 Lincoln Tower, 1800 Cannon Dr. Columbus, Ohio, 43210
Yan.Zhang@osumc.edu

KUN HUANG PhD,
Dept. Biomedical Informatics, Ohio State University,
250 Lincoln Tower, 1800 Cannon Dr. Columbus, Ohio, 43210
Kun.Huang@osumc.edu

Mouse brain transcriptomic studies are important in the understanding of the structural heterogeneity in the brain. However, it is not well understood how cell types in the mouse brain relate to human brain cell types on a cellular level. We propose that it is possible with single cell granularity to find concordant genes between mouse and human and that these genes can be used to separate cell types across species. We show that a set of concordant genes can be algorithmically derived from a combination of human and mouse single cell sequencing data. Using this gene set, we show that similar cell types shared between mouse and human cluster together. Furthermore we find that previously unclassified human cells can be mapped to the glial/vascular cell type by integrating mouse cell type expression profiles.

* This work is partially supported by RGP0053 of the Human Frontier Science Program

1. Introduction

Mouse models are an important part of biomedical research and are routinely used as a stepping-stone towards treatments for humans – gleaning knowledge from high-throughput low risk experiments. Translating this knowledge requires a firm understanding of similarities between these two species [1-2]. Homologous genes exist between these species and these genes often play similar roles in the brain [3]. However, the biochemical pathways within each species have subtle to extreme differences leading to subsets of homologous genes without exact mechanistic overlap in the brain [4]. To address the issue of identifying functionally similar homologous genes we propose the concept of concordant genes defined as gene homologs that mechanistically behave similarly between two species [5]. Specifically, we hypothesize that concordant genes between mouse and human exist and that those genes can be algorithmically derived from combined mouse-human data. We also hypothesize that based off of these concordant genes we can determine cell type matching between mouse and human. Specifically in this study we focus on the comparison of brain cell gene expression profiles between mouse and human to identify concordant gene expression patterns in the brain tissue associated with different cell types taking advantages of recent development in single cell transcriptomics for brain cells. We hope that the single cell granularity of these comparisons will augment the tissue level comparisons of the human and mouse brain transcriptome [6].

RNA sequencing (RNA-Seq) in the past has been used to study brain structure, development, and disease [7]. Recently RNA-Seq has become more granular in the form of single cell RNA sequencing (scRNA-Seq) which is an important tool in the study of tissue heterogeneity due to its unique ability to characterize transcriptomes at the cellular level [8]. Recent advances in single cell transcriptomics in the brain have provided researchers with an influx of new data spanning different brain regions, diseases, and species [9]. Specifically, the Linnarsson group amassed a large single cell dataset from the mouse cortex and hippocampus which was clustered into multiple cell types based expression profiles [10]. Subsequent to the mouse single cell transcriptomic study, the Zhang group created a large human brain scRNA-Seq dataset from postmortem brain tissue and clustered the cells into unique cell types based on expression profiles [11]. Because of the availability of both datasets we believe that in-depth comparative analyses of these two datasets is fundamental to our understanding of neuronal cell types, the distribution of these cell types, and the evolution of brain anatomy in these two species. Furthermore a clear understanding of concordant genes in both human and mouse provides valuable information on how mouse studies can be translated to human research. We provide a methodology and gene set that can be used for these comparative studies and hopefully for future translational research. We demonstrate the method by not only identifying concordant cell types between mouse and human brains with the same set of concordant feature genes, but also matching un-categorized cells in the human brain to a salient cell type based on mouse brain information.

2. Methods

2.1. *Data normalization and cleaning*

The mouse scRNA-Seq unique molecular identifier (UMI) counts [12] were downloaded from the Data section of the Linnarsson lab website (http://linnarssonlab.org/) and human scRNA-Seq transcripts per million (TPM) data was downloaded from the Links section of the SCAP-T website (scap-t.org). Since these data files contain various numbers of genes with different order, we preprocessed the files by scanning matching gene symbols between files then sorting the gene symbols so that the orders were consistent. While this process may not be able to identify all homologous genes, it provides a large list for us to extract concordant genes. The shared gene symbols in the human and mouse datasets were retained for further study (Figure 1). Within the human dataset there were genes that were originally left out of analysis by the original authors due to low expression, resulting in some cells with low number of expressed genes. Because of this, such human cells as well as human cells without annotation in the metadata were also removed from further analysis, resulting in 3,086 human cells each containing 13,355 genes. The mouse dataset resulted in 3,005 cells each containing expression values from 13,355 genes. Both human and mouse data then were transformed into comparable units. Each dataset was log2 transformed and the expression values converted into the within cell z-scores.

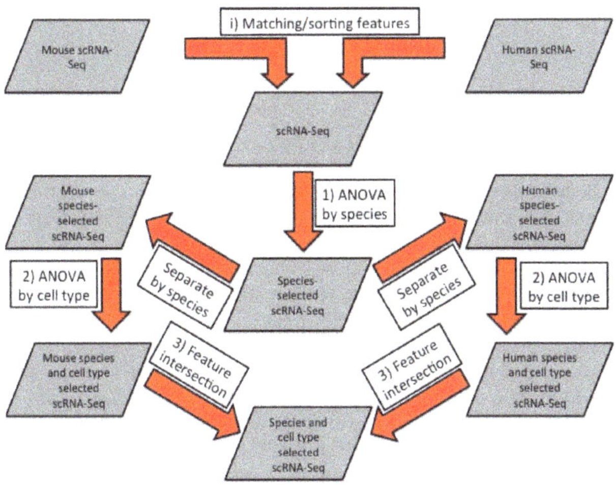

Figure 1. Workflow of data normalization (i) and three step feature selection method (1-3).

2.2. *Feature selection*

We developed a three-step approach to find concordant genes between mouse and human based on gene expression profiles (Figure 1). This feature selection was performed to identify genes that were informative at separating cell type but uninformative at separating mouse from human cells. Genes that meet this criterion would be more useful at identifying similar cell types across species.

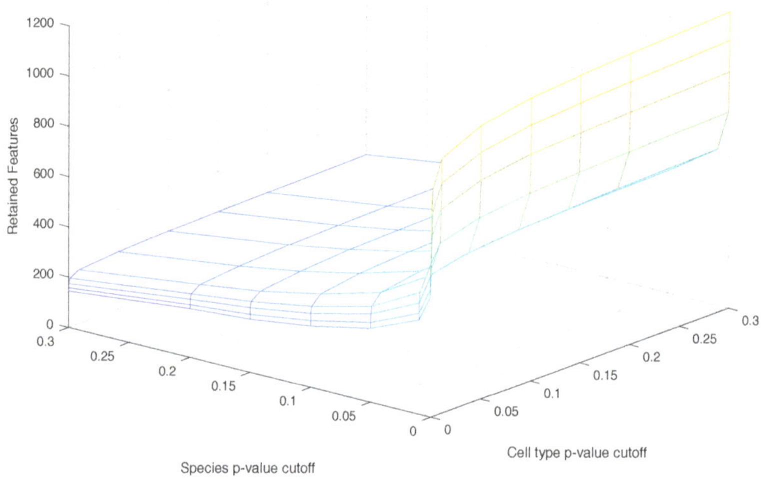

Figure 2. Number of retained features as a function of p-value cutoffs.

First, the human and mouse data matrices were concatenated such that the first 3086 columns consisted of human cells and the last 3005 columns consisted of mouse cells. For each gene in the data matrix, a one-way ANOVA was performed grouped by species to detect genes with significantly different expression level between human and mouse. Only genes with p-values larger than 0.1 were kept. This was done to remove genes that would separate cells by species. Because we are removing the significant genes from our gene set in Step 1, a greater threshold makes our criterion for retaining genes more strict than using a standard significance level. Second, the human and mouse matrices were separated and in each separate matrix a one-way ANOVA was performed on the remaining genes grouped by cell type label and using a threshold p-value of 0.01 – any genes found with a p-value of 0.01 or less were retained. The 0.01 threshold was used to provide stricter criteria for retained genes that were informative about cell type. The 0.1 and 0.01 p-value cutoffs used in the feature selection method are near the inflection point of retained features as a function of cutoff p-value (Figure 2). Third, the intersection of retained genes from human and mouse were retained in the final dataset such that genes that existed in both human and mouse gene sets after Step 2 were retained in the final combined mouse-human gene set.

To compare the differences between cell types and in concordance with previous single cell studies [13], principal component analysis (PCA) was applied to the human and mouse datasets prior to feature selection. The first 2 principal components were then plotted to visually show the differences in cell types and species (Figure 3). After feature selection, principal cross-species cell-type clusters can be viewed in the PCA of the first two principal components colored by species (left) and cell type (right) (Figure 4).

2.3. *Functional annotation of retained concordant genes*

When selecting features, it is important to study the relation of these feature/gene sets to the functional, anatomic, and phenotypic relationships that are being selected for. If there are functional relationships related to a phenotype, then the feature selection method targeting that phenotype is likely more robust. The retained genes from the feature selection step were used as input for the DAVID functional annotation software [13-14]. The functional annotation clusters were reviewed for over represented terms that can be attributed to neural pathways and cell types. We display the three most highly enriched terms within the three most highly enriched clusters from the DAVID functional annotation clustering (Table 1).

2.4. *Clustering cells using Gaussian mixture models*

Gaussian mixture models are effective in clustering microarray expression profiles [16]. We apply Gaussian mixed models (GMMs) in the mouse and human scRNA-Seq data to cluster the cells into principal cell types and to compare the relative proportions of human and mouse cells within each cluster. To perform the GMM we used the first two principal components, the same components used in the PCA plot of cell types. Four GMMs were fit to the data with two, three, four and five components respectively. The cells were clustered into three major cluster using the three component GMM fit in concordance with the three major cell types present in the human dataset. The remaining GMM fits were used in comparison against the three-component GMM fit.

Principal cell types of the mouse and human labels were compared in the PCA space to determine the most similar cell types between both species. To quantitate the mouse-human overlap the mouse and human data were split into three groups from the three major cell types in the original publications. Human cells were split into 3 major groups from their original labels [11]. All "Int" labeled cells were considered Interneuron. All "Ex" labeled cells were considered pyramidal. All "NoN" (No Nomenclature) labeled cells from a C1 Fluidigm chip with reduced mapping rates were without a biologically derived label but were considered a singular group. Similarly, mouse cells were also split based on cell type label mapping to GMM clusters [10]. All cells labeled Interneurons were still considered Interneurons. All S1 Pyramidal and CA1 Pyramidal were considered Pyramidal. All Oligodendrocytes, Microglia, Endothelial, Astrocytes, Ependymal and Mural were considered Glial/Vascular cells. All human and mouse cells that were contained within each GMM cluster were compared by the their original cell type labels to the labels of the GMM cluster. For each cluster a fisher exact test was conducted to calculate the odds ratios and confidence intervals between published cell type labels and GMM predicted cell types.

The VennX package in MATLAB was used to convert the cell type labels into Venn Diagrams to show overlap with both three component GMM predicted cell types and original mouse/human cell type labels from their original publications.

3. Results

3.1. *Feature selection*

Prior to feature selection the human and mouse cells created two clusters separated by species. The mouse cells formed sub-clusters within the major mouse clustering of cells. The human cells formed one main cluster with little differentiation (Figure 3).

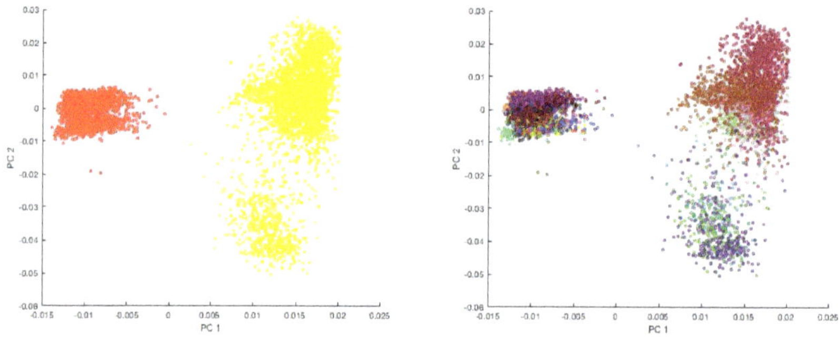

Figure 3. PCA of all human and mouse cells after normalization/cleaning. Left is colored by species, mouse (yellow) and human (red). Right is colored by cell type (36 cell types).

After feature selection, 358 concordant genes were retained, which are informative in terms of distinguishing cell types and uninformative in terms of separating species. As a result, human and mouse cells were no longer completely separate from each other. The mouse cell types still have more variability than the human cell types in the PCA space but cells from both species are contained within the same major clusters of cells (Figure 4).

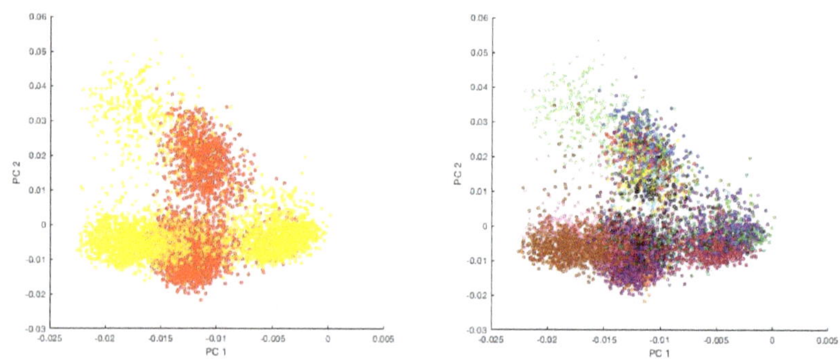

Figure 4. PCA of all human and mouse cells after normalization/cleaning and feature selection. Left is colored by species, mouse (yellow) and human (red). Right is colored by cell type (36 cell types).

3.2. *Functional annotation of concordant genes*

Functional annotation analysis of the concordant gene set revealed GO terms related to binding, ion transport and neural cells. The third most highly enriched annotation cluster was that of the GO terms axon, cell projection and neuron projection with an enrichment score of 1.57 (Table 1). Cluster 7 (not displayed) also contained many neuron related ontology terms.

Table 1. Functional annotation clustering using DAVID. Shown below are the three most highly enriched clusters and three most highly enriched terms within each cluster.

Category	Term	PValue	Fold Enrichment	Bonferroni
Annotation Cluster 1	Enrichment Score: 1.670			
SP_PIR_KEYWORDS	atp-binding	0.008	1.573	0.939
SP_PIR_KEYWORDS	nucleotide-binding	0.010	1.478	0.969
GOTERM_MF_FAT	GO:0032559~adenyl ribonucleotide binding	0.012	1.463	0.997
Annotation Cluster 2	Enrichment Score: 1.594			
GOTERM_BP_FAT	GO:0006826~iron ion transport	0.002	8.868	0.979
SP_PIR_KEYWORDS	iron transport	0.007	10.076	0.919
GOTERM_BP_FAT	GO:0000041~transition metal ion transport	0.012	4.347	1.000
Annotation Cluster 3	Enrichment Score: 1.568			
GOTERM_CC_FAT	GO:0030424~axon	0.010	3.027	0.946
GOTERM_CC_FAT	GO:0042995~cell projection	0.036	1.611	1.000
GOTERM_CC_FAT	GO:0043005~neuron projection	0.056	1.877	1.000

3.3. *Clustering cells using gaussian mixture models*

Gaussian mixture models showed major patterns within the cell profiles. Interneurons from both human and mouse (red and yellow respectively)(Figure 5) clustered in the same GMM. Whereas human pyramidal/projection neurons clustered (green) clustered with the remaining 2 cell types in mouse (S1 pyramidal, CA1 pyramidal). It is also worth consideration that the non-biologically labeled "NoN" human cell types in purple are mapped to a third cluster that begins to appear at 3 GMM components that contains the remaining 6 mouse cell types (mural, endothelial, microglia, ependymal, astrocytes, oligodendrocytes) (Figure 5).

The GMM clustering using three components (BIC = -9.08×10^4) split the cells into three groups that can be roughly defined as Interneurons (red), Pyramidal cells (green) and Glial/Vascular cell types (blue) (Figure 6: Top left). After identifying these three groups and comparing the mouse and human labels the GMM labels it was found that these three groups, Interneurons, Pryamidal cells, and Glial/Vascular cells are very closely mapped between both mouse and human. Also the "NoN" cell type cluster found in the human scRNA-Seq paper were clearly and uniquely clustered with the mouse Glial/Vascular cells (Figure 6 bottom right) with no significant difference between Glial/Vascular mouse cells and "NoN" human cells on PC 1 p-value = 0.41.

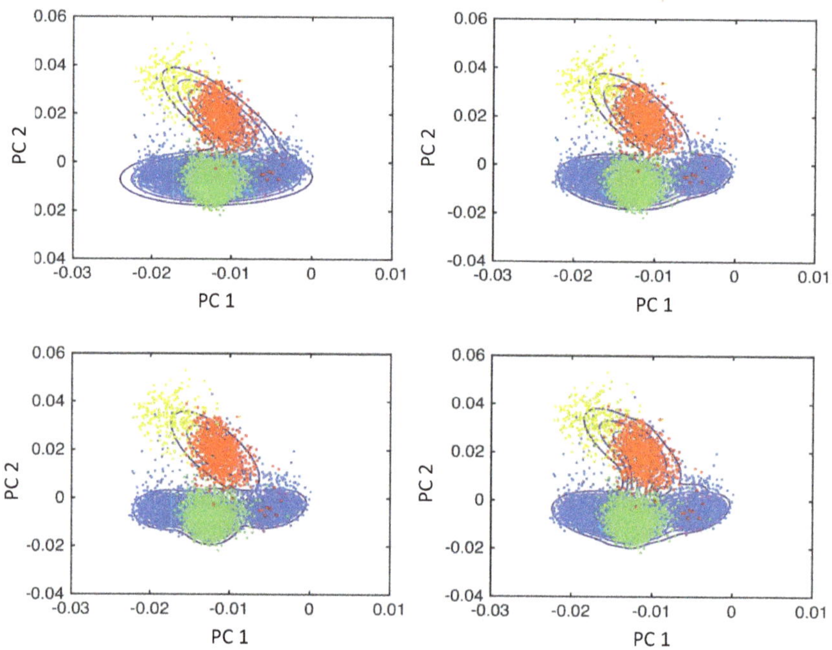

Figure 5. Gaussian mixture model clustering of human and mouse cell types where top left: two components, top right: three components, bottom left: four components and bottom right: five components.

The cell types predicted by the three component GMM were representative of the original cell type labels. The interneuron GMM had an odds ratio of 2.00×10^3 and confidence interval of $(1.16\times10^3, 3.46\times10^3)$, the pyramidal GMM had an odds ratio of 9.93×10^2 and a confidence interval of $(6.84\times10^2, 1.44\times10^3)$, and the glial/vascular GMM had an odds ratio of 1.15×10^2 and a confidence interval of $(91.34, 1.44\times10^2)$ (Figure 6). The GMM cluster for glial/vascular cells had a higher false negative rate than the other GMM clusters due to incorrect clustering of glial/vascular labeled mouse cells.

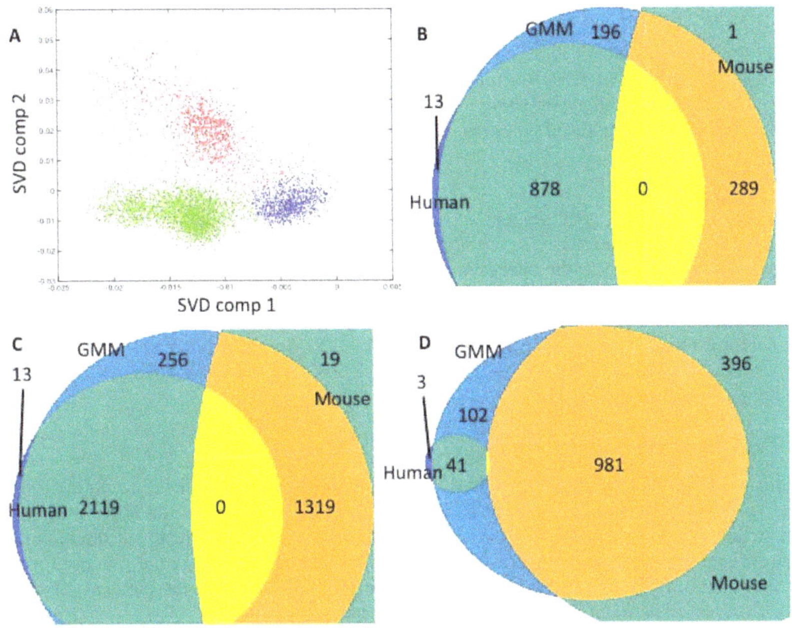

Figure 6. Comparing GMM clustering of human and mouse cells versus reported cell types. A) SVD components colored by GMM predicted clusters red (interneurons), green (pyramidal) and blue (mural/vascular). B-D are Venn diagrams comparing reported human and mouse cell types with GMM predicted cell types. The following superscripts represent if the point was included + or excluded – from species and GMM cluster. The colors from left to right consist of $Human^{+}$-GMM^{-} (blue), $Human^{+}$-GMM^{+} (green), $Human^{-}$-$Mouse^{-}$-GMM^{+} (blue), Null set (yellow), $Mouse^{+}$-GMM^{+} (orange), $Mouse^{+}$-GMM^{-} (green). B) GMM Interneurons cluster (red in panel A) with mouse and human interneuron labeled cells. C) GMM Pyramidal (green in panel A) with mouse and human pyramidal labeled cells ("CA1, S1" and "Ex" respectively). D) GMM Glial/Vascular (blue in panel A) with mouse glial/vascular labeled cells and human "NoN" labeled cells.

4. Discussion

4.1. *Insights*

In this study we found that through feature selection it is possible to find informative gene sets that can be used across species. This feature selection of "concordant gene sets" is an important application of single cell data that has multiple downstream applications in relation to cross species modeling, especially in translation of preclinical studies. It is important to note that the data used to find the concordant genes cannot be paired by sample which makes correlation matrices impossible to generate. Without correlation matrices to discover concordant genes, the gene sets must be derived from ulterior methods such as minimizing redundant gene sets through machine learning [17] or grouped statistical tests like ANOVA.

4.1.1. *Scalability*

The feature selection method is based on ANOVA which is calculated across multiple groups. Unlike t-tests, this facet of ANOVA makes the feature selection method scalable in relation to number of species and cell types being studied. Because of this, finding concordant gene sets between many organisms and cell types simultaneously is possible and should be pursued.

4.1.2. *Functional relevance*

The annotated concordant gene set had a clear relationship to the brain through gene ontology which is an important control due to the tissue origin [18]. It is important to note that gene sets with no functional overlap to the phenotype being selected for could potentially be selecting for unknown associated phenotypes. The functional ontology analyses of this concordant gene set shows that there is selection of genes with direct relation to neuronal phenotypes. Because of the enrichment of phenotypically similar ontology terms, a case can be made that seemingly phenotypically dissimilar ontology terms are more likely to have an unknown but direct relationship to our concordant gene set.

4.1.3. *Evolutionary potential*

Concordant gene sets also contain unique evolutionary information. Gene homologs which express differently between two species (Discordant genes) potentially do not share exactly the same functionality. Discordant genes may have the same down-stream effects but the biological mechanism may have changed [6] such that the same quantity of mRNA is not produced across species. Concordant genes are informative because they could represent pathways that are relatively conserved between through the evolution of species.

4.1.4. *Medical and research potential*

In the medical realm concordant gene sets could be of use in translational research. Much of research is conducted in model organisms and using concordant gene sets gives the user an ability to distinguish between transcriptional changes that likely cause similar phenotypes or likely do not between the model and human. Though we do not immediately condone the clinical use of concordant genes at the present these concordant gene sets could help to quickly and efficiently integrate cross-species knowledge to improve translational research.

4.1.5. *Future work*

The scalability of cell type and species number should be tested upon the arrival of comparable data in other species. Aside from the direct feature selection of concordant genes multiple comparisons could be carried out to create hierarchical concordant gene sets for higher granularity. Another option to improve granularity would be to test models that include interaction variables between species, brain location, and cell type. With the generation of concordant gene sets cross-species deconvolution could become more accurate than with more heuristic approaches. Also concordant gene sets can be used in classification of cell types across species. With further refinement of the procedure human cell types could be classified using mouse expression profiles which would require refinement of feature selection and of classification algorithms and validation of such methods on another dataset.

4.1.6. *Importance of single cell granularity*

Single cell technologies in the form of fluorescence-activated cell sorting (FACS) and flow cytometry have been effectively used to model cell heterogeneity [19] before the advent of single cell transcriptomics. Through FACS sorting [20] and flow cytometry [21] deriving the transcriptome of a single cell is much higher throughput than original methodologies that required manual isolation of single cells [22]. Without the single cell granularity of these techniques, it would be impossible to study concordant genes effectively at the cellular level and acquire the sample sizes large enough to properly study concordant gene sets, especially when many species and phenotypes are involved. Only through these recent advances in scRNA-Seq is it possible to properly glean enough information about cell types to model across species.

4.2. *Limitations*

There are some limitations to this study, which included the use of zscores as the measurement of expression. This measurement makes the assumption that the data has a normal distribution. Because of the nature of scRNA-Seq data the distribution is negative binomial. It was important to use zscores because other normalization techniques would not be effective. Quantile normalization introduced artificats in the data that made it unrepresentative. Conversion of UMI counts to TPM alos posed a problem because TPM is based on aligned reads opposed to tag counts from UMIs.

Aside from normalization, the diversity of cell types in each dataset also potentially introduced bias. The human dataset consisted of fewer major cell types than the mouse dataset. The mouse dataset contained more glial cell types while the human dataset had higher granularity within interneurons and pyramidal cells.

5. Conclusion

We were able to find a concordant gene set between mouse and human brain cells that had direct functional ontology relationships to the brain. The concordant gene set allowed us to reduce the distance between cell types of different species allowing separation of cell type regardless of each cell's species. Through the study of these aggregate cell types the biologically unresolved human cell type "NoN" (No Nomenclature) was able to be categorized as Glial/Vascular. Furthermore we show that our methodology is scalable to multiple species and cell types to find concordant gene sets between multiple species and these concordant genes sets are important stepping stones toward evolutionary and translational research goals.

6. Acknowledgements

This work is partially supported by the Human Frontier Science Program (RPG0053/2014) and the NLM T15 training grant. The Ohio Supercomputer Center provided computing support.

References

[1] S. Lin, Y. Lin, J. R. Nery, M. A. Urich, A. Breschi, C. A. Davis, A. Dobin, C. Zaleski, M. A. Beer, W. C. Chapman, T. R. Gingeras, J. R. Ecker, and M. P. Snyder, *Proc. Natl. Acad. Sci.*, **111**, 17224 (2014).

[2] P. P. C. Tan, L. French, and P. Pavlidis, *Front. Neurosci.*, **7**, 1 (2013).

[3] K. Taeho, G. S. Vidal, M. Djurisic, C. M. William, M. E. Birnbaum, C. K. Garcia, B. T.

Hyman, and C. J. Shatz, **341**, 1399 (2013).
[4] S. Matsuda, M. Katane, K. Maeda, Y. Kaneko, Y. Saitoh, T. Miyamoto, M. Sekine, and H. Homma, *Amino Acids* **47**, 975 (2015)
[5] J. W. Rowley, A. J. Oler, N. D. Tolley, B. N. Hunter, E. N. Low, D. a Nix, C. C. Yost, G. a Zimmerman, and A. S. Weyrich, *Blood* **118**, 101 (2011).
[6] J. a Miller, S. Horvath, and D. H. Geschwind, *Proc. Natl. Acad. Sci.* **107**, 12698 (2010).
[7] S. a. Fietz, R. Lachmann, H. Brandl, M. Kircher, N. Samusik, R. Schroder, N. Lakshmanaperumal, I. Henry, J. Vogt, a. Riehn, W. Distler, R. Nitsch, W. Enard, S. Paabo, and W. B. Huttner, *Proc. Natl. Acad. Sci.* **109**, 11836
[8] E. Z. Macosko, A. Basu, R. Satija, J. Nemesh, K. Shekhar, M. Goldman, I. Tirosh, A. R. Bialas, N. Kamitaki, E. M. Martersteck, J. J. Trombetta, D. A. Weitz, J. R. Sanes, A. K. Shalek, A. Regev, and S. A. McCarroll, *Cell* **161**,1202 (2015).
[9] A. P. Patel, I. Tirosh, J. J. Trombetta, A. K. Shalek, S. M. Gillespie, H. Wakimoto, D. P. Cahill, B. V Nahed, W. T. Curry, R. L. Martuza, D. N. Louis, O. Rozenblatt-Rosen, M. L. Suvà, A. Regev, and B. E. Bernstein, *Science* **344**, 1396 (2014).
[10] A. Zeisel, A. B. M. Manchado, S. Codeluppi, P. Lönnerberg, G. La Manno, A. Juréus, and S. Marques, *Science* **347**, 1138 (2015).
[11] B. B. Lake, R. Ai, G. E. Kaeser, N. S. Salathia, Y. C. Yung, R. Liu, A. Wildberg, D. Gao, H.-L. Fung, S. Chen, R. Vijayaraghavan, J. Wong, A. Chen, X. Sheng, F. Kaper, R. Shen, M. Ronaghi, J.-B. Fan, W. Wang, J. Chun, and K. Zhang, *Science* **352**, 1586 (2016).
[12] S. Islam, A. Zeisel, S. Joost, G. La Manno, P. Zajac, M. Kasper, P. Lönnerberg, and S. Linnarsson, *Nature methods* **11**, 163 (2014).
[13] D. Ramsköld, S. Luo, Y.-C. Wang, R. Li, Q. Deng, O. R. Faridani, G. a Daniels, I. Khrebtukova, J. F. Loring, L. C. Laurent, G. P. Schroth, and R. Sandberg, *Nat. Biotechnol.* **30**, 777 (2012).
[14] D. W. Huang, R. a Lempicki, and B. T. Sherman, *Nat. Protoc.* **4**, 44 (2009).
[15] D. W. Huang, B. T. Sherman, and R. A. Lempicki, *Nucleic Acids Res.* **37**, 1 (2009).
[16] P. D. McNicholas and T. B. Murphy, *Bioinformatics* **26**, 2705 (2010).
[17] C. Ding and H. Peng, *Journal of Bioinformatics and Computational Biology* **3**, 185 (2003).
[18] N. A. Twine, K. Janitz, M. R. Wilkins, and M. Janitz, *PLoS One* **6,** e16266.
[19] P. Qiu, E. F. Simonds, S. C. Bendall, K. D. Gibbs, R. V Bruggner, M. D. Linderman, K. Sachs, G. P. Nolan, and S. K. Plevritis, *Nat. Biotechnol.* **29**, 886–891 (2011).
[20] N. K. Wilson, D. G. Kent, F. Buettner, M. Shehata, I. C. Macaulay, F. J. Calero-Nieto, M. Sanchez Castillo, C. A. Oedekoven, E. Diamanti, R. Schulte, C. P. Ponting, T. Voet, C. Caldas, J. Stingl, A. R. Green, F. J. Theis, and B. Gottgens, *Cell Stem Cell* **16**, 712 (2015).
[21] D. A. Jaitin, E. Kenigsberg, H. Keren-Shaul, N. Elefant, F. Paul, I. Zaretsky, A. Mildner, N. Cohen, S. Jung, A. Tanay, and I. Amit, *Science* **343**, 776 (2014).
[22] F. Tang, C. Barbacioru, Y. Wang, E. Nordman, C. Lee, N. Xu, X. Wang, J. Bodeau, B. B. Tuch, A. Siddiqui, K. Lao, and M. A. Surani, *Nat. Methods* **6**, 377 (2009).

A SPATIOTEMPORAL MODEL TO SIMULATE CHEMOTHERAPY REGIMENS FOR HETEROGENEOUS BLADDER CANCER METASTASES TO THE LUNG

KIMBERLY R. KANIGEL WINNER[1,2], JAMES C. COSTELLO[1,2,3]
[1]*Computational Bioscience Program,*
[2]*Department of Pharmacology,*
[3]*Univeristy of Colorado Cancer Center*
University of Colorado Anschutz Medical Campus
12801 E. 17th Ave. MailStop 8303,
Aurora, CO 80045, USA
Email: kimberly.kanigelwinner@ucdenver.edu, james.costello@ucdenver.edu

Tumors are composed of heterogeneous populations of cells. Somatic genetic aberrations are one form of heterogeneity that allows clonal cells to adapt to chemotherapeutic stress, thus providing a path for resistance to arise. *In silico* modeling of tumors provides a platform for rapid, quantitative experiments to inexpensively study how compositional heterogeneity contributes to drug resistance. Accordingly, we have built a spatiotemporal model of a lung metastasis originating from a primary bladder tumor, incorporating *in vivo* drug concentrations of first-line chemotherapy, resistance data from bladder cancer cell lines, vascular density of lung metastases, and gains in resistance in cells that survive chemotherapy. In metastatic bladder cancer, a first-line drug regimen includes six cycles of gemcitabine plus cisplatin (GC) delivered simultaneously on day 1, and gemcitabine on day 8 in each 21-day cycle. The interaction between gemcitabine and cisplatin has been shown to be synergistic *in vitro*, and results in better outcomes in patients. Our model shows that during simulated treatment with this regimen, GC synergy does begin to kill cells that are more resistant to cisplatin, but repopulation by resistant cells occurs. Post-regimen populations are mixtures of the original, seeded resistant clones, and/or new clones that have gained resistance to cisplatin, gemcitabine, or both drugs. The emergence of a tumor with increased resistance is qualitatively consistent with the five-year survival of 6.8% for patients with metastatic transitional cell carcinoma of the urinary bladder treated with a GC regimen. The model can be further used to explore the parameter space for clinically relevant variables, including the timing of drug delivery to optimize cell death, and patient-specific data such as vascular density, rates of resistance gain, disease progression, and molecular profiles, and can be expanded for data on toxicity. The model is specific to bladder cancer, which has not previously been modeled in this context, but can be adapted to represent other cancers.

1. Introduction

1.1. *Tumor heterogeneity and drug resistance*

Intratumoral heterogeneity is increasingly recognized as a major contributor to cancer progression, metastatic potential, and drug resistance.[1,2] Metastatic tumors that arise from the primary site are generally established from single clones, but may also display initial genetic heterogeneity.[3,4,5] Sub-clonal cell phenotypes with varying metastatic potential and drug resistance have also been shown to develop in 90% of lung metastases within weeks of establishment in mice.[5] This heterogeneity can lead to differential drug response within or among metastases, with newly arising clones developing additional resistance.[5] After the death of sensitive cells and continuing replication of resistant survivors, the spatial dynamics of drug diffusion and accumulation during later drug delivery cycles may change.

A bottleneck in clinical research studies of drug resistance is the lack of tumor sample measurements over the course of treatment from the same patient that can be used to explore the relationship between tumor polyclonality and drug resistance.[6] By building explicit computational models with evolving dynamics, we can manipulate, visualize, and quantitatively analyze patterns of resistance that emerge in a growing tumor. Here, we have created a spatiotemporal model of bladder cancer metastasis to the lung that includes cycles of drug delivery, tumor vascularity, and clumped clonal populations with different drug sensitivities. We model how a heterogeneous tumor responds to the standard first-line regimen of gemcitabine plus cisplatin (GC). Results show that a 100 cell simulated tumor, composed of four clonal populations ranging from highly sensitive to highly resistant cells will not be completely killed by this regimen, and will grow while gaining cross-resistance to both gemcitabine and cisplatin. In this work we aim to model drug response in bladder cancer metastases and establish a baseline set of results that can be extended to model additional visceral sites, determine how varying tumor composition affects drug response, and determine how altering drug scheduling will affect drug response.

1.2. *Prior spatiotemporal models of drug delivery, tumor heterogeneity, and resistance*

Our model is a cellular Potts model, which represents cells and chemical fields on a spatial lattice, interacting and evolving over time. Spatiotemporal models have been used to represent disease development and drug delivery in a variety of cancers, and have generated observations that are not easy to measure in real biological systems.[7–9] They have incorporated parameters such as response to oxygen, information sources provided to the cell such as nutrients and toxicity, and distance from the information source.[8] Spatiotemporal cancer therapy models have used cell cycle, chemotherapy, and radiation data to predict changes in tumor size during treatment. Some have included more specialized events and data, such as bystander effects (in which tumor cells assist in killing damaged cells) resulting from radiotherapy[10] and patient data from CT scans in models of brain cancer.[11,12] These models have successfully produced qualitatively and semi-quantitatively comparable results to *in vitro* studies, mouse models, and patient outcomes, showing the promise of spatiotemporal modeling for *in silico* oncology. To our knowledge, there are no existing spatiotemporal models of drug delivery to lung metastases arising from bladder cancer.

Tumor heterogeneity and resistance have been explored with spatiotemporal methods, including two agent-based models (one incorporating game theory for trade-offs between proliferation and migration), field theory, a cellular automaton/cellular Potts model, and a pure cellular automaton. Interestingly, in three of these models,[13,14,15] slowing of the cell cycle was an important predictor of resistance, whether due to cells being driven into quiescence by drugs, by a shortage of oxygen and nutrients, or from initial heterogeneity between clonal populations in their endogenous cell cycles; cells with inherently slow growth were reservoirs for survival during therapies that depend on cell division.[14,15] This last model is the most similar to ours, and is part of a comparison of spatiotemporal implementations, showing that there are trade-offs between performance and resolution for different model types, but that similar types parameterized to the same system will produce cross-validating results. The simulated tumor in ref. 15 was composed of cell populations having heterogeneous cell cycles that changed in response to oxygen, chemotherapy, and radiation (in a 300×300 cellular Potts model). Our model similarly includes cell cycles and chemotherapy, but is different in that it creates a site-specific tumor environment incorporating vascular density specific to metastases to the lung, with *in vivo* concentration curves for drug delivery, and initial and gained resistance modeled using bladder cancer cell lines. In both models, the spatial arrangement of vessels creates a drug concentration unique to each cell in a simulation, allowing spatially driven phenomena to emerge.

1.3. *Bladder cancer drug regimen and cell response*

Annually, it is estimated that there will be nearly 77,000 new cases of bladder cancer with over 16,000 succumbing to the disease.[16] Overall survival has not improved since 1989.[16] The most aggressive form, muscle-invasive bladder cancer, occurs in 30% of patients.[17] Treatment is radical cystectomy, requiring removal of the bladder and sometimes surrounding tissues, followed by chemotherapy. The 5-year survival rate varies from 25-50%. Failure is likely due to occult metastases present before treatment, with the most common visceral metastatic sites in the liver and lungs.[17,18] Patients with inoperable locally advanced or metastatic cancer who undergo GC or methotrexate/vinblastine/doxorubicin/cisplatin (MVAC) regimens have a 5-year overall survival of 13%, but a progression-free survival of 9.8%.[19] Those with lung, liver, or bone[18] metastases have a 5-year overall survival rate of 6.8%.[19] Here, we model this last group of patients, with aggressive metastatic disease localized to the lung.

The standard regimen defined by the National Comprehensive Cancer Network (NCCN) for metastatic bladder cancer includes six 21-day cycles, with GC delivered simultaneously on day 1 (or cisplatin instead on day 2) and gemcitabine alone on day 8.[20] For patients with muscle-invasive or metastatic cancer, who cannot receive cisplatin, monotherapy regimens without cisplatin produce no long-term disease-free survival, with a median survival of six to nine months.[17] This was reflected in initial runs of the model, with rapid acquisition of resistance during cisplatin or gemcitabine monotherapy regimens. Reported efficacy of such regimens is derived from clinical trials. Computational models of drug delivery can additionally be used to generate hypotheses at a small scale where we can explore mechanisms of drug action and drug resistance, as well as adjust the regimen in a consequence-free environment where results for 18 weeks of time course data can be obtained in just hours.

Cisplatin and gemcitabine are genotoxic agents, damaging DNA and causing a cell to undergo apoptosis during cell division. Cisplatin incorporates into DNA as platinum-DNA adducts,[21] whereas gemcitabine is a nucleoside analog that interrupts DNA synthesis and triggers apoptosis.[22] The 50% inhibitory concentration (IC50) is a concentration of drug that inhibits a cellular process by 50%. IC50 for cytotoxicity and drug accumulation in cells are linearly correlated for both cisplatin and gemcitabine, especially at clinically relevant concentrations, which tend to be at the lower end of cytotoxicities measured *in vitro*.[23–25] There is also a linear relationship between tissue platinum concentration and tumor size reduction.[26] These relationships were used to parameterize cellular accumulation of the two drugs.

Synergy between gemcitabine and cisplatin occurs during pre-treatment with gemcitabine or co-treatment with GC in ovarian and neuroblastoma cells.[27,28] In these studies, one in four and one in five cell lines did not respond synergistically. Patients with non-small-cell lung cancer also responded better to a day 1 combination of gemcitabine and cisplatin than to day 1 cisplatin alone (30.4% response compared to 11%, $p<1e\text{-}4$), with improved median time to progression and improved overall survival.[29] Synergy in cisplatin during the GC regimen is an important dynamic that we include in the model.

2. Methods

2.1. *Summary of model design*

Our model represents a partially drug-resistant lung metastasis that arose from a primary bladder tumor, containing four clonal cell patches with different sensitivities to gemcitabine and cisplatin. The drugs are delivered through vasculature in the tumor at levels found in patient plasma based on the regimen dosages. Drugs diffuse from vessels with effective diffusion coefficients measured in tumor tissue, and accumulation is a cell-type-specific proportion of drug concentration at the cell site. Synergy between the drugs causes increased intracellular cisplatin accumulation. If cells attempting to replicate have accumulated enough drug to reach their IC50 or greater, they will either die with 50% probability or increase their resistance. Finally, when a cell divides, its accumulated drug is halved between the two child cells. Drug delivery frequency and dosage are from the basic GC drug regimen for metastatic bladder cancer (see Fig. 1 for model).

Tumor and vessel cell types are represented, along with cell division, cell death, and clearance of dead cells as a proxy for the immune system. Vascular density for lung metastases is equal to the ratio of microvessel density between primary and lung metastases in non-clear cell renal cell carcinoma.[30,31] Further biometric parameters, derivations, fits for drug concentrations in patients, and their sources can be found in Table 1. Model permutations include runs with and without synergy, variations on the drug regimen, and variations in rates of resistance gain in the cells.

The modeling platform is Compucell3D (CC3D),[32] an integrated programming and visualization environment for cellular Potts models. Cellular Potts models couple mobile, single-cell agents to a cellular automaton process at the cells' surfaces. Cell agents live on their own 2-D or 3-D lattice, and chemical fields can be layered on in other lattices. Partial differential equations for drug diffusion are solved using the Forward Euler method. For more explicit descriptions of the cellular Potts model for modeling drug delivery in tumors, please see Kanigel Winner, et al.,[33] and Extended Methods are available at https://synapse.org/MetHet. In short, pre-defined biological rules comprise an energy function that drives the behavior of the cellular automaton process at the cell surface during each Monte Carlo time step (MCS). Meeting the rules (by convention) lowers this energy or keeps it the same, allowing biologically reasonable cellular events contributing to growth, division, and death (though stochasticity can be added). Cell death, cell type switches due to drug accumulation, and drug delivery calculated from continuous functions (fits to patient plasma drug concentrations) are expansions of the basic CC3D model coded in a Python wrapper. These processes are non-stochastic. More modeling methods, details of parameter acquisition, and source code that is plug-and-play in CC3D can be found at https://synapse.org/MetHet.

2.2. *Specifics of biological parameters and model dynamics*

- IC50 data for gemcitabine and cisplatin sensitivity in 18 bladder cancer cell lines were acquired from the Genomics of Drug Sensitivity in Cancer (GDSC) database.[34]
- Cell growth and division occurred in all cancer cells. Replication rate was approximated from the averages of 14 cancer cell lines varying in metastatic capacity (31 to 33 hrs.).[22,36,37]
- Cisplatin and gemcitabine in normal cells (lung and phagocytic cells) were given accumulation rates for the bladder cancer cell line (SW780) closest to the middle of the range for both drugs.

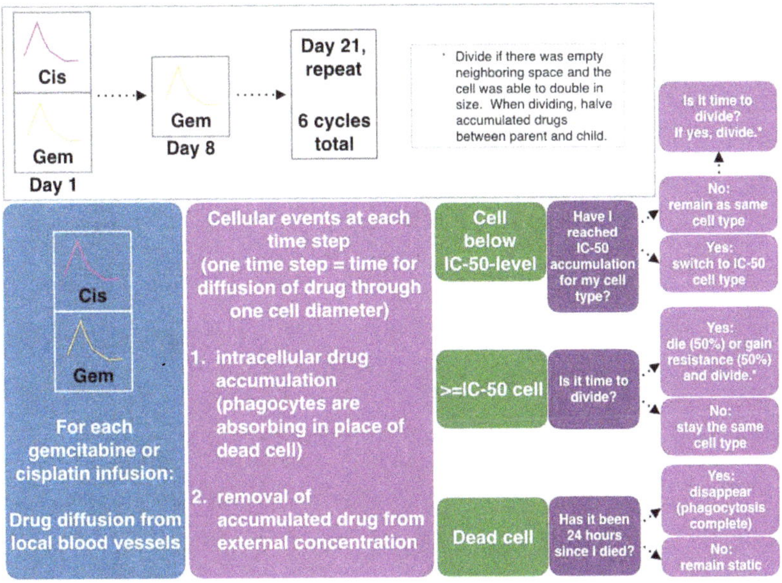

Figure 1. Flow chart of events in the model at each time step, reflecting body- and cellular-scale processes

- Acquired resistance was modeled as an increase in the IC50 of any cell that survived an IC50 accumulation of gemcitabine or cisplatin at division time, increasing the chances of being below IC50 and another gain in resistance at the next division time. The quantity to be added to the IC50 for each gain in resistance (Table 1) was derived from bladder cancer cell lines, passaged to increase resistance, as the increase per division required to acquire maximum resistance over one year ("quick") or two years ("slow") for cells with a 30-hour cell cycle.[35]
- Cell accumulation rate and peak of gemcitabine is linearly correlated with concentration *in vitro* and *in vivo*.[38] In bladder cancer cells, cytotoxicity is linearly correlated with gemcitabine concentration,[25] and accumulation is correlated with IC50.[36] Cisplatin DNA lesion counts are linearly correlated with concentration.[27] We therefore fit cellular accumulation rates for both gemcitabine and cisplatin linearly to the IC50 of each cell type, with some modifications.[36,39]
- Cells at IC50 for both gemcitabine and cisplatin at division underwent two chances at death.
- Gemcitabine and cisplatin were modeled with the same effective diffusion coefficient as sodium fluorescein.[40] For details on this choice, please see ref. 33. Both molecules diffused at the same rate in all cell types except for blood vessel, which either took away molecules, ostensibly into flowing blood, or delivered them from the vessel surface.

Table 1. Model parameters and fits to data

Parameter	Value	Units	Source
Cell diameter (BC* T24 line, aggressive/invasive)	30	μm	[42]
Eff. diffusion coefficient sodium fluorescein	6.40E-06	cm^2/s	[40]
Division time (mean, S.D.)	30, 1	h	[22,36,37]
Time from death to complete phagocytosis	24	h	[43]
Fraction cross-sectional microvessel area in metastasis from urinary system cancer to lung	0.146		[30]
Pixel dimension	1	cell	
Cisplatin resistance gain per survived division	0.125 – 0.25	+ IC50	[35]
Gemcitabine resistance gain per survived division	0.05 – 0.1	+ IC50	[35]
IC50 cis. accumulation for initial cell lines. Seed gem. & cis. sensitive, Seed res. gem./sens. cis., Seed res. cis./sens. gem., Seed gem. & cis. resistant	0.8106177157, 3.774888444, 6.586828431, 5.923917064	μM per cell	calculated using fit from [44]
IC50 gem. accumulation for initial cell lines. Seed gem. & cis. sensitive, Seed res. gem./sens. cis., Seed res. cis./sens. gem., Seed gem. & cis. resistant	0.000017923, 270.913928515, 0.145644144, 46.134163935	μM per cell	calculated using fit from [36]
Accumulation rates of cis. in initial cell lines. Seed gem. & cis. sensitive, Seed res. gem./sens. cis., Seed res. cis./sens. gem., Seed gem. & cis. resistant	7.98701E-05, 6.82909E-05, 7.42347E-06, 5.46716E-05	* cis. (μM) at cell site per MCS	fit from [44]
Accumulation rates of gem. in initial cell lines Seed gem. & cis. sensitive, Seed res. gem./sens. cis., Seed res. cis./sens. gem., Seed gem. & cis. resistant	4.41575E-04, 2.68443E-04, 4.41518E-04, 4.22858E-04	* gem. (μM) at cell site per MCS	fit from [36]
Fit for cisplatin plasma concentrations during 3h infusion (top) and decay (bottom)	$= 0.11*hrs^3 - 0.83*hrs^2 + 2.2*hrs - 2.6E\text{-}16$ $= 57.4124 * e^{(-1.0927 * hrs)}$	μM	[46]
Fit for gemcitabine plasma concentrations during 30m infusion (top) and decay (bottom)	$= 6.8*(min/15 - 1) + 7.3$ $= 101.3452 * e^{(-0.0676 * min)}$	μM	[45]
Synergy multiplier for cisplatin accumulation	2.5		[27,28]
Total Monte Carlo (simulation) Steps (126 days)	11,916,800		
Time in one Monte Carlo Step (MCS)	0.914	s	

* Bladder Cancer

3. Results

3.1. *No standard or alternate regimen prevents regrowth of a drug resistant tumor*

In preliminary simulations containing only GC dual-sensitive cells, cells declined over time and the population was killed late on day 48, five days after the third round of GC. However, for an initial tumor with three additional cell types that had increased resistance to gemcitabine, cisplatin, or both, neither the standard GC regimen (Figs. 2,3), nor an unrealistically high rate of delivery of gemcitabine could kill all cells.

The initial 2-D tumor of 100 cells consistently quadrupled to 400 cells in 14 to 15 days. At simulation end, 0 to 18 days after the last round of drug (depending on the simulation) the domain was completely filled with drug-resistant tumor cells, primarily cisplatin-resistant and GC dual-resistant cells, as well as a sub-population of the most GC dual-resistant seed population. Within six hours of regimen start, the two most sensitive cell types reached the IC50 for gemcitabine accumulation. While some of these sensitive cells died, some went on to propagate as sub-clones with gained resistance.

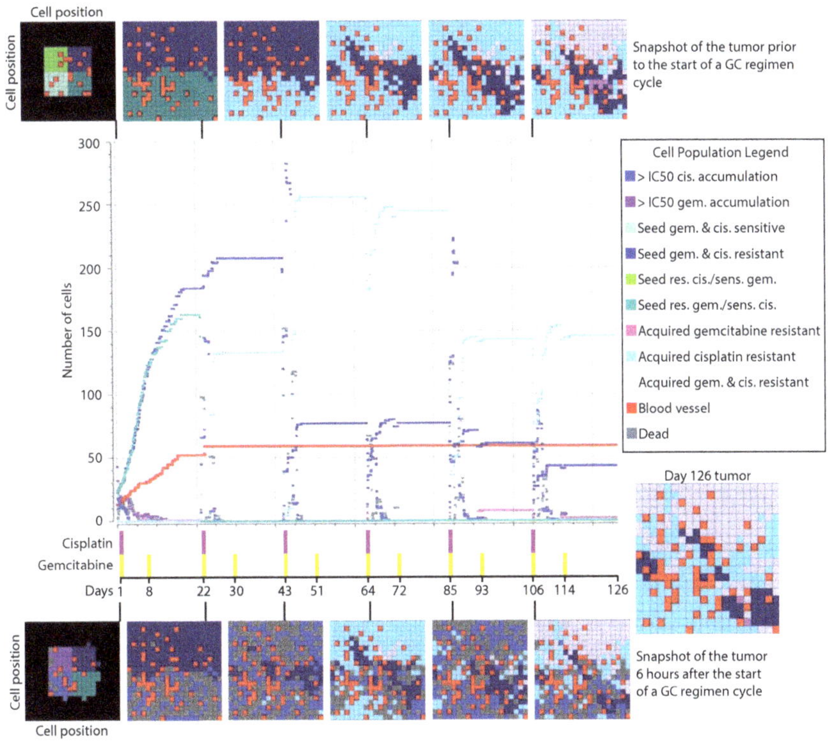

Figure 2. A simulation with random uniform "slow" to "quick" acquired resistance for all cells, and drug synergy in all cells. Pulses of gemcitabine and cisplatin or gemcitabine alone for the first-line chemotherapy regimen are displayed and matched to the simulation. Cells in the simulation were seeded in 100-cell tumors shown in the top-leftmost simulated tumor diagram. The row of simulated tumors on the top represent the state of the tumor before the start of a chemotherapy cycle; the row of simulated tumors on the bottom represent the state of the tumor 7 hours after a GC cycle. Resistant seed cells, cells with dual resistance, and cells with cisplatin resistance composed the final population as shown as the final simulated tumor after 126 days of treatment.

3.2. *Effects of acquired resistance*

3.2.1. *Ability of cells to gain resistance increases likelihood of dual resistance*

When acquired resistance was allowed to arise in the cell populations, cells with acquired resistance comprised the majority of the final tumor (Figs. 2, 3). GC dual-resistant sub-clones arose at day 43, after the third cycle of GC, suggesting that increased dosage or delivery rate prior to this time point may help to keep cross-resistant strains from arising. Interestingly, the fastest rate of acquired resistance for both gemcitabine and cisplatin drove cells with acquired resistance to cisplatin to dominate the population.

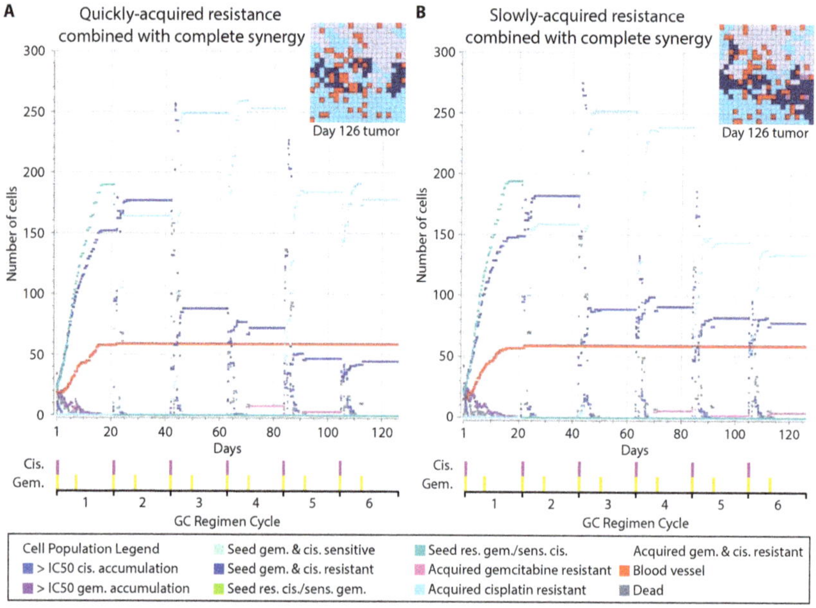

Figure 3. (A) Quickly-acquired resistance and (B) slowly-acquired resistance resulted in tumors composed primarily of cells with newly acquired resistance, with a smaller population of highly GC dual-resistant seed cells. Quickly-acquired resistance drove the tumor toward greater cisplatin resistance.

3.2.2. *Simulated tumors show complete resilience to even intense treatment*

In simulations with an added pulse of gemcitabine at day 18 during each cycle (we mirrored the timing of the 28-day regimen, which has an additional gemcitabine infusion on day 18), we found an earlier rise of the GC dual-resistant phenotype, and more gemcitabine-resistant cells. We also applied single-drug regimens with cisplatin or gemcitabine alone at standard frequencies. Cells with resistance to the treatment drug were the majority of the final population.

To try treatment prior to all cells entering a new cell cycle (30 hrs) while using a potentially tolerable regimen, we shifted the three pulses of gemcitabine to the first three days of each 21-day cycle, at every 24 hours, in addition to the standard cisplatin every 21 days. This caused the end state to be dominated by GC dual-resistant cells. Finally, we pulsed gemcitabine every 24 hours for 126 days, with cisplatin every 21 days. The tumor was not killed, and the simulated tumor area was fully repopulated, primarily with cisplatin-resistant cells after 126 pulses of gemcitabine reduced the gemcitabine-resistant populations.

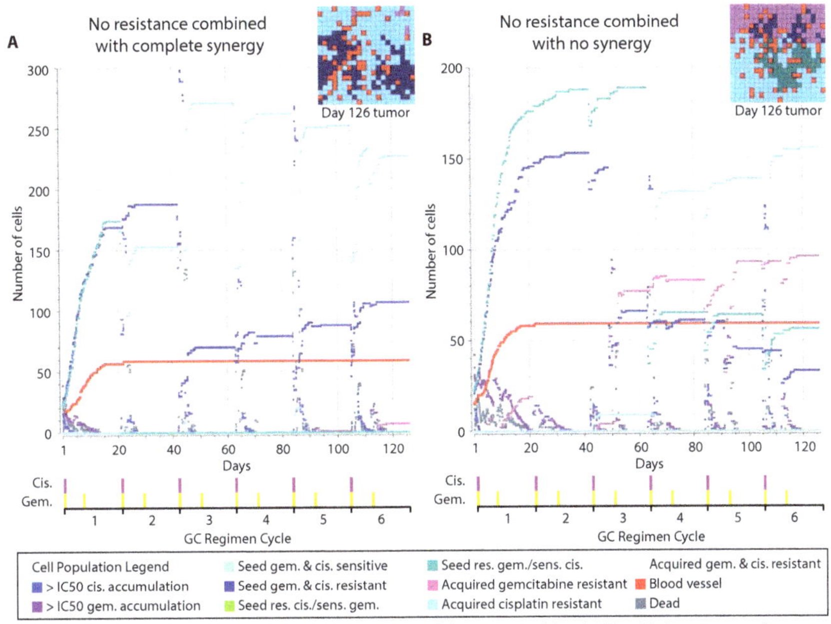

Figure 4. In simulations without acquired resistance, models were considered (A) with drug synergy between gemcitabine and cisplatin and (B) without synergy. The final tumor was composed of cells that survived division after reaching either cisplatin IC50 accumulation, or gemcitabine IC50 accumulation. Cells that survived both gemcitabine and cisplatin IC50 levels did not arise. When there was no synergy in cisplatin (2.5× normal accumulation rate, B), an extra cell type (teal-colored) derived from the seed populations remained.

3.3. *In the absence of acquired resistance, diffusion of drug via cell division allows survival*

In simulations where cells did not acquire resistance, populations primarily composed of cells that randomly survived an IC50-cisplatin division (Fig. 4A) repopulated the simulation space. A subpopulation of initial highly GC dual-resistant seed cells also survived. Because of the division of drug equally between two progeny cells, acquired resistance was not required for tumor repopulation, suggesting that cells reaching IC50 accumulation may survive *in vivo* without newly acquired resistance. In simulations with synergy and resistance (Figs. 2, 3), one clone in the original tumor died during the second round of GC at day 21 (teal-colored; $IC50_{cisplatin}$ = 14.0μM in range 2.6μM to 225.2μM). When synergy and the ability to gain resistance were absent, this cell type comprised a substantial portion of the final tumor, the most heterogeneous final tumor in our models (Fig. 4B). Hence, for the most resistant seed cells, and in less resistant seed cells in which synergy may not be active, no acquisition of extra resistance was required for repopulation.

4. Discussion

In this work, we were able to capture population-level responses to chemotherapy stress in a model of lung metastasis arising from the bladder. Unless the initial tumor was comprised of highly sensitive cells, the *in vivo* concentration and timing of the standard first-line regimen did not kill the metastasis. Cells were then able to proliferate and fill the simulation space after completion of treatment. A striking result was that in tumors without any ability to acquire resistance, some cells survived the IC50 threshold and were able to repopulate the space. When tumors were allowed to acquire resistance, there was consistent emergence of cells that had coordinately increased resistance to both gemcitabine and cisplatin around 43 days. This occurred after the third cycle of GC, suggesting that early aggressiveness in treatment may be important in avoiding cross-resistant sub-clones. In terms of drug-directed cell selection, when cells were given the ability to acquire resistance, even at slower rates described *in vitro*, final tumors were composed of a majority of cells with acquired resistance. Because metastases starting from single clonal populations in the lung have been shown to develop sub-clones within weeks of establishment,[5] and because cell lines and living tumors are known to gain resistance mutations over time, metastases with large proportions of cells with acquired resistance is a likely scenario in a patient, and the model likely reflects selection *in vivo*.

Qualitative comparisons can be made between prior data and model outcomes. Overall, the acquired resistance model produced rounds of cell death under drug concentrations in patients, showing that the parameters are biologically reasonable. The results are consistent with survival data for patients with inoperable locally advanced or metastatic bladder cancer undergoing a GC or MVAC regimen; those who had lung, liver, or bone metastases had a 5-year overall survival rate of 6.8%.[19] The likelihood of a patient presenting with a completely drug-sensitive metastatic population is low, creating low likelihood of complete cell killing in the tumor. Similarly, in the model, we saw only the most sensitive populations being eradicated by the standard regimen. A patient's metastatic population might have been completely sensitive if metastasis was recently established from a sensitive primary cell and lacked the time to gain genetic heterogeneity. Less likely still, several weeks or more after establishment, the metastases may have either not gained new genetic heterogeneity, or simply not acquired resistance through genetic aberrations. Finally, cells in the model had IC50s derived from cell lines, and some cells died at *in vivo* drug concentrations, suggesting that cell line data reasonably reflects the range of resistances found in patients' tumor cells. While these comparisons to patients and cells are speculative, they are valuable observations for generating hypotheses and represent opportunities for empirical validation as we develop the model further.

When acquisition of resistance was removed, some cells that had initial resistance survived and propagated. This "resistance" occurred because accumulated drug was divided in half between offspring, giving both sensitive and resistant primary sub-clones more time to grow and replicate before reaching IC50, with proliferation outpacing the delivery of drug. This result, in which cells randomly evade death without incorporating new resistance mechanisms, emphasizes the importance of considering growth rate in an aggressive metastatic population.

To estimate the number of doses required to actually kill a metastasis, we simulated delivery of gemcitabine every 24 hours over 126 days, along with synergistic cisplatin every 21 days. Even this unrealistic regimen did not kill the tumor, and drove it to gain cisplatin resistance. Increasing gemcitabine dosage, in combination with increasing the frequency of cisplatin at lower doses should be explored in the future. Additionally, drug regimens that incorporate other drugs besides cisplatin and gemcitabine will be explored in future iterations of the model.

There are caveats to this approach that we must consider. Our model is small (20×20×1 cells) for relatively fast computation so that many scenarios could be explored. Although this size still allowed differential effects to emerge between different drug scenarios, and computational costs scaled proportionally to the number of cells during growth from 100 to 400 cells, larger grids will be part of future work, hopefully approaching the clinical detection limit for lung metastases. The system modeled is specific to bladder cancer; however, lung is a common metastatic site for many other cancers. This and the available data on vascularity at urogenital metastatic sites helped justify the choice of the system modeled. Additionally, the model is simple and general, in part because a GC regimen is used in a variety of cancers, and can be relatively easily adapted to other metastatic or primary sites by replacing parameters in the code. The primary bottleneck to adaptation to other cancers will be the availability of empirical data to derive model parameters.

The model may be allowing consistent tumor survival despite an aggressive drug regimen due to a cell cycle time of 30h +/-1h (S.D.); slower- (or even faster-) cycling cells may create different dynamics. Drug is not delivered from vessels outside of the tumor, inherent cell death rates are not included, and the immune system is not directly considered. Most importantly, although the model can be manipulated unrealistically, useful hypothetical regimens must include practical considerations for regimens given to patients. If a new regimen kills more cells, perhaps the immune system will have a greater chance to reduce a smaller residual population. A simple increase in cell kill under an organizational and dosing scheme reasonable for patients is therefore a goal for this modeling process and the subject of future studies.

Finally, our model results recapitulate prior work by Powathil *et al.*[15] regarding the importance of accounting for cell cycle in drug delivery. Also our results concur with aspects of Waclaw *et al.*,[41] showing that after cell kill opens up space in the tumor, it takes only one or two cells to repopulate the vacant area with a new more resistant sub-clone. Such cell behavior is extremely difficult to track through time in a patient, and even in experimental models such as mice. Therefore, the importance of spatiotemporal models incorporating realistic parameters, with behavior that can be tracked over time to clinically relevant outcomes, cannot be underestimated.

Acknowledgments

We would like to thank members of the Costello lab, in particular Andrew Goodspeed, and Anis Karimpour-Fard for constructive comments. This work is supported by the Boettcher Foundation (J.C.C.) and NIH grant 2T15LM009451 (K.R.K.W.).

References

1. Saunders, N. A. *et al. EMBO Mol. Med.* **4,** 675–84 (2012).
2. Tabassum, D. P. & Polyak, K. *Nat. Publ. Gr.* **15,** 1–11 (2015).
3. Yamamoto, N. *et al. Cancer Res.* **63,** 7785–90 (2003).
4. Talmadge, J. E. & Zbar, B. *J Natl Cancer Inst* **78,** 315–320 (1987).
5. Poste, G. *et al. Proc. Natl. Acad. Sci. U. S. A.* **79,** 6574–8 (1982).
6. Burrell, R. A. *et al. Mol. Oncol.* **8,** 1095–111 (2014).
7. Andasari, V. *et al. J. Math. Biol.* 1-31–31 (2010).
8. Mansury, Y. *et al. J. Theor. Biol.* **219,** 343–370 (2002).
9. Martins, M. L. *et al. Phys. Life Rev.* **4,** 128–156 (2007).
10. Powathil, G. G. *et al. J. Theor. Biol.* **401,** 1–14 (2016).
11. Tracqui, P. *et al. Cell Prolif.* **28,** 17–31 (1995).
12. Stamatakos, G. & Antipas, V. *IEEE Trans.* (2006).
13. Mansury, Y. *et al. J. Theor. Biol.* **238,** 146–156 (2006).
14. De La Cruz, R. *et al.* arXiv:1607.01449v1 (2016).
15. Powathil, G. G. *et al. Semin. Cancer Biol.* **30,** 13–20 (2015).
16. Siegel, R. L. *et al. CA. Cancer J. Clin.* **66,** 7–30 (2016).
17. Piergentili, R. *et al. Curr. Med. Chem.* **21,** 2219 (2014).
18. Sternberg, C. N. *Ann. Oncol.* **17,** 23–30 (2006).
19. von der Maase, H. *et al. J. Clin. Oncol.* **23,** 4602–4608 (2005).
20. National Comprehensive Cancer Network. Bladder Cancer v.1.2016, accessed 5/3/2016
21. Johnstone, T. C. *et al. Philos. Trans. A. Math. Phys. Eng. Sci.* **373,** (2015).
22. van Moorsel, C. J. *et al. Br. J. Cancer* **80,** 981–90 (1999).
23. Mistry, P. *et al. Cancer Res.* **52,** 6188–6193 (1992).
24. Henderson, P. T. *et al. Int. J. cancer* **129,** 1425–34 (2011).
25. Torres, M. P. *et al. PLoS One* **8,** e80580 (2013).
26. Kim, E. S. *et al. J. Clin. Oncol.* **30,** 3345–52 (2012).
27. Moufarij, M. *et al. Mol. Pharmacol.* **63,** 862–9 (2003).
28. Besançon, O. G. *et al. Cancer Lett.* **319,** 23–30 (2012).
29. Sandler, A. B. *et al. J. Clin. Oncol.* **18,** 122–130 (2000).
30. Fukata, S. *et al. Cancer* **103,** 931–942 (2005).
31. Papadopoulos I. *et al. J. Clin. Pathol.* **57,** 250 (2004).
32. Swat, M. H. *et al. Methods Cell Biol.* **110,** 325–66 (2012).
33. Kanigel Winner, K. *et al. Cancer Res.* 0008-5472.CAN-15-1620- (2015).
34. Yang, W. *et al. Nucleic Acids Res.* **41,** D955-61 (2013).
35. Vallo, S. *et al. Transl. Oncol.* **8,** 210–216 (2015).
36. Damaraju, S. *et al. Biochem. Pharmacol.* **79,** 21–9 (2010).
37. Ning S. *et al. Int. J. Oncol.* (2004).
38. Grunewald, R. *et al. Cancer Chemother. and Pharmacol.* **27,** 258 (1990).
39. Köberle, B. & Piee-Staffa, A. *Bladder Cancer - From Basic Science to Robotic Surgery.* **Chapter 13**, 265 (2012).
40. Nugent, L. J. & Jain, R. K. *Cancer Res.* **44,** 238–244 (1984).
41. Waclaw, B. *et al. Nature* **525,** 261–264 (2015).
42. Gilloteaux, J. *et al. Anat. Rec. A. Discov. Mol. Cell. Evol. Biol.* **288,** 58–83 (2006).
43. Wagner, B. J. *et al. J. Cell. Sci.* **124,**1644 (2011).
44. Koberle, B. *et al. Biochem. Pharmacol.* **52,** 1729–1734 (1996).
45. Fan, Y. *et al. Acta Pharmacol. Sin.* **31,** 746–52 (2010).
46. De Jongh, F. E. *et al. J. Clin. Oncol.* **19,** 3733–3739 (2001).

SCALABLE VISUALIZATION FOR HIGH-DIMENSIONAL SINGLE-CELL DATA

JUHO KIM

Department of Electrical and Computer Engineering, University of Illinois at Urbana-Champaign
Urbana, Illinois, 61801, USA
Email: juhokim2@illinois.edu

NATE RUSSELL

Institute of Genomic Biology, University of Illinois at Urbana-Champaign
Urbana, Illinois, 61801, USA
Email: ntrusse2@illinois.edu

JIAN PENG

Department of Computer Science, University of Illinois at Urbana-Champaign
Urbana, Illinois, 61801, USA
Email: jianpeng@illinois.edu

Single-cell analysis can uncover the mysteries in the state of individual cells and enable us to construct new models about the analysis of heterogeneous tissues. State-of-the-art technologies for single-cell analysis have been developed to measure the properties of single-cells and detect hidden information. They are able to provide the measurements of dozens of features simultaneously in each cell. However, due to the high-dimensionality, heterogeneous complexity and sheer enormity of single-cell data, its interpretation is challenging. Thus, new methods to overcome high-dimensionality are necessary. Here, we present a computational tool that allows efficient visualization of high-dimensional single-cell data onto a low-dimensional (2D or 3D) space while preserving the similarity structure between single-cells. We first construct a network that can represent the similarity structure between the high-dimensional representations of single-cells, and then, embed this network into a low-dimensional space through an efficient online optimization method based on the idea of negative sampling. Using this approach, we can preserve the high-dimensional structure of single-cell data in an embedded low-dimensional space that facilitates visual analyses of the data.

1. Introduction

Many traditional biological experiments have been conducted on bulk-cell populations[1] with an assumption that cells in the same group share homogeneous properties. However, some evidence[1-3] shows that heterogeneity can exist even within a small group of cells. The assumption based on homogeneity of each cell group can mislead averages and does not properly explain small but critical changes in individual cells. Each cell can have different biological properties such as cell sizes, gene expression levels, RNA transcripts, and bio marker expressions. These variations can be very important to answer previously unsolved questions in stem cell research, cancer biology, and immunology. Single-cell data analysis has contributed to understand the various and important behaviors of individual cells[1-15].

The recent development of single-cell technologies has also improved the analysis to be more reliable and reasonable. For example, mass cytometry[4,16] can measure up to 60 parameters at the same time for tens of thousands of individual cells. In addition, single-cell RNA sequencing techniques[17,18] also have been widely used, which deal with hundreds of or thousands of parameters per cell.

Even though the advanced single-cell technologies can provide quality data, such data sets are still difficult to analyze. Traditionally, single-cell data are analyzed in a biaxial scatter plot for two variables at once[19]. However, this method requires the order of dimension squared to represent all pairwise relationships between variables, which is computationally expensive. In addition, scatter plots cannot capture multivariate relationships between more than two variables. Thus, new computational methods have been developed for analyzing single-cell data. For instance, SPADE[6] tries to find hierarchies of high-dimensional single-cell data showing cellular heterogeneity by clustering of down-sampled cytometry data, constructing minimum spanning trees, and up-sampling. However, this method considers not each cell itself but cell groups and their behaviors on average. X-shift[12] is recently developed to discover cell subsets and visualize them based on a weighted k-nearest neighbor density estimation.

Another approach to deal with the high-dimensionality of single-cell data is to use dimensionality reduction techniques. Some researchers applied principle component analysis (PCA)[20] to find low-dimensional projections of single-cell data[21,22]. Although PCA is possibly the most popular method of dimensionality reduction, it is a linear projection method. Thus, it cannot capture nonlinear structures in single-cell data. In order to address this issue, advanced methods based on nonlinear dimensionality reduction have been developed. Both viSNE[8] and ACCENSE[10] are based on an algorithm called t-Distributed Stochastic Neighbor Embedding (t-SNE)[23]. viSNE applies t-SNE to mass cytometry data and reveals biologically meaningful relationships from bone marrow and leukemia data. ACCENSE combines the results of t-SNE with kernel-based density estimation and finds subpopulations of given single-cell data sets. However, the runtime complexity of t-SNE is $O(n^2)$, and that of its accelerated version, Barnes-Hut-SNE[24] is $O(n \log n)$ where n is the number of cells. Thus, both methods require excessive computational time for large-scale single-cell data sets with hundreds of thousands or millions of cells.

In this paper, we propose a scalable embedding-based visualization method for large-scale and high-dimensional single-cell data based on a new graph embedding algorithm, LargeVis[25]. The proposed method constructs a k-nearest neighbor (k-NN) network to find the structure of similarities between high-dimensional single-cell data. This process is accelerated by an approximate k-NN construction method based on random projection trees[26] and neighbor exploring[30]. This approach optimizes a probabilistic utility function to embed the high-dimensional single-cell data into a low-dimensional space (2D or 3D). For efficient training, the utility function is approximated using negative sampling[28] that was introduced in word2vec[28]. The runtime complexity of our method is linear with regard to the number of cells, which is faster than previous single-cell visualization tools such as viSNE[8] and ACCENSE[10].

2. Methods

We propose a new approach for visualizing high-dimensional single-cell data via efficient dimensionality reduction based on LargeVis[25]. The algorithm consists of two steps: constructing an approximate k-NN network to find the similarity structure between high-dimensional single-cell data and embedding the constructed network into a 2D or 3D space while preserving the high-dimensional structure in an easily visualized low-dimensional space. Pairwise similarity between single-cell data points is determined by the distance between them in their marker expression representation space. The core assumption is that numerical proximity in the marker space is proportional to cell similarity.

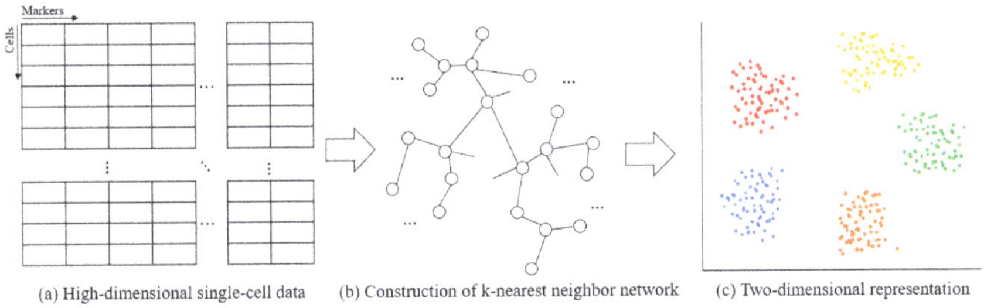

Figure 1. Outline of high-dimensional single-cell data visualization: constructing a k-nearest neighbor network and embedding the network into a 2D space.

2.1. *Notation*

We denote a set of high-dimensional single-cell data as $\mathcal{X} = \{x_i | x_i \in \mathbb{R}^p,\ i \in [n],\ p > 3\}$, where p is the dimension of measurements and n is the number of cells in the data; and the embedded representations of cells are denoted as $\mathcal{Y} = \{y_i | y_i \in \mathbb{R}^2\ or\ \mathbb{R}^3,\ i \in [n]\}$ in a low-dimensional space.

2.2. *Construction of a k-nearest neighbor network*

Constructing a k-nearest neighbor (k-NN) network is a very crucial step in many applications of machine learning such as a distance-based similarity search, manifold learning, and topological data analysis. Finding the exact k-NN network for large-scale single-cell data is time-consuming because it requires $O(n^2)$ time to compute all pairwise distances between all cells in the data set. Approximate methods for constructing a k-NN network have been developed, all of which have a tradeoff between speed and accuracy. Common approaches include locality sensitivity hashing[29], neighbor exploring methods[27], and partitioning methods based on random projection trees[26], k-d trees[31] and k-means trees[31].

As suggested by LargeVis[25], we develop a fast method to construct an approximate k-NN network. We first partition the whole high-dimensional space into two subspaces and generate a tree having only a root node. A set of single-cells in each partitioned subspace belongs to child nodes of the root node. Then, for the two subspaces that each set of single-cells in the child nodes belongs to, we partition each subspace into two sub-subspaces and generate two child nodes for each child node of the root node. The single-cells in each sub-subspace are assigned to each generated child nodes' child node. By continuing to partition the space iteratively, we can build a tree that assigns a group of single-cells belonging to partitioned small subspaces to its nodes. When the number of cells in a certain node is equal to or less than a predefined threshold, we stop the iterations. The single-cells in each leaf node are considered to be a candidate of approximate nearest neighbors. The generated tree is called a random projection tree.

By generating many random projection trees, we can increase the accuracy of the construction of a k-NN network, but it is time consuming. Instead of building many random projection trees, we use a neighbor search method in order to enhance both the accuracy and the efficiency. Specifically, we search the neighbor j of the neighbor of each node i assuming that its neighbor's neighbor is likely to be its neighbor also[30]. If the number of neighbors of node i is less than k, the method pushes some searched neighbor's neighbor j into the set of nearest neighbors of the node i. By iteratively doing this procedure, we can improve the accuracy of the construction and finally find our approximate k-NN network. Regarding the accuracy of the k-NN network construction, one can refer to the paper of largeVis[25], which dealt with several benchmark tests for the accuracy. The k-NN network construction process has linear time complexity because we build only a few random projection trees and because searching a certain node's neighbor's neighbor requires just a few iterations.

We then calculate the weight of each pairwise edge that represents the similarity structure of the constructed network using the Gaussian kernel, which was also used by t-SNE[23,24]. The conditional probability that the edge from data x_i to x_j is observed is first computed by:

$$p_{j|i} = \frac{exp\left(-\|x_i - x_j\|^2/2\sigma_i^2\right)}{\sum_{(i,k)\in E} exp\left(-\|x_i - x_k\|^2/2\sigma_i^2\right)} \qquad (1)$$
$$p_{i|i} = 0$$

where the parameter σ_i is determined by setting the perplexity, and E is the set of all edges in the k-NN network. To make the network symmetric, the weights are defined as:

$$w_{ij} = \frac{p_{j|i} + p_{i|j}}{2n} \tag{2}$$

where n is the number of input single-cell data. Since the number kn is much smaller than the number of all pairs (n^2), the constructed k-NN network is sparse. The sparsity of the k-NN network can make us compute w_{ij} within linear time complexity. Through the steps, our method can find the similarity structure of high-dimensional single-cell data within linear time complexity $O(kn)$.

2.3. *Network embedding into a low-dimensional space*

Embedding the constructed k-NN network is intended to preserve local and global network topology such that neighbors in the network are near each other in a low-dimensional space. First, for two nodes v_i and v_j, LargeVis[25] defines the probability that they come from the same neighborhood, i.e. the probability that we can observe the edge between two nodes in the k-NN network, as:

$$p(e_{ij} = 1|y_i, y_j) = f(dist(y_i, y_j)) \tag{3}$$

where f is a transformation function to map the distance between y_i and y_j into a probability value.

The function f satisfies the idea that when the distance between two low-dimensional points is small, the probability observing the connection between them is high. After considering some candidates like a multinomial logistic model and a sigmoid function, we chose $f(x) = \frac{1}{1+\alpha x^2}$ ($\alpha > 0$) due to its computational simplicity. The selected function f does not require any normalization across the data set, thus only $O(n)$ runtime is needed for objective evaluation and gradient calculation in the embedding optimization (see below). In addition, we can control the thickness of the tail of the function f by controlling α. When α becomes smaller, its tail gets thicker. When $\alpha = 1$, f is Student's t-distribution with degree of freedom one except a scaling factor $\frac{1}{\pi}$. On the other hand, t-SNE[23] uses the Gaussian kernel p_{ij} of (1) and a t-distributed kernel $q_{ij} = \frac{(1+\|y_i-y_j\|^2)^{-1}}{\sum_{k\neq l}(1+\|y_k-y_l\|^2)^{-1}}$ to measure its high-dimensional and low-dimensional similarity, respectively. By minimizing the Kullback-Leibler divergence between two similarities through gradient descent, t-SNE finds its low-dimensional embedding. The gradient of its cost function contains the normalization term of q_{ij}. Computing the term requires $O(n^2)$. To avoid inefficiency, accelerated t-SNE[24] uses Barnes-Hut algorithm[32] and reduces its time complexity from $O(n^2)$ to $O(n \log n)$. Two versions of t-SNE are more expensive than our approach.

Like LargeVis[25], we chose Euclidean distance as a distance metric in a low-dimensional space because computing Euclidean distance between embedded single-cell data is simple. In addition, we can map each calculated distance to one of the various probability function values since the range of Euclidean distance is $[0, \infty)$.

To embed the high-dimensional data, we define a log likelihood utility function (4) that considers both the probabilities of all edge connections E of the constructed k-NN network and the probabilities of all negative edges E^C. Negative edges mean that pairwise single-cell connections that are not observed in the k-NN network. This idea originally comes from noise-contrastive estimation (NCE)[33], which considers estimation that differentiates its observed data from noise using nonlinear logistic regression. Using the idea of NCE, we want to discriminate the same type of cells from different types of cells. Specifically, by maximizing the first term of (4), we can make similar single-cells become closer to each other in a low-dimensional space, and by maximizing the second part of (4), we can make dissimilar single-cells move away from each other.

$$J = \sum_{(i,j)\in E} w_{ij} \log p(e_{ij} = 1|y_i, y_j) + \sum_{(i,j)\in E^C} \gamma \log(1 - p(e_{ij} = 1|y_i, y_j)) \quad (4)$$

However, considering all negative edges is computationally expensive or even intractable when input data are very large. Thus, instead of using all negative edges, we use the idea of negative sampling[28]. This approach considers only a few samples drawn from a noise distribution. We assumed $P_n(j) \sim d_j^{3/4}$ as the noisy distribution where $d_j$is the degree of node j, which was used in word2vec[28]. By letting M the number of negative samples, we can redefine the utility function as:

$$J = \sum_{(i,j)\in E} w_{ij} \log p(e_{ij} = 1|y_i, y_j) + \sum_{k=1}^{M} \mathbb{E}_{j_k \sim P_n(j)} \gamma \log(1 - p(e_{ij_k} = 1|y_i, y_k)) \quad (5)$$

Then, we optimized (5) by applying asynchronous stochastic gradient descent (ASGD)[34]. It is a powerful optimization technique which can be efficiently parallelized and can make our algorithm more scalable. ASGD can be used in this context because the network constructed by the first step is sparse and there are few memory access conflicts between the threads we used. The learning rate is determined by $\rho_t = \rho(1 - t/T)$ where T is the total number of edge samples[25], and the initial learning rate ρ_0 is determined by considering the properties of input single-cell data. The time complexity of each SGD step of (5) is $O(M)$. For a large number of data set, the number of SGD iterations is usually proportional to the number of the given data set, n. Thus, the time complexity of the optimization is $O(Mn)$, which is linear with respect to the number of samples.

3. Experiments and Discussion

3.1. *Data and data processing*

We used mass cytometry data that are provided by X-shift[12]. They consist of 10 data sets that contain mice bone marrow samples stained with surface markers, and each of them has 51 parameters. Instead of using all of them, we used 39 surface marker expressions[12,35] that were utilized for mass cytometry experiments of the immune system reference framework[35]. In addition, the data was processed through noise thresholding and asinh transformation, i.e. $y = \text{asinh}(\max(x - 1, 0)/5)$ like X-shift[12] and viSNE[8] applied. The data sets also offer 24 gating annotations of each cell, which were used to distinguish cells in visualization and compare the clustering performance of viSNE and our method.

3.2. *Experimental setting*

We compared our method with viSNE[8] because it is a state-of-the-art method of single-cell visualization based on nonlinear embedding like our approach. Before implementing both algorithms, we set the parameters of each method. viSNE is based on Barnes-Hut-SNE[24], which has two parameters: perplexity and theta that controls the tradeoff between speed and accuracy. In our experiments, we set the two as 30 and 0.5, respectively. Our method allows for more control and therefore has more parameters: number of trees, number of neighbors, perplexity, number of negative samples, rho, gamma, and alpha. We set the parameters considering our input data set. The first three parameters are related to constructing a k-NN network. The number of trees and neighbors can determine the shapes of a k-NN network, and perplexity is related to computing edge weights of section 2.2. The other parameters are related to network embedding. The number of negative samples is M of (5), rho is the initial learning rate, gamma is the weight of negative edges, and alpha determines the thickness of the tail of f. Table 1 shows the parameters we tuned for our visualization.

Table 1. Parameters for constructing a k-NN network and for network embedding

Parameters for constructing a k-NN network			
Number of trees	Number of neighbors		Perplexity
20 – 100	20 – 150		10 – 50
Parameters for network embedding			
Number of negative samples	Rho	Gamma	Alpha
5 – 10	1 – 10	1 – 10	< 1

All experiments for measuring the computation time were performed on a machine with Intel Xeon E5-2650 CPUs running at 2.30GHz. 40 threads were used except the experiments about the effectiveness of multiple threads.

3.3. *Results*

3.3.1. *Visualization*

Figure 2 represents the visualization for mice bone marrow replicate 7 data set[12]. Overall, the same type of cells forms a dense subset. The number of a certain class of cells such as HSC in the data set was so small that they were difficult to distinguish from other cells and to find in our visualization. Except these cells, we can see clearly that the same type of cells gathers together and different types of cells move away from each other in a two-dimensional space. In addition, we can find some similar cells to stay together in Figure 2. For example, similar cell types like Intermediate Monocytes (red) and Classical Monocytes (yellow) appear close to each other. Two types of B cells (purple and light green) are also stay near each other.

In addition, we applied viSNE to the same data set. viSNE also represented cell subpopulations very well. The same type of cells was grouped together, and it can clearly distinguish different types of cells. In the experiments, our method tended to form denser and rounder clusters than viSNE but to have more randomly scattered samples. Due to the space limit, the visualization results of viSNE are shown through our web-based visualization tool (see section 4). We also compare our method with other embedding methods such as PCA in the tool.

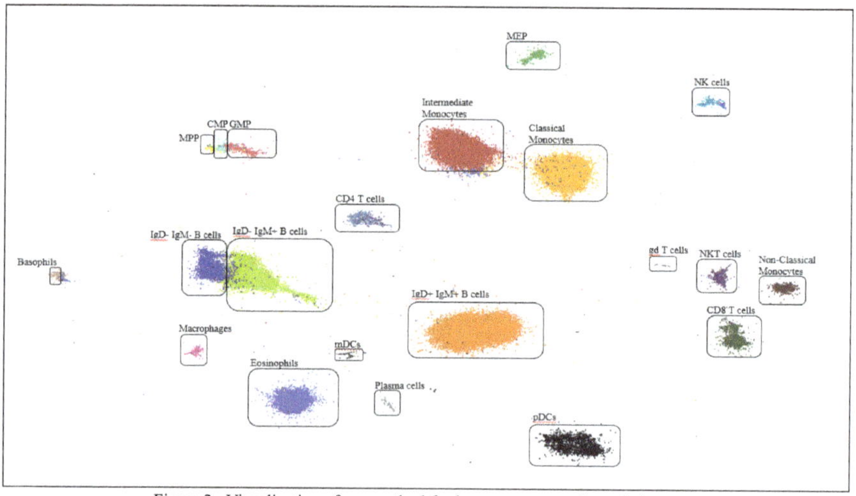

Figure 2. Visualization of our method for bone marrow replicate 7 data set.

3.3.2. *Computation time*

One of the main goals of our method is to make visualization of high-dimensional single-cell data be faster and more scalable. Thus, we compared the computation time between viSNE[8] and our method for various cases. In addition, to test the scalability and parallelizability, we measured the effectiveness of speedup with respect to the number of threads.

To measure the computation time and evaluate the scalability with respect to the size of the data set, we constructed 8 single-cell data sets that contained 5,000, 10,000, 25,000, 50,000, 75,000, 100,000, 250,000, and 500,000 data, respectively. For each data set, cells were uniformly sampled from the union of 10 data sets (total number: 841,644). Each data set contained 39 parameters and were preprocessed by noise thresholding and asinh transformation before sampling. Figure 3(a) shows that our method was faster than viSNE for all 8 sampled data sets and our method is easier to make scalable. The total computation time of our method consists of two computation times: one is for constructing a k-NN network and the other is for embedding the network. Figure 3(b) shows how much time we needed for each step.

In addition, we tested the parallelization of our method in the multi-core setting. Since our method uses asynchronous stochastic gradient descent (ASGD)[34] for training, it can be more accelerated by using multiple threads. We measured the computation time of our method when dealing with the union of all 10 single cell data sets with respect to the number of threads. By increasing the number of threads from 1 to 8, we measured the effectiveness of the multiple threads for our method. When we used 8 threads simultaneously, the speedup rate was 4.1 times faster than single-thread implementation in Figure 3(c). The results show that our method can be easily parallelized and can be made more scalable through a multi-core system.

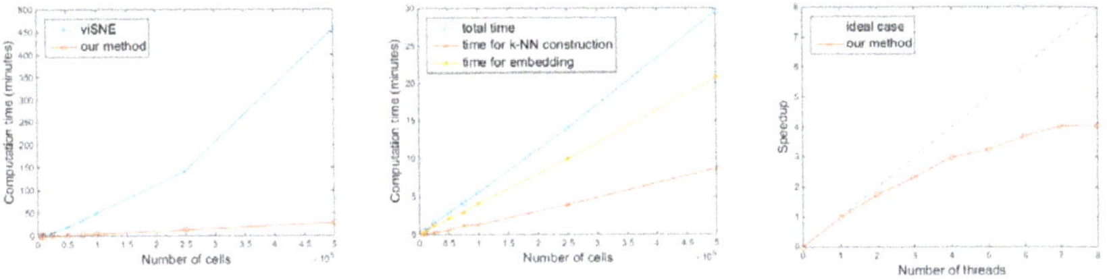

(a) (b) (c)

Figure 3. (a) Comparison of the computation time of viSNE and our method with respect to the number of single-cell data samples. (b) Separate analysis of the computation time for constructing a k-NN network and for embedding with regard to the number of single-cell data samples. (c) Effectiveness of the multiple threads for speedup of our method.

3.3.3. *Clustering*

In this section, we compared the quality of embedding by comparing the performance of clustering. In our experiments, we first applied one of the off-the-shelf clustering algorithms, k-means clustering[20] to the embedded vectors by viSNE[8] and those by our method. Next, we measured the performance of clustering using hand-gated annotations of each cell. Specifically, we followed the process of X-shift[12] to compare the clustering result and hand-gated labels and to calculate F1-measures. As the number of clusters changed from 2 to 100, we computed F1-measures for each cluster that a label was assigned to by the Hungarian algorithm[36]. This process was applied to our 10 data sets, and we obtained an average F1-measure sum. As another performance measure, we obtained maximum F1-measures for each data set across all the number of clusters and took a median.

As the input of clustering, we used the two-dimensional vectors embedded by viSNE[8] and our method. We compared an average F1-measure sum of both methods and a median of maximum F1-measures. Figure 4(a) shows that the clustering performance of our method was better than that of viSNE across all the number of clusters with respect to an average F1-measure sum. In addition, we compared a median of maximum F1-measures of viSNE and our method. Our two-dimensional embedding obtained 14.68 while viSNE obtained 13.23 as its median. Our method also outperformed viSNE for this metric.

Since our method is developed mainly for visualization, two or three dimensional vectors are usually used as a result of embedding. However, the algorithm can embed high-dimensional single-cell data into another arbitrary low-dimensional space other than a two- or three-dimensional space. The vectors embedded in a higher-dimensional space than a space for visualization can lose less intrinsic information about original high-dimensional single-cell data. Thus, they can be used to enhance the performance of clustering. We clustered the data by using 5-, 10-, 15- and 20-dimensional representations obtained by our embedding.

Figure 4(b) shows that the performance of clustering was improved when we used the vectors with higher dimensions than two. The performances as we used 10-, 15-, and 20-dimensional vectors are similar to each other and better than the performance as we used two- or 5-dimensional vectors.

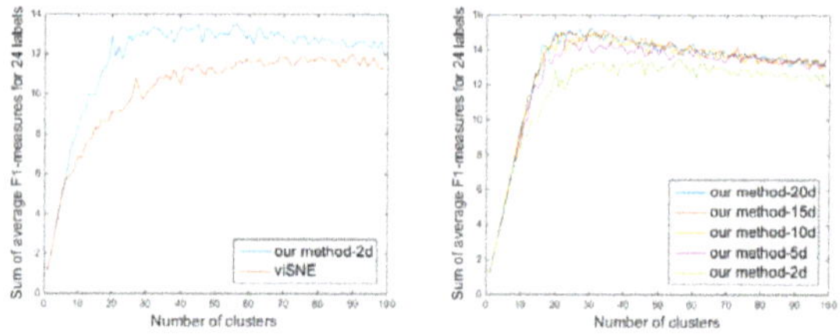

(a) (b)

Figure 4. (a) Comparison of the clustering performance of viSNE and our method when using two-dimensional vectors with respect to the number of clusters. (b) Comparison of the clustering performances when we changed the dimension of our embedding.

4. Interactive Visualization

To better aide analysis, we also introduce an interactive web browser based visualization tool featured in Figure 5. It allows researchers to examine their own data quickly by enabling functionality like mouse-over, zoom, pan, brushing, and linking on the embedded data. Users can color data by quantities of marker values as well as qualitative gate information. One can select arbitrary groups of single-cell data points, tag them, and save them for downstream analysis. We provide code, documentation, and video demonstrations to reproduce experiments and apply our methods to new single-cell data through the link[a]. All code is made available under an MIT license.

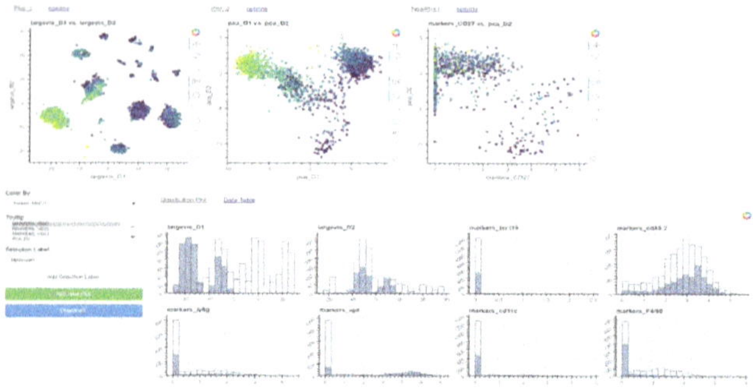

Figure 5. Screen shot of web browser based visualization developed in python. The left scatter plot depicts the result of our proposed method and on the middle, a PCA projection of the data. The right plot describes embedded expressions

[a] https://github.com/nate-russell/SVHD-Single-Cell

of a specific marker with respect to a certain projection. Color assignment and data selection labeling are also available through widgets at the bottom left. Some data statistics and the table to the right show all provided marker data and meta data regarding the single-cell data.

5. Conclusion

In this paper, we introduced a new visualization method for large-scale and high-dimensional single-cell data based on LargeVis[25], which consists of two parts: constructing an approximate k-NN network and embedding the constructed network into a low-dimensional space. Since the both steps have linear time complexity, our method is scalable and readily for analyzing large-scale single-cell data sets with hundreds of thousands or even millions of single cells. Specifically, our experiment results showed that the proposed method is much faster than viSNE[8], a state-of-the-art single-cell visualization method. In addition, through the experiments about clustering, we showed that the quality of our embedding is better than that of viSNE on cell identity mapping with respect to F1-measures. We also provide a web based interactive visualization tool and all necessary code and documentation to extend this approach to new data.

Acknowledgments

This study was supported by a Sloan Research Fellowship and a National Center for Supercomputing Applications (NCSA) Fellowship of University of Illinois at Urbana-Champaign.

References

1. O. Stegle, S. A. Teichmann, and J. C. Marioni, *Nat. Rev. Genet.* **16**, 133-145 (2015).
2. F. Buettner, K. N. Natarajan, F. P. Casale, V. Proserpio, A. Scialdone, F. J. Theis, S. A. Teichmann, J. C. Marioni, and O. Stegle, *Nat. Biotechnol.* **33**, 155-160 (2015).
3. C. Trapnell, D. Cacchiarelli, J. Grimsby, P. Pokhare, S. Li, M. Morse, N. J. Lennon, K. J. Livak, T. S. Mikkelsen, and J. L. Rinn, *Nat. Biotechnol.* **32**, 381-386 (2014).
4. O. Ornatsky, D. Bandura, V. Baranov, M. Nitz, M. A. Winnik, and S. Tanner, *J. Immunol. Methods.* **361**, 1-20 (2010).
5. S. C. Bendall, E. F. Simonds, P. Qiu, E. D. Amir, P. O. Krutzik, R. Finck, R. V. Bruggner, R. Melamed, A. Trejo, O. I. Ornatsky, R. S. Balderas, S. K. Plevritis, K. Sachs, D. Pe'er, S. D. Tanner, and G. P. Nolan, *Science.* **332**, 687-696 (2011).
6. P. Qiu, E. F. Simonds, S. C. Bendall, K. D. Gibbs Jr, R. V. Bruggner, M. D. Linderman, K. Sachs, G. P. Nolan, and S. K. Plevritis, *Nat. Biotechnol.* **29**, 886-891 (2011).
7. S. C. Bendall, G. P. Nolan, M. Roederer, P. K. Chattopadhyay, *Cell.* **33**, 323-332 (2012).
8. E.-A. D. Amir, K. L. Davis, M. D. Tadmor, E. F. Simonds, J. H. Levine, S. C. Ben dall, D. K. Shenfeld, S. Krishnaswamy, G. P. Nolan, and D. Pe'er, *Nat. Biotechnol.* **31**, 545-552 (2013).
9. S. C. Bendall, K. L. Davis, E.-A. D. Amir, M. D. Tadmor, E. F. Simonds, T. J. Chen, D. K. Shenfeld, G. P. Nolan, and D. Pe'er, *Cell.* **157**, 714-725 (2014).
10. K. Shekhar, P. Brodin, M. M. Davis, and A. K. Chakraborty, *Proc Natl Acad Sci.* **111**, 202-207 (2014).
11. M. Setty, M. D. Tadmor, S. Reich-Zeliger, O. Angel, T. M. Salame, P. Kathail, K. Choi, S. C.

Bendall, N. Friedman, and D. Pe'er, *Nat. Biotechnol.* **34**, 637-645 (2016).
12. N. Samusik, Z. Good, M. H. Spitzer, K. L. Davis, and G. P. Nolan, *Nat. Methods.* **13**, 493-496 (2016).
13. B. Anchang, T. D. P. Hart, S. C. Bendall, P. Qiu, Z. Bjornson, M. Linderman, G. P. Nolan, and S. K. Plevritis, *Nat. Protocols.* **11**, 1264-1279 (2016).
14. A. P. Patel, I. Tirosh, J. J. Trombetta, A. K. Shalek, S. M. Gillespie, H. Wakimoto, D. P. Cahill, B. V. Nahed, W. T. Curry, R. L. Martuza, D. N. Louis, O. Rozenblatt-Rosen, M. L. Suvà, A. Regev, and B. E. Bernstein, *Science.* **344**, 1396-1401 (2014).
15. Q. Deng, D. Ramskold, B. Reinius, and R. Sandberg, *Science.* **343**, 193-196 (2014).
16. D. R. Bandura, V. I. Baranov, O. I. Ornatsky, A. Antonov, R. Kinach, X. Lou, S. Pavlov, S. Vorobiev, J. E. Dick, and S. D. Tanner, *Anal. Chem.* **81**, 6813-6822 (2009).
17. F. Tang, C. Barbacioru, Y. Wang, E. Nordman, C. Lee, N. Xu, X. Wang, J. Bodeau, B. B. Tuch, A. Siddiqui, K. Lao, and M. A. Surani, *Nat. Methods.* **6**, 377-382 (2009).
18. C. Trapnell, *Genome Res.* **25**, 1491-1498 (2015).
19. L. A. Herzenberg, J. Tung, W. A. Moore, and D. R. Parks, *Nat. Immunol.* **7**, 681-685 (2006).
20. C. Bishop, *Springer.* (2006).
21. H. C. Fan, G. K. Fu, and S. P. A. Fodor, *Science.* **347**. 1258367 (2015).
22. D. A. Lawson, N. R. Bhakta, K. Kessenbrock, K. D. Prummel, Y. Yu, K. Takai, A. Zhou, H. Eyob, S. Balakrishnan, C. Wang, P. Yaswen, A. Goga, and Z. Werb, *Nat.* **526**, 131-135 (2015).
23. L. J. P. van der Maaten, and G. E. Hinton, *J. Mach. Learn. Res.* **9**, 2579-2605 (2008).
24. L. J. P. van der Maaten, *J. Mach. Learn. Res.* **15**, 3221-3245 (2014).
25. J. Tang, J. Liu, M. Zhang, and Q. Mei, *Proc. 25th Int. Conf. WWW.* (2016).
26. S. Dasgupta and Y. Freund, *Proc. 40th ACM STOC.* 537-546 (2008).
27. W. Dong, M. Charikar, and K. Li, *Proc. 20th Int. Conf. WWW.* 577-586 (2011).
28. T. Mikolov, I. Sutskever, K. Chen, G. S. Corrado, J. Dean, *Proc. 26th Adv. NIPS.* 3111-3119 (2013).
29. M. Datar, N. Immorlica, P. Indyk, and V. S. Mirrokni, *Proc. 20th ACM SoCG.* 253-262 (2004).
30. J. Tang, M. Qu, M. Wang, M. Zhang, J. Yan, and Q. Mei, *Proc. 24th Int. Conf. WWW.* 1067-1077 (2015).
31. M. Muja and D. G. Lowe, *IEEE Trans Pattern Anal Mach Intell.* **36**, 2227-2240 (2014).
32. J. Barnes and P. Hut, *Nat.* **324**, 446-449 (1986).
33. M. U. Gutmann and A. Hyvarinen, *J. Mach. Learn. Res.* **13**, 307-361 (2012).
34. B. Recht, C. Re, S. Wright, and F. Niu, *Proc. 24th Adv. NIPS.* 693-701 (2011).
35. M. H. Spitzer, P. F. Gherardini, G. K. Fragiadakis, N. Bhattacharya, R. T. Yuan, A. N. Hotson, R. Finck, Y. Carmi, E. R. Zunder, W. J. Fantl, S. C. Bendall, E. G. Engleman, G. P. Nolan, *Science.* **349**, 1259425 (2015).
36. J. Munkres, *J. Soc. Ind. Appl. Math.* **5**, 32-38 (1957).

HARNESSING BIG DATA FOR PRECISION MEDICINE: INFRASTRUCTURES AND APPLICATIONS

KUN-HSING YU
Biomedical Informatics Training Program, Stanford University
3165 Porter Dr., Room 2270, Palo Alto, CA 94304
Email: khyu@stanford.edu

STEVEN N. HART
Center for Individualized Medicine, Mayo Clinic
200 First Street SW, Rochester, MN 55905
Email: Hart.Steven@mayo.edu

RACHEL GOLDFEDER
Biomedical Informatics Training Program, Stanford University
870 Quarry Rd, Stanford, CA 94305
Email: rlg2@stanford.edu

QIANGFENG CLIFF ZHANG
School of Life Sciences, Tsinghua University
Medical Science Building, B-1002, Tsinghua University, Beijing, China 100084
Email: zhang.lab@biomed.tsinghua.edu.cn

STEPHEN C. J. PARKER
Computational Medicine and Bioinformatics, University of Michigan
100 Washtenaw Ave, 2049B, Ann Arbor, MI 48109
Email: scjp@umich.edu

MICHAEL SNYDER
Department of Genetics, Stanford University
300 Pasteur Dr., M344 MC 5120, Stanford, CA 94305
Email: mpsnyder@stanford.edu

Precision medicine is a health management approach that accounts for individual differences in genetic backgrounds and environmental exposures. With the recent advancements in high-throughput omics profiling technologies, collections of large study cohorts, and the developments of data mining algorithms, big data in biomedicine is expected to provide novel insights into health and disease states, which can be translated into personalized disease prevention and treatment plans. However, petabytes of biomedical data generated by multiple measurement modalities poses a significant challenge for data analysis, integration, storage, and result interpretation. In addition, patient privacy preservation, coordination between participating medical centers and data analysis working groups, as well as discrepancies in data sharing policies remain important topics of discussion. In this workshop, we invite experts in omics integration, biobank research, and data management to share their perspectives on leveraging big data to enable precision medicine. Workshop website: http://tinyurl.com/PSB17BigData; HashTag: #PSB17BigData.

1. Introduction

Throughout medicine's history, disease prevention and treatment has been based on the expected outcome of an average patient[1]. Data from patients with the same disease were often pooled together for statistical analysis, and clinical guidelines derived from the aggregated analysis informed health and disease management for billions of patients. Although this approach achieves some success, it ignores important individual differences, which can result in different treatment responses[2].

Precision medicine aims to tailor clinical treatment plans to individual patients, with the goal of delivering the right treatments at the right time to the right patient[3]. Recent advances in omics technologies provide clinicians with more complete patient profiles[4,5]. The decreasing cost of sequencing and associated data storage[6] and the development of effective data analysis methods make it possible to collect and analyze big biomedical data for various human diseases at an unprecedented scale[7]. These advancements can improve the diagnostic accuracy of complex diseases, identify patients who will benefit from targeted therapeutics, and predict diseases before their occurrence[3,8].

Nevertheless, many challenges still remain. Conventional methods for data storage, database management, and computational analysis are insufficient for the petabytes of biomedical data generated every year. In addition, as datasets become larger and more diverse, advanced distributed file storage and computing methods are needed to make the data useful. Furthermore, data-sharing policies and result reproducibility continue to be vigorously debated issues[9-10].

In this workshop, world-renowned experts in personal omics profiling, biobanks, biomedical databases, and medical data analysis will describe recent advancements in these areas and discuss associated challenges and potential solutions.

2. Workshop presentations

This section provides a brief summary for each presentation. The full abstracts could be found at the workshop website http://tinyurl.com/PSB17BigData.

2.1. DeepDive: A Dark Data System

Dr. Christopher Ré, Department of Computer Science, Stanford University, CA, USA

Many pressing questions in science are macroscopic, as they require scientists to integrate information from numerous data sources, often expressed in natural languages or in graphics; these forms of media are fraught with imprecision and ambiguity and so are difficult for machines to understand. Here I describe DeepDive, which is a new type of system designed to cope with these problems. It combines extraction, integration and prediction into one system. For some paleobiology and materials science tasks, DeepDive-based systems have surpassed human volunteers in data quantity and quality (recall and precision). DeepDive is also used by scientists in areas including genomics and drug repurposing, by a number of companies involved in various forms of search, and by law enforcement in the fight against human trafficking. DeepDive does not allow users to write algorithms; instead, it asks them to write only features. A key technical challenge is scaling up the resulting inference and learning engine, and I will describe our line of work in computing without using traditional synchronization methods including Hogwild! and DimmWitted. DeepDive is open source on github and available from DeepDive.Stanford.Edu.

2.2 Results of the VariantDB Challenge

Dr. Steven Hart, Department of Health Sciences Research, Mayo College of Medicine, MN, USA

The current standard formats for storing genomics data is the VCF and gVCF, but manipulating these large files is an imperfect and impractical long-term solution. Scalability, availability, consistency, are all important drawbacks to the file-based approach. Multiple pieces of metadata are often required to interpret genomic data, but there is no specification for how to tie sample level data (e.g. smoking status, disease status, age of onset, etc.) with variant-level data. The motive of the VariantDB Challenge is to identify a scalable, robust framework for storing, querying and analyzing genomics data in a biologically relevant context. The contextual focus is a central theme in the challenge since it is relatively easy to optimize simple database lookups, but forming queries with multiple predicates becomes a much more complicated task. The VariantDB_Challenge is a 100% open source project, meaning that all code and solutions used must be made publically available via GitHub. In this session, we will present an overview of the challenge and summarize the results from all submitters.

2.3. ADAM: Fast, Scalable Genome Analysis

Mr. Frank Austin Nothaft, Department of Computer Science, UC Berkeley, Berkeley, CA, USA

The detection and analysis of rare genomic events requires integrative analysis across large cohorts with terabytes to petabytes of genomic data. Contemporary genomic analysis tools have not been designed for this scale of data-intensive computing. This talk presents ADAM, an Apache 2 licensed library built on top of the popular Apache Spark distributed computing framework. ADAM is designed to allow genomic analyses to be seamlessly distributed across large clusters, and presents a clean API for writing parallel genomic analysis algorithms. In this talk, we'll look at how we've used ADAM to achieve a 3.5× improvement in end-to-end variant calling latency and a 66% cost improvement over current toolkits, without sacrificing accuracy. We will also talk about using ADAM alongside Apache Hbase to interactively explore large variant datasets.

2.4. Personalized Medicine: Using Omics Profiling and Big Data to Understand and Manage Health and Disease

Dr. Michael Snyder, Department of Genetics, Stanford University School of Medicine, CA, USA

Understanding health and disease requires a detailed analysis of both our DNA and the molecular events that determine human physiology. We performed an integrated Personal Omics Profiling (iPOP) on 70 healthy and prediabetic human subjects over periods of viral infection as well as during controlled weight gain and loss. Our iPOP integrates multiomics information from the host (genomics, epigenomics, transcriptomics, proteomics and metabolomics) and from the gut microbiome. Longitudinal multiomics profiling reveals extensive dynamic biomolecular changes occur during times of perturbation, and the different perturbations have distinct effects on different biomolecules in terms of the levels and duration of changes that occur. Overall, our results demonstrate a global and system-wide level of biochemical and cellular changes occur during environmental exposures.

2.5. Statistical and Dynamical Systems Modeling of Real-Time Adaptive m-Intervention for Pain

Dr. Jingyi Jessica Li, Departments of Statistics and Human Genetics, University of California, Los Angeles, CA, USA

Nearly a quarter of visits to the Emergency Department are for conditions that could have been managed via outpatient treatment; improvements that allow patients to quickly recognize and receive appropriate treatment are crucial. The growing popularity of mobile technology creates new opportunities for real-time adaptive medical intervention, and the simultaneous growth of "big data" sources allows for preparation of personalized recommendations. We present a new mathematical model for the dynamics of subjective pain that consists of a dynamical systems approach using differential equations to forecast future pain levels, as well as a statistical approach tying system parameters to patient data (both personal characteristics and medication response history). We combine this with a new control and optimization strategy to ultimately make optimized, continuously-updated treatment plans balancing competing demands of pain reduction and medication minimization. A workable hybrid model incorporating both mathematical approaches has been developed. Pilot testing of the new mathematical approach suggests that there is significant potential for (1) quantification of current treatment effectiveness for pain management, (2) forecast of pain crisis events, and (3) overall reduction of pain without increased medication use. Further research is needed to demonstrate the effectiveness of the new approach for each of these purposes.

2.6. Integrated Database and Knowledge Base for Genomic Prospective Cohort Study: Lessons Learned from the Tohoku Medical Megabank Project

Dr. Soichi Ogishima, Tohoku Medical Megabank Organization, Tohoku University, Japan

The Tohoku Medical Megabank project is a national project to revitalize medical care and to realize personalized medicine in the disaster area of the Great East Japan Earthquake. In our prospective cohort study, we recruited 150,000 people at Tohoku University, its satellites health clinics, and Iwate Medical University. We collected biospecimen, questionnaire, and physical measurement during baseline and follow-up investigations. Along with prospective genome-cohort studies, we have developed integrated database and knowledge base, which will be the foundation for realizing personalized medicine and disease prevention.

3. Conclusion

Big data in biomedicine presents a great opportunity to understand health and disease states at an unprecedented level. This workshop will highlight landmark achievements in integrative omics studies, biobank research, and novel data mining methods for large datasets. With the growing number and size of biomedical datasets worldwide, we envision that approaches discussed in this workshop will facilitate the development of precision medicine.

4. Acknowledgments

K.-H. Y. is supported by a Howard Hughes Medical Institute (HHMI) International Student Research Fellowship and a Winston Chen Stanford Graduate Fellowship. R.G. is supported by a National Science Foundation (NSF) Graduate Research Fellowship. M.P. is partially supported by National Institutes of Health grants 1U54DE02378901, 5P50HG00773502, and 5U24CA16003605.

5. References

1. Collins FS. Exceptional opportunities in medical science: a view from the National Institutes of Health. *JAMA.* **313**:131-2 (2015).
2. Shastry BS. Pharmacogenetics and the concept of individualized medicine. *Pharmacogenomics J.* **6**:16-21 (2006).
3. Collins FS, Varmus H. A new initiative on precision medicine. *N Engl J Med.* **372**:793-5 (2015).
4. Chen R, Mias GI, Li-Pook-Than J, et al. Personal omics profiling reveals dynamic molecular and medical phenotypes. *Cell.* **148**:1293-307 (2012).
5. Yu KH, Snyder M. Omics Profiling in Precision Oncology. *Mol Cell Proteomics.* **15**:2525-36 (2016).
6. Stein LD. The case for cloud computing in genome informatics. *Genome Biol.* **11**:207 (2010).
7. Wichmann HE, Kuhn KA, Waldenberger M, et al. Comprehensive catalog of European biobanks. *Nat Biotechnol.* **29**:795-7 (2011).
8. Ashley EA. The precision medicine initiative: a new national effort. *JAMA.* **313**:2119-20 (2015).
9. Longo DL, Drazen JM. Data Sharing. *N Engl J Med.* **374**:276-7 (2016).
10. Ioannidis JP. Expectations, validity, and reality in omics. *J Clin Epidemiol.* **63**:945-9 (2010).

THE TRAINING OF NEXT GENERATION DATA SCIENTISTS IN BIOMEDICINE[1]

LANA X GARMIRE[2†], STEPHEN GLISKE[3†], QUYNH C NGUYEN[4†], JONATHAN H. CHEN[5†], SHAMIM NEMATI[6†], JOHN D. VAN HORN [7†], JASON H MOORE[8], CAROL SHREFFLER[9], MICHELLE DUNN[10]

[2]*Epidemiology Program, University of Hawaii Cancer Center, Honolulu, HI, 96813, USA. Email: LGarmire@Hawaii.edu*
[3]*Department of Neurology, University of Michigan, Ann Arbor, MI 48109-5322, USA*
[4]*Department of Health, Kinesiology and Recreation, University of Utah, 84112, USA*
[5]*Department of Medicine, Stanford University, Stanford, CA, 94305, USA*
[6]*Department of Biomedical Informatics, Emory University, Atlanta, GA, 30322, USA*
[7]*Mark and Mary Stevens Neuroimaging and Informatics Institute, University of Southern California, Los Angeles, CA, 90032, US*
[8]*Institute of Biomedical Informatics, University of Pennsylvania, Philadelphia, PA, 19104, USA*
[9]*National Institute of Environmental Health Sciences, Research Triangle Park, NC, 27709, USA*
[10]*Office of the Associate Director for Data Science (ADDA), National Institute of Health, Bethesda, MD, 20892, USA*

With the booming of new technologies, biomedical science has transformed into digitalized, data intensive science. Massive amount of data need to be analyzed and interpreted, demand a complete pipeline to train next generation data scientists. To meet this need, the trans-institutional Big Data to Knowledge (BD2K) Initiative has been implemented since 2014, complementing other NIH institutional efforts. In this report, we give an overview the BD2K K01 mentored scientist career awards, which have demonstrated early success. We address the

[1*] This work is supported by BD2K K01 program

2†3†4†5†6†7† Work partially supported by grant NIH Big Data 2 Knowledge Award K01ES025434 (to LXG), K01ES026839 (to SG), K01ES025433 (to QCN), K01ES026837 (to JHC), K01ES025445 (to SN), U24 ES026465 (to JDV), by the National Institute of Environmental Health Sciences through funds provided by the trans-NIH Big Data to Knowledge (BD2K) initiative,

2† work is also partially supported by P20 COBRE GM103457 awarded by NIH/NIGMS, NICHD R01 HD084633, NLM R01 LM012373, and Hawaii Community Foundation Medical Research Grant 14ADVC-64566.

specific trainings needed in representative data science areas, in order to make the next generation of data scientists in biomedicine.

1. Biomedical science as data intensive science

There is little doubt that biomedical science has become data intensive science. In the last decades, we have witnessed the booming of new biomedical technologies which generated massive amount of bio-data. In the genomics realm, next generation sequencing (NGS) has produced various types of omics-data. It is now a reality to sequence patients' genomes to seek personalized medication. In the medical imaging field, petabytes of imaging data are stored, processed and analyzed in institutions[1]. Sensor-based wearable devices monitor daily exercise and other life-style routines, and generate real-time physiological data. With the adoption of Electronic Health Record (EHR) data by hospitals, it is now feasible to access and mine the massive amount of clinical and phenotypic data.

For junior researchers, the timing has never been better to seek a career in data science. Given the global "open-data" movement, many of the data types mentioned above are available publically, significantly saving the time and cost to conduct large-scale data mining and discoveries. We perceive that secondary data analysis would empower a whole new level of knowledge discoveries and hypothesis generation, which will reciprocally benefit other fields of biomedical research. Facility-wise, high-performance-computing (HPC) environments are well set-up in many major research universities; moreover, private sectors such as Google and Amazon offer cloud-computing as an alternative to the localized (thus restrictive) HPC access. Additionally, advancing in mathematical and statistical modeling, machine learning and the new derivatives of deep learning, is playing an increasingly important role in biomedical and healthcare industries.

2. The increasing needs to train the next generation data scientists in biomedicine

Compared to the prolific amount of biomedical data, developing computational methods and algorithms and training data scientists with domain expertise in biomedicine are major limiting factors to understanding the complex interactions in human health and disease. Unlike many other disciplines, data science in biomedicine is very interdisciplinary and requires training in domains including computer science, statistics, mathematics, and biomedicine. This interdisciplinary nature requires that data science in biomedicine be adaptive and involve constant learning and training by all, from undergraduate, graduate, postdoc to faculty levels.

Recognizing such needs, National Institute of Health (NIH) spearheaded The trans-NIH Big Data to Knowledge (BD2K) Initiative in 2014. The mission of the BD2K initiative is to support training in and research and development of innovative and transformative new approaches and tools, in order to maximize and accelerate the utility of the Big Data being generated. Since its inception, training has been one of the major thrust areas of the BD2K program. The term "training" is meant to include training, education, and workforce development that provides learners, no matter what career level, either foundational knowledge or skills for immediate use. Training currently accounts for 15% of the BD2K budget. There are two main goals for training: (1) to increase the number of people trained in developing the tools, methods, and technology to maximize the information which can be obtained by biomedical Big Data, and (2) to elevate the data science competencies of all biomedical scientists.

3. Funding mechanisms of National Institute of Health to train next generation data scientists

To accomplish these goals, a diverse set of grants and grants types have been developed (see the complete report: https://datascience.nih.gov/bd2k/funded-programs/enhancing-training). The work being showcased in this paper relates to the goal of increasing the number of biomedical data scientists. Although the establishment of biomedical data science as a career requires a complete career pipeline, from undergraduate training on up, the focus here is on the latter end of the pipeline, at the postdoc and junior faculty level. To support junior faculty, the NIH developed the K01 Career Development program. K01s in Biomedical Big Data Science are designed to facilitate the career transition of research oriented interdisciplinary investigators who are significantly altering their research focus. Candidates can enter the mentored experience from any of the three major scientific areas of Big Data Science: (1) Computer science or informatics; (2) Statistics and Mathematics; or (3) Biomedical Science. At the end of the program, awardees are expected to have competence in all three areas, as well as depth in one area. Competence is gained through course work as well as through a mentorship from a team that includes all of the expertise listed above. In 2014 and 2015, BD2K awarded 21 K01 projects. The PIs come from diverse backgrounds, including: (1) 9 physicians with specialties in hematology/oncology, neurology, neuroradiology, surgery, urologic surgery, pulmonary and critical care medicine, and internal medicine; (2) 7 PhDs with primarily quantitative or computational backgrounds, with degrees in Electrical Engineering and Computer Science, Physics, Nuclear Physics, and Biomedical Engineering; (3) 3 interdisciplinary scientists with backgrounds in fields that blend the biomedical and computational sciences (Molecular Genetics, Bioinformatics and Computational Biochemistry); and (4) 2 behavioral or social scientists (Social Epidemiology, Quantitative Psychology). These awardees represent 18 unique institutions from 12 states, among whom 9 awardees are female. .The expectation of the program is that the K01 awardees will be, by the end of the project period, competitive for new research grants (e.g. R01) in the area of Big Data Science. Many K01 awardees have moved on to faculty positions, and some have already obtained competitive NIH grants (e.g. R01s).

4. Areas of biomedical data science demanding new workforce

Data science in biomedicine includes, but is not limited to, the categories: translational bioinformatics and computational biology, clinical informatics, consumer health informatics and public health informatics[2]. Maximal success can be obtained by the biomedical data scientists trained in not only the technical aspects of data science (computer science, signal processing, math, statistics, etc.), but also the specific area of biomedicine of application. This, in part, sets apart the biomedical data scientists from general data scientists. Below we focus on a few representative categories funded by the current BD2K K01 program.

4.1. *Translational Bioinformatics*

Large national and international consortia and data repositories have formed, significantly increasing the sample sizes and discovery powers for many diseases. Training in translational bioinformatics needs to rapidly adapt to the global environment by emphasizing broad, interdisciplinary training in computer science, statistics, bioinformatics, and biology. Good suggestions on bioinformatics training courses have been made earlier[3]. Here we put more focus on multi-omics areas, beyond single-omics data analysis and pipeline construction. At the input data level, the trainees will be expected to deal with missing values and normalizing data within and across various technical platforms. The trainees should be able to creatively transform data, by taking advantage of prior biological knowledge such as pathway or network information[4,5]. The trainees should have courses in statistics to thoroughly understand issues such as sample size, power, multiple hypothesis testing, classification (unsupervised learning), and generalized regression techniques (supervised learning)[6,7]. Training in multi-omics data integration (from the same population cohort) and meta-omics data integration (from heterogeneous populations) will be paramount to derive meaningful discoveries on molecular subtypes of diseases[8]. The trainees will also learn about omics-clinical/phenotypic data integration, using methods such as correlational and survival analysis.

Two new areas of translational bioinformatics are microbiome and single cell genomics. Both fields have measurement uncertainty. While the microbiome has the unknown variables of microbe numbers and strains, single cell genomics has the unknown variable of noise due to complicated batch effects, cell cycle and stress states, amplification biases etc[9]. In additional to the skills noted above, data visualization and tools to enhance reproducibility should be required for trainees, to enable efficient exploratory analysis and hypothesis generation. Last but not least, the trainees should go through rigorous training in HIPPA compliance to protect the private (including genetic) information of study subjects.

4.2. *Clinical Informatics*

The generation and dissemination of medical knowledge towards the practice of modern medicine arose in the past century, when there were relatively few effective interventions that the discipline had to offer for patient care. However, such norms now collide with the current reality of an explosive growth in biomedical knowledge[10]. Fortunately, with the new era of biomedical informatics, the clinicians are presented with great opportunities, along with challenges. The meaningful use of electronic health records (EHR)[11] presents the big data opportunity with the widespread routine capture of real-world clinical practice data, further augmented by high volume clinical data streams from claims, registry data, genomics, sensor systems, to patient generated content forms. Such digitized records offer new approaches to generate medical knowledge and to synthesize it into usable tools that can affect real-world clinical practice by assimilating and managing the increasing complexity of medical information. Principled, data-driven approaches are critical to unlocking the potential of large-scale healthcare data sources to impact clinical practice, compared to the otherwise limited and preconceived concepts manually abstracted out of patient chart reviews.

The current clinical practice force needs a paradigm shift. The next generation of data scientists will have the technical capability to generate useful insights from large complex data sources (machine learning, statistical analysis methods). They should have the tenacity to tackle enormous noisy and unstructured data that was not generated for precise research purposes (data wrangling, software engineering). Training in the appreciation of the applied subject domain is particularly important, in order to transcend the data-information-knowledge-wisdom hierarchy (translational inquiry). For physician scientists, complementary knowledge is needed to bridge the evolving practice of medicine from one that is traditionally apprenticeship, heuristic, pattern based learning to the new approach of using big data analytics creatively to inform decision making. Meanwhile, clinician scientists will need to gain experience on meaningfully informing practice, including recognizing pitfalls and limitations of data science.

4.3. *Public Health Informatics*

The curriculum for public health graduate students typically includes classes on population health, research methods, ethics of scientific research, and applications in public health. Other courses covering research methods are usually on study design, data analyses, and causal inference. However, training is generally lacking on how to fully utilize larger and nontraditional data sources. Public health investigators are usually trained to implement and analyze health surveys and clinical trials. However, training on processing large unstructured text data is lacking. Clinical text is the most pervasive data type in EHR[11]. Leveraging techniques in data mining, machine learning and natural language processing will enable the extraction of information on patient characteristics and clinical outcomes. Mining EHR allows us to better understand longitudinal patterns in treatment outcomes[12], treatment heterogeneity, and drug interactions. In addition, social media text has been useful for outbreak detection, tracking health conditions, and monitoring social influences on human health[13,14]. New public health training with data science concentration may include additional course in computer science, including database systems, data mining, machine learning, advanced algorithms, and visualization. More specialized training in natural language processing, image processing, high performance computing, and network security would be beneficial, too. These courses would increase expertise in the creation and maintenance of database structures for efficient storage and processing, and also increase the incorporation of large, emerging data sources such as text, images and videos in health research. The addition of training in database management and analytics would further enhance the understanding of drivers of health and disease, by incorporating novel and integrated data sources to account for disease complexities.

4.4 Exemplary Emerging Area of Informatics

In neuroscience, one area in need of data scientists that is only beginning to be recognized involves electroencephalogram (EEG). For example, the US Brain initiative funded many projects focused on acquiring high resolution EEG data, yet little attention has been focused ensuring that there is a sufficiently trained work force to analyze such data. Even with current technology, there is great need for more data scientists related to EEG analysis, both intracranial EEG (e.g., in epilepsy research) [15] and extracranial EEG (e.g., sleep medicine). The training needs for these individuals are similar to other fields: fluent programming skills, a strong understanding of machine learning, statistics and applied mathematics, and an understanding of the application of focus. One training method that has worked quite well for this applications is for students to get a PhD in either a technical field (e.g., biomedical engineering) or an applied field (e.g., neuroscience), and augment their coursework in order to obtain the needed breadth of subject matter. Some universities, such as the University of Michigan, has created a graduate certificate in data science, which can be paired with a PhD in specific discipline. Additionally, coursework needs to be matched with appropriate "hands on" research activities at the graduate and post-doctoral levels. One main challenge facing the next generation of data scientists is to establish the culture of interactions between disciplines. In addition to the challenges common to upcoming biomedical data scientists, these students face the extra barrier of EEG analysis being an emergent application area.

5. Conclusion

The golden era of big data science in biomedicine has just begun[2]. Many fields, such as EMR mining, mobile health and community-based health data mining are very new, and clearly challenges exist.

However, the data volume will only increase, thus "more is more, less is bore". The need for data scientists specialized in bio-medicine will continue to drive the market. On the other hand, while the paradigm shift towards data intensive biomedical science is happening, we must also bring to the attention that the "brain drain" from academia to private sectors is likely, and it is critical for institutions to create tenure-track career paths for the new generation of biomedical data scientists after their training programs end.

6. Acknowledgement

We would like to thank all BD2K K01 awardees for their support to make this workshop a reality.

References

1. Van Horn JD. Opinion: Big data biomedicine offers big higher education opportunities. Proc Natl Acad Sci U S A. 2016 Jun 7;113(23):6322–6324. PMID: 27274038
2. Moore JH, Holmes JH. The golden era of biomedical informatics has begun. BioData Min. 2016;9:15. PMID: 27069509
3. Greene AC, Giffin KA, Greene CS, Moore JH. Adapting bioinformatics curricula for big data. Brief Bioinform. 2016 Jan;17(1):43–50. PMID: 25829469
4. Huang S, Yee C, Ching T, Yu H, Garmire LX. A novel model to combine clinical and pathway-based transcriptomic information for the prognosis prediction of breast cancer. PLoS Comput Biol. 2014;10(9):e1003851. PMID: 25233347
5. Huang S, Chong N, Lewis NE, Jia W, Xie G, Garmire LX. Novel personalized pathway-based metabolomics models reveal key metabolic pathways for breast cancer diagnosis. Genome Med. 2016;8(1):34. PMID: 27036109
6. Ching T, Huang S, Garmire LX. Power analysis and sample size estimation for RNA-Seq differential expression. RNA New York N. 2014 Sep 22; PMID: 25246651
7. Menor M, Ching T, Zhu X, Garmire D, Garmire LX. mirMark: a site-level and UTR-level classifier for miRNA target prediction. Genome Biol. 2014;15(10):500. PMID: 25344330
8. Wei R, De Vivo I, Huang S, Zhu X, Risch H, Moore JH, Yu H, Garmire LX. Meta-dimensional data integration identifies critical pathways for susceptibility, tumorigenesis and progression of endometrial cancer. Oncotarget. 2016 Jul 9; PMID: 27409342
9. Poirion OB, Zhu X, Ching T, Garmire L. Single-Cell Transcriptomics Bioinformatics and Computational Challenges. Front Genet. 2016;7:163. PMID: 27708664
10. Bastian H, Glasziou P, Chalmers I. Seventy-five trials and eleven systematic reviews a day: how will we ever keep up? PLoS Med. 2010 Sep;7(9):e1000326. PMID: 20877712
11. Jensen PB, Jensen LJ, Brunak S. Mining electronic health records: towards better research applications and clinical care. Nat Rev Genet. 2012 Jun;13(6):395–405. PMID: 22549152
12. Jensen AB, Moseley PL, Oprea TI, Ellesøe SG, Eriksson R, Schmock H, Jensen PB, Jensen LJ, Brunak S. Temporal disease trajectories condensed from population-wide registry data covering 6.2 million patients. Nat Commun. 2014;5:4022. PMID: 24959948
13. Eysenbach G. Infodemiology and infoveillance tracking online health information and cyberbehavior for public health. Am J Prev Med. 2011 May;40(5 Suppl 2):S154–158. PMID: 21521589
14. Nguyen QC, Kath S, Meng H-W, Li D, Smith KR, VanDerslice JA, Wen M, Li F. Leveraging geotagged Twitter data to examine neighborhood happiness, diet, and physical activity. *Applied Geography* 2016;73:77-88.
15. Gliske SV, Stacey WC, Lim E, Holman KA, Fink CG. Emergence of Narrowband High Frequency Oscillations from Asynchronous, Uncoupled Neural Firing. Int J Neural Syst. 2016 Jul 14;1650049. PMID: 27712456

NO-BOUNDARY THINKING IN BIOINFORMATICS

JASON H. MOORE
Institute for Biomedical Informatics
University of Pennsylvania
Philadelphia, PA 19104, USA
Email: jhmoore@upenn.edu

STEVEN F. JENNINGS
Sector3 Informatics
Marana, AZ 85658, USA
Email: drsfjennings@gmail.com

CASEY S. GREENE
Department of Systems Pharmacology and Translational Therapeutics
University of Pennsylvania
Philadelphia, PA 19104, USA
Email: csgreene@upenn.edu

LAWRENCE E. HUNTER
Computational Bioscience Program,
University of Colorado School of Medicine
Aurora, CO 80045, USA
Email: larry.hunter@ucdenver.edu

ANDY D. PERKINS
Department of Computer Science and Engineering,
Mississippi State University
Jackson, MS 39762, USA
Email: perkins@cse.msstate.edu

CLARLYNDA WILLIAMS-DEVANE
Department of Biological and Biomedical Sciences,
North Carolina Central University
Durham, NC 27707, USA
Email: clarlynda.williams@nccu.edu

DONALD C. WUNSCH
Department of Electrical and Computer Engineering
Missouri Science and Technology University
Rolla, MO 65409, USA
Email: dwunsch@mst.edu

ZHONGMING ZHAO
Center for Precision Health
University of Texas Health Science Center
Houston, TX 77030, USA
Email: zhongming.zhao@uth.tmc.edu

XIUZHEN HUANG
Department of Computer Science
Arkansas State University
Jonesboro, AR 72467, USA
Email: xhuang@astate.edu

1. Bioinformatics is a Mature Discipline

Bioinformatics had its origins in the 1970s with the convergence of DNA sequencing, personal computers, and the internet. The field rapidly evolved as biotechnology improved making it critical to store, process, retrieve, and analyze bigger and bigger data to address important questions in the biological and biomedical sciences. Bioinformaticians throughout the 1980s and 1990s were often seen as consultants that provided a data service that represented one step in the process of asking a question, formulating a hypothesis, carrying out an experiment, analyzing the results, and making an inference. Much of bioinformatics at that time was about developing the capacity for providing this service. As the discipline matured in the 2000s it quickly became apparent that bioinformaticians were needed as collaborators and not just consultants. This facilitated the integration of bioinformatics into every aspect of a research project. We are at yet another turning point in the evolution of bioinformatics that will see in the coming years bioinformaticians transition from collaborators to leaders that bring interdisciplinary teams together to solve a complex problem. In other words, bioinformaticians will be able to ask the questions, define the hypotheses, and orchestrate the scientific study. This is the natural result of interdisciplinary training, the public availability of data, open-source software, the widespread availability of core facilities for conducting experiments and, importantly, the ability to integrate and synthesize knowledge sources to ask more impactful questions.

2. The Golden Era of Bioinformatics Has Begun

The turning point in the maturity of bioinformatics as a discipline has led some to speculate that we are entering a golden where the focus on computational approaches to biomedical research will be front and center[1]. There are several reasons for this speculation. First, big data is now the norm rather than the exception and computational methods are critical for successful storage, management, retrieval, analysis, and interpretation for answering scientific questions. Bioinformatics has never been so important for moving research forward. Bioinformatics areas such as databases, machine learning, and visualization are in high demand. Second, high-performance computing (HPC) is inexpensive and widely available in different technologies such as cloud computing and parallel computing using graphic processing units (GPUs) that bring thousands of compute core to a single desktop computer. Third, artificial intelligence and machine learning have matured and are now routinely being used to solve complex problems in the biomedical sciences. This is the result of decades of research on intelligent algorithms and software and the HPC resources necessary to apply them to big data. Fourth, the power of combining computational intelligence with statistical methods has emerged in the form of data science that allows the integration of different philosophical and quantitative schools of thought to solve biomedical problems. Fifth, visual analytics that brings visualization technology together with data science and human-computer interaction is maturing quickly with areas such as virtual reality, augmented reality, and 3D printing. Visual analytics will be essential for allowing human to interact with and understand data and research results that are too big and too complex to understand. Sixth, data and knowledge integration is maturing quickly as we have seen with electronic health records, data warehouses, and knowledge resources such as PubMed. Seventh, there is an increasing recognition of the importance of bioinformatics by federal funding agencies, biotechnology and pharmaceutical companies, and academic institutions. Investment in bioinformatics personnel and technology has never been greater and is expanding quickly. Now is the time for bioinformatics to have a substantial impact on biological and biomedical research.

3. No-Boundary Thinking in Bioinformatics

The purpose of this workshop is to introduce and discuss the future of bioinformatics as a mature discipline. We have previously defined this evolution and its impact as No-Boundary Thinking (NBT) in Bioinformatics[2,3]. The NBT philosophy provides bioinformaticians with the unique opportunity to move past being service providers to asking and answering research questions. This is because they are in the best position to integrate and synthesize knowledge across many disciplines to articulate a question that might have broader impact than one formulated from the knowledge of a single discipline. This allows them to be an equal contributor to the motivation and design phases of research studies. NBT puts the emphasis on knowledge-based question definition with big data serving a secondary and supporting role. This is counter to the current philosophy of letting big data drive the questions that are asked[3]. The workshop will introduce and define the NBT approach and will provide several scientific examples. An important component the workshop is providing examples of how NBT can be moved into the classroom to prepare bioinformatics students for a future where they are leading scientific studies. Panel discussions around NBT in science and education will allow for a robust discussion about these new ideas.

References

1. J.H. Moore, J.H. Holmes, *BioData Mining* **9**, 15 (2016).
2. X. Huang, B. Bruce, A. Buchan, C.B. Congdon, C.L. Cramer, S.F. Jennings, H. Jiang, Z.Li, G. McClure, R. McMullen, J.H. Moore, N. Nanduri, J. Peckham, A. Perkins, S.W. Polson, B. Rekepalli, S. Salem, J. Specker, D. Wunsch, D. Xiong, S. Zhang, Z. Zhao, *BioData Mining* **6**, 19 (2013).
3. X. Huang, S.F. Jennings, B. Bruce, A. Buchan, L. Cai, P. Chen, C.L. Cramer, W. Guan, U.K. Guan, U.K. Hilgert, H. Jiang, Z. Li, G. McClure, D.F. McMullen, B. Nanduri, A. Perkins, B. Rekepalli, S. Salem, J. Specker, K. Walker, D. Wunsch, D. Xiong, S. Zhang, Z. Zhao, J.H. Moore, *BioData Mining* **8**, 7 (2015).

OPEN DATA FOR DISCOVERY SCIENCE

PHILIP R.O. PAYNE[1], KUN HUANG[2], NIGAM H. SHAH[3], JESSICA TENENBAUM[4]

[1]*Washington University Institute for Informatics, Washington University in St. Louis School of Medicine, St. Louis, MO 63130, United States of America*

[2]*Department of Biomedical Informatics, The Ohio State University College of Medicine, Columbus, OH 43210, United States of America*

[3]*Center for Biomedical Informatics Research, Stanford University, Stanford, CA 94305, United States of America*

[4]*Department of Biostatistics and Bioinformatics, Duke University, Durham, NC 27710, United States of America*

Email: prpayne@wustl.edu, kun.huang@osumc.edu, nigam@stanford.edu, jessie.tenenbaum@duke.edu

The modern healthcare and life sciences ecosystem is moving towards an increasingly open and data-centric approach to discovery science. This evolving paradigm is predicated on a complex set of information needs related to our collective ability to share, discover, reuse, integrate, and analyze open biological, clinical, and population level data resources of varying composition, granularity, and syntactic or semantic consistency. Such an evolution is further impacted by a concomitant growth in the size of data sets that can and should be employed for both hypothesis discovery and testing. When such open data can be accessed and employed for discovery purposes, a broad spectrum of high impact end-points is made possible. These span the spectrum from identification of *de novo* biomarker complexes that can inform precision medicine, to the repositioning or repurposing of extant agents for new and cost-effective therapies, to the assessment of population level influences on disease and wellness. Of note, these types of uses of open data can be either primary, wherein open data is the substantive basis for inquiry, or secondary, wherein open data is used to augment or enrich project-specific or proprietary data that is not open in and of itself. This workshop is concerned with the key challenges, opportunities, and methodological best practices whereby open data can be used to drive the advancement of discovery science in all of the aforementioned capacities.

1. Rationale for Workshop

There are significant realized and potential benefits associated with the use of open data for discovery science. Unfortunately, despite such opportunities, the computational and informatics tools and methods currently used in most investigational settings to enable such efforts are often labor intensive and rely upon technologies that have not be designed to scale and support reasoning across heterogeneous and multi-dimensional data resources (1-3). As a result, there are significant demands from the research community for the creation and delivery of data management and data analytic tools capable of adapting to and supporting heterogeneous analytic workflows and open data sources (4-7). This need is particularly important when researchers seek to focus on the large-scale identification of linkages between bio-molecular and phenotypic data in order inform novel systems-level approaches to understanding disease states. In these types of situations, the scalar nature of such data exacerbates almost all of the aforementioned challenges. In this context, it is of interest to note that while the theoretical basis for the use of knowledge-based systems to overcome such challenges have evolved rapidly, their use in "real world" context remains the domain of experts with specialized training and unique access to such tools (1, 8, 9).

All of the preceding issues are further amplified when considering the nature of modern approaches to hypothesis discovery and testing when exploring biological and clinical open data, which are often based on the intuition of the individual investigator or his/her team to identify a question that is of interest relative to their specific scientific aims, who then carry out hypothesis testing operations to validate or refine that question relative to a targeted data set (10, 11). This approach is feasible when exploring data sets comprised of hundreds of variables, but does not scale to projects involve data sets with magnitudes on the order of thousands or even millions of variables (1, 8). An emerging and increasingly viable solution to this particular challenge is the use of domain knowledge to generate hypotheses relative to the content of such data sets. This type of domain knowledge can be derived from many different sources, such as complementary and contextualizing databases, terminologies, ontologies, and published literature (8). It is important to note, however, that methods and technologies that can allow researchers to access and extract domain knowledge from such sources, and apply resulting knowledge extracts to generate and test hypotheses are largely developmental at the current time (1, 8).

Finally, even when the major hurdles to the regular use of open data for discovery science as noted above are adequately addressed, there remains a substantial reliance on the use of data-analytic "pipelining" tools to ensure the systematic and reproducible nature of such data analysis operations. These types of pipelines are ideally able to support data extraction, integration, and analysis workflows spanning multiple sources, while capturing intermediate data analysis steps and products, and generating actionable output types (12, 13). Using data-analytic pipelines provide a number of potential benefits, including: 1) they support the design and execution of data analysis plans that would not be tractable or feasible using manual methods; and 2) they provide for the capture meta-data describing the steps and intermediate products generated during such data analyses. In the case of the latter benefit, the ability to capture systematic meta-data is critical to ensuring that such *in-silico* research paradigms generate reproducible and high quality results (12, 13). Again, while there are a number of promising technology platforms capable of supporting such data-analytic "pipelining", their widespread use is not robust, largely due to barriers to adoption related to data ownership/security, usability, scalability, and socio-technical factors (7, 14).

Given the aforementioned challenges and opportunities and the current state of knowledge concerning the use of open data across and between types and scales for the purposes of discovery science, this workshop addresses the following major topic areas:

- The state-of-the-art in terms of tools and methods targeting the use of open data for discovery science, including but not limited to syntactic and semantic standards, platforms for data sharing and discovery, and computational workflow orchestration technologies that enable the creation of data analytics "pipelines";
- Practical approaches for the automated and/or semi-automated harmonization, integration, analysis, and presentation of "data products" to enable hypothesis discovery or testing; and
- Frameworks for the application of open data to support or enable hypothesis generation and testing in projects spanning the basic, translational, clinical, and population health research and practice domains (e.g., from molecules to populations).

3. Workshop Speakers

Philip R.O. Payne, PhD: Dr. Payne is the founding Director of the Institute for Informatics (I2) at Washington University in St. Louis, where he also serves as a Professor in the Division of General Medical Sciences. Previously, Dr. Payne was Professor and Chair of the Department of Biomedical Informatics at The Ohio State University. Dr. Payne's research primarily focuses on the use of knowledge-based methods for in silico hypothesis discovery. He received his Ph.D. with distinction in Biomedical Informatics from Columbia University, where his research focused on the use of knowledge engineering and human-computer interaction design principles in order to improve the efficiency of multi-site clinical and translational research programs.

Kun Huang, PhD: Dr. Kun Huang is Professor in Biomedical Informatics, Computer Science and Engineering, and Biostatistics at The Ohio State University. He is also the Division Director for Bioinformatics and Computational Biology in OSU Department of Biomedical Informatics and Associate Dean for Genomic Informatics in the OSU College of Medicine. He has developed many methods for analyzing and integrating various types of high throughput biomedical data including gene expression microarray, next generation sequencing (NGS), qRT-PCR, proteomics and microscopic imaging experiments. Dr. Huang received his BS degree in Biological Sciences from Tsinghua University in 1996 and his MS degrees in Physiology, Electrical Engineering and Mathematics all from the University of Illinois at Urbana-Champaign (UIUC). He then received his PhD in Electrical and Computer Engineering from UIUC in 2004 with a focus on computer vision and machine learning.

Nigam Shah, MBBS, PhD: Dr. Nigam Shah is associate professor of Medicine (Biomedical Informatics) at Stanford University, Assistant Director of the Center for Biomedical Informatics Research, and a core member of the Biomedical Informatics Graduate Program. Dr. Shah's research focuses on combining machine learning and prior knowledge in medical ontologies to enable use cases of the learning health system. Dr. Shah was elected into the American College of Medical Informatics (ACMI) in 2015 and to the American Society for Clinical Investigation (ASCI) in 2016. He holds an MBBS from Baroda Medical College, India, a PhD from Penn State University and completed postdoctoral training at Stanford University.

Jessica Tenenbaum, PhD: Dr. Tenenbaum is Assistant Professor in the Division of Translational Biomedical Informatics, Department of Biostatistics and Bioinformatics at Duke University, and Associate Director for Bioinformatics for the Duke Translational Medicine Institute. Her primary areas of research include infrastructure and standards to enable research collaboration and integrative data analysis; informatics to enable precision medicine; and ethical, legal, and social issues that arise in translational research, direct to consumer genetic testing, and data sharing. After earning her bachelor's degree in biology from Harvard, Dr. Tenenbaum worked as a program manager at Microsoft Corporation in Redmond, WA for six years before pursuing a PhD in biomedical informatics at Stanford University.

2. Acknowledgements

The authors wish to acknowledge the contributions of Drs. Gustavo Stolovitzky (IBM) and Josh Swamidass (Washington University in St. Louis) to the preparation of this workshop summary.

References

1.Ruttenberg A, Clark T, Bug W, Samwald M, Bodenreider O, Chen H, et al. Advancing translational research with the Semantic Web. BMC Bioinformatics. 2007;8 Suppl 3:S2.

2.Fridsma DB, Evans J, Hastak S, Mead CN. The BRIDG project: a technical report. J Am Med Inform Assoc. 2008;15(2):130-7.

3.Kush RD, Helton E, Rockhold FW, Hardison CD. Electronic health records, medical research, and the Tower of Babel. The New England journal of medicine. 2008;358(16):1738-40.

4.Payne PR, Johnson SB, Starren JB, Tilson HH, Dowdy D. Breaking the translational barriers: the value of integrating biomedical informatics and translational research. J Investig Med. 2005;53(4):192-200.

5.Maojo V, García-Remesal M, Billhardt H, Alonso-Calvo R, Pérez-Rey D, Martín-Sánchez F. Designing new methodologies for integrating biomedical information in clinical trials. Methods of information in medicine. 2006;45(2):180-5.

6.Casey K, Elwell K, Friedman J, Gibbons D, Goggin M, Leshan T, et al. Broken Pipeline: Flat Funding of the NIH Puts a Generation of Science at Risk
. 2008. p. 24.

7.Ash JS, Anderson NR, Tarczy-Hornoch P. People and Organizational Issues in Research Systems Implementation. Journal of the American Medical Informatics Association : JAMIA. 2008.

8.Payne PR, Mendonca EA, Johnson SB, Starren JB. Conceptual knowledge acquisition in biomedicine: A methodological review. J Biomed Inform. 2007;40(5):582-602.

9.Richesson RL, Krischer J. Data standards in clinical research: gaps, overlaps, challenges and future directions. Journal of the American Medical Informatics Association : JAMIA. 2007;14(6):687-96.

10.Erickson J. A decade and more of UML: An overview of UML semantic and structural issues and UML field use. Journal of Database Management. 2008;19(3):I-Vii.

11.Butte AJ. Medicine. The ultimate model organism. Science. 2008;320(5874):325-7.

12.van Bemmel JH, van Mulligen EM, Mons B, van Wijk M, Kors JA, van der Lei J. Databases for knowledge discovery. Examples from biomedicine and health care. International journal of medical informatics. 2006;75(3-4):257-67.

13.Oster S, Langella S, Hastings S, Ervin D, Madduri R, Phillips J, et al. caGrid 1.0: An Enterprise Grid Infrastructure for Biomedical Research. Journal of the American Medical Informatics Association : JAMIA. 2008;15(2):138-49.

14.Kukafka R, Johnson SB, Linfante A, Allegrante JP. Grounding a new information technology implementation framework in behavioral science: a systematic analysis of the literature on IT use. J Biomed Inform. 2003;36(3):218-27.

www.ingramcontent.com/pod-product-compliance
Lightning Source LLC
Chambersburg PA
CBHW071247180926
51613CB00002B/159
9789813207806